The Neurobiology of an Insect Brain

The Neurobiology of an Insect Brain

Malcolm Burrows

Department of Zoology,
University of Cambridge

Oxford New York Tokyo
OXFORD UNIVERSITY PRESS
1996

Oxford University Press, Walton Street, Oxford OX2 6DP

Oxford New York
Athens Auckland Bangkok Bombay
Calcutta Cape Town Dar es Salaám Delhi
Florence Hong Kong Istanbul Karachi
Kuala Lumpur Madras Madrid Melbourne
Mexico City Nairobi Paris Singapore
Taipei Tokyo Toronto

and associated companies in
Berlin Ibadan

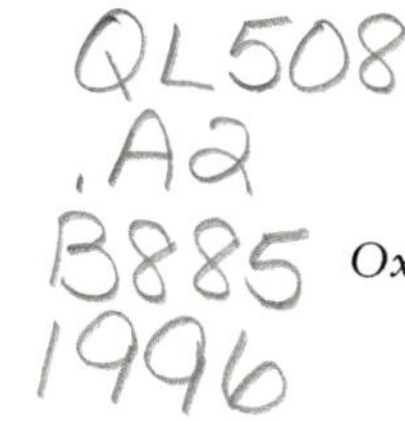

Published in the United States
by Oxford University Press Inc., New York

A catalogue record for this book is available from the British Library

Library of Congress Cataloging in Publication Data
Burrows, Malcolm.
The neurobiology of an insect brain / Malcolm Burrows.
Includes bibliographical references.
1. Locusts–Nervous system. 2. Insects–Nervous systems.
3. Brain. I. Title.
QL508.A2B885 1996 595.7′26–dc20 96–17651
ISBN 0 19 852344 0 (Hbk)

Typeset by EXPO Holdings, Malaysia

Printed in Great Britain by
Bookcraft (Bath) Ltd
Midsomer Norton, Avon

'Protect me from knowing what I don't need to know.

Protect me from even knowing that there are things to know that I don't know.

Protect me from knowing that I decided not to know about the things that I decided not to know about.'

Douglas Adams (1992)
Mostly Harmless, William Heinemann Ltd.

PREFACE

THE NEED FOR SYNTHESIS IN A TIME OF CHANGE

This book is written at a time of major change in the concepts and techniques in neurobiology and, moreover, in the areas that are thought fashionable to pursue. Such are generally not times for synthesis, but it is precisely because of these shifts that we need to establish and consolidate the present state of our knowledge as a foundation for future growth and development. Neurobiology looks set to grow massively in two areas driven by the extraordinary power of even our current molecular biological techniques on the one hand and computational approaches on the other. To have any meaning, both of these lines of advance need a firm foundation in the way the nervous system is constructed and functions; the detailed knowledge of the structure and action of molecules needs to be placed in the context of how they might work in a functioning brain, and the computational models need to be based on the properties of real neurons, the way that they are actually connected and the way they normally work. Without this structure, the molecular biology is likely only to provide an enormous wealth of interesting detail without the explanations at higher levels for how neurons and groups of neurons might work, and the computational approach is likely to lead to a series of interesting designs of machines that can perform some of the functions of the brain, without necessarily shedding much light on the mechanisms used by the brain. Integrating the results of these new approaches into the base of functional information that we have, but which still needs to be advanced strongly to fill many voids in our knowledge, offers great hope for providing explanations at the many levels that will be necessary.

OBJECTIVES

My objectives are to define and analyse our current knowledge of the functioning of the brain of one animal, the locust, and to show how this contributes to our understanding of brains in general. The analysis is at all times functional, by trying to relate the processes at molecular receptors, channels, synapses, neurons and networks to the functioning of the nervous system in generating behaviour. The overriding question is always how the brain processes the gamut of sensory signals, integrates them and then produces movements that are behaviourally appropriate. The emphasis is, therefore, on how movements (behaviour) are produced, to the exclusion of a description of how the major senses work.

I have deliberately been selective in both the isssues I analyse, the data I present and the reference I give to the published literature. This is to restrict the size of the book and to make the text flow more easily, but inevitably means that there are omissions and that work will sometimes be described without precise references to all the key papers. No slight is intended to the authors that I do not cite, but reference to the thousands of papers that I have attempted to digest would be unwieldy and in itself indigestible. I have always worked from the original research articles and have avoided reviews which tend to tell simplified stories often using data that I have failed to find in the original literature.

THANKS

The space and the time to think about and write this book was made possible by a leave of absence from Cambridge, by a generous award from the Alexander von Humboldt Stiftung, Bonn, Germany, and by the hospitality and tolerance of Prof. H.-J. Pflüger and Prof. R. Menzel at the Institut für Neurobiologie, Freie Universität Berlin, Germany and Prof. W. Rathmayer at the Fachbereich Biologie, Universität Konstanz, Germany.

I am particularly grateful to Ali Cooper who copy-edited the manuscript and to Ken Baker for his help in preparing the figures. Many thanks are also due to the numerous people who have read drafts of parts of the book, offered suggestions, discussed some of my more absurd ideas and told me when I was daft; Peter Bräunig, Peter Evans, Berthold Hedwig, Eleni Kalogianni, Wolfram Kutsch, Tom Matheson, Phil Newland, Dave Shepherd, Paul Stevenson, Alan Watson, and Harald Wolf. My dog also maintained some semblance of balance in my life by unfailingly reminding me when it was time to stop writing and eat breakfast. The greatest thanks nevertheless are due to the people who have performed with such verve the exciting experiments and analyses that I discuss.

Acknowledgements

Full details and bibliographic references to all material reproduced are to be found in the Figure legends and in the reference section.

Figure 9.3 is reprinted from *J. Insect Physiol.* 37, 129–138, ©1991, Fig. 11.26 is reprinted from *J. Insect Physiol.* 19, 763–772, ©1973 and Fig. 12.13 is reprinted from *J. Insect Physiol.* 28, 53–60. ©1982, with kind permisssion from Elsevier Science Ltd, The Boulevard, Langford Land, Kidlington OX5 1GB, UK. Figures 9.7 and 11.21 are partly or wholly reproduced from the *European Journal of Neuroscience* by permission of Oxford University Press, Walton St., Oxford OX2 6DP, UK.

The following are thanked for permission to reproduce or reprint copyright material: Academic Press, Inc., Orlando, Florida 32887–6777, USA, for material from *Biological neural networks in invertebrate neuroethology and robotics* used in Fig. 8.12; The American Physiological Society, 9650 Rockville Pike, Bethesda, MD 20814–3991, USA, for material from the *Journal of Neurophysiology* used in Figs 3.13, 6.1, 7.15, 7.18, 8.10, 9.10, 11.7, and 11.8; Blackwell Science, Osney Mead, Oxford OX2 OEL, UK, for material from *Physiological Entomology* used in Figure 8.14; Chapman & Hall Ltd., Cheriton House, North Way, Andover, Hampshire SP10 5BE, UK, for material from the *Journal of Neurocytology* used in Figures 3.11 and 3.19; The Company of Biologists Ltd., Bidder Building, 140 Cowley Road, Cambridge CB4 4DL, UK, for material from the *Journal of Experimental Biology* used in Figs 3.2, 3.7, 4.5, 4.9, 7.1, 8.1, 8.4, 8.7, 8.8, 8.9, 8.12, 8.14A, 8.19, 8.23, 9.1, 9.2, 9.3, 9.4, 9.7, 9.8, 9.9, 9.13, 9.14, 10.7, 11.4, 11.12, 11.14, 11.15, 11.16, 11.17, 11.19, 11.23, 11.24, 12.5, 12.6, 12.8, 12.9, 12.10, 12.11, 12.12 and also for material from the *Journal of Embryology and Experimental Morphology* (now *Development*) used in Fig. 4.3; Elsevier Trends Journals, 68 Hills Road, Cambridge CB2 1LA, UK, for material from *Trends in Neurosciences* used in Figure 8.2; Gustav Fischer Verlag, Wollgrasweg 49, D–70577, Stuttgart, Germany, for material from *Fortschritte der Zoologie* used in Figure 11.3; National Research Council of Canada, Research Jourrnals, Ottawa, Ontario K1A 0R6, Canada, for material from the *Canadian Journal of Zoology* used in Fig. 11.16; The Physiological Society, Shaftesbury Road, Cambridge CB2 2BS, UK for material from the *Journal of Physiology* used in Figures 3.14, 6.2, and 7.17; The Royal Society of London, 6 Carlton House Terrace, London SW1 5AG, for material from *Philosophical Transactions of the Royal Society of London* used in Figures 2.5, 2.6, 2.7, 3.18, 11.1, 11.20 and also for material from *Proceedings of the Royal Society of London* used in Figures 3.8 and 12.1; The Society for Neuroscience, 11 Dupont Circle, NW Suite, Washington DC 20036, USA, for material from the *Journal of Neuroscience* used in Figures 3.13, 4.4, 4.6, 5.13, 7.2, 7.5, 7.6, 7.10, 7.11, 7.21, 7.22, 7.24, 7.25, 8.22, 8.24, 10.5 and 11.14; Dr K. Seymour for material used in Figs 4.1 and 4.2, Professor D.C. Sandeman for material used in Figure 11.3, Dr C. Hill-Venning for material used in Figures 12.2, 12.4, 12.8, 12.12, and Drs P.A. Stevenson, R.M. Robertson, H.-J. Pflüger and L.H. Field for permission to use unpublished drawings.

The following have been applied to for permission to reproduce copyright material: CIRARD, Paris, France, for material from *Les acridiens des formations l'herberses d'Afrique de l'ouest* used in Figure 1.1; Gustav Fischer Verlag, Stuttgart, Germany, for material from *BIONA report 1:Insektenflug 1* used in Figure 11.14; Alan R. Liss, New York, USA, for material from *Cytochemical methods in neuroanatomy* used in Figure 4.3; Springer-Verlag, Heidelberg, Germany, for material from *Cell Tissue Research* used in Figs 2.4, 2.8, 5.5, 5.8, 5.9, 5.11, 5.12, 5.13, 8.15, 8.22, 11.7, 11.8, 11.9, for material from the *Journal of Comparative Physiology A* used in Figs 5.7, 7.1, 7.14. 8.2, 8.16, 8.18, 8.20, 8.21, 9.4, 9.5, 9.6, 9.8, 9.9, 9.11, 9.12, 9.15, 10.1, 10.2, 10.4, 10.6, 10.7, 11.2, 11.5, 11.6, 11.9, 11.10, 11.14, 11.16, 11.18, 11.19, 11.20, 11.21, 11.22, 11.25, 12.2, 12.3, 12.6, for material from *Histochemistry* used in Fig. 5.13, for material from *Biological Cybernetics* used in Fig. 8.14, for material from *Zoomorphology* used in Fig. 8.22, for material from *Naturwissenschaften* used in Fig. 11.20 and for material from *Behavioural physiology and neuroethology: roots and growing points* used in Fig. 3.5; John Wiley & Sons, Inc., New York, USA, for material from the *Journal of Comparative Neurology*, used in Figs 2.3, 2.8, 3.1, 3.4, 3.6, 3.9, 3.10, 3.11, 3.12, 3.16, 3.17, 4.7, 4.8, 5.2, 5.3, 5.4, 5.10, 5.14, 7.3, 7.4, 7.7, 11.21 and for material from the *Journal of Neurobiology* used in Figs 8.4, 8.11, 10.7, 11.7 and 11.15; Verlag Paul Parey, Berlin, Germany for material from *Insect Locomotion* used in Fig. 11.15, and The Zoological Society of London, London, UK for material from the *Journal of Zoology* used in Fig. 2.2.

Although every effort has been made to trace and contact copyright holders, in a few instances this has not been possible. If notified, the publishers will be pleased to rectify any omissions in future editions.

Contents

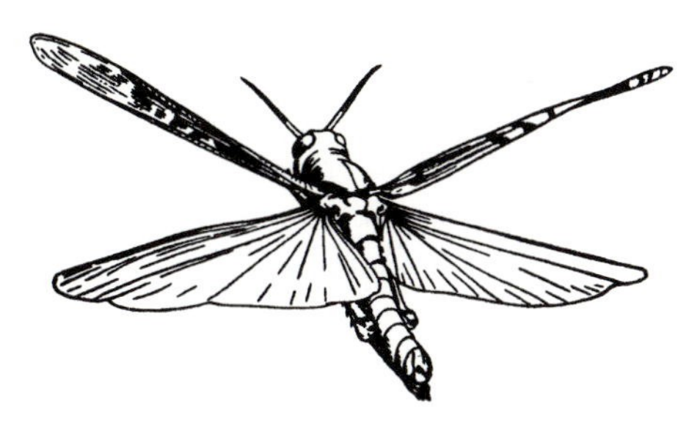

1

Biology of locusts and grasshoppers

1.1. LOCUSTS AND GRASSHOPPERS

Locusts belong to a large group of orthopteroid insects in the superfamily **Acridoidea**, with most in the family **Acrididae**. This family also includes many grasshoppers. Locusts are similar to grasshoppers in their morphology and differ from them only in their behaviour in which they show a strong tendency to group together and become gregarious, and then to migrate in large swarms. The behavioural distinction is not complete, for there is really a spectrum of types; at one extreme are species that readily become gregarious and swarm – typical locusts, and at the other are those that always live solitary lives, never aggregate and never swarm – typical grasshoppers. In between, there are species of locusts that live for many generations without aggregating and similarly, there are species of grasshoppers that may aggregate but do not migrate in swarms. Locusts are therefore those species that aggregate and swarm more commonly and grasshoppers are those which rarely aggregate. The terms are thus often used loosely and interchangeably, but for most of what we know about the central nervous system this does not matter.

Locusts consist of about 12 species which can change their behaviour and appearance when they occur in dense groups. When they live singly, or in small groups, they behave like grasshoppers and are called **solitary phase**, but in large groups they assume a **gregarious phase**. As immature larvae that have no wings and cannot fly, these groups are called **bands**. When adult, the large groups are called **swarms** which may then take to the air and migrate for large distances. The simultaneous occurrence of a number of swarms in one area becomes a **plague**. The number of locusts in a particular region fluctuates enormously with time so that there can be many years where the numbers are small, before a critical number and density is reached for the next occurrence of bands and swarms.

The species of grasshoppers are more numerous and more widespread so that in the UK alone there are 21 species. These animals do not form swarms and are generally seasonal. Their general morphology is similar to that of locusts and many parallels can be found in the structure of the nervous system, even at the level of the structure and connections of neurons that can be recognised from species to species (Wilson *et al.*, 1982). Despite this wide variety of species, the number of locusts and grasshoppers that have been studied by neurobiologists (Table 1.1) is nevertheless relatively limited.

1.1.1. *Schistocerca* and *locusta*

Most of what is known about the neurobiology of locusts has, however, been derived from just two of these species ***Schistocerca gregaria*** and ***Locusta migratoria migratoriodes*** (Fig. 1.1 and Table 1.2).

The most obvious differences between these two species are their external appearance, apparent as different coloration, hairiness and shape of the head and compound eyes. Most of the behaviour of the two species that is studied by neurobiologists is, however, the same; the legs are moved in the same way during walking and the wings are moved in the same way, but at slightly different frequencies, during flight. All the indications so far are that the neural organisation of the two is very similar in that the same identified neurons with the same actions and the same connections can be found in both. No differences have been observed

Table 1.1. Locusts and grasshoppers studied by neurobiologists

Name	Common name	Country
Schistocerca gregaria (Forskål)	Desert locust	North Africa, Arabia, India
Locusta migratoria migratoriodes (Reiche and Fairmaire)	African migratory locust	Africa (south of Sahara)
Locusta migratoria migratoria (Linnaeus)	Asian migratory locust	Asia
Nomadacris septemfasciata (Serville)	Red locust	Africa
Locustana pardalina (Walker)	Brown locust	South Africa
Schistocerca americana (Fabricius)	–	Central and North America
Schistocerca nitens (Thunberg) (= S. vaga (Scudder))	Gray bird locust	Central and North America
Chortoicetes terminifera (Walker)	Australian plague locust	Australia
Melanoplus differentialis (Thomas)	Rocky mountain locust	North America
Dissosteira carolina (Linnaeus)	Carolina locust	North America
Romalea microptera (Beauvois)	Lubber grasshopper	North America

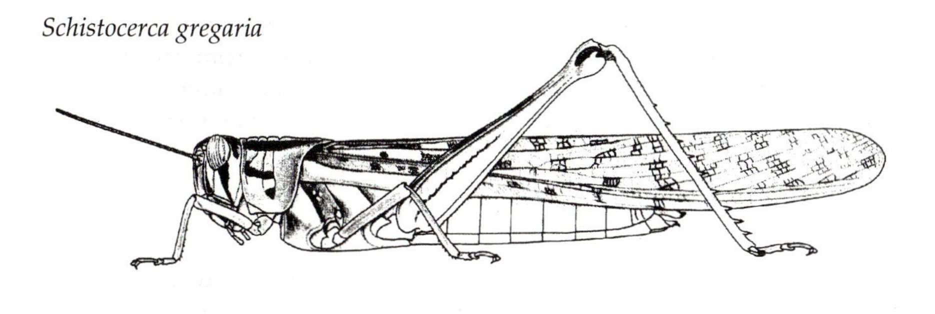

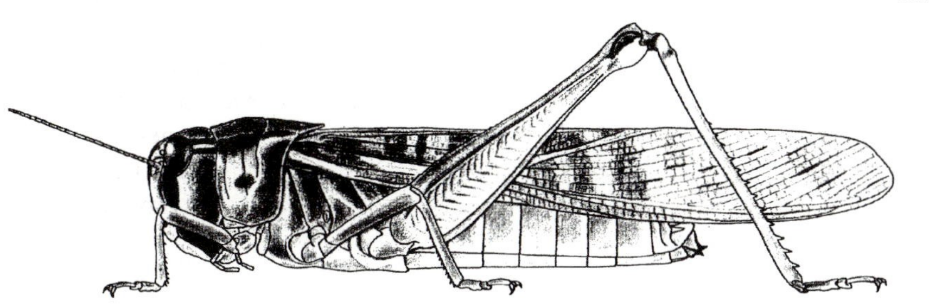

Fig. 1.1 Drawings of the two species of locusts most commonly used by neurobiologists. Based on Mestre (1988).

Table 1.2. Classification of Schistocerca gregaria and Locusta migratoria migratoriodes

Phylum	Superorder	Order	Superfamily	Family	Subfamily	Genus	Species	Subspecies
Arthropoda								
	Insecta							
		Orthoptera						
			Acridoidea					
				Acrididae				
					Cyrtacanthacridinae	Schistocerca	gregaria	
					Oedipodinae	Locusta	migratoria	migratoriodes

at this level in the organisation, but nothing can be said about the weighting of the apparently homologous synaptic connections, or the numbers of synapses involved. Differences are to be expected in, for example, the sensory processing, where more tactile exteroceptors are present in *Locusta migratoria* than in *Schistocerca gregaria*. Throughout the text, these two species are used interchangeably with attention drawn to the particular species only where possible differences may be critical. It is recognised that some of the more complex behaviour of the two species is different and that this must have a neuronal correlate. This invites a broader caution in extrapolations from one species to the other.

1.1.2. Life cycle

Eggs are laid in moist ground during the rainy season, with embryonic development taking place below ground. The first instar larva then crawls to the surface as a miniature version of the adult but without moveable wings. It then grows postembryonically by a series of moults so that there are five larval instars of progressively increasing size, with these immature locusts called larvae, nymphs or hoppers. Fully moveable wings are formed only at the final moult to the adult. The period of development varies from a few weeks to 6 months depending on prevailing conditions, but in the controlled conditions of the laboratory usually takes 6 weeks (*see Chapter 4*).

1.1.3. Adult appearance

The adults are initially pink in colour and their cuticle is soft, but as they become sexually mature about 2 weeks after the final moult they turn yellow and their cuticle becomes harder. The adults are sexually dimorphic with the most obvious difference being their size; females finally attain a length of about 90 mm and a mass of 2.5–3.5 g, while the males are slightly smaller at 75 mm and 1.5–2.0 g.

The distinctive features of both sexes are the well-developed wings and hind legs. The stiff front pair of wings is normally folded to cover the hind wings and most of the abdomen. The massive hind legs are considerably larger than the other two pairs of legs and are used in jumping and kicking. The antennae on the head are unbranched and short, and the cerci that act as wind detectors at the tip of the abdomen are so short that they are barely visible.

The body is bilaterally symmetrical with very few visible structural or behavioural asymmetries. The mandibles on the left and right sides are of slightly different shape and because they must move past each other to chew food, the movements can be either left- or right-handed. A similar asymmetry occurs during copulation when the male mounted on the back of a female has the choice to move his abdomen down and under either the left or right side of the female and usually chooses the left (Kutsch, unpublished). One pair of neck muscles (muscle 54) run diagonally from one side of the body to the other and the same one is always dorsal to the other. Few asymmetries have been sought or described in the organisation of the nervous system, and it is generally assumed that symmetry prevails. There are, however, unpaired neurons that arise from an unpaired precursor cell at the midline, but these neurons have symmetrically arranged projections to both sides of the nervous system or body (*see Chapter 3*). It remains a real possibility that there will be structural differences between neurons on both sides of the body, or differences in the expression of some of their putative neuromodulatory or neurosecretory contents. Some hints of these sorts of asymmetries are emerging from immunocytochemistry that cannot readily be dismissed as artefacts of the staining method.

1.1.4. Phases

In size, colour and behaviour the solitary and gregarious phases can be quite different, to the extent that they were once thought to represent distinct species. Gregarious larvae (hoppers) of *Schistocerca gregaria* are black and yellow, while solitary ones are green or yellow-green according to the colouration of the background where they are raised. Gregarious adults are different in size and shape with a broader head, larger compound eyes, a shorter pronotum (neck shield) and shorter hind legs than solitary individuals. Behaviourally, gregarious hoppers develop more quickly and then march in a concerted fashion in bands for up to 2 km a day, whereas the solitary ones do not aggregate, and walk for only short distances. Gregarious adults fly readily and mostly by day, whereas solitary ones fly more reluctantly and by night. Solitary locusts lay more eggs than gregarious ones, but gregarious nymphs can survive for longer without food. The phase differences in colouration, and probably other features, are under hormonal control (Pener, 1991): a hormonal deficiency resulting from a mutation leads to a strain of albino locusts even when they are crowded together (Tanaka, 1993). Juvenile hormone from the corpora allata (*see Chapter* 2), for example, is a major factor in causing the green colouration of hoppers and a peptide from the corpora cardiaca can induce the dark colouration of adults (Tanaka and Pener, 1994).

The transition from solitary to gregarious phase depends on a number of factors including visual, chemical and tactile stimuli, and the diet. Of particular importance are the volatile substances present in the faeces that act as aggregation pheromones, bringing the locusts together for protection, reproduction and feeding.

1.1.5. Swarms

A number of species have the ability to form vast swarms that migrate, while there are others that are intermediate between locusts and grasshoppers in that they can, under particular circumstances, multiply rapidly and form large bands. The size of the swarms and the numbers of insects involved can be quite staggering, and reports of them have occurred throughout recorded history. Carved images of locusts date from at least 2400 B.C. and their occurrence is recorded in both the Koran and the Bible '*... and the Lord brought an east wind upon the land all that day, and all that night; and when it was morning, the east wind brought the locusts. And the locusts went up over all the land of Egypt, and rested on all the coasts of Egypt; very grievous were they; before them there were no such locusts as they, neither after them shall be such. For they covered the face of the whole earth, so that the land was darkened; and they did eat every herb of the land, and all the fruit of the trees which the hail had left; and there remained not any green thing in the trees, or in the herbs of the field, through all the land of Egypt.*' (Exodus, 10). '*And the Lord turned a mighty strong west wind, which took away all the locusts, and cast them* *into the Red Sea; there remained not one locust in all the coasts of Egypt'* (Exodus, 19). In 1865 a ship's captain, some 2400 km west of the African coast, reported that *'the air and the sails were full of locusts for 2 days'* (Selys-Longchamps, 1878). In 1986, during 1 week in September, 360 000 hectares of Senegal were sprayed with malathion to try to control *Oedalus senegalensis*, a grasshopper that aggregates in large numbers but does not swarm. In 1988, a swarm of locusts in Algeria was estimated to cover 390 km^2 and to contain more than 20 billion locusts at densities of up to 150 per m^3 when on the ground. These locusts consumed more than 35 000 tonnes of vegetation in a day (each locust eating about 1.5 g per day), which would have been sufficient to feed 20 000 humans for a year. Most recently, in 1995 desert locusts ate their way through 25 000 acres of sorghum in Gitena province in Sudan.

1.1.6. Control of locusts

Despite these recent reports of swarms, it is easy nowadays to underestimate the economic and agricultural importance of locusts, given the widespread, but decreasingly effective, use of insecticides. Nevertheless, swarms of *Schistocerca gregaria* are estimated to be able to affect some 20% of land in the world and thus some 10% of its population. The presence of locusts occasionally thrusts itself upon western consciousness as when, for example, during an England cricket tour to Australia in the 1980s, an up-country pitch had to be cleared of them before play could start.

The cost of controlling locusts and grasshoppers is high. During the years 1986–89 some $US400 million was spent on spraying with insecticides in the Sahel and north-west Africa alone. The problem of control has been compounded because the environmental impact of persistent insecticides has meant that the use of these is no longer desirable. The cost of the nonpersistent pesticides that are now used is higher because they must contact a locust during application if they are to have any effect; this means that the young larvae must first be detected•before it is sensible to spray and that applications must be repeated. An alternative to chemical sprays is only just appearing and that is the use of a pathogenic fungus *Metarhizium flavoviride* (Goettel *et al.*, 1995) which, when applied just once, may be able to spread through the population (Thomas *et al.*, 1995).

1.2. WHY ANALYSE THE NERVOUS SYSTEM OF AN INSECT?

The attractions of an insect brain for analysis of the construction and action of the nervous system are many.

1. In comparison with the nervous system of a mammal, it contains a relatively small number of neurons. While the numbers involved in visual processing

remain large, those involved in controlling movement are small; a leg is controlled by a population of only 70 motor neurons.
2. Access to the neurons is not impeded by a pulsating blood supply.
3. Many of the neurons are large enough to allow electrodes to be placed in their various parts to analyse the processing of signals that occurs within them.
4. It is possible to treat all of the motor neurons, many of the larger interneurons and some of the sensory neurons as identified individuals, allowing the properties and connections of particular neurons, rather than simply members of a population, to be determined in successive experiments.
5. The motor neurons innervate muscles that move levers, in the same way as those of vertebrates, so that the design of the nervous system must be fitted to solve similar problems in controlling movement in both groups of animals.
6. The actions of many of the neurons can be related directly to the behaviour in which they are involved and which they may control.

1.2.1. Why the nervous system of a locust?

The impetus for the analysis of the nervous system of locusts stemmed originally from their promotion by the Anti-Locust Research Centre in London, founded in the colonial days of Britain to solve the problem with locusts in Africa. The particular advantages of this large and robust insect then soon became apparent to neurobiologists and developmental biologists.

Locusts can be easily bred and their embryos can be manipulated and the large size of the neuronal precursor cells gave the initial impetus to the analytical work on developmental processes. Postembryonic development through five larval instars means that the neuronal organisation of certain motor patterns can be analysed before the locust is able to express these patterns as behaviour; moveable wings, for example, are only present in adults, yet the motor pattern for flying can be expressed in early larvae. As adults, locusts show a wide variety of behaviour that is worthy of analysis, both at the ethological and neurobiological levels. As the information about the nervous system and its actions in controlling and generating behaviour has increased, any of the reasons for continuing with such studies become self-sustaining; the detailed base of knowledge continually allows more and more detailed questions to be asked, with a greater probability of obtaining useful information and the solutions to more general problems.

The intent behind most of the analyses of the nervous system of the locust is to reveal mechanisms and design principles that apply to more general issues of the operation of neurons and nervous systems. Of course, there is detail that relates only to the locust or to other insects, and this is still a proper reason for analysis; insects are a major element in the biotype and their lives impinge upon almost all aspects of ours, often with drastic effects. Understanding how they work is an essential step to our living in harmony with then. From a purely neurobiological standpoint, it is possible to say with certainty that many of the mechanisms revealed and analysed in an insect also occur in other animals. Clues, and even an understanding of, the operation of more complex nervous systems can be obtained from simpler nervous systems.

1.2.2. Drawing information from other insects

Descriptions of other locusts, grasshoppers or other insects are included either because of their special attributes or because of the specific contribution that they make to our understanding. Thus, for example, most of our information about the use of giant fibres in eliciting escape movements comes from crickets and cockroaches. The default descriptions are, however, always from locusts.

1.3. ANALYSING THE NERVOUS SYSTEM OF A LOCUST

It is increasingly seen as fashionable to denigrate the contribution made by analyses of the relations between neurons and behaviour and the likely contribution that such an approach will make in the future. This is an attitude that I reject entirely, and I hope that this book will be a testament to the past, continuing and future value of these approaches, and to the enormous progress that they have enabled. The question of how a brain works is far more difficult to answer than how a spike is conducted, how signals are transmitted across a synapse, or how the activation of a receptor allows an ion to move across a membrane. Our failure to answer this question is at the heart of much of the disillusionment with past approaches, despite the fact that it is the most complex problem in modern biology. Moreover, it is uncertain in what form we would like the answer to be framed and whether we would recognise the answer were we to be presented with it. Perhaps it is simply a human trait to castigate failures to solve seemingly intractable problems, rather than celebrating the wondrous advances that have been made. If there should be any doubt of the real advances that have been made, then consider the diagrams in books of only 30 years ago – roughly the start of the active period of research covered in this book – that sought to explain how the nervous system of an insect worked. No neurons were identified, recordings from the central nervous system were virtually unknown, and while anatomy from 60 years earlier had indicated the likely complexity of neuronal structure it had, at least for insects, largely been ignored.

Of course, it is not surprising to find that satisfying explanations for all the questions that have been tackled are sometimes lacking, or that some of the original questions were posed with unreasonable optimism and expectations, but rather than abandoning them from frustration, what is needed is to incorporate them with other approaches designed to give explanations at many levels. Any understanding of how a nervous system works is likely to be given at many different levels; molecules, synapses, neurons, circuits, brain, and behaviour. The real excitement and challenge that lies ahead is to harness the power of the new techniques of molecular biology and computer modelling to provide functional explanations in terms of the actions of real neurons and the real networks that our present approaches have defined in such exquisite detail.

1.3.1. Ethics of experiments on locusts

To understand how behaviour is generated and controlled by the nervous system inevitably involves analysis of the action of individual neurons in an alert animal. It is therefore appropriate to raise ethical questions about this form of analysis and to ask whether insects have any sensation equivalent to our concept of pain.

1.3.1.1. *Defining pain*

Pain is generally understood, in human terms, as a subjective experience that varies with a huge range of different prevailing factors in one individual and between individuals. Our inference of pain in other animals comes from an extrapolation of our own experiences coupled with our interpretation of their overt behaviour. We readily accept that our dog or cat experiences pain, but have no means of finding out from them what their experience might be and whether it equates with our own experience of pain. Defining the pain that might be experienced by an insect requires even greater extrapolations and no hope of sharing those experiences (Eisemann *et al.*, 1984).

1.3.1.2. *Neural organisation*

The perception of pain is an interpretation by the brain of particular sensory signals. In mammals, much of the experience of pain results from the activation of so-called nociceptors whose fine fibres make up a large percentage of the fibres entering the dorsal roots of the spinal cord. These receptors are activated by several types of stimuli that are likely to damage the tissues in which they are embedded. Insects have no such receptors and no fibres that fit any of the characteristics of those of the nociceptors and hence no parts of the nervous system that are dedicated to the decoding of such information. Any experience of pain by an insect must therefore be based on a different type of neural organisation than in mammals. Nevertheless, this by itself does not exclude the experience of pain, which might be coded in particular frequencies or patterns of spikes in receptors that would normally respond to other modalities. Similarly, were nociceptors to be found in insects, their activation would not necessarily imply an experience of pain.

1.3.1.3. *Behaviour*

Behaviour provides no evidence for a general experience of pain caused by a wide range of different stimuli that are either likely to, or actually do cause damage. Both insects and mammals will withdraw a limb from a potentially harmful stimulus and both, to different degrees, will learn from such experiences. The occurrence of these reactions does not necessarily indicate pain because they are adequately explained by activation of exteroceptors that lead to appropriate motor responses. A locust will, for example, withdraw its leg when a few tactile hairs on its leg are lightly touched, a stimulus that would not, in human terms, be considered as one likely to engender pain.

There are no reports of insects showing protective behaviour toward damaged parts of their body, but there are many examples of insects continuing with apparently normal behaviour despite injury. A locust with a damaged tarsus will continue to walk and will use it to support the body weight. A locust will continue to feed while being eaten by a mantis (Eisemann *et al.*, 1984), and a male mantis will continue to copulate while being devoured by its mate. A cockroach stung by a wasp shows none of the responses of a mammal that has been stung. The grooming of wound sites is no different to the grooming of the site of mechanical stimulation, or the site where a foreign piece of material has adhered. Similarly, the struggling movements when grasped, the release of alarm pheromones or repellent secretions are appropriate motor responses to potentially harmful or threatening stimuli, and do not necessarily imply a sense of pain. The uncoordinated movements or writhings that occur upon poisoning with insecticides can be explained by the unnatural excitation of neurons, and again do not imply a sense of pain.

1.3.1.4. *Are insects aware of their behaviour?*

Inextricably bound with the concept of an experience of pain is whether an insect is aware of its behaviour, or has any consciousness. The central issue is when we infer from an overt behaviour that the brain of an insect may be 'thinking' about a problem rather than simply computing it. Does the flexibility of their behaviour and their ability to learn imply a conscious comprehension of the problem that is being solved? Two examples of the behaviour of bees, which are generally thought to be amongst the most capable of insects at learning, indicate that it is not necessary to invoke any comprehension.

A foraging bee learns about the colour of a food source only in the last few seconds before it lands (Menzel and Erber, 1978). In the preceding approach time, in the time that it spends feeding, and in the time it spends circling the source after feeding, it must be aware of the source but it does not learn its colour. To learn the colour of the food source requires several visits and yet the odour of the food is learnt on just one visit. Information about the food source seems to be stored as a set, so that if one element of the set is altered, the bee has to relearn the set at the characteristic rates of each of the elements (Bogdany, 1978). Everything in this behaviour suggests programmed learning rather than flexibility or awareness.

Many insects release chemical messages in the form of pheromones, alarm substances and repellent odours in response to particular stimuli. The release, for example, of oleic acid by a dying bee is a signal for its corpse to be removed from the hive by other bees, but oleic acid painted on a live bee will also result in removal of this healthy and moving bee from the hive. Everything in this behaviour suggests a specific response to a particular sensory stimulus and no comprehension of the chemical message that is used.

The overwhelming conclusion from this sort of evidence suggests that there is no indication of conscious experience in insects (Gould and Gould, 1982). These authors are forceful in their conclusion and '*fail to see how the hypothesis of animal consciousness has any major value at this stage for furthering our understanding of insect behaviour*'. I share their view that it is better to proceed with a fuller exploration of the complex neuronal mechanisms to see how far beyond our present limits of understanding this will take us before being forced to invoke notions of consciousness to account for the behaviour of insects that we observe and analyse.

1.3.1.5. *Does a locust experience pain?*

Whether insects experience pain is essentially an unanswerable question, but both the organisation of the nervous system and observations of behaviour do not suggest that they experience pain in a way comparable to humans. If they do experience pain then the underlying mechanisms must differ from those in humans. This, of course, does not in any way free us from the obligation to treat insects as experimental animals with the care and respect demanded of any living animal. But can we envisage a world where many of the devastating insect pests are not killed deliberately, or where many insects are not killed accidentally as we proceed with our lives in the world that we have largely manipulated for our own ends?

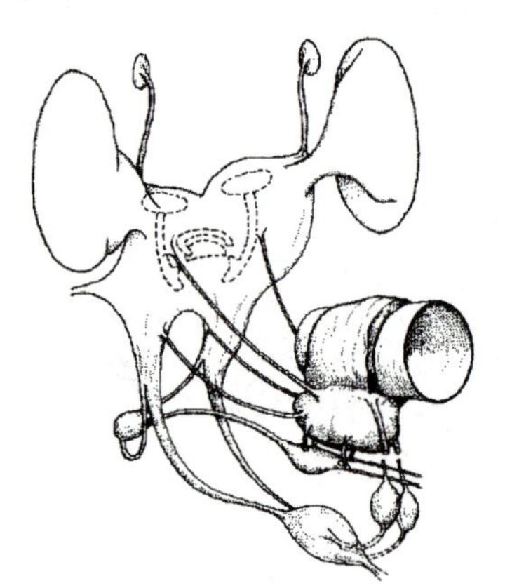

Anatomy of the nervous system

Exquisite descriptions of the structure of the nervous system are now possible by coupling classical staining methods with more modern techniques that link structure to the action and chemistry of individual neurons, or small ensembles of neurons. The neurons can be described in terms of their anatomical relationships with other neurons at both the light and electron microscopical level. This means that the processes of an individual neuron can be attributed to known tracts, commissures and areas of neuropil, allowing some restrictions to be set on inferences about its likely set of connections. The synapses that a neuron makes can also be described in terms of their spatial distribution, likely transmitters, and contacts with other known neurons. The use of antibodies raised against particular chemical components of the neurons, most frequently putative neurotransmitters and neuromodulators, can reveal the structure of individual neurons, the distribution of certain groups of somata, and patterns of neuronal projections, so giving a chemical mapping of the nervous system. Antibodies can also be used to reveal the distribution of different types of ion channels in the nervous system and even within different parts of an individual neuron. There seems every reason to expect that these methods will continue to be extended to provide information about the expression of particular genes in specific areas and at specific times. This combination of anatomical, physiological and chemical methods allows such good characterisation of the neurons that they can often be treated as identified individuals.

Classically, the nervous system of insects has been considered to consist of three parts (Snodgrass, 1935), although as more is learnt about the component neurons these distinctions become blurred.

The somatic nervous system consists of a ventral chain of bilaterally symmetrical segmental ganglia, including the brain, that are linked by paired bundles of axons called connectives running the length of the locust (Fig. 2.1). In each ganglion, the cell bodies (somata) of the neurons form a cortex with associated glial cells surrounding the central core of tracts, commissures and regions of neuropil formed chiefly by the fine branches of neurons and interspersed glial cells. Paired

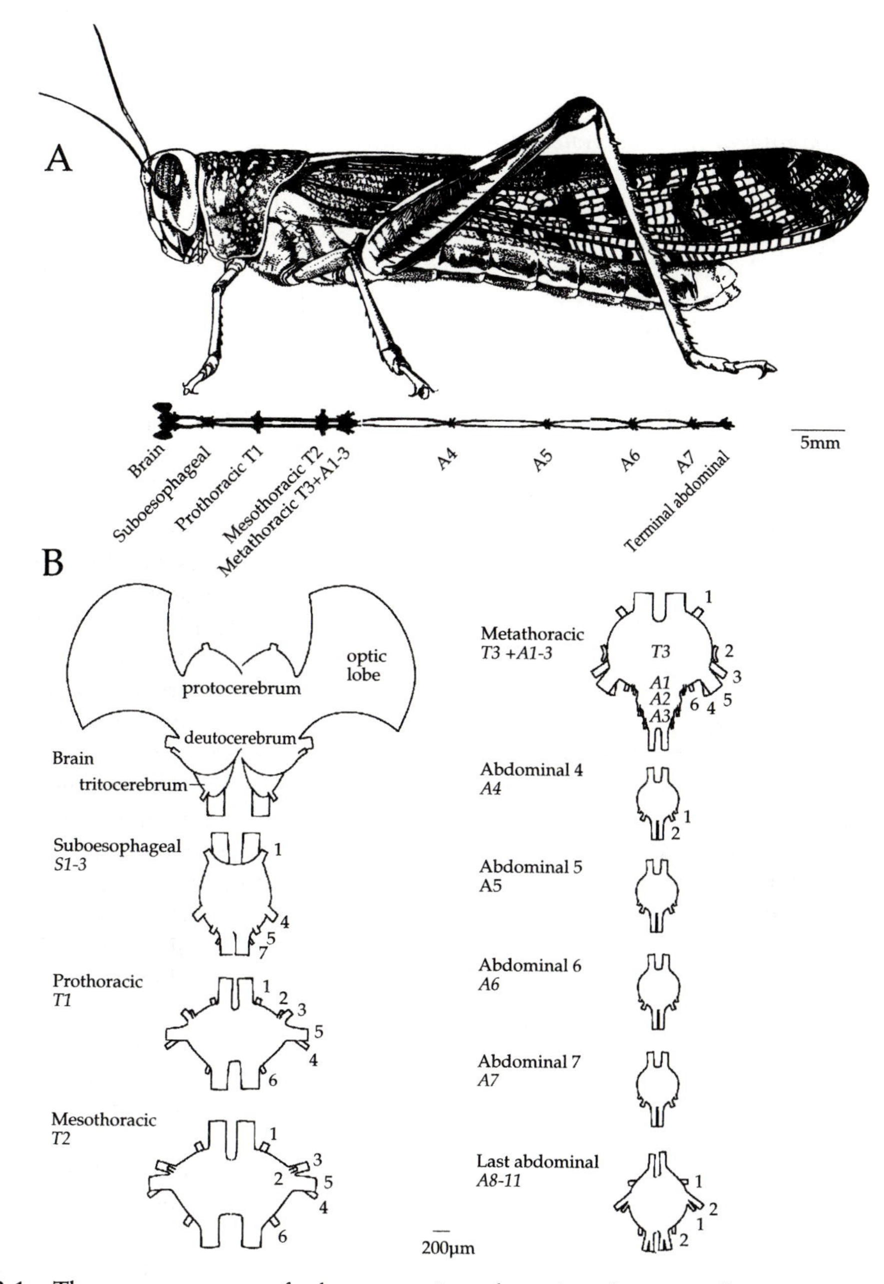

Fig. 2.1 The nervous system of a locust consists of a series of segmentally arranged ganglia linked by paired connectives. A. A drawing of *Schistocerca gregaria* and a diagram (at the same scale) of its nervous system. B. Drawings made from the ventral surface of the brain, suboesophageal, thoracic and abdominal ganglia. Each segment contributes a neuromere and this may form a separate ganglion, or several neuromeres may be fused. The neuromeres of each segment are numbered according to the following scheme: suboesophageal S1-3; thoracic T1-3; abdominal A1-11.

lateral nerves and median unpaired nerves contain the axons of sensory neurons projecting into the central nervous system and the axons of effector neurons to the periphery, although not every nerve contains both afferent and efferent axons.

The peripheral nervous system consists of the axons and terminals of the motor and other efferent neurons, and the axons, terminals and cell bodies of sensory neurons, most of which have their cell bodies in the periphery close to their receptor. Some peripheral nerves may have cell bodies of sensory and neurosecretory neurons along their surface and some of these may be aggregated into diffuse peripheral ganglia.

The visceral nervous system consists of a series of small ganglia distinct from, although linked to, the main segmental ganglia of the somatic nervous system. Close to the brain is the **retrocerebral complex**, which consists of the paired corpora cardiaca and corpora allata that have a neurosecretory function, and the unpaired hypocerebral ganglion. A series of small ganglia called either the **stomodeal** or **stomatogastric** ganglia innervate the anterior region of the gut (stomodeum), whereas the posterior portion of the gut is innervated by nerves from the terminal abdominal ganglion in the somatic chain. The median nerves from most segmental ganglia, their associated neurohaemal areas and nerves that link them in a chain are sometimes considered as a separate part of the visceral nervous system. Neurons are also associated with the gut, and the dorsal blood vessel or heart.

There is a great disparity of information on the different parts of the nervous system. Most is known about the thoracic ganglia, and the metathoracic ganglion in particular, reflecting the large amount of work that has been carried out on the control of the large pair of hind legs, on the control of the wings in flight and on the control of ventilation. A consequence is that although most of the neurons in the whole central nervous system occur in the brain, we only know how they are organised in certain parts and overall know few of them as individuals. The exceptions are the early stages in processing visual signals from the ocelli, and from the compound eyes in the lamina of the optic lobes, and odour signals in the deutocerebrum (antennal lobes). This means that there is still woefully little that can be said either about the organisation of the rest of the brain or about the processing that is performed there.

2.1. BLOOD–BRAIN BARRIER

Each ganglion is surrounded by a sheath, or neural lamella, consisting largely of connective tissue, under which is a thin specialised layer of glial cells called the perineurium. This arrangement provides an ionic barrier, which acts like a blood–brain barrier, so that the ionic concentrations of the fluid surrounding the membranes of the neurons are different from those in the haemolymph, which are high in K^+ and low in Na^+ (Treherne and Schofield, 1979) (see Fig. 3.2). The result is the maintenance of a constant internal milieu for the largely Na^+-mediated spikes and K^+-dependent resting potentials, in the face of considerable daily fluctuations in the ionic composition of the haemolymph.

The brain and ganglia are bathed in the haemolymph, but no vessels penetrate the sheath so that any substances in the haemolymph, including circulating hormones, must cross the barrier and then diffuse considerable distances before they can affect the neurons.

Gaseous exchange is effected by tracheae that push deep into all parts of the brain and segmental ganglia to form an extensive network of blind-ending tracheoles throughout. A thoracic ganglion, for example, is invested with a complex arrangement of air sacs and is supplied by two pairs of tracheae (see Fig. 12.1). Inside the ganglion, these tracheae branch profusely so that neuronal processes are, at most, only a few microns away from the tip of a tracheole where gaseous exchange can take place by diffusion.

2.2. BRAIN

The head of a locust is oriented so that the mouthparts point downwards (a hypognathous arrangement) and, as a consequence, the brain is at right angles to the long axis of the ventral nerve cord (Fig. 2.2A). Descriptions of the brain are confusingly given according to two axes; at first they were given according to the coordinates of the body (Williams, 1975), but more recently are given according to the neuraxis (Boyan *et al.*, 1993). The confusion arises because in early development the entire nervous system is a flat sheet with all parts oriented along the same axis as the body. The ventral surface of the brain then flexes dorsally so that it faces forward in the head of a hatchling and thereafter in the larval instars and adults. The ability to make comparisons between the organisation of the brain and the other ganglia in the locust, between the brains of different species that may be oriented differently, and between the central nervous system of embyros and adults, demands that a uniform set of coordinates be used. This can best be met by relating the planes of section to the neuraxis.

The brain develops from about 130 neuroblasts on each side, suggesting the involvement of neuromeres of 3–4 segments (*see Chapter 4*) (Zacharias *et al.*, 1993). The fusion of the neuromeres is so complete, however, that almost nothing can be recognised of the segmental origins in the adult pattern of tracts and commissures, even in embryos as early as 30% development. In *Drosophila*, however, molecular markers indicate the presence of four pregnathal segments (**labral, ocular, antennal** and **intercalary**), and three gnathal (**mandibular, maxillary** and **labial**) segments in the head (Schmidt-Ott and Technau, 1992) that are each assumed to contribute a neuromere to the central nervous system. The neuroblasts of the deutocerebrum suggest that this part of the brain is the neuromere of the antennal segment, and this is in keeping with the processing it performs. It is unclear how the neuroblasts of the protocerebral structures can be related to the labral and ocular segments. The neuroblasts of the intercalary segment seem to become divided between the brain, where they form the tritocerebrum, and the suboesophageal ganglion, where they form the anterior part of the mandibular neuromere, with others migrating further to

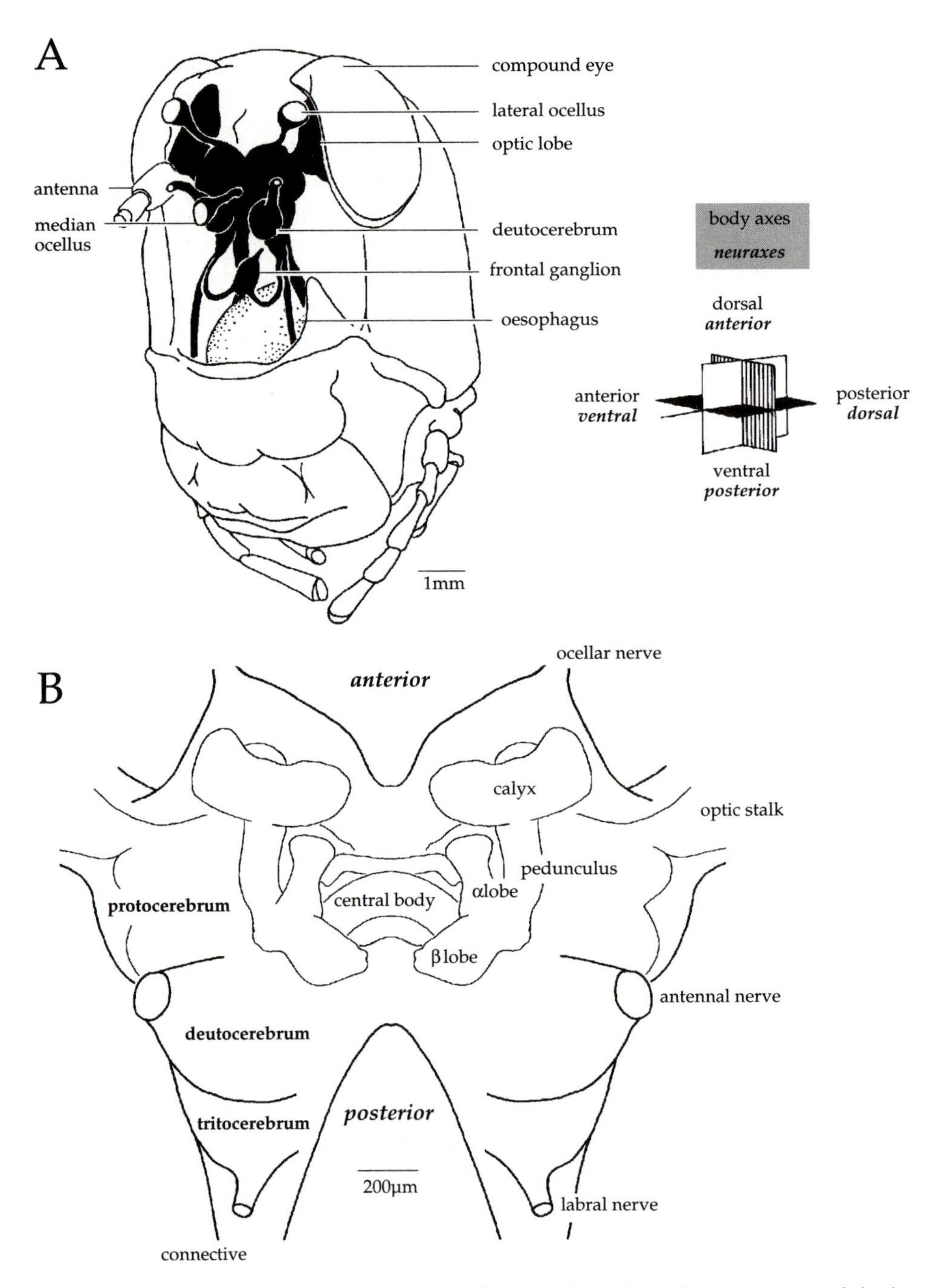

Fig. 2.2 The brain of a locust. A. Drawing of the head to show the orientation of the brain. Based on Williams (1975). B. A drawing of the brain from the ventral surface to show the major regions. The optic lobes have been omitted.

form the mandibular ganglion (Doe and Goodman, 1985) (*see section 2.4*). As the gut enlarges during development, the neuromeres of the three gnathal segments become separated from the brain and form the suboesophageal ganglion.

The brain of a locust has a volume of almost 6mm^3, to which the optic lobes contribute the most. It is bilaterally symmetrical and consists of three main regions, the **protocerebrum, deutocerebrum** and **tritocerebrum** (Fig. 2.2B), with the optic lobes forming prominent anterior and lateral lobes. Each of these divisions consists of a cortex of cell bodies surrounding a core of tracts, commissures and neuropils (Fig. 2.3). Some of the neuropils are distinctive, have a glomerular arrangement and are named, but the overall complex organisation is relatively poorly understood compared with, for example, the brain of a fly (Strausfeld, 1976). Some idea of the complexity is indicated by the description of 73 named commissures in the central part of the brain (Boyan *et al.*, 1993), compared with 10 for a thoracic ganglion (*see section 2.5.1*). Calculating the numbers of neurons in the brain is difficult and estimates consequently vary widely. Common suggestions that it contains some 360 000 neurons may, for example, underestimate the contribution of the small neurons in the mushroom bodies and optic lobes. Nevertheless, even the most conservative estimates suggest that it contains about 20 times the total number of neurons in the rest of the nervous system.

2.2.1. Protocerebrum

This is the main part of the brain with several distinctive areas of neuropil and the following nerves.

The **paired lateral ocellar nerves** containing the axons of photoreceptors in the two lateral ocelli.

The **median ocellar nerve** containing the axons of the photoreceptors in the median ocellus.

Two small paired nerves linking with the corpora cardiaca (*see section 2.2.4*) called **nervus corporis cardiaci I,II (NCCI,II)**.

2.2.1.1. *Mushroom bodies*

The mushroom bodies are prominent, bilateral, essentially neuropilar regions in the anterior of the brain, with a shape that, at least in some insects, reflects their name. They are involved in the processing of olfactory signals and are implicated in olfactory learning. There are two major regions, the cap-shaped calyx and the stalk-shaped peduncle, that contain many tens of thousands of small, intrinsic neurons. These neurons are eponymously named Kenyon cells, but more rarely they are also called intrinsic neurons or globuli cells. The Kenyon cells are thought to consist of at least three morphological types, but virtually nothing is known of the physiological distinctions between them. For the locust, the best estimate of the number of these neurons is 50 000, which is similar to that in a fly. By contrast, a drone bee has some 1.2 million neurons in its mushroom bodies, of which 300 000 are Kenyon cells.

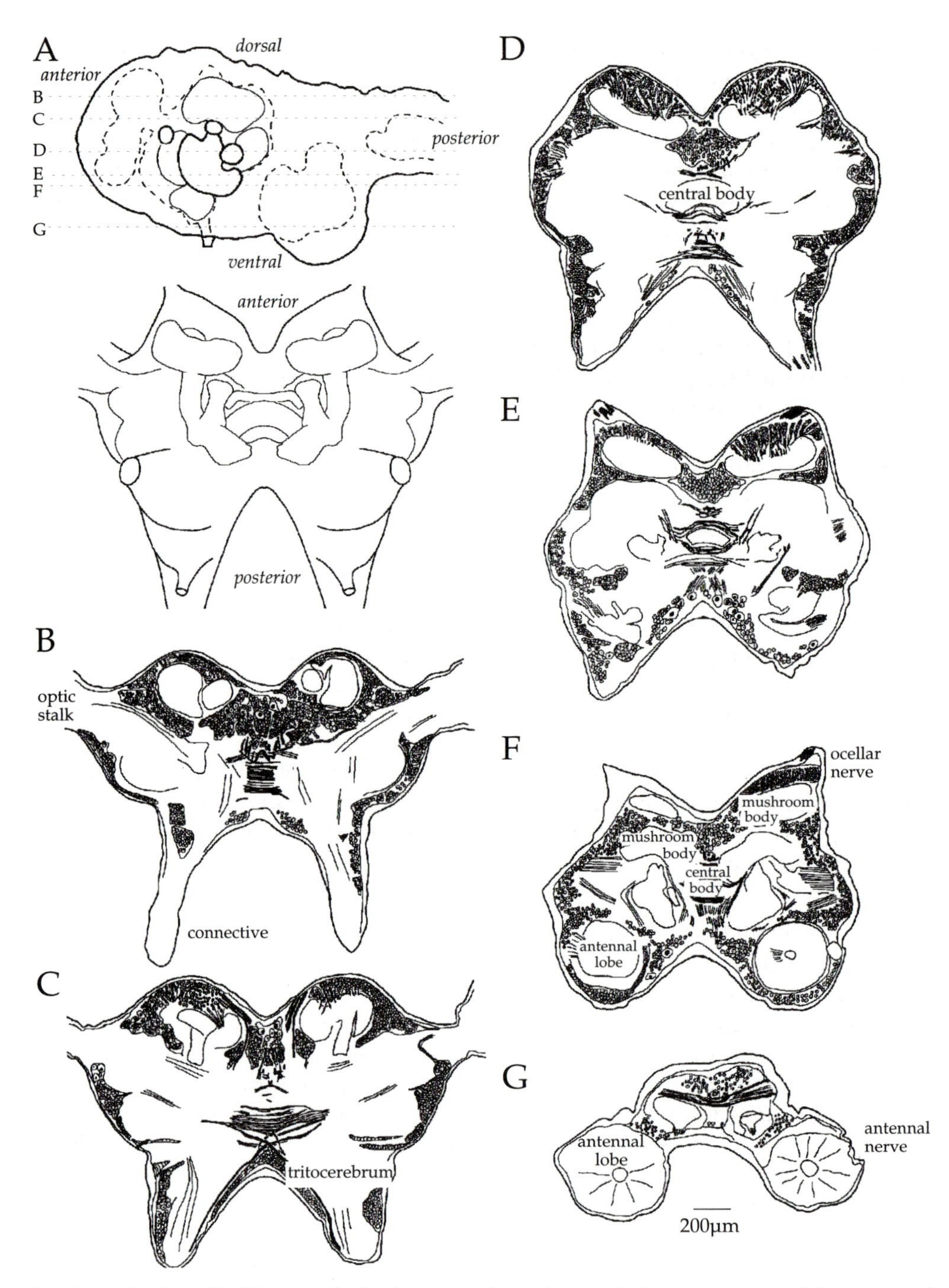

Fig. 2.3 Sections (B–G) through the brain to show the cortical arrangement of the neuronal cell bodies, the major neuropil areas, and the commissures. The orientation of the sections is shown in the diagrams in A. Based on Boyan *et al.* (1993).

Calyx The **calyx** in the locust consists of a concave primary calyx and a bulbous accessory calyx (Weiss, 1981).

The primary calyx is bilayered, with two anterior arms and a special central zone. The layer (zona interna) closest to the anterior cell bodies of the Kenyon cells consists of the processes of these neurons and hence appears fibrous. The other layer (zona externa) closest to the pedunculus and furthest from the cell bodies, is a neuropil region. The central zone (or central ring) is a distinctive neuropil surrounding the columns of the pedunculus, through which pass many of the processes of neurons from the zona interna on their way to the pedunculus.

The accessory calyx is an unlayered, uniform neuropil surrounded by the cell bodies of the Kenyon cells that send their processes into it at particular points, before projecting toward the junction with the pedunculus where they coalesce into one of the peduncular columns. It also has two anterior extensions beneath the primary calyx.

Inputs to both regions come from the TOG (tractus olfactorio globularis; *see section 2.2.2*) that originates in the deutocerebrum, connects with the calyx near the pedunculus and continues on to the lateral protocerebral neuropil and the tritocerebral tract.

Pedunculus The pedunculus consists of one barrel with three major columns of fibres, two from the primary and one from the accessory calyx. It forks at its lower end into an anteriorly directed α lobe and posteriorly directed β lobe.

This arrangement is thought to have originated from an ancestral pattern that had two equivalent bilayered calyces and a double-barrelled pedunculus. The primary calyx of modern locusts may have been formed by the coalescence of the two calyces, and the pedunculus from the fusion of the two barrels. The accessory calyx may have originated from intrinsic neurons at the base of the primary calyx, perhaps in response to a larger input from the tritocerebral tract.

2.2.1.2. *Central body complex*

This central region of the brain consists of four parts: the protocerebral bridge, the upper and lower (which may be homologous to the ellipsoid body of other insects) divisions of the central body, and paired noduli that are interconnected by small field interneurons and with the surrounding protocerebrum by large field interneurons. The upper division of the central body is a layered structure consisting of an anterior lip and layers I-III (Homberg, 1991). The anterior lip and part of layer I are linked to the lateral accessory lobes of the protocerebrum by a small finger of neuropil. The central body is a region where fibres pass from one hemisphere of the brain to the other, but despite the importance that this implies, there is little information on the types of neuron present or the sort of processing that occurs. The central body is implicated in visual processing because it is linked by pathways from the compound eyes and ocelli. It receives a large input from neurons that show serotonin-like and octopamine-like immunoreactivity which may indicate that its processing is highly modifiable (*see Chapter 5*).

2.2.1.3. *Lateral accessory lobes*

These are paired regions of neuropil, lateral and posterior to the central body and in front of the antennal lobes. They appear to consist of several subregions because individual neurons that project into them do not occupy all the area of neuropil. Local interneurons link an accessory lobe to other regions of the brain and projection interneurons with axons projecting to the thoracic ganglia also have branches here. Many of these neurons spike during flight, but they also spike in response to mechanosensory stimulation of particular parts of the body and to visual stimuli (Homberg, 1994).

2.2.1.4. *Optic lobes*

The optic lobes process the visual signals from the compound eyes in a series of neuropil layers; lamina, medulla, accessory medulla and lobula, in which the neurons are further arranged in a retinotopic array of columns. The photoreceptors themselves act as rather generalist receptors, signalling light intensity and wavelength of the light so that the complex features and quality of the visual world have to be extracted by the successive layers of neurons in the optic lobes. A single compound eye consists of about 3000 ommatidia, each of which contains eight receptor (retinula) cells.

The columnar organisation of the neuropils is such that, in the lamina and medulla, there are the same number of columns of neurons as there are optical units or ommatidia. In the lamina, the columns are where the retinula cells make their first synaptic connections. They are segregated into cartridges that are tightly bounded by glial cells and consist of repeating sets of neurons; the input neurons are the retinula cells and the output cells with axons to the medulla are the lamina monopolar cells. Tangential cells run between different cartridges and feedback neurons send signals back from the medulla. In the medulla itself, the number of neurons contributing to the repeating elements of the columns is now much greater, but little is known of the processing performed by these neurons. In the lobula, the number of columns is reduced and large interneurons provide outputs to the main part of the brain. Between the lamina and medulla the optic chiasma reverses the visual world, but the orientation is then restored by neurons in the lobula.

2.2.2. Deutocerebrum (antennal lobes)

The posterior part of the brain is indented where the gut protrudes between the connectives, so that the deutocerebrum is separated ventrally into left and right parts and from the tritocerebrum dorsally. The commissures linking these regions are thus displaced anteriorly.

Each part of the deutocerebrum consists of an antennal lobe that has a dome-shaped anterior (ventral) protrusion on either side of the oesophagus and a smaller dorsal lobe, with the antennal nerves entering laterally. Cortically arranged cell bodies of neurons are clustered into particular groups.

There are three nerves from each half of the deutocerebrum (Figs 2.2B and 2.4).

The **antennal nerve** contains the axons of the many sensory receptors on the antennae and the axons of motor neurons innervating the antennal muscles.

The **dorsal tegumentary nerve** contains the axons of sensory neurons from the apical part of the head capsule.

The **ventral tegumentary nerve** contains the axons of sensory neurons from the lateral part of the head capsule.

An antennal lobe is the main processing region for the primary olfactory signals conveyed to it from an antenna by many thousands of small diameter axons in the antennal nerve. The olfactory receptors in cockroaches and moths can be specialised for a particular odour, such as the sex pheromones released by females, or may respond best only to a limited range of odours. The receptors themselves thus extract particular features of a stimulus and it is assumed that those in a locust operate in a similar way. The dorsal lobe, often called the mechanosensory and motor centre, receives inputs from antennal mechanoreceptors generally concentrated at the base of an antenna and contains the motor neurons to muscles moving an antenna. This

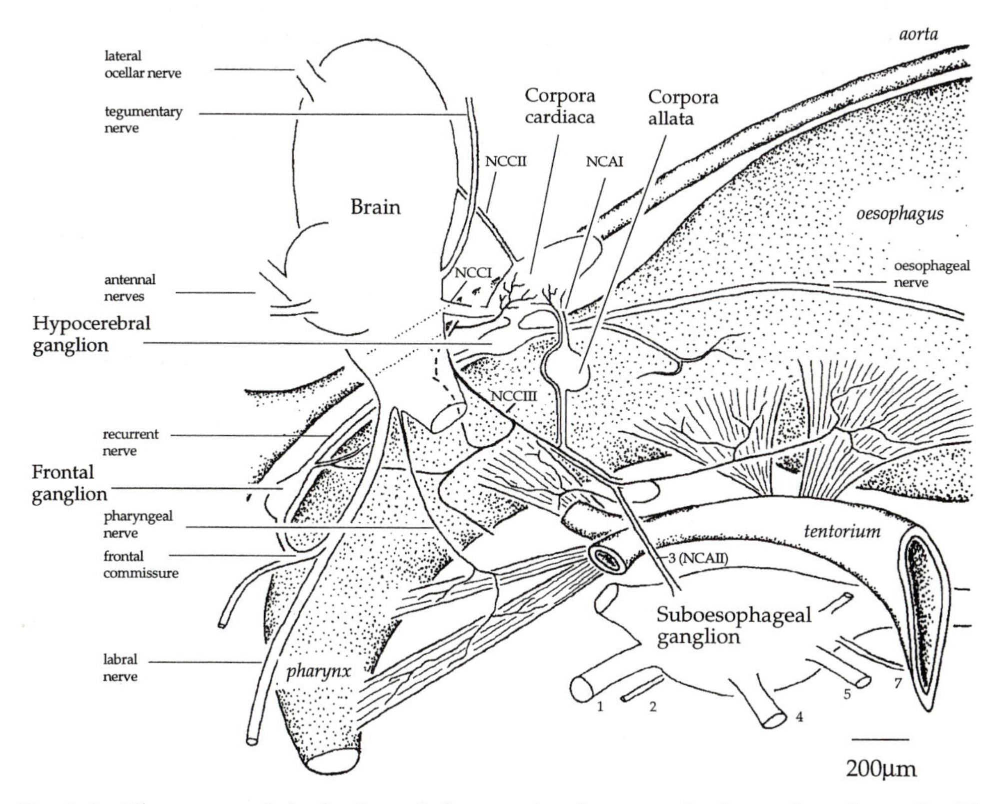

Fig. 2.4 The nerves of the brain and the associated retrocerebral complex of ganglia. The drawing is made from a side view and shows the aorta and gut passing between the brain and the suboesophageal ganglion. The muscles linking the anterior part of the gut to an internal strut of the exoskeleton (the tentorium) are also shown. Based on Bräunig (1990a).

organisation suggests that much of the processing of antennal mechanosensory and olfactory signals is performed separately, but there are many neurons that respond to both modalities of stimuli.

Each antennal lobe receives inputs from perhaps 100 000–200 000 olfactory neurons and projections from centrifugal neurons of the protocerebrum. It contains fewer than 750 local neurons and only a few hundred output neurons. The axons of the output neurons project to the mushroom bodies and other regions of the protocerebrum in a number of tracts, the most prominent of which is the tractus olfactorio globularis (TOG), which contains about 1200 axons. Some 300 axons from the tritocerebrum may also run in this tract.

The structure of an antennal lobe is characterised by a series of spherical regions of neuropil called glomeruli, the number of which varies in different species of insects from 10–200. Such spherical glomeruli are also characteristic of brain regions that process olfactory signals in vertebrates and crustacea. These glomeruli are islands of synaptic neuropil, largely isolated from each other by glia but linked by the processes of interneurons. They consist of an inner core of fine neuronal branches surrounded by the terminals of the sensory neurons and are the sites of most synaptic interactions and of the convergence of the many input neurons onto the few output neurons in this region of the brain. It is not known how an antenna is mapped onto the antennal lobe, although many suggestions are made that particular glomeruli process related odours, rather than particular spatial arrays of receptors on the different annuli of an antenna.

In moths and cockroaches, the structure of the deutocerebrum is sexually dimorphic; in the male, it receives an influx of axons from the many thousands of additional receptors specialised for the detection of female pheromones. The signals in these neurons are processed in an enlarged glomerulus, called the macroglomerulus or macroglomerular complex, and the integrated output is conveyed to the protocerebrum by a small number of specific projection neurons. Neither of these two specialisations are present in a female brain.

2.2.3. Tritocerebrum

There are few insights into the structure of the tritocerebrum. It consists of two symmetrical lobes bounded anteriorly by the deutocerebrum and posteriorly tapering into the circumoesophageal connectives. Each lobe consists of a cortex of cell bodies surrounding longitudinal tracts, commissures and areas of neuropil. It has connections with the mushroom bodies and with the optic lobes, and provides a link with the stomatogastric nervous system and the retrocerebral complex.

There are four pairs of nerves (Figs 2.2B and 2.4).

The **frontal connectives** (nerves) arise close to the connectives and link to the unpaired frontal ganglion (*see section 2.3.1*). Branches of this nerve also innervate muscles of the foregut.

The **labral nerves** contain sensory and motor neurons of the labrum (mouthpart). They fuse with the frontal connectives close to their entry into the tritocerebrum.

The **pharyngeal nerves** are much thinner and supply muscles of the foregut.

Nervus corporis cardiaci III (NCCIII) contains the axons of neurosecretory cells in the tritocerebrum that run to the corpora cardiaca (*see section 2.2.4*). Other neurosecretory neurons with cell bodies in the tritocerebrum have axons in NCCI.

The tritocerebral commissure links the two circumoesophageal connectives together a short distance posterior to the tritocerebrum. It contains only a small number of axons, including those of two identified projection interneurons (*see Chapter 11*).

The tritocerebrum contains motor neurons that innervate muscles in the labrum and pharyngeal dilator muscles (Aubele and Klemm, 1977) and receives projections from sensory neurons associated with the mouthparts (moths: Kent and Hildebrand, 1987). It also contains the cell bodies of a number of well-characterised intersegmental interneurons with axons that project to the thorax (*see Chapter 11*). Close to these cell bodies are neurons which contain octopamine and have elaborate projections in other regions of the brain (*see Chapter 5*). This paucity of information on the tritocerebrum, first pointed out almost 20 years ago, clearly still needs to be rectified.

2.2.4. Corpora cardiaca

The corpora cardiaca are paired secretory structures situated just behind the brain and in front of the oesophagus (Fig. 2.4) that consist of storage lobes and glandular (secretory) lobes. The storage lobes contain the terminals of neurosecretory cells originating in the brain. The glandular lobes consist of intrinsic neurosecretory cells that are unipolar and with long thin processes, and that synthesise, store and release into the haemolymph a number of peptides, most notably the adipokinetic hormones (AKH) and a peptide that has yet to be characterised that controls colouration of the body. The corpora cardiaca are innervated directly by two nerves from the brain (Mason, 1973) and indirectly by a third (Bräunig, 1990a). These nerves (nervus corporis cardiaci) are not numbered in an obvious anatomical sequence (Fig. 2.4), but instead the names derive from the order in which they were discovered.

NCCI runs directly to the corpora cardiaca and contains axons from about 400 cell bodies at the contralateral side of the midline of the pars intercerebralis of the protocerebrum, and from a group of about 12 cell bodies in the tritocerebrum. They probably convey peptidergic material to the storage lobe for release into the haemolymph (Konings *et al.*, 1989a), while some of these neurons may be dopaminergic and may modulate rather than effect release from the glandular lobe. Some axons in this nerve also project to the hypocerebral ganglion.

NCCII also runs directly to the corpora cardiaca and contains axons from about 30 cell bodies in the lateral protocerebrum (pars lateralis) that regulate the release of AKH hormones from the glandular lobes, and some axons to the storage lobes. In cockroaches, some of these neurons contain the peptide corazonin, which may be involved in controlling the heart (Veenstra and Davis, 1993). The nerve also contains the axons originating from about 15 cell bodies in the pars lateralis near the calyx that project through the corpora cardiaca and into the nervus corporis allati I (NCAI) to the corpora allata.

NCCIII takes a circuitous route to the corpora cardiaca, the posterior branch first anastomosing with NCAII (nervus corporis allati, sometimes called N3) of the suboesophageal ganglion to reach the corpora allata, and then with NCAI to reach the corpora cardiaca. Another branch loops dorsally to reach the corpora cardiaca more directly and to innervate the aorta. It also innervates dilator muscles of the pharynx and oesophagus, the auxiliary hearts at the base of the antennae and has a connection to the frontal ganglion (*see section 2.3*). The whole nerve contains axons of neurons with cell bodies in the suboesophageal ganglion.

The corpora cardiaca are also linked by the paired nerves NCAI with the corpora allata and by two paired nervi stomatogastrici (NCSI and II) to the hypocerebral ganglion. They also receive inputs from neurons with cell bodies in the suboesophageal ganglion and axons in NCAII which bypass the corpora allata (Bräunig unpublished). These neurons have many of the characteristics of neurosecretory neurons and apparently contain a number of peptides.

2.2.5. Corpora allata

The corpora allata are paired neurosecretory structures behind the brain that are each linked to the corpora cardiaca by NCAI and to the suboesophageal ganglion by NCAII (also called suboesophageal N3) (Fig. 2.4). In the pars lateralis of the brain is a group of about 13 cell bodies that have axons projecting in NCCII to the corpora cardiaca and then by NCAI to the ipsilateral corpora allata, and a further two that have axons to both corpora allata (Virant-Doberlet *et al.*, 1994). Each half of the corpora allata is thus innervated by 17 neurons, 13 with ipsilateral cell bodies and axons, and four with bilateral axons. In the cockroach, some of these neurons show allatostatin-like immunoreactivity (Stay *et al.*, 1992), and the two neurons with bilateral projections show some resemblance to the similarly placed neurons in moths that contain prothoracicotropic hormone (PTTH) (*see Chapter 5*). This hormone controls the level of ecdysone production by the prothoracic gland in larvae, and probably has additional roles in adults. Whether the corpora allata are a target for these neurons or simply a release site is unresolved.

The corpora allata secrete juvenile hormone (JH3) which plays an important role in development and in maturation (*see Chapter 4*), under the control of the neurons from the brain and various neuropeptides such as allatostatins. Electrical stimulation of NCAI inhibits the production of juvenile hormone in adult females, while section of the nerve leads to an increase in its production (Horseman *et al.*, 1994).

2.3. STOMATOGASTRIC GANGLIA

The stomatogastric or stomodeal ganglia are a series of small aggregations of neurons that are closely associated with the brain, the corpora cardiaca and allata, and the anterior portion of the gut. Virtually nothing is known of their physiological actions, in stark contrast to their counterparts in crustacea. More detailed descriptions of the gross anatomy of these ganglia and their nerves are available for crickets (Kirby *et al.*, 1984).

2.3.1. Frontal ganglion

The frontal ganglion is an unpaired structure on the dorsal surface of the pharynx anterior to the brain (Fig. 2.4). It is linked to the anterior surface of the tritocerebrum by paired frontal connectives that fuse with the labral nerves (Willey, 1961). A small recurrent nerve (nervus recurrens) passes backwards along the dorsal surface of the pharynx to link it with the hypocerebral ganglion just behind the brain. Two pairs of nerves arise laterally to innervate the gut. The ganglion contains the cell bodies of about 100 neurons in locusts, and in cockroaches many have axons in the nervus recurrens and in the frontal connectives (Gundel and Penzlin, 1978), but only a few have axons in the nervus connectivus (Jagota and Habibulla, 1992). What does the frontal ganglion do? Most of the suggestions for its function have come from experiments where nerves of the stomatogastric ganglia were cut, or whole ganglia removed, with the usual attendant difficulties in the interpretation of any effects resulting from such procedures (Clarke and Langley, 1963). Removal of the frontal ganglion results in cessation of growth and failure to moult, with the body mass remaining constant despite continued feeding. This suggests an interruption in the pathway linking sensory information from the gut, or from any systems monitoring levels of substances in the haemolymph, to the neurosecretory neurons of the brain or corpora cardiaca. In cockroaches, the levels of serotonin (5-HT) in the corpora cardiaca and allata show a circadian rhythm with a maximum at midnight and a minimum at noon; this rhythm is disrupted when the frontal ganglion is destroyed (Jagota and Habibulla, 1992). The frontal ganglion itself also contains 5-HT and some of its neurons show a circadian rhythm in their spike activity.

2.3.2. Hypocerebral (occipital) ganglion

This is a small unpaired ganglion lying behind the brain, dorsal to the pharynx and ventral to the corpora cardiaca (Fig. 2.4). It is linked anteriorly by the nervus recurrens to the frontal ganglion, dorsally to the corpora cardiaca and posteriorly forms the paired oesophageal nerves. It gives rise to a pair of nerves that run to the crop. Its neurons are undescribed in the locust.

2.3.3. Ingluvial (ventricular or paraventricular) ganglia

The paired oesophageal nerves that arise posteriorly from the hypocerebral ganglion run along the lateral surface of the oesophagus and then broaden to form the paired ingluvial ganglia on either side of the crop. In cockroaches, each ingluvial ganglion contains about 60 neurons with small cell bodies, some of which may contain biogenic amines (Aloe and Levi-Montalcini, 1972). The nerves from these ganglia innervate the anterior parts of the gut and it has to be assumed that they are involved in controlling movements of the foregut.

2.4. SUBOESOPHAGEAL GANGLION

The suboesophageal ganglion consists of the fused mandibular, maxillary and labial neuromeres, with some neurons at the anterior derived from the intercalary neuromere. The internal structure of the three main neuromeres of the ganglion in terms of tracts, commissures and neuropils is very similar to that of a thoracic ganglion (*see section 2.5.1*), presumably reflecting the homologies in the structure of the mouthparts and the legs (Snodgrass, 1928). It contains some 5000 neurons that are involved in controlling movements of the mouthparts associated with the particular segments (mandibles, maxillae and labium) and processing sensory signals from them. It is also involved in controlling the action of the neck muscles, secretions of the salivary glands and the corpora cardiaca (Altman and Kien, 1987). The suboesophageal ganglion also contains many interneurons with axons projecting to the thoracic ganglia and others projecting to the brain that are involved in processing auditory signals (Boyan and Altman, 1985) and in controlling, or at least influencing, the motor patterns involved in walking, flying and breathing (*see Chapters 8, 11 and 12*) and stridulation in some grasshoppers. Its action therefore extends more widely than the segments from which it is derived and for which it provides the local innervation. There are eight pairs of lateral nerves and one posterior median nerve.

N1 of the mandibular, **N4** of the maxillary and **N5** of the labial neuromere all innervate the mouthparts. N4, for example, contains the axons of about 50 motor neurons.

N2 the hypopharyngeal nerve, arises ventrally from the mandibular neuromere.

N3 (NCAII) arises dorsally and runs to the corpora allata, with some axons projecting to the corpora cardiaca.

N6 and N7 from the labial neuromere innervate the neck and salivary glands. N7 contains the axons of two neurons (salivary neurons 1 and 2, SN1,2) with large cell bodies in the labial and mandibular neuromeres, respectively, which innervate the salivary glands (*see Chapter 5*).

N8 runs posteriorly close to a posterior connective and innervates muscles in the neck.

2.5. THORACIC GANGLIA

There are three thoracic ganglia and of all the parts of the nervous system these have been the most intensively studied, both anatomically and physiologically. Consequently, much is known about the organisation of these ganglia and of the structure and action of a large number of their component neurons. Given the similar way that the ganglia in all segments are formed embryologically (*see Chapter 4*), there seems good reason to believe that the principles underlying their design and operation will be broadly applicable to the rest of the central nervous system.

The prothoracic and mesothoracic ganglia (Thoracic = T, hence, T1 and T2) contain only the neuromeres of their own segment, but the metathoracic ganglion contains neurons belonging to four neuromeres; the metathoracic segment itself (T3) and the first three abdominal neuromeres (Abdomen = A, hence, A1-3). The fusion of the neuromeres of these segments begins at 45–50% development (*see Chapter 4*) for definition of embryonic stages) with the fusion of A1 to T3, followed at 65% with the fusion of A2, and finally at 70% with the fusion of A3. The fused ganglionic mass then expands in volume, as does the rest of the central ganglia, due to the enlargement of existing neurons and the addition of further ones, countered by the loss of others by cell death. Each thoracic ganglion contains the cell bodies of some 2000 neurons. These cell bodies belong to the motor neurons of the muscles in a particular segment, to local interneurons, to intersegmental interneurons, to a few sensory neurons and to neuromodulatory interneurons (*see Chapter 3*). The complement of neurons in each ganglion is sufficient to organise many of the movements of the legs (*see Chapters 7 and 8*) or wings (*see Chapter 11*) of that segment, but of course these movements are coordinated with those of the rest of the body by signals in neurons with axons in the linking connectives. Each of these connectives in the thorax contains about 4000 axons, of which only about 3% have a diameter greater than 5μm (Rowell and Dorey, 1967). These axons belong mostly to interneurons with cell bodies in various parts of the nervous system, to a few sensory neurons from thoracic receptors, and to neuromodulatory neurons which release different substances that alter the processing within the ganglia or the performance of effectors in the periphery (*see Chapters 5 and 6*).

2.5.1. Organisation of a thoracic ganglion

Each thoracic ganglion has the same basic structure, and the principles of organisation also extend to the suboesophageal ganglion and to the abdominal ganglia. The thoracic ganglia are known in the greatest detail, due largely to the high standards set first by the descriptions of cockroach ganglia (Gregory, 1974, 1984) and then locust ganglia (Tyrer and Gregory, 1982; Pflüger *et al.*, 1988). Comparable descriptions have also been made for ganglia in stick insects (Kittmann *et al.*, 1991), a moth (Suder and Wendler, 1993), a cricket (Wohlers and Huber, 1985), a bee (Rehder, 1989) and *Drosophila* (Power, 1948) and show that the organisation may be a common feature of insects. Many similarities also extend to the organisation of ganglia in crustacea (Skinner, 1985a,b), suggesting a general ground plan for arthropods.

Each thoracic ganglion is bilaterally symmetrical and consists of a cortex of some 2000 cell bodies surrounding a series of longitudinal tracts, transverse commissures, smaller vertical tracts, aggregations of axons into bundles that form the nerves, and interspersed areas of neuropil (Tables 2.1 and 2.2 and Figs 2.5–2.7). Most of the cell bodies of the neurons are in a ventral cortical layer, but some wrap around the sides towards the dorsal surface and others are in prominent groups at the dorsal midline.

Into this elaborate framework can be woven the shapes of individual neurons; their cell bodies can be ascribed to particular groups (Siegler and Pousman, 1990a,b) and their positions mapped, their axons to particular tracts, their main branches to particular commissures and tracts, and their fine branches to particular areas of neuropil. This detail is essential for an adequate description of the morphology of a neuron and for comparison of its shape with what may be suspected to be the same neuron in another animal, or with other neurons in the same animal. These anatomical descriptions, while of enormous value in themselves, can also be of predictive value in indicating the sort of processing that a neuron might perform and the sorts of connections that it might make. Such predictions can only suggest connections, or exclude the possibility of them, because the close apposition of so many fine branches in the neuropils makes it virtually impossible to define connectivity at the level of the light microscope. This can be achieved at the level of the electron microscope for individual neurons that have been labelled with electron-opaque markers, and is further aided by the detailed anatomy of the ganglion, which allows the observed synapses to be ascribed to particular known areas of neuropil.

2.5.1.1. *Nerves*

Each ganglion has six pairs of lateral nerves and a median nerve, and while the targets that each innervate are much the same in each segment, there are some differences because only the meso- and metathoracic segments have wings; the prothoracic ganglion has an anterior and a posterior median nerve; N6 fuses with N1 of the next anterior segment, and the metathoracic segment contains sensory

Table 2.1. Planes of sections used to describe the structure of segmental ganglia

Plane of section	Definition
Transverse	at right angles to the longitudinal axis of the nerve cord.
Longitudinal	parallel to the longitudinal axis of the nerve cord in the vertical plane. When the section is directly through the midline it is called **sagittal** and when off the midline but still in the same plane it is called **parasagittal**.
Horizontal	parallel to the longitudinal axis of the nerve cord in the horizontal plane.

Table 2.2. Organisational features of a thoracic ganglion

Feature	Abbreviation
Longitudinal tracts	
Dorsal intermediate tract	DIT
Dorsal median tract	DMT
Median dorsal tract	MDT
Median ventral tract	MVT
Lateral dorsal tract	LDT
Lateral ventral tract	LVT
Ventral intermediate tract	VIT
Ventral lateral tract	VLT
Ventral median tract	VMT
Commissures	
Dorsal commissures	DCI–DCVI
Ventral commissures	VCI, VCII
Supramedian commissure	SMC
Posterior ventral commissure	PVC
Neuropils	
Ventral neuropils	
Ventral association centre	VAC
anterior	aVAC
lateral	lVAC
medial	mVAC (aRT)
ventralmost	vVAC
Dorsal neuropils	
Lateral association centre	LAC
anterior	aLAC
posterior	pLAC
Vertical and oblique tracts	
C-Tract	CT
I-Tract	IT
Oblique tract	OT
Perpendicular tract	PT
Ring tract	RT
T-tract	TT
Deep DUM tract	DDT
Superficial DUM tract	SDT

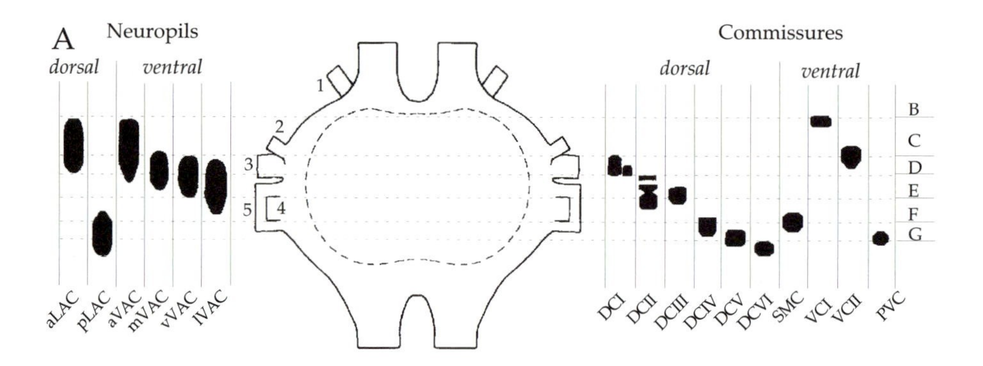

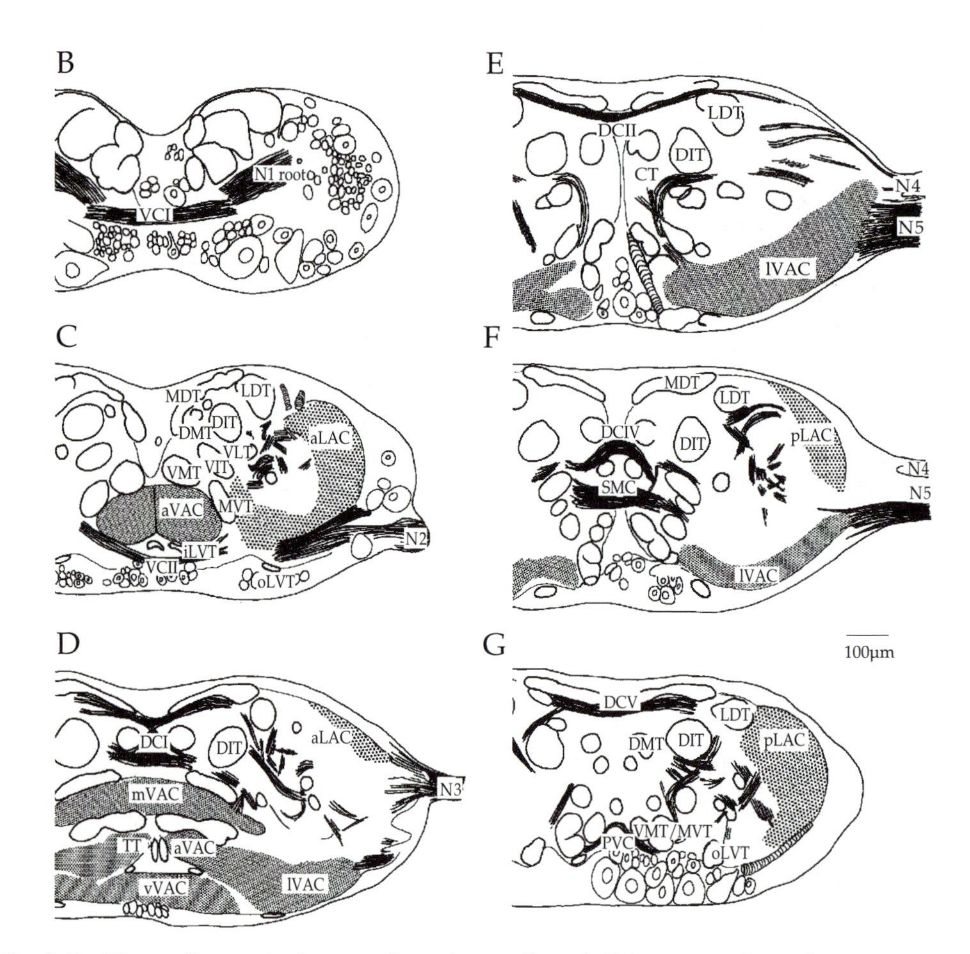

Fig. 2.5 Neuropil areas in the mesothoracic ganglion. A. Diagram to show the area occupied by a number of named areas of neuropil and their position relative to the dorsal and ventral commissures. B-G. Drawings of transverse sections at the planes indicated in (A) show the areas of neuropil (shaded) relative to the major longitudinal tracts and commissures. See Table 2.2 for a list of the abbreviations used. Based on Pflüger *et al.* (1988).

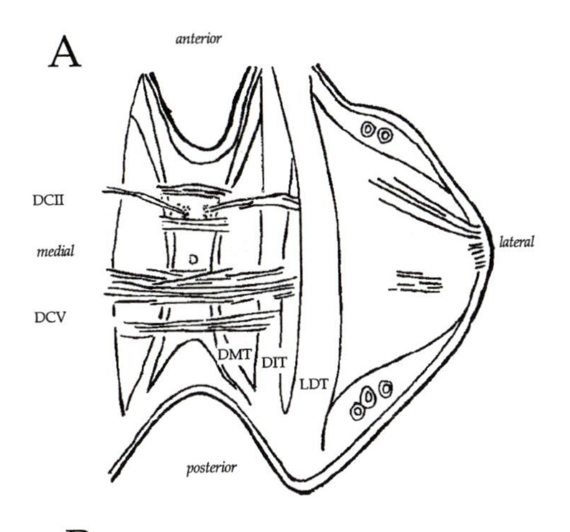

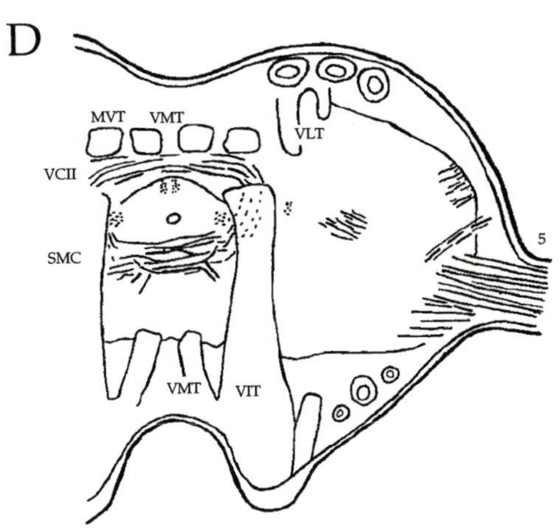

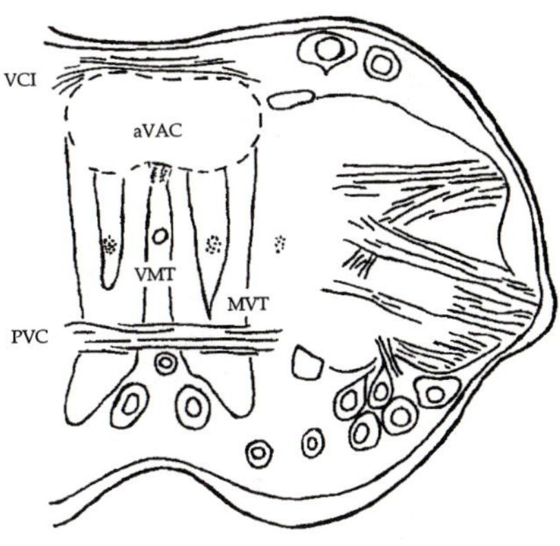

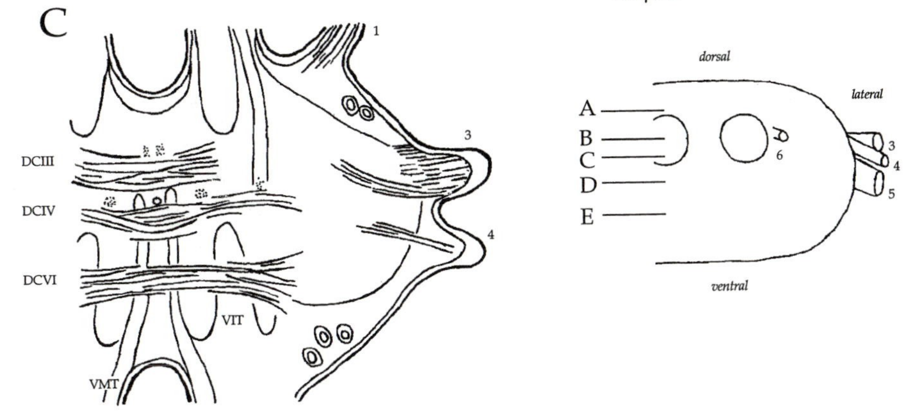

Fig. 2.6 The major longitudinal tracts and commissures of the mesothoracic ganglion as seen in drawings of a series of horizontal sections. Based on Tyrer and Gregory (1982).

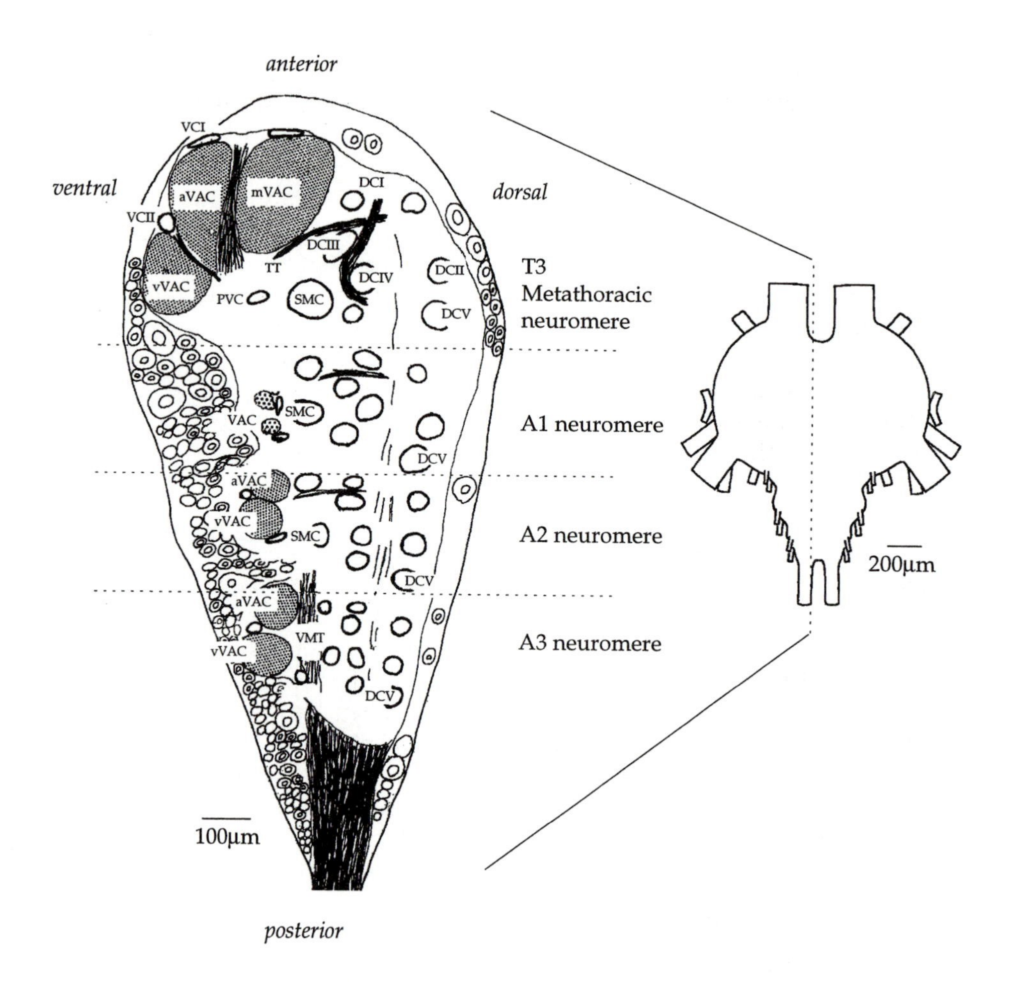

Fig. 2.7 Longitudinal section through the metathoracic ganglion to show the fusion of the neuromeres of segments of T3, A1, A2 and A3. The repeating structural organisation of each neuromere is exemplified by the aVAC neuropils and commissures DCV. Based on Pflüger *et al.* (1988).

neurons from the tympanum (ear). All nerves except N2, which is purely sensory, contain both afferent and efferent axons. The lateral nerves in all segmental ganglia are numbered from anterior to posterior rather than being named. Their first order branches are lettered A, B, etc., second order branches are numbered 1,2, etc., third order branches are given lower case letters a,b, etc., and fourth order branches are numbered; hence, for example, **N5B1b2**.

Mesothoracic nerves In the mesothoracic segment, the nerves innervate the following muscles and sensory receptors (for further details of the constituent motor axons and sensory neurons see Tables 3.1, 8.3 and 11.1).

N1 contains sensory neurons from the wing and wing hinge, and motor neurons with contralateral cell bodies to the main dorsal longitudinal muscle and a few other smaller thoracic muscles.

N2 is an entirely sensory nerve innervating thoracic receptors.

N3 contains the axons of motor neurons that innervate dorso-ventral flight muscles in the anterior part of the segment and some leg muscles of the proximal parts. It also contains sensory neurons from these regions.

N4 contains the axons of motor neurons that innervate dorso-ventral flight muscles in the posterior part of the segment and leg muscles at the thoraco-coxal and coxo-trochanteral joints. One of its branches (N4D) contains sensory neurons from the posterior part of the ventral wall of the thorax (the epimeron), and N4A contains sensory neurons from hairs on the coxa. The remaining two branches (N4B,C) contain motor neurons but no sensory neurons.

N5 contains the axons of motor neurons that innervate leg muscles and contains sensory neurons. N5A contains only two motor neurons and sensory neurons from hairs on the coxa. N5B innervates the rest of the leg.

N6 joins with N1 of the metathoracic ganglion and contributes motor neurons to the dorsal and ventral longitudinal muscle of the metathoracic segment.

Median nerve is an unpaired nerve that arises from the dorsal midline. It soon divides into transverse nerves that run to the left and right sides of the body. Sometimes the distinction is maintained between the median and transverse nerves, but commonly only the term median nerve is used. It innervates the spiracles on either side of the mesothoracic segment and contributes neurosecretory neurons to a neurohaemal area that forms a swelling on the nerve itself.

2.5.1.2. Longitudinal tracts

The few thousand axons that make up a connective split as they enter a ganglion into a series of nine separate longitudinal tracts that run throughout the ganglion before coalescing to reform the connective at the opposite end (Fig. 2.6). Most remain as distinct bundles of fibres, but the lateral ventral tract (LVT) is usually split into a more lateral outer bundle (oLVT) and a more medial inner bundle (iLVT), the dorsal median tract (DMT) is split into dorsal and ventral parts (dDMT, vDMT) by DCIV, and the median ventral tract (MVT) can be diffuse.

2.5.1.3. *Commissures*

Six dorsal commissures, labelled from anterior to posterior as DCI-DCVI, and four ventral commissures VCI, VCII, SMC and PVC link the two symmetrical halves of a ganglion (Figs 2.5–2.7).

2.5.1.4. *Other tracts*

Many small tracts run vertically or obliquely in lateral regions of a ganglion, but some such as the C, I, R and T tracts are large enough to stand out as landmarks.

The Ring tract, for example, is a horizontal ring of fibres mingling anteriorly with VCII and posteriorly with SMC and is more properly to be considered a neuropil area. Other tracts are named because they contain the processes of identified sets of neurons. The DDT and SDT contain the processes of dorsal unpaired median (DUM) neurons (*see Chapter 3*), and the PT contains the neurites linking the dorsal and ventral fields of neurites of the midline spiking local interneurons (*see Chapters 3 and 7*). The roots of the lateral nerves also form obliquely running tracts at the edge of the ganglion that fan out and become less distinct towards the middle.

2.5.1.5. *Neuropils*

The neuropils are the regions where the vast majority of synaptic contacts occur between the fine branches of neurons (Fig. 2.5A–G). Only a very few synaptic contacts between neurons occur in the tracts and no synapses are made on the cell bodies. Most descriptions have admirably avoided functional names for these neuropils but some have slipped into terms such as sensory (ventral), motor (dorsal) and intermediate regions, or even into elaborate descriptions of functional layers. While it is true that a large number of sensory neurons from exteroceptors do project only to ventral regions of neuropil and that flight motor neurons, for example, have branches in dorsal neuropils, such general functional names are both misleading and meaningless. The production of behaviour, which is the measure of the output of the these ganglia, results from complex network interactions of neurons with branches in different regions of the neuropil. What is more useful, therefore, is to describe the branches of identified neurons in the different anatomical regions of neuropil relative to the framework of other features of a ganglion. Thus, it is helpful to describe the projections of sensory neurons from hairs as being in the ventral neuropils, and the branches of flight motor neurons in the dorsal neuropils, but not to ascribe a sensory processing function to the ventral neuropils and a motor organisational one to the dorsal neuropils.

The ventral neuropil area or ventral association centre (VAC) consists of four regions: anterior (aVAC), lateral (lVAC), ventralmost (vVAC) and medial [mVAC, formerly called the anterior Ring tract (aRT) because of its encirclement by the Ring tract (RT)]. The processes in these regions are generally of small diameter and are closely packed to give a dense appearance. Many of the processes are from sensory neurons of hairs that project to the first three areas, and from chordotonal organs that project to the fourth. They also contain the processes of many local and intersegmental interneurons. The more dorsal areas of neuropil, called the lateral association centres (LAC), are not as clearly defined as the ventral ones and appear to contain fibres of larger diameter. An anterior region (aLAC) is present close to the roots of N1 and contains the projections of some of the sensory neurons from the wings, while the posterior region (pLAC) is close to the roots of N3 and N5. These regions contain the processes of motor neurons and of many different types of interneuron.

2.6. ABDOMINAL GANGLIA

The locust has 11 abdominal neuromeres. The first three (A1-3) are fused to the metathoracic neuromere (T3) and form the metathoracic ganglion (*see section 2.5*), and the last four (A8-11) are fused to each other to form the last (terminal) abdominal ganglion. The four intervening neuromeres (A4-7) each form a separate ganglion, with A4-6 being very similar and innervating repeating structures in the abdomen; in females, A7 contains additional neurons that innervate the oviducts. Each neuromere of segments 4–6 contains the cell bodies of some 500–600 neurons (Sbrenna, 1971).

2.6.1. Organisation of an abdominal ganglion

2.6.1.1. *Nerves*

Each of the neuromeres A1-9 has two pairs of lateral nerves: nerve 1, the dorsal or tergal nerve, and nerve 2, the ventral or sternal nerve. In the terminal abdominal ganglion, the nerves of A10 and A11 form the epiproct and the cercal nerves (Seabrook, 1968).

Each unfused abdominal ganglion has two median nerves, one of which arises anteriorly and the other posteriorly. These nerves anastomose with each other and with branches of particular lateral nerves to form a complex ladder of nerves in the abdomen (Schmitt, 1965) (Fig. 2.8).

The anterior median nerve of a posterior ganglion swells to form a neurohaemal organ and then anastomoses with the posterior nerve of the next anterior ganglion. From this fused median nerve arise the transverse nerves to the left and right sides of the body, both of which swell to form neurohaemal release sites. On each side of the nerve cord, these transverse nerves branch to form the paired paramedian nerves which run parallel to the nerve cord and which are also joined to N1 by a small link nerve. A transverse nerve then anastomoses with a link nerve that joins N2 of its own ganglion to N1 of the next posterior ganglion.

Close to the dorsal midline, N1 from each of the abdominal ganglia anastomose to form the lateral heart nerves. These paired nerves, lying on either side of the dorsal tube that forms the heart, thus run the length of the abdomen in parallel to both the nerve cord and the paired paramedian nerves.

2.6.1.2. *Tracts and commissures*

The internal structure of the unfused abdominal ganglia (A4-7) resembles that of the thoracic ganglia in that the same longitudinal tracts (with the possible exception of a distinct LVT), commissures (six dorsal and two ventral) and areas of neuropil can be recognised (Watson and Pflüger, 1987). The number of neurons with cell bodies in a particular ganglion is reduced, and because different structures are innervated by the abdominal ganglia there is no guarantee that the same sets of neurons occur as in the thorax. For example, so far, no groups of motor neurons equivalent to those that innervate the legs have been found in these ganglia.

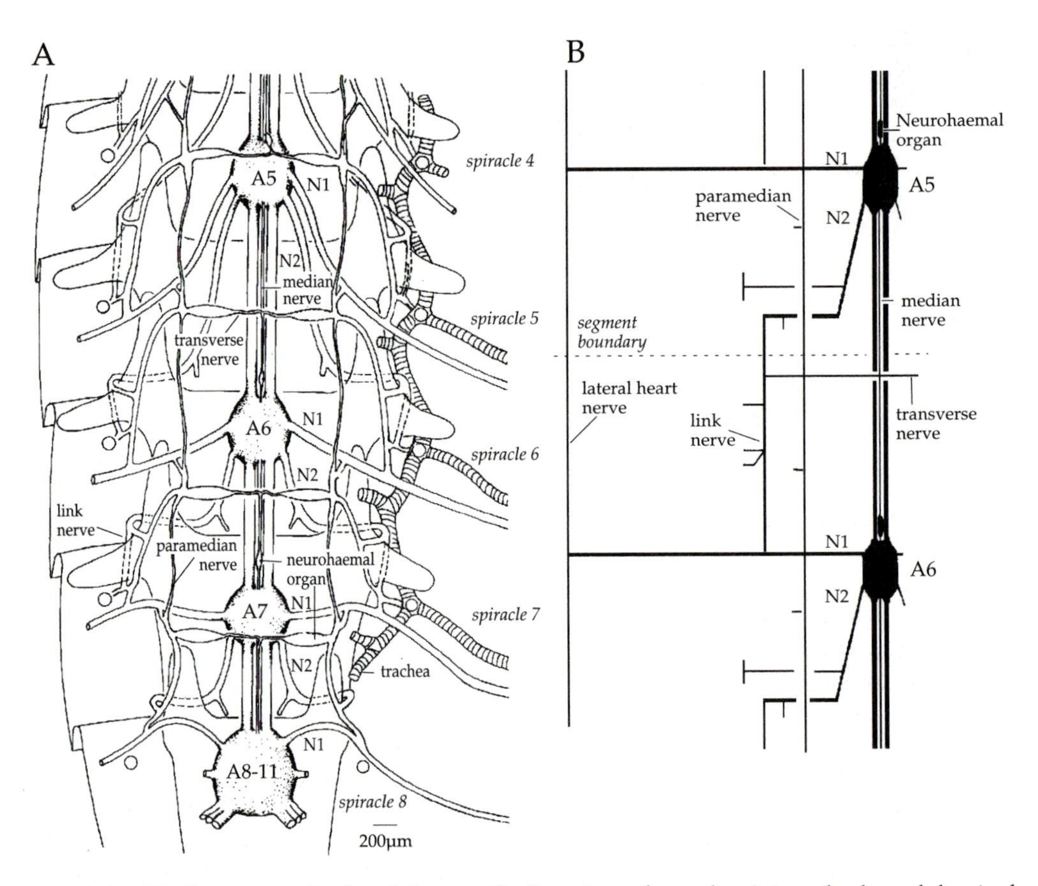

Fig. 2.8 Median nerves in the abdomen. A. Drawing of ganglia A5 to the last abdominal ganglion (containing neuromeres of segments 8–11) viewed dorsally against the ventral wall of the thorax. One of the longitudinal tracheae is drawn on the right. B. A diagrammatic representation of the anastomoses between the lateral nerves of the ganglia and the median, paramedian and lateral heart nerves. A is based on Dircksen *et al.* (1991), B on Ferber and Pflüger (1990).

In the last abdominal ganglion, the 8th neuromere has a similar structure to that of the unfused abdominal ganglia, the 9th lacks DCI, TT and one ventral commissure, and in the 10th only DCII and DCIII of the commissures can be recognised. The existence of the 11th neuromere is not obvious from the adult anatomy.

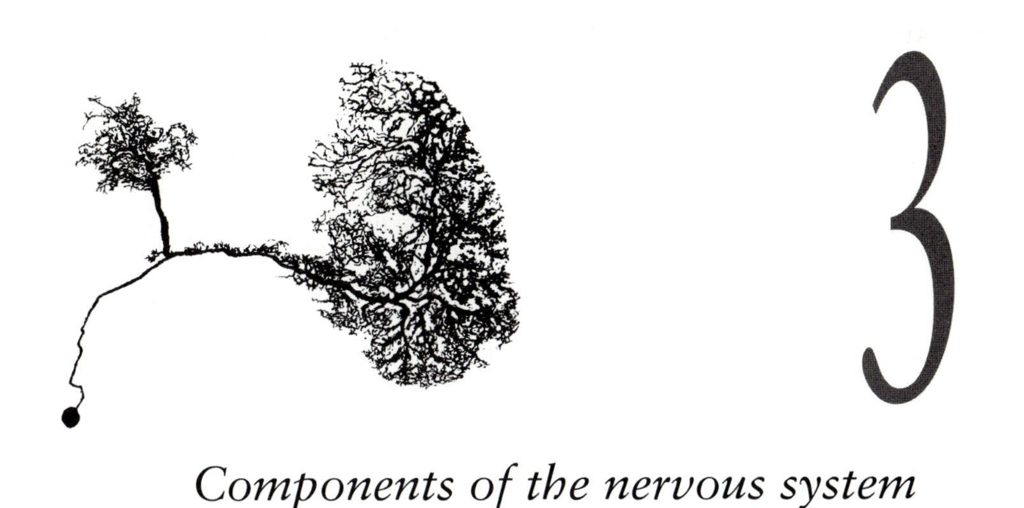

3 Components of the nervous system

3.1. GLIAL CELLS

Estimates of the number of glial cells in the central nervous system vary so widely that it is safe to conclude only that we do not know how many there are. In crickets, for example, they are suggested to outnumber neurons by as much as 8:1 in the central nervous system and to account for as much as half of its volume (Gymer and Edwards, 1967), whereas in the brain of bees they are suggested to represent only 15% (~19 000 glial cells in a worker bee) of the total number of cells within the neuropil (Witthöft, 1967). Hope for providing more accurate estimates of the numbers of glial cells comes from the use of antibodies against specific gene products. In *Drosophila*, the gene *repo* is expressed in most glial cells but not in neurons of the developing nervous system (*see Chapter* 4). In an abdominal ganglion, it is present in only 60 glial cells (Halter *et al.*, 1995).

Our lack of knowledge of their numbers is matched only by the paucity of information about the structure of the different types of glial cell and of the role that they might be playing in the functioning of the central nervous system. The shining exception to this is the considerable amount of information on the glial cells of the perineurium that surrounds the central nervous system and forms a barrier with the haemolymph.

3.1.1. Glia in the peripheral nervous system

The nerves and sense organs have two major types of glial cells associated with them, called the **peripheral glia**, which are located along the nerves, and the **exit glia**, which are located at the junctions of the nerves with the segmental ganglia. Both have processes that ensheath the nerves and may serve some sort of insulating function equivalent to that of vertebrate Schwann cells, although individual axons are not ensheathed. Glial cells are also associated with sensory neurons. In hairs, campaniform sensilla and chordotonal organs, these glial cells express the glycoprotein glionexin that may aid adhesion or act as an anionic matrix (Field *et al.*, 1994).

3.1.2. Glia in the central nervous system

The glia in the central nervous system have been classified on their positions (Saint Marie *et al.*, 1984; Hoyle, 1986) rather than on the basis of their developmental origin (*see Chapter* 4), their structure, or their suspected actions. This essentially pragmatic scheme may therefore cut across the more important functional roles of these cells, and lends itself to endless subdivisions that may have little relevance. The ultrastructure of these glial cells is known from electron microscopy, but information is largely lacking about the overall shapes of individual cells and the extent of their ramifications. Some reconstructions of shapes have, however, been made from what is seen in the electron microscope (Fig. 3.1). Glial cell bodies can have diameters of 10-50 μm, which is within the range of sizes of neuronal processes that can be penetrated with microelectrodes and stained by the intracellular injection of dye. Intracellular recordings have occasionally been made from insect glial cells (Schofield *et al.*, 1984) but we do not know their overall shape from intracellular staining. This is even more surprising given the extensive intracellular recordings that have been made from neurons in particular parts of the central nervous system. It might have been expected that a probing microelectrode would periodically penetrate and stain a glial cell, but nothing that matches our conception of a glial cell has been reported from such studies.

3.1.2.1. *Perineurial cells*

Sheath cells form the outermost layer or neurolemma of the central nervous system. These cells are probably of mesodermal origin, unlike the other glial cells, and

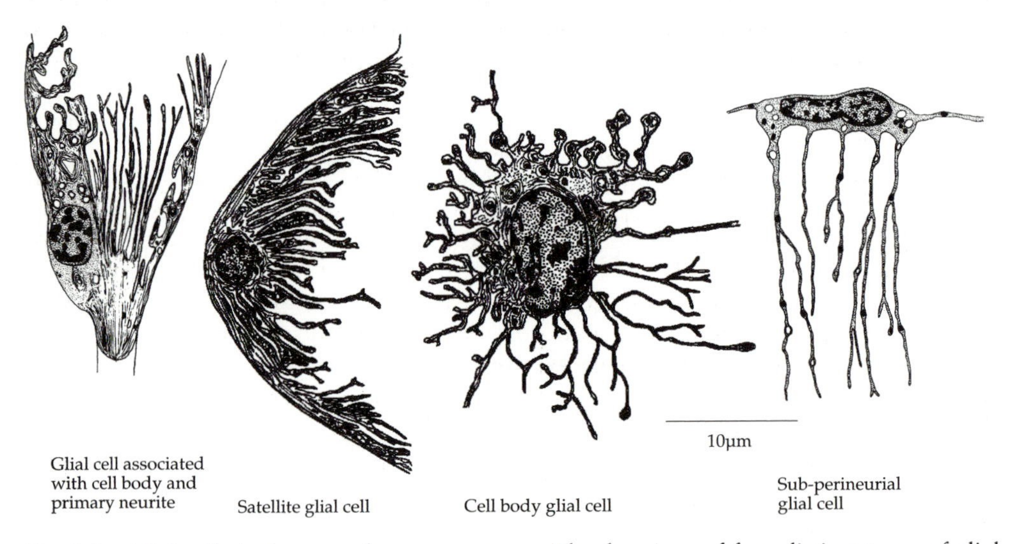

Fig. 3.1 Glial cells in the central nervous system. The drawings of four distinct types of glial cells from different regions of a segmental ganglion are reconstructions made from serial electron microscope sections. Based on Hoyle (1986).

probably secrete the acellular outer coating of the lamella that consists of collagen embedded in a matrix of glycosaminoglycans. The distinction between the sheath cells and the other glia in the central nervous system is further emphasised during neural repair (*see section 3.1.3.7*) and during development (Shepherd and Bate, 1990). Sheath cells are only found in the embryonic central nervous system after 75% development (*see Chapter 4*), probably arising from a remote site and migrating towards the developing nervous system.

Below the sheath cells are the perineurial cells, which are distinct from the glial cells that are more intimately associated with the neurons. These cells are basically cuboid in shape and have a dense granular cytoplasm containing many mitochondria and stores of lipid and carbohydrate. They form a continuous layer that varies in different places from 1–4 cells in thickness with the membranes of adjacent cells linked by many junctions to form a series of narrow and occluded channels. Beneath these perineurial cells are the subperineurial cells that also have a dense cytoplasm but are distinguishable by their more irregular shapes and longer processes.

3.1.2.2. *Class I*

These are multipolar cells that extend from the outer perineurial layer and send long thin, finger-like processes between the neurons. Each cell body of a motor neuron in a thoracic ganglion may have several of these glial cells associated with it, some of which it shares with neighbouring cell bodies. The glial cells form layers that encircle individual cell bodies, while others wrap particular groups (*see section 3.3.2.1*) and these, in places, can form layers as thick as 20 μm with as many as 100 parallel membranes. Motor neurons appear to be heavily wrapped but the dorsal unpaired median (DUM) neurons are less heavily wrapped. There is extensive interdigitation of the processes of the glial cells and the neuronal cell bodies, particularly where a cell body narrows to form a primary neurite that will enter the neuropil.

3.1.2.3. *Class II*

These cells at the boundary between the cortical layer of cell bodies and the neuropil form extensive invaginations into the neurons, particularly at the point where the primary neurite is emerging. The invaginations of the glia into the neurons are numerous and deep and form a trophospongium. The neuron may also extend processes into the glia so that there are reciprocal interdigitations. The invaginations of the glia into the primary neurite may be so numerous and deep as to almost obliterate the cytoplasm of the neuron. The primary neurites of the neurons are also wrapped in glia so that few synapses are made onto the initial part of these processes, but the invaginations of the glia into the neurons are now shorter and stouter. These glial cells do not extend processes elsewhere in the neuropil.

3.1.2.4. *Class III*

These cells, which may not be a uniform class, surround the neuropil and send thin processes with sparse cytoplasm between the neuronal branches in the neuropil.

Another type of cell is associated with the tracheae that form extensive branches within the central nervous system (*see Chapter 12*). The nuclei of these glial cells are close to the largest tracheae and enter the neuropil along with the tracheae and other glial cells.

3.1.2.5. *Microglia*

It seems probable that there is also a distinct class of cells comparable to the microglial cells of vertebrates which, in response to trauma, migrate and change their shape to amoeboid and assume phagocytic activity, mopping up any damaged fragments of tissue (Howes *et al.*, 1993; Sonetti *et al.*, 1994). They are probably closely related to haemocytes and are often so named. These cells are mobile, capable of phagocytosis and are able to transform into glial cells following injury (*see section 3.1.3.7*). These characteristics and their inhibition by morphine are shared with mobile immune cells of vertebrates.

3.1.2.6. *Junctions associated with glia cells*

The glial cells make a number of different sorts of junctions with each other and with neurons. The characterisation of these junctions is based almost entirely on their ultrastructure, with extrapolations that particular junctions are associated with particular functions.

Glia–glia junctions **Desmosomes** and **hemidesmosomes** staple the perineurial cells to the neural lamella.

Septate junctions are adhesive and keep the glial cells together.

Gap junctions provide for metabolic and ionic interdependence between the glial cells.

Tight junctions between the perineurial cells may be responsible for forming the barrier that prevents the entry of certain ions and molecules into the central nervous system.

Glia–neuron junctions Desmosomes, tight and septate junctions occur between glia and neurons but gap junctions are rare. Moreover, neurons may sometimes form synapses with glial cells and these are known in the initial stages of visual processing in the lamina of the optic lobes and in the thoracic ganglia from some identified motor neurons.

3.1.3. Contributions of glia to the functioning of the central nervous system

3.1.3.1. *Blood–brain barrier*

The fluid that bathes the neurons in the restricted extracellular space within the central nervous system differs in composition from the haemolymph. Its ionic composition is held constant, while in the haemolymph the concentration of various ions such as K^+ and Na^+ may vary considerably depending on the diet. The neurolemma itself is leaky to ions, but the perineurium severely restricts their access. The superficial layer of perineurial cells and the septate and tight junctions that they make with each other constitute an intercellular diffusion barrier, but an extracellular matrix and cation transport by the glial cells also play a part (Treherne and Schofield, 1981) (Fig. 3.2). The cells of this layer have tortuous narrow clefts between them that open at one end below the neurolemma and at the other in the cortical layer of neurons in a ganglion, or axons in a connective. The inner ends of these clefts are occluded by junctions, so that both the length and bore of the clefts and the junctions themselves restrict access. The perineurial glial cells are also linked by gap junctions to glial cells with processes between the neurons. Markers such as

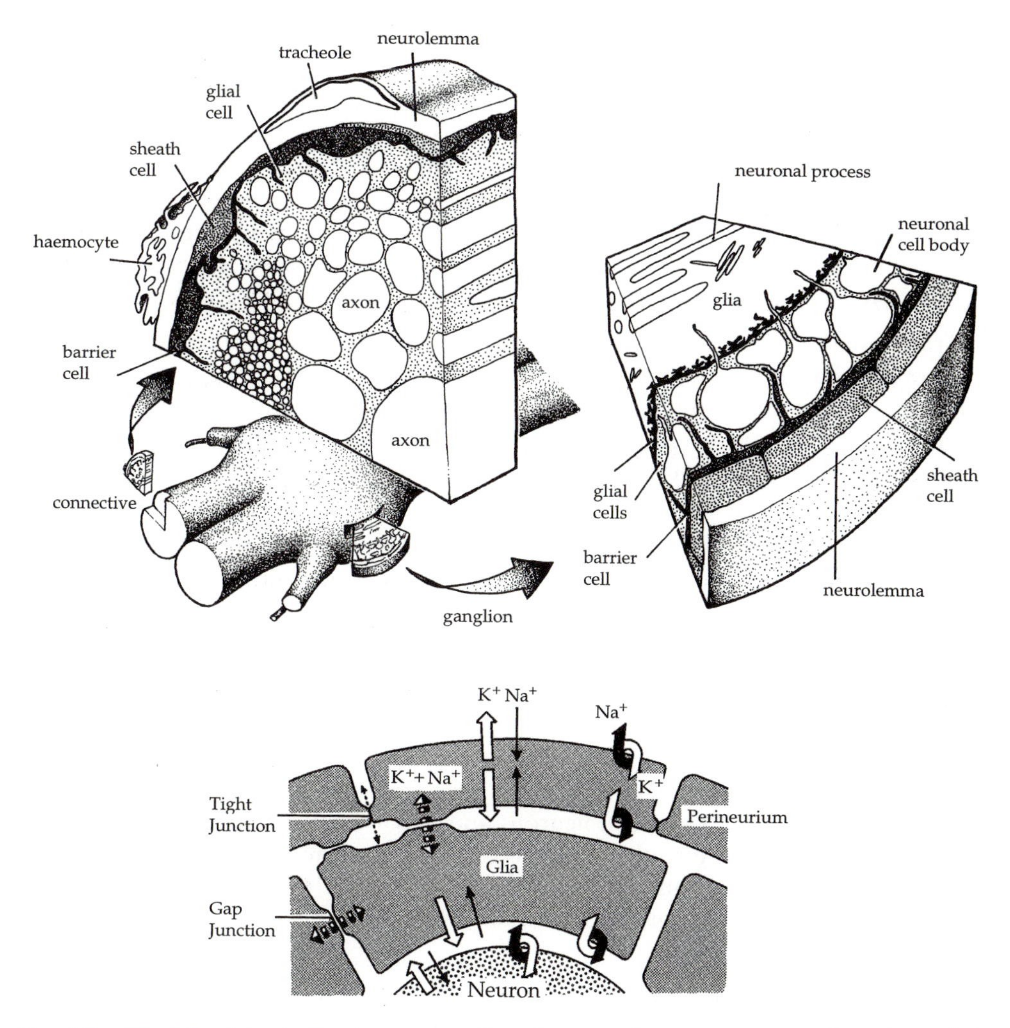

Fig. 3.2 The blood–brain barrier surrounding the central nervous system. Drawings of the arrangement of glial cells surrounding a connective and a segmental ganglion. The diagram below shows the exchange of ions between the perineurial cells, glial cells and the neurons within. Based on Treherne and Scofield (1981).

lanthanum added to the haemolymph pass through the neurolemma and accumulate at these junctions in the perineurium, but this does not necessarily imply that the junctions are the barrier to smaller ions. Ionic transfer across the barrier is thus largely through the perineurial cells by means of Na^+/K^+ pumps and by movement down electrochemical gradients. A consequence of the separation of fluids with different ionic concentrations is that the barrier formed by the perineurial cells offers an electrical resistance and maintains a positive potential in the central nervous system relative to the surrounding haemolymph.

The effectiveness of the barrier, the lack of blood vessels in the central nervous system and the small amount of extracellular space also means that the neurons are potentially susceptible to the accumulation of ions that result from their own signalling. Glial cells within the neuropil and around the cell bodies may therefore also be involved in regulating the immediate ionic environment of the neurons, particularly the concentration of K^+ when spike activity has been high. The anionic matrix associated with some of these glial cells may also serve as a reservoir of certain ions, releasing Na^+ in particular circumstances and absorbing K^+ in others.

3.1.3.2. *Structural support*

The traditional view is that a major function of glial cells is to provide support for the neurons. They do this by virtue of the junctions that they make with each other and with the neurons, and by the orientation of the microtubules, which should provide some rigidity in the neuropil. Allied with this structural role, they may maintain neuron spacing and orientation in the neuropil.

3.1.3.3. *Metabolism*

Many glial cells store glycogen and lipids, which are presumably capable of being given up to the neurons. Sugars also pass from the haemolymph to the central nervous system through the perineurial cells. The extensive invaginations of the glia into the neurons, particularly at the cell bodies, may play some role in maintaining the synthetic processes that occur here.

3.1.3.4. *Compartmentalisation*

The glia wrap particular groups of neurons and may thus restrict the flow of current, channelling it in certain directions and thus playing an important role in signalling by neurons. This function is clear in the lamina of the optic lobes, but its consequences in other parts of the nervous system have not been explored.

3.1.3.5. *Neuronal guidance*

During development, and perhaps also during regeneration, the glial cells may be one of a number of influences on the direction in which the growth cones of neurons will grow (*see Chapter 4*). They may also limit the proliferation of the neuroblasts in dividing to form neurons.

3.1.3.6. *Signalling*

The possibility that the glial cells may be involved in signalling and therefore in the processing of information is only beginning to be explored. Clues that they may play such a role come from the demonstration that they have receptors for transmitters such as acetylcholine and that neurons may make contacts with them that have all the ultrastructural features of synapses. Presumably, their membranes also have channels that operate in the same ways as those on neurons. It must be hoped that the ability to dissociate glia from neurons and then maintain and recognise them in culture will be used to determine the nature of the receptors and channels that they possess.

3.1.3.7. *Repair*

The ability of an insect to regenerate may be intimately linked to the microenvironment provided by the glial cells of the central nervous system. Certain glial cells may phagocytose degenerating neurons, but more extensive injuries require more elaborate mechanisms. The sorts of processes involved can be observed by applying the DNA intercalating agent ethidium bromide to the connective linking the abdominal ganglia of the cockroach (Smith *et al.*, 1984). Only the glia have cell bodies in this region so that they and not the neurons are damaged. The blood–brain barrier soon breaks down after such treatment, but is then soon re-established by the invasion of haemocytes that form a layer that becomes impermeable to small ions. After 4–6 days there is a large proliferation of cells at the site of the lesion, probably as a result of the division of perineurial cells that have migrated from neighbouring undamaged ganglia. The haemocytes then disappear, most probably because they transform into functional components of the blood–brain barrier and become the sheath cells. Regeneration may thus reflect the awakening after damage of developmental potential in mature animals.

3.2. NEURONS

3.2.1. Concept of identified neurons

The small number of neurons in the nervous system of insects relative to that in mammals makes it possible to treat many as individuals that can be recognised from animal to animal. Identification of the neurons is an important asset in the analyses of this nervous system as it allows information to be accumulated in different experiments about the properties and actions of a particular neuron. Without identification, the accumulated knowledge could be applied only to a like class of neuron from a particular part of the nervous system, with the accompanying uncertainty that the properties revealed may not belong to all. The impetus for identification has clearly come from the initial analyses that dealt only with the larger neurons in the nervous system. Inevitably, there are fewer of these neurons and many are the motor neurons or intersegmental interneurons with large cell bodies or large diameter axons. For these neurons identification is clear and unambiguous, but for many of the smaller neurons identification may be rendered uninformative or even impossible by the need to use many identifying criteria that cannot be easily met during the course of a particular experiment. Attempting to identify the small neurons in repeating structures such as the optic lobes, or the huge numbers that occur in some regions such as the mushroom bodies, is not a useful or attainable goal. It has to be recognised, therefore, that the limitations of the identified neuron are met when dealing with local neurons, and that for many of these small neurons the pursuit of identity offers little aid to understanding. It is hard to see how identification will help understanding when the number in a population exceeds 100, let alone when it exceeds thousands, as in some regions of the brain. The task becomes self-defeating.

Identification of a neuron is essentially an open-ended process whereby information is accumulated about the various structural, chemical and physiological properties of a particular neuron that allow it to be distinguished unequivocally from all other neurons in the nervous system. The identification also extends to characterising the synaptic inputs that it receives and the output connections that it makes with other neurons, and its actions during different behaviour. Thus, neurons once thought to be equivalent may be distinguishable by further criteria. To be useful in the course of experiments, the neurons must be readily identifiable by criteria that can be used when applying the technical methods of those experiments.

3.2.1.1. *Naming neurons*

Identifying neurons as individuals inevitably means that they must be given names so that information about them is readily communicable. This is no trivial problem (Rowell, 1989), particularly as more is learnt through disparate techniques about an increasing number of neurons both in one species and in species that are often only distantly related. It is important to have schemes that are unambiguous, expandable and give clear information about the identity and type of neuron to which the short label or name applies. So far, no scheme has successfully met all these needs for all types of neurons.

Naming motor neurons Motor neurons can be easily named according to the muscle they innervate, or numbered after the scheme devised for the muscles of the body by Snodgrass (1929). It is thus possible to say that there is a set number of motor neurons that innervate a particular muscle, and that these numbers are invariant from animal to animal. Thus, the extensor tibiae motor neurons are obviously those motor neurons that innervate this muscle, numbered 136 in a hind leg. The members of a motor pool that innervate one muscle can then be further subdivided according to the type of contraction that they cause, or the position of their cell body in the central nervous system. On this basis, the two extensor tibiae motor neurons are called the fast and the slow.

Naming intersegmental interneurons Intersegmental interneurons can also be readily labelled because staining the tracts containing their axons can reveal the total number of neurons of a particular type in a specific region of the nervous system. Thus, we already have a good idea of the number of interneurons with cell bodies in a thoracic ganglion that have ascending or descending axons. This has prompted the introduction of a three digit numbering scheme in which the first digit represents the course taken by the axon from the cell body and the second two arbitrary numbers represent the sequence in which the neurons were discovered by a particular author(s) (Fig. 3.3) (Robertson and Pearson, 1983). While this scheme has the enormous advantage of avoiding the functional names that at one time were favoured, and which always gave a partial and therefore distorted view of the processing a neuron

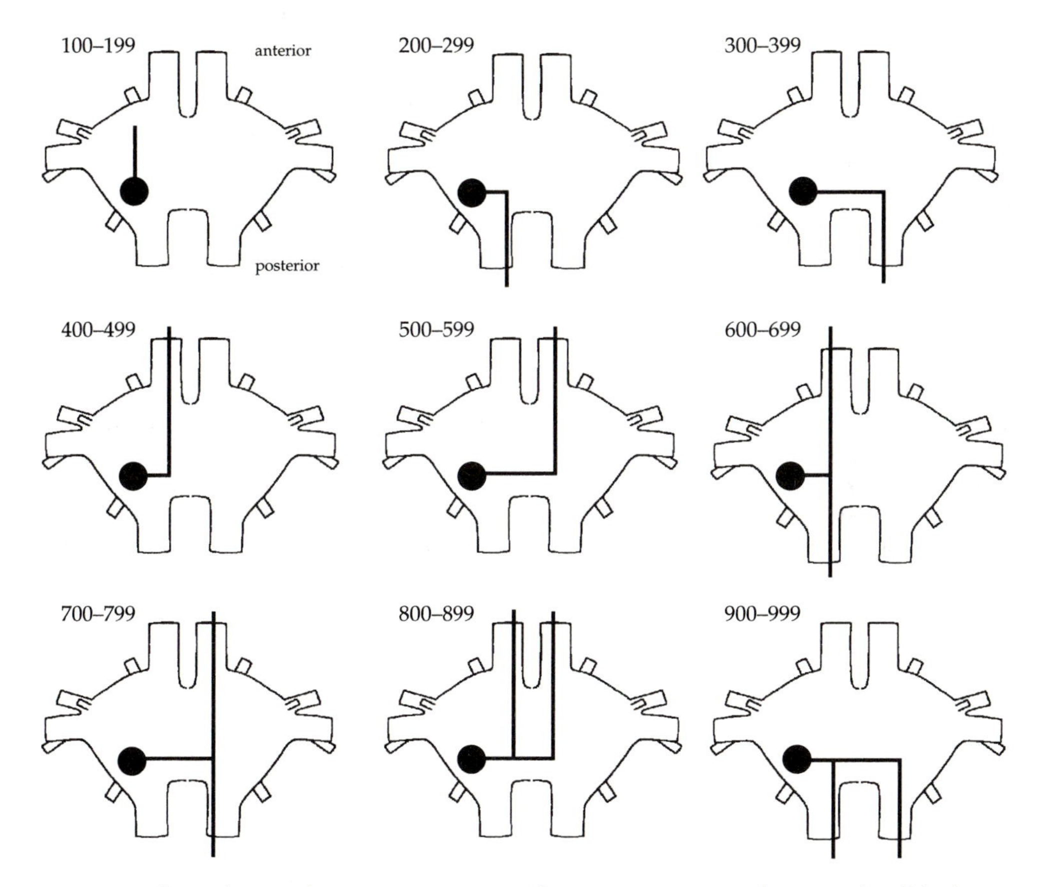

Fig. 3.3 Scheme for numbering interneurons. The various types are shown with cell bodies in the mesothoracic ganglion, but the scheme also holds for neurons in other segmental ganglia. Axonless neurons are numbered from 100. For other neurons, the first digit indicates the route taken by their axon. The second two numbers are arbitrary and are assigned on precedence. The scheme was devised by Robertson and Pearson (1983).

performed, it still has considerable drawbacks. First, the numbering allows no grouping of the interneurons into populations of similar shapes or similar actions and thus does not maintain natural relationships. Second, it gives no indication of homology even between the different segments of one species. Third, a different number may be ascribed to neurons that are clearly the same, simply because they occur in different species. Fourth, the same neuron from the same species may be given different numbers by different laboratories.

Naming local interneurons Identification and a numbering scheme face their biggest challenge in trying to deal with local neurons. Unlike the motor neurons and the intersegmental interneurons there is no easy way to determine the total number of local interneurons in a population; the only available method is to count the number of cell bodies in a region and trust that they all belong to the same population of local neurons. Even for the well-characterised local neurons in the thoracic ganglia, there is uncertainty as to whether any can be treated as individuals. It is probable that they can be treated as identified individuals but the procedures that can currently be used are not sufficient to establish identity. Names and numbers have thus been avoided in dealing with local interneurons in the locust, although some of the apparently equivalent neurons have been labelled (using a different scheme to the one outlined above) in stick insects. This has added rather more confusion than clarity because a name refers not to an individual but to a few interneurons that have different actions and different connections (Büschges, 1990; Driesang and Büschges, 1993). The three digit numbering scheme that separates the intersegmental interneurons into many different categories lumps all the local interneurons into one, despite the fact that they are the most numerous neurons in the central nervous system.

Towards an effective nomenclature The ideal solution would be to name neurons according to their lineage so that the name would reflect their segment, neuroblast of origin, and the timing in the divisions of the neuroblast and their ganglion mother cells (*see Chapter 4*) (Rowell, 1989). Such a method would preserve homologies both within one species and between species, and would maintain natural relationships between neurons. It seems that such a desirable scheme may be difficult to achieve, given the difficulty of defining the lineages of neurons; so far it has been possible to trace the lineages of only a few of the neurons about which so much is known of their morphology and physiology. To work effectively, a system of names will need similar monitoring to that for the taxonomy of animals, and the creation of a large database similar to that for protein sequences against which new neurons can be compared. Initially, these developments could provide no more than a partial solution, but eventually they should lead to a rational system once more substantial numbers of neurons are well characterised. There is little cause for optimism that the situation will change markedly in the near future. Since the impassioned, plausible and widely accepted plea for rationalisation by Rowell (1989) more than 5 years ago, little has changed. We are no closer to a scheme based on lineage and must continue to rely on the partial pictures of a neuron that emerge from anatomical and physiological studies, coupled with the sometimes illuminating but often confusing input provided by immunocytochemistry (*see Chapter 5*). The problems have, however, been well defined and the goals to which we must strive are clearly set out.

3.2.1.2. *Parts of a neuron and what to name them*

Cell body or soma The cell bodies of the different types of neurons are packed together in a multilayered cortex that surrounds the central areas of neuropil, tracts and commissures where the processing of signals takes place. The position of the soma of a particular neuron is relatively constant from locust to locust with some of the larger cell bodies acting as landmarks for the position of others. Sensory neurons, in general, have their cell bodies in the periphery, with a notable exception being the strand receptors that have cell bodies intermingled with those of the motor neurons and interneurons in the central nervous system (*see Chapter 8*). Some sensory neurons and some neurosecretory neurons have cell bodies along particular peripheral nerves.

A cell body forms no dendrites and receives no synapses, although it has receptors linked to ion channels for many of the transmitters that are used at synapses, and for many neuromodulators (*see Chapter 5*). The position of a cell body is thus of little functional significance in the processing performed in the central nervous system since no synaptic connections are involved. Its contribution to integrative processing seems to be limited to the electrical loading that it provides as both synaptic and spike events are reflected into it from a distance. Only the cell bodies of DUM neurons (*see Chapter 5*) are able to support active spikes, while in all motor and interneurons the spikes are simply electrotonic remnants conducted passively from distant excitable membranes.

Each cell body is wrapped in glia, which indent its surface with many finger-like processes to form an extensive trophospongium, particularly where its single process (primary neurite) arises. Particular groups of cell bodies may be further enwrapped by layers of glia.

Primary neurite The single process that emerges from a cell body is called the primary neurite. It enters the neuropil and forms a complex array of branches that are characteristic of each type or even of each individual neuron, before forming an axon that carries spikes to its targets in either the periphery (motor neurons and efferent neurosecretory neurons) or other parts of the central nervous system (interneurons and other neurosecretory neurons). The diameter of the primary neurite is enlarged in the region of neuropil where it forms the array of branches, often to reach a diameter of 10–15 μm, and this is sometimes called the integrating zone, implying a facet of its function that is not established. Integration occurs throughout a neuron and is not restricted to one large branch, although this may be the region where the final sum of the synaptic inputs determines whether spikes should be generated, if the neuron is of the type that normally signals with spikes. Where input and output synapses are intermingled on fine branches, much integration and signalling to other neurons can occur at many distributed sites. For some neurons, at least, this raises the possibility of local processing and the compartmentalisation of action.

In motor neurons and many interneurons with long axons, spikes are generated at a zone in the enlarged primary neurite and are actively conducted orthodromically along the axon and passively conducted antidromically along the primary neurite to the soma. Most neurons have a single spike-initiating zone, but some intersegmental interneurons may be able to generate spikes at different sites in response to different sets of synaptic inputs.

Neurites The primary neurite gives rise to many fine branches, the distribution of which is characteristic of an individual neuron to the extent that many can be identified on the pattern of these branches alone. Deciding what to call these branches has been a persistent problem. The term dendrite has often been used, but a strict definition from its vertebrate origin indicates a process of a cell body that has only input synapses, and thus on both counts fails to define the branches of an insect neuron; they arise from the main process and can both receive input synapses and can make output synapses. The term neurite, branch and arborisation are used interchangeably, with neurite being my preferred one. It carries no implication as to whether a branch receives input synapses alone, makes only output synapses, or has a mixture of input and output synapses.

In many neurons there is not an obvious distinction between the branches of the primary neurite and the axon because one is, after all, merely the extension of the other, and their branches are intermingled. At the other extreme are neurons whose axonal branches are in different ganglia, or different parts of the brain, from those of the primary neurite, and in between are neurons whose axonal branches are close to, but distinct from, those of the primary neurite. Often, the axonal branches of these neurons have a beaded appearance when stained with intracellularly injected dyes, and are more sparse than the smoother and more numerous branches from the primary neurite. The inference is often therefore drawn, and predictive extrapolations made from light microscopic appearance alone, that the beaded branches are output regions of a neuron and the smoother branches the input regions. A further extrapolation that is often made is that the varicosities represent the sites of the output synapses. For some neurons, these generalisations may hold and can be backed by the necessary evidence from electron microscopy, but even then the output region may have input synapses and the input region may have output synapses, thus raising the likelihood that local processing is widely distributed. Indeed, in some intersegmental interneurons, where the axonal branches may be in distant ganglia, the branches from the primary neurite can have an equal number of input and output synapses, so that to call this the input region would be misleading.

3.3. MOTOR NEURONS

3.3.1. General features of motor neurons

The motor neurons represent the output channels by which the processing performed in the central nervous system is conveyed to the effectors, but the design of the insect nervous system is to keep these channels to a minimum. In a neuromere of the metathoracic segment there are the cell bodies of about 2000 neurons but no more than about 200 motor neurons, or 100 to be divided between the control of one hind leg, one hind wing and the muscles on one side of the thorax. The motor neurons therefore represent only a very small percentage of the overall neuronal complement of the central nervous system, and only a few are responsible for any particular movement.

3.3.1.1. *Excitatory and inhibitory motor neurons*

Insect motor neurons are of two basic types; excitatory ones that act in a conventional way to depolarise muscle fibres and make them contract, and inhibitory ones that usually hyperpolarise muscle fibres and reduce the amount of force produced by a muscle and increase its rate of relaxation. They have either a postsynaptic action on muscle fibres or a presynaptic action on some, usually the slow, excitatory motor neurons. In contrast to excitatory motor neurons that innervate just one muscle or functional group, the inhibitory motor neurons innervate several muscles, some of which have antagonistic actions during normal movements. Inhibitors with an effect on several muscles are called common inhibitory motor neurons, originally to distinguish them from specific inhibitors that supply a single muscle. The inhibitors that supply leg muscles are all common inhibitors (Hale and Burrows, 1985), but two of the inhibitors to the small intersegmental thoracic muscles may be specific and innervate just one muscle. The number of inhibitory motor neurons is very limited in comparison to the number of excitatory motor neurons. A leg, for example, is innervated by just three common inhibitory motor neurons compared to 70 excitatory motor neurons (see Table 8.3). The inhibitors innervate only those muscle fibres that are also innervated by slow motor neurons and this may explain their absence from the power producing muscles used in flight that are innervated only by fast motor neurons.

3.3.1.2. *Motor pools*

Each muscle is innervated by only a few motor neurons (Table 3.1, and *see Tables 8.3 and 11.1*), often only one or two, but sometimes as many as 11. The constancy in the number of motor neurons in a pool is only occasionally breached by the occurrence of supernumerary neurons electrically coupled to each other so that their spikes usually occur together (Siegler, 1982). Our knowledge of the sizes of the pools has come from physiological recordings of the synaptic (junctional) potentials in muscle fibres and stains of motor nerves that reveal the number of cell bodies of neurons associated with a particular muscle in the central nervous system. Both methods have their limitations and there are often discrepancies between them, particularly since the use of the current sophisticated morphological methods increasingly shows larger numbers of neurons; typically a small group of large diameter cell bodies, that match those predicted from physiological studies, and a larger group of small cell bodies. These small neurons are most probably either efferent neuromodulatory neurons, distinct from the dorsal unpaired median (DUM)

Table 3.1. Nerves, excitatory motor neurons and the muscles they innervate in the metathoracic segment

Nerve	Number of motor neurons	Muscles innervated	
		Hind wing	Hind leg
1	10,4 meso 6 meta	112 116 (ventral longitudinal) other small thoracic muscles	–
RNa	8,5 meso 3 meta	thoracic muscles?	–
2	0	–	–
3A	14	113 118 127 128	125 126
3B	8	–	131 135
3C	9	133b/c*	121 130 133d
4AB	10	115 (pleurosternal)	132 CxTrMRO
4C	7	–	122 123 124
4D	11	114 119 120* 129	–
5A	2	–	133a
5B	24	–	135 136 137 138 139

Muscles are numbered according to Snodgrass (1929)
* Suggested to be bifunctional muscles, used in flying and walking
CxTrMRO = coxo trochanteral muscle receptor organ
meso = mesothoracic ganglion
meta = metathoracic ganglion
RNa = recurrent nerve

neurons (*see sections 3.9.3 and 5.12.2*), or sensory neurons with central cell bodies like those of the strand receptors (*see section 8.11*).

3.3.1.3. *Actions of the members of a motor pool*

Within these pools, the motor neurons may differ in their behavioural actions, the sequences of spike traffic that they carry, the type of junctional potentials that they evoke in the muscle fibres, and the contractions that they elicit from a muscle. On this basis, the motor neurons can often be divided into fast (phasic) and slow (tonic) motor neurons representing the extremes of a spectrum of types. These names often become muddled as to which features of the action of the neurons they are describing, although many are interrelated. Strictly, the terms 'fast' and 'slow' refer to the speed of the muscle contraction that is evoked and not to the speed with which the motor spike is conveyed to the muscle, while the terms 'phasic' and 'tonic' refer to the sequences of spikes that a neuron produces during a movement. The two excitatory motor neurons innervating the extensor tibiae muscle of a hind leg well illustrate the use of the terminology. One motor neuron evokes a fast twitch contraction of the muscle, has a large diameter axon so that the spikes are conducted rapidly, and is used only intermittently in moving the leg. This is the fast motor neuron. The other, slow motor neuron produces only small and slow twitch contractions, the amplitude of which depend on the pattern of the motor spikes. It has a smaller diameter axon, so that its spikes are conducted more slowly, and is used continuously, often in sequences of high frequency spikes, to vary the force generated by the muscle in a graded manner. In larger motor pools, some of the members have intermediate properties so that the contractions they produce differ in amplitude and the rate of onset.

The definition of motor neuron types is further complicated by differences in the mechanical, structural and chemical properties of individual fibres within a muscle which are intrinsically capable of generating different amplitudes and time courses of force. Most of the muscle fibres differ markedly in the enzymes that they contain (Muller *et al.*, 1992), but, in contrast to crustacean muscles, differ little in the lengths of their sarcomeres. Some can generate force rapidly while others at the opposite end of the spectrum of types generate force more slowly. These intrinsic properties may, in turn, be influenced during development by the types of motor neuron that innervate them and the patterns of spikes that these carry. The force generated by a muscle is thus the result of a complex interplay between the pattern of innervation by the different types of motor neurons and the different types of muscle fibres that can lead to many permutations of actions. There are, however, few examples where the type of motor innervation has been matched precisely to the properties of individual muscle fibres. The general pattern appears to be that the fast motor neurons preferentially innervate the fast muscle fibres and the slow motor neurons the slow muscle fibres. This would imply an organisation into motor units, which may well hold for the extremes of the types of muscles fibres, but for many there is overlap in the innervation so that a particular muscle fibre is polyneuronally innervated.

3.3.2. Morphology

Motor neurons are all built to the same basic plan from which a large variety of shapes have been elaborated. An individual motor neuron can be characterised, usually at least to the level of the pool that innervates a particular muscle, by the position of its cell body, the path of its primary neurite, the three-dimensional array of its fine branches in the neuropil, and the nerve in which its axon emerges (Fig. 3.4). Most can then be identified individually by a combination of their morphological and physiological properties.

The cell bodies (somata) of the motor neurons are some of the largest in the nervous system, with diameters of 30–100 μm, but none are pigmented in a way that makes them immediately visible. The cell bodies of particular groups of motor neurons are packed together in the multilayered cortex of the central nervous system in such a way that the position of an individual varies only within restricted limits from animal to animal.

The axon usually emerges from the side of the nervous system that contains the cell body and the majority of the branches so that the whole organisation is unilateral. All the leg motor neurons conform to this pattern (Fig. 3.5).

Other motor neurons may have branches on both sides of a ganglion and an axon that emerges from the side contralateral to the cell body. Only motor neurons with axons in N1 have contralateral cell bodies and although the reasons for this are obscure, the distribution probably arises during development. It is a highly conserved feature that is seen in all species of insects that have been examined. Thus, the dorsal longitudinal muscles of the wings are innervated by four neurons with ipsilateral cell bodies and axons in N6 of an anterior ganglion and one with a contralateral cell body and an axon in N1 of the next posterior ganglion. This pattern is also repeated for the abdominal dorsal longitudinal muscles.

Still other motor neurons may have their cell bodies and branches in one ganglion and an axon that runs anteriorly or posteriorly to emerge from an adjacent ganglion. Some of the motor neurons controlling abdominal muscles have this shape.

The majority of motor neurons have a single axon, but the axon of certain common inhibitory motor neurons divides into several branches before or just after leaving the central nervous system to innervate several muscles on one side of the body (Fig. 3.5, and see Fig. 8.6) and the axons of excitatory spiracular motor neurons divide in the median nerve to innervate muscles on both sides of the body (*see Chapter 12*). Despite these differences in shape, all the motor neurons exist as bilaterally symmetrical pairs.

3.3.2.1. *Grouping of the motor neurons*

Is there an underlying organisation, either developmental or functional, to the way the cell bodies of the motor neurons are packed in the cortex of a ganglion, the way their primary neurites course through the neuropil, or the distribution of nerves in which their axons emerge to the periphery? An organisation according to the nerve containing the axon does not hold because, for example, the cell bodies of neurons with axons in N3B or N4A,B occur in different regions of a ganglion (Fig. 3.6). Similarly, there is no organisation according to their action in particular movements such as walking or flying because the cell bodies of neurons involved in these movements are widely distributed and intermingled. Such an arrangement would be unlikely because many motor neurons participate in various different behaviours. Finally, distribution according to the muscle that is innervated has its anomalies in that the cell bodies of

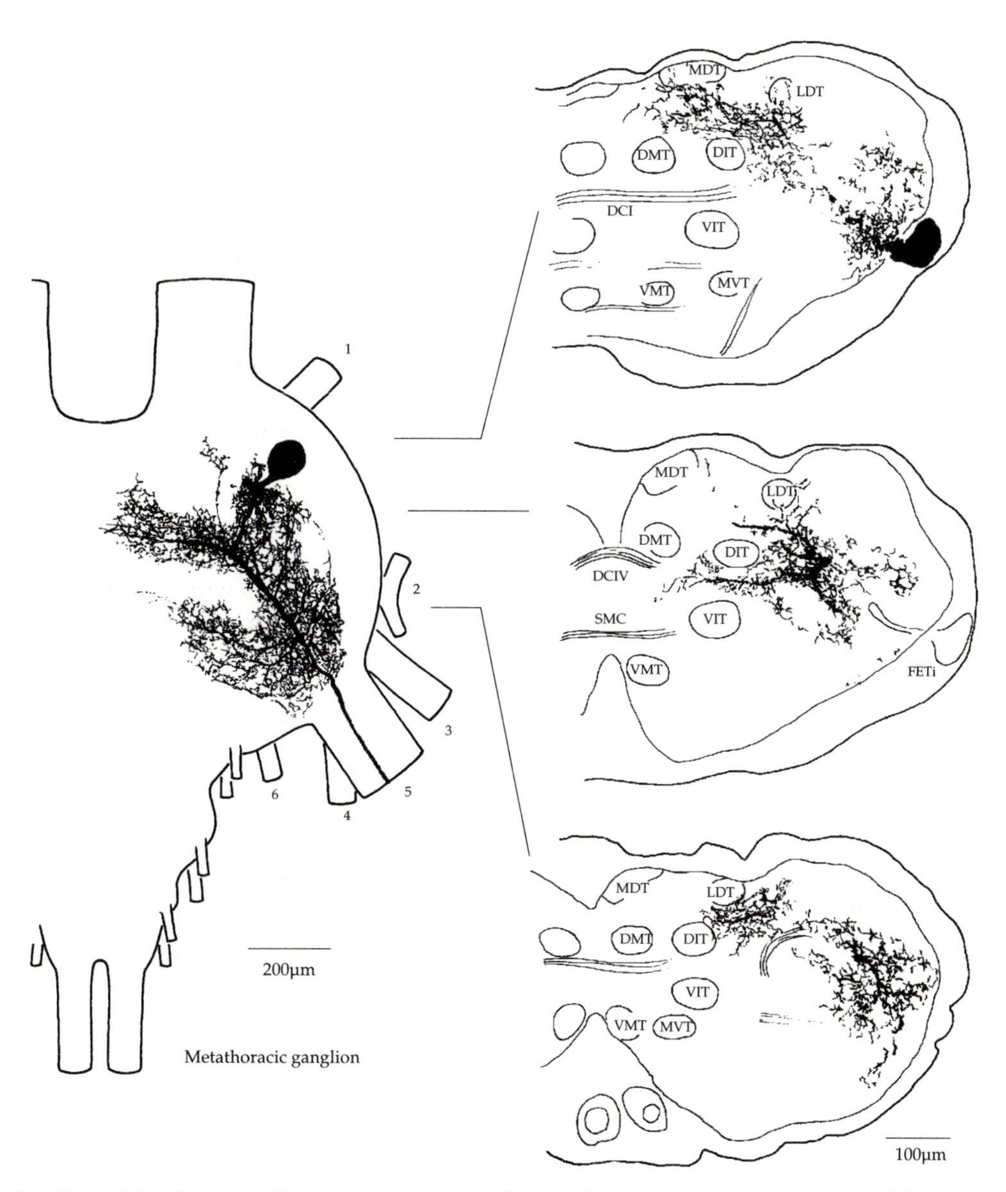

Fig. 3.4 Morphology of a motor neuron; a flexor tibiae motor neuron of a hind leg. The drawing on the left is of a whole mount of half of the metathoracic ganglion after the motor neuron was identified physiologically and then injected intracellularly with cobalt (method of Pitman *et al.*, 1972b; Brogan and Pitman, 1981). The stain was then enhanced by the Timm's silver method (Bacon and Altman, 1977). The transverse sections on the right through the ganglion at the levels indicated by the horizontal lines relate the branches of the neuron to the known tracts, commissures and areas of neuropil (for definitions of the abbreviations see Table 2.2). The unstained cell body of the fast extensor tibiae motor neuron (FETi) is drawn on the middle section. Based on Watkins *et al.* (1985).

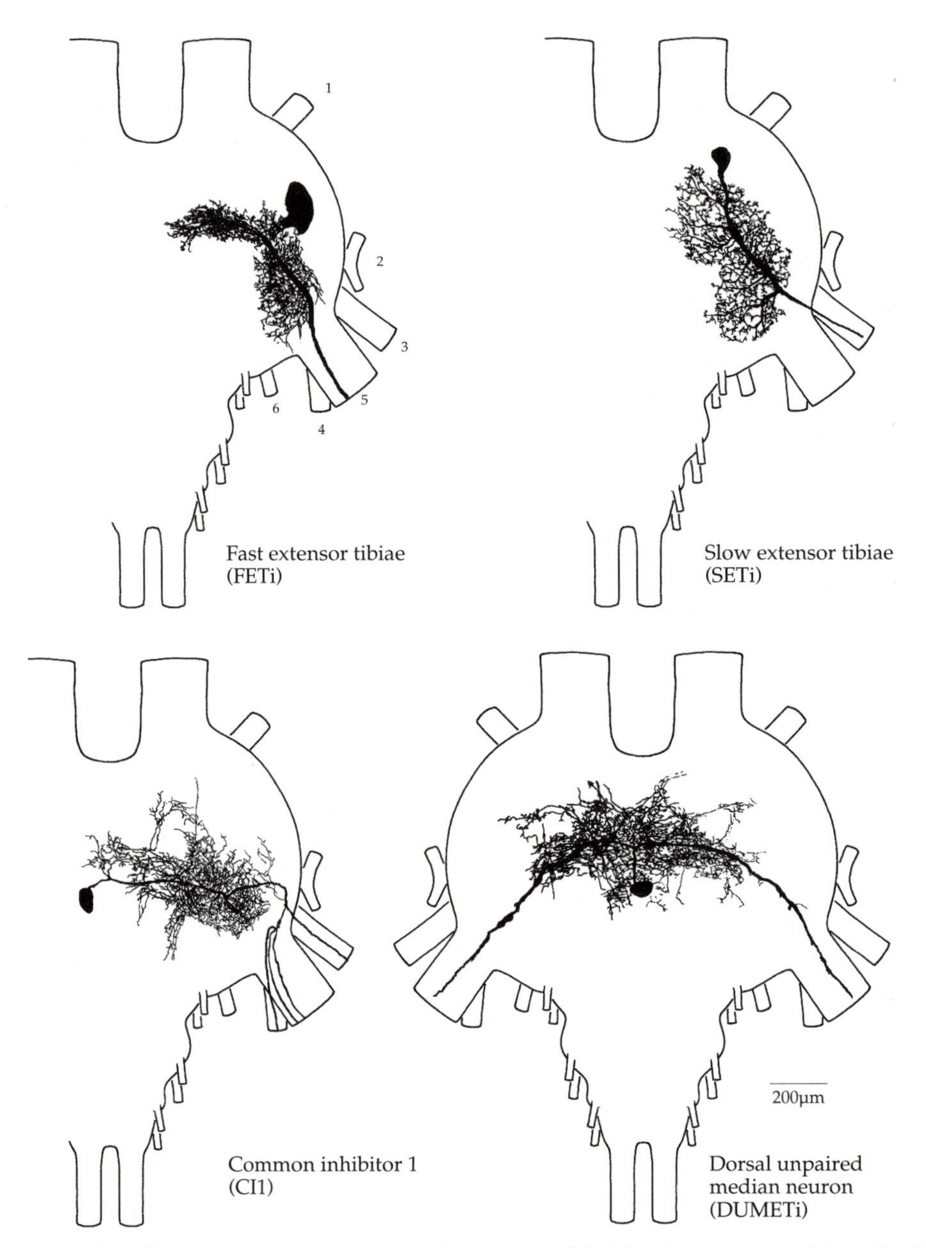

Fig. 3.5 The efferent innervation of a muscle as exemplified by the extensor tibiae of a hind leg. This muscle is innervated by four identified neurons. Two are excitatory motor neurons; a single spike in the fast extensor tibiae (FETi) motor neuron produces a rapid twitch contraction and in the slow extensor tibiae (SETi) a slower and less powerful contraction. The third neuron is a common inhibitory neuron (CI1). All these neurons are paired neurons with cell bodies in the ventral cortex of the metathoracic ganglion. The fourth neuron is an unpaired median neuron with its cell body at the dorsal midline; it is called DUMETi (dorsal unpaired median neuron to the extensor tibiae muscle). Based on Burrows (1983).

the two motor neurons innervating the extensor tibiae muscles are in different places and their axons run in different nerves (Fig. 3.5).

The organisation that promises to advance understanding is the finding that 82 out of the 100 excitatory motor neurons in one half of the metathoracic neuromere occur in eight groups (Fig. 3.6 and Table 3.2) (Siegler and Pousman, 1990a,b). These eight groups are intermingled with the cell bodies of local and intersegmental interneurons in a densely packed lateral region of the cortex and are distinct from the medial common inhibitory neurons. The cell bodies of motor neurons in one group are packed together and surrounded by cortical glial cells with fibrous regions up to 20 μm thick separating adjacent groups. Their primary neurites are also bundled together as they follow the same course through the neuropil, so that neurons within a group have the same general shape, with the divergence of their fine neurites imparting a characteristic and recognisable appearance to individuals. Within an individual group, the position of specific cell bodies may vary, particularly the smaller ones, but the number of neurons contributing is constant from animal to animal. This use of the word 'group', implying a distinct and constant assembly of particular motor neurons, may thus be more restrictive than in common practice.

Each group may represent the progeny of an individual neuroblast, although the progeny of one neuroblast may be further subdivided into separate groups (*see Chapter 4*). The members of a group may thus be related linearly. The groupings are not according to function, as motor neurons to the same muscle may be in different groups (e.g. extensor tibiae), although all members of some motor pools may be within one group (e.g. flexor tibiae). Similarly, the groupings are not according to the nerve containing the axons of the motor neurons, as the axons of the members of a group may diverge to enter several different nerves. Some, but not all, of the apparent lack of functionality of the groupings may result from the functional names that are used for the muscles (Snodgrass, 1929), and hence for the motor neurons, which take little account of the subdivisions of action that can occur within the muscles themselves, the developmental history of the muscles, or the changes in organisation that have occurred during evolution (Alsop, 1978).

Such groupings of neurons nevertheless provide the best conceptual organisation that we have and may well provide a framework for understanding the organisation of ganglia into which can eventually be placed the local and intersegmental interneurons, and the DUM neurons. They also provide a spur to the expansion of further linkages between developmental and functional studies of the nervous system.

3.3.3. Structure and distribution of synapses

Synapses associated with motor neurons are similar in basic structure to those found throughout the thoracic nervous system (see Table 3.3). In the presynaptic neuron, they are characterised by an accumulation of a discrete population of vesicles around a bar-shaped density, which often has a thin plate at one end while the other end is close to, or in contact with, the synaptic membrane. The majority of vesicles are

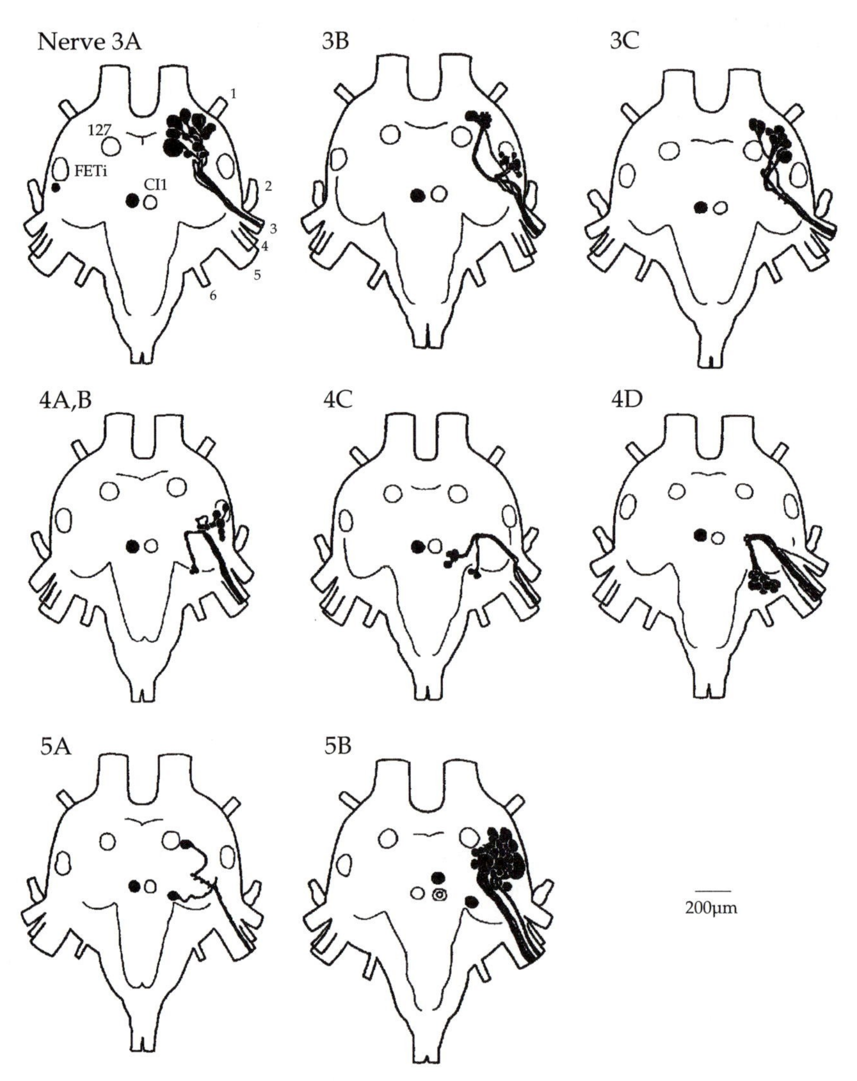

Fig. 3.6 The distribution of cell bodies in the metathoracic ganglion of motor neurons that innervate leg muscles, according to the nerve in which their axons run. The top row shows the motor neurons with axons in the three primary branches of N3, the middle row of N4, and the bottom row of N5. Three cell bodies are drawn in each to serve as landmarks; FETi, CI1 and the motor neuron innervating the basalar muscle (127) of the hind wings. The cell bodies were stained by allowing cobalt to diffuse along a selected nerve. This also stains the central cell bodies of sensory neurons of strand receptors (see the small cell bodies in the anterior of the ganglion after N3B was stained). Based on Siegler and Pousman (1990a).

either round or more irregular in shape (pleomorphic) and agranular, but with occasional larger granular vesicles nearby. There are few functional implications that can be drawn from the shapes and sizes of the vesicles seen in these or other neurons.

Table 3.2. Grouping of the excitatory motor neurons in the metathoracic segment

Group	Number of motor neurons in the group	Nerve	Muscle: Wing	Muscle: Leg
1	7	3A	118	125
			128	126
2	6	3A	113	
			118	125
		3C		121
				133b/c*
3	6	3A	127	125
				126
		3C		130
				133d
4	5	3B		135
		3C	133b/c*	
		5B		138 or 139
5	24	4A		132
		5A		133a
		5B		136
				137
				138
				139
6	13	3B		131
		4A		132
		4B	115 (pleurosternal)	
		5B		135
7	16	4A		132
				CxTrMRO
		4C		122
				123
				124
		4D	114	
			119	
			120*	120*
			129	
8	5	4C		122
				123
				124

Data from Siegler and Pousman (1990a,b)
CxTrMRO = coxo-trochanteral muscle receptor organ
* Suggested to be bifunctional muscles, used in flying and walking
Muscles are numbered according to Snodgrass (1929).
See also Tables 8.2 and 8.3 for details of muscles of a front, middle and hind leg.

Table 3.3. Ultrastructural features of synapses of the component neurons in the central nervous system

Character	Motor neuron	DUM neuron	Nonspiking local interneuron	Spiking local interneuron	Projection interneuron	Sensory neuron
presynaptic specialisation	bar-shaped density often with a plate	density	bar-shaped density 0.3 μm	bar-shaped density 1–2 μm	bar-shaped density	bar-shaped density and row of discrete densities
vesicle number	200–500	few	500–1600	?	2000	–
vesicle features	round agranular	round agranular	round agranular	round agranular	round agranular	round agranular
vesicle diameter	40 nm	40–60 nm	47 nm	50 nm	34 or 48 nm	37 nm
ratio of input and output synapses	most lack outputs	very few output synapses	intermingled inputs and outputs	different ratio of inputs and outputs in two fields	varies between different neurons	unknown

The pleomorphic vesicles are frequently associated with neurons that show immunoreactivity to γ-aminobutyric acid (GABA), and the larger granular vesicles are usually associated with the release of neurosecretory products. This means that the vast majority of substances used in synaptic transmission or neuromodulation must be contained in similar vesicles. The gap between the pre- and postsynaptic membranes is wider than usual and may be filled with an electron-opaque material. The postsynaptic membrane may be dense and may sometimes have further cytoplasmic specialisations.

3.3.3.1. *Input synapses*

Input synapses are made onto all the processes of a motor neuron in the neuropil; the primary neurite, the large secondary neurites that arise from it and the many further levels of branches, including the short and thin (0.2–0.5 μm diameter) branches, or spines, that arise from all diameters of processes (Watson and Burrows, 1982; Watson *et al.*, 1985). The spines, which are also found on many other types of neurons, are not comparable in structure to those on vertebrate neurons for they are merely short blind-ending processes. They do not consist of a swollen head at the end of a narrow neck, they do not have a spine apparatus, they can have both input and output synapses, and there is no evidence that particular inputs are restricted just to the spines. A motor neuron is usually one of two postsynaptic processes at these synapses which are therefore called diadic. Often, the primary neurites of all the neurons within a particular pool, such as the flexor tibiae, are bundled tightly together for part of their route through the neuropil so that they do not participate in synaptic interactions. Where the glial sheath of a primary neurite is lacking, or where the packing is less tight, input synapses occur either directly onto the neurite or onto its spines. The input synapses are not necessarily evenly distributed along a branch but can be clustered; the highest density of synapses occurs on the smaller diameter branches and the lowest on the primary neurite. A single presynaptic neuron may make several contacts along one neurite, often within the space of 1 μm. These repetitive contacts can be with the neurite itself or with one of its spines so that the effectiveness of these synapses cannot be assumed to be the same.

3.3.3.2. *Output synapses*

A few motor neurons, for example the fast but not the slow extensor tibiae, also make output synapses that are intermingled on the same neurites as their input synapses. Commonly, only one postsynaptic process is associated with the presynaptic specialisation, so forming a monodic arrangement, but diadic arrangements involving two postsynaptic processes also occur. Each output synapse has a population of 200–500 round, agranular vesicles with a diameter of about 40 nm, and is not necessarily associated with the varicose swellings that occur along the length of a neurite and at the ends of fine neurites. These swellings frequently have no vesicles and show no specialisations that are associated with synapses, but instead are the sites where several mitochondria accumulate. They are also present in motor neurons such as the common inhibitors that make no output synapses at all within the central nervous system. For some neurons viewed only with the light microscope, a varicose appearance to an array of neurites is often taken to indicate that it represents an output region, whereas those with smoother neurites are assumed to be input regions. Moreover, swellings on the ends of branches are often assumed to be synaptic boutons. Electron microscopy shows that these are not always safe assumptions.

The intermingled input and output synapses may be spaced no more than 0.5 μm apart, leading to complex serial and reciprocal arrangements between a motor neuron and other neurons. This can result in recurrent loops in which a motor neuron is presynaptic to one neuron which then is either directly presynaptic to this motor neuron (a reciprocal synapse), or is presynaptic to a third neuron (a serial synapse), which then synapses back onto the motor neuron. Where one of the postsynaptic neurons (e.g. a flexor tibiae motor neuron) is also labelled, the estimate is that several hundred anatomical synapses underlie the direct physiological connection from the fast extensor tibiae motor neuron (Burrows *et al.*, 1989) (*see Chapter 9*). The message from these sorts of arrangements is that the anatomical representation of a physiological connection may be extremely complex. A further complexity of as yet unappreciated functional significance is that some of the output synapses are with glial cells.

3.3.4. Membrane properties

The axons of motor neurons carry spikes that are primarily dependent on an inward Na^+ current, but the rest of the neuron, including the primary neurite, the mass of fine neurites and the cell body, does not normally support spikes. Spikes recorded in a cell body are thus the electronic remnants of spikes reflected from the axon. Similarly, synaptic potentials recorded in a cell body are also reflections of distant events because synapses are only made onto the neurites. The variety of shapes of the motor neurons would suggest that these potentials will decrement to different extents in different neurons. Thus, in the fast extensor motor neuron, the spikes recorded in the soma are about 20 mV in amplitude and 3 ms in duration at half-amplitude, those in a spiracular closer motor neuron are about 15 mV in amplitude but of much shorter duration, while those in most other wing and leg motor neurons are only a few millivolts in amplitude and sometimes difficult to distinguish from the synaptic potentials generated at sites closer to the cell body. All of these differences can be explained by the geometry of a neuron, but intrinsic differences in the membranes of different motor neurons, even those within the same motor pool, result in different frequencies of spikes in response to the same injected or synaptic currents. For example, the slope of the relationship between current and frequency of evoked spikes is about 13spikes.s^{-1}.nA^{-1} for the slow coxal depressor motor neuron of the cockroach but only 2.7spikes.s^{-1}.nA^{-1} for the fast motor neuron to the same muscle (Meyer and Walcott, 1979). This appears to reflect real differences in input resistance, conductances associated with the spikes, and voltage thresholds for spikes between the motor neurons and is not a result of experimentally applying the current to different parts. These differences may also reflect the effectiveness of synaptic currents and may therefore determine the order of recruitment of the motor neurons and their response to common synaptic inputs.

3.3.4.1. *Excitability of the cell body*

Normally, spikes invade the cell body passively from the distant spike-initiating zone in the neuropil. Over most voltages, the membrane behaves linearly, but at depolarised potentials, the membrane rectifies, probably because of the activation of a rapid transient and a delayed rectifying K^+ current. The cell body of an identified cockroach motor neuron shows an N-shaped current–voltage relationship at membrane potentials of about +100 mV that indicates the activation of a voltage-sensitive Ca^{2+}-activated K^+ current with a reversal potential close to normal resting potential (Thomas, 1984). This current will ensure that any inward current carried by Ca^{2+} will be countered by the rapid activation of an outward current carried by K^+ and thus result in the soma being inexcitable. It follows that if these outward currents are suppressed experimentally, the soma membrane should be able to support spikes.

A few days after the axon is cut or colchicine (which disrupts intracellular tubules) is applied, the cell body starts to generate spikes predominantly dependent on Na^+ that can still be recorded 1–2 weeks later (Pitman *et al.*, 1972a; Pitman, 1975; Goodman and Heitler, 1979). Anoxia can also cause the delayed appearance of Na^+-dependent spikes that persist for several days, and which are not caused by changes in intracellular pH (Pitman, 1988). The ionic dependence of these spikes is thus similar to those in the axons. Cockroach motor neurons can also produce Ca^{2+}-dependent spikes when calcium chelating agents such as citrate ions or EGTA are injected intracellularly, or tetraethylammonium ions are present extracellularly (Pitman, 1979). Similarly, after axotomy, the membrane can carry Ba^{2+} spikes so divalent cation channels must be present. When K^+ currents are blocked, the locust fast extensor tibiae motor neuron can generate overshooting spikes that are carried by Ca^{2+} and Na^+. Either of these two currents is sufficient as spikes can still occur in the presence of cobalt, or in the absence of external Na^+. Long-term recordings from the isolated somata of motor neurons show that their excitability changes with time and without pharmacological intervention (Hancox and Pitman, 1992). Initially, depolarising currents can produce only small oscillations in the membrane but after some time they evoke Ca^{2+}-dependent spikes with an amplitude of about 18 mV and a duration at half-amplitude of 3.5 ms. These spikes can also be driven by induced plateau potentials (*see section 3.3.4.3*) and do not reflect a gradual deterioration in the neurons.

Together, these results indicate that voltage-sensitive Na^+ and Ca^{2+} channels can either be unveiled in the soma membrane when they are normally masked by a much larger K^+ conductance, or induced if they are not normally present. The generation of spikes by the soma membrane is not the important consequence of the presence of these channels because no output synapses are made from this membrane. The channels may, however, be activated by the voltages generated by synaptic inputs and thus alter the response of the neuron in a nonlinear way. Furthermore, these properties may not be restricted to the soma but may also reflect the properties of the 'passive' membrane of the neurites that make and receive synapses. The existence of these voltage-sensitive channels also opens the possibility that they can be activated by neuromodulators and thus selectively alter the integrative processing under certain circumstances. These properties of a motor neuron add a further nonlinear component to their operation, indicating that they must be regarded as active elements in the shaping of the motor output and not merely the passive distributors of patterns generated elsewhere in the networks.

3.3.4.2. *Spike initiation*

A motor neuron receives a wealth of synaptic inputs to its many branches in the neuropil. These can sum to evoke spikes which then invade the soma passively and are propagated actively along the axon that leaves the central nervous system to innervate a muscle. Where in the neuron are the graded synaptic signals converted into overshooting spikes? The best answers to this question have come from an analysis of the largest motor neuron innervating a hind leg, the fast extensor tibiae motor neuron (FETi) (Gwilliam and Burrows, 1980) (Fig. 3.7), and while most motor neurons seem to conform to the mechanisms revealed, the hazards of extrapolation from one example must be recognised. In the axon, the spike overshoots zero, has a prominent after-hyperpolarisation and at half-height has a duration of about 1ms. At recording sites along the primary neurite, the spike progressively changes in shape, so that by the time it reaches the soma its rate of rise is slowed, the amplitude is about 20 mV, the after-hyperpolarisation is reduced and prolonged, and the duration at half-height is now 3 ms. Active membrane has clearly given way to passive membrane. Three experimental strategies indicate that the transition between these two types of membrane and the spike-initiating zone

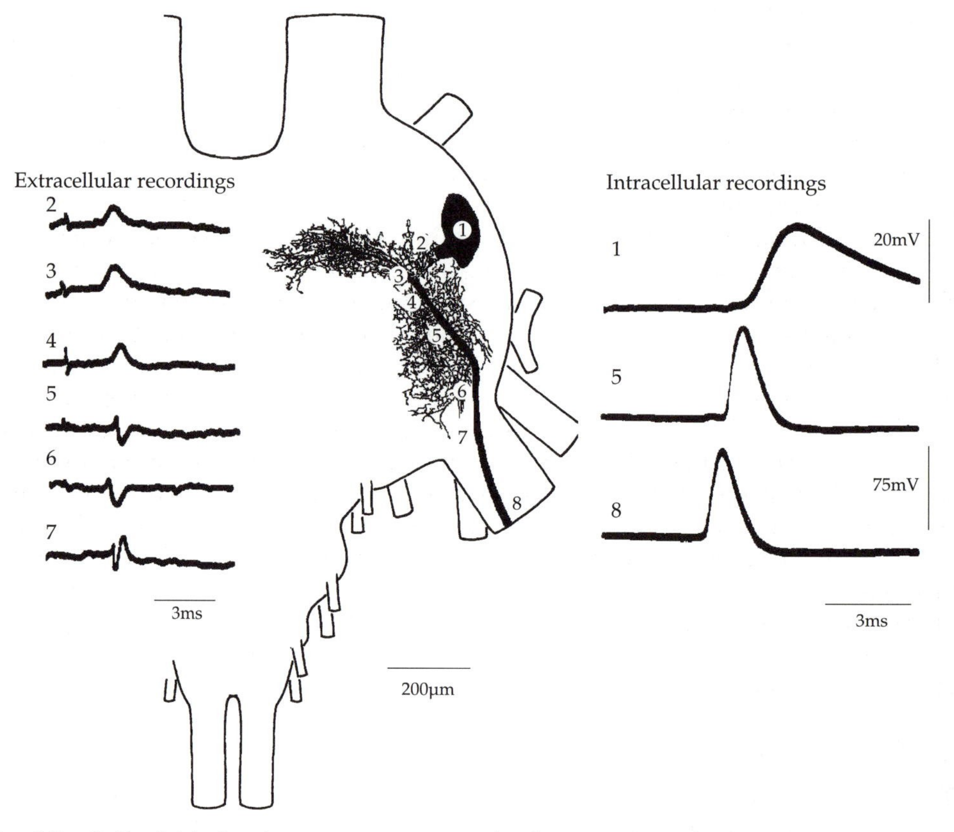

Fig. 3.7 Spike initiation in a motor neuron, the fast extensor tibiae in the metathoracic ganglion. The extracellular recordings of antidromic spikes evoked by stimulation in the muscle in the left-hand column were made at the sites numbered on the drawing in the middle. The change in the waveform suggests that spikes are initiated where the primary neurite emerges from the neuropil. The recordings in the right-hand column were made simultaneously at the three sites indicated. The spike is delayed and attenuated as it propagates passively into the cell body. Based on Gwilliam and Burrows (1980).

corresponds to a zone of the primary neurite within the neuropil that gives rise to many side branches. First, intracellular recordings chart the progressive change in the rise time, amplitude and duration of the spike at different points along the neurite. Second, the time taken for an antidromic as compared to an orthodromic spike to travel between the cell body and a recording site can be measured. Third, extracellular recordings made just before the electrode penetrates the neurite show that the field potential of the spike is monophasic in the neurite leading from the cell body but changes to a bi- or triphasic waveform further along the neurite where it becomes capable of actively propagating a spike. It seems likely that the initiation is from a zone rather than a point and that the actual initiation can shift within this zone. Moreover, the spikes do not actively propagate along the fine side branches. Within the ganglion, the spike travels at about 0.8 $m.s^{-1}$ while in the axon, both ortho- and antidromic spikes travel five times faster at about 4.1 $m.s^{-1}$.

This design of a motor neuron may allow faster initiation of spikes because the large size, and hence large capacitance of the cell body is kept distant from the spike-initiating zone and will thus act as less of a shunt for the synaptic currents. The inexcitability of the main neurite and the cell body also prevents the spikes actively reaching the cell body and perhaps then being reflected back into the main neurite and thus possibly into the axon.

The distribution of voltage-sensitive Na^+ channels can be revealed by antibodies raised against specific sequences of peptides in the channel proteins. When applied to a thoracic ganglion of a cockroach, such an antibody reveals intense staining of axons in tracts, commissures and nerves but little staining in the neuropil, or on the membranes of the cell bodies (French *et al.*, 1993). The density of channels in the membrane is about 90 per μm^2 which is thus similar to that in the membrane of lobster nerves (Ritchie *et al.*, 1976). The distribution is thus in accord with the physiological evidence about the location of the spike-initiating zone and with the expectation that the spikes spread only passively into the neurites within the neuropil and into the cell body. Nevertheless, intense staining also occurs in the cytoplasm of the cell bodies representing either the channels themselves, their precursors or their breakdown products in transit to the axon membranes. It should now be possible to extend this method to identified neurons and therefore relate the distribution of these channels more directly to their known physiological functioning. It would be particularly informative to compare spiking local neurons with nonspiking local neurons.

3.3.4.3. *Plateau potentials*

The active participation of the membrane of a motor neuron in integration is further emphasised by its ability to produce plateau potentials in response to an applied or a synaptically driven depolarisation (Hancox and Pitman, 1991, 1993) (Fig. 3.8). Bistable membrane properties could therefore contribute to the shaping of motor patterns. A brief depolarisation of an identified cockroach motor neuron above a certain threshold produces a prolonged depolarisation and spikes that can be terminated by a brief hyperpolarising pulse. The membrane of an isolated soma can also generate these potentials, indicating that they are an inherent property of

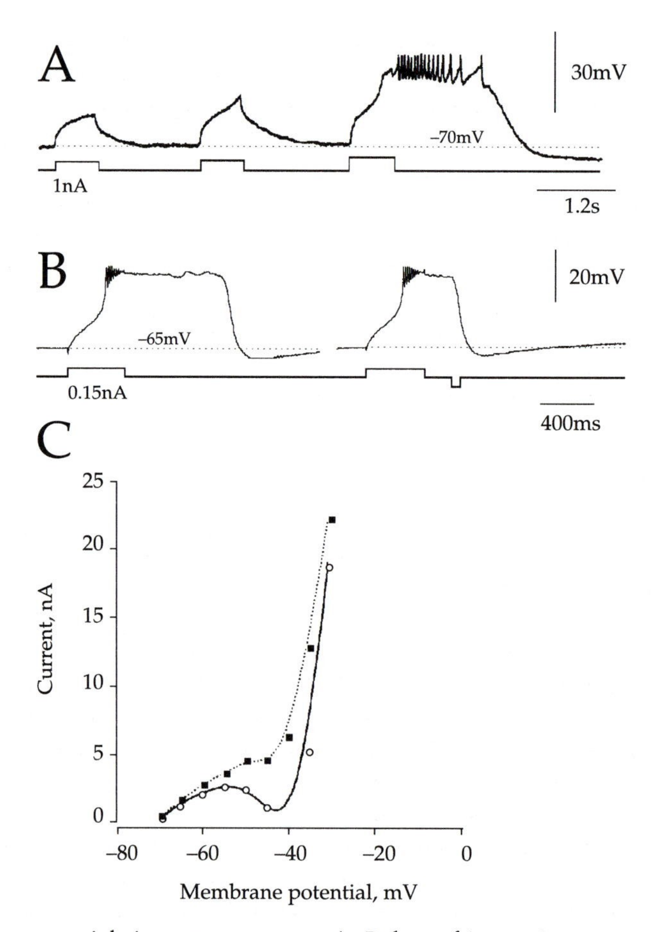

Fig. 3.8 Plateau potentials in motor neurons. A. Pulses of increasing current injected into a leg motor neuron of a cockroach cause a depolarisation that may lead to spikes and outlast the current pulse. B. A plateau potential induced by a depolarising current can be terminated abruptly by a brief hyperpolarising current. C. Current voltage relationship in a motor neuron showing plateau potentials (open circles) and in one that does not (filled squares). Based on Hancox and Pitman (1991).

the membrane independent of the spike-initiating zone. The potentials are accompanied by a large increase in the conductance of the membrane, indicating the activation of a large inward current probably carried by Ca^{2+}. The voltage-dependent current also appears as a region of negative slope in the current-voltage relationship, especially if the currents are measured 600 ms after the imposed change. How widespread these properties are amongst the different motor neurons has yet to be tested, but it has to be recognised that some can produce these potentials in response to voltage changes caused by synaptic inputs, and without the experimental addition of drugs. Potential neuromodulators such as octopamine can also induce these currents in other neurons (*see Chapter 5*), and some locust motor neurons will show them in the presence of picrotoxin. This property of the membrane means that a neuron will respond in a nonlinear way to the synaptic inputs that it receives, depending on whether the voltage changes are sufficient to activate or deactivate the relevant currents; a brief excitatory synaptic input may, if the conditions are right, generate a prolonged response, and a brief inhibitory input will terminate a prolonged depolarisation. The neuron can thus be switched suddenly between two excitatory states by brief synaptic input, obviating the need for sustained synaptic inputs to maintain a sustained shift in the membrane potential. These plateau potentials are a further example of the way that the nonlinear membrane properties of neurons contribute substantially to integrative processing and to the shaping of motor patterns.

3.3.5. Controlling the actions of a motor pool

Members of a motor pool are not coupled to each other by electrical junctions or by synaptic pathways, and with one notable exception (a connection between the fast extensor motor neuron and flexor motor neurons in the metathoracic ganglion – *see Chapter 9*), antagonistic motor pools show no direct coupling. This is in marked contrast to the organisation in crustacea, where electrical coupling between motor neurons is widespread (e.g. Heitler, 1978). It has been suggested that particular motor neurons of a pool may be reserved for specific actions. For example, particular slow motor neurons may be used to control limb muscles for reflex responses while others are used for generating locomotory movements (Zill and Moran, 1982). Particular fast motor neurons may only be used for certain explosive movements. There is little evidence for such complete segregation. Instead, most pools of motor neurons appear to be driven by common sources of synaptic inputs with their inherent membrane properties determining their spike response and hence their order of recruitment. This form of control is widespread and extends to motor neurons involved in most movements, so that particular pools of motor neurons innervating muscles involved in flying, breathing, jumping and walking are driven by common synaptic inputs in their respective movements. For example, the slow and fast extensor tibiae motor neurons receive many synaptic inputs in common so that both are depolarised together during most movements, but with the lower threshold of the slow motor neuron determining that it alone spikes during certain movements such as walking. Both motor neurons do, however, receive specific synaptic inputs from sensory neurons and projection interneurons (*see Chapter 9*), so that, for example, the excitation of the fast can be increased above that of the slow in certain movements that require rapid actions of the extensor tibiae muscle. Some nonspiking interneurons excite all members of a motor pool, recruiting them in accordance with the rapidity of their action, while other interneurons excite only particular motor neurons.

The pattern of common synaptic inputs, different inherent membrane properties and spike thresholds that are modifiable, coupled with specific inputs to selected members of a motor pool impart great flexibility on the use of individual motor neurons for different tasks. Nevertheless, the small size of most motor pools means that gradation of contraction must be achieved by varying the frequency code in a small number of channels rather than by recruiting additional channels. Slow movements can be produced by slow motor neurons exciting the slow muscle fibres and faster movements by the recruitment of fast motor neurons exciting fast muscle fibres. The small number of motor neurons in a pool is compensated by a shift in control to the neuromuscular junctions and to the muscles themselves. This shift is further emphasised by the action of the inhibitory motor neurons whose action at the muscles can reduce the residual tension in slow muscle fibres so that faster movements can proceed.

3.3.6. Controlling the force produced by a muscle

The force that is generated by a muscle depends on a complex interplay between the actions of the excitatory motor neurons, the inhibitory motor neurons and the neuromodulatory DUM neurons. The force also depends on the history of the activity of these neurons as the muscles themselves have catch-like properties, the substances released by the neuromodulatory neurons can act over long periods, and there may be long-term cycles in the concentrations of various neurohormones and neuromodulators circulating in the haemolymph. These nonlinearities in the response of a muscle to a particular pattern of spikes in the excitatory motor neurons means that the force generated cannot easily be predicted; this is a prime reason why the locust, along with most other animals, must monitor its actions with sensory neurons.

3.3.6.1. *Catch-like tension*

The ability of a muscle to produce different levels of force in response to a set motor command depends on the history of the motor signals that it has received (Blaschko *et al.*, 1931). The phenomenon is most easily demonstrated by stimulating a single motor neuron to a muscle at a low constant frequency until the force has reached a plateau. The interjection of a few stimuli at higher frequency then cause the force to rise and to remain at a higher level when the original frequency of stimulation resumes. There is thus a mismatch or 'catch' between the force produced before and after the high frequency stimulation. The ability of a muscle to produce catch-like tension is also dependent on the presence of neuromodulators such as octopamine released by DUM neurons (*see Chapter 5*). Little is known of the effects of this form of force generation and its modulation in behaviour, but where precise control is needed, such as the control of steering during flight, it seems that it might play an important role. There remains reasonable doubt, however, as to whether it is generated in normal behaviour, because the occurrence of a few spikes in a common inhibitory motor neuron that innervates many of these muscles might be sufficient to abolish all potential catch-like effects.

3.3.6.2. *Common inhibitory motor neurons*

All the inhibitory motor neurons innervate several muscles, often those that are used antagonistically, so that they cannot be used to alter the force generated by one muscle. Their action is instead to reduce the force that is generated by slow motor neurons and to accelerate the rate of muscular relaxation. Through these actions they can abolish basal tension in muscles, or reduce the tonic forces generated by spikes in slow motor neurons so that a rapid movement is not impeded by any lingering force. A prime role is thus the control of the rate of relaxation of the muscles. The implication behind this is that the motor patterns are organised to ensure the correct temporal relationship in the action of the inhibitory and excitatory motor neurons. A clear example of this action is the simultaneous excitation of inhibitory neurons and the inhibition of excitatory motor neurons to a particular muscle when the locust is about to jump (*see Chapter* 9).

The organisation of the inhibitory motor neurons has many parallels with that in crustacea (Wiens and Wolf, 1993). In both, just three inhibitory motor neurons innervate a limb. Their cell bodies are in similar places in the respective ganglia, their primary neurites follow similar paths through the neuropil, the distribution of their innervation of limb muscles is comparable, and they probably all use GABA as their transmitter. Common inhibitor 1 (CI1) in the locust innervates many muscles that move the proximal joints of the leg (*see Fig.* 8.6), some of which are normally antagonistic (Hale and Burrows, 1985; Wolf, 1990a). The other two inhibitors (CI2 and CI3) innervate fewer of the more distal muscles. Similar patterns of innervation hold for the three crustacean inhibitors, although one innervates all leg muscles and another only the dactyl opener muscle. These parallels are an example of neurobiology, backed by molecular biology (Ballard *et al*., 1992), giving further insights into evolutionary ancestries and relationships than have been apparent from the seemingly endless debate about the origin of uniramous and biramous limbs.

3.3.6.3. *Transmitters*

Excitatory motor neurons Most of the excitatory motor neurons are thought to use glutamate as their transmitter, based on evidence obtained from a few. Many show immunoreactivity to glutamate in their cell bodies and in their terminals on the muscles (Bicker *et al*., 1988). There is much evidence for glutamate being released at their neuromuscular junctions and causing a depolarisation by opening K^+ and Na^+ channels (Usherwood, 1994). Some excitatory motor neurons of the cockroach may co-release a transmitter with a fast action that is presumed to be glutamate, and one with a slower action that is thought to be the peptide proctolin (Adams and O'Shea, 1983) (*see Chapter* 5 for more detail). The muscle fibres also have extra-junctional receptors, for glutamate at least, but these are associated with Cl^- channels that result in a hyperpolarisation of the membrane. It seems unlikely that the action of the excitatory motor neurons has any effect on these receptors and channels.

Inhibitory motor neurons The inhibitory motor neurons evoke hyperpolarising potentials in the muscle fibres by releasing GABA which causes a transient increase in the permeability of the muscle membrane to Cl^- (Usherwood and Grundfest, 1964, 1965) (*see Chapter 5*). The reversal potential for these GABA-evoked inhibitory postsynaptic potentials (IPSPs) is close to resting potential, so that under some experimental circumstances they can appear as depolarising potentials. This was one of the reasons that led, for a short period, to these neurons being called inhibitory conditioning neurons (Hoyle, 1966). Further evidence that GABA is the transmitter at these inhibitory synapses is provided by the GABA-like immunoreactivity of the three inhibitory motor neurons that innervate leg muscles, but by none of the other excitatory leg motor neurons (Watson, 1986).

3.4. SENSORY NEURONS

3.4.1. General features

Sensory neurons provide a large number of input channels to the central nervous system, in stark contrast to the small number of output channels of the central nervous system as represented by the motor neurons. The most numerous sensory neurons are the hundreds of thousands that signal odours detected by receptors on the antenna, and the numerous other small neurons that function as contact chemoreceptors and are associated with receptors distributed all over the body. Exteroceptors on each leg also outnumber leg motor neurons by almost 40 times. The few sensory neurons involved in hearing are an exception to the general mass of sensory neurons that signal the other modalities.

The majority of sensory neurons have cell bodies in the periphery close to the receptor with which they are associated and axonal projections to the central nervous system. An exception are the strand receptors that act as mechanoreceptors at the joints and which have cell bodies in their segmental neuromeres. Spikes are the usual means of intracellular signalling because the distances that must be covered are large and the diameters of the axons are small. Notable exceptions are the visual receptors, the retinula cells, which use graded signals. Each type of receptor has a characteristic pattern of central branches, typically restricted to the ipsilateral half of the neuromere of the segment in which it is located, but some have projections to more distant neuromeres. Most of the central branches have input synapses that modify the effectiveness of the signals and thus control the information that is delivered to their postsynaptic target neurons (*see Chapter 7*). The effectiveness of signalling may also be altered by the action of neuromodulators that act on the coding of a stimulus at the receptor and perhaps also on the output of the signals at the central synapses (*see Chapter 5*). This implies that the peripheral cell bodies or dendrites have receptors for particular neuromodulators but this facet of their functioning is largely unexplored. There is, however, no direct efferent control of most receptors and their associated sensory neurons.

3.4.2. Morphology

The sensory neurons from the different types of receptors form characteristic patterns of projections in the central nervous system and are described in detail elsewhere (*see Chapters 8, 9 and 11*). Three generalisations can be made about the projections of mechanosensory neurons.

First, the sensory neurons from receptors on the distal parts of a limb project only to the ipsilateral half of their segmental ganglion, but more proximal receptors may have both a contralateral projection and a projection to the next anterior ganglion. For example, the single sensory neuron of a tactile hair on a femur, and the sensory neurons from the femoral chordotonal organ, all project to the ipsilateral half of their segmental ganglion, whereas those from hairs on the sternum of the thorax or abdomen, and those from thoracic chordotonal organs have an additional projection to the next anterior ganglion. Sensory neurons from hairs on the prosternum (ventral neck) have either ipsilateral or contralateral projections, depending on the position of the receptor relative to the midline. In the cricket, a pair of long hairs on a thoracic sternum project only to the ipsilateral half of their segmental ganglion, whereas those from hairs in the same position on the abdomen project to both halves of their segmental ganglion and to the next three anterior ganglia (Hustert, 1978), suggesting that they are involved in coordinating the movements of the abdominal segments. Sensory neurons from hairs on the head that detect wind project to the brain, suboesophageal ganglion and some as far as the prothoracic ganglion.

Second, the sensory neurons from different classes of receptor project to different regions of neuropil. For example, sensory neurons from exteroceptive mechanoreceptors project to the most ventral regions of neuropil whereas those from joint proprioceptors project to separate and more dorsal areas of neuropil. Mechanoreceptors from the tympanum (ear) project to a further distinct region of neuropil. This segregation presumably simplifies the organisation of the neurites of the postsynaptic interneurons that must sample and integrate these signals.

Third, the central projections of the sensory neurons are highly ordered according to some feature of the signal extracted by the particular type of receptor which they supply. The projections of visual receptors of the compound eye are arranged retinotopically in an array of columns that correspond to the number of ommatidia in the eye, auditory receptors are arranged tonotopically depending on the frequency characteristics of the receptors, and sensory neurons from exteroceptors are arranged topographically according to the spatial positions of the receptors on the body. This means that the vast numbers of sensory terminals are arranged as maps of some feature of the stimulus so that the task of delivering their signals to the correct postsynaptic neurons is greatly simplified. The terminals of the exteroceptors, for example, form a spatial map of a leg that corresponds to the branches of particular postsynaptic interneurons that read specific parts of that map.

3.4.3. Structure and distribution of synapses

The characteristic feature of the terminals of most types of sensory neurons in the central nervous system of virtually all animals, locusts included, is that they receive input synapses close to the output synapses that they make with other neurons. A consequence of this arrangement is that there is no guarantee that the spikes they have carried from their sense organ will be reliably delivered to all their output synapses with different neurons.

3.4.3.1. Input synapses

The terminals in the prothoracic ganglion of the sensory neurons from hairs on the neck illustrate the usual synaptic arrangements of the input synapses on mechanosensory neurons (Watson and Pflüger, 1984). Diadic input synapses can occur within 0.5 μm of an output synapse and the two types of synapses are intermingled on all the neurites, although their distribution may often be clumped rather than uniform. About half the input synapses show GABA-like immunoreactivity associated with pleomorphic vesicles, 40% show glutamate-like immunoreactivity, whilst the putative transmitters in the remaining synapses are unknown. At the terminals of other mechanosensory neurons the proportions of the presynaptic neurons with these immunoreactivities may be different, but it is uncertain what this might imply about the presynaptic control of the sensory signals (*see Chapter* 7). Processes that are presynaptic to the terminals of a sensory neuron and that have GABA-like immunoreactivity may also be presynaptic to other unlabelled processes, and these, in turn, may be presynaptic to the sensory neuron. This suggests that GABA is a major, but by no means the only, transmitter to change the properties of the terminals of the sensory neurons. The only available evidence points to presynaptic alterations being effected by GABA, but the morphology clearly shows that other mechanisms, or different manifestations of the same inhibitory mechanism, must exist.

From all the mechanosensory afferents that have been examined so far, the following features of the presynaptic input synapses can be extracted (Watson, 1992a). First, the synapses made by a presynaptic neuron usually involve two postsynaptic processes that can either belong to two sensory neurons, or to one sensory neuron and another type of neuron. Second, a presynaptic process can synapse on other presynaptic processes close to where they all synapse with a sensory neuron. Third, the same presynaptic neuron may make several successive contacts with the same branch of a sensory neuron, and the sensory neuron may make output synapses with this presynaptic neuron, thus forming reciprocal synapses. These arrangements indicate that complex interactions take place between the neurons that exert a presynaptic control over the signalling of sensory neurons.

None of the neurons responsible for this presynaptic control have yet been identified, although they can be characterised from the synaptic potentials recorded in identified sensory neurons in particular circumstances. Identifying these neurons will be essential for any understanding of sensory signalling, and for understanding the way these signals might influence or control the expression of motor patterns. At present, there are only a few clues as to the identity of these neurons, so that finding them promises to be a tedious, if ultimately rewarding, process.

3.4.3.2. *Output synapses*

The output synapses of mechanosensory neurons, as illustrated by those from prosternal hairs, have a presynaptic bar, a row of densities spaced 85 nm apart and a population of round agranular vesicles 37 nm in diameter. The number of these output synapses that a sensory neuron makes with a particular central neuron can rarely be measured, but the most accurate figures indicate that fly retinula cells make 200 synapses with a lamina monopolar neuron (Nicol and Meinertzhagen, 1982). The stretch receptor of a locust wing is estimated to make 600 synapses with a wing motor neuron (Peters *et al.*, 1985), and a cercal hair afferent in a cockroach to make 85 synapses with a giant interneuron (Blagburn *et al.*, 1985). These values are estimates, based on the unfounded assumption that the density and distribution of synapses will be the same throughout the array of neurites as in the few region(s) that are sampled.

3.4.4. Integrative properties

The diversity of the different sorts of sensory receptors means that few generalisations are possible about the actions of their sensory neurons. The initial coding of the sensory signals will depend on the properties of the receptor and the membrane properties of the sensory neurons. The reliable delivery of these signals, usually coded as sequences of spikes, will depend on events at their terminals in the central nervous system. Here, the input synapses, receptors for neuromodulators, the membrane properties of the terminal branches and the transmitter(s) that is released will determine the signals that are delivered to postsynaptic neurons.

3.4.4.1. *Membrane properties*

The combination of the properties of the receptor and the inherent membrane properties of the sensory neuron is particularly apparent in determining the signalling by mechanoreceptors. These sensory neurons may adapt with varying degrees of rapidity to a maintained stimulus because of the mechanics of the receptor and the currents that flow across the membrane of the sensory neuron. The stiffness of a hair and the rigidity of its mounting in its socket clearly contribute to their response; filiform hairs with long, thin shafts and flexible mountings respond to air currents, whereas hairs with short, stout shafts and more rigid mountings respond only to tactile stimulation. In cockroach tactile spines, the spikes of the single sensory neuron adapt completely within a second to a maintained stimulus;

this is caused mainly by the processes that lead to the encoding of the spikes rather than transduction of the stimulus (French and Torkkeli, 1994). The neuronal mechanism is probably the slow inactivation of the Na^+ current that underlies the spike, because if this is prevented then the rapid adaptation is also prevented. At least three K^+ currents also play a part. First, a Ca^{2+}-activated K^+ current carries almost half the outward current and determines much of the time course of adaptation. Second, a rapid 'A' current is activated by the hyperpolarisation at the end of a spike, with a threshold of about –75 mV, and causes a progressive rise in spike threshold during a burst of spikes (Torkkeli and French, 1994). It is specifically blocked by 4-aminopyridine which therefore prolongs the spike response to a step stimulus by lowering the threshold for the generation of spikes without altering the duration of individual spikes. The activation and inactivation of this current are, however, too fast to contribute much to normal adaptation. Instead, its action is to prolong interspike intervals by opposing depolarisation and to increase the threshold for generating spikes. Third, a 'delayed rectifier current' controls the duration of the action potentials but does not contribute much to adaptation.

3.4.4.2. *Coupling between receptors*

The signalling of some sensory neurons is influenced by their electrical or synaptic coupling to other receptors. Retinula cells in the eye are electrically coupled to each other and this may improve the signal-to-noise ratio in their signalling by graded potentials to the interneurons in the first layer of the optic lobes. Electrical coupling has not been demonstrated for other receptors. Some mechanosensory neurons from a chordotonal organ in the leg of *Drosophila* are thought to be synaptically coupled (Shanbhag *et al.*, 1992). These interactions occur between the axons on their route to the central nervous system so it is unclear what effect to expect on the patterning of the sensory spikes and therefore on the information coded at the receptor in the different sensory channels. In crustacea, electrical coupling between mechanosensory neurons seems to be widespread (Wildman and Cannone, 1991; Marchand and Leibrock, 1994), but has not been demonstrated in locusts. The reasons for the different design solutions adopted by a crayfish and a locust to essentially the same problem of monitoring the movements of the body remain as elusive as they are intriguing.

3.4.4.3. *Transmitters*

There is no universal transmitter for sensory neurons or universal polarity of the effect caused in postsynaptic neurons. Retinula cells in the compound eyes of flies, and ocellar photoreceptors in locusts make inhibitory connections with interneurons and probably use histamine as their transmitter (Hardie, 1987). A generalisation which so far seems to hold is that all mechanosensory neurons make excitatory synapses in the central nervous system, probably by releasing acetylcholine (*see Chapter 5*). It is noteworthy that in *Drosophila*, but not some other flies, the sensory neurons of the hair sensilla are immunoreactive to histamine

(Buchner *et al.*, 1993). Sensory neurons from campaniform sensilla or scolopidial organs show no histamine-like immunoreactivity. Whether this indicates that the real transmitter in most insects is more closely related to histamine than acetylcholine, or that *Drosophila* is unusually different is an interesting conjecture. In a leg chordotonal organ, some sensory neurons are immunoreactive to 5-HT, suggesting that this may be a possible transmitter or even a co-transmitter (Lutz and Tyrer, 1988).

Most of the evidence that acetylcholine is the transmitter used by mechanosensory neurons comes from the cercal hair sensory neurons in cockroaches and their connections with giant interneurons (Sattelle, 1985). In locusts, this conclusion is backed by the presence of acetyltransferase in mechanosensory neurons (Emson *et al.*, 1974) and by the immunoreactivity to acetyltransferase shown by many mechanosensory neurons in legs (Lutz and Tyrer, 1988). Moreover, pharmacological evidence suggests that sensory neurons from campaniform sensilla on the tibia also appear to use acetylcholine as their transmitter in the direct connections that they make with particular leg motor neurons (Parker and Newland, 1995) (*see Chapter 5* for the pharmacological evidence, and *Chapter 9* for the role that these receptors play in influencing leg movements).

3.4.4.4. *Release properties on different target neurons*

A further variable in the signalling of a sensory neuron is that the release properties of its terminals on its various target neurons may differ so that the effect on each is different. The sensory neuron from a particular hair on a locust hind leg makes synapses with different effective gains on different spiking local interneurons so that a deflection of the same hair may be able to evoke spikes in one interneuron, but only a depolarisation in another interneuron (Burrows, 1992; *see Chapter 7*). The cause is probably the different quantal contents of the different synapses rather than their different placement relative to the spike-initiating zone in the different interneurons. The synapses made by the same cercal hairs of a cricket may generate synaptic potentials that facilitate in one projection interneuron, and potentials that depress in another, suggesting that the release properties are determined locally in the branches of the sensory neuron, possibly under the influence of the postsynaptic neurons themselves (Davis and Murphey, 1993).

3.4.4.5. *Regulation by presynaptic receptors*

Muscarinic receptors are present on the terminals of many sensory neurons close to their output synapses and close to the input synapses on postsynaptic neurons and can alter synaptic transmission mediated by acetylcholine through pre- and postsynaptic actions that probably involve different types of receptors (*see Chapter 5*). Most of these effects are known for mechanosensory hairs in *Manduca sexta*, but similar events also occur in the wing hinge stretch receptor of the locust (*see Chapter 11*) and the sensory neurons from tibial campaniform sensilla. The normal action of the presynaptic receptors is to reduce the release of acetylcholine and hence reduce the amplitude of an excitatory postsynaptic potential (EPSP) evoked in a postsynaptic neuron so that if these receptors are blocked, the EPSP becomes larger. By contrast, the postsynaptic receptors increase the excitability of the target neurons by causing a parallel, but slower depolarisation to the released acetylcholine, so that there are two opposing effects.

3.5. LOCAL INTERNEURONS

3.5.1. What are local interneurons?

The term local interneuron is an arbitrary but nevertheless useful definition based on the morphology of a neuron. In the segmental chain of ganglia, it can be applied rigorously to a neuron whose processes are restricted to one neuromere or segment. The term then has functional meaning, for it implies that the interneuron is concerned with processing signals within that segment and generating outputs only to neurons in that segment, although it may be influenced by signals in intersegmental interneurons that originate elsewhere. In the brain, the term becomes less useful because distinctions between regions may not be so anatomically clear and such divisions may have little functional meaning. It may therefore define an interneuron whose branches are restricted to one region of neuropil such as the lamina (lamina monopolar cells) or mushroom bodies (Kenyon cells), but also a neuron that projects from one region to another. In the rest of the nervous system, it is unhelpful to extend this use of the term to describe neurons that project from one ganglion to an adjacent one (Marquart, 1985) when more meaning is conveyed if these are called intersegmental interneurons. Inevitably, the definition becomes blurred, particularly where segmental neuromeres are fused into one ganglion, and a local neuron has branches extending to more than one of these neuromeres. The justification for using the term in these circumstances is if the neuron belongs to a population of other, more obviously local, interneurons. The use of this terminology is designed as an aid to understanding and concern over the precision of its use should only become paramount when this objective is obscured.

3.5.2. How many local interneurons are there?

Estimating the number of local interneurons is fraught with difficulties, even in a segmental ganglion. The only method available to date is to count the number of cell bodies in each ganglion [a thoracic ganglion is estimated to contain the cell bodies of 2000 neurons, and an abdominal ganglion 500–600 neurons (Goodman and Bate, 1981; Sbrenna, 1971)] and from this, subtract the number of known motor neurons, other efferent neurons and the intersegmental neurons that can be identified by various means. The cell bodies that are left are assumed to belong to local interneurons. Such crude estimates indicate that perhaps 60% or 1200 neurons in a thoracic and 360 in an abdominal ganglion are local interneurons. These are imprecise figures and should be taken as no more than the rough estimates that they are, but they do indicate the enormous contribution that local interneurons must make to central processing.

3.5.3. Types of local interneurons

The local interneurons are subdivided, using their ability or inability to generate spikes during the normal course of their integrative actions, into spiking local interneurons and nonspiking local interneurons. This does not imply that **all** the intercellular communications of spiking local interneurons must necessarily depend on their generating spikes, or conversely that the nonspiking neurons are restricted only to a graded action. It is, however, a working assumption that these are the respective modes of operation of these distinct types of local neurons. Within these two major categories are many distinct groups of neurons with characteristic morphologies which perform distinctive roles in the integrative processes in different parts of the nervous system.

3.5.4. Are local interneurons identifiable?

The simple answer is that most are not, but a few are. Motor neurons and sensory neurons can be readily labelled in the sense that they can be traced to the effector or receptor that they innervate. Often it is possible to identify these neurons on the basis of their actions or their morphology alone, but for interneurons, the identification almost always has to rely on a combination of criteria. For projection interneurons, the destination of their axons generally allows the number of members in a population to be determined, and from this can eventually come the morphological characterisation of an individual. For local interneurons it is almost impossible to determine the number in a population either by anatomical or physiological methods. If anatomy is to be used then the population must be defined either as an aggregation of cell bodies in the cortex, or neurites in particular tracts. At present, it appears that the cell bodies of different types of interneuron are intermingled for there are no descriptions that allow coherent groupings as are possible for the much larger cell bodies of the motor neurons. If physiological methods are used, then the best that can be said is that interneurons with similar actions can be found in different animals. If anatomical and physiological methods are combined by staining an individual interneuron after its actions have been characterised, the best that can then be said is that interneurons with similar actions and similar shapes can be found in different animals. This does not identify the interneuron because there could be more than one with the same action and the same morphology. It is common to stain, often fortuitously, two or more interneurons with the same shape, which, according to current information, cannot be separated. The need then is to know from the numbers of physiological and anatomical types how many of each could be represented in the whole population. For example, the midline population of spiking local interneurons in each thoracic ganglion consists of perhaps 50 interneurons to each leg and at least 30 individual types are known. Most types cannot be represented more than once and thus almost certainly represent identified individuals, while some can only be ascribed to a small group of like individuals. For these, identification is an open-ended procedure so that the more that is learnt about them, the more it becomes possible to separate and perhaps identify them.

3.6. SPIKING LOCAL INTERNEURONS

3.6.1. General features

Spiking local interneurons occur in all parts of the central nervous system but are most numerous in the optic neuropils, mushroom bodies and antennal lobes where they have distinctive roles in processing particular modalities of sensory signals. There are also extensive groups of these neurons in the thoracic ganglia and much is known about the morphology and integrative actions of these neurons.

In a thoracic ganglion, three bilaterally paired groups of spiking local interneurons have so far been identified, but there are almost certainly more. The known groups are named after the location of their cell bodies in the ventral cortex; midline, antero-medial, and antero-lateral groups. These groups of interneurons seem to be repeated in each of the thoracic neuromeres but most have not been found in the abdominal neuromeres, perhaps because they are concerned with controlling the movements of the wings and the legs. A fourth group consists of unpaired neurons with cell bodies at the dorsal midline intermingled with the unpaired intersegmental and efferent neurons. These are the DUM (dorsal unpaired median) neurons. In each group, all the interneurons have the same basic shapes and similar actions, but individual interneurons have characteristic shapes and specific connections. Only rough estimates of the numbers of interneurons in each group can be given, because of the problems of identifying local interneurons and because the cell bodies in a particular group are often close to those of other neurons. The arrangement of the groups of cell bodies and the properties of the individual members suggest that one group is the progeny of a particular neuroblast, and for one of the groups that have been analysed this has proved to be true (*see Chapter 4*).

3.6.2. Morphology

The diversity of shapes is illustrated by the local interneurons in three of the four groups in a thoracic ganglion (Fig. 3.9). The first three groups are bilaterally symmetrical and occur in each thoracic ganglion, with striking similarities in the morphology and action of individual interneurons (Burrows and Watkins, 1986).

3.6.2.1. *Ventral midline interneurons*

These interneurons have their 10–15 μm diameter cell bodies amongst a group at the ventral midline, primary neurites in ventral commissure II and two extensive fields of neurites in ventral (VAC) and more dorsal neuropil linked by a process in

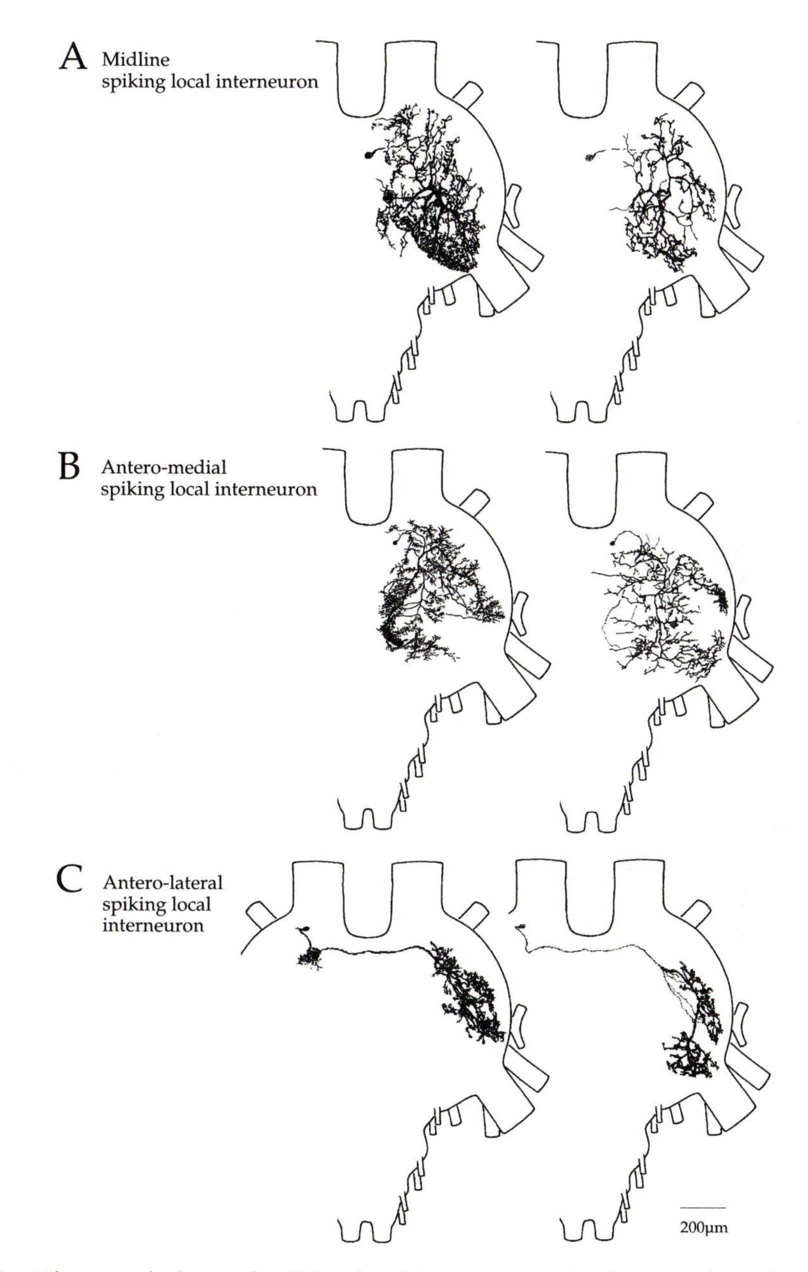

Fig. 3.9 The morphology of spiking local interneurons in the metathoracic ganglion. Individual neurons belonging to three groups are illustrated. A. An interneuron with its cell body in a ventral midline group. B. An interneuron with its cell body in a ventral, antero-medial group. C. An interneuron with its cell body in a ventral, antero-lateral group. All the interneurons have two fields of branches; one in ventral neuropil, shown on the left, and the other in more dorsal areas of neuropil. A is based on Burrows and Siegler (1984), B on Nagayama (1989), C on Siegler and Burrows (1984).

the perpendicular tract (Figs 3.9A and 3.10) (Siegler and Burrows, 1984; Burrows and Siegler, 1984). The cell bodies are in the contralateral half of a ganglion with respect to the main arrays of fine neurites. The cell bodies of interneurons that have neurites in the left side of the ganglion are intermingled with those which have neurites in the right half of the ganglion and, in total, about 100 neurons belong to each half. Not all of the cell bodies in this group belong to spiking local interneurons, for some are nonspiking interneurons and some are intersegmental interneurons with axons that project anteriorly. The ventral neurites can be profuse with a uniform appearance and occur in the most ventral and lateral regions of neuropil to which certain mechanosensory neurons from a leg also project. By contrast, the dorsal branches are sparser, often have a varicose appearance, and are in regions of neuropil to which the leg motor neurons and other interneurons have their main array of neurites. Each interneuron in this group has a shape that is an elaboration on this basic plan and that is correlated with its receptive field of receptors on a leg that provide its mechanosensory input. Interneurons with large receptive fields have large arrays of ventral branches and those with restricted fields have similarly restricted ventral branches. Moreover, some of the interneurons that receive inputs only from proprioceptors in a leg have only a small ventral field with no neurites in the most ventral regions of neuropil where there are no projections from the proprioceptors. At least 30 individuals are known in this group and their role in the processing of sensory information is well defined (*see Chapter 7*).

3.6.2.2. *Antero-medial interneurons*

These interneurons have their cell bodies in a cluster at the medial edge of an anterior connective and just medial to the somata of some of the large flight motor neurons in groups 2 and 3 that have axons in N3A. Amongst the 80 cell bodies in this cluster are both local and intersegmental interneurons, but the cell bodies of the local interneurons are more anterior and have diameters of only 10–15 μm compared with the 20–30 μm cell bodies of the intersegmental interneurons (Gramoll and Elsner, 1987). The primary neurites of the local interneurons enter the neuropil together in a bundle and then form extensive fields of branches in both ventral and dorsal neuropil linked by processes in an oblique tract (Fig. 3.9B). The ventral field of an interneuron consists of numerous fine branches in dense arrays and often in two distinct layers, whereas the more dorsal branches are less numerous. The ventral branches project to the ventral and lateral areas of neuropil where there are terminals from exteroceptors, while the dorsal branches can extend through neuropil where there are neurites of leg motor neurons and into neuropil occupied by neurites of wing motor neurons. Fifteen types of these local interneurons have been distinguished and placed in three subgroups according to their morphology and their physiology (Nagayama, 1989). Interneurons in the first subgroup have the most extensive and the most ventral branches and are excited by exteroceptors on a hind leg, those in the second subgroup have branches in more intermediate and lateral neuropil and are excited by proprioceptors signalling

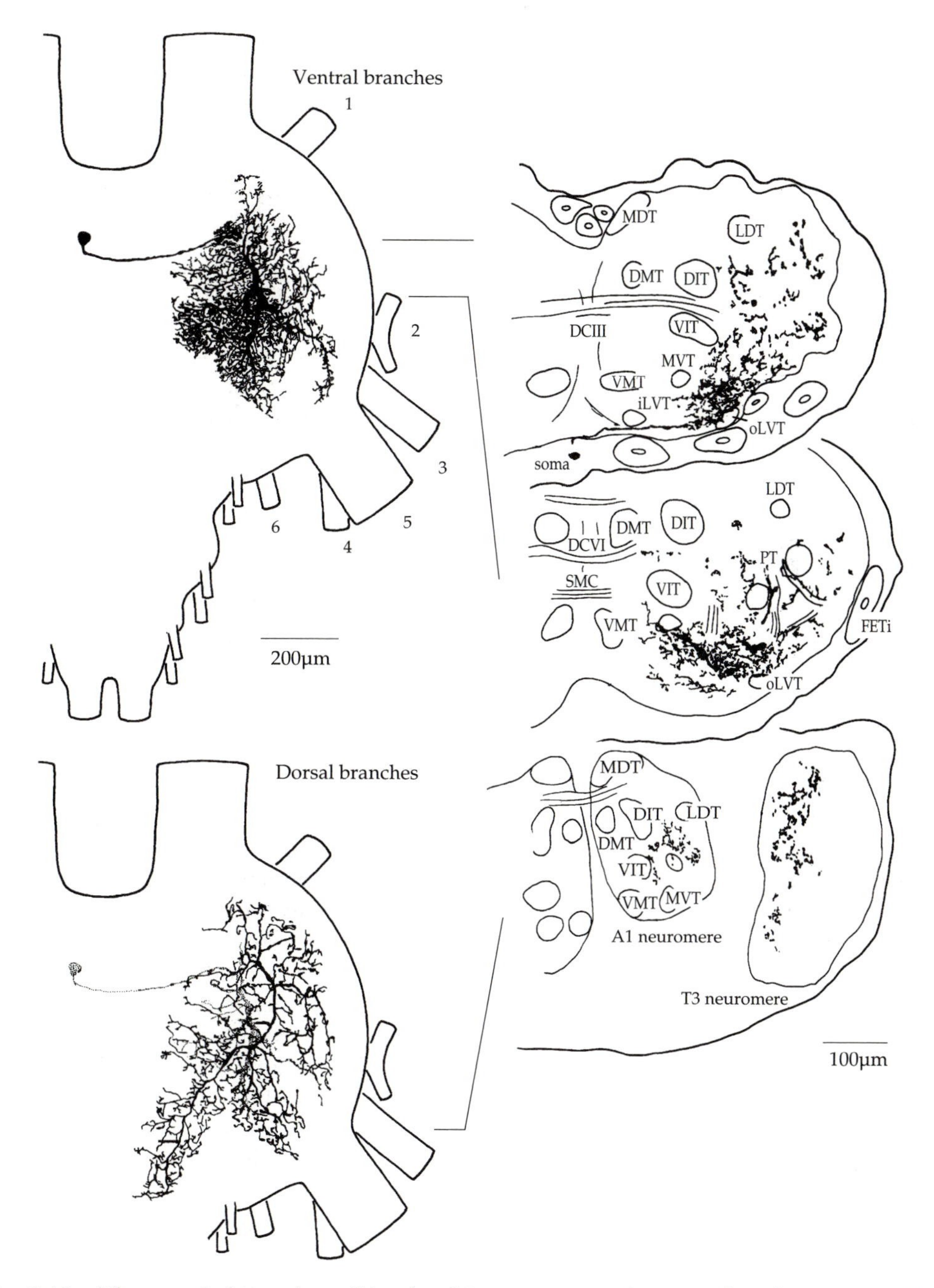

Fig. 3.10 The morphology of a spiking local interneuron in the ventral midline group. The upper drawing on the left shows the branches in ventral neuropil areas, and the lower one the branches in more dorsal neuropil. The transverse sections at the levels indicated relate the branches of this neuron to known structural features of the metathoracic ganglion (for definitions of the abbreviations see Table 2.2). Based on Burrows and Siegler (1984).

movements of the joints, and those in the third have only a few ventral branches and are inhibited by input from a hind leg. By contrast, the array of dorsal branches in each group is similar.

3.6.2.3. *Antero-lateral interneurons*

These interneurons have their 10–30 μm diameter cell bodies amongst a group of about 40 cell bodies close to the outer edge of an anterior connective of a thoracic ganglion, primary neurites in ventral commissure I and two extensive fields of neurites in the dorsal and ventral neuropil in the contralateral half of the ganglion linked by a process in the oblique tract (Fig. 3.9C) (Siegler and Burrows, 1984). Some of these interneurons may also have neurites in dorsal neuropil ipsilateral to the cell body so that they have potential sites for input and output synapses in both halves of a ganglion. The contralateral ventral branches are profuse and project into lateral and ventral regions of neuropil where sensory neurons from a leg terminate. The dorsal branches are sparser and project to neuropil where leg motor neurons and other interneurons have branches. Few of these interneurons are known in any detail, but they do spike during active or imposed movements of a leg and when exteroceptors are stimulated.

3.6.2.4. *Dorsal midline (local DUM) neurons*

The characteristic feature of these interneurons is that they have branches in both halves of the ganglion, usually with a clear symmetry to their arrangement (Fig. 3.11A) (Thompson and Siegler, 1991). They are, nevertheless, unpaired neurons that arise embryonically from the median neuroblast. The primary neurites enter the neuropil in a tract (the posterior DUM tract, PDT), seemingly composed entirely from DUM interneurons, and then branch twice to form two prominent neurites that run laterally in the supramedian commissure (SMC) in each half of the ganglion. Most of the neurons have branches in neuropil to which the auditory sensory neurons project (mVAC), with some neurites extending into more lateral and more posterior regions of neuropil. Many of the interneurons are excited by sound, but none of their input or output connections are known. These interneurons show GABA-like immunoreactivity, suggesting that they may have inhibitory actions, but when an individual interneuron is made to spike at a high frequency it does not cause any observable change in posture or movements.

The progeny of the median neuroblast were originally thought to consist of neurons with a spectrum of membrane properties, with the firstborn neurons all spiking, and all of the lastborn nonspiking (Goodman *et al.*, 1980). The local DUM neurons were thus considered as nonspiking, because no evidence could be found for voltage-sensitive Na+ channels. When outward rectifying currents were blocked, Ca^{2+}-dependent spikes could, however, be produced. Subsequent recordings from these neurons show that they produce conventional spikes in response to particular sensory inputs such as sound, or when depolarised with injected current (Thompson and Siegler, 1991). The assumption is that these are conventional Na^{+}-dependent spikes.

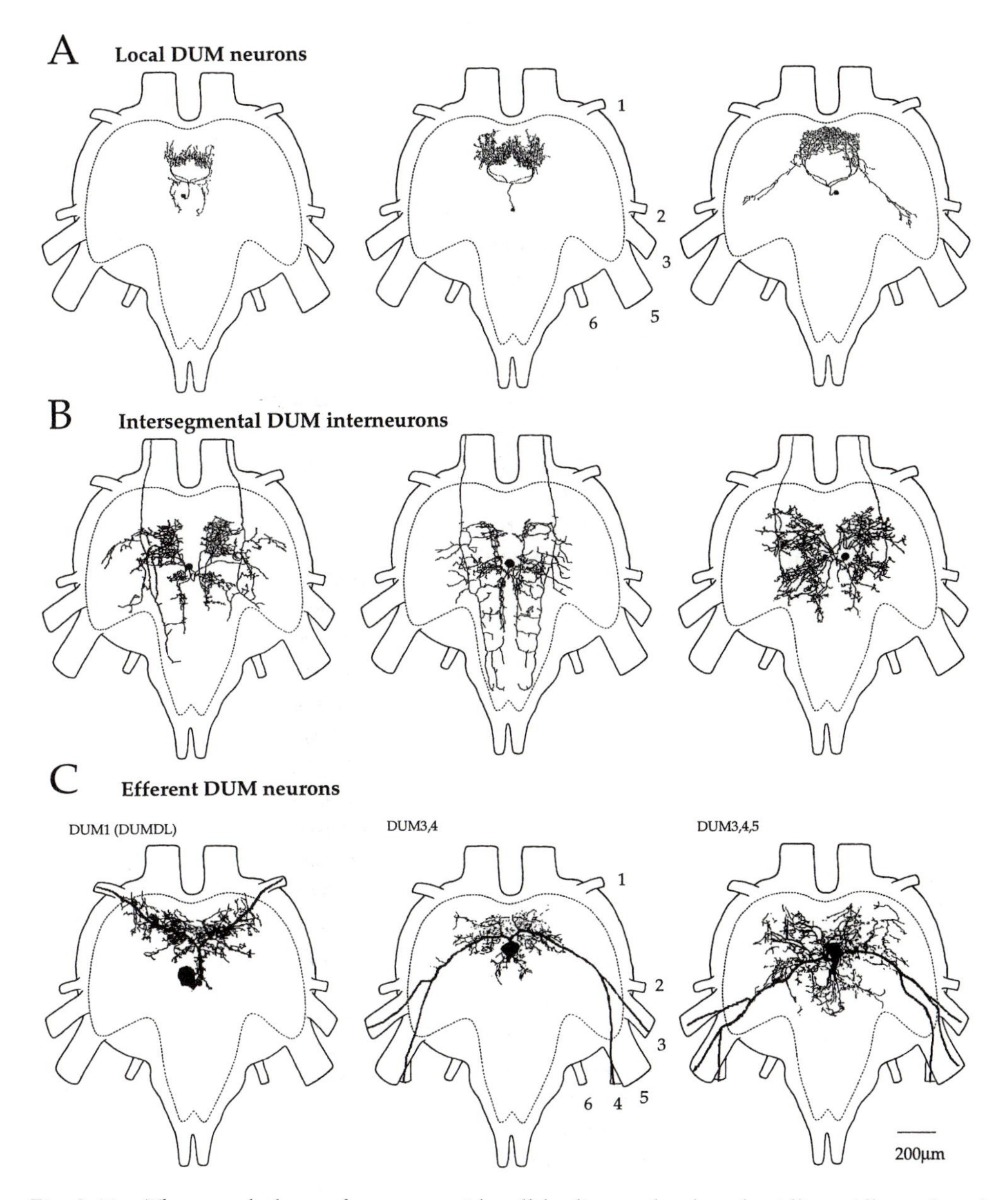

Fig. 3.11 The morphology of neurons with cell bodies at the dorsal midline. All are dorsal unpaired median (DUM) neurons, but they are of three basic types. A. Local DUM neurons. Three neurons are illustrated, all of which have no axons. B. Intersegmental DUM interneurons. Three interneurons are shown that each have axons in the two anterior connectives linking the meta- and mesothoracic ganglia. C. Efferent DUM neurons. These neurons have axons that emerge in particular lateral nerves to innervate effectors on both sides of the body. The neurons are labelled according to the nerves in which their axons run. A and B are based on Thompson and Siegler (1991), C on Watson (1984).

3.6.3. Structure and distribution of synapses

The main information on the distribution of synapses in spiking local interneurons comes from those in the ventral midline group of the metathoracic ganglion (Watson and Burrows, 1985). This distribution may, of course, be specific to these types of interneuron, but it does illustrate the complexities of the processing performed by local neurons and the insights that can be gained into their organisation. The extensive neurites of these interneurons are in two fields. The neurites of the ventral field are in neuropil to which the axons of mechanoreceptors from a leg project, and receive predominantly input synapses, whereas the neurites of the dorsal field are in neuropil containing the processes of motor neurons and other interneurons, and make predominantly output synapses. The single process linking these two regions of an interneuron is tightly packed with processes of other interneurons and wrapped in glia so that it neither receives input synapses nor makes output synapses. This apparent polarisation of the two regions is not, however, complete, for output synapses are made from the ventral neurites and dorsal neurites receive input synapses. Local interactions are therefore to be expected, whereby the input synapses modify the outputs made from either field. The main flow of information must be the integration of sensory information from exteroceptors in the ventral field and its conversion into a spike code that is transmitted to the output synapses in the dorsal field.

In the ventral field, the secondary neurites form a dense array of fine branches with diameters of 0.1–0.3 μm that receive most of the input synapses, while the smaller number of output synapses are made from larger branches or from varicosities in the finer branches. The presence of varicosities is therefore not a good guide to the distribution of output synapses. The distribution of the input and output synapses is not uniform throughout the field. In the most ventral fine branches, input synapses predominate over output synapses in a ratio of 4.5:1, but in the more dorsal branches of this same region, output synapses can outnumber the inputs by a ratio of 2:1. These inputs and outputs are intermingled and can be separated by no more than 1μm. This arrangement means that the majority of the input synapses are some distance from the output synapses and there is not the widespread intermingling of synapses that occurs on the neurites of nonspiking local interneurons. About 40% of all the input synapses to the ventral field have GABA-like immunoreactivity and pleomorphic vesicles, and are therefore probably inhibitory (Leitch and Laurent, 1993). This suggests that considerable integration occurs in this field between these inputs and the excitatory inputs from the sensory neurons.

In the dorsal field, the relative distribution of synapses is reversed so that the outputs exceed the inputs by a ratio of 6.5:1. The majority of output synapses are made from varicosities that are separated by long lengths of intervening neurite devoid of synapses. A varicosity may contain many mitochondria and may make several output synapses, each with two postsynaptic processes that have diameters of less than 0.4 μm. The output synapses are characterised by 50 nm diameter, round, agranular vesicles sometimes accompanied by a few larger 100–130 nm diameter, granular vesicles. An input synapse may also occur on these varicosities, seemingly to exert a maximal effect on the nearby output synapses, but only a very few of these input synapses are GABAergic, suggesting that they are not involved in presynaptic inhibitory control of the spike outputs.

The distribution of input and output synapses changes during the course of embryonic development (Leitch *et al.*, 1992). Recognisable synapses first appear at 70% development, at the time when the axons of the majority of sensory neurons from the leg arrive in the central nervous system, and when synaptic potentials and spikes can first be recorded in these interneurons. The ventral branches then have more output synapses than input ones, in the ratio 7.5: 1, but as development proceeds, more input synapses are added to arrive at the adult ratio. In the dorsal branches, output synapses always predominate as they do in the adult.

3.6.4. Membrane properties

The occurrence of local neurons that spike and others that do not spike poses the question of why neurons with similar gross morphologies should use such different mechanisms of signalling. Clearly, the need for a spike cannot be because signals have to be transmitted over different distances in the two types, but instead must lie in some integrative action determined, at least in part, by the arrangement of the input and output synapses. In the midline spiking local interneurons, developmental studies (*see Chapter 4*) would seem to suggest that the process linking the dorsal and ventral fields can be regarded as an axon linking input and output regions, but there is no evidence as to where the spikes are initiated. The spikes are larger in recordings from neurites than from the soma and their peaks can occur up to 0.7 ms earlier, but this only implies that spikes are initiated somewhere other than close to the cell body. The simplest scheme would have the inputs in the ventral field initiating spikes close to the linking process which would then channel these signals to the main concentration of output synapses in the dorsal field. The reason why this arrangement should require the generation of spikes must lie either in the need to flood all the output synapses with the same signal at about the same time, or with some developmental or evolutionary derivation of these neurons from ones that had long axons and which were therefore imbued with a sufficient density of voltage-sensitive Na^+ channels to enable them to spike.

3.6.4.1. *Transmitters*

There is almost certainly no universal transmitter used by spiking local interneurons, so diverse are their actions and connections and so widespread their occurrence. Many do, however, have inhibitory actions and occur in places as distinct as the glomeruli of the antennal lobes, and the unpaired and paired groups at the dorsal and ventral midline of the thoracic ganglia. Some, or even all, of the interneurons in each of these groups show GABA-like immunoreactivity. For example, particular interneurons in the midline group of a thoracic ganglion show GABA-like immunoreactivity, and their effects on postsynaptic neurons can be blocked by drugs which block chloride channels that are usually opened by GABA (Watson and Burrows, 1987). Their actions can be mimicked by the application of GABA. The inhibitory potentials evoked by these interneurons in postsynaptic nonspiking interneurons reverse at around −80 mV so that the potentials are normally hyperpolarising (Laurent and Sivaramakrishnan, 1992). Their effects are thus probably caused by the release of GABA. Other local interneurons, such as those in the antero-medial groups in the thoracic ganglia, have excitatory effects on postsynaptic neurons and most probably therefore use a different transmitter, although, as yet, this is not identified.

3.6.4.2. *Transmitter release onto different target neurons*

Each spiking local interneuron synapses onto many postsynaptic neurons but the effectiveness of these output synapses is not necessarily the same (Table 3.4). In the thoracic midline spiking local interneurons, the quantal content at output synapses onto different neurons can range from 2 to 10 (Laurent and Sivaramakrishnan, 1992). These differences arise because of differences in the probability of release at the sites associated with different synapses. The properties of the synaptic connections from one of these interneurons cannot therefore be extrapolated from the study of one connection. The implication is that the effectiveness of connections with various neurons can be very different and may also depend on the frequency and patterning of presynaptic spikes and on other events that can change the probability of release selectively at the different synapses.

Table 3.4. Average values for synaptic transmission from spiking local interneurons to nonspiking interneurons

Amplitude of IPSPs (mV)	1.9
Quantal amplitude (μV)	290
Quantal content	6.25
Number of release sites (size of releasable pool)	13
Probability of release	0.45

3.7. NONSPIKING LOCAL INTERNEURONS

3.7.1. General features

3.7.2.1. *What is a nonspiking interneuron?*

The use of the negative name 'nonspiking interneuron' is inappropriate as it does not convey the richness of the integrative actions of these interneurons, but years of searching have failed to provide a more definitive and positive descriptive name. The name applies to those neurons that do not produce spikes during normal processing and that cannot be made to spike by experimentally activating a particular input pathway to them, or by injecting depolarising current directly into them. The nonspiking state is real and is not the result of damage inflicted by the recording electrode, or caused by anoxia in preparing the central nervous system for access to these neurons. The important feature of these interneurons from the standpoint of integration is their ability to communicate with other neurons without the intervention of spikes; they exert graded control over their postsynaptic neurons. A simple division of neurons into 'spiking' and 'nonspiking' might therefore lead to the erroneous conclusion that there is a similar dichotomy of synaptic function, whereas all neurons may be capable of graded interactions, at least at some of their synapses.

Nevertheless, such a negative definition does not, in practice, identify them, because spiking neurons often do not spike and nonspiking interneurons can sometimes be made to spike. Spiking neurons can be resistant to attempts to make them spike, or can be converted to nonspiking neurons by damage or anoxia. Some of the large interneurons in the lobula of the fly that respond to particular movements of the whole visual field were once thought to be truly nonspiking, but it now appears that this state is induced by anoxia during the procedures used to study them and that normally they are spiking neurons (Hengstenberg, 1977). Similarly, some of the progeny of the median neuroblast (the so-called DUM neurons; *see sections 3.6.2.4 and 3.9.3*) were formerly thought to be nonspiking, but capable of generating spikes carried by an inward Ca^{2+} current when outward currents were blocked (Goodman *et al.*, 1980). They are now, however, regarded as conventional spiking interneurons (Thompson and Siegler, 1991). On the other hand, some nonspiking interneurons can, under extreme conditions that may not normally be met, generate what may be Ca^{2+} spikes (*see section 3.7.4.2 Inward current*). For unequivocal identification, therefore, the definition of nonspiking has to be supplemented by a demonstration of the ability of the neuron to release transmitter onto postsynaptic neurons in a graded fashion and without the intervention of spikes. They must also be shown morphologically to be axonless local interneurons, but the morphological definition alone is not sufficient as many axonless local interneurons do generate spikes (*see sections 3.6.1 and 3.6.2*). It is also possible that part of a spiking intersegmental interneuron may act without spikes in influencing postsynaptic targets. A combination of criteria is thus needed to identify a nonspiking local interneuron, and there has to be a willingness to accept that some may be misnamed and that revision may be necessary as more information about them is accumulated.

3.7.1.2. The discovery of nonspiking interneurons

Nonspiking interneurons were first discovered in crustacea as part of the networks that control ventilatory movements (Mendelson, 1971) and then subsequently in cockroaches as part of the networks that control walking (Pearson and Fourtner, 1975). They are now known to be common in brains of insects, where the lamina monopolar cells process signals from the retinula cells, in the thoracic ganglia where they are involved in the control of most limb movements (except curiously, as yet, in control of the wings in flight), and in the last abdominal ganglion where they are part of the networks that process signals from wind-sensitive hairs on the cerci. Graded transmission from neurons, some of which may also spike, also occurs widely in crustacea, where it is involved in many diverse integrative processes, for example, in the stomatogastric networks (Graubard, 1978; Graubard *et al.*, 1980).

3.7.2. Morphology

Nonspiking interneurons are local neurons with all their branches restricted to one region of the nervous system, often a segmental ganglion. None of their basic morphological features (Fig. 3.12) immediately set them apart from some of the spiking local interneurons, so that it is impossible to predict from structure whether a local interneuron will be spiking or nonspiking. None have a process which can be called an axon that projects either within or beyond a restricted region, but they nevertheless have a diverse array of shapes (Siegler and Burrows, 1979; Watkins *et al.*, 1985), and homologous neurons seem to be repeated in the different thoracic ganglia (Wilson, 1981).

In the metathoracic ganglion, most seem to be unilateral interneurons with cell bodies and branches confined to one side, but some have branches contralateral to their cell bodies, and a few others have branches on both sides. The cell bodies are distributed in the dorsal and ventral cortices of a ganglion, but interneurons with a similar shape have their major neurites in the same tracts, suggesting that groups of like individuals occur. At present, however, the interneurons can only be assembled into loose and perhaps unnatural groupings: unilateral interneurons with either ventral or dorsal cell bodies, and interneurons with branches contralateral to their ventral cell bodies. These groupings do not reflect particular features of their actions, as interneurons with output connections to the same motor neurons can have their cell bodies in different positions. The apparent diversity results from the accumulation of information derived from microelectrode recordings made in the neuropil, where the probability of encountering different types is greater. Such recordings, nevertheless, reinforce the idea that there is an underlying organisation, not yet understood, because many have processes in the same tracts.

The cell bodies range in diameter from 15–30 μm with most of their volume occupied by the nucleus and have relatively smooth outlines because glial cells form few trophospongial invaginations. The larger neurites are often grouped together and are woven among regions of neuropil and bundles of fibres that contribute to the roots of the lateral nerves. They cannot, however, be assigned to any of the known tracts or commissures. The fine neurites are profuse and of uniform appearance with extensive projections into intermediate and dorsal areas of neuropil and some in lateral and ventral regions. The neurites thus overlap with the neurites of motor neurons, local and intersegmental interneurons and some sensory neurons and are the basis for the widespread role of these interneurons in the

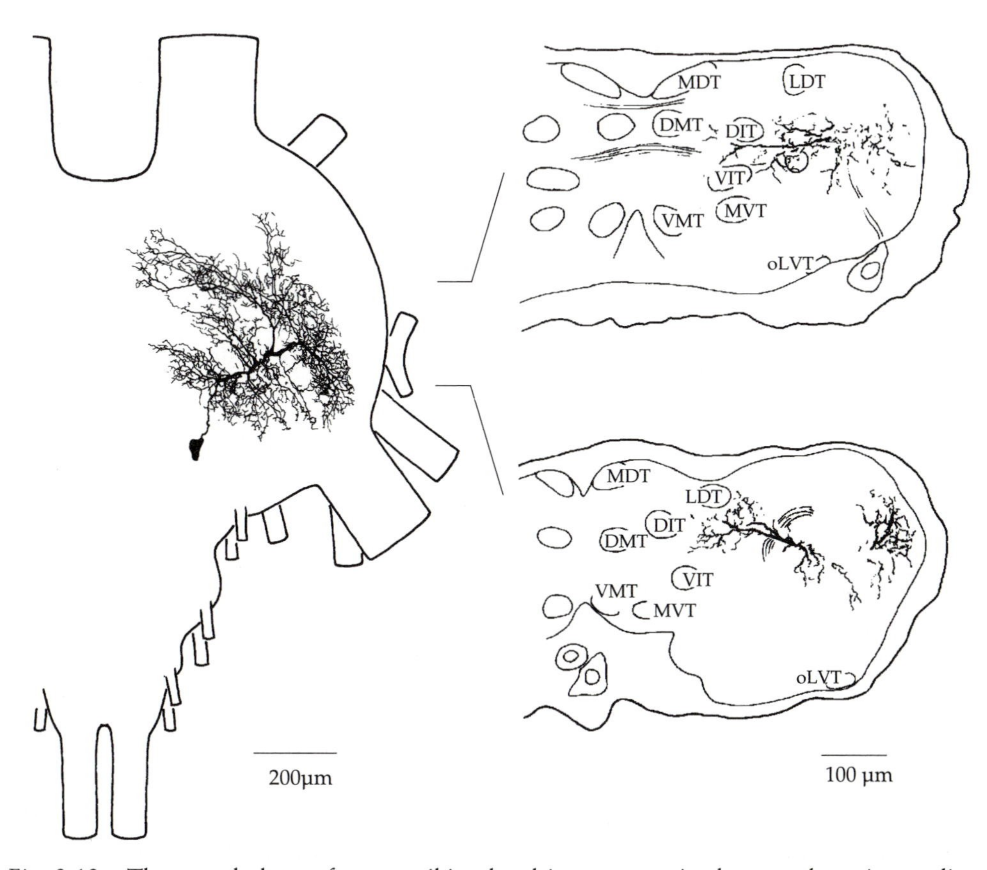

Fig. 3.12 The morphology of a nonspiking local interneuron in the metathoracic ganglion. The drawing on the left shows the overall shape of the interneuron in a wholemount of the ganglion. The transverse sections at the levels indicated relate the branches of this interneuron to known structural features of the metathoracic ganglion (for definitions of the abbreviations see Table 2.2). Based on Watkins *et al.* (1985).

integration of mechanosensory signals and the moulding of motor patterns for movement.

In the last abdominal ganglion of the cockroach, three groups of nonspiking interneurons process mechanosensory signals from the cerci (Kondoh *et al.*, 1991a, 1993). One group has neurites ipsilateral to the cell body. The other two groups have cell bodies in either the dorsal or ventral cortex with bilaterally arranged neurites that are linked by 1–3 processes that can have diameters of up to 30 µm, or the same size as their cell bodies, in different commissures (see Fig. 10.5). These shapes are quite remarkable and pose as yet unanswered questions about how the neurons might integrate signals. The interneurons with single commissural processes are similar in shape to nonspiking interneurons in the last abdominal ganglion of the crayfish (Reichert *et al.*, 1982, 1983).

3.7.3. Structure and distribution of synapses

Output synapses in the thoracic nonspiking interneurons are characterised by a 3 μm long presynaptic bar associated with an average of 1000 round, agranular vesicles, 47 nm in diameter (Watson and Burrows, 1988). In the larger processes, the vesicles are scattered with only slight clustering around the presynaptic bar. No large granular vesicles are present. The output synapses are almost always made with two postsynaptic neurons in a diadic arrangement, although a few synapses with only one postsynaptic process do occur. In nonspiking neurons of the dragonfly eye (Armet-Kibel *et al.*, 1977), the presynaptic bar is similar to that in the locust neurons, but in the fly eye it is distinctively T-shaped (Burkhardt and Braitenberg, 1976). About 75% of the input synapses contain round agranular vesicles 35–40 nm in diameter often accompanied by 90–150 nm granular vesicles, and some 25% contain pleomorphic agranular vesicles. A single presynaptic process can make several synapses with a neurite of a nonspiking interneuron within a few microns.

The initial part of the primary neurite leading from the cell body of a thoracic nonspiking interneuron is either bundled with other fibres, or wrapped in glia so that it does not participate in any synaptic interactions. Deeper in the neuropil the primary neurite gradually loses its glial sheath, although both it and some of the secondary neurites may still be partially wrapped. It then both makes output synapses and receives input synapses. All the secondary and finer neurites have both input and output synapses that are intermingled with each other, so that whole regions of a neuron are not devoted to inputs and others to outputs. One 50 μm long neurite that was reconstructed from serial sections received 22 input synapses and made 64 output synapses. In neurites like these, the output synapses therefore outnumber the input synapses by a ratio of 3:1, whereas in some of the fine branches almost all the synapses may be inputs. The relative distribution of input and output synapses seems to differ both within neurites of one interneuron and between different neurons.

3.7.4. Membrane properties

Two problems dominate the functioning of these neurons. First, how do the voltages generated by synaptic inputs at one point spread through the numerous small branches and affect the output synapses? Second, does the membrane operate linearly over the range of voltages that it normally experiences, or are voltage-sensitive channels involved? These factors will be crucial in determining how signals from different inputs will be integrated to produce outputs that can affect other neurons. Moreover, the magnitude of the voltage changes generated by the synaptic inputs determines the release of transmitter onto postsynaptic neurons, and because input and output synapses are intermingled on the branches, function could be compartmentalised; a synaptic input at one point could activate neighbouring synapses without influencing more distant synapses.

To understand how these neurons integrate synaptic inputs and influence the actions of their postsynaptic targets, it is necessary to know their membrane constants, the channels that govern the voltage changes, and the distribution of these channels. All the physiological data on these neurons have come from recordings with intracellular electrodes placed in the branches of these interneurons, where they should be close to both input and output synapses.

3.7.4.1. *Membrane constants*

The determination of the membrane constants (Table 3.5) relies heavily on the accuracy of the methods used to reveal them, and for an interneuron that is embedded deeply in a network of connections with other neurons, the pattern and level of inputs that it receives. This problem can be reduced somewhat by making measurements from neurons in an isolated ganglion which is therefore separated from the signals generated in other parts of the nervous system, and from those supplied by sensory neurons (Laurent, 1990). Nevertheless, a resting potential really has no meaning for these cells because it is solely determined by the context in which it is measured and therefore the synaptic inputs that it receives at that time.

3.7.4.2. *The channels*

The membrane does not act as a linear and passive resistance over the range of voltages in which it normally operates, so that there must be voltage-sensitive channels carrying both outward and inward currents. Two K^+ channels and possibly a voltage-sensitive Ca^{2+} channel have been identified.

Outward currents The membrane operates in a linear and passive way at potentials more negative than about –70 mV, but at more positive potentials shows a strong outward rectification, so that applied currents of more than 3 nA cannot depolarise the membrane by more than 10–20 mV above resting potential (Fig. 3.13) (Laurent, 1990). Hyperpolarising the membrane potential from rest results in an increase of the membrane resistance whereas a depolarisation results in a substantial reduction, indicating the presence of an outward current that is active at the resting potential. At the same time, the membrane time constant changes from 33 ms at hyperpolarised potentials in the linear part of the current-voltage curve, to 26 ms at resting potential. The whole outward current is activated at –60 mV, reaches a peak at –30 mV, where it involves a large conductance change of about 500 pS, and reverses at –75 mV (Laurent, 1991). It consists of two components, a large but

Table 3.5. Membrane constants of thoracic nonspiking interneurons

Resting potential (mV)	–58
Input resistance (MΩ)	16
Membrane time constant (ms)	33
Equalising time constant (ms)	3
Specific membrane resistance (kΩ.cm^2)	33

rapidly inactivating transient current carried by K^+ with characteristics of an 'A' current, and a later slowly inactivating current also probably carried by K^+ with characteristics of a 'delayed rectifier current'. The ratio between the transient and the late current varies between 1.6 and 5.4 in different interneurons. The transient current is activated at –60 mV with a time-to-peak of 11 ms, and is inactivated with a time constant of 3 ms that does not depend on membrane potential. Recovery from inactivation occurs with a time constant of 100 ms at –80 mV. At resting potential, this current will be half inactivated.

Inward current Fast Na^+ currents do not seem to be present in nonspiking interneurons as they cannot be revealed in the presence of tetraethylammonium (TEA) and 4-aminopyridine (4AP) even when the membrane is depolarised to –30 mV. When the resting potential of a cockroach interneuron is hyperpolarised from its apparently normal level of –40 mV to about –60 mV, single spikes are generated in response to an abrupt stimulus (Bodnar *et al.*, 1991). At its normal resting potential, spikes are not generated. Nonspiking interneurons with similar shapes in crayfish also show similar membrane responses. Some locust nonspiking interneurons produce resonant oscillations and regenerative potentials when depolarised to voltages less than –40 mV (Fig. 3.13) (Laurent *et al.*, 1993). These voltages are outside the normal operating range of these neurons and thus these potentials are not seen when the membrane is depolarised only by natural arrays of synaptic inputs. These regenerative potentials are slow and have a half-width of 20–30 ms and an amplitude of only 25–35 mV, often with a pronounced after-hyperpolarisation. The inward current generating these potentials has not been characterised in the interneurons imbedded in their circuits, but instead a voltage-dependent Ca^{2+} current has been identified in the somata of neurons taken from dissociated embryonic ganglia and maintained in culture for a few days. The assumption is that the cell bodies in the culture which do not generate tetrodotoxin (TTX)-sensitive spikes belong to nonspiking interneurons, but this may not be valid.

Fig. 3.13 Membrane properties of a nonspiking local interneuron. A. Outward rectification. Pulses of current injected into a neuropilar branch reveal a large outward rectification when the interneuron is depolarised from its resting potential, whereas the membrane behaves linearly at more hyperpolarised potentials. B, C. Currents. Voltage clamp reveals a transient, rapidly inactivating outward current and also a late, slowly inactivating outward current. Whole cell patch-clamp recording also reveals an inactivating inward current. D. Active membrane properties determine the response to synaptic inputs. A sinusoidally varying current is injected into a neuropilar branch of an interneuron held at three different membrane potentials. At –59 mV the membrane behaves linearly; at –51 mV, K^+ currents are activated, the membrane resistance falls and depolarisations are rectified; at –42 mV, a depolarising conductance is activated that accentuates the depolarisations. E. Effects of membrane potential on depolarising synaptic potentials. The more the membrane is depolarised, the more the potentials are reduced in amplitude and duration. F. Spikes can sometimes be induced in a nonspiking interneuron by large imposed depolarisations. A and E are based on Laurent (1990), B on Laurent (1991), and C,D,F on Laurent *et al.* (1993).

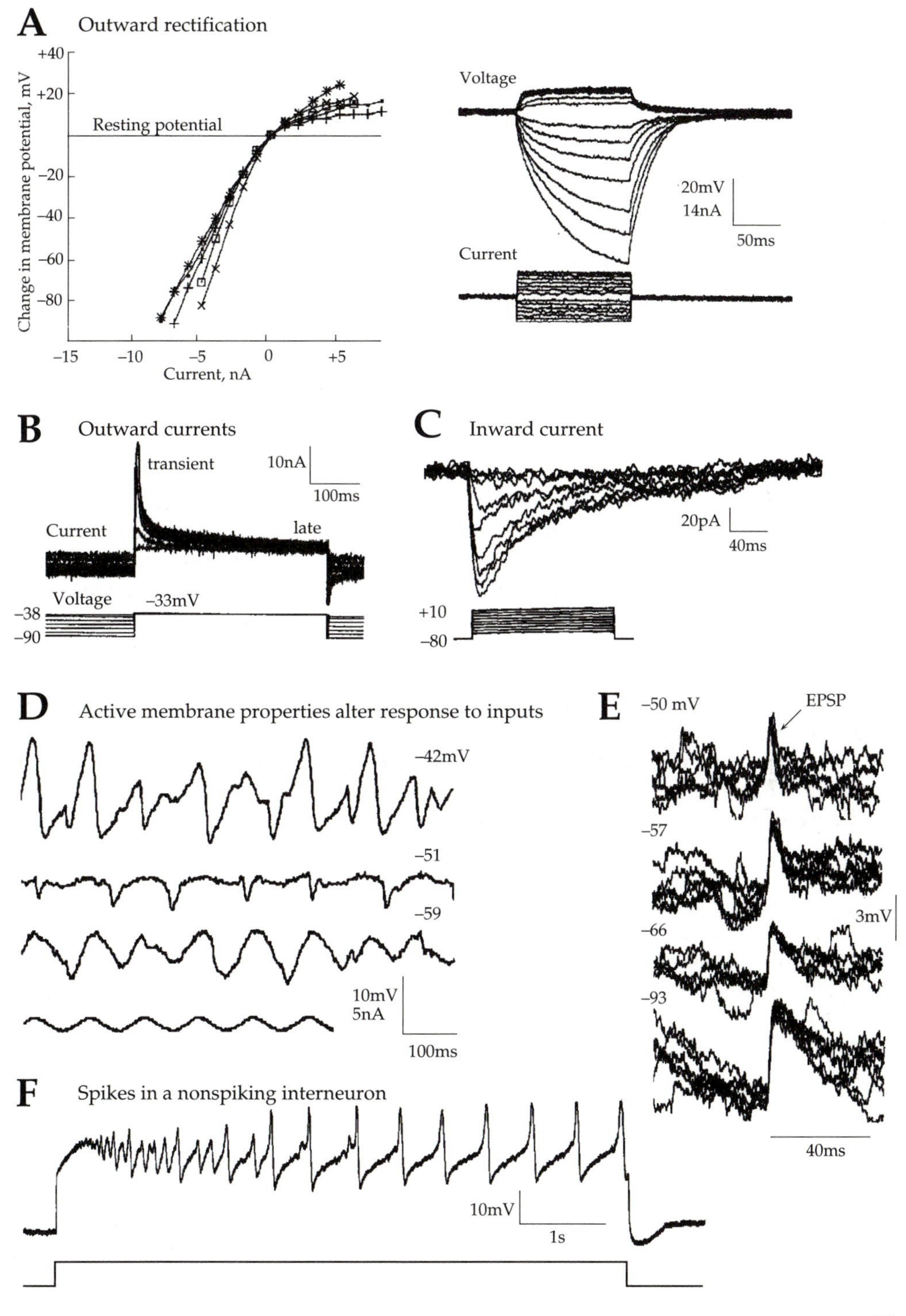
A
Outward rectification
+40
+20
Resting potential
-20
-40
-60
-80
Change in membrane potential, mV
-15
-10
-5
0
+5
Current, nA
Voltage
20mV
14nA
50ms
Current
B
Outward currents
transient
10nA
100ms
Current
late
Voltage
-33mV
-38
-90
C
Inward current
20pA
40ms
+10
-80
D
Active membrane properties alter response to inputs
-42mV
-51
-59
10mV
5nA
100ms
E
-50 mV
EPSP
-57
-66
-93
3mV
40ms
F
Spikes in a nonspiking interneuron
10mV
1s

The characteristics of this current are that it is resistant to TTX, can be carried by Ba^{2+} and is blocked by Cd^{2+} and Co^{2+}, suggesting that it is normally carried by Ca^{2+}. It is activated at –40 mV and is voltage dependent, reaching a peak at zero. The current inactivates rapidly, and is half-inactivated at –25 mV so that at resting potential some 90% of the current is available for activation.

3.7.4.3. *The distribution of the channels*

There is no precise definition of the distribution of the channels, but one observation suggests that it may not be uniform. The regenerative potentials evoked at membrane potentials more positive than –40 mV can be of different amplitudes in the same interneuron, and may have several peaks. The simplest explanation is that the potentials are generated in different regions of the interneuron, which would, in turn, imply a nonuniform distribution of the putative Ca^{2+} channels in the different branches.

3.7.4.4. *Transmitters*

There is no information on the transmitters that are likely to be used by nonspiking interneurons. This is an enormous lacuna in our knowledge and is a substantial impediment to understanding the actions of these interneurons in the control of behaviour. The difficulties in filling this void should not, however, be underestimated because there are few clues that might direct this search for putative transmitters. Most of the actions of these interneurons can be explained by the release of fast acting transmitters that evoke conductance changes in postsynaptic neurons (*see section 3.7.5*), but the release of other substances with slower actions through second messengers cannot be excluded. What is needed is to combine the physiological characterisation of these interneurons with intracellular staining followed by labelling with the gamut of antibodies that are available against neuroactive substances known to be present in the nervous system of the locust.

3.7.5. Graded synaptic transmission

The absence of spikes in these neurons during normal behaviour means that the only intracellular signals that are available to effect communication with other neurons are the graded changes in membrane potential caused by the synaptic inputs. There is no evidence that any of these neurons are electrically coupled to other neurons, so that transmission must be chemical and the voltage changes caused by synaptic inputs must be sufficient to alter the rate of release of transmitter. When a locust is not moving, a nonspiking interneuron receives a continuous barrage of depolarising and hyperpolarising synaptic inputs, but when it moves its legs, the membrane potential of some thoracic interneurons can be depolarised by 10 mV and hyperpolarised by as much as 20 mV. These observations during normal movements define the voltages which must affect the release of transmitter.

Injecting current into a nonspiking interneuron to simulate, in a controlled way, such voltage changes, leads to changes in the membrane potential of postsynaptic neurons (Fig. 3.14) (either motor neurons or other nonspiking interneurons) (Burrows and Siegler, 1978; Burrows, 1987). The voltage changes in the postsynaptic neurons are continuously graded with respect to the voltage of the presynaptic nonspiking interneuron. The more the nonspiking interneuron is depolarised, the greater is the change in voltage in the postsynaptic neuron, so that the relationship between the voltage of the pre- and the postsynaptic neurons is sigmoidal. This indicates that the release mechanism has a voltage threshold and that it saturates at more positive membrane potentials. The saturation is also due to properties of the postsynaptic neuron, such as the membrane approaching the reversal potential for the ions carrying the evoked potential. The transmission occurs without spikes being detected in the presynaptic neuron and without abrupt discontinuities in the membrane potential of the postsynaptic neuron that might indicate the presence of otherwise unseen presynaptic spikes. The transmission thus contrasts strongly with that from spiking presynaptic neurons that have output connections with the same postsynaptic neurons. Under voltage-clamp, the changes in voltage of the nonspiking neuron that are needed to cause changes in postsynaptic neurons are within the range of the voltage fluctuations that occur during normal behaviour. Depending on the nonspiking interneuron, the effects on the postsynaptic neuron can be either excitatory or inhibitory, although those with inhibitory actions seem to dominate in the circuitry controlling movements of the legs.

3.7.5.1. *Chemical synaptic transmission*

The synaptic transmission from a nonspiking interneuron to other neurons is considered to be chemical for the following reasons. First, nonrectifying electrical junctions do not exist between them because injecting current into a postsynaptic neuron has no effect on the membrane potential of the presynaptic nonspiking interneuron. Second, there is a delay of 1–2 ms between the application of brief current pulses to a nonspiking neuron and the appearance of a voltage change in a postsynaptic neuron, even if both electrodes are in neuropilar processes. The delay is difficult to measure as there is no reference point such as a presynaptic spike, and in some pairs of neurons the synaptic delay may appear to be several milliseconds. Third, the voltage changes in a postsynaptic neuron are associated with a conductance change. When a nonspiking interneuron produces a hyperpolarisation of a postsynaptic neuron, a reversal potential can be demonstrated. When the postsynaptic effect is a depolarisation with a short latency, a dependence on membrane potential and a projected reversal potential can be demonstrated. Fourth, chemically mediated synaptic potentials generated in a postsynaptic neuron by other inputs can be modulated by the transmission from a nonspiking interneuron.

The compact nature of the neuropil and the tight packing of the neuronal processes has precluded showing that the release is Ca^{2+}-dependent or that it is quantal, although in some postsynaptic neurons, the evoked potential is accompanied by a change in small voltage fluctuations that may represent synaptic noise. All the evidence points to a transmitter release mechanism that is essentially

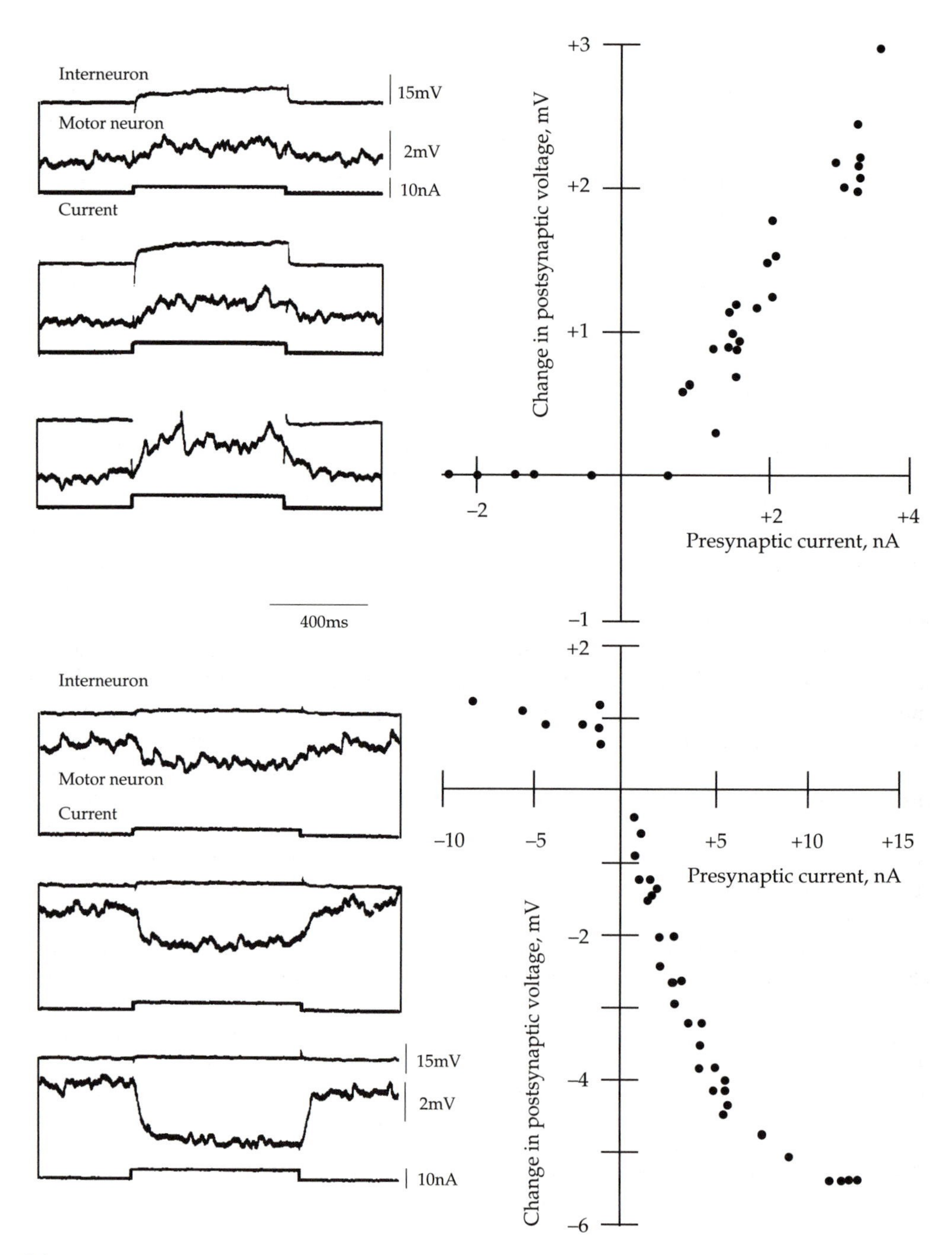

the same as that in spiking neurons, except that the linear part of the input-output curve is matched to the voltage changes caused by synaptic potentials rather than the much larger voltages caused by spikes.

3.7.5.2. *Properties of graded transmitter release*

On many occasions, nonspiking interneurons must be depolarised by a few millivolts before an increase in the release of transmitter can be detected as a voltage change in a postsynaptic neuron. This might imply that they have a voltage threshold for release, but this has little meaning in the context of the continuous fluctuations in the membrane potential that occur as the result of changes in the locust's actions. The membrane potential is not stable, but fluctuates according to the pattern of synaptic inputs that are generated in particular circumstances. At certain times, therefore, transmitter will be released tonically, so that depolarising inputs will increase release and hyperpolarising ones will decrease it (Fig. 3.15). Postsynaptic neurons may therefore be kept continuously hyperpolarised by this tonic release of transmitter. They will be further hyperpolarised when the nonspiking interneuron is depolarised, and will be allowed to depolarise through cessation of release when the nonspiking interneuron is itself hyperpolarised. All the nonspiking interneurons are capable of releasing transmitter tonically, depending on their present membrane potential, and there is not a subgroup with this action as has been suggested (Wilson and Phillips, 1982).

The clear ability of small voltage changes to effect the release of transmitter is most dramatically illustrated by the ability of some single synaptic potentials to cause the release of transmitter and evoke synaptic potentials in postsynaptic neurons by themselves (see Fig. 7.21B) (Burrows, 1979a). Each of these depolarising synaptic potentials in a nonspiking interneuron can, in appropriate circumstances, directly evoke an inhibitory synaptic potential in a postsynaptic motor neuron. If the membrane of the nonspiking interneuron is hyperpolarised then the ability of this synaptic input to generate a postsynaptic effect is reduced, implying that other synaptic inputs can modulate the effectiveness of the transmission of this particular input. This arrangement means that the sequence of events in a single input pathway can be represented in the output and that this is then conveyed directly to at least one of the postsynaptic neurons. The juxtaposition of input and output synapses further indicates that those output synapses close to this particular input will be more strongly affected than more distant ones.

If the nonspiking interneurons are to act tonically or be driven by long-lasting changes in synaptic inputs, then there must be a mechanism for sustaining the release

---

Fig. 3.14 Graded synaptic transmission. The left-hand column of records shows simultaneous paired recordings from a nonspiking neuron and a leg motor neuron. Increasing currents were used to depolarise the interneurons. The graphs show the relationship between the current injected into an interneuron and the resulting change in the membrane potential of the motor neurons. The interneuron in the upper panels caused a depolarisation of the motor neuron, while that in the lower panels caused a hyperpolarisation. Based on Burrows and Siegler (1978).

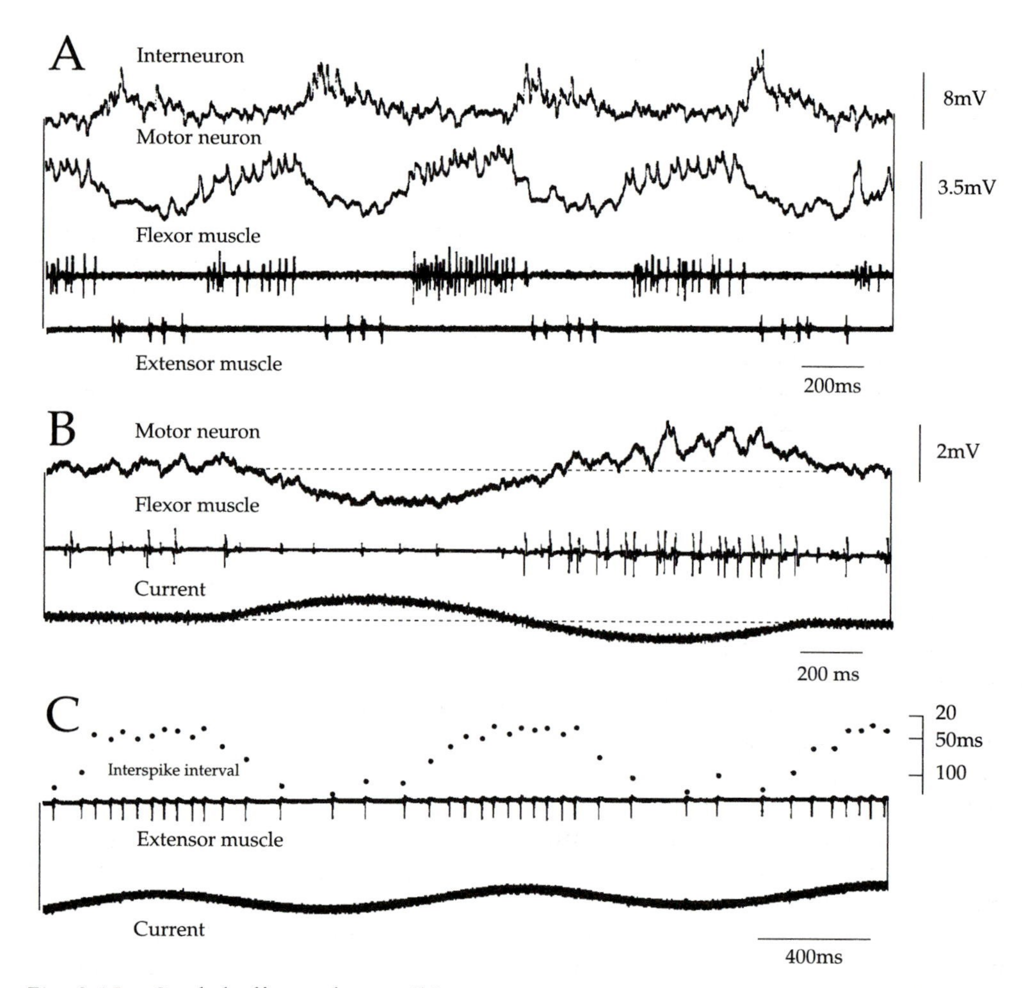

Fig. 3.15 Graded effects of nonspiking neurons on motor neurons. A. During rhythmic movements of the femoro-tibial joint of a hind leg, a nonspiking neuron is depolarised alternately with a flexor tibiae motor neuron. The interneuron has an inhibitory effect on this motor neuron. B. Sustained release of transmitter by a nonspiking interneuron. Depolarisation of the interneuron causes a graded hyperpolarisation of a flexor tibiae motor neuron. Hyperpolarisation of the interneuron allows the motor neuron to repolarise above the potential when the interneuron is at its resting potential. C. Graded control of the force generated by a muscle by a nonspiking interneuron. Sinusoidally varying current is injected into a single nonspiking interneuron. It causes a sinusoidal change in the frequency of spikes in the only slow motor neuron innervating the extensor tibiae muscle of a hind leg. This modulation of the spike frequency is transformed by the muscle into a sinusoidal variation of force.

of transmitter, and moreover, the postsynaptic receptors must show little or no desensitisation. Transmitter release and its consequent effects on postsynaptic neurons can be sustained for several minutes while a nonspiking interneuron is held depolarised with a continuous current. This suggests that there is no barrier to the idea that the nonspiking interneurons exert continuous effects on their postsynaptic neurons by the graded release of transmitter.

3.7.5.3. *Excitatory and inhibitory output connections*

Each nonspiking interneuron in the circuits that control leg movements appears to make output connections with several motor neurons, with other nonspiking neurons, and probably with other interneurons. A depolarising input to a nonspiking neuron will thus spread widely within the network and alter the membrane potential of many neurons. Given these widespread effects of a nonspiking interneuron and the graded nature of the synaptic transmission, it becomes difficult to determine the real extent of the connections made by an individual nonspiking interneuron.

Some nonspiking interneurons have an excitatory effect on their postsynaptic neurons whereas others have an inhibitory effect. Interneurons with these different effects do not obviously fall into distinct populations. Moreover, given the graded nature of the transmission process it is difficult to establish with physiological criteria whether the connections between a nonspiking interneuron and another neuron impaled simultaneously with a second microelectrode is direct. The most convincing evidence is for the inhibitory output connections. Here, short synaptic latencies compatible with a direct connection can be demonstrated, and clear conductance changes underlie the evoked postsynaptic hyperpolarisations. Some excitatory connections are also effected with a short latency but the majority are longer and show no conductance changes. The simplest explanation of these effects is that they are disynaptic and caused by disinhibition; the impaled nonspiking interneuron makes an inhibitory connection with a second nonspiking interneuron which, in turn, makes an inhibitory connection with the other neuron that is impaled. The overall conclusion is that many nonspiking interneurons have direct connections with other neurons that cause inhibition. Some other nonspiking interneurons also make direct connections that cause excitation, but the majority of excitatory effects are probably to be attributed to disinhibition.

3.7.6. Integrative actions

The presence in a network of nonspiking interneurons that release transmitter in a graded and sustained way has profound implications for how integration is accomplished. The cellular consequences are considered below, and the network consequences in *Chapter 7*.

3.7.6.1. *Effects of membrane properties on integration*

Membrane time constant The changes in membrane resistance and its time constant caused by the activation of the outward K^+ currents affect the frequency response of the membranes of nonspiking interneurons and the time course of their synaptic potentials (Fig. 3.13) (Laurent, 1990). The frequency response of the membrane to an applied voltage decreases by a factor of nine as the membrane is depolarised from –85 to –35 mV, with consequent effects on its ability to transmit repetitive signals. Similarly, the half-width of a postsynaptic potential (PSP) decreases by a factor of 7.5 as the membrane is depolarised from –90 to –50 mV. Moreover, the effect on the PSP also depends on the sign and the rate of change of an imposed voltage signal, because PSPs are shunted on the depolarising phase of an imposed voltage change and reappear at the same potential on the hyperpolarising phase. This means that the PSPs become faster as the membrane is depolarised, and in order to maintain a constant level of summation they must occur at a higher frequency. The integrative capacity of an interneuron will thus be highly dependent on its membrane potential. In some nonspiking interneurons of stick insects, the membrane seems to behave in a more linear way, although the evidence for this conclusion is much weaker, so the synaptic inputs that result from activation of a particular sense organ are not dependent on the membrane potential (Driesang and Büschges, 1993).

Ca^{2+} currents The possible inward Ca^{2+} current that has so far been identified is not responsible for regulating transmitter release, as the threshold for release is at more negative voltages than those needed to activate this current. Moreover, transmitter release can be sustained but this current is rapidly inactivated. The current may, however, have two effects on integration.

First, the regenerative responses of the membrane might boost the synaptic potentials in the fine branches, thus counteracting the shortening of the space constant caused by activation of the K^+ currents. An effect of this might be to enable a small input signal to activate nearby output synapses on the same or adjacent branches.

Second, it may help to equalise the synaptic gain over different voltage ranges (Laurent, 1993). From –55 mV to –45 mV the membrane is least responsive to a synaptic depolarisation or current injected through a microelectrode because the membrane resistance is lowered by the activation of the K^+ currents. At potentials more negative than –55 mV, the response is strong because the membrane is linear, and at potentials more positive than –45 mV, the response is also strong because the Ca^{2+} current is activated. Transmitter release from an interneuron, however, starts at potentials more positive than about –65 mV and follows a sigmoidal curve so that it is maximal at about –50 mV. This means that the output of an interneuron is greatest at the potentials where its input synapses generate the least changes in voltage. Modelling suggests that a consequence of this arrangement is that the effectiveness of an input signal in releasing transmitter is largely independent of the voltage of the nonspiking interneuron. Nevertheless, such a mechanism would leave little scope for an input signal to interact with the many others that impinge on an interneuron and would thus seem to undermine the value of these interneurons in integrating signals from many sources. This they can be clearly demonstrated to do.

3.7.6.2. *Graded control of postsynaptic neurons*

The graded release of transmitter is a mechanism for the precise control of the membrane potential of postsynaptic neurons. This analogue control mechanism is free of the constraints that result from digital control, in which the coding with spikes can operate only over a limited range of frequencies determined by the ability of the membrane to carry spikes. Moreover, converting the analogue signal in one neuron directly to an analogue signal in another eliminates the variability associated with the intervening step of generating spikes. The precision of this mechanism is well illustrated by the way a nonspiking interneuron controls its postsynaptic motor neurons and hence the force generated in the extensor tibiae muscle (Fig. 3.15). Depolarising a nonspiking interneuron with a sinusoidally varying current causes a sinusoidal change in its membrane potential that effects a graded release of transmitter onto the single slow motor neuron. The evoked changes in the membrane potential of the motor neuron lead to a sinusoidal change in the frequency of its spikes, and this motor pattern, in turn, leads to a sinusoidal variation in the force that the muscle produces. Finally, this leads to smooth and sinusoidally varying changes in the angle of the femoro-tibial joint, with phase lags imposed by the mechanical components. Overall, therefore, the graded changes in membrane potential of the nonspiking interneuron are reflected directly in the fine movements of a leg. Such a graded control mechanism should also allow the same precision of control to be achieved with fewer neurons than if spiking neurons were always used.

3.7.6.3. *Are nonspiking interneurons compartmentalised*

The juxtaposition of input and output synapses on the numerous fine branches, and the absence of a spike that could sweep rapidly to all parts, immediately suggests that different regions of a nonspiking interneuron could act with some autonomy. The neuron could thus perform several computations simultaneously, and with the independence of different regions under continuous control by different patterns of synaptic inputs. Thus at certain times processing of certain signals would be performed independently of other inputs but at other times would be affected by these other inputs.

The idea that processing could be compartmentalised is based on the seemingly reasonable assumption that an input synapse will have a stronger influence on nearby output synapses than on more distant ones, particularly as these synapses need only to experience a small voltage change to alter their release properties. The working of such an arrangement will be determined first by the spatial arrangement of input synapses on the branches on a nonspiking interneuron, and second by the way in which the potentials they generate are attenuated as they spread from their site of initiation.

Distribution of input synapses Almost nothing is known about the distribution of the synapses made by a particular neuron with a nonspiking neuron, a statement that also holds true for other neurons in the central nervous system. Just how many synaptic contacts does one neuron make with another and are these restricted to a particular set of branches or are they widely distributed over all the branches? Some clues can be gained from double labelling of two neurons thought by physiological criteria to connect, but viewing of such stained neurons is usually only possible at the level of the light microscope so that areas of overlap rather than synaptic contacts are all that can be described. What these methods do show, however, is that the input synapses from certain neurons can only be made on particular regions of a postsynaptic neuron and cannot be distributed over all of its branches. Serial electron microscopy can only realistically be undertaken for small areas of branches and not for a whole neuron so that it can only point to the likely synaptic arrangements and density of synapses. It is a dangerous assumption to extrapolate from samples of such small areas to the whole neuron because there is no reason to suppose that the synaptic arrangements are uniform. The estimates that have been made suggest that a physiological synaptic connection may be represented by many anatomical synaptic contacts.

Attenuation of synaptic potentials It has not been possible to measure directly the attenuation of synaptic potentials in a nonspiking neuron because of the small diameter of the branches and because the distribution of synapses from a given presynaptic neuron is not known (*see above section 3.7.6.3*). Estimates of likely attenuation can be made from descriptions of the anatomy and assuming linear cable properties of the membrane (Rall, 1981). The difficulty here is that the histological treatment of the neurons after they have been stained may give false figures for important parameters such as the diameter of the processes, and furthermore the membrane is known to behave in a nonlinear fashion with the possible uneven distribution of voltage-activated channels. These estimates suggest that synaptic potentials generated in the small branches will be severely attenuated by the time they reach output synapses on other small branches.

Evidence suggestive of compartmentalisation Direct experiments have not been possible to test the idea of compartmentalisation but some observations on the actions of nonspiking interneurons can be most easily interpreted by invoking this idea (Laurent and Burrows, 1989a). The observations are that only inputs to a nonspiking interneuron from particular intersegmental interneurons and not those from others are able to alter the effectiveness of direct sensory inputs to the same nonspiking interneuron in causing a reflex motor response (*see also Chapter 7*). This can be explained by proposing that the inputs from one intersegmental interneuron and the sensory neurons occur on the same branches so that their conductance or voltage changes can interact, but that the inputs from the other intersegmental interneurons are anatomically and electrically separate.

3.8. INTERSEGMENTAL INTERNEURONS

3.8.1. Morphology

An intersegmental interneuron has an axon that projects from one part of the central nervous system to another. The typical structure is of a cell body and an array of associated neurites in one part of the central nervous system and an axon that ends in a series of branches in another, often forming side branches in intervening regions through which it passes (Figs 3.16 and 3.17). They are a diverse group, with

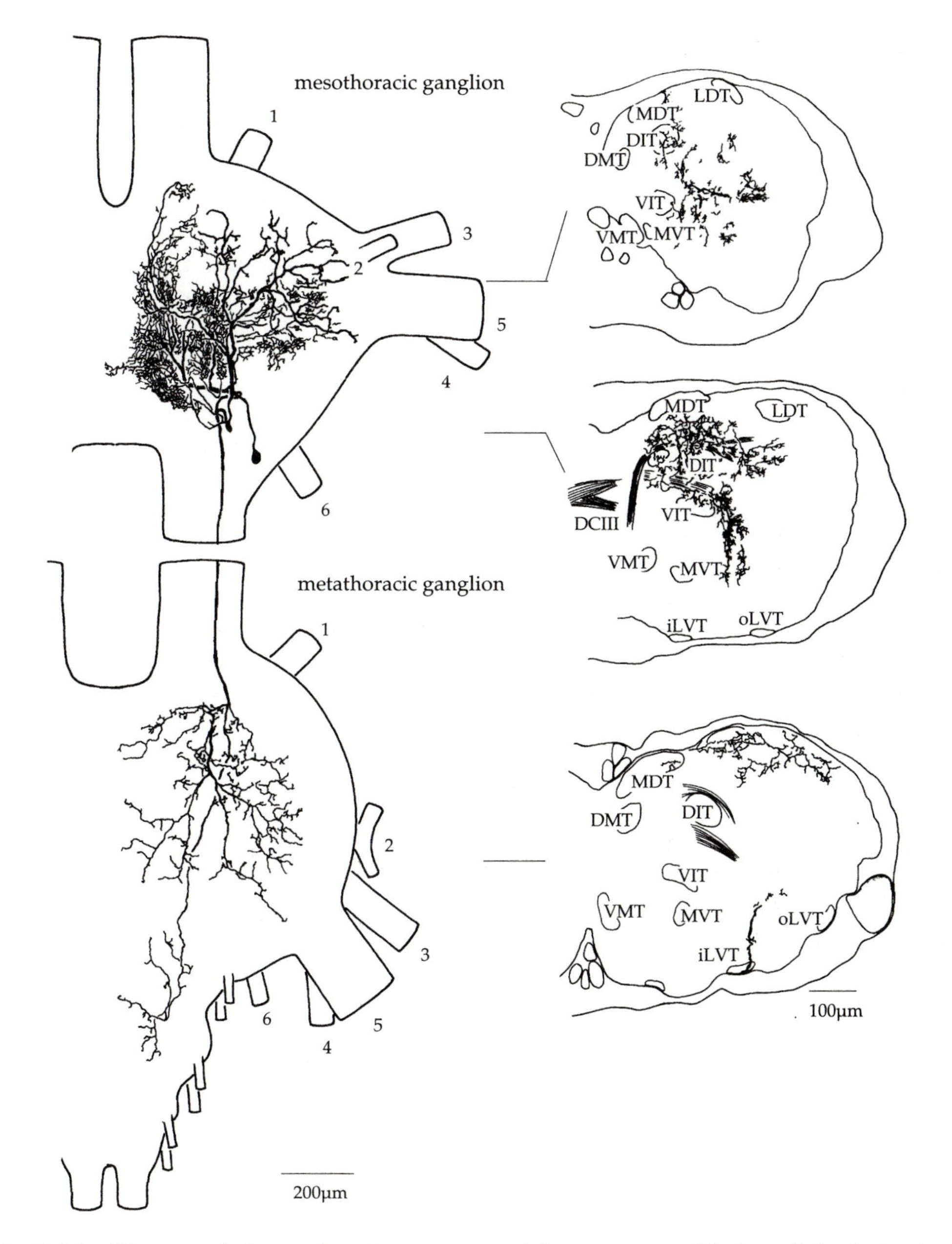

Fig. 3.16 The morphology of an intersegmental interneuron with its cell body in the mesothoracic ganglion (upper left-hand drawing) and an axon that projects to the metathoracic ganglion (lower left-hand drawing). The transverse sections at the levels indicated relate the branches of this interneuron to known structural features of the meso- and metathoracic ganglia (for definitions of the abbreviations see Table 2.2). Based on Laurent (1987).

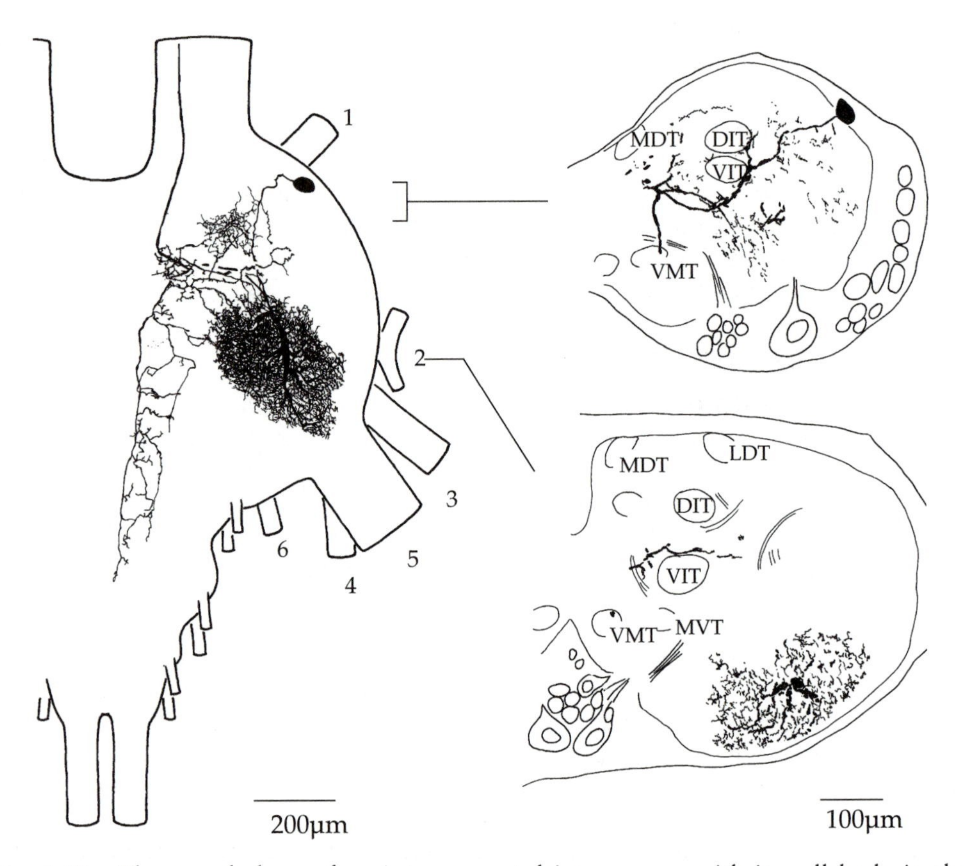

Fig. 3.17 The morphology of an intersegmental interneuron with its cell body in the metathoracic ganglion and an axon that projects to more anterior ganglia. The transverse sections at the levels indicated relate the branches of this interneuron to known structural features of the metathoracic ganglia (for definitions of the abbreviations see Table 2.2). Based on Laurent and Burrows (1988).

different patterns and extents of projections and hence different morphologies. Some project from one segment to the neighbouring one, largely collating inputs in one segment and delivering output signals to the other. Others have axons that project to many ganglia and are thus often distinguished by the name plurisegmental interneurons. These can have cell bodies in the brain and axons that extend to the abdomen, others have cell bodies in the terminal abdominal ganglion with their axons extending to the brain, while others may have cell bodies in the suboesophageal, thoracic or abdominal ganglia and axons that either ascend or descend. The axons of many of these interneurons run in the connective contralateral to the cell body so that the majority of outputs are made on the side opposite to that on which they receive the majority of their inputs, but others have ipsilateral axons. A small number of intersegmental interneurons have bilateral axons. In the thoracic and abdominal ganglia of embryos, H neurons, named after the arrangement of their axons, are known from their morphology if not their actions (Bate *et al.*, 1981; Goodman *et al.*, 1981).

Overall, therefore, the intersegmental interneurons provide routes for the transfer of signals between different parts of the central nervous system. Despite the diversity, it is possible to recognise populations of these interneurons that have similar structures and for some it is even possible to say that they perform similar roles in integration. Nevertheless, few generalisations can be made about the morphology of these interneurons as an overall class.

3.8.1.1. *Intersegmental DUM interneurons*

One distinctive group of intersegmental interneurons are those that have their cell bodies at the dorsal midline of the thoracic ganglia in the same group as the local DUM interneurons (*see section 3.6.2.4*) and the efferent DUM neurons (*see section 3.9.3*). They have a distinctive structure and are probably not octopaminergic like the efferent DUM neurons, but are still unpaired (Fig. 3.11B) (Thompson, and Siegler, 1991). Their primary neurites enter the neuropil bundled together with those of the local DUM interneurons, before bifurcating to form symmetrical bilateral branches in the supramedian commissure (SMC), which then form thin axons of about 1μm in diameter that enter both anterior connectives. The fine neurites, which vary in extent from interneuron to interneuron, are in regions of dorsal neuropil occupied by the processes of flight motor neurons, but are more dorsal than the neurites of leg motor neurons, or the auditory sensory neurons that excite the local DUM interneurons. Their input and output connections are unknown and little is known about their action during behaviour.

In the suboesophageal ganglion, DUM interneurons form some of the most amazing structures in the whole nervous system (Fig. 3.18A) (Bräunig, 1991a). Seven of these neurons of one group have been identified. They have cell bodies at the midline of the anterior part of the suboesophageal ganglion, bilateral branches in its anterior neuromeres, and bilateral axons that ascend to the brain where they may form incredibly complex, bilateral arrays of branches in major neuropils such as the antennal lobes, mushroom bodies and the lobula. A second group of six neurons with cell bodies and bilateral branches towards the posterior of the ganglion have bilateral axons that descend to the thorax and may again be octopaminergic (Fig. 3.18B). Other DUM neurons in the suboesophageal ganglion are local neurons (Fig. 3.18C).

3.8.2. Structure and distribution of synapses

The diversity in the structure of intersegmental interneurons is so great that extrapolations from the few for which ultrastructural information is available can only be undertaken with caution. Three examples illustrate the complexity of the interrelationships that must be considered.

First, a 404 interneuron that may be involved with the initiation of flight (*see Chapter 11*) has its cell body and a large array of neurites in the mesothoracic

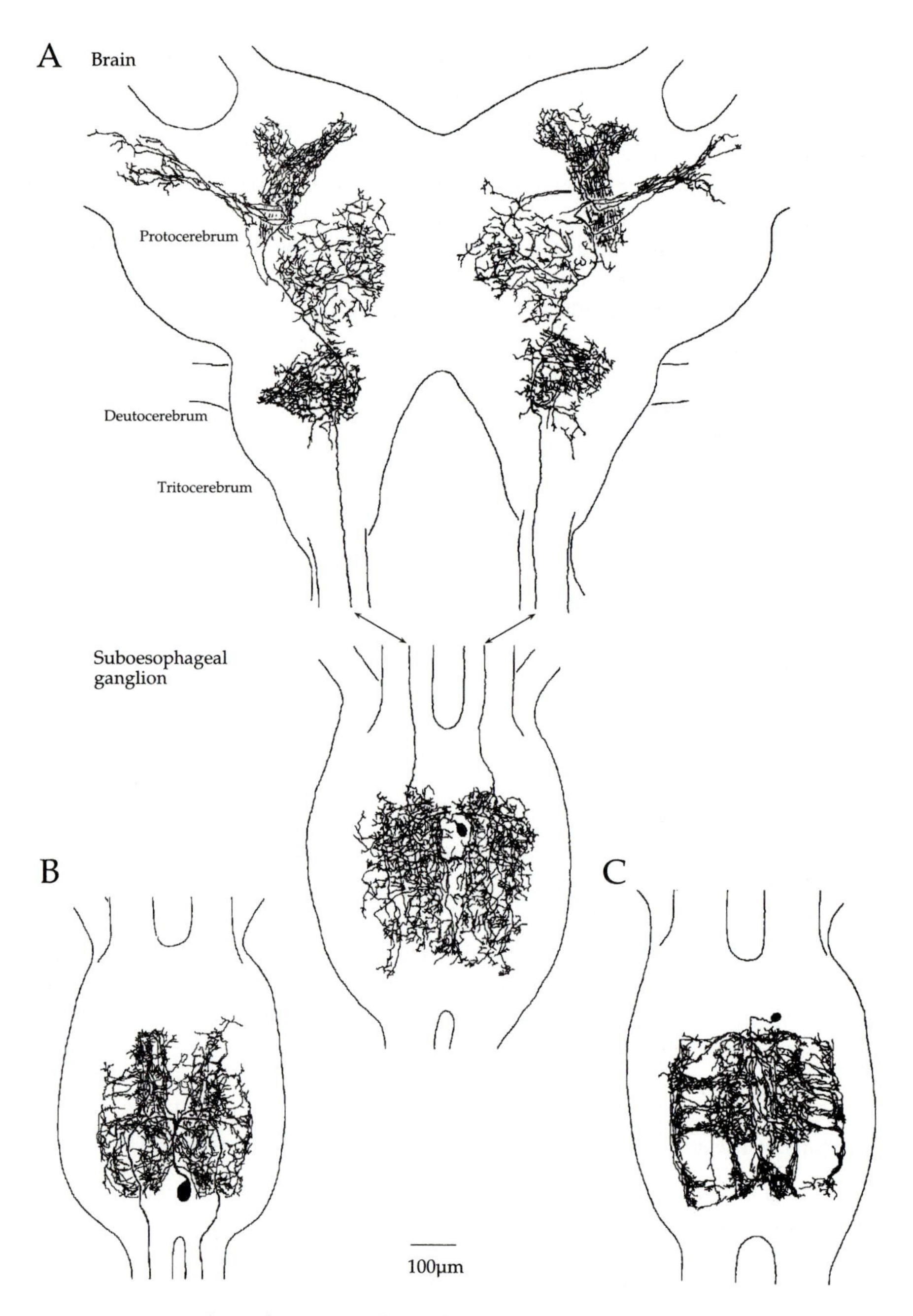

Fig. 3.18 DUM (dorsal unpaired median) interneurons with cell bodies in the suboesophageal ganglion. A. An interneuron with axons in both anterior connectives that form extensive, bilaterally symmetrical branches in the deutocerebrum and protocerebrum of the brain. B. An intersegmental interneuron with axons in the posterior connectives. C. A local DUM interneuron. Based on Bräunig (1991a).

ganglion and an axon that projects anteriorly (Watson and Burrows, 1983). The primary neurite and the initial parts of the secondary neurites are wrapped in glia and so have no synapses, but where the glia is lacking, synapses are made. All the other mesothoracic branches have an approximately equal number of input and output synapses, amounting to a total of between 10^5 and 10^6 synapses, often spaced within 0.5 µm of each other. There is thus no segregation of these neurites into input and output regions. Many of the synapses are associated with spines, the finest of which (0.1–0.2 µm in diameter) have only input synapses, while the thickest (0.2–0.4 µm) may be packed with 48 nm round agranular vesicles that are related to output synapses. About 2000 vesicles are grouped around the presynaptic density of each output synapse.

Second, the A4I1 interneuron that processes signals from wind-sensitive hairs for probable use in flight steering (*see Chapter 11*) has its cell body in the fourth abdominal ganglion and an axon that projects to the brain, forming branches in the intervening ganglia (Watson and Pflüger, 1989). In this interneuron there is regional specialisation, with ventral branches in the prothoracic ganglion (the main site of its sensory input and spike initiation) and fourth abdominal ganglion receiving predominantly input synapses, and dorsal branches in the other thoracic ganglia making predominantly output synapses. Other branches have a mixture of input and output branches. The output synapses have round agranular vesicles with a diameter of 34 nm and a few granular 90–110 nm vesicles, whereas the input synapses have either round or pleomorphic vesicles.

Third, an interneuron (TN1) that responds to sound (Peters *et al.*, 1986) also has a segregation of its input and output synapses. Branches that occur in the ventral neuropil (mVAC) where the primary sensory neurons from the tympana terminate have input synapses, whereas branches in other regions have predominantly output synapses. This distribution also correlates with the appearance of the branches as seen with the light microscope; the branches with input synapses are fine and smooth whereas those with mostly output synapses are thicker and more varicose.

3.8.3. Membrane properties

3.8.3.1. *Spike initiation*

Intersegmental interneurons may have a single spike-initiating zone in the primary neurite near the cell body, or they may have spike-initiating zones distributed along their axon(s) in different parts of the nervous system. Many of the interneurons with cell bodies in the brain and axons that extend through much of the rest of the body probably only initiate spikes in the brain. Even for some of these interneurons, however, the spike-initiating zone may shift, depending on the modality of the stimulus that activates them. The evidence is indirect and comes largely from an interneuron whose branches extend from the optic lobes to the protocerebrum. This interneuron, the lobula giant movement detector (LGMD), is primarily excited by movement anywhere in the visual field of one eye, but can also be excited by loud, transient sounds (O'Shea, 1975). The spikes initiated by these two modalities arise from sites about 600μm apart, and may thus allow inhibitory inputs to be directed at a particular site and thus modality.

Many spike-initiating zones Many intersegmental interneurons have distinct areas of neurites in different parts of the nervous system. Some of these areas have both input and output synapses so that each could be a potential site for spike initiation. For example, spikes in an interneuron called A4I1 (*see Chapter 11*) that has its cell body in the fourth abdominal ganglion (A4) and an axon that projects to the brain can be initiated at two sites, one close to the cell body and the other in the prothoracic ganglion. This means that spikes can potentially be conducted in both directions through the thorax, with the consequent possibility of collisions. Interneurons that collate tactile information from different parts of the body have spike-initiating zones in each ganglion through which their axon passes (Hustert, 1985). A hair anywhere in their extensive receptive fields has the ability to evoke spikes in these interneurons, but its branches project only to its own segmental ganglion. This arrangement would inevitably lead to congestion of spike traffic when a stimulus affects a large area and suggests that the action of these interneurons is not to collate information from a huge area and convey it to one output site, but instead to act as an integrating channel for the exchange of information between local regions.

3.8.3.2. *Transmitters*

For interneurons of such diversity, it is not surprising that there should be no generalisations about the transmitters that they use. Only for a few intersegmental interneurons, such as the tritocerebral dwarf (Tyrer *et al.*, 1988) (*see Chapter 11*), that contain GABA has it been possible to pinpoint the putative transmitter. For the vast majority of the other intersegmental interneurons there are few clues as to the identity of the transmitter. Some predictions have, however, been made on the basis of the shape of an interneuron as to whether it is likely to have inhibitory or excitatory effects on its postsynaptic neurons. Thus, some of the interneurons involved in flying that have a cell body close to the midline and branches taking particular routes through a ganglion may make inhibitory output connections, whereas those with more laterally placed cell bodies may make excitatory output connections (Pearson and Robertson, 1987). Some of the interneurons with inhibitory outputs may use GABA but others do not (Robertson and Wisiniowski, 1988). This correlation does not necessarily hold for all interneurons.

3.9. NEUROSECRETORY NEURONS

3.9.1. What are neurosecretory neurons?

Neurosecretion is part of a continuum between neurotransmission on the one hand and hormonal secretion on the other so that the distinction between these different processes can easily become blurred. Neurosecretory neurons, in general, show nonfocal release of various chemicals at sites other than the usual classical synapses. There must be a real doubt as to whether it is worth trying to separate a neurosecretory neuron from a neuromodulatory neuron, or from a neuron releasing a substance with a neurohormonal action. These qualifying descriptions of actions are fuzzy because they try to subdivide a continuous spectrum of actions, filled with the concepts imposed by the experimentalist. Both peptides and some amines can have all three actions, and because many descriptions of neurons presently rely solely on immunocytochemical identification of particular substances, it is necessary to deal with intersecting schemes of neuron groups. It seems pointless to worry too much about the validity of the present definitions until we know much more about the identity, connectivity and action of these important neurons.

3.9.2. Neurosecretory interneurons

These neurons have cell bodies and release sites within the central nervous system. In general, not much is known about them save for the projections of their axons and the long duration spikes that are recorded from their cell bodies. There are a few notable exceptions where more information is available, such as the pair of interneurons that contain vasopressin-like peptide and with cell bodies in the suboesophageal ganglion and axons that project to the last abdominal ganglion. These and other interneurons containing material that may serve a neurosecretory or neuromodulatory role are described in *Chapter 5*.

3.9.3. Efferent neurosecretory neurons

These neurons have cell bodies in the central nervous system and release sites in the periphery, either at neurohaemal sites, or close to a particular effector. The substances must be released close to their targets because of the poor circulatory system and because they may be broken down quickly once they are in the haemolymph. Neurons projecting from the pars lateralis to the corpora cardiaca and allata would be numbered among these (*see Chapter 2*). The best known example of these neurons is, however, the efferent DUM (dorsal unpaired median) neurons of the segmental ganglia and they illustrate the enormous impact that such neurons can have on the manifold aspects of neuronal action (*see Chapter 5*).

At the dorsal midline of each segmental ganglion is a group of cell bodies that was first described by Plotnikova (1969) using the Methylene Blue staining methods so successfully promulgated by A. Zawarzin. Neurons in this area were later found to produce overshooting spikes, unlike most other cell bodies of insect neurons, and to have axons in many nerves (Crossman *et al.*, 1971a, 1972). They are unpaired neurons, unlike other neurons in the central nervous system, and from this feature and the dorsal midline position of their cell bodies they were called Dorsal Unpaired Median (DUM) neurons (Hoyle *et al.*, 1974). They are distinct from all the other neuronal cell bodies and are probably the progeny of a single median neuroblast (Bate, 1976a) (*see Chapter 4*).

In the meso- and metathoracic ganglia, each dorsal midline group contains about 90 cell bodies, in the prothoracic ganglion, somewhat fewer cell bodies, and in the abdominal ganglia only about 60. In the metathoracic ganglion, the group consists of three distinct types of neurons that are all unpaired but otherwise have different anatomical and physiological characteristics (Fig. 3.11). It is not known whether all three groups are represented in each ganglion of the central nervous system.

First, there is a group of **efferent DUM neurons** with large cell bodies 50–90 µm in diameter and with axons that emerge in particular lateral nerves from both the left and right sides of the ganglion (Fig. 3.19). These neurons have neuro- and myomodulatory effects and show octopamine-like immunoreactivity. Their properties and their actions are described in detail in *Chapter 5*.

Second, there is a group of **intersegmental DUM interneurons** that have axons in both anterior connectives (*see section 3.8.1.1*).

Third, there is a group of **local DUM interneurons** with bilateral arrays of branches that are all restricted to one ganglion (*see section 3.6.2.4*).

The latter two groups have cell bodies 10–25 µm in diameter with some showing GABA-like immunoreactivity (Thompson and Siegler, 1993) and are therefore thought to act as conventional interneurons for which an action has yet to be established. All the neurons within these groups are spiking interneurons. The cell bodies of the efferent DUM neurons have long duration, overshooting spikes whereas those of the interneurons have only small spikes that are electrotonically conducted from distant spike-initiating sites.

3.9.4. Peripheral neurosecretory neurons

These neurons have cell bodies attached to peripheral nerves, and release sites also in the periphery. Some of the peripheral nerves have small cell bodies along their length, and while this is currently seen as a particular feature of abdominal median nerves and the nerves with which they anastomose, and of some of the nerves to the mouthparts, it seems probable that such neurons are more widespread and may even be regularly associated with skeletal muscles. Some of these neurons contain neurosecretory granules and have spikes of long duration, and on this basis are reasonably assumed to have a neurosecretory action, but others located in the same regions act as conventional sensory neurons signalling stretch of the nerve to which they are attached.

3.9.4.1. *Neurons with cell bodies on peripheral nerves*

On some of the peripheral nerves in the abdomen there are cell bodies of multipolar neurons, the processes of which lie in or close to the surface of the nerve (Finlayson and Osborne, 1968; Fifield and Finlayson, 1978). In the stick insect, for example, there are about 12 such neurons associated with the nerves of each abdominal ganglion. Some four neurons associated with the nerve linking the lateral branch of the median nerve to a lateral nerve (the link nerve) are packed with neurosecretory

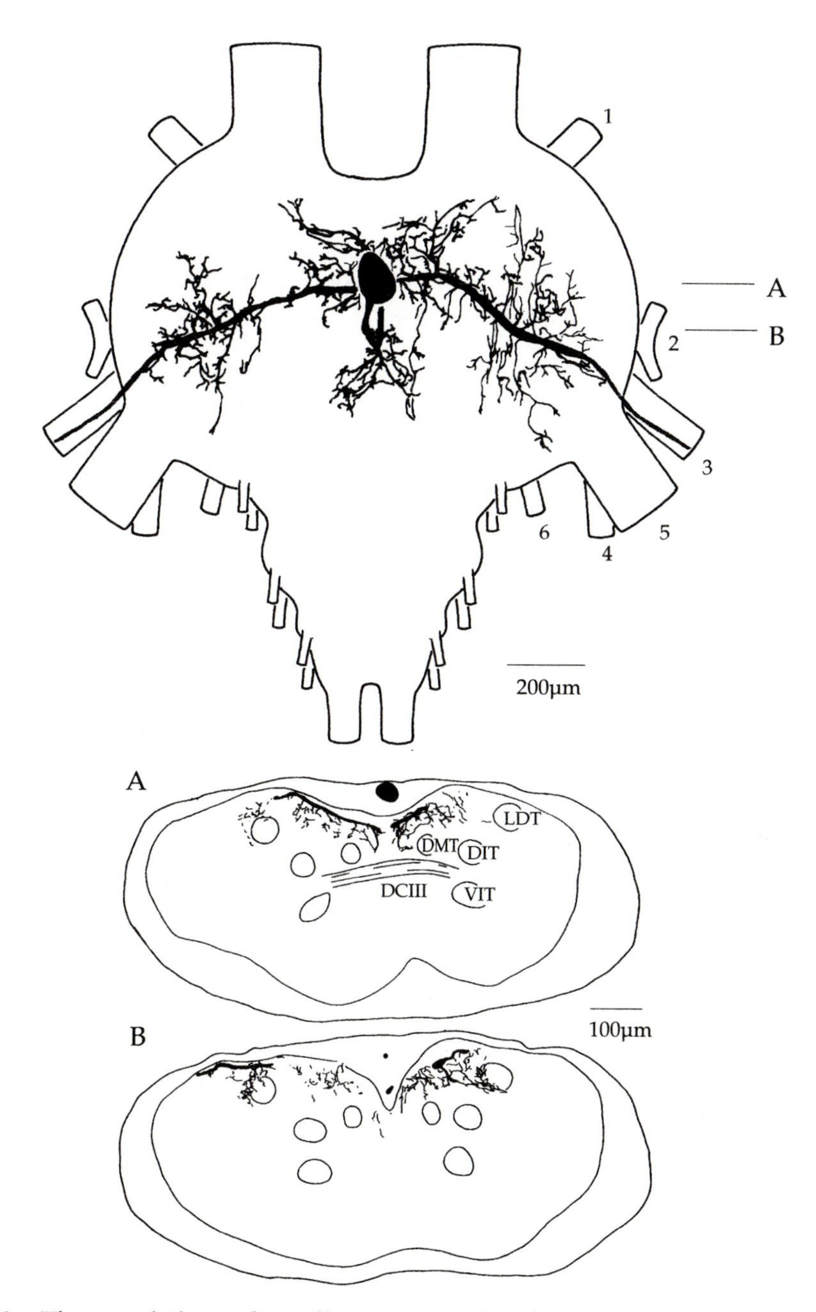

Fig. 3.19 The morphology of an efferent DUM (dorsal unpaired median) neuron in the metathoracic ganglion. This neuron is called DUM 3 because it has axons in N3 to the left and to the right sides of the body. The transverse sections at the levels indicated (A and B) relate the branches of this neuron to known structural features of the metathoracic ganglion (for definitions of the abbreviations see Table 2.2). Based on Watson (1984).

granules and have terminals at the surface of the nerve, suggesting that they release their contents into the haemolymph. The remaining neurons are sensory neurons that conduct spikes towards the central nervous system in response to stretch of a strand closely associated with the nerve. The supposed neurosecretory neurons have spikes with a duration at half-height of up to 20 ms, the inward current of which may be carried by Ca^{2+} (Orchard and Osborne, 1977; Orchard and Finlayson, 1977). The spikes are initiated near the electrically excitable cell bodies, usually occur at low frequencies, and are conducted slowly at velocities of 0.6–0.8 $m.s^{-1}$. The role(s) that such neurons might be playing is unknown.

3.9.4.2. *Median neurohaemal organs*

The neurohaemal organs are swellings on the median nerves that represent the release sites into the haemolymph of neurons with cell bodies in the central nervous system and of neurons with cell bodies associated with the organs or neighbouring nerves. Paired, segmentally arranged neurohaemal organs are associated with the median nerves of each ganglion, but the endings of the neurons associated with these structures can be widely distributed along the nerves. In the mesothoracic segment, the median nerve and the neurohaemal organ itself have three groups of cell bodies (Myers and Evans, 1988). In abdominal neuromeres 1–3, there are several groups of ventrally located cell bodies that project to the neurohaemal organs.

3.9.4.3. *Suboesophageal neurons*

Mandibular ganglion These paired groupings of cells are attached to the mandibular nerve (N1) of the suboesophageal ganglion and each contains the elongated cell bodies (10×50 µm) of only 22–25 neurons that are either clustered together or more widely distributed along the nerve (Bräunig, 1990b). These neurons do not form any branches within the mandibular ganglion itself and there are no projections from other neurons into it. Furthermore, recordings from the cell bodies reveal no spikes or synaptic potentials, indicating that they are remote from integrative actions. Most of the neurons have axons that project to the suboesophageal ganglion, form neurites in the mandibular neuromere and then have axons in both connectives to the brain. This structure thus sets them apart from the neurosecretory neurons found elsewhere on peripheral nerves, and although their function remains obscure, some neurosecretory action seems a possibility.

Four of the larger neurons are, however, sensory neurons that have a meshwork of branches associated with a strand proprioceptor of a mandible, and an axon that returns to the suboesophageal ganglion to form an extensive array of fine branches in the mandibular neuromere with collaterals in the maxillary neuromere. The axon of at least one of these neurons then projects into the tritocerebrum where it again forms fine branches before emerging in the labrofrontal nerve. These neurons are presumed to be exclusively sensory but their peculiar anatomy raises obvious questions about how they might work.

Other neurons In addition to the DUM neurons that have axons which project into the NCCIII nerves of the brain are a group of three paired neurons with cell bodies contralateral to their axons (Bräunig, 1990a). Two of these neurons probably contain CCAP and the third 5-HT, but their action is unknown, and it is assumed that they contribute to neuromodulatory or neurohormonal control mechanisms.

3.9.4.4. *Abdominal neurons*

Each abdominal ganglion has groups of neurons with cell bodies at the midline and with bilateral axons that project to the neurohaemal structures associated with the heart (Ferber and Pflüger, 1990). These neurons are distinct from the efferent DUM neurons in that they do not show octopamine-like immunoreactivity, their primary neurites do not run in the DUM tracts, and their fine branches are often asymmetric. Moreover, they are not the progeny of the median neuroblast as they occur in A11 where this neuroblast is absent. In A7 of females, some of these neurons may also innervate the oviducts. Some of these neurons may belong to the midline group of neurons that shows bovine pancreatic polypeptide-like immunoreactivity (*see Chapter 5*).

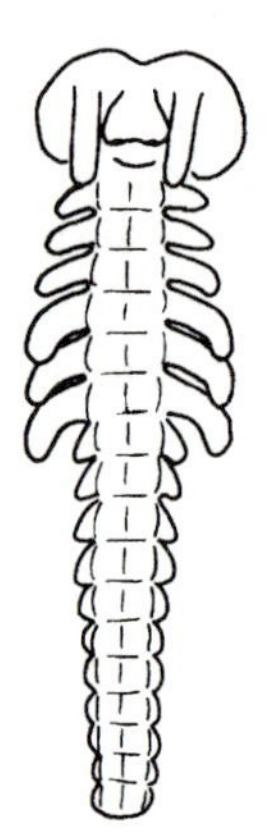

Development of the nervous system

4.1. DEVELOPMENT OF A LOCUST

Locusts are holometabolous insects that undergo a period of embryonic growth and development within the egg before hatching as an insect that has most of the characteristics of an adult, but is much smaller in size. There then follows a postembryonic period of growth and moulting through five larval stages, or instars, of progressively increasing size, lasting some 6 weeks in controlled laboratory conditions. Only after the final moult to the adult are the wings functional, and only after a further 2 week period of maturation, in which the cuticle progressively hardens and changes in colour, is sexual maturity reached.

A functional nervous system must be laid down embryonically and be capable of producing adaptive movements in response to particular sensory stimuli as soon as the diminutive larval locust emerges from the ground. During embryonic development, the majority of central neurons are formed and have established their connections, as have the majority of proprioceptors at joints. For example, all the motor neurons innervating the leg muscles have formed, all the proprioceptors at the leg joints are in place, and all the peripheral nerve pathways are established. Many exteroceptors are also formed and have established connections in the central nervous system, but others will develop postembryonically as the surface area of the body increases.

4.1.1. Development of locusts and flies

Many of the early processes in development are the same in *Drosophila* and locusts (Thomas *et al.*, 1984) so it is possible to use the vast amount of information that is available on the genetics of *Drosophila*, and on the developmental processes themselves, to illuminate the control of these common events. Of course there are differences, particularly in the postembryonic development of *Drosophila,* which is hemimetabolous; the larva, which is quite different in appearance to an adult when it first emerges from the egg, pupates after a certain period. The adult eventually emerges from the pupa. In locust embryos, the cells are larger and thus more easily recognised and development can proceed for a substantial proportion of its time in culture, allowing the divisions and growth of individual cells to be observed (Myers and Bastiani, 1993). Furthermore, expression of specific genes in selected lineages of neurons can be inhibited by the injection of antisense oligodeoxynucleotides into individual precursor cells (Condron *et al.*, 1994). These clear attributes of locusts for developmental studies are, to some extent, offset by their relatively long generation time and by their relatively large genome.

In the following account, information is drawn from both insects but the default description is always of the locust.

4.2. EGG LAYING

The site where the eggs are to be laid is first selected by sweeping movements of the valves of the ovipositor on the tip of the abdomen as they open and close across the surface of the ground. Tactile, chemo- and hygroreceptors are present on the exposed valves of the ovipositor and are presumably involved in the selection process. The favoured substrate is moist sand.

Once a suitable site is selected, the ovipositor then burrows into the ground over a period of 30–45 min to excavate a hole 100 mm deep. The burrowing movements of the ovipositor are produced by rhythmic movements of its shovel-shaped valves, while the front two pairs of legs grip the ground firmly and the hind legs are held ready to deliver defensive kicks to protect the locust in its now vulnerable state. The valves are modified during postembryonic development from the paired ventral appendages originally present on all abdominal segments of the embryo. They are powered by 10 abdominal muscles consisting basically of openers and closers, retractors and protractors (Snodgrass, 1935). The two dorsal valves belong to segment A8 and the ventral valves to A9; this segmentation is maintained in the origin of the nerves from the neuromeres of the fused last abdominal ganglion which innervate them (Thompson 1986a). The ovipositor is pushed downwards with the valves closed, and when they open, soil is forced sideways and upwards so that it is compacted into the sides of the hole. Protraction and retraction movements then elongate the abdomen to enable the hole to be advanced. These opening and closing movements of the valves during digging are periodically interrupted by tamping movements in which the valves remain closed and the ovipositor is rotated to pat and smooth the sides of the hole. During this process of digging a hole, the abdomen is stretched from its normal length of 30 mm to about 130 mm (*see section 12.9.5* for a description of how the muscles and cuticle achieve this extension).

Egg laying starts once the hole is complete. The ovipositor is withdrawn slightly and the rhythmic, neurogenically mediated and anteriorly directed contractions of the oviduct, which have retained the eggs close to the ovaries during digging, are replaced by myogenically mediated and posteriorly directed contractions that transfer an egg to the ovipositor. The eggs are extruded from the ovipositor and are laid in pods, each containing 30–90 eggs in *S. gregaria*. The accessory glands then produce a white froth on top of an expelled egg. The process is repeated until the whole clutch of eggs is laid and capped by more froth which then hardens and darkens and glues the eggs together into a cylinder that fills most of the excavated hole. The whole process can be repeated at intervals of 7–10 days, when more eggs have matured sufficiently within the abdomen.

4.2.1. The motor pattern for digging

The adult female digs a hole in the ground, by movements of her ovipositor, into which she then deposits her eggs. The muscles of the ovipositor valves contract in a particular sequence to produce these digging movements (Fig. 4.1). The contractions are repeated with a periodicity of 1–4 s depending on the consistency of the substrate, and last for 15–20 s. This pattern alternates with a second in which the opener muscles are inactive and results in tamping movements that compact the side of the hole (Fig. 4.2). This pattern occurs at the same frequency as that for digging but lasts for periods of 20–25 s.

The whole motor pattern for digging can be released in mature females by sectioning the connectives posterior to T3, suggesting that expression of the pattern is normally prevented by descending inhibition (Thompson, 1986b). The same phase relationships between the muscles are maintained if all sensory feedback is denied, although the period of the rhythm increases to 3–8 s. This indicates that, as for many of the other rhythmical movements that the locust generates, the properties and connections of the neurons in the central nervous system are sufficient to generate the basic motor pattern (*see Chapters 8, 11 and 12*). The essential neurons are contained in A7 and in A8 (the terminal abdominal ganglion), but if the connections between these two ganglia are cut both the reliability with which the rhythm is expressed and its form are disturbed. Nevertheless, the variability of the rhythm under different external conditions points to the influence that sensory feedback would normally have on its production.

4.2.2. Hatching

Hatching occurs underground after about 20 days at 35°C in *S. nitens*, with embryos emerging from the cluster of eggs laid by one female within 5h of each other. Emergence is brought about by a series of powerful peristaltic contractions that sweep forwards along the body and rupture the egg but leave the embryo still covered by the second embryonic cuticle. The newly hatched larva must then dig upwards through the layer of froth and pehaps sand that covers the buried egg pod. The digging movements begin as soon as the larva has freed itself from the egg shell and the movements used to rupture the egg case grade into those used for digging (Bernays, 1971). The head of these newly hatched and so-called vermiform larvae is pointed, and the neck contains two bladders underneath the embryonic cuticle that can be inflated by increases in blood pressure and pulled back into the body by muscles in the neck. The digging movements result from a wave of contractions in the abdomen that moves forwards rhythmically. The abdomen, followed by the thorax, first

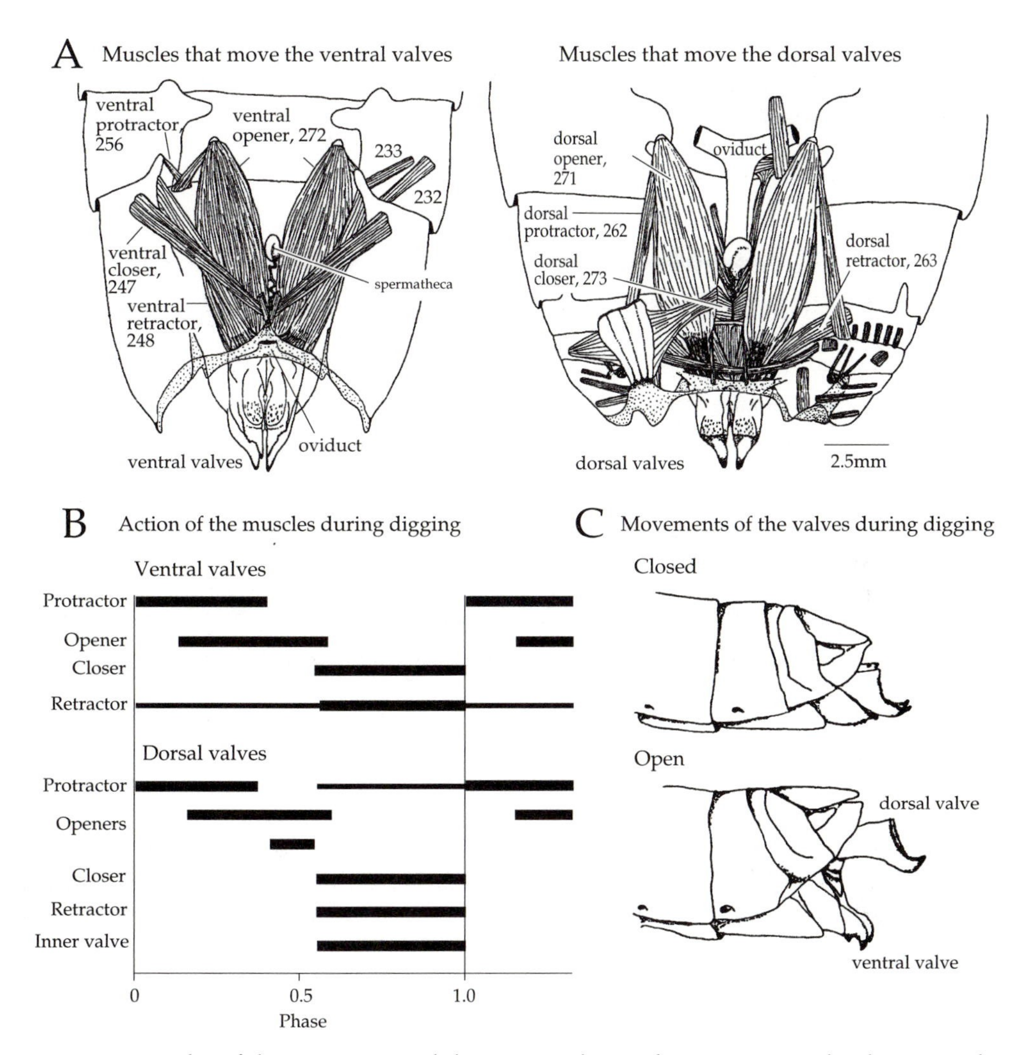

Fig. 4.1 Muscles of the ovipositor and their action during digging. A. Muscles that move the ventral and dorsal valves of the ovipositor. The posterior part of the abdomen is opened ventrally. B. Summary of the action of the muscles during one cycle of the digging rhythm. The solid bars indicate when a particular muscle is active. C. The resultant movements of the valves, shown first in the closed position and then open. Muscles are numbered according to Snodgrass (1935). Based on Seymour (1990).

shortens quickly and the ampullae in the neck are withdrawn. The abdomen then begins to lengthen, the ampullae are protruded to make a hole through which the rest of the body can pass, and the head is thrust forward. After a variable pause the cycle is repeated, with each cycle lasting 2–3 s. Some of the neck muscles involved in generating these hatching and digging movements break down during the first larval instar and are thus not present in an adult (Bernays, 1972).

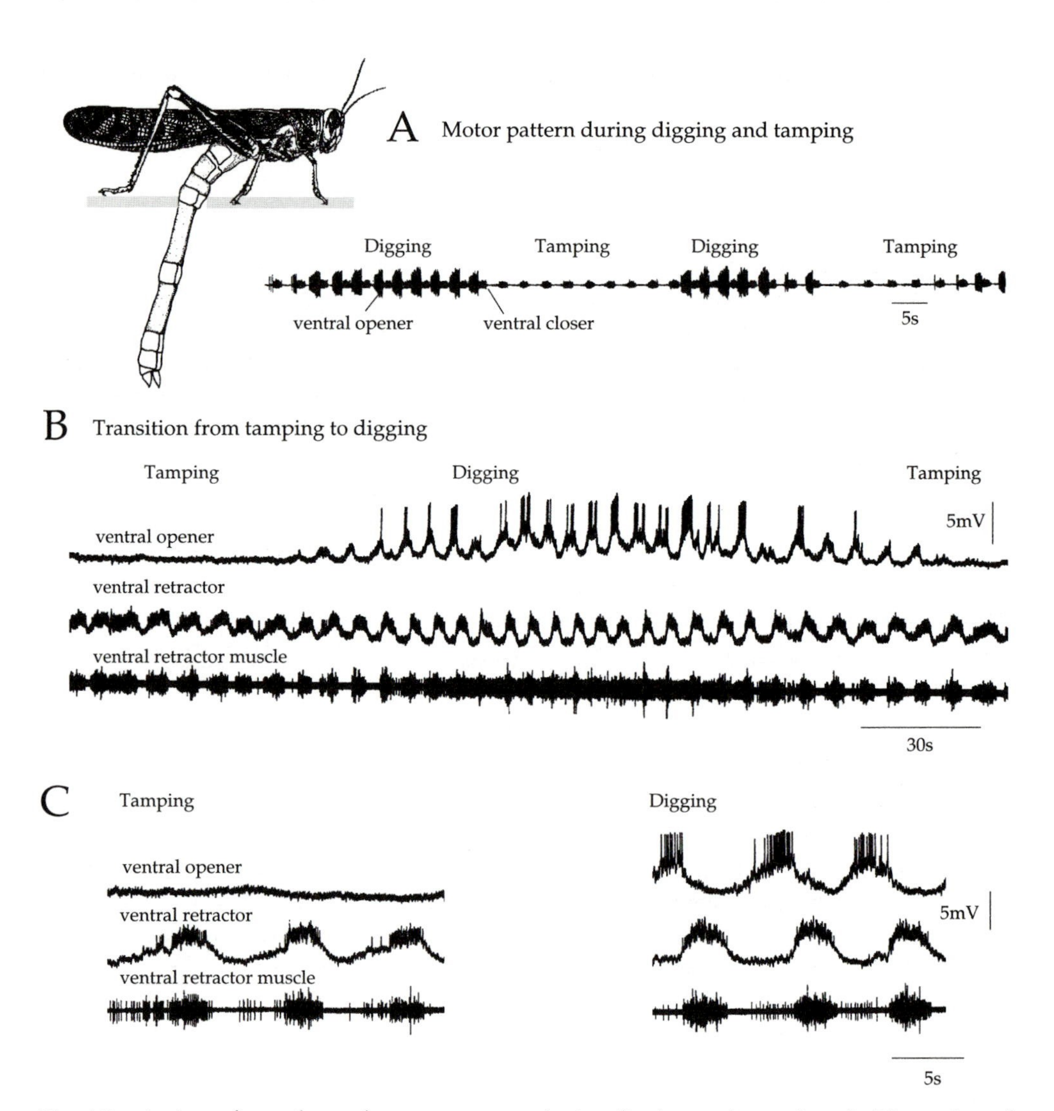

Fig. 4.2 Action of muscles and motor neurons during digging and tamping. A. The action of the ventral opener and ventral closer muscle during a continuous sequence in which digging and tamping alternate. B. Action of the motor neurons during the transition between digging and tamping. At the start of digging, the amplitude of the rhythmic depolarisations gradually increase in a ventral opener motor neuron. C. In tamping, the ventral opener motor neuron does not spike, but during digging it is rhythmically depolarised before the ventral retractor spikes. Based on Seymour (1990).

Within 10 min of reaching the surface, the larva begins to moult the second embryonic cuticle by splitting it along the dorsal midline to allow the first instar larva to emerge. The remnants of the abdominal limb buds are lost, the mouthparts close medially for the first time, and the hairs become erect. After a few minutes of quiescence, this larva is ready to take its first steps.

4.3. MOULTING

The larval locust that emerges from an egg and climbs to the surface of the ground is a miniature version of the adult, but lacks wings. Its hard exoskeleton means that it can grow only by periodically shedding the cuticle, and then expanding the body before the new cuticle hardens (*see section 12.9.4* for a description of the movements that bring about moulting). This process is repeated through five larval instars of progressively increasing size, but even after the final moult into adulthood there are still maturational changes in both appearance and behaviour. The cuticle changes in rigidity, and accompanying this is a progressive increase in the probability of jumping (*see Chapter 9*) and an increase in the frequency of wing beats during flying (*see Chapter 11*). Sexual maturity is achieved some 2 weeks after the final moult and with this comes copulatory and then egg laying behaviour so that the life cycle is repeated. Most of these processes of periodic growth and maturation are under the control of hormones that show a temporal series of changes in concentration in the haemolymph.

4.3.1. Juvenile hormone

Juvenile hormone is secreted by the corpora allata and has many functions both in larval and adult life. Its secretion in larvae determines that the next moult will be to a larva and its absence during the last (fifth) larval stage is necessary for the moult to adulthood. Secretion of juvenile hormone is controlled by neurons in the brain that are immunoreactive for allatostatin (*see Chapters 2 and 5*). If its secretion is prevented in early larval stages, precocious moults to adult-like stages result. If juvenile hormone is injected into fifth instars then a supernumerary (sixth) larval stage is formed, which may then subsequently moult into an adult (seventh stage). Secretion of juvenile hormone by the corpora allata of adults is essential for ovarian development and maturation, and if prevented results in sterility.

Juvenile hormone was first identified in the moth *Manduca sexta* and then subsequently synthesised artificially. Some natural plant products affect metabolic pathways normally activated by juvenile hormone. Extracts from the balsam fir tree mimic the effects of juvenile hormone, while the precocenes, extracted from *Ageratum houstonianum*, have an anti-juvenile hormone action, and when applied to larvae can induce precocious moulting to an adult (Fridman-Cohen and Pener, 1980), and when applied to adults renders them sterile. It acts by disrupting the synthesis of juvenile hormone in the corpora allata through a cytotoxic action on the secretory cells (Bowers, 1981). These adultiform larvae have a mixture of larval and adult characteristics, including miniature, moveable wings that can be flapped in a pattern that resembles flight even though they are not large enough to generate sufficient lift for free flight.

Azadirachtin (a natural extract from the Indian neem tree, *Azadirachta indica*) prolongs larval instars and prevents locusts moulting to adulthood by interfering with the neuroendocrine control mechanisms. In the fly *Lucilia cuprina*, it causes degeneration of cells in the corpora cardiaca, corpora allata and prothoracic gland, which are all fused together in this species (Meurant *et al.*, 1994), thereby disrupting the secretion of juvenile hormone, among many other effects.

4.3.2. Eclosion hormone

The stereotyped movements that are used to shed the cuticle during moulting are triggered by the secretion of an eclosion hormone. In the moth *Manduca sexta* this peptide hormone is secreted for a period of only some 20–90 min at the end of each moult by a group of four cells located in the brain (Reynolds *et al.*, 1979). During this time, almost 95% of the peptide stored in these cells is released. The secretion itself is triggered by the falling levels of ecdysone (ecdysteroid) in the haemolymph. In the intermoult period, these neurons receive a continuous synaptic input but do not generate action potentials (Hewes and Truman, 1994). About an hour before a moult they show a marked increase in excitability that is not accompanied by a change in membrane resistance or synaptic input, so that they spike tonically and have spikes in their somata that are of longer duration. A few hours after the moult, the membrane properties return to those seen during the remainder of the intermoult period. These transient changes are caused by changes in the level of ecdysone which appears to act only slowly and with a time course that suggests an action on the genome. The ecdysone could act either directly on the neurons, or on other interposed neuromodulatory neurons, by mechanisms that require the synthesis of new RNA and protein.

4.3.2.1. *Eclosion hormone neurons*

In locusts, little is known of the processes that lead to the induction of moulting (ecdysis), and it has to be assumed that the mechanism is similar to that characterised in other insects. In moths, for example, larval and pupal ecdyses are triggered by the release of an eclosion hormone contained largely in two pairs of neurons with cell bodies in the brain and with axons that project along the entire length of the nervous system to the proctodeal nerves of the last abdominal ganglion (Truman and Copenhaver, 1989). There are also five other paired neurons in the brain that have axons innervating the corpora cardiaca and corpora allata that release eclosion hormone (Copenhaver and Truman, 1986).

4.4. EMBRYONIC DEVELOPMENT

Each egg is oval in shape, about 8 mm long and 2 mm wide, and has a tough protective shell enclosing a yolky interior. The initial pronucleus lies in cytoplasm within the yolk, and after fertilisation it divides into several thousand cleavage nuclei each surrounded by cytoplasm, but not separated by cellular membranes. These nuclei migrate to the periphery of the egg to form a syncytial blastoderm, which gradually becomes cellular as membranes form around each nucleus and its associated cytoplasm. A part of the blastoderm becomes the 'membrane' or serosa surrounding the yolk, while another part becomes the germ band, which then elongates to form a segmentally repeating strip of cells. The blastoderm gastrulates to form three layers: a ventral ectoderm, a dorsal endoderm and a mesoderm in between.

The eggs in a pod hatch at much the same time and with an accuracy that represents no more than 1% of the total time for embryonic development (Bentley *et al.*, 1979). This provides a foundation from which the developmental stage of an embryo can be described by reference to the percentage of development that has elapsed and with an accuracy that can be increased from the usual 5% to 1% intervals if required (Table 4.1). The 20 days of development means that a single day corresponds to 5% of the total developmental time. Each stage can be described by particular features of its appearance.

4.5. FORMATION OF NEURONS

The mature central nervous system is a highly differentiated three-dimensional structure but it arises from a flat longitudinal strip of ectoderm that runs the length of the embryo and which is separated dorsally from the mesoderm by a basement membrane. In each segment, interactions between about 300 (150 on each side of the midline) undifferentiated ectodermal cells result in about 60 of them (30 on each side of the midline) differentiating and enlarging to form precursor cells, called **neuroblasts** and **midline precursor cells**. These cells become spherical, achieving diameters of 20–30 μm, and then separate from the ventral surface of the ectoderm by gradually displacing the surrounding smaller cells, which then either die or become nonneural cells. The size and shape of neuroblasts and midline precursor cells clearly distinguishes them from the other epithelial cells. It is these few cells that must give rise to the numerous and enormously diverse neurons of the adult nervous system. The spatial and temporal sequence of neuroblast formation is controlled by the neurogenic genes such as *notch*, and by the *achaete–scute* complex of proneural genes. Loss of a neurogenic gene results in all cells in the ventral strip of ectoderm becoming neuroblasts, whereas loss of all the *achaete–scute* complex can result in embryos that lack only a single neuroblast, but which then show much neuronal degeneration.

4.5.1. Numbers of neuroblasts and midline precursor cells

In each segment along the length of the body there is a similar set of neuroblasts and midline precursor cells and this arrangement is constant from embryo to embryo (Table 4.2). The arrangement of these precursor cells is conserved across a wide range of different types of holo- and hemimetabolous insects, including flies, bugs and moths, although the numbers of cells may differ (Thomas *et al.*, 1984). There

Table 4.1. Developmental stages of an embryo

Stage (%)	General features	Nervous system
0	Egg light yellow when laid, gradually tanning to brown: posterior pole marked by opaque cap, anterior pole at pointed end of egg	
5	Egg brown: cytoplasm at posterior pole: colourless island of cytoplasm	
10	Single layer of cells forming rudimentary disc-shaped blastoderm at posterior pole: cytoplasm throughout the yolk	
15	Embryo as a one-layered band on ventral surface of egg and attached dorsally to the yolk: differentiated head toward anterior pole	
20	Embryo on dorsal surface, has rotated so that the head faces posterior pole: distinct head and posterior region: cleavage along two thirds of egg	Neuroblasts visible in thoracic segments
25	Segmentation complete to third abdominal segment, with mouthpart and larger limb buds visible: cleavage complete	Neuroblasts visible in anterior abdominal segments: division of neuroblasts starts in brain and anterior ganglia
30	Segmentation complete: limb bud on T3 larger than T1,2 and limb buds present on A1,2: lobes on head contain presumptive retina	Neuroblasts visible in last abdominal segment: progeny of some thoracic midline precursor cells are formed: all neuroblasts dividing to form ganglion mother cells and undifferentiated neurons: Ti1 pioneers in legs
35	Limb buds on all abdominal segments: clear film of cuticle covering all appendages: invagination of proctodeum	Pioneer fibres in antennae. Cx1,2, Tr1, Fe1, and FeCO neurons in legs
40	Embryo occupies half length of egg: hind leg with clear tibia and femur	Fibres in the CNS extend intersegmentally and commissures formed intrasegmentally. Ti2, Ta1,2 and SGO neurons in legs: first leg pioneers have reached CNS

Table 4.1. Continued

Stage (%)	General features	Nervous system
45	Embryo starts moving around posterior pole producing a sequence of anteriorly directed waves of contraction at frequency of 10–20 min^{-1}: tarsus on hind leg: retinula cells in eye: small white spots on either side of midline that become fat bodies	Fibres in CNS form ladder-like arrangement with two commissures in each segment: fusion of A1 to T3
50	Movement about posterior pole complete so that head now faces anterior pole and ventral surface is against ventral side of egg: proctodeum reaches as far anteriorly as A7: tibia of hind leg flexed and parallel to femur: eye plate pigmented: median ocellus present: separation of primary embryonic cuticle from underlying epidermis	
55	Blood sinus along dorsal midline with peristaltic contractions at 10–20 min^{-1}: herring-bone array of muscle fibres in hind leg: antennae divided into annular segments	Major longitudinal tracts recognisable: 5-HT present in intersegmental interneurons: GABA present in neuropil
60	Embryo almost fills egg: rotation about longitudinal axis so that ventral surface is adjacent to dorsal side of egg: white line on eye plate	GABA present in CI motor neurons and other cell bodies: all types of leg receptor now present: exteroceptive sensory neurons from leg first reach CNS
65	Tibia of hind leg has double curve: leg muscles twitch but movements not co-ordinated: peristaltic waves replaced by synchronous contractions along length of median sinus (heart)	Fusion of A2 to T3+A1: asymmetrical contacts between neurons: major areas of neuropil recognisable: second wave of ingrowth of sensory neurons: octopamine present in thoracic DUM neurons
70	Embryo pale green: red pigmentation of posterior half of compound eye: hind tibia straightens	Fusion of A3 to T3+A1,2: synapses with mature features first seen: spikes and synaptic potentials in local interneurons: majority of sensory neurons from exteroceptors reach CNS

Table 4.1. Continued

Stage (%)	General features	Nervous system
75	Rows of brown spots along length of hind femur: facets of compound eyes outlined in white: second embryonic cuticle separates from epidermis but is not shed until hatching, third cuticle starts to form: this will be cuticle of first instar larva	Sensory neurons from leg continue to reach CNS
80	Head pressed against anterior end of egg: brown pigment spots on femora of all legs: second embryonic cuticle separates from epidermis leaving a third layer to be the cuticle of the first instar larva	
85	Embryo bright green: vertical stripes and pseudopupil in eye	Hair sensilla on cerci
90	Three dark green stripes on hind femur: tarsal claws black: eyes brown: occurrence of indentations of cuticle due to contractions of the muscles	
95	Cuticle opaque white against which brown and black markings are prominent: eye stripes fade: blue colour inside hind legs	Black hairs on head, mouthparts, legs and cerci
100	Embryo hatches from egg	

A = Abdomen
A1 = Ganglion of first abdominal segment
CI = Common inhibitory motor neuron
CNS = Central nervous system
Cx = Coxa
Fe = Femur
FeCO = Femoral chordotonal organ
SGO = Subgenual organ
T = Thorax
T1-3 = Ganglia in pro-, meso- and metathoracic segments
Ta = Tarsus
Ti = Tibia
Tr = Trochanter

are also small differences in the number of neuroblasts in different segments and this is under the control of homeotic genes.

4.5.1.1. *Thoracic segments T1-3*

Our detailed knowledge of the development of the thoracic ganglia provides a model against which the small variations that occur in other segments can be compared. In each thoracic segment there are 61 neuroblasts arranged in two bilaterally symmetrical flat plates, each containing 30 neuroblasts. Within these plates, the neuroblasts are arranged in seven rows of between two and six cells (Fig. 4.3) (Bate, 1976a). A single neuroblast, called the median neuroblast, is unpaired and lies close to the posterior boundary of the segment. In the prothoracic, mandibular and abdominal segments 9 and 10 there is an additional anterior median neuroblast.

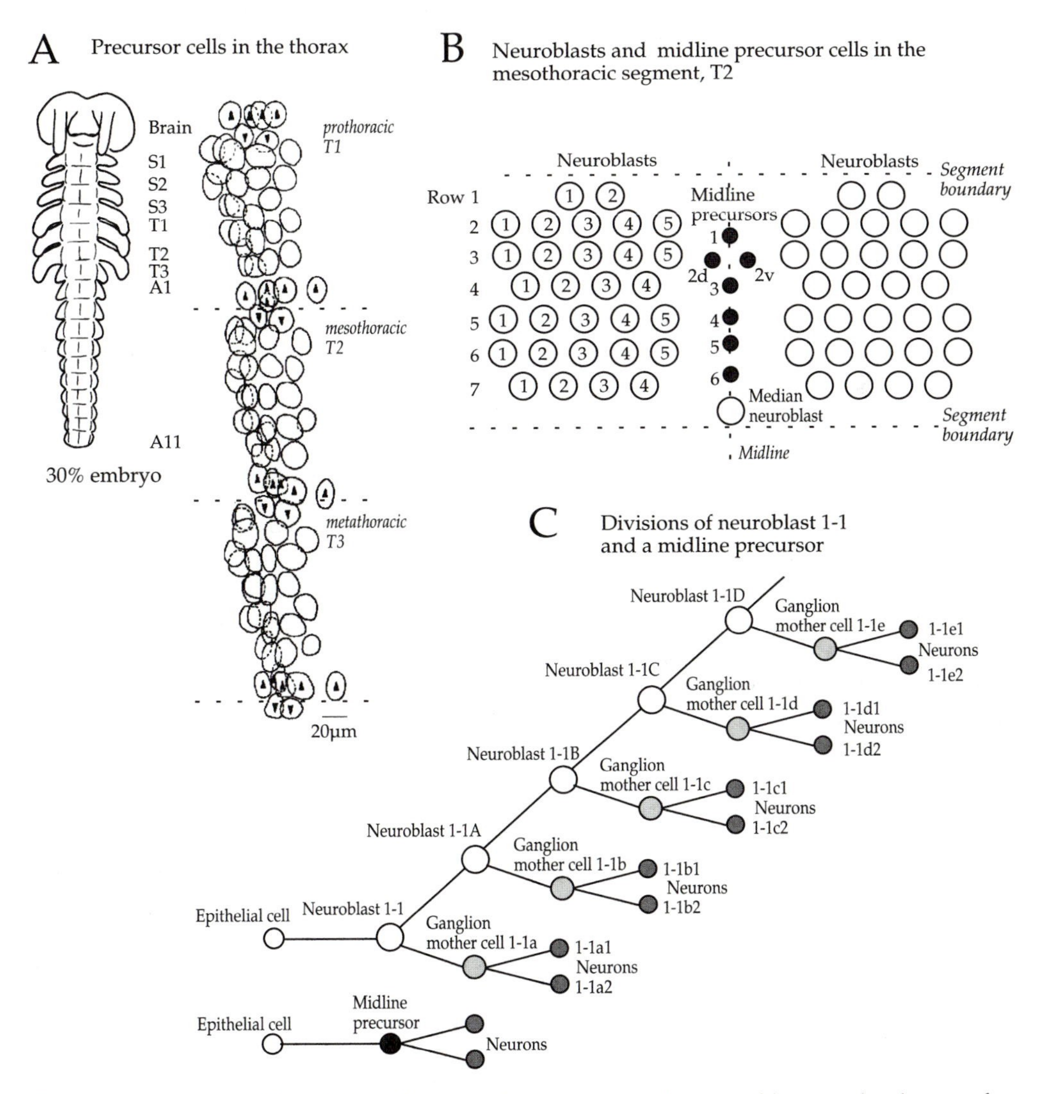

Fig. 4.3 Precursor cells and their divisions. A. Drawing of the neuroblasts in the thorax of a 30% development embryo. B. Diagram to show the arrangement of the neuroblasts and midline precursor cells in the mesothoracic segment. All but the median neuroblast are paired and they are arranged in seven rows. The midline precursor cells are also unpaired and are arrranged at the midline. C. A neuroblast divides asymmetrically to form a ganglion mother cell, which then divides symmetrically to form two neurons. A midline precursor cell divides symmetrically to form two neurons. A and B are based on Bate (1976a) and Goodman *et al.* (1982).

4.5.1.2. *Abdominal segments A1-7*

In each of the first seven abdominal segments there are 59 neuroblasts (neuroblast 5-5 is missing), even though the final tally of neurons in an adult abdominal neuromere is only about a quarter (500–600 neurons) of that in a thoracic neuromere (Doe and Goodman, 1985). The smaller number of neurons in an adult abdominal ganglion results from the earlier degeneration of certain neuroblasts, from the slower rate of division of those that survive, and from the larger number of progeny that subsequently die. For example, the median neuroblast produces about 100 progeny in the metathoracic ganglion but only 90 in A1, because its divisions start later and finish earlier, but many of these offspring die so that only 45 remain in the adult.

4.5.1.3. *Suboesophageal segments S0-3*

In the intercalary segment that forms the anterior part of the suboesophageal ganglion there are only 26 bilateral neuroblasts arranged in a different pattern to the other segments. The mandibular segment has 48 bilateral neuroblasts, but the maxillary and labial segments have 58 bilateral neuroblasts and one median neuroblast and are thus similar to A1-7 in lacking neuroblast 5-5.

4.5.1.4. *Brain*

In the brain of 30–45% development embryos, 260 neuroblasts are arranged in two symmetrical plates and are grouped in ways that reflect the main regions (Zacharias *et al.*, 1993). These are the neuroblasts that divide asymmetrically in the same way as those in the more posterior segments and are distinct from plate-like aggregations of smaller precursor cells that divide symmetrically. The difficulty of defining the segmentation of the brain also means that it is difficult to ascribe the neuroblasts to particular segments. The most widely accepted estimates assign 86 pairs of neuroblasts to the protocerebrum, (excluding the optic lobes for which no estimate is available), 32 to the deutocerebrum and 12 to the tritocerebrum (Table 4.2). The number for the deutocerebrum is similar to that for the more posterior segments and adds further weight to the idea that it is formed from the antennal segment. The small number for the tritocerebrum suggests that these neuroblasts belong to the intercalary segment, and have become separated by the growth of the gut into one group of 12 pairs that form the tritocerebrum and a second group of 13 pairs that form the anterior part of the suboesophageal ganglion.

4.5.1.5. *The neuroblasts can be identified*

The reliable arrangement of the neuroblasts in each segment means that an individual neuroblast can be identified by its position in a row and numbered accordingly at different stages during early development. Molecular markers identify different overlapping sets of neuroblasts, and individual neuroblasts can be identified by the particular genes that they express (Doe, 1992). The segment polarity genes could provide unique identities in the antero-posterior axis. In *Drosophila*, for example, neuroblast 5-2 expresses *wingless* and *seven-up*, and 7-4 expresses *seven-up*, *achaete*

Table 4.2. Numbers of precursor cells in each segment

Segment	Neuroblasts in each half segment	Median neuroblasts	Midline precursor cells
Brain			
Protocerebrum			
Pars intercerebralis	20	?	?
Medial anterior	41	?	?
Lateral	25	?	?
Deutocerebrum	32	?	?
Tritocerebrum	12	?	?
Suboesophageal			
Intercalary, S0	13	1	?
Mandibular, S1	24	2	4
Maxillary, labial, S2,3	29	1	7
Thoracic			
Prothoracic, T1	30	2	7
Meso-, metathoracic, T2,3	30	1	7
Abdominal			
A1–7	29	1	7
A8	30	1	8
A9	29	2	7
A10	25	2	7
A11	6	0	?

Data from Doe and Goodman (1985); Zacharias et al. (1993).

and *engrailed*. This pattern of expression changes with time and may, in part, determine cell lineages. The *ming* gene is expressed in some neuroblasts from the time of their formation (e.g. 6-1), in others only after they have divided a number of times (e.g. 1-1), but it is not transcribed in any of the neurons that are subsequently formed. Similarly, the gene *prospero* is expressed in only some neuroblasts and its absence results in them failing to express certain other genes and leads to the production of abnormal progeny, some of whose axons follow incorrect paths (Doe *et al.*, 1991). Thus, although the position and morphology of a neuroblast do not change as it goes through its cycle of division, changes are clearly taking place that may determine the fate of its progeny. As a consequence of the different times of expression of the different genes, the ganglion mother cells will differ in the products that they inherit and this could influence the fate and differentiation of their neuronal progeny.

The arrangement of the midline precursor cells is also regular, so that they can be treated as identified individuals, which, like the neuroblasts, form an invariant family of neurons and glial cells. Identification of the neuroblasts and midline precursor cells is an essential prerequisite to tracing their progeny during development. The fate of an individual neuron is thus determined by the neuroblast of origin, its place in the lineage, and its interactions with sibling neurons.

4.5.2. Divisions of neuroblasts and midline precursor cells

The division of these cells begins just before 30% development in anterior segments, moves progressively posteriorly and is complete by 90% development, whereupon the neuroblasts die (Shepherd and Bate, 1990). Each division of a neuroblast is asymmetric and produces **a ganglion mother cell** while at the same time replicating itself (Bate, 1976a), with the complete cell cycle lasting about 5 h. A ganglion mother cell divides symmetrically once more to produce **two ganglion cells** that do not divide again but differentiate into neurons (Fig. 4.3). The neuroblast itself divides again so that the whole process is repeated, with the result that different neuroblasts each produce between 10 and 100 progeny arranged as inwardly directed radial columns that add the third dimension to the developing central nervous system. The progeny of a single neuroblast are typically wrapped as individual, separate packets by layers of glia so that they are distinguishable from other lineages. By contrast, each midline precursor cell divides just once to form two neurons, but if they are simply ganglion mother cells then they are merely following the established pattern. This means that the repeated divisions of the 61 neuroblasts in a thoracic segment produce 1000 ganglion mother cells which then divide once to form most of the 2000 neurons of an adult thoracic neuromere. Single divisions of the seven midline precursor cells provide the rest. The first division of a neuroblast in *Drosophila* can occur only 10–20 min after the neuroblast has enlarged and moved dorsally from the epidermis, with subsequent cell cycles lasting 40–50 min (Hartenstein *et al.*, 1987). The divisions of the neuroblasts may be regulated by neighbouring glial cells. In *Drosophila*, the *anachronism* locus encodes a glycoprotein that is expressed in glial cells but not in the neuroblasts, and in mutants in which it is lacking, the neuroblasts proliferate earlier, suggesting that their divisions are normally held back by the influence of the glial cells (Ebens *et al.*, 1993).

4.5.2.1. *Timing of the neuroblast divisions*

The division of the neuroblasts in the different ganglia follows a strict temporal pattern, beginning first in the presumptive brain and then spreading posteriorly. By 45% development, some of the neuroblasts in the segmental ganglia have completed their divisions and are either degenerating or regressing in size so that they can no longer be distinguished from their offspring that are increasing in size. By 50% development, some 14 of the original complement of neuroblasts in a thoracic ganglion have finished dividing and are missing, but in an abdominal ganglion more than half of the neuroblasts have already stopped dividing. By 70% development, half of the thoracic neuroblasts continue but only five to six abdominal neuroblasts continue to divide. In the brain, however, only three of the 130 pairs of neuroblasts have stopped dividing at this stage. The implication is that the neuroblasts in the brain have a longer life span than those in the segmental ganglia and must therefore contribute to the large final number of neurons. By 85–90% development, division of all neuroblasts in the segmental ganglia has ceased, although some in the brain may still retain their capability for division. This must mean that new neurons are formed well into late embryonic life, and because full differentiation of individual neurons to the point where they can express electrical activity takes approximately 25% of development (Goodman and Spitzer, 1979), they will not be fully differentiated at hatching. It is, of course, possible that neurons born later develop more rapidly that those born earlier so that all are at the same stage of maturation at hatching. Particular neuroblasts start to die at the same time as the epidermis becomes folded and periodically bulged to form the ganglia. This means that as development progresses, the number of neuroblasts present depends on a balance between those ectodermal cells that are being committed to form neuroblasts and the number of neuroblasts that are dying.

Within this overall pattern of neuroblast divisions, individual neuroblasts have their own particular life histories determined by the time of onset of their divisions, their rate of division, and the time when they stop dividing. The individual neuroblasts can be recognised in different locusts and their individual patterns of divisions appear to be repeated by homologous neuroblasts in each thoracic segment. For example, the various neuroblasts in the mesothoracic ganglion begin dividing at different times between 23 and 30% development, but neuroblasts 6-3 and 7-3 are always the first to complete their divisions and have disappeared by 45%. During the early part of their lineages, the neuroblasts in a thoracic ganglion divide at the same rate of four to five times in 24h, but as the lineage comes to an end, the rate declines. In an abdominal ganglion, by contrast, the rate of division can be much lower (two divisions in 24h), and the neuroblasts stop dividing earlier than do those in thoracic ganglia. These factors may contribute to the smaller final complement of neurons in the abdominal neuromeres.

One of the striking features of Table 4.3 is that one of the lineages is known in great detail (the median neuroblast in a thoracic segment, *see sections 4.6 and 4.11.4*), but that most of the lineages of the other neuroblasts are unknown. We know the origins of only a very few of the bigger motor neurons and interneurons in the central nervous system even though large numbers of them are well characterised both anatomically and physiologically in the adult. It would seem that the initial enthusiasm for tracing the development of known neurons that emerged in the early eighties waned once the general idea of a lineage had been established, and has been replaced by the obvious attractiveness of the genetic approaches offered in *Drosophila*. This now means that there is a large lacuna in our knowledge on the origins of identified neurons that is ready to be filled. The genetic information from *Drosophila*, which provides so many useful genetic and molecular markers, combined with the use of many new dyes that can be injected into neuroblasts and precursor cells, should permit identification of the cells in the early stages of a lineage. Finally, this information can then be combined with our detailed knowledge of the identity of different neurons in the adult nervous system. Such an analysis would be invaluable in providing insights into the organisation and relationships of these neurons that elude physiological or anatomical studies on the mature neurons.

In *Drosophila*, about 20 neurons in a thoracic ganglion can be recognised in the early embryo, but the final identity of these neurons is generally not known. Exceptions are a group of five motor neurons given the unlikely names RP (Raw Prawn) 1-5 even though the thoracic muscles that they innervate are known. It is time that these nicknames were replaced with names appropriate to their patterns of innervation, as for all other motor neurons. As an example of the power of this

Table 4.3. Lineages of neuroblasts and midline precursor cells in a thoracic ganglion

Neuroblast number	Cells produced in its lineage
1-1	aCC motor neuron: pCC intersegmental interneuron
1-2	
2-1	
2-2	leg motor neurons including 133a
2-3	
2-4	
2-5	
3-1	antero-medial spiking local interneurons: intersegmental interneurons
3-2	
3-3	
3-4	
3-5	
4-1	midline spiking local interneurons: intersegmental interneurons
4-2	local and intersegmental interneurons; RP1-3 neurons
4-3	
4-4	
5-1	CI2: motor neurons: intersegmental interneurons
5-2	
5-3	
5-4	
5-5	CI1, CI3: motor neurons: intersegmental interneurons with GABA-like immunoreactivity
5-6	
6-1	
6-2	
6-3	
6-4	
7-1	
7-2	
7-3	S1-3, serotonin immunoreactive interneurons
7-4	C (314), 311, 319 and G (714), 531, 529 intersegmental interneurons: Q1-6 neurons
Median neuroblast, MNB	MP4-6: midline glial cells: efferent, local and intersegmental DUM neurons

(*continued on facing page*)

Table 4.3. Continued

Midline precursor	Cells produced in its lineage
1	pioneers posterior longitudinal tracts
2L	pioneers anterior longitudinal tracts
2R	pioneers posterior longitudinal tracts
3	H cell, H cell sibling
4 Ganglion mother cell of median neuroblast	help establish anterior commissure: superficial DUM tract SDT
5 Ganglion mother cell of median neuroblast	Deep DUM tract, DDT
6 Ganglion mother cell of median neuroblast	help establish posterior commissure: posterior DUM tract, PDT

aCC, pCC = anterior and posterior corner cells
C = 'cocking' interneurons given number 314 (see Chapter 9)
G = identified interneuron responding to sound, given number 714
CI1-3 = common inibitory motor neurons 1-3
MP = midline precursor
Q1-6 = no name given
RP1-3 = 'raw prawn' neurons

approach and what could be achieved by its wider application, the growth cone of the RP3 neuron in *Drosophila* can be traced through its pathfinding stages in the central nervous system and in the periphery to the establishment of synapses with a particular abdominal muscle, so that almost the complete life history of this neuron can be described, even though its neuroblast of origin is still uncertain (Broadie *et al.*, 1993).

4.5.3. Midline precursor cells

At the midline of a thoracic segment are seven midline precursor (MP) cells in addition to the median neuroblast, and this pattern is repeated in most segments, although the segments that form the suboesophageal ganglion may lack specific cells. These are the precursor cells that will each divide symmetrically just once to form two neurons, and this pattern of division seems to set them apart from the neuroblasts that form the vast majority of cells in the central nervous system. This pattern of divisions is similar to those of the ganglion mother cells of a neuroblast, and it is probable that this is what at least some of them are. By tracing their lineage into early development, midline precursor cells 5 and 6 (MP5, MP6) and probably also MP4 can be shown to be ganglion mother cells of the median neuroblast and are not cells formed separately from the midline epithelium (Condron and Zinn, 1994). So embedded are these names in the literature that they have been retained, even though they give a false impression about the nature of these cells. The similarity in the pathways pioneered by these neurons to the axonal morphology of the later progeny of this neuroblast, however, greatly simplifies and rationalises our understanding of this early phase of development at the midline. It seems probable, however, that the remaining midline precursors, MP1-3, are distinct precursor cells that are not the progeny of other neuroblasts. The processes of some of these cells act as pioneers that establish the basic ground plan of a ganglion by forming the longitudinal tracts and some median tracts, and contributing some of the first fibres to two of the commissures. A glial precursor cell is also present at the midline.

4.5.4. Determination of cells in the lineages of neuroblasts

The neuroblasts arise in a continuous sequential process, with NB3-5 being the first to appear followed by 2-5 and with the others being added to form a series of four columns, with 7-3, or 5-5 if present, being the last to be formed (Doe *et al.*, 1985). The formation of these neuroblasts does not appear to be according to chemical gradients in any direction. The individual epithelial cells are not initially specified to form a particular neuroblast because if an enlarging epithelial cell is removed by laser ablation then another undifferentiated epithelial cell takes over its role. The new neuroblast assumes the fate of the one that has been removed, even though it has been formed by an epithelial cell from a new position. Thus, the fate of the enlarging neuroblast is determined by its position in the epithelial sheet. This mechanism ensures that an intact embryo is formed even if there are small losses in the epithelial cells early in development, and it is only when all the undifferentiated epithelial cells at a particular position are removed that a specific neuroblast is not formed. When a neuroblast enlarges, it appears to inhibit nearby epithelial cells so that only one neuroblast develops in a particular position and the other epithelial cells are left to form nonneural cells.

The ganglion mother cells are determined by their neuroblast of origin and by their sequence in the lineage of divisions. If, therefore, a ganglion mother cell formed early in the lineage is removed, the neurons it would produce are not formed by later ganglion mother cells and are thus lost. By contrast, if a neuroblast is removed after it has already divided to produce the first neurons in its lineage, then the neuroblast that is formed in its place will also divide to produce duplicate sets of the same neurons, although the delay may result in the growth cones of these neurons following different routes. These ganglion mother cells express a number of genes amongst which are *fushi tarazu* and *even-skipped*, loss of whose function results, respectively, in abnormal neuronal determination and in neurons with axon morphologies that differ from the normal. Such genes may be involved in the many processes that lead to the determination of the neurons.

The neurons that are the progeny of a ganglion mother cell are initially equivalent and are not committed to a particular fate. They become determined before there is any overt sign of differentiation, such as the formation of an axonal growth cone, with one of the pair of phenotypes being dominant so that it is always formed if either of the two offspring of the ganglion mother cell is ablated. The neuron that has been deleted is not replaced by the division of any other cells. This means that a neuron is determined by at least three factors.

1. Its lineage from a particular neuroblast.
2. The position of its particular ganglion mother cell in the sequence of divisions of a neuroblast.
3. Its interactions with its sibling from the same ganglion mother cell, but no genes are yet known which could control or mediate these interactions. Interactions with the progeny of neighbouring precursor cells seem to play little part.

4.6. FORMATION OF GLIAL CELLS

Glial cells are formed by the division of distinct precursor cells called glioblasts whose progeny are solely glial cells, and by particular divisions of certain neuroblasts. In the locust, the median neuroblast forms glia, and in *Drosophila*, neuroblast 1-1 also forms glia (Udolph *et al*., 1993). The glia arise in three different regions of the developing nervous system and many are formed before the first neurons begin to extend their axonal growth cones.

First, on each side of each segment where the paired longitudinal connectives will form are six longitudinal glial cells that are the progeny of a lateral and anterior glioblast which expresses the gene *fushi tarazu*. In contrast to the neuroblasts, this glioblast divides symmetrically into two cells which then migrate anteriorly across the segment boundary before dividing to form these glial cells.

Second, on each side of the midline where the intersegmental nerves will form (in the adult these are the nerves formed by the fusion of N6 of an anterior thoracic ganglion and N1 of the next posterior ganglion) are two bilateral segmental boundary cells that become glial cells.

Third, at the midline of each segment where the anterior and posterior commissures will form are six midline glial cells. These are the progeny of cells at the midline that express the gene *single-minded* and stain with an antibody raised against annulin. In locusts, these midline glial cells are all the progeny of the median neuroblast (Condron and Zinn, 1994). Previously, the origin of these midline glial cells had been ascribed simply to midline cells of the sort which also form the median neuroblast and midline precursors, but tracing their lineage shows that they, like some of the numbered midline precursor cells themselves, are actually the ganglion mother cells of the median neuroblast. The midline glia in locusts are further subdivided into six classes in each segment.

1. One or two posterior midline glial cells with processes extending to the medial edges of the longitudinal fascicles.
2. About six sheath glia that envelop the median neuroblast and its progeny.
3. Two anterior commissural midline glia that wrap the anterior commissure.
4. Two posterior commissural midline glia that wrap the posterior commissure.
5. About six intercommissural midline glia with processes that wrap the commissures and the progeny of the MP3 precursors.
6. About 10 anterior midline glia whose processes also wrap the commissures.

In *Drosophila*, the homeobox gene *repo* is required for the differentiation and maintenance of glial function but not for the initial determination of glial cells (Halter *et al.*, 1995). The protein encoded by this gene is expressed in most embryonic glial cells (not, however, in the midline glia and some of the nerve root glial cells) but not in neurons, unlike many other genes that are expressed only in subsets of glia and in certain neurons.

4.7. CENTRAL PIONEER FIBRES

The basic scaffold of the central nervous system appears to be constructed according to three rules. First, a small number of neurons that are born and differentiate early pioneer an array of pathways that are repeated in each segment. Second, these pathways are labelled, most probably by molecules on the surfaces of their cells that may be expressed only at certain times. Third, the growth cones of neurons that are born and differentiate later follow these established routes but have to make choices that are determined by their reading of the appropriate labels. These labels may often need to contain information on polarity, perhaps in the form of gradients, so that a growth cone turns anteriorly or posteriorly. The filopodia of the growing neurons must therefore sample their immediate environment and at successive choice points make the correct decision about the direction in which to turn. A succession of these choices leads to the complex shape and appropriate connections of a neuron.

4.7.1. The basic scaffold of a ganglion

The first pioneers are formed by division of midline precursor cells 1, 2L and 2R to form two groups of three cells on either side of the midline (Table 4.3) (Fig. 4.4) (Bate and Grunewald, 1981). MP1 contributes a cell to each side, and the two MP2 cells contribute a ventral and a dorsal cell to each side. A single growth cone process from the ventral progeny of an MP2 cell runs anteriorly along the basement membrane, and the processes from MP1 and the dorsal MP2 run posteriorly, to establish the first longitudinal tracts extending anteriorly and posteriorly beyond the segmental boundaries on either side of the midline. In *Drosophila*, MP1 and both the MP2 cells express the gene *fushi tarazu,* in contrast to the neuroblasts which do not.

The progeny of MP4 extend processes anteriorly to establish the superficial DUM tract (SDT, originally called the median fibre tract, MFT) at the midline, but then bifurcate and turn laterally to form some of the first fibres in the anterior commissure in a 35–40% development embryo. This arrangement means that each progeny of MP4 has an axon in this anterior commissure to both the left and the right sides of the body.

The progeny of MP5 have axons that bifurcate to enter the anterior commissure but follow a more ventral route than that pioneered by MP4 so that they pioneer what will eventually be the deep DUM tract (DDT) of adult ganglia.

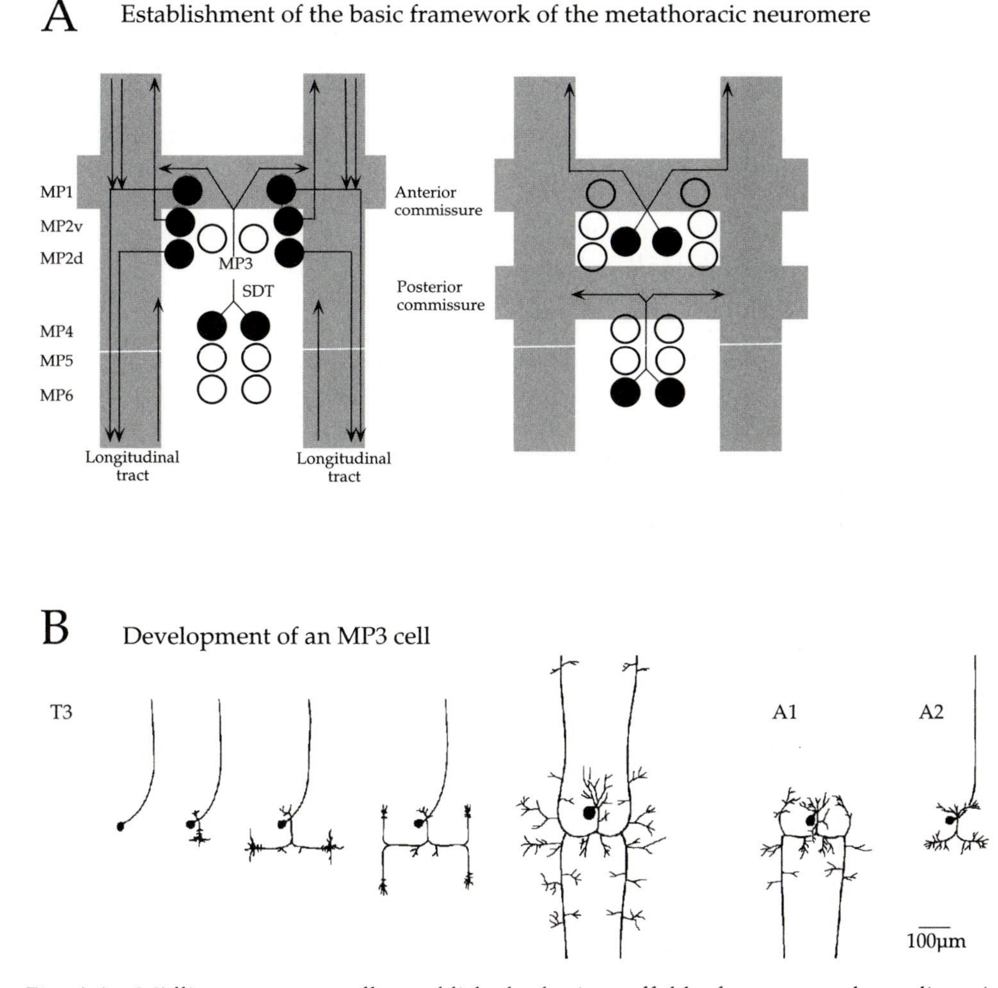

Fig. 4.4 Midline presursor cells establish the basic scaffold of a segmental ganglion. A. Growth cones of particular midline cells establish the longitudinal tracts and the anterior and posterior commissures. The particular midline cells contributing are shown as filled circles in the two diagrams. B. The MP3 cell in the metathoracic (T3) and abdominal (A1, A2) neuromeres and its differentiation during development into particular neurons. In the metathoracic ganglion it retains posterior and anterior axons to become the H neuron; in A1 only the posterior axons persist, and in A2 only the contralateral anterior axon persists. Based on Goodman *et al.* (1981) and Bate *et al.* (1981).

The progeny of MP6 have axons that pioneer the posterior DUM tract (PDT) of adult ganglia before bifurcating to contribute some of the first fibres to the posterior commissure.

The progeny of the remaining midline precursor cells then follow these established routes. Thus, each of the progeny of MP3 has an axon that crosses the midline following the anterior commissure before turning anteriorly and joining the axons of MP1 in the first longitudinal tract. Somewhat later, at about 40% development, a further transformation of one of these cells occurs in the meso- and metathoracic ganglia, with the appearance of a posterior growth cone that bifurcates at the posterior connective. When these growth cones reach the longitudinal tracts they again branch so that four axons are formed, one in each of the anterior and posterior longitudinal tracts. By 60% development, the initial anterior axon is lost, leaving a neuron with four axons in a pattern that has led to its being called the H cell (Goodman *et al.*, 1981). It is thus an unpaired but symmetrical neuron. The other progeny of MP3 retains its original shape but is called the H cell sibling, and in 20% of embryos dies.

The same cells in each segment divide and send out axons in the same way, thus forming a ladder-like arrangement of tracts and commissures along the length of the embryo. The processes of these neurons thus form the scaffold that links together the repeating pattern of neurons into a chain of ganglia that, in turn, represents the complete central nervous system. Other commissures and longitudinal tracts are then established with the help of the progeny of the neuroblasts, so forming the basic structure around which the subsequently generated neurons will differentiate.

4.7.1.1. *Scaffold of glial cells*

The neurons establishing the first commissures and tracts have no other neurons to guide their growth cones, so they must derive clues either from the epithelial cells or from primitive glial cells. They also use extracellular clues present in the basement membrane and basal lamina. Three classes of glial cells are in position before any of the neurons start to produce growth cones in both the locust and *Drosophila* (Jacobs and Goodman, 1989). These three types of cell, and perhaps more, provide an ordered structure of narrow channels over which the neurons can then erect the first neuronal scaffold. In locusts, the channels are wider and the glial covering is less complete than in *Drosophila*, so that the developing neurons have to navigate over exposed regions of basement membrane. The developing neurons make specific contacts with particular members of this glial population. The role of these glial cells becomes especially apparent in the formation of the intersegmental nerve. The growth cones of the motor neurons extending axons into this nerve show a high affinity for the segmental boundary cell, whereas other nearby neurons forming the fascicles of the connectives do not (Bastiani and Goodman, 1986). This glial cell continues to divide and its progeny become glial cells that wrap the axons in this nerve.

All of this suggests that the first neuronal scaffold is erected on a substrate of oriented glia. Once the neuronal scaffold is formed, later neurons follow the more elaborate and specific clues established by these early neurons.

4.7.1.2. *Scaffold for the final structure of a ganglion*

These early structures form the basis for the elaboration of the final adult structures. Some of the major longitudinal tracts of a thoracic ganglion can be recognised as early as 42% development, and virtually all can be seen by 55% (Leitch *et al.*,

1995). Some of the longitudinal fascicles that form first are clearly the forerunners of particular longitudinal tracts (*see section 4.7.2.2*), and there seems no real impediment to describing the developmental formation of all the tracts and the neurons that contribute to them, although this has not yet been attempted. The major commissures are also established by 55% development although the separation of some into distinct parts is not complete until 65%. The major areas of neuropil develop later so that they are not distinct until 65–70% development, at the time when synapses are first established (*see section 4.12*), when the processes of local neurons increase in number, and when the majority of sensory neurons from the legs and the body arrive at the central nervous system. The emergence of the neuropil areas is thus, in large part, due to the expansive growth of neurons within the central nervous system and the arrival of many sensory neurons.

4.7.2. Choice of routes by the growth cones

The growth cones of neurons continually have to make choices about in which direction to turn and which bundles of axons to join. In general, the routes that are followed by particular neurons are the same in each locust, although mistakes are made and then corrected. These mistakes are most commonly manifested as the extension of filopodia or small branches along incorrect routes before the final axonal extension along the correct route is established. The decisions that must be made are well illustrated by two particular sibling neurons and in the establishment of the longitudinal fascicles that link segmental ganglia.

4.7.2.1. *Choices made by identified interneurons*

The C (=314) and G (=714) intersegmental interneurons arise from the second ganglion mother cell of neuroblast 7-4 in the mesothoracic ganglion (Raper *et al.*, 1983) (Fig. 4.5). The C interneuron, which may be involved in the control of jumping (*see Chapter 9*), has an axon contralateral to its cell body that turns posteriorly to run to the metathoracic ganglion. The sibling G interneuron, which responds to sound and makes output connections with metathoracic motor neurons, also has an axon contralateral to its cell body, but this must turn anteriorly and produce a smaller diameter axon that runs posteriorly. Clearly, the two siblings must make distinct choices when they meet axons in the 25 fascicles that comprise the contralateral connective at about 40% development. The path of both interneurons across the ganglion is pioneered in the posterior commissure by neurons Q1 and Q2, which are the first progeny of neuroblast 7-4, and these meet their counterparts from the opposite side of the ganglion at the midline. Q1 then turns posteriorly close to the growth cones of cells MP1 and MP2 that form a medial fascicle of the longitudinal tract, but then both it and Q2 appear to die. The G interneuron is the next neuron to follow this route, followed soon by the C interneuron, but they both pass the turning point of the Q1 neuron. In this more lateral position, they meet the anteriorly growing axons of the A1 and A2 neurons, which have cell bodies in the anterior part of the metathoracic ganglion

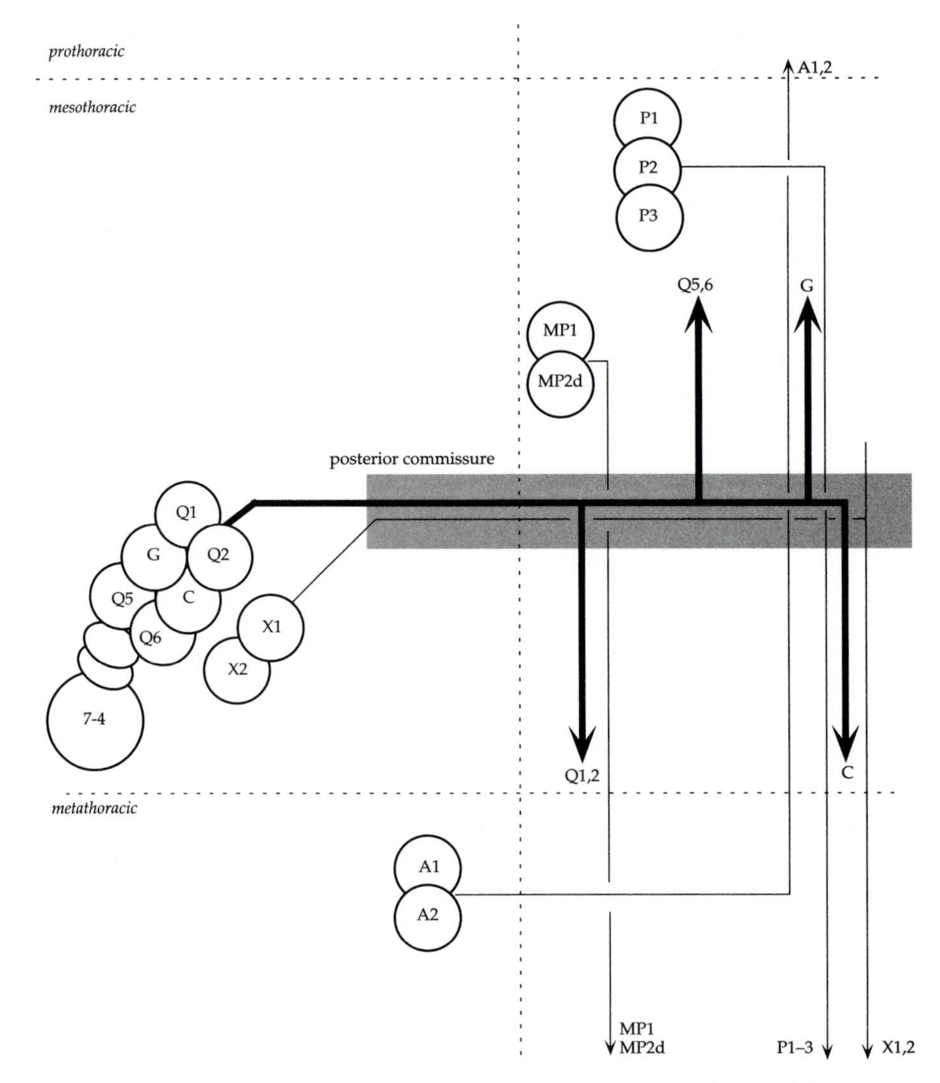

Fig. 4.5 Choices made by the growth cones of the progeny of neuroblast 7-4 in the mesothoracic segment. The growth cones may turn anteriorly or posteriorly within a longitudinal tract, and within that tract join particular fascicles pioneered by other neurons. Two of the progeny are the identified C (314) and G (714) neurons (see Fig. 4.10 for morphology of the G interneurons). Based on Bastiani *et al.* (1984).

contralateral to their axons, and the posteriorly growing axons of the P1 and P2 neurons, which have cell bodies in the anterior part of the mesothoracic ganglion ipsilateral to their axons. It is at this point that the growth cone of the C interneuron turns posteriorly and that of the G interneuron anteriorly. The filopodia of the G interneuron are most closely associated with the P neurons, which presumably express some antigen on their surface that is not present on the other axons in the longitudinal tracts. If these P neurons are ablated then the growth of the G interneuron is abnormal and the wrong choice can be made, but ablation of the A neurons has no effect (Raper *et al.*, 1984). The conclusion is that the choice of route taken by the G interneuron is determined by its contact and interaction with a particular small set of neurons, even though its growth cone comes into contact with most of the axons forming some 25 different fascicles of axons in the longitudinal tracts in the neuropil, and that these play an active role in guidance rather than just providing a convenient substrate on which they can grow. Other remaining axons in the longitudinal fascicles could provide such a substrate, but obviously they cannot provide all the required clues for the correct decision to be made by the growth cone of the G interneuron. This implies that the neighbouring fascicles have different surface markers that allow them to be recognised by the growth cones of other neurons.

4.7.2.2. *Choices made by pioneers of longitudinal fascicles*

The pioneering of the longitudinal fascicles by the progeny of the MP1 and MP2 cells also requires that choices be made about whether to turn anteriorly or posteriorly. The posterior turn of the MP1 neuron is made after its filopodia have become associated with a cell called the posterior corner cell (pCC). A junction is established through which dye can pass and the filopodia of MP1 interdigitate deeply with pCC, inducing the formation of coated vesicles (Bastiani and Goodman, 1984).

The first three longitudinal fascicles in the neuropil that have been pioneered by the progeny of the MP1 and MP2 cells consist of seven neurons at about 35% development: the most medial vMP2 fascicle contains the vMP2 axon alone, the middle MP1/dMP2 fascicle contains the axons of MP1, dMP2 and pCC, while the most lateral U fascicle contains the axons of the anterior corner cell (aCC) and of the U1 and U2 neurons that are the progeny of an unidentified neuroblast. By 40% development, the number of longitudinal fascicles has risen to 25 within the neuropil, but close to the connectives these are fused into 8-10 and within the connectives to just three fascicles, so that at 42% development there are about 140 axons in each thoracic connective. The fusion of these longitudinal fascicles eventually forms the nine major longitudinal tracts of the adult nervous system and even at these early embryonic stages many can be recognised and the axons they contain attributed to particular identified interneurons. For example, the vMP2 fascicle becomes part of the ventral median tract (VMT) and the MP1/dMP2 fascicle becomes part of the ventral intermediate tract (VIT).

Anterior and posterior corner cells (aCC and pCC) are sibling neurons produced by the first ganglion mother cell of neuroblast 1-1 and which express the pair-rule gene *even-skipped*. They first guide the route of particular MP cells and then themselves extend axons, but as these grow they make independent choices and join two distinct longitudinal fascicles; the pCC neuron has an absolute preference for the fascicle formed by the axons of MP1 and dMP2 and turns anteriorly (Bastiani *et al.*, 1986), whereas aCC has an affinity for the U fascicle so that it turns posteriorly (du Lac *et al.*, 1986). If the neurons in the fascicles are late in forming, the growth cones of either the aCC or pCC neurons do not simply join a pathway that is already established, but instead they wait for the arrival of the correct neurons, which can be from any segment. Similarly, if neuroblast 1-1 is ablated so that the birth of the aCC and pCC neurons is delayed, the same paths are still followed by their growth cones although they must now navigate in an older environment that contains more axons and more pathways. The neurons in specific fascicles thus bear particular surface labels irrespective of their segment of origin and provide clues about the correct direction in which subsequent neurons should turn and a substrate along which axons can grow. This implies that pathways, rather than individual neurons, have specific labels.

4.8. PERIPHERAL PIONEER FIBRES

The majority of sensory neurons arise from peripheral epithelial cells, groups of which also contribute to the associated sensory structures. A few sensory neurons that belong to the strand receptors of the limb joints have cell bodies within the central nervous system instead of the periphery and although their origin is obscure, the assumption must be that they arise from central neuroblasts like the other central neurons. The sensory neurons arise in two waves of cell divisions that correspond to the formation of the first and second embryonic cuticle (Kutsch, 1989). In the first wave at about 35% development, the first sensory neurons are born and these act as pioneer cells, establishing by about 58% development a pattern of internal receptors that is characteristic of the first instar larva. The second wave begins at about 55% development and represents the formation of the exteroceptors, but their external structure is not completed until the formation of the third and final embryonic cuticle. The deposition of these three layers of cuticle is also correlated with peaks in the levels of ecdysteroids (Lagueux *et al.*, 1981), which may thus have some influence on the formation of the sensory neurons and their sensory structures. More exteroceptors are progressively added until the number seen in a first instar larva is reached at 65–70% development, but in postembryonic development still more exteroceptors are added.

The major pathways from most sense organs that have sensory neurons with peripherally located cell bodies are established by the sensory neurons themselves which send axons towards the central nervous system. The earliest sensory neurons have no established pathways to follow, so these neurons must pioneer the route to the central nervous system. These established routes are then followed by the neurons that are born later. The first sensory neurons migrate from the epithelium where they are formed at particular times and sites, and by 45% development their ingrowing axons have established the main scaffold of peripheral nerves. Soon after the arrival of the growth cones of the sensory neurons at the central nervous system, the growth cones of motor neurons begin to grow outwards, some following the routes established by the sensory neurons, but others pioneering their own routes. The similarity in the pattern of nerves from animal to animal indicates that these initial pioneers must always arise at the same place and time and respond to the same clues as they navigate towards the central nervous system. Sensory neurons that arise later, either embryonically or postembryonically, join these established pathways and grow along them to the central nervous system, although they may have to pioneer their way to the nearest established nerve route. This all points to some sort of labelling of the cells according to their position in addition to the type of sensory neuron that they represent and their time of birth, and this labelling is also reflected in the pattern of branches that their axons will make within the central nervous system. For example, the position of a hair receptor on a leg is reflected in a topological map in a ventral region of neuropil (*see Chapter* 7).

Cells that pioneer the establishment of peripheral nerves have been recognised in the antennae and cerci, but it is those in the leg that are known in the greatest detail. The way that these leg pioneers establish the peripheral nervous system illustrates the way that all nerve routes are formed.

4.8.1. Pioneers in a leg

The routes of the nerves in a leg are pioneered by a number of specific cells that appear in a particular sequence from particular positions. Some of the axons of these pioneer neurons growing towards the central nervous system meet the outgrowing axons of motor neurons and their axons then grow past each other; the sensory neurons on their way to the central nervous system and the motor neurons on their way to the muscles. At the tip of a limb bud of a hind leg of a 30% development embryo, a single cell moves to the inside of the epithelium and then divides to form two sibling offspring called the Ti1 cells (Table 4.4) which then migrate proximally into the tibia. At this stage, the limb bud is about 300 μm long and consists of an external layer of epithelial cells and an inner layer of mesodermal cells separated from each other by a basal lamina. Thus, the Ti1 cells are about 200 μm from the central nervous system, but as the leg elongates, they must grow for some 300 μm before they establish contact with the central nervous system. Once contact has been made these neurons are anchored both peripherally and centrally, and further axonal growth occurs, with the result that in a 55% development embryo the axons are 1000 μm long and in an adult, 35 000 μm.

The Ti1 cells (see Table 4.4 for explanation of names) start to grow by first sending out filopodia that advance 1–2 $\mu m.h^{-1}$, but may be suddenly withdrawn (Keshishian and Bentley, 1983a). Each growth cone has an array of about 20 filopodia of less than 0.5 μm in diameter extending up to 100 μm so that they span some 10% of the ectoderm of the limb bud. A particular filopodium can either persist for more than 10 min or can be retracted within a minute and forges a particular path on the epithelial side of the basal lamina. The growth cones of these neurons have no other cells to follow, and are thus the first cells to pioneer this route

Table 4.4. Cells establishing the major nerves in a hind leg

Cell or receptor	Synonym	Nerve formed	Time of appearance (% development)
Coxa			
Cx1			
+ CI2	CT1	5B	35
Cx2			
+ SETi	CT2	3B	35
Trochanter			
Tr1	F2, 3B	5B1	30–35
Tr2	F4, 1A1	lateral nerve	35–38
Femur			
Fe1	F1, 2B	5B1	30–35
Fe2	F3, 1C	lateral nerve	35–38
FeCO	1B1	5B1	35
Tibia			
Ti1	1A	5B1	30
Ti2	1D	5B2	40
SGO	4B	5B1	40
Tarsus			
Ta1	1E	5B2a	36–38
Ta2	1F	5B2b	38–40

The first column gives the names currently in use; the synonyms are those used previously.

CI = common inhibitory motor neuron
CT = coxa/trochanter
Cx = coxa
F, Fe = femur
FeCO = femoral chordotonal organ
SETi = slow extensor tibiae motor neuron
SGO = subgenual organ
Ta = tarsus
Ti = tibia
Tr = trochanter

(Fig. 4.6). In a hind leg, they are the first peripheral cells to reach the central nervous system and establish contact by 34% development. The route that they have established becomes N5B1 and the axons of subsequently formed ingrowing sensory neurons from certain specific locations on the leg, and of specific outgrowing motor neurons join this route (Bate, 1976b; Keshishian, 1980). The axons of the sensory neurons then become invested with layers of glial cells and become detached from the epithelium that has served as the substrate for the formation of the nerve pathway. As a consequence, the established nerve moves to the interior of the developing leg and must thus move through the basal lamina. The Ti1 cells do not leave behind a ciliary dendrite in the epithelium, as do subsequent sensory neurons,

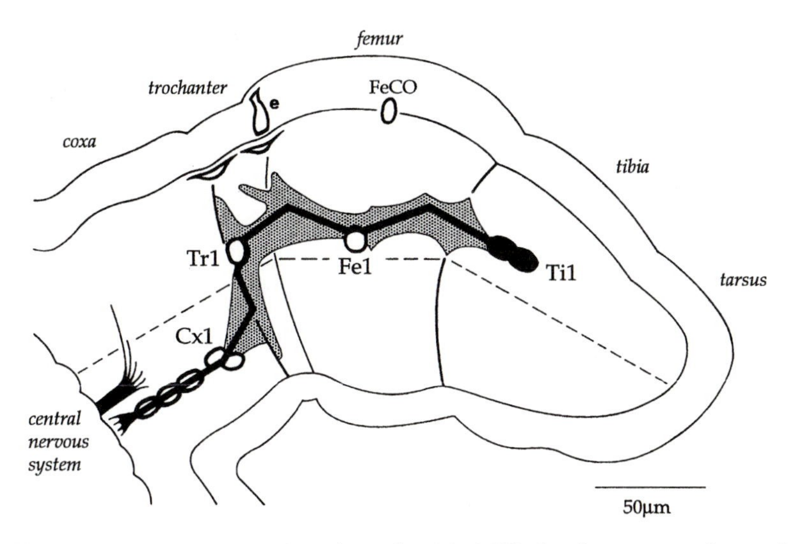

Fig. 4.6 Pioneer sensory neurons in a leg of a 33–35% development embryo. A pair of sensory neurons in the tibia (the Ti1 cells) encounters guidepost cells Fe1, Tr1 and Cx1 when establishing the route of nerve 5 into the central nervous system. See Table 4.4 for explanations of names. Based on Caudy and Bentley (1986).

they have no associated support cells, and they die at 55–60% development without forming a sensory structure.

Two Cx1 pioneer cells arise at the same time as the Ti1 cells (35% development), but unlike them are not siblings. They are formed individually from the anterior and the posterior epithelium of the limb and then migrate to contact each other before extending their axons together (Bentley and Toroian-Raymond, 1989). In the pro- and mesothoracic segments, the Cx1 pioneers are the first to reach the central nervous system and start axogenesis before they are contacted by the Ti1 cells (Keshishian and Bentley, 1983b). By contrast, in the metathoracic segment they begin axogenesis only after they have been contacted by the Ti1 cells.

To this scaffold of the main leg nerves are added the axons of sensory neurons that are born later. They merge with a particular nerve, depending on their position in the epithelium. Adhesive connections between the sensory neurons themselves are established rapidly and contribute to the guidance of a growth cone over an axon or bundle of axons of other sensory neurons.

4.8.1.1. *Formation of the main leg nerves*

Other pioneer cells arise in the same way as the Cx1 and Ti1 neurons, and in a characteristic sequence, to establish the routes of all the major nerves (Table 4.4).

Nerve 5B1 from the tibia to the central nervous system is established by the Ti1 pioneer cells, and is then joined by the axons of other sensory neurons in an ordered sequence. The first neurons of the femoral chordotonal organ appear at 35% development and their axons project along the ventral ectoderm to reach Fe1 at

38%, whereupon their axons follow the route of N5B1 established by the Ti1 pioneers. The first neurons of the subgenual organ in the tibia appear at 40% development and their axons project to the cell bodies of the Ti1 cells, their axons again joining the N5B1 pathway.

Nerve 5B2 in the femur is formed by Ti2 pioneers that arise at 40% development. Their axons grow into the femur where they meet the outgrowing axon of the CI2 motor neuron that has established the main N5B branch. The axons of these pioneers then grow past each other, the peripheral pioneer following the route to the central nervous system established by the motor neuron and *vice versa*.

Nerves 5B2a,b in the tibia and tarsus are formed, respectively, by tarsal neurons Ta1 and 2. Ta1 arises first at 36–38% development and its axon grows along the ventral ectoderm of the tarsus, finally reaching the Ti2 neurons at 40%. The Ta2 neuron arises a little later, at 38–40% development, and its axon pioneers a more posterior route in the tibia before converging on Ti2 at 43%. Other tarsal pioneers (Ta3,4) pioneer different initial routes before converging on either Ta1 or Ta2.

The lateral nerve in the femur is formed by femoral pioneers Fe2 and Tr2 that appear at 35–38% development and whose axons run together but parallel to N5B, joining it only when proximal to the Cx1 cells.

Nerve 3B is formed by both motor and sensory pioneers. The Cx2 cells arise at 35% development when the motor pioneer, the slow extensor tibiae motor neuron, emerges from the central nervous system. The axons grow over each other, with the sensory neurons reaching the central nervous system in the metathoracic segment by 38% development and the motor pioneer reaching Tr1 by 40%.

4.8.1.2. *Ablation of pioneer cells*

If the Ti1 cells are ablated just after they become visible at 30% development, the first cell that they would normally encounter, Fe1, is still able to extend filopodia and contact the second cell, Tr1 (Keshishian and Bentley, 1983c). Similarly, Tr1 can successfully contact Cx1, and the FeCO cells can find Fe1 as normal, so that together they can take over the pioneering role of the ablated cells and establish the normal route of N5B1. Deletion of the normal pioneer cells in a cercus results in the fasciculation of the axons of the sensory neurons born later into abnormal paths caused by the spatial arrangement between the different neurons (Edwards *et al*., 1981). This suggests that there is nothing particularly unique about the abilities of the early pioneers and that other neurons can read the same clues and establish the same routes. Moreover, it also indicates that contact from the first pioneers is not essential for other cells to begin their own axogenesis. The first pioneer neurons have unique opportunities that are afforded solely by the time and place where they were born. This view is strengthened by the ability of sensory neurons that are born later to navigate their way a short distance to the major routes that are already established. The conclusion is that all neurons have to do some pioneering of their paths, but that the first pioneers must navigate without clues from other sensory neurons growing along the same route.

4.9. GUIDANCE OF GROWTH CONES

The initial direction taken by a growth cone is determined by the polarity of the cell following its division. In a leg, the direct paths then taken by the growth cones of pioneer cells towards the central nervous system suggest directed growth, even though their filopodia extend in a wide arc and for distances of at least 100 μm. The axons of neurons that are born later follow the routes established by the pioneers, so that they have to navigate only short distances on their own. In the presence of cytochalasin B or D, which disrupt actin microfilaments, pioneer neurons such as Ti1 do not extend filopodia, and their growth cones are frequently distorted and follow the wrong paths by making incorrect turns so that their routes are often circular (Bentley and Toroian-Raymond, 1986). They are, however, still able to elongate their axons along what must be an adhesive substrate and to recognise the appropriate guidepost cells (*see section 4.9.1*) if they contact them. This indicates that the filopodia are necessary for correctly oriented growth but not for elongation of the axon. Microtubules are present in axons as tightly packed linear arrays and in growth cones as a rich network of single tubules or small bundles with some branches, but they do not occur in the filopodia (Sabry *et al.*, 1991). When turns are made towards guidepost cells, the microtubules enter the filopodia that have actually contacted the guidepost cell. The microtubules enter several filopodia but are then withdrawn from all but the one that is oriented in the appropriate direction. This selective alignment of the microtubules could be influential in steering the growth cone along its correct route.

The morphology of the filopodia is also an indicator of any guidance clues that might be present at different times during development and at different sites within a limb (Caudy and Bentley, 1986). At the boundaries between different parts of a leg, such as the femur and tibia, the growth cones extend more lateral filopodia that grow around the circumference of the limb bud, unless a pioneer has previously crossed this region. The filopodia selectively wrap around the guidepost cells but do not extend in the same region of the leg if the guidepost cells are absent. The extent of filopodial growth is also dependent on the position of the growth cone so that, for example, more filopodia are produced in the proximal femur than in the distal femur. All of these observations indicate that the neuron is responding to external clues rather than expressing an internal program of growth. The clues are from at least three types of cell: developing neurons that serve as guidepost cells, muscle pioneer cells of mesodermal origin, and epithelial cells at the boundaries between different parts of a leg. All of these cells appear to have at least some surface molecules in common, as all can be stained by the same anti-HRP (horseradish peroxidase) antibody. The cells respond to these clues in a hierarchical way, with the guidepost cells being dominant, by extending a filopodium toward the clue of highest affinity and then withdrawing the others. The filopodia do not simply adhere best to certain cells and pull the growth cone in a particular direction because differences in adhesiveness are too small. Instead, more active processes must be involved, perhaps involving the transmission of second messengers from the tip of the filopodium that has contacted the clue of highest affinity to the actomyosin filaments at its base. These internal signals thus amplify the small differences in extrinsic clues so that the filopodia respond selectively to an array of clues and that the growth cone does not simply advance in a way that is determined by the sum of the forces acting on all its filopodia.

4.9.1. Guidepost cells

The route taken by the initial pioneer cells often involves a series of sharp turns and is not the shortest route to the central nervous system. This implies that the growth cones are not moving along a diffusible gradient of chemical clues, but instead suggests that they are guided by local clues that are either bound to particular substrates or secreted locally. For the sensory neurons from a limb, no clues extrinsic to that limb are involved in determining their route to the central nervous system, but rather it is their encounters with particular cells on their way that is important. The cells that act as guideposts for the pioneer neurons are distinguishable from surrounding cells by particular surface molecules, but eventually themselves become neurons. The growth cones of the Ti1 pioneers are associated with two cells in the femur, Fe1 and Tr1, and a third at the junction of the trochanter and coxa, Cx1 (Fig. 4.6). These cells are positioned such that the Ti1 pioneers have to grow 50 μm before encountering Fe1 and then the same distances to encounter the other two. These distances as thus about the same as the length of the filopodia extended by a pioneer cell. Abrupt turns are made at cells Tr1 and Cx2, indicating that choices must be made at these points. As a pioneer cell grows past each of these guidepost cells, it makes a junctional contact with it that allows dye to pass between the two cells. Thus, a pioneer cell grows from one guidepost cell to another on its route to the central nervous system with the extension and exploration by the filopodia of the growth cone spanning the gaps in between. Once a particular guidepost cell has been contacted, the leading filopodia have already been extended further and may even be close to contacting the next guidepost cell. It is therefore the position of the guidepost cells and the ability of a pioneer to recognise certain of these cells that determines the pathway that is established. At the time that the guidepost cells have been contacted, they too begin axogenesis and their axons follow along the route established by the pioneer that has contacted them. The guidepost cells thus become neurons but their identity is generally unknown.

Nevertheless, routes are still established if some of the guidepost cells are removed, but then the growth cones of the pioneer neurons grow more slowly and extend more filopodia. Fe1 is often detectable only after the Ti1 pioneers have grown past it, and if it is removed the pioneer growth cones still find the next guidepost cell, but sharp turns are not involved in navigating this part of the route. If Cx1 is removed, however, the growth cones either do not enter the coxa or follow an incorrect route, failing to make the abrupt turn that is required (Bentley and Caudy, 1983). In some places, therefore, the guidepost cells provide the only clues for correct pathfinding, but in others different clues must also be present.

4.9.2. Boundaries

Boundaries between the different parts of a leg cause reorientation of the growth cones, probably because they offer an interface between a high affinity epithelial substrate distally and a low one proximally. No physical boundaries in the form of membranes are present that explain the different behaviour of the growth cones at these local regions. At these boundaries, the growth cones extend more filopodia than they do in regions of the leg in between, perhaps indicating that the substrate proximal to the boundary is less favourable for migration. Thus, at the boundary between the femur and trochanter where the growth cone of Ti1 encounters guidepost cell Tr1, it also meets two orthogonal bands of boundary cells; a distal band of cells that have high adhesiveness and a proximal band with lower adhesiveness. The growth cone extends branches dorsally and ventrally on the first band of cells but always eventually takes the ventral route. The pioneer neurons seem to establish a bridge across these boundaries over which the axons of later sensory neurons can navigate. For example, when the Ti1 cells cross the border between the tibia and the femur, the boundary cells have not yet differentiated and thus do not provide a barrier, but when they have differentiated, the later developing sensory neurons are able to follow this route through the established barrier. If the Ti1 neurons are prevented from differentiating, however, then the axons of sensory neurons, such as those of the subgenual organ, that are formed later and that would normally join these pioneers in N5B1, are unable to cross this boundary (Klose and Bentley, 1989). The second route across this boundary pioneered by Ti2 and forming N5B2 is unaffected, and sensory neurons that follow this route are still able to cross.

4.9.3. Gradients of cell adhesiveness

Local clues in a leg are sufficient for successful navigation, because the same path is followed by a sensory pioneer even when that leg is removed from the rest of the embryo. Some of these clues may be provided by a gradient of particular molecules on the surface of the epithelial cells and this may be related to the different adhesiveness of cells along the proximo-distal axis of a limb: adhesion between the cells is higher at the distal end and this may provide the clue for the peripheral pioneers to grow proximally and the motor pioneers to grow distally. The clues appear to be present in the epithelium, as the same routes are pioneered by the sensory neurons even when the mesoderm is disrupted, when the basal lamina is enzymatically removed, or when some guidepost cells are absent. This means that the basal lamina is not necessary for the growth cones to be extended, for the interactions of the growth cones with guidepost and boundary cells, or for correct guidance. The growth cones of peripheral pioneers nevertheless preferentially contact and adhere to the basal lamina, in contrast to the few contacts that they make with epithelial cells and the lack of contacts with mesodermal cells (Anderson and Tucker, 1988). If the basal lamina is removed before the Ti1 pioneer neurons have made contact with other neurons then the growth cones are withdrawn into their cell bodies, but this retraction is prevented by cytochalasin D, suggesting that the growth cones are normally under tension generated by microfilaments (Condic and Bentley, 1989a,b). The growth cones also adhere to particular guidepost cells and cells at the boundaries between different parts of a leg, so that if the basal lamina is removed after these contacts have been made, the adhesion is maintained but the cell bodies become unstuck and move proximally.

4.9.4. Surface molecules

A pervasive conclusion is that the growth cones are reading signals on the surface of neurons, glial cells or mesodermal cells encountered by their filopodia and that these are expressed only by particular sets of these cells and at particular times during development. Several glycoproteins, such as TERM-1, fasciclin I-IV, annulin, semaphorin, lachesin and neuroglian act as adhesion molecules and are expressed at the time of axonal outgrowth on cell surfaces of particular subsets of central neurons, peripheral neurons, glia and some other cells. These molecules are expressed transiently and on different subsets of neurons, but annulin is only expressed on glial cells after 40% development (Bastiani *et al.*, 1992). TERM-1 is expressed only in a pair of neuroblasts in the brain whose progeny form projection interneurons with axons in the connectives (Meier *et al.*, 1993). Fasciclin II is expressed in neuroblasts 1-2 and 5-4, and in midline precursor cells MP1 and 2, where it may determine the ability of these neurons to fasciculate and follow their particular route in longitudinal fascicles (Harrelson and Goodman, 1988). Fasciclin IV, by contrast, is expressed on the axons of MP4–6 cells in the median fibre tract when they form the first commissures, in some motor neurons with axons in lateral nerves, and in bands of epithelial cells in the limb buds, some of which correspond to the positions of guidepost cells for the first sensory pioneers (Kolodkin *et al.*, 1992). If these pioneers grow in embryos where antibodies have bound to this protein they are more likely to follow incorrect routes, suggesting that fasciclin IV plays some role in guiding growth cones. Inactivation of fasciclin I disrupts the fasciculation of the Ti1 cells but does not affect their growth and orientation in the leg, probably because it is expressed in the Ti1 cells themselves but not in the substrate cells (Jay and Keshishian, 1990). If the anchorage of certain cell surface proteins to the membrane is cleaved with a specific phospholipase C (PI-PLC), the early sensory pioneer neurons then lose their ability to grow directly towards the central nervous system and fail to make the appropriate turns at the boundaries between different parts of a leg (Chang *et al.*, 1992).

The implication is that these proteins are involved in guiding growth cones and in enabling specific axons to fasciculate with each other, either by promoting cell adhesion, or by acting as receptors in pathways mediating intercellular signalling. Clearly, proteins by themselves cannot explain the choices made by individual neurons, so that the further implication is that they must act with other molecules. Nevertheless, the molecular heterogeneity necessary to provide appropriate guidance clues does not have to be vast and could be met by a relatively small number of molecules present in different combinations and expressed at different times. These different molecules will almost certainly be identified soon by powerful molecular and genetic techniques, but it will still remain a difficult task to show what role these play in guiding the development of specific pathways in the nervous system.

4.10. SEGMENTATION AND THE SPECIALISATION OF SEGMENTS

It is chiefly from *Drosophila* that we know something of the events that lead to the normal pattern of segmentation and the specialisation of individual segments. Given the similarity in the early events of development in both *Drosophila* and the locust it may well be that what is learnt from *Drosophila* can be extrapolated directly to the locust. In locusts, however, the majority of segments arise sequentially from a budding zone which gradually adds more segments to the developing germ band until the final number is reached. By contrast, in *Drosophila*, the primordia for all segments are laid down at the blastoderm stage so that a complete fate map is established before gastrulation. Many genes are involved in the processes that lead to segmentation and to the development of specific features of those segments, but they fall into the following categories.

4.10.1. Segmentation genes

These genes interact to establish the polarity and number of segments in the embryo and to control the identity of the cells within each segment. They can be divided into three groups based on the defects seen in mutants lacking a particular gene.

1. **Zygotic gap** genes such as *Krüppel* and *hunchback* produce a deletion between contiguous segments.
2 **Pair-rule** genes such as *fushi tarazu* and *even-skipped* loci produce homologous defects in alternate segments.
3. **Segment polarity** genes such as *engrailed* are needed for pattern formation within each segment.

The segmental divisions of the body of a *Drosophila* embryo are formed rapidly by division of a sheet of cells and without growth, but are not visible as morphological specialisations until gastrulation has occurred. By contrast, in locusts, segments are added one at a time, so that the first stripe reflecting *engrailed* expression in T1 occurs shortly after gastrulation at about 17% development and the pattern of 14–15 stripes marking each segment is complete at 31% (Patel *et al.*, 1989). The segmentation of the abdomen appears in a clear posteriorly extending sequence, but the anterior sequence of the appearance of the suboesophageal and cephalic segments is less precise. The segmentation results from the sequential action of a number of different types of genes: the zygotic gap and pair-rule genes establish the repeating pattern of the segmentation genes, and the homeotic genes give each segment its own particular identity.

At the blastoderm stage of *Drosophila* embryos, which consists of a single layer of cells enclosing the yolk, the pair-rule segmentation genes are initially represented as a series of seven transverse stripes that correspond to alternate segments, while the segment polarity genes such as *engrailed* are expressed in 14–15 stripes corresponding to each segment. Thus, the *fushi tarazu* gene seems to be responsible for determining the formation of alternate segments and recessive lethal *fushi tarazu* mutants have only half the number of normal segments (Laughon and Scott, 1984). When the nervous system forms and neuroblasts are being generated the pair-rule genes are expressed in every neuromere, suggesting different mechanisms controlling expression at the different stages and in different tissues. Certain genes are expressed in several neuroblasts as they enlarge and others in the progeny of certain neuroblasts just as they are born, but this does not appear to determine the features of these neurons as there are no common characteristics amongst those expressing the product of a particular gene. More than one gene can be expressed in a single neuron, suggesting that interactions between many gene products are necessary to influence differentiation. For example, *fushi tarazu* controls the expression of *even-skipped* so that when either it or *even-skipped* is absent, particular motor neurons that normally express both send their axons to the wrong place by following the route of their sibling (Doe *et al.*, 1988a). Similarly, inactivation of the *even-skipped* protein also results in particular neurons with aberrant axonal pathways (Doe *et al.*, 1988b). The aCC motor neuron, which normally pioneers the intersegmental nerve, shows no change in the absence of *fushi tarazu* expression, but in the absence of *even-skipped* it no longer sends its axon into this nerve, which is nevertheless still established. Its sibling, the pCC interneuron is, however, unaffected by the loss of either gene. The implication is that these genes are influential in determining the fate of particular neurons and that they may interact in a hierarchical way with each other and with many other genes.

4.10.2. Homeotic genes

These genes do not affect segment boundaries, but instead, the loss of one of these genes results in the transformation of a segment or part of a segment into the pattern characteristic of another. Many of these genes occur in two clusters

1. **Bithorax cluster** containing, for example, *ultrabithorax* and *bithoraxoid.*
2. **Antennapedia complex** containing, for example, *antennapedia*, *deformed* and *sex combs reduced.*

These genes are all expressed in both the body and in the embryonic nervous system, with particular ones expressed most strongly in a particular region consisting of the posterior part of one segment and the anterior part of the next posterior segment – a parasegment. Mutants lacking *ultrabithorax* and *bithoraxoid* develop legs on segment A1 and a corresponding leg neuromere in the central nervous system, with the actions on the epidermis and central nervous system occurring in parallel. This specific expression in only certain segments contrasts with the expression of the segmentation genes in each neuromere and implies that the homeotic genes are not involved in the early processes of neuronal differentiation, which are the same in all neuromeres, but must instead be involved in events that are specific to particular regions. Their early expression could influence the neurogenic genes in determining the different numbers of neuroblasts in the various segments, and their later actions could influence the different structures and connections of homologous neurons in different neuromeres. A giant interneuron normally makes output connections only in T2, but when T3 is transformed into another T2 by *bithorax* mutations it extends its axon into this neuromere and duplicates its synapses (Thomas and Wyman, 1984). Again, it is the interactions between many of these and other types of genes that regulate the action of each and, in turn, influence the differentiation of the neurons.

4.10.3. Fusion of ganglia

The metathoracic ganglion of the locust consists of neurons belonging to four neuromeres; the metathoracic segment itself (T3) and the first three abdominal neuromeres (A1-3). The fusion of the neuromeres of these segments begins at 45–50% development with the fusion of A1 to T3, followed at 65% with the fusion of A2, and finally at 70% with the fusion of A3. It is presumed that the fusion of the neuromeres of the suboesophageal and terminal abdominal ganglion proceeds in a similar fashion, but at different stages in development.

4.10.4. Differentiation of limbs

The different anterior segments have five types of limbs; the labrum, antenna, mandible, maxilla, labium (fused second maxilla) and the thoracic legs. These are presumed to be derived from ancestors that had similar limbs on each segment. In *Drosophila*, the differentiation of these limbs is under the control of homeotic genes and limb patterning genes such as *distal-less*. The latter gene is expressed in all the developing limb buds of the butterfly *Precis coenia*, except those of the mandibles (Panganiban *et al.*, 1994), and in *Drosophila* its expression in the abdomen is excluded by homeotic genes so that no abdominal prolegs are ever formed. All of this is taken to indicate that common mechanisms are responsible for the formation of appendages in different insects.

4.11. DIFFERENTIATION OF NEURONS

The differentiation of a ganglion cell into a neuron involves many processes that result in its characteristic morphological features, the connections that it receives and makes with other neurons, the transmitter(s) or neuromodulators that it uses, its own intrinsic membrane properties, and the numerous molecules, particularly on its surface membrane, that may be markers by which it is recognised by other neurons. Many of these processes are determined by the segment in which the ganglion cell resides, so that neurons with different properties emerge in different segments, despite the apparent repetitive pattern of precursor cells. The process has to be sufficient to specify most of the neurons in a segmental ganglion as individuals; we know that all the motor neurons and most of the large intersegmental interneurons are individually specified, and we strongly suspect that the numerous local interneurons are also individuals. For many neurons, only a small subset of these criteria are necessary for an experimenter to establish the identity of an individual. For many groups of similar neurons, the use of further criteria might establish individual identity.

The specification of a neuron is determined by the identity of the parent neuroblast and by the rank order in the lineage of the neuroblast in which it is produced. Within this framework, the following factors are important. First, positional clues within the ventral region of ectoderm specify individual neuroblasts. Second, interactions between neighbouring neuroblasts and ectodermal cells lead to differentiation of the ectodermal cells as specific neuroblasts. Third, each neuroblast produces a particular and invariant sequence of ganglion mother cells from which neurons are formed in a strict sequence. Fourth, interactions between the two sibling neurons of a ganglion mother cell further determine their fate.

The fate of the cells produced by the neuroblasts may be determined by the sequential action of a number of genes. First, certain segmentation genes provide positional clues within the ectoderm in both the antero-posterior and dorso-ventral axes that determine the formation of the neuroblasts. Second, genes such as *prospero* translate these positional clues into a unique identity for the individual neuroblasts. Thus, each neuroblast will express a unique combination of these genes at particular times. Third, certain homeotic genes are expressed in subsets of ganglion mother cells under the control of the genes that have given the neuroblasts their identity, and themselves impose an identity on the ganglion mother cells and hence the neurons that they produce.

The development of the different types of neurons has not always been studied by equivalent methods; some have been labelled and examined with the electron microscope whereas others have only been examined with the light microscope. Thus, only for some types of neurons do we have a clear picture of when synapses start to form, and when spikes and synaptic potentials start to be generated. It is, however, not possible to extrapolate from one type of neuron to another about the development of these features.

4.11.1. Development of motor neurons

A few of the motor neurons that innervate a leg can be recognised in embryos at 50% development, but most can be recognised by the positions of their cell body and the paths of their axons only in a 60% embryo (Whitington and Seifert, 1981). The routes of their axons from the central nervous system into a leg are pioneered by specific motor neurons (Whitington, 1989). In the metathoracic ganglion of a 33%

development embryo, five large motor neurons send axonal growth cones dorsally to the basal lamina and then posteriorly and laterally for 100 μm to the edge of the ganglion, forming a parallel row of axons. One of these neurons, which can eventually be identified as the slow motor neuron to the extensor tibiae muscle (SETi), pioneers a route into the anterior of the coxa where there are no other sensory or motor axons and this will become N3B. A second motor neuron, so far not identified, extends a little later into the coxa where it meets sensory pioneer fibres Cx1 and Ti1 from the leg to form N5B. It has access to the route pioneered by SETi, but does not follow this route. The axons of the other motor neurons leave the thoracic ganglia along these two routes by 60–65% development. One of the first to follow the N5 route is common inhibitor 2 (CI2). Motor neurons that will eventually have axons in N4 initially follow the N5 route.

The cell bodies of the three motor neurons to the extensor tibiae muscle of a hind leg (slow and fast excitors, SETi and FETi, and common inhibitor 1, CI1) can all be recognised in a 35% development embryo. The large array of branches that characterises these and other motor neurons in an adult and gives many an identity based on morphology alone starts to be formed as a series of short side branches from the primary neurite, so that their basic adult shape is recognisable by 55% development. These branches then increase in length while the cell body gradually becomes separated from them by a length of primary neurite that has no branches. The development of the adult form of these neurons does not depend on the axon establishing contact with the muscle (Whitington *et al.*, 1982), although when its target muscle is removed, FETi may sometimes develop an extra anterior and medially directed neurite. This same unusual structure is also occasionally seen in these neurons of adults.

CI1 and CI3 are siblings of an early ganglion mother cell from neuroblast 5-5 (Wolf and Lang, 1994) which is present only in the thoracic segments. We can therefore expect that these inhibitory neurons will be absent in abdominal segments. The CI3 neuron has an axonal growth cone that grows towards the ipsilateral N5, whereas the growth cone of CI1 crosses the midline in the posterior commissure to send axons into several contralateral nerves. This neuroblast starts dividing at about 30% development and the two inhibitors are some of its earliest progeny, but later progeny, numbering about 60, are intersegmental interneurons with contralateral axons that cross the midline in either the anterior or posterior commissures. GABA-like immunoreactivity first appears in CI1 at about 60–65% development and then appears in many of the interneurons derived from the 5-5 neuroblast. The remaining common inhibitory neuron to the leg, CI2, which in its action and innervation pattern is similar to CI3, is not clonally related to the other two, and appears to be formed by neuroblast 5-1.

4.11.2. Development of sensory neurons

The separation or delamination of cells from the ectoderm which will form the sensory structures occurs after the segregation of the neuroblasts. These sensory organ mother cells then differentiate into the various types of sense organs, such as tactile or chemosensory sensilla, campaniform sensilla and chordotonal organs. As for the neuroblasts, these processes depend on the action of proneural genes and probably many others yet to be characterised.

The development of only a few individual sensory neurons can be followed during embryonic development, mostly because they are small neurons and because many have similar shapes. An exception is the single sensory neuron from a stretch receptor at the hinges between the wings of the body. These neurons have large (10 µm) diameter axons that follow a unique route into the central nervous system, so that they can be readily identified (*see Chapter 11*). One of these sensory neurons can be identified in a 40% development embryo when its peripheral cell body starts to produce an axon that grows towards the central nervous system (Fig. 4.7) (Heathcote, 1981). In a 60% development embryo, the sensory neuron becomes able to produce action potentials, but the basic pattern of arborisations in the central nervous system is not formed until 80%. It is not known whether its connections with interneurons and motor neurons are established embryonically, but they certainly are present in the early instars before the wings are functional.

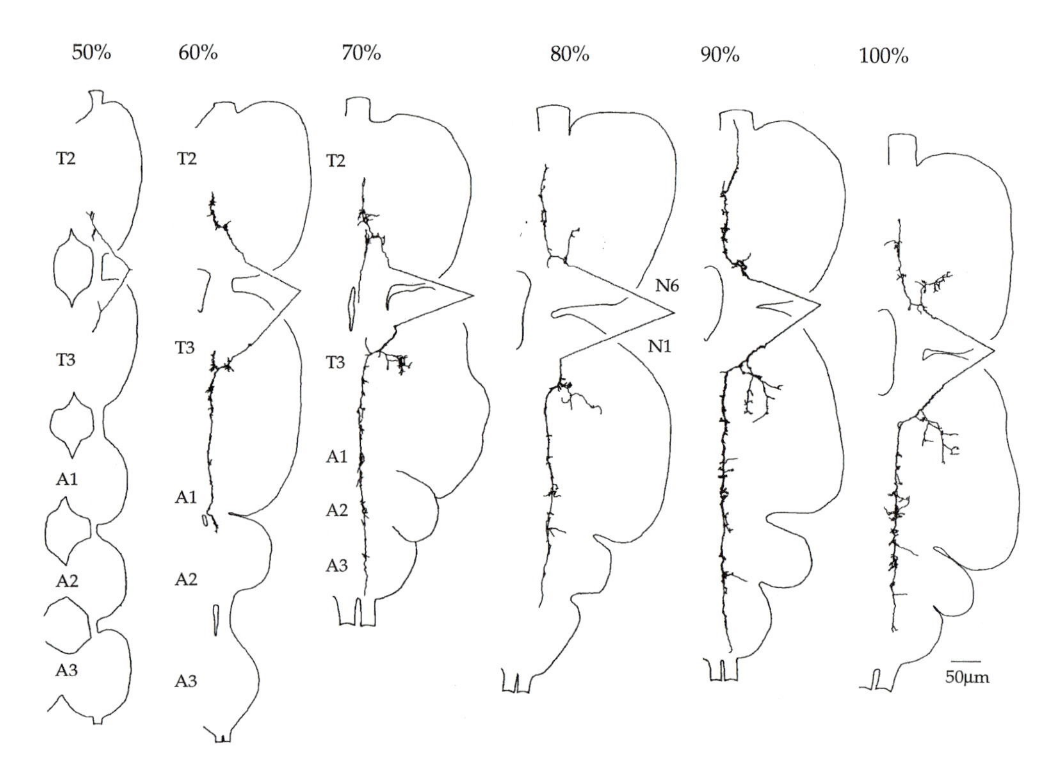

Fig. 4.7 Embryonic development of the single sensory neuron of a hind wing stretch receptor. The projections of this sensory neuron into the meso- and metathoracic neuromeres are shown at various stages of development. The projections in a mature adult are shown in Fig. 11.14C. Note the progressive fusion of neuromeres A1, A2 and A3 with the metathoracic neuromere as development progresses. Based on Heathcote (1981).

4.11.3. Development of interneurons

For a few of the large well-characterised interneurons and for particularly coherent groups of smaller interneurons it is possible to trace their lineages and chart the stages in their embryonic development at which they differentiate and assume certain of their diagnostic features.

4.11.3.1. *Intersegmental interneurons*

One of the largest neurons with its cell body in the brain and with a descending axon is the descending contralateral movement detector (DCMD) (*see Chapter 9*). In the adult, it can be readily characterised by its morphology and by its responses to small moving objects in the field of one eye. This interneuron can be first seen in a 45% development embryo, implying that it must have arisen from a ganglion mother cell in a 40% embryo early in the sequence of division of its parent neuroblast (Bentley and Toroian-Raymond, 1981). At this stage, the cell body is about 10 μm in diameter and is already in its adult location. The primary neurite has established its path through the brain, but it is of uniform diameter and bears only a few short side filopodia. As development proceeds, these branches lengthen and themselves branch further, but are always restricted to the region of the brain occupied by the branches of the interneuron in the adult. This implies that the final characteristic shape of the interneuron is achieved more by directed growth of particular branches than by the overproduction of branches and then selective loss. After hatching, the interneuron continues to grow in step with increases in the size of the body, but its adult form has already been established during embryogenesis.

4.11.3.2. *Spiking local interneurons*

Midline spiking interneurons in a thoracic ganglion are produced during the latter stages of the lineage of neuroblast 4-1 (Shepherd and Laurent, 1992). Some of the first of the 81 progeny of this neuroblast are intersegmental interneurons, but all the later progeny appear to be spiking local interneurons in the midline group. This group of spiking local interneurons may thus be a clone. Of the other nearby neuroblasts, 4-2 also forms local interneurons with cell bodies contralateral to their main neurites, but these do not have separate dorsal and ventral fields of branches as do the midline group. Neuroblast 3-1 forms local neurons that belong to the anteromedial group (*see Chapter 3*) and have neurites ipsilateral to their cell bodies, and some intersegmental interneurons. On the other hand, neuroblast 5-1 forms common inhibitory motor neurons CI1, CI3, some other motor neurons, and intersegmental interneurons, but no local interneurons.

The spiking local interneurons in the midline group can be recognised as a group from 55% development and their embryonic growth and differentiation occurs in three stages. First, from 55–70% development, the basic pattern of primary and secondary neurites is formed so that the array of dorsal and ventral neurites is established (Fig. 4.8). The ventral branches, which will provide the main sites for synaptic inputs, develop from outgrowths of the primary neurite, while the dorsal branches that will provide the main sites for synaptic outputs are formed from branches of the advancing primary growth cone. The process linking these two

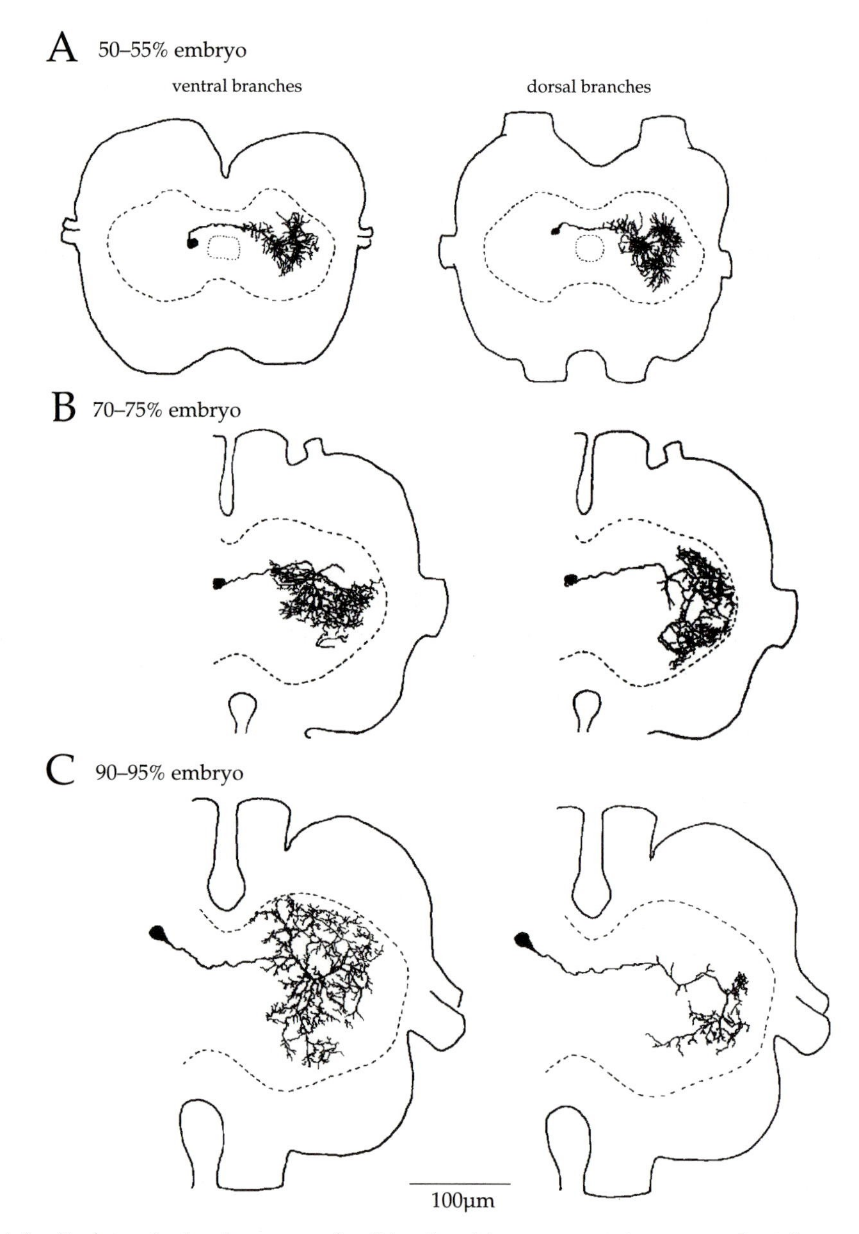

Fig. 4.8 Embryonic development of spiking local interneurons in a ventral midline group of the metathoracic ganglion. A. 50–55% development. B. 70–75% development. C. 90–95% development. Each neuron has two fields of branches that are recognisable as early as 50–55% development. Based on Shepherd and Laurent (1992).

fields is formed directly by the primary growth cone, and is thus developmentally equivalent to an axon. Filopodia extend from the ends of many of its neurites. Second, from 70–80% development, this basic pattern is elaborated by the rapid growth in both the volume and number of neurites. At this time, the ventral branches have more output synapses than input ones, but as development proceeds, more input synapses are added. In the dorsal branches, output synapses always predominate, as they do in the adult. Third, from 80–100% development, the number of branches is reduced and reorganised and varicosities appear in the dorsal branches, so that essentially the interneurons now have their final structure.

The components of the neural circuits in which the spiking local interneurons participate appear to form in an ordered sequence. The motor neurons and intersegmental interneurons that represent the output elements are formed first, followed by the local interneurons. Output synapses of the midline local interneurons are initially in greater abundance than input synapses. Sensory neurons from the exteroceptors arrive later and finally those from developing proprioceptors reach the central nervous system and establish synaptic contacts with the spiking local interneurons and other interneurons.

4.11.4. Development of DUM neurons

Dorsal unpaired median (DUM) neurons are formed by repeated divisions of the median neuroblast. The first 3-4 divisions of this neuroblast form neuronal precursor cells, but at about 34% development in the thoracic ganglia, the neuroblast switches to producing midline glial cells dividing 8-10 times (*see section* 4.6) (Condron and Zinn, 1994). At 40% development it switches back to producing neurons, dividing 50 times, or approximately every 5h, to produce 100 DUM neurons that remain tightly clustered at the dorsal midline, before it dies at 80% development.

The neuronal precursors that result from the early divisions are the cells that have been numbered MP5 and 6, but earlier divisions probably produce the MP4 cells. All these cells, which are really ganglion mother cells of the median neuroblasts, rather than separately derived cells, are the pioneers of the routes to be followed by neurons that are produced later in the lineage of this neuroblast. The transition to the production of glial cells requires the activity of the *engrailed* gene for when this is inhibited by injection of antisense oligodeoxynucleotides into the median neuroblast, no glial cells are produced (Condron and Zinn, 1994).

The transition back to the production of neurons is triggered by an elevation in the level of cAMP and the activation of a cAMP-dependent protein kinase (PKA) (Condron and Zinn, 1995). The transition to the production of neurons can be advanced by increasing the activity of PKA by injecting one of its subunits (PKA-C) directly into the median neuroblast during the time it is producing glial cells. Moreover, nuclear translocation of PKA-C normally occurs at the time when the switch to the production of neurons occurs. In the later divisions that form the DUM neurons, the ganglion mother cells are pushed forwards and start to divide symmetrically. The first divisions result in the formation of efferent DUM neurons that can be identified by the paths taken by their axons in particular commissures and then nerve roots (Table 4.5). Thus, the first four neurons follow the route pioneered by MP4, the next four neurons follow the route pioneered by MP5, and the third four follow the posterior commissure pioneered by MP6. In the adult, these routes are associated with the superficial (SDT), the deep (DDT) and the posterior (PDT) DUM tracts, respectively. The four neurons in each quartet are initially dye-coupled to one other, but not to other developing DUM neurons.

A possible developmental sequence for these different groups of neurons thus envisages the efferent DUM cells being formed first, followed by the local DUM interneurons, and finally by the intersegmental DUM interneurons. This would require the expression of genes at different times during development to influence the formation of the different types, not least because the efferent DUM neurons contain octopamine and the local DUM neurons contain GABA. In fact, the efferent DUM neurons do not express *engrailed* whereas many, perhaps all, of the local and intersegmental DUM neurons do (Condron and Zinn, 1994). It also means that the median neuroblast produces neuronal progeny that can be diverse in several of their features, in addition to the variety of glial cells that it forms earlier.

There was once thought to be further diversity in the membrane properties of the neuronal progeny of the median neuroblast, with the firstborn neurons all spiking, and all of the lastborn nonspiking (Goodman *et al.*, 1980). In this scheme, the local DUM neurons were thought to be born last and were considered as nonspiking, because no evidence could be found for voltage-sensitive Na^+ channels. Subsequent recordings from

Table 4.5. Divisions of the median neuroblast

Division	Progeny		Tract
Until 34%			
	MP4?		SDT
	MP5		DDT
	MP6		PDT
34–40%			
	midline glial cells		
After 40%			
1st	DUM3,4,5	DUM3,4	SDT
2nd	DUM5a (DUMETi)	DUM5b	SDT
3rd	DUM3a	DUM3b	DDT
4th	DUM1a (DUMDL)	DUM1b	DDT
5th, 6th	?	intersegmental interneurons	PDT
7th	DUM3,4	DUM3,4	SDT
8th	DUM3,4	DUM3,4	
9th	DUM3,4	DUM3,4	SDT
10th	DUM3,4	DUM3,4	
11th and more	4 DUM	interneurons	PDT

these neurons showed that they produce conventional spikes (Thompson and Siegler, 1991) and lineage studies show that they are not the last to be born.

4.11.4.1. *Differentiation of individual DUM neurons*

The individual DUM neurons differentiate in a way that is exemplified by DUMETi, which innervates the extensor tibiae muscle, but the timing of events depends on when they are born in the lineage. The median neuroblast in the metathoracic ganglion begins dividing at about 30% development and, at its second division in the second neuronal wave of its lineage, gives rise to the ganglion mother cell that forms DUMETi. Axonal outgrowths begin from DUMETi at 35% development, and by 40% its growth cone has reached the anterior commissure, where it branches to form a bilaterally symmetrical process on both sides of the midline. By 50% development, the axon has entered N5 and by 60% has reached the extensor tibiae muscle in the limb. The cell body and the neurites then enlarge so that the neuron has the characteristic shape of that found in the adult and, as for the motor neurons that innervate these muscles, this differentiation is not dependent on the presence of the target muscles. It is first sensitive to applied acetylcholine or GABA at 40% development but is still electrically inexcitable, even though it is electrically coupled to others born at the same time (Goodman and Spitzer, 1981). Thereafter, nonlinearities appear in the response of the membrane caused by the activation of a K^+ current that causes rectification, followed at 60% development by the appearance of action potentials. At this stage it is no longer electrically coupled to other DUM neurons. At the time that the axon reaches the extensor tibiae muscle, octopamine can be detected in its cell body. By 70% development synaptic potentials can be recorded, indicating the establishment of input synapses.

4.11.5. Segment-specific differentiation

Homologous neuroblasts and midline precursor cells can have progeny that differentiate into neurons with different structures and actions in different segments.

In the mesothoracic ganglion, the second ganglion mother cell of neuroblast 7-4 forms the sibling G (=714) and C (=314) interneurons (*see Chapter* 2 for an explanation of neuron numbering, and *Chapter* 9 for a description of these interneurons). The same division in the metathoracic ganglion leads to a pair of sibling interneurons: one called, without much logic, B1 (=531), the other called 311. In A1, the division gives rise to one interneuron called B2 (=529) and another called 319 (Pearson *et al.*, 1985a; Boyan, 1992). These three interneurons are homologous but undergo segment-specific differentiation. The three have a similar shape in both 60% development embryos and adults, with a cell body contralateral to an ascending axon and bilateral arrays of fine neurites (Fig. 4.9). There are differences in morphology that define each, most notably that the mesothoracic G neuron has an additional contralateral descending axon and the fine neurites of 531 are markedly asymmetric. All three also share some input connections and responses in common, but even here there are differences; interneuron 531, but not the other

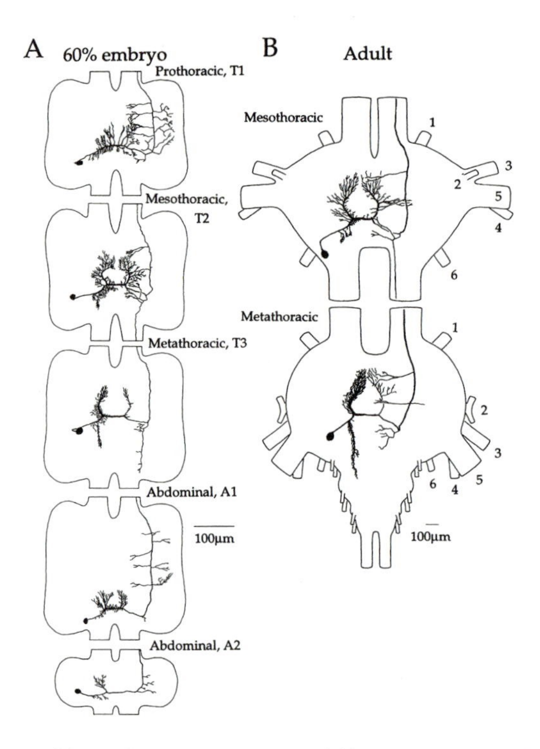

Fig. 4.9 Differentiation of homologous neurons in different segments. A. The structure of the G neurons in the pro-, meso- and metathoracic neuromeres, and in abdominal neuromeres A1 and A2 of a 60% development embryo. B. The structure of the neurons in a meso- and metathoracic neuromere of a mature adult. Based on Bastiani *et al.* (1984).

two interneurons, is hyperpolarised by sound delivered to the ear ipsilateral to its somata. Auditory sensory neurons apparently make direct connections with G (714) and 529, but not 531. The descending contralateral movement detector (DCMD) makes connections with G (714) and 529. Finally, the three differ in their output connections with 529 but not G (714) and 531, synapsing on specific motor neurons that innervate muscles that move the wings.

In the thoracic neuromeres, one of the two progeny of median precursor 3 initially has a single axon ascending in the contralateral connective, but secondarily differentiates into a neuron with four axons, one in each of the anterior and posterior connectives, giving it a characteristic H shape (Bate *et al.*, 1981). In other ganglia this neuron assumes different morphologies; in A1 only the two posterior axons are formed, and in A2-7 the cell dies in more than 90% of embryos, whilst in the others the original contralateral ascending axon is not replaced (Fig. 4.4).

Thus, even in ganglia that control similar sets of appendages and muscles, the local environment can lead to differentiation of neurons from the homologous precursor cells that have different structures, connections and presumably different actions during the processing of sensory signals and the generation of movements. Inferences about homology based on similarities in structure or performance in an adult are thus difficult, unless backed by evidence derived from their embryonic origins.

4.11.5.1. *Cell death*

The developing structure of the nervous system is sculpted not simply by the birth and differentiation of cells but also by their death, which can be segment specific, even within the lineage of a particular neuroblast. This enhances the inequality in the final numbers of cells in different neuromeres that arise from only slightly different numbers of neuroblasts. For example, the final neuronal progeny of the median neuroblast in the metathoracic segment T3 number about 90, but in the first abdominal segment A1 number only about 60 (Thompson and Siegler, 1993). These differences arise primarily from the selective death of certain progeny, and from the different periods during which the neuroblasts in the two segments are actively dividing: the median neuroblast in A1 is born after that in T3 and yet dies earlier. The neuroblasts continue to divide at the same rate throughout their lineage, but the number of progeny in the metathoracic neuromere increases at a steady rate until declining just before the median neuroblast dies, whereas the number of progeny of the A1 median neuroblast increases steadily at the beginning and end of its lineage with a plateau in the middle. The decline at the end of the T3 lineage and the plateau in the A1 lineage are caused by cell death. The fivefold greater incidence of cell death despite the longer period of division in A1 accounts for the smaller final tally of progeny. The timing of these different phases of cell death corresponds to changes in the primary and secondary cuticles, respectively, and suggests a possible hormonal involvement, perhaps by ecdysteroids. The different times of cell death also means that there are different proportions of the types of cells produced in this lineage within the two segments; intersegmental interneurons formed late in the lineage are removed from T3, whereas efferent DUM neurons (*see Chapter 3*) and local interneurons are removed from A1.

Presumably, similar events in the lineages of other neuroblasts explain the difference in cell numbers in segments which either possess or lack an appendage. For example, in an abdominal segment with no appendages, cells with the same lineage as those of an unidentified neuroblast that forms the fast extensor tibiae motor neuron in T3 send axons to the edge of the neuromere, but then die (Whitington and Seifert, 1981). A consequence is that whole sets of neurons may die (in this example, the leg motor neurons) emphasising how quickly identical sets of precursor cells differentiate into segment-specific neuronal populations.

4.12. DEVELOPMENT AND FORMATION OF SYNAPSES

The ability to identify a population of midline spiking local interneurons at different stages in development allows the formation of their input and output synapses to be studied at the level of the electron microscope and gives insight into the ways that synapses form in the developing nervous system. What appears as a morphological sequence of synapse development may, in reality, not be one, because the earlier structures may not necessarily be the forerunners of the later ones. Instead, they may merely be devices for holding neurons together while synapses are formed. At 55–60% development, when these midline interneurons can first be recognised, junctions with other neurons are apparent only as simple filopodial contacts (Leitch *et al.*, 1992). At 60–65% development, punctate contacts are formed in which the gap between the two membranes is filled by an amorphous dense material also present intracellularly close to the membranes of both neurons. At 65–70% development, vesicles are present close to the regions where the membranes of two neurons are closely apposed and symmetrically thickened, but these symmetrical contacts still do not function as synapses because the interneurons at this stage are not able to generate spikes and synaptic potentials are not recorded. At 70–75% development, recognisable asymmetrical synapses with vesicles and presynaptic bars are first seen, and, at this time, synaptic potentials and spikes can first be recorded in these interneurons. Many of the sensory neurons from the leg, particularly exteroceptors, are also arriving at the central nervous system at this time and establishing functional synaptic connections. At 95% development, these synapses are the only types of contacts to remain.

4.12.1. Expression of putative transmitters and neuromodulators

Many of the potential transmitters and neuromodulators that are used by neurons in the central nervous system can be recognised by immunocytochemical methods at the time when contacts are being made between the developing neurons, but before synapses with a mature structure have formed on at least many of the neurons. The following examples give an indication of the development of these neurotransmitters and neuromodulators.

GABA can first be detected in the developing neuropil of the thoracic ganglia at 55% development but does not appear in the cell bodies of neurons until about 60–62% (O'Dell and Watkins, 1988). Common inhibitor 2 (CI2) in the thoracic ganglia is one of the first neurons expressing GABA that can be identified (Wolf and Lang, 1994). The number of immunoreactive neurons then increases in an antero-posterior pattern of expression until, at 85% development, the adult pattern of neuronal distribution is reached. The cell bodies of the common inhibitory motor neurons and of the midline spiking local interneurons can be identified at 60–65% development.

5-HT (serotonin) can first be detected in some thoracic interneurons at 55% development and it is then found progressively later in the abdominal ganglia. In an embryo that is about to hatch, three pairs of interneurons in the prothoracic ganglion (neurons S1-3), two in the mesothoracic and each of the abdominal ganglia A4-6 (neurons S1,2), and one in the metathoracic and A1 ganglia (neuron S1) show immunoreactivity to 5-HT (Taghert and Goodman, 1984). In A4, all three interneurons have both ascending and descending axons contralateral to their cell bodies, but in the prothoracic ganglion, two of the neurons have contralateral ascending axons whereas the third is a local interneuron. It seems probable that the same set of interneurons is present in each ganglion but that serotonin is expressed only in certain neurons in particular ganglia. Nothing is known of the actions or connections of any of these interneurons in adults, but the presence of 5-HT might suggest a neuromodulatory role (*see Chapter 5*). All the neurons are the progeny of neuroblast 7-3, S1 and 2 arising from its first division and S3 from its second division. Later divisions give rise to neurons that do not contain 5-HT.

Octopamine is first detected in efferent DUM neurons in the thoracic ganglia at about 60% development, although they are sensitive to applied acetylcholine or GABA at 40% development, indicating that receptors are present before their own putative neuromodulator is ready for release.

Proctolin can also first be detected in 50–60% embryos, but the first neurons to stain with an antibody raised against proctolin are seen in the suboesophageal ganglion at 65–70% development (Keshishian and O'Shea, 1985a). It then appears in neurons in more posterior ganglia in a progressive sequence so that the adult pattern of staining is present in the metathoracic ganglion at 70% development.

Myomodulin is first detected in a DUM neuron that is probably the H neuron at 50% development, and after 60% appears in a group of cells in the posterior and lateral region of the thoracic ganglion (Swales and Evans, 1994) (*see Chapter 5*).

4.13. DEVELOPMENT OF MUSCLES

4.13.1. Muscle pioneers

The muscles develop from mesodermal cells, thus providing an immediate contrast with neurons, which develop from ectodermal cells. Certain mesodermal cells enlarge to form muscle pioneer cells with a flattened shape and a tendency to adhere to the ectoderm. They form a basic scaffold by extending large growth cones that insert into the ectoderm and mark the future insertion points of the muscles, and into the ectodermal invaginations that form muscle apodemes or tendons. Other mesodermal cells then aggregate around a pioneer and fuse with it to form a multinucleate syncytium from which the muscle fibres divide and differentiate. A simple muscle consisting of one bundle of fibres may thus be formed by a single muscle precursor cell, but more complex muscles with many bundles of fibres are formed from many. These different mechanisms of muscle development are well illustrated by the ways that the extensor and flexor tibiae muscles of a leg are formed.

4.13.1.1. *Development of the extensor tibiae muscle*

The extensor tibiae muscle in all the legs begins as a single mesodermal cell that becomes visible at 35% development spanning the ectoderm of the invaginating apodeme and the wall of the femur where the muscle fibres will eventually insert (Ball and Goodman, 1985a). It then fuses with surrounding mesodermal cells to form a giant, horseshoe-shaped syncytial cell called the supramuscle pioneer around the developing apodeme. This multinucleate pioneer provides the framework for the whole extensor tibiae muscle. At about 55% development, the pioneer splits into a series of thin cytoplasmic bridges, containing many nuclei and surrounded by many small mesodermal cells, that insert on the wall of the femur and link to the apodeme. By 60% development the bridges have started to divide and have lost syncytial contact with each other so that they form an array of smaller multinucleate precursors which provide the framework for the individual bundles of muscle fibres. These precursors then divide again to form the individual muscle fibres, replete with internal filaments, sarcoplasmic reticulum and T tubules. In the hind leg, but not in the other legs, an accessory pioneer separates at about 45% development, eventually to form the two distal accessory bundles of the muscle.

4.13.1.2. *Development of the flexor tibiae muscle*

By contrast, the flexor tibiae muscle, which is more complex and consists of an array of muscle bundles, develops by the sequential recruitment of pioneer cells from the surrounding undifferentiated mesoderm along the lengthening apodeme (Ball and Goodman, 1985b). Initially, at 37% development, two pioneer cells appear at the ends of the invaginating apodeme linking it to the ectoderm. These are soon joined by a third pioneer in between, but this soon dies. The number of pioneers then increases by the symmetrical addition of new ones along each side of the apodeme and by 50% development these cells have enlarged and begin to divide. This topological array of pioneer cells thus defines the arrangement of the bundles of muscle fibres that constitute the final muscle. Each precursor becomes surrounded by a cluster of undifferentiated mesodermal cells that fuse with it so that it is now multinucleated, and at 60% development it divides into the individual muscle fibres of a particular bundle. The differences in the development of the flexor and extensor tibiae muscles suggest a different evolutionary origin of the two, which may also be reflected in their markedly different patterns of motor innervation: the extensor has only two excitatory motor neurons, the flexor has at least nine.

4.13.1.3. *Innervation of the muscles*

The growth cones of most of the motor neurons leave the central nervous system along routes established by the first pioneer motor neurons, but at some point have to leave these routes to establish contact with the muscle pioneer cells of their particular muscle. This implies that particular motor neurons read clues in particular muscle pioneer cells and make the appropriate choices about the direction in which they should grow.

The SETi motor neuron that has pioneered the route of N3B is the first to arrive at the developing extensor tibiae muscle at about 42% development and by 50% has spread along the length of the supramuscle pioneer. The FETi motor neuron, following the route of N5 established by other motor neurons and by peripheral pioneers, arrives soon after this stage, so that both excitatory motor neurons of this muscle are present before the muscle bundles have been formed or the muscle cells have differentiated. It is possible, therefore, that the different types of muscle fibres that characterise the adult muscle arise from interactions between the motor neurons and the precursor cells.

On their way to the extensor muscle, these and at least 18 other motor neurons pass the muscle pioneer cells of muscles in the coxa and yet grow past them and into the femur. At the same time, the appropriate motor neurons of these coxal muscles leave the main routes and establish contact with their correct muscle pioneers but go no further. The two excitatory motor neurons that innervate the depressor trochanteris muscle (133a) (*see Table 8.2*) follow the N5 route into the coxa but then abruptly leave it to contact the muscle pioneer cell of this muscle at 40% development (Ball *et al*., 1985). The sibling motor neuron of one of the 133a motor neurons that is formed by division of one of the first ganglion mother cells of neuroblast 2-2 takes the same route into the coxa but does not deviate from it, despite contact with the133a muscle pioneer, and continues on into the tibia. These different choices are made at a time when the two siblings are still dye coupled to each other. If the 133a muscle pioneer is ablated at about 37% development it is not replaced and then the motor neurons of this muscle also fail to leave the N5 route as they pass through the coxa. This implies that specific motor neurons read signals, probably on the surface membranes of specific muscle pioneer cells and not of surrounding mesodermal cells, and make turns according to these signals to establish specific contacts.

4.14. POSTEMBRYONIC FORMATION OF NEURONS

Most neurons are born during embryonic life: the neuroblasts then degenerate and die and their offspring are postmitotic. While this is the general rule, a small number of cells in the central nervous system remain undifferentiated and can resume their cycles of divisions after the embryo hatches. This occurs particularly in the brain during the larval stages, but may also persist into adulthood. Moreover, many sensory neurons are added postembryonically.

4.14.1. Addition of neurons to the central nervous system

The compound eyes continue to grow postembryonically by the addition of ommatidia at the anterior margin and by a parallel addition of new neurons to the anterior margin of the lamina and medulla. In the fourth instar, for example, some 10 new rows of ommatidia are added anteriorly and mitotic division of neuroblasts occurs in the lamina (Anderson, 1978). These divide in the characteristic asymmetrical way to form another neuroblast and a ganglion mother cell, which then divides symmetrically to form two cells that then differentiate into neurons of the lamina and medulla. Thus, the addition of sets of neurons in the two most distal layers of the optic lobes keeps pace with the addition of receptor cells.

In holometabolous insects such as flies, which undergo complete metamorphosis from a crawling larva through a pupal stage before finally emerging as an adult, the division of the neuroblasts is prolonged through postembryonic development. In embryos of *Drosophila*, the 500 neuroblasts (which are arrayed in a pattern of about 40 per segment and are thus similar to those in a locust) undergo only a few cycles of division to form the larval nervous system of some 300 cells in each neuromere before becoming smaller so that they can no longer be recognised easily. Once the larva has hatched, the neuroblasts enlarge again so that they appear in a repeating, segmental pattern, and resume their cycle of division postembryonically, adding many more undifferentiated neurons to each neuromere (Prokop and Technau, 1991). Thus, the postembryonic neuroblasts are either simply the embryonic ones that resume proliferation after a quiescent period or are the daughter cells of the embryonic neuroblasts. The differentiation of these neurons is delayed until pupation, when the division of the neuroblasts also ends, and is controlled by the levels of ecdysteroids. Without these hormones the neurons remain undifferentiated, but in their presence they start to grow axons, some are respecified, and many of them also die, particularly those formed embryonically.

In the mushroom bodies in the brain of adult, but immature, female crickets, a group of undifferentiated neuroblast cells persist during the first 2 weeks of adult life (Cayre *et al.*, 1994). These divide in the usual asymmetric way to form smaller ganglion mother cells that then divide symmetrically to form neurons, and which finally differentiate as Kenyon cells (intrinsic interneurons, *see Chapter 2*). The rate of division of the cells is stimulated by increases in the level of juvenile hormone in the haemolymph which, at the same time, also controls maturation of the gonads and oviposition behaviour. The formation of new neurons under hormonal control implies considerable plasticity in the adult brain accompanied by the probable formation of new connections that may be associated with changes in the egg-laying behaviour of females. These changes are similar to the structural changes that occur in the brains of adult canaries (Alvarez-Buyalla, 1990) under the influence of steroid hormones, although these hormones do not induce neurogenesis of the type seen in the cricket (Brown *et al.*, 1993). The wider question that is raised by these observations is whether the divisions of the neuroblasts in embryos are also under some sort of hormonal control, rather than the internal factors within the neuroblasts and their interactions with immediate neighbouring cells that are currently thought to determine their actions.

4.14.2. Addition and maturation of sensory neurons

As the surface area of the body increases during postembryonic growth so too does the number of sensory neurons. The most striking increase in locusts comes from the wings, which are present only as small, immoveable wing buds in the later instars, finally becoming fully formed and moveable in the adult (*see Chapter 11*). In a first instar larva, the nerve that will innervate the wing has fewer than 50 axons but this number increases gradually to about 200 by the third instar, and then abruptly to almost 900 at the fourth instar as the wing buds are first formed (Altman *et al.*, 1978). In a young adult, the number increases further to 2000 before being reduced to 1000 in mature adults. In cockroaches, there is an even more dramatic increase:

the number of sensory neurons from an antenna increases from 14 000 in a first instar larva to 270 000 in an adult (Schaller, 1978).

4.14.2.1. *Sensory neurons on the neck*

The ventral part of the neck in an adult *Locusta migratoria* has about 300 long flexible hairs that respond to low velocities of air currents (*see Chapter 11*). A first instar larva hatches with only about 30 of these hairs; new hairs are added at each moult with the majority appearing at the last two moults, and those already present increase in length and in the pigmentation of their bases (Pflüger *et al.*, 1994). There is thus a correlation between the final morphology of one of these hairs and its birth date. In the adult, the ventral hairs project to the contralateral half of the prothoracic ganglion where they synapse directly with the axonal branches of an identified projection interneuron called A4I1 (it has its cell body in abdominal ganglion A4). In early instars, however, each of these sensory neurons has a projection in both ipsilateral and contralateral neuropil and makes synaptic connections with both the ipsilateral and the contralateral A4I1 interneurons. In later instars, the ipsilateral projection is retracted and at the same time the connection with the ipsilateral interneuron disappears. The final pattern of projections and connections is therefore established by segregation of fibres and loss of synaptic contacts after an initial overgrowth. This process can be blocked if some of the surrounding hairs are immobilised so that they no longer respond to air currents, with the result that some lateral hairs retain both their ipsilateral and contralateral projections into adulthood. The final structure and connections made by a sensory neuron may thus depend on sensory activity as in this example neither the position of the receptor nor the hormonal environment was altered. The functional consequences of the changes in receptive fields of the A4I1 interneurons have not been explored. In adults, these interneurons appear to play a role in adjusting the movements of the wings in response to changes in air currents, but as functional wings are not present until the adult, their role in larvae, in which they will have a larger receptive field, is unknown.

4.14.2.2. *Sensory neurons on a cercus*

In cockroaches only two filiform hairs are present on a cercus at the time of hatching but this number increases to about 220 in an adult; in crickets the increase is from 50 to 2000, but no values are available for the locust. Accompanying these changes in number are increases in the diameter and length of the hair shaft and thus concomitant changes in their stiffness and hence response properties. The changes allow a spectrum of different hair shafts to be formed, each with their own characteristic response properties, so that at different times during postembryonic development, equivalent stimuli will be signalled to the central nervous system by different patterns of sensory spikes. Nevertheless, the sensory neurons present in the embryo have already made functional synapses so that the sensory neurons formed postembryonically must fit into the pattern of connections already established on the expanding numbers of branches of the interneurons and motor neurons. Moreover, the integration performed in the central nervous system must compensate for the change in the number of inputs it receives and for the changing response properties of these neurons. Competition for synaptic sites and the increase in the number of branches of an interneuron compensates for the increased number of sensory neurons, so that during periods when the number of cercal sensory neurons doubles, the properties of the giant interneurons on which they synapse show only small changes, remaining more or less constant in those properties that are thought to be important for coding the escape behaviour (Chiba *et al.*, 1992). Further compensation may also involve changes in the strength (defined by the amplitude of the synaptic potential) of a connection made by a sensory neuron with different interneurons (Chiba *et al.*, 1988).

In the face of these progressive changes, the behaviour of the insects remains much the same: a first instar cockroach can still move away from a wind stimulus of a certain direction in a way comparable to the adult (Dagan and Volman, 1982) (*see Chapter 10*) but there must be subtle changes that await elucidation. Stability of overall response during the progressive increase in sensory neurons seems to occur because the changes at every level from the transduction properties of the sensory neurons to their synaptic connections with their target interneurons produce matching and compensatory alterations.

4.14.3. Changes in volume of the central nervous system

The volume of the nervous system increases progressively during postembryonic development and this is attributable to a number of factors. It is generally assumed that most neuroblasts do not continue to divide, but the example given above indicates that this can still occur. Sensory neurons are still added so that their projections into the central nervous system and the formation of connections will lead to an increase in volume. The largest changes must come from the expansion in volume of individual neurons, either by growth or by the increase in the number of their branches. Such changes do not necessarily occur evenly throughout all the different parts of the nervous system and may be associated with particular aspects of behaviour. In beetles, flies and bees, the most dramatic changes occur in the protocerebrum. In the first 20 days of adult life, the corpora pedunculata of the beetle *Aleochara curtula* increase in volume by about 70% (Bieber and Fuldner, 1979). In *Drosophila*, the number of cells in the mushroom bodies increases by 15% before declining as the fly ages (Technau, 1984). More cells are present in females than in males and the total number can depend on the environment in which the flies are raised. In bees, changes that occur in the mushroom bodies can be directly related to behaviour and to the division of labour amongst different members of a hive (Withers *et al.*, 1993). Adult worker bees spend the first 3 weeks as nurses in the hive before spending the rest of their 4–7 week life as foragers. During the first 3 weeks, the total volume of the brain remains stable, but the volume of the olfactory glomeruli of the nurses increases. In foragers, the volume of the Kenyon cell bodies in the mushroom bodies decreases by about 30% but the volume of neuropil in the mushroom bodies increases by about 15% so that the ratio between neuropil and cell body is almost double that of the 1 day old bee. The volume of the olfactory glomeruli also declines, reflecting a lesser dependence on the olfactory signals that are a major means of communication in the dark hive, and greater dependence on the associations made between sight and smell in the outside world. Some of the changes in the neuropil are probably attributable to increases in spines compared to those of the nurse bees (Coss *et al.*, 1980). The changes in the brains of foragers appear to be linked to changes in behaviour rather to ageing because precocious foragers only 4 days old have mushroom bodies similar to those of the older foragers.

What all of these changes emphasise is the potential plasticity of the adult central nervous system. The structure that is formed during embryonic development is not immutable, but instead may be altered, not just with age but also to facilitate specific changes in behaviour. These alterations can involve the addition or deletion of neurons, but mostly are assumed to encompass changes in either the number or strength of the synaptic connections made by the neurons that are already present. The picture to emerge is one that should forever dispel the notion that the central nervous system of an insect is always fixed in its structure and in the functional responses that it produces.

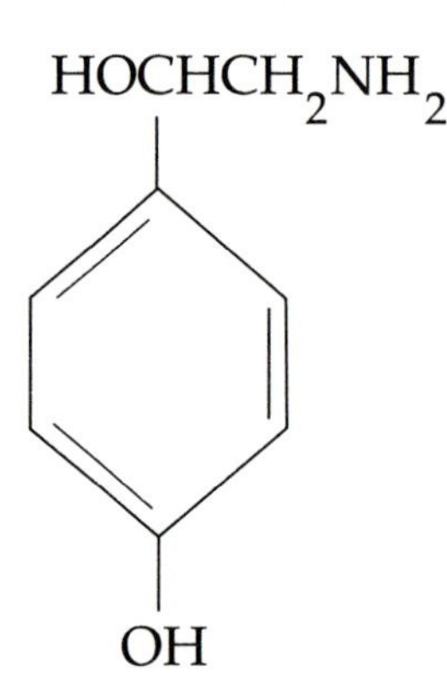

Neurotransmitters, neuromodulators and neurohormones

Defining neurotransmitters, neuromodulators and neurohormones

A **neurotransmitter** is a messenger released from a neuron at an anatomically specialised junction and that diffuses across a narrow cleft to affect one or sometimes two postsynaptic neurons, a muscle cell, or other effector cell. Typically, a neurotransmitter acts directly on a postsynaptic neuron to cause a change in its membrane potential, although it may sometimes act through second messengers.

A **neuromodulator** is a messenger released from a neuron in the central nervous system, or in the periphery, that affects groups of neurons, or effector cells that have the appropriate receptors. It may not be released at synaptic sites, often acts through second messengers and can produce long-lasting effects. The release may be local so that only nearby neurons or effectors are influenced, or may be more widespread, which means that the distinction with a neurohormone can become very blurred. The act of neuromodulation, unlike that of neurotransmission, does not necessarily carry excitation of inhibition from one neuron to another, but instead alters either the cellular or synaptic properties of certain neurons so that neurotransmission between them is changed.

A **neurohormone** is a messenger that is released by neurons into the haemolymph and which may therefore exert its effects on distant peripheral targets. It may differ only in degree from a neuromodulator in the extent of its action. Those with restricted actions may sometimes be called local neurohormones to emphasise that their effects are localised.

These terms do not define rigid categories but rather the peaks in a continuum of effects. Some substances, such as nitric oxide (NO), can act as a transmitter but by virtue of their diffusibility might act on many cells, while substances such as octopamine can fulfil the requirements of all three definitions. A restricted but workable definition of neuromodulation (Kaczmarek and Levitan, 1987) suggests that it is '*the ability of neurons to alter their electrical properties in response to intracellular biochemical changes resulting from synaptic or hormonal stimulation*'. On this basis, neuromodulation can result from the actions of a substance defined in any of the three categories.

Neurotransmitters

5.1. RECEPTORS FOR PUTATIVE TRANSMITTERS

5.1.1. Extrasynaptic receptors

Just as the muscle fibres have receptors distributed along their membranes away from the synaptic sites, so too do the central neurons. In particular, the cell bodies, which have no dendritic branches and neither receive nor make synapses, have receptors for many putative transmitters and neuromodulators (Table 5.1), but the reason for their presence and whether they serve any function is unknown. Does their presence merely indicate that receptors are made in the cell body for transport to the distant synaptic sites in the neuropil, or does it indicate that they respond to substances which have gained access through the sheath of the ganglion, or by some other unknown route? Are they involved in any signalling with the surrounding glial cells? If they do function in these ways, the changes they effect could contribute to integration by altering the electrical loading of the cell body and hence the time and space constants of the membrane. Alternatively, these possible actions could influence the expression of genes that might be involved in controlling neuronal growth, or those that are only expressed at particular times. It should be borne in mind that, within the central nervous system, the cell bodies of the neurons are wrapped with many layers of glial cells. In interpreting the results of tests for receptors on the cell bodies of neurons, even those isolated from the central nervous system, the possibility should be considered that the applied drugs may be acting on the glial membranes rather than directly on the neuronal membranes.

The properties of the receptors on a cell body are known in more detail than any of the synaptic receptors because the cell body offers a large recording site that can be voltage-clamped. Moreover, the use of individual cell bodies that are mechanically and/or enzymatically dissociated from the nervous system (Usherwood *et al.*, 1980) allows the ionic environment to be changed readily, and allows agonistic and antagonistic drugs to be applied easily to the receptors. The identity of the isolated cell body cannot, however, readily be determined, but the large diameter of those that are usually used and their inability to support spikes indicates that they are probably motor neurons. The overriding issue that haunts these studies is whether receptors at the central synapses have the same properties as those characterised on the isolated cell bodies. The characterisation of these receptors on cell bodies (Benson, 1993) has concentrated on those either thought to be the most widespread or the most important in the nervous system, and is thus biased by our

Table 5.1. Properties of receptors on the cell bodies of neurons

Receptor	Responses	Antagonists	Agonists
Glutamate	Cl^--dependent hyperpolarisation	picrotoxin	L-aspartate
	kainate and quisqualate cause a depolarisation	–	–
$GABA_A$-like	reverses at –65 to –85 mV in different neurons	picrotoxin	muscimol
$GABA_B$-like	reverses at –95 mV	–	3-aminopropylphosphonic acid 3-aminopropylphosphonous acid
Acetylcholine			
Nicotinic	Na^+, reverses at + 20mV	α-bungarotoxin> mecamylamine> D-tubocurarine> hexamethomium> tetraethylammonium> picrotoxin	nicotine
Muscarinic	decreases to zero at –80 to –90 mV	QNB> scopolamine> atropine> pirenzepine	oxotremorine> nicotine> ACh
5-HT			
1	K^+	Cs^+, Rb^+ >4-AP, TEA, ketanserin, ritanserin	2-methyl-5-HT
2	Na^+	Na^+-free saline	2-methyl-5-HT
3	K^+ ?	4-AP	–
Octopamine	depolarisation, maximal at –70 mV	phentolamine, mianserin> yohimbine, metoclopramide	–
Dopamine	depolarisation; current reverses at 0 mV; region of negative slope conductance in current/voltage curve	fluphenazine, flupenthixol, holperidol	ergotamine, apomorphine, bromocryptine
Taurine	hyperpolarisation, reverses at same potential as GABA-evoked response	picrotoxin	–

4-AP = 4-aminopyridine
QNB = quinuclidinyl benzilate
TEA = tetraethlyammonium

current thinking. It thus omits receptors for some candidate transmitters and neuromodulators that have recently emerged, and for some agents that are well known in other nervous systems, but which have so far been found either not to occur in, or not to be effective in locust nervous systems.

5.1.2. Synaptic receptors

Synaptic receptors are more difficult to study for many reasons. First, synapses are made between the fine processes of neurons that are usually too small to be penetrated by intracellular electrodes. Electrical events at synapses can thus only be observed by electrodes placed at a distance, with inevitable distortions caused by the interposed membrane. Second, the synapses lie within a dense neuropil and the central nervous system is surrounded by a blood-brain barrier that restricts access even of small ions. It is thus difficult to gain access to a synapse with agonists and antagonists, or even to change its ionic environment. Third, neurons exist only in a web of connections with other neurons. This means that it is difficult to establish that an experimentally applied change acts primarily on the neuron being studied and not secondarily on other neurons to which it is connected. All of these reasons probably explain the paradox of why so little is known about the pharmacology of signal transmission in the central nervous system of a locust, when so much is known about the pathways and so many of the neurons have been identified. Nevertheless, the lack of knowledge is all the more surprising given the enormous effort that there has been to kill pest insects with substances that interfere with synaptic transmission.

5.2. GLUTAMATE

The distribution of glutamatergic neurons is revealed by antibodies raised against glutamate that do not at the same time label neurons containing GABA, for which glutamate is a precursor. Glutamate-like immunoreactivity occurs in some 360–400 cell bodies in the pro- and mesothoracic ganglia and in about 600 in the metathoracic ganglion, including the A1-3 neuromeres (Watson and Seymour-Laurent, 1993). Each of the free abdominal ganglia, A4-7, contains 80–100 cell bodies with glutamate-like immunoreactivity and the last abdominal ganglion contains about 250 immunoreactive cell bodies. Caution is necessary in drawing firm conclusions from these numbers as the methodology relies on deciding whether a neuron shows immunoreactivity by ascribing an arbitrary threshold level for detection. Moreover, it is unclear whether the method might also stain aspartate.

5.2.1. Identity of glutamatergic neurons

In the thoracic ganglia, the cell bodies that are labelled have diameters ranging from 10 to 100 μm and thus probably include all the excitatory motor neurons. Some of these labelled neurons can be clearly identified as specific motor neurons, for example, the fast motor neuron to the extensor tibiae muscle of a hind leg (Bicker *et al.*, 1988). In the median nerve of the mesothoracic ganglion there are two large stained axons that presumably belong to the spiracular closer motor neurons (*see Chapter 12*). The motor neurons to the dorsal longitudinal muscle of a hind wing, one of which has its cell body in the metathoracic ganglion contralateral to its axon, and the other four with cell bodies in the mesothoracic ganglion can all be recognised as showing glutamate-like immunoreactivity. In the abdominal ganglia, most of the glutamatergic neurons are likely to be motor neurons, but the corollary of whether all excitatory abdominal motor neurons are glutamatergic is unproven. Some of the motor neurons that control the movements of the ovipositors can be identified as glutamatergic.

If all the excitatory motor neurons are glutamatergic, this must mean that the number of glutamatergic interneurons in segmental ganglia is small, assuming that the stained neurons represent the true number of those that contain glutamate. The only thoracic interneurons that have so far been recognised as glutamatergic are a group called the 404 interneurons that may be involved in initiating flight (*see Chapter 11*). This attribute of their identity is ascribed on the basis of their unusual anatomy and not according to their physiological connectivity. In the brain of the honeybee, motor neurons of the antennal muscles appear to be glutamatergic and particular groups of interneurons also stain with an antibody raised against glutamate (Bicker *et al.*, 1988). In the optic lobes, lamina monopolar cells stain, as do some of the large interneurons receiving information from the ocelli. Many interneurons with axons that descend to the thorax also stain.

5.2.2. Glutamate receptors

There are three broad classes of receptors in the locust that can be activated by L-glutamic acid and with 1000 times less potency by D-glutamic acid (Usherwood, 1994).

1. Receptors which gate anion-selective channels, typically Cl^- and therefore most commonly cause a hyperpolarisation.
2. Receptors which gate cation-selective channels and therefore cause a depolarisation. Within this category are subtypes that can be distinguished by their sensitivity to the agonists quisqualate, kainate and aspartate.
3. Metabotropic receptors which act *via* second messengers to cause a change in membrane potential.

The receptors are located at synapses within the central nervous system and at neuromuscular junctions. They also occur on the cell bodies, at other extrasynaptic sites in the central nervous system and at extrajunctional sites along the membranes of muscle fibres. Much of what we know about insect glutamate receptors comes from pharmacological studies of the extrajunctional receptors on muscles fibres and by their analysis with patch-clamp methods. It seems clear that a fuller understanding of the synaptic receptors will emerge only when they have been cloned and expressed in a suitable membrane. A kainate-sensitive subunit from the *Drosophila* central nervous system has already been cloned (Ultsch *et al.*, 1992).

5.2.2.1. *Receptors in the central nervous system*

The typical response of a locust or cockroach cell body to the application of L-glutamate is a hyperpolarisation that reverses at the chloride equilibrium potential, but without any cross-sensitisation to the GABA response (Wafford and Sattelle, 1989). The response is ibotenate-sensitive and is blocked noncompetitively by picrotoxin but at higher concentrations than are needed to block responses to GABA. Kainate and quisqualate, however, produce a depolarisation with a slower time course but again involving a conductance increase. It seems probable, therefore, that the membranes of cell bodies contain at least two types of receptors, one gating anions and the other cations.

Glutamate applied directly to the desheathed surface of a meso- or metathoracic ganglion inhibits flight motor neurons, as recorded in their neuropilar processes, by activating a chloride conductance (Dubas, 1991). The resulting voltage change can be depolarising or hyperpolarising depending on the resting potential, but it is uncertain whether it results from the activation of extrajunctional receptors rather than those associated with the synapses. Much remains to be done to sort out the properties of these central receptors, but the hyperpolarising response suggests that these receptors are different from the synaptic receptors where a depolarisation and excitation are the usual responses to synaptic inputs from neurons that are thought to release glutamate.

5.2.2.2. *A central glutamatergic synapse*

One of the leg motor neurons that is immunoreactive to antibodies raised against glutamate makes glutamatergic synapses on the muscle fibres that it innervates. Moreover, it also makes central output synapses which again seem to use glutamate (Sombati and Hoyle, 1984a; Parker, 1994). The fast extensor tibiae motor neuron in the metathoracic ganglion makes a direct excitatory synapse with many of the motor neurons that innervate the flexor tibiae muscle (*see Chapter* 9 for details of this connection and its functional role). The EPSP in the flexor motor neurons caused by spikes in FETi is blocked by the glutamate antagonist glutamic acid diethyl ester and can be mimicked by injecting glutamate or quisqualate, but not by kainate or NMDA, into the neuropil containing the processes of the flexor motor neurons. Both the natural EPSP and the depolarisations evoked by glutamate have the same reversal potential close to zero. Thus, glutamate can cause a depolarisation when acting on receptors at the synapse and in the neuropil, but a hyperpolarisation when acting on the membrane of the cell body with a reversal potential close to −90 mV, or a depolarisation. Glutamate thus has similar central and peripheral effects because at muscle fibres it can cause a depolarisation at the synapses but a hyperpolarisation at extrajunctional receptors. FETi itself is also depolarised by glutamate injected into the neuropil but its receptors appear different from those on the flexors in that kainate and not quisqualate activates them. Clearly, the receptors that are activated by the synaptic release of glutamate produce different effects to those on the soma that are activated by the experimental application of glutamate. These results invite extreme caution in any extrapolations that might be made from studies on the receptors on the soma, where no function has yet been clearly demonstrated, to the likely action of receptors at the functioning synapses in the neuropil.

5.2.2.3. *Junctional and extrajunctional receptors on muscles*

The receptors for glutamate at the synapses made by excitatory motor neurons with muscle fibres are a mixed population with sensitivities to quisqualate, aspartate and ibotenate. Receptors with similar properties to the ibotenate-sensitive and quisqualate receptors also occur at extrajunctional sites, and metabotropic receptors may occur here as well. The quisqualate-sensitive receptors have a high conductance of some 150pS with at least three open and four closed states but with a restricted ion selectivity for Na^+ and K^+ but not Ca^{2+}. The quisqualate-sensitive receptor can be blocked by polyamide-containing toxins such as argiotoxin (extracted from the venom of certain spiders) and philanthotoxin (extracted from the venom of the wasp *Philanthus triangulum*). These substances are normally used by these predators to paralyse their insect prey by blocking neuromuscular transmission of the excitatory motor neurons. The extrajunctional metabotropic receptors produce slow depolarisations of the muscle membrane that seem to be mediated by cAMP and to involve a conductance decrease (Robinson, 1982). Receptors sensitive to NMDA may also occur on the terminals of the motor neurons but, as yet, there is little detail of their properties or of their likely action.

5.3. GABA

GABA is present in all parts of the central nervous system (Breer and Heilgenberg, 1985) and the evidence so far suggests that it has an inhibitory action. This is most clearly demonstrated for the common inhibitory neurons that innervate the leg muscles (Usherwood and Grundfest, 1965) and for some of the local interneurons in a thoracic ganglion. The distribution of cell bodies containing GABA has been well described by the use of antibodies raised against GABA that appear to be very selective. Moreover, the synapses made by GABAergic neurons can be seen in the electron microscope after immunogold labelling (Watson, 1988). The vesicles associated with their output synapses are pleomorphic in shape, whereas those of glutamatergic neurons in the same region of neuropil are round.

5.3.1. Identity of GABAergic neurons

5.3.1.1. *Brain*

Information about the distribution of neurons in the brain that might contain GABA comes from bees (Schäfer and Bicker, 1986) and the moth *Manduca sexta*

(Homberg *et al.*, 1987), but as yet no systematic study of locust brain for GABAergic neurons has been performed. In each half of the moth brain, about 18 000 neurons in the optic lobes, 1000 in the protocerebrum, 550 in the deutocerebrum, and 100 in the tritocerebrum show GABA-like immunoreactivity. This means that about 40 000 neurons or 5% of the population of the whole brain may contain GABA and emphasises the widespread importance of this amino acid as a transmitter. The neurons in the optic lobes are centrifugal neurons of the lamina and tangential neurons of the medulla and lobula. Some are large projection interneurons from the lobula plate to the protocerebrum. In the protocerebrum itself, about 300 labelled neurons supply the calyces and lobes of the mushroom bodies, and others supply the lower division of the central body. In the deutocerebrum, GABA-like immunoreactivity is present throughout the neuropilar regions and is probably associated with the local interneurons. Only a few axons stain in any of the brain nerves and these probably belong to inhibitory motor neurons.

5.3.1.2. *Thoracic ganglia*

In each of the thoracic neuromeres, about 250 cell bodies arranged in five groups, four on the ventral and one on the dorsal surface, show GABA-like

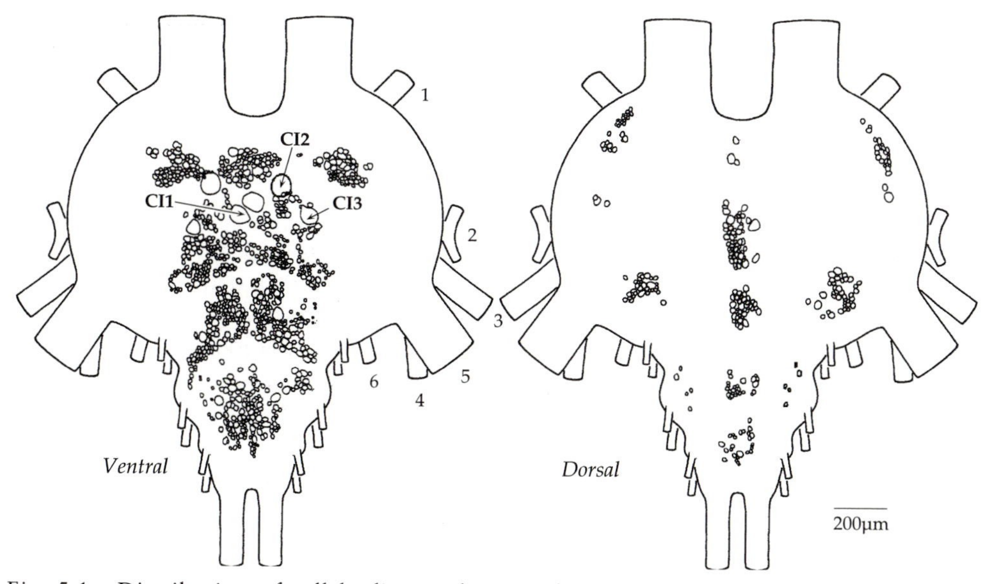

Fig. 5.1 Distribution of cell bodies in the metathoracic ganglion that show GABA-like immunoreactivity. Three of the cell bodies belong to the common inhibitory neurons (CI1–3) that innervate leg muscles. Some of the cell bodies at the ventral midline belong to midline spiking local interneurons (see Fig. 3.10) and some at the dorsal midline belong to local DUM interneurons (see Fig. 3.11A). The other neurons are not yet identified. Figure courtesy of P.A. Stevenson.

immunoreactivity (Watson, 1986) (Fig. 5.1). A few of the neurons in these groups can be identified as individuals, some can be ascribed to populations of interneurons with characterised actions, but the properties and actions of most have yet to be explored.

The three common inhibitory motor neurons all show GABA-like immunoreactivity in their cell bodies and in some of their larger neuropilar processes. In the peripheral nerves to the legs, only three axons stain and these belong to the common inhibitory motor neurons. This is a clear demonstration of the specificity of the staining and that none of the excitatory motor neurons or any of the sensory neurons contain GABA. Eight of the identified intersegmental interneurons (interneurons 301 and 302 in the mesothoracic ganglion and 501, 507, 511, 513, 520, and 538 in the metathoracic ganglion) (*see Chapter 11*) that are involved in controlling the movements of the wings in flight are GABAergic (Robertson and Wisiniowski, 1988). Of these, six are known to make inhibitory output connections, but some other interneurons involved in flying that are thought to have inhibitory output connections are not GABAergic, emphasising that GABA is not the only transmitter with an inhibitory action.

Some of the remaining cell bodies can be assigned to particular populations of interneurons. Many of the cell bodies in the ventral midline group belong to a population of spiking local interneurons (*see Chapters 3* and 7) that make inhibitory output connections with motor neurons and other interneurons. Some of the cell bodies in the dorsal midline group belong to local DUM neurons, but the action of these neurons is not defined (Thompson and Siegler, 1993). In the lateral and posterior group in the mesothoracic ganglion, some 15% of a population of intersegmental interneurons with axons that project posteriorly and ipsilaterally to the metathoracic ganglion show GABA-like immunoreactivity (Watson and Laurent, 1990). This fits with the observation that a similar percentage of these interneurons make inhibitory output connections with metathoracic nonspiking interneurons and motor neurons. The branches of these interneurons also receive GABA-like immunoreactive input synapses, and this again correlates well with the inhibitory inputs that they receive from GABAergic midline spiking local interneurons. The other interneurons in this population of cell bodies make excitatory output connections, emphasising how little can be predicted about function from cell body position and axon trajectories (*see Chapter 11)*.

5.3.1.3. *Abdominal ganglia*

In the abdominal ganglia, about 150 cell bodies, or some 25% of the total cell body population, show GABA-like immunoreactivity (Watson and Pflüger, 1987). About half of these neurons are in three recognisable groups; first, a lateral and anterior group of about 15 cell bodies of which two are large; second, a posterior lateral group of about 15 small cell bodies; third a medial posterior group of about 35–40 cell bodies with one much larger than the rest. None of the interneurons in these groups have yet been identified so it remains uncertain whether any are homologues of those identified in the thoracic ganglia. In N1, two large diameter (>4 μm) axons show GABA-like immunoreactivity, while in N2 there are no stained axons with comparable diameters but a variable number (1–4) of smaller stained axons. It is assumed that these are the axons of inhibitory motor neurons.

5.3.2. Receptors on cell bodies

The general view of GABA receptors envisages a GABA recognition site, a Cl^- channel and modulatory sites for benzodiazepines and barbiturates. All our knowledge of these receptors in the central nervous system comes from extrajunctional sites on cell bodies, backed by pharmacological evidence that is more difficult to interpret on synaptic receptors. On unidentified cell bodies, activation of these receptors evokes a fast current that is carried by Cl^- ions and which reverses at potentials from –65 mV to –85 mV, the variation presumably being a reflection of the differing internal Cl^- concentrations in different neurons. At the normal resting potential, this current results in a hyperpolarisation, but because the reversal potential is so close to resting potential it can, at slightly more negative potentials, cause a depolarisation. Its pharmacological profile resembles the vertebrate $GABA_A$ receptor, in that it is agonised by muscimol and blocked by picrotoxin, but it differs in that bicuculline has no blocking action. Only in the moth antennal lobe does bicuculline have any consistent effect on pathways that might be mediated by GABA (Waldrop *et al.*, 1987). Both benzodiazepines and barbiturates can enhance the currents activated by GABA.

The existence of a receptor with properties akin to the vertebrate $GABA_B$ receptor has been doubted for a long time, largely because baclofen, the common agonist of this receptor in vertebrates, is without any action. In the cell body of an identified leg motor neuron of the cockroach, some agonists of vertebrate $GABA_B$ receptors, such as 3-aminopropylphosphonic acid and 3-aminopropylphosphonous acid, can induce hyperpolarisations with a slow onset and slow time course which are insensitive to picrotoxin (Bai and Sattelle, 1995). Baclofen is, however, again without effect so that the receptor must differ from those in vertebrates. The potential reverses at an estimated value of –95 mV, leading to the suggestion that the current is carried by K^+.

In the central nervous system, many of the synapses made on the terminals of sensory neurons (Burrows and Laurent, 1993) and by a particular group of thoracic spiking local interneurons onto nonspiking neurons (Laurent and Sivaramakrishnan, 1992) appear to operate by chloride conductance mechanisms that are sensitive to picrotoxin, although the reversal potential in the nonspiking neurons is more negative, presumably reflecting a different internal chloride concentration. All of these effects are mimicked by the application of GABA itself.

5.4. ACETYLCHOLINE

The transmitter most commonly thought to be used by mechanosensory neurons in insects is acetylcholine (ACh), with the bulk of the evidence for this assertion coming from the cercal receptors of cockroaches (Callec *et al.*, 1971) and the tactile hairs of moth prolegs (Trimmer and Weeks, 1989). In locusts, many sensory neurons stain with antibodies raised against choline acetyltransferase (Lutz and Tyrer, 1988), and contain acetyltransferase (Emson *et al.*, 1974), but pharmacological evidence has only been explored in any detail for the direct connections made by the sensory neurons from tibial campaniform sensilla (for their action during jumping, *see Chapter 9.*) with an identified leg motor neuron (*see section 5.4.2*). The receptors for acetylcholine appear to fall into three broad classes. First, nicotinic receptors which are antagonised by α-bungarotoxin. Second, muscarinic receptors which are not sensitive to the nicotinic antagonists. Third, receptors with mixed properties that can be activated by both nicotinic and muscarinic agonists but which are not antagonised by α-bungarotoxin.

5.4.1. Receptors on cell bodies

Cell bodies appear to have nicotinic receptors with fast kinetics and receptors with a mixed nicotinic and muscarinic pharmacology [unidentified *Locusta* neurons (Benson, 1993); identified cockroach motor neuron (David and Pitman, 1993)]. Activation of the nicotinic receptor at the normal resting potential produces a rapid depolarisation accompanied by a decrease in membrane resistance with an inward current carried by Na^+ that reverses at a projected potential of +20 mV. The receptors appear to operate a nonspecific anion channel and to desensitise. By contrast, activation of the mixed receptor (there may be several types, with a pharmacology that differs from the known vertebrate types) generates little or no current at normal resting potential, but an inward current that increases at more positive potentials in a voltage-dependent way. This current declines to zero at –80 to –90 mV. The inward current generated by these receptors may be caused by the closure of voltage-dependent Ca^{2+} channels and consequent reduction of a current mediated by Ca^{2+}-activated K^+ channels. The resulting depolarisation is slow to develop and to decline.

Isolated cell bodies of DUM neurons from the cockroach respond to muscarinic agonists with an initial hyperpolarisation followed by a slower depolarisation and the two components can be separated by different muscarinic antagonists, although it is uncertain that all these actions are mediated by muscarinic receptors (Lapied *et al.*, 1992). It is probable that the muscarinic receptors are not directly coupled to an ion channel but instead to a G protein and therefore affect ion channels either by the action of a G protein subunit on the ion channels, or through the action of second messengers such as cAMP or inositol phosphates.

The two types of receptors have different agonist and antagonist profiles, with a distinguishing feature being that the nicotinic but not the mixed receptor is blocked by α-bungarotoxin. The muscarinic receptors, however, cannot be fitted into any of the conventional vertebrate classes (Table 5.1). The pharmacological profiles of the receptors also differ between different species of insects.

5.4.2. Synaptic acetylcholine receptors

The presence of receptors for acetylcholine is most readily demonstrated on some of the large motor neurons. For example, the fast extensor tibiae (FETi) has cholinergic receptors that can be activated by ionophoresis of ACh, or its agonists nicotine and muscarine, into the region of neuropil containing its processes (Parker and Newland, 1995). The depolarisations evoked by nicotine are associated with a conductance change whereas those evoked by muscarine are not. This motor neuron receives a direct synaptic input from sensory neurons of particular campaniform sensilla on the tibia (for the action of these neurons *see Chapter 9*) and the excitatory potentials evoked by them can be blocked by the nicotinic antagonists mecamylamine and decamethonium bromide. The muscarinic antagonist scopolamine, by contrast, leads to an increase in the depolarisation evoked by the sensory neurons, suggesting it acts presynaptically to increase transmitter release. This evidence suggests that ACh is released from the sensory neurons and acts on nicotinic receptors on the motor neuron leading to an increase in the conductance of the membrane, but that the release is normally reduced by the presence of muscarinic receptors on the presynaptic terminals of the sensory neurons, or on interposed interneurons.

5.4.2.1. *Nicotinic receptors*

There are very limited data on whether the synaptic receptors have the same properties as those on the cell bodies, but two types of nicotinic receptors, which may possibly represent synaptic and extrasynaptic receptors, have been isolated from the central nervous system and reconstituted in lipid bilayers (Tareilus *et al.*, 1990).

5.4.2.2. *Muscarinic receptors*

Muscarinic receptors are now thought to be widespread, both on the terminals of neurons close to their output synapses and close to the input synapses on postsynaptic neurons. In general, they can alter synaptic transmission in two ways, by pre- and postsynaptic actions that probably involve different types of receptors.

First, they can reduce the release of ACh from mechanosensory neurons. Blocking these receptors increases the amplitude of an EPSP evoked by a mechanosensory neuron in postsynaptic neurons, presumably because the spikes now release more transmitter (Trimmer and Weeks, 1989). The receptors are apparently on the presynaptic terminals, where they can act as autoreceptors with a negative feedback action. The effect of these muscarinic receptors on the presynaptic terminals must be inhibitory.

Second, postsynaptic receptors can increase the excitability of neurons in addition to the excitation caused by the nicotinic receptors. For example, sensory neurons from hairs on the prolegs of *Manduca sexta* caterpillars evoke fast EPSPs in certain proleg motor neurons that appear to result from the activation of nicotinic receptors, and a slower depolarisation caused by the activation of muscarinic receptors on the motor neurons themselves (Trimmer and Weeks, 1993). The current underlying the slow depolarisation appears to be carried by Na^+ but is insensitive to tetrodotoxin, and its major effect is to reduce the spike threshold and thus increase the probability that spikes will be generated. If the mixed receptor present on cell bodies also occurs at synaptic membranes then its reduction of a K^+ current will increase the membrane resistance and hence increase the excitability of the postsynaptic neuron.

The release of acetylcholine from the sensory neurons can thus alter the excitability of postsynaptic neurons through parallel fast and slow pathways, and at the same time will regulate its own release through the presynaptic receptors. The overall excitability of the neurons postsynaptic to the sensory neurons will also be increased depending on the number of sensory neurons that are active; the more that are active, the greater will be the increase. This positive feedback form of gain control could counter the homosynaptic depression of EPSPs that occurs as the frequency of spikes in a sensory neuron increases. If such positive feedback effects operate when proprioceptive sensory neurons are active, they will work against the negative feedback control caused by presynaptic inhibition (*see Chapter 7*). It would also work against the negative feedback control of transmitter release by the presynaptic muscarinic receptors.

A combination of effects on these primary actions of the muscarinic receptors probably explains the potent action of the muscarinic agonist pilocarpine in releasing rhythmic motor patterns from an isolated metathoracic ganglion (Ryckebusch and Laurent, 1993) (*see Chapter 8*). The rhythmicity must result from the intrinsic membrane properties of particular neurons and from the way that they are connected. However, only rhythms that might represent walking and none that might represent the many other motor patterns that this ganglion is involved in producing have so far been elicited.

5.5. HISTAMINE

Histamine is distributed widely in the locust nervous system, where it can be both synthesised and inactivated (Elias and Evans, 1983). The greatest concentrations occur in the retina and in the lamina of the optic lobes where the terminals of the photoreceptors show histamine-like immunoreactivity (Nässel *et al.*, 1988). Histamine H_1-like receptors that, like some octopamine receptors, bind [3H]mianserin, are also abundant (Roeder, 1990; Roeder *et al.*, 1993).

5.5.1. Identity of the histaminergic neurons

In flies, there is strong evidence to indicate that histamine is the transmitter of retinula cells and that the receptor is coupled to a chloride channel in postsynaptic lamina monopolar cells (Hardie, 1987, 1989; Sarthy, 1991). In locusts, there is evidence that it is also the transmitter used by photoreceptors in the ocelli (Simmons and Hardie, 1988).

There is little information about the neurons in the rest of the locust nervous system that might contain histamine. In the brain of the moth *Manduca sexta*, however, there are 10 pairs of neurons with histamine-like immunoreactivity, and there is one pair in the suboesophageal ganglion (Homberg and Hildebrand, 1991). The morphology of some of the neurons in the brain is known (Fig. 5.2). In the brain of *Drosophila* there is a similar number of neurons to those in *Manduca* (Pollack and Hofbauer, 1991; Nässel and Elekes, 1992). In the thoracic and abdominal neuromeres of flies there are nine pairs of segmentally arranged neurons with histamine-like immunoreactivity that have processes both in the neuropil and in the neural sheath of the ganglion where they could possibly act as neurohaemal release sites (Nässel *et al.*, 1990). The brain neurons have extensive branches, often arranged bilaterally in the same regions of neuropil, but none extend into the mushroom bodies. This anatomy gives few clues about the role that histamine might play, but the superficial similarities to some neurons containing 5-HT suggest an action akin to the modulatory one proposed for those neurons.

In the prothoracic ganglion of crickets are a pair of local spiking interneurons called the omega (ON1) neurons, each of which receives direct inputs from sensory neurons of one ear and makes inhibitory output connections with its contralateral partner (Selverston *et al.*, 1985) and with some ascending intersegmental interneurons that carry auditory signals. The effectiveness of the reciprocal inhibitory connection between the two interneurons is reduced by pyrilamine, a histamine antagonist, and is taken as evidence for the histaminergic nature of these interneurons (Skiebe *et al.*, 1990). However, these neurons also stain with an antibody raised against 5-HT, so there must remain some doubt about the transmitter that mediates their inhibitory action.

In *Drosophila* but not some other flies, the sensory neurons of the mechanosensory hairs show histamine-like immunoreactivity that is absent in mutants unable to synthesise histamine (Buchner *et al.*, 1993). In locusts, the available evidence still points to mechanosensory neurons using acetylcholine as their transmitter.

In other arthropods, histamine is now implicated as the transmitter used by barnacle photoreceptors (Callaway and Stuart, 1989), by particular neurons in the stomatogastric nervous system of lobsters (Claiborne and Selverston, 1984), in the olfactory pathway of lobsters (Orona and Ache, 1992), and by some neurons mediating presynaptic inhibition of mechanosensory neurons in crayfish (El Manira and Clarac, 1994). This diversity of action indicates that a re-examination of its role in locusts would be rewarding.

5.5.2. Receptors for histamine

Photoreceptors in both the compound eye and ocelli depolarise in response to increases in light intensity and hyperpolarise their postsynaptic neurons in a chloride-dependent manner by the graded release of transmitter. Ionophoresis of histamine onto fly lamina monopolar cells, or L neurons of the locust ocelli, causes a rapid hyperpolarisation that mimics the normal response to light in its kinetics, its

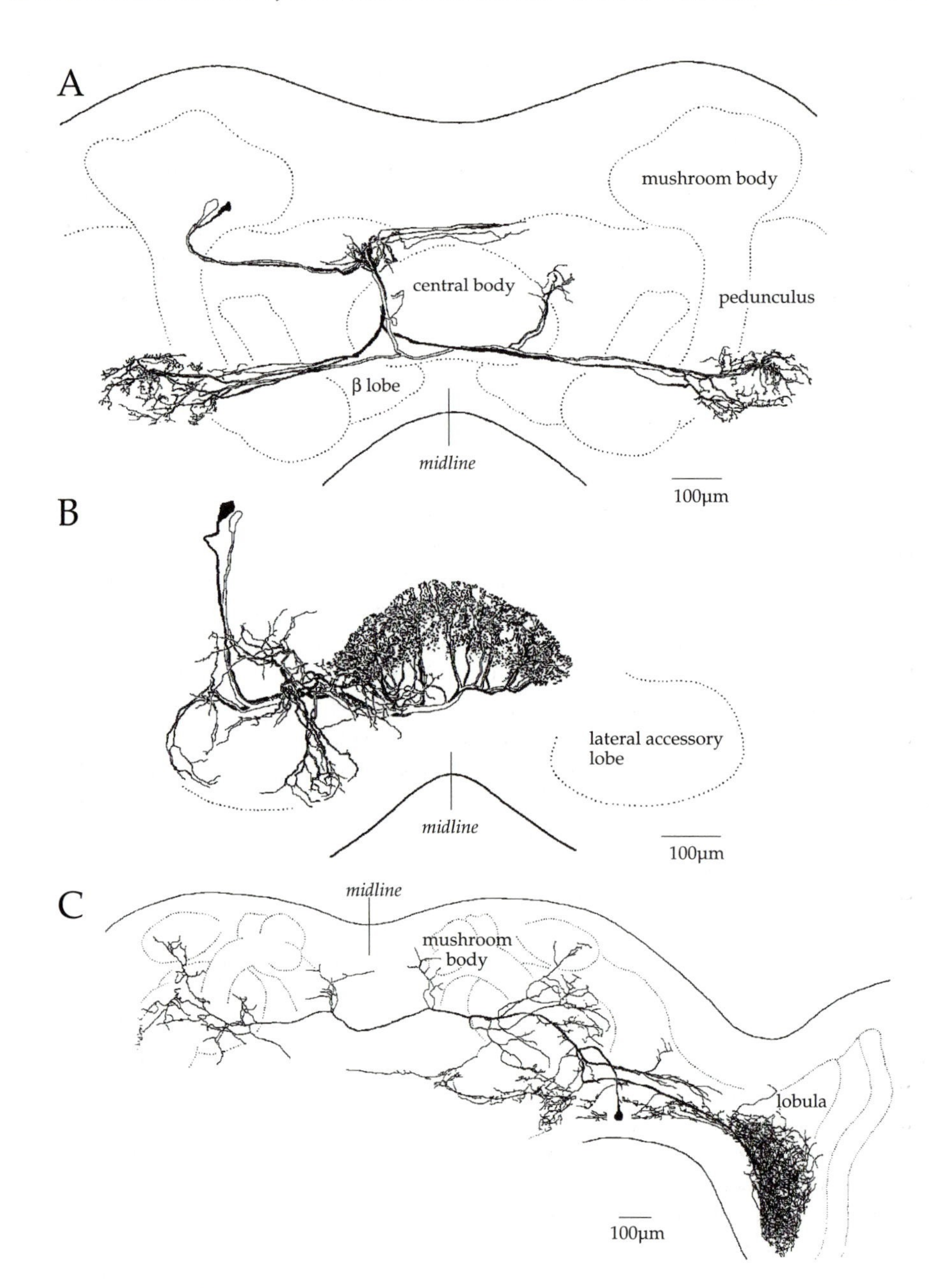

Fig. 5.2 Morphology of neurons in the brain of the moth *Manduca sexta* that show histamine-like immunoreactivity. A. Reconstructions of two neurons with branches in the protocerebrum and bilateral branches lateral to the calyces of the mushroom bodies. B. Two neurons with branches in the lateral accessory lobe of the protocerebrum and more extensive branches in the central body. C. A neuron with a cell body near the optic stalk, extensive branches in the lobula of an optic lobe and branches in the mushroom bodies on both sides of the brain. Based on Homberg and Hildebrand (1991).

dependence on chloride and its reversal potential (Hardie, 1989). The chloride channels that are coupled to this histamine receptor have a conductance of about 60 pS and an open time at physiological membrane potentials of about 0.5 ms. The response of these neurons to light is blocked by the same array of antagonists with the same potencies as those that block the response to histamine, with benzoquinonium the most effective (Hardie, 1988). The receptor is different from any of the vertebrate histamine receptors that act through second messengers, and is blocked more effectively by certain nicotinic cholinergic ligands than by histaminergic ones.

5.6. NITRIC OXIDE

The possibility that nitric oxide (NO) may be used as an intercellular signalling molecule seems to be turning from an unexpected possibility to an almost certain reality. Being such a small and reactive molecule, it readily diffuses through the plasma membrane to adjacent neurons where, amongst other actions, it may stimulate soluble guanylate cyclase, so that its effects may be mediated through the cGMP pathways. Its presence is detected by identifying the enzyme nitric oxide synthase (NOS) that is activated by Ca^{2+}/calmodulin to generate NO from L-arginine, or the enzyme nicotinamide adenine dinucleotide phosphate diaphorase (NADPHd) which is probably identical with NOS. Biochemical assays can demonstrate the presence of NOS in the nervous system (Elphick *et al.*, 1993) and histochemical staining of NADPHd can localise the cells that are likely to be releasing NO (Müller and Bicker, 1994). Moreover, isolated neurons extracted from regions of the brain rich in neurons that stain for NADPHd show a Ca^{2+}-dependent release of NO effected by the application of acetylcholine.

In the brain, NADPHd staining is restricted to a group of about 45 cell bodies in the dorsal part of each antennal lobe and a few scattered cell bodies in the tritocerebrum and optic neuropils (Müller and Bicker, 1994; Elphick *et al.*, 1995). The neurons in the antennal lobe have extensive branches in the glomeruli but there is no staining in the tractus olfactorio globularis that leads to the protocerebrum or in the antennal nerve. The stained neurons are therefore local interneurons that may be part of the networks processing olfactory signals. The lobes of the mushroom bodies stain only lightly. In the thoracic ganglia, a pair of neuronal cell bodies at the anterior of each ganglion medial to the entry of the anterior connectives and approximately 20 other cell bodies also stain, but as yet none of these neurons can be equated with the many neurons that have been identified in these ganglia. In A7 of both males and females, a pair of neurons with extensive neuropilar branches and ascending axons in the contralateral connectives also stain.

In the brains of *Apis mellifera* and *Drosophila melanogaster* the staining is largely restricted to the antennal lobes, the central complex and the mushroom bodies, and is absent from the primary visual areas (Müller, 1994). In the antennal lobes, the central regions of the glomeruli and the chemosensory, but not the mechanosensory, axons are stained. Somata and the lower division of the central body are stained but the ellipsoid body is not. In the mushroom bodies, the cell bodies of the Kenyon cells do not stain but the lip region of neuropil that receives input from the olfactory pathway of the antennal lobe is heavily stained. The staining in these regions and the mutations that interfere with cAMP, and the expression in them of these gene products, has led to speculation that NO may be important as a signalling molecule in olfactory learning. Applying the same logic, and coupling it with the lack of staining in the mushroom bodies of locusts, would lead to the conclusion that locusts are relatively poor at olfactory learning or use a different mechanism for signalling.

Neuromodulators and neurohormones

5.7. DECIDING WHETHER A SUBSTANCE IS A NEUROMODULATOR

An initial problem is to decide what substances are likely to be neuromodulators. Substances that have emerged as neuromodulators have frequently been recognised first as biologically active compounds in other contexts before their presence and effects in the nervous system have been established. Only then has the more important question been addressed of the identity of neurons that contain these substances and their action during normal behaviour. Nevertheless, our understanding is made difficult because demonstrating a natural role for neuromodulators is quite different to producing an effect by the application of high concentrations to many neurons simultaneously. The analyses are usually led by the prevailing technology of the time. Initial studies of amines, for example, demonstrated their presence by fluorescent methods on homogenates of the nervous system. The advent of the Falck–Hillarp histochemical method in the 1960s allowed the localisation of some amines to particular regions of the central nervous system. The technology of the moment allows the use of antibodies raised against particular molecules or parts of molecules and this has revealed new classes of neuroactive compounds by demonstrating their presence, or that of their synthetic enzymes, directly in specific neurons. As with all techniques, there are limitations to what can be learnt. For the Falck–Hillarp fluorescence method, the distinction between different classes of amines was uncertain, whereas with immunocytochemistry, confusion may arise when dealing with families of peptides if the specificity of the antibody is unrecognised. The introduction of each new technique thus provides additional information but some of this is subsequently seen to be inaccurate in the light of still further advances. At best, the current methods which combine the use of antibodies and sensitive assay methods allow some substances to be localised to individual and often identified neurons, and the amount present to be measured.

This implies that the current balance of our information may not reflect the real importance of particular substances, or particular sets of neurons, but merely our ability to identify and study some of them. For example, much more is known about the actions of octopamine than tyramine (the precursor of octopamine, but which may play a role in its own right) in the locust, even though there is ten times as much tyramine as octopamine in the thoracic nervous system (Table 5.2). Measurements of the concentrations of particular substances do not, however, give an indication of their likely functional role as many other factors must also be taken into account. A description of the current knowledge may thus give an unbalanced picture of the way the neural circuits are likely to be modulated and emphasises how much more we need to know. The surface has only been scratched as far as peptides or small molecules such as nitric oxide (NO) are concerned.

To establish a substance as a neuromodulator requires that the same rigid set of criteria that are set down for transmitters are met, but, as for most transmitters, it is usually only feasible to fulfil a subset of these criteria. Octopamine is a good example where these criteria have been met in the locust, at least for its action on muscles. It is usually possible to show the presence of these substances; harder to show the presence of specific receptors,when most receptors have an affinity for a range of substances; and harder still to show that they are released during the normal activity of the nervous system. But even if a satisfying number of the requisite criteria are met, it is still a difficult step to establish that an experimentally observed effect following the application of a substance is the explanation of a similar phenomenon during normal behaviour.

5.8. AMINES AS MODULATORS

Neurons containing various biogenic amines are widely distributed throughout the nervous system (Table 5.2) and have a variety of actions. Curiously, we know very little about tyramine, which is the most abundant, but considerably more about the distribution and action of octopamine, 5-HT and dopamine, in large part because the neurons containing these substances have been located and, at least for the octopaminergic ones, can be relatively easily studied.

Table 5.2. Biogenic amines in the thoracic nervous system

Amine	Amount in thoracic nervous system (ng)
Catecholamines	
dopamine	6
noradrenaline	0.15–0.2
Indolalkylamines	
5-HT	1.3–1.5
Phenolamines	
p-tyramine	28–32
p-octopamine	2–3

Data from Macfarlane *et al.* (1990)

5.9. DOPAMINE

In the protocerebrum of the **brain**, about 110 pairs of neurons show dopamine-like immunoreactivity, with cell bodies in eight groups, many of which have branches in both halves (Fig. 5.3) (Wendt and Homberg, 1992). The most intense immunoreactivity occurs in the central body from neurons with cell bodies below the calyces. The pedunculus of a mushroom body is supplied by neurons with cell bodies close to the mechanosensory area of the deutocerebrum, and the lobes by two groups of neurons from the median protocerebrum. Other neurons have branches in the protocerebrum and in the optic tubercles. Some of the neurons with cell bodies in the pars intercerebralis and axons in NCCI to the storage lobes of the corpora cardiaca may be dopaminergic. Neurons with similar projections may also contain 5-HT and octopamine (*see below, sections 5.10 and 5.12*). In an optic lobe, there are about 3000 immunoreactive neurons, or about one neuron for each ommatidium, with cell bodies around the chiasma and branches in all layers of the medulla. The vast majority are thus intrinsic and columnar neurons of the medulla, but a small number are tangential neurons linking the medulla to the posterior edge of the lamina. Staining is absent in the neuropils of the antennal lobes, the calyces of the mushroom bodies and large parts of the lobula and lateral protocerebrum. In the brain of the bee (Schäfer and Rehder, 1989) and of flies (Nässel and Elekes, 1992) there are similar numbers of dopamine-containing neurons, but in bees there is no staining in the optic lobes.

The SN1 neurons that supply the salivary glands from the mandibular neuromere of the **suboesophageal ganglion** contain dopamine (Ali *et al.*, 1993). Stimulation of the nerve containing these axons gives a hyperpolarisation of the acinar cells with a latency of 1 s, and dopamine also causes a hyperpolarisation mediated by specific dopamine receptors.

In the **prothoracic ganglion**, three pairs of neurons show dopamine-like immunoreactivity, whereas the mesothoracic ganglion and abdominal ganglia A5-7 and the eighth neuromere of the last abdominal ganglion each have only one pair. The metathoracic ganglion and abdominal ganglia A1–4 and neuromeres 9–11 of the last abdominal ganglion have none (Watson, 1992b). About 7–10 immunoreactive axons enter each ganglion through each anterior connective and there are numerous immunoreactive profiles in the neuropil, indicating that the few dopaminergic neurons have widespread, possibly neuromodulatory, effects.

5.9.1. Receptors for dopamine

Dopamine depolarises the cell body of the prothoracic common inhibitory motor neuron (CI1) of the cockroach accompanied by a decrease in membrane conductance and with the current markedly voltage dependent (Davis and Pitman, 1991). At resting potentials the inward current is very small, but increases dramatically between –120 and –150 mV, and again between –20 to –10 mV, before reversing at 0 mV, so that there is a region of negative slope conductance in the relationship between membrane potential and the current induced by dopamine. The current is

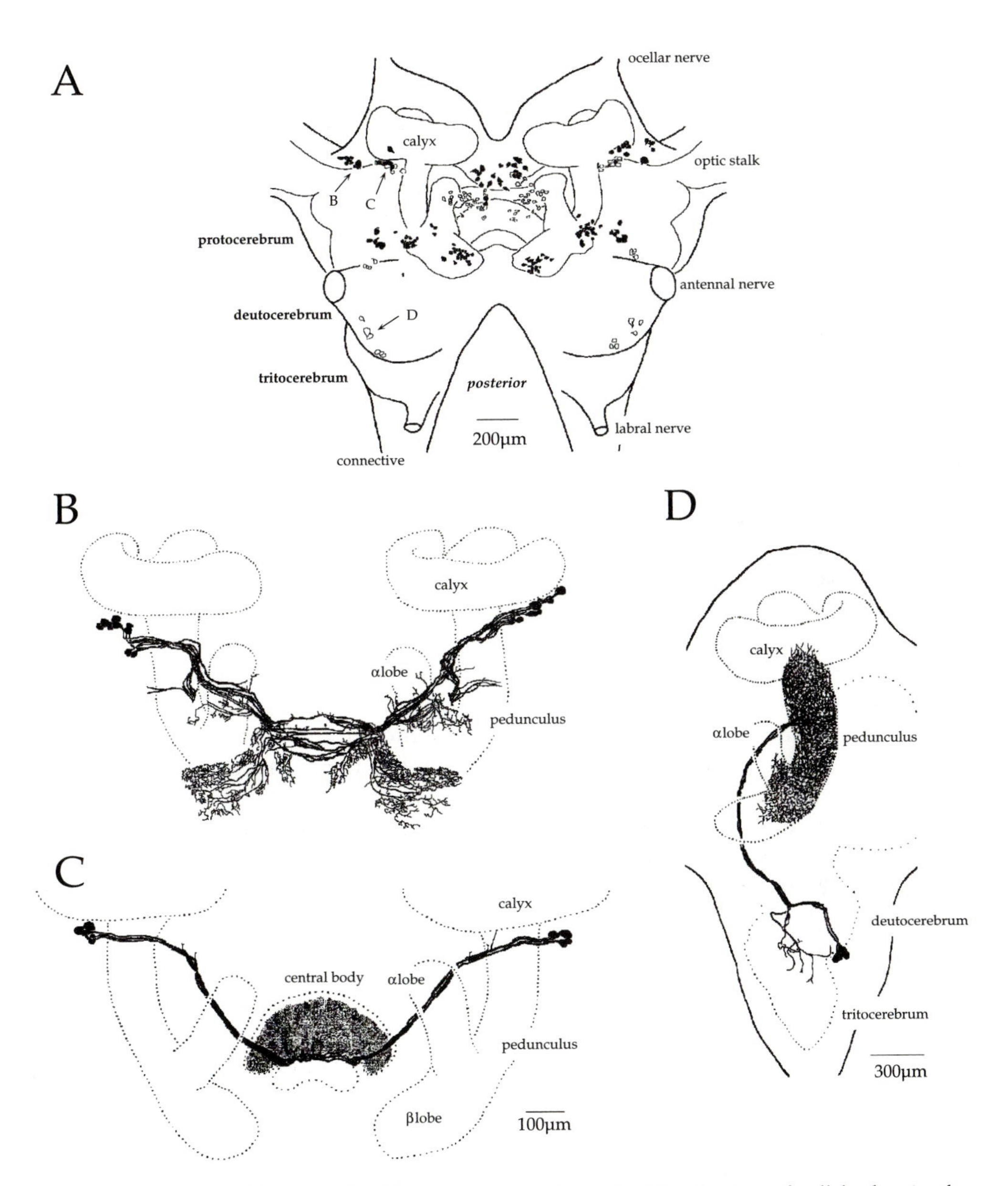

Fig. 5.3 Neurons with dopamine-like immunoreactivity. A. Distribution of cell bodies in the brain. B-D. Reconstructions of the morphology of three groups of neurons in the brain. Nothing is known of the physiology of these neurons. Based on Wendt and Homberg (1992).

abolished in the absence of external Ca^{2+} or in the presence of Cd^{2+} and so may be carried by Ca^{2+}, but it is also dependent on the external concentration of Na^{+}, K^{+} and perhaps Cl^{-}. The response is mimicked by apomorphine and ergotamine and blocked by flupenthixol, which has no effect on the octopamine receptor, so that the pharmacological profile of the receptor is thus different from that of known vertebrate dopamine receptors. In *Drosophila*, two types of receptor have been cloned which have D_1- and D_2-like properties. There are no comparable data on possible synaptic receptors, apart from the demonstration that dopamine-sensitive adenylate cyclases are present. If such receptors exist at synapses, they would be activated only when the membrane was depolarised by other sources and might therefore influence the production of plateau potentials in response to brief synaptic inputs, or the ability of a neuron to generate bursts of spikes. Such effects, however, have yet to be demonstrated.

5.10. 5-HT (SEROTONIN)

Only a small number of neurons in the central nervous system contain 5-HT but these may have extensive branches characterised by numerous varicosities. In flies, it is estimated that there are only 154 neurons that contain 5-HT, and the number in locusts is also likely to be small (Fig. 5.4). Most of the neurons that show 5-HT-like immunoreactivity are interneurons, but a few might be sensory neurons and an even smaller number efferent neurons (Tyrer *et al.*, 1984). The terminals of these neurons, at least as seen with the pre-embedding methods used, contain large granular vesicles about 100 nm in diameter and smaller agranular vesicles about 60 nm in diameter, both of which show immunoreactivity (Nässel and Elekes, 1984), so that it is unclear in what form the 5-HT is packaged. There is similar uncertainty over the mechanism of release. The terminals of 5-HT neurons in the optic lobes of flies do not appear to form conventional synaptic structures with other neurons, but instead appear to release nonsynaptically. Nevertheless, conventional diadic synaptic structures occur in the equivalent neurons of other species. 5-HT-immunoreactive terminals on the muscles of the gut lie in close contact with the sarcolemma of the muscle fibres for up to 1 μm, but show no synaptic specialisations, although they contain both the large and small vesicles characteristic of contacts in the central nervous system. These observations suggest that 5-HT may act either as a neuromodulator or as a conventional transmitter.

5.10.1. Interneurons

The characteristic feature of the few interneurons that contain 5-HT is that each supplies a large area of neuropil. The mushroom bodies of the **brain** are innervated by four extrinsic groups of 5-HT-immunoreactive neurons (Klemm and Sundler, 1983). One group, with cell bodies near an optic lobe, innervates the anterior and apical region of the calyx. A second group of 10–12 postero-lateral cell bodies innervates the peduncle and calyx. A third group, with cell bodies in the ventral protocerebrum, innervates the α and β lobes. A fourth group, the cell bodies of which have not been located, innervate the peduncle and an antennal lobe. The Kenyon cells themselves do not appear to contain 5-HT.

In the central body, about 60 columnar neurons with cell bodies in the pars intercerebralis have branches in the protocerebral bridge, layer III of the upper division, and the noduli (Homberg, 1991). A group of 15–20 bilateral neurons connects the optic tubercle with the protocerebral bridge, and these have widespread branches in particular layers of the upper division and cell bodies in the protocerebrum. A further 10 pairs of neurons have cell bodies between the β lobe of a mushroom body and an antennal lobe and branches in layer I of the upper division. Some large field neurons also branch in layer I and the lateral accessory lobes of the protocerebrum. The lower division of the central body has no neurons with 5-HT-like immunoreactivity.

In the protocerebrum, six pairs of neurons with cell bodies 40–60 μm in diameter have axons that descend in DMT of a connective, at least as far as the metathoracic ganglion, forming branches in each of the intervening ganglia [the PI(2) neurons of Williams (1975)]. One interneuron from the protocerebrum also sends branches into an antennal lobe, while others send branches into the central body.

In an antennal lobe of the deutocerebrum some 5–8 intrinsic neurons send branches to ramify among the glomeruli, but there are no stained afferents entering a lobe, or projection neurons leaving it. In the deutocerebrum of the bee most, if not all, of the 5-HT immunoreactivity comes from a pair of large neurons, the deutocerebral giants (DCG), that have extensive branches in both the olfactory and mechanosensory parts, cell bodies located laterally and axons that project to the suboeosphageal and more posterior ganglia (Rehder *et al.*, 1987). Similarly, in the moth *Manduca sexta*, each lobe is supplied by just one neuron containing 5-HT (Sun *et al.*, 1993). The cell body of this neuron is in one lobe and its terminals are in the centre of the glomeruli of the opposite lobe where they make mostly output synapses but also receive some input synapses. They may therefore modulate the action of the interneurons rather than the cortically arranged terminals of the sensory neurons.

The stained processes in the optic lobes originate from separate groups of large and small cell bodies in the medulla and ramify through the columns of the medulla and lamina, with some corresponding to the wide field centrifugal neurons. In the flies *Calliphora erythrocephala* and *Drosophila melanogaster* there are two large neurons that supply the lamina on both sides of the brain, suggesting that they modulate visual processing in a symmetrical and widespread fashion. In addition, there are between 10 and 20 wide field amacrine neurons in parts of the medulla. This arrangement of neurons suggests a feedback or feedforward role for 5-HT in regulating visual processing.

The storage lobes of the corpora cardiaca in locusts are supplied by about 20 neurons with 5-HT-like immunoreactivity, axons in NCCI and cell bodies in the pars intercerebralis (Konings *et al.*, 1988) which may modulate the release of AKH hormones from the glandular lobe caused by neurons with axons in NCCII. Neurons with similar projections may also contain dopamine and octopamine (*see sections 5.9 and 5.12*).

In each of the **thoracic ganglia** there are two pairs of neurons with 20–40 μm diameter cell bodies lateral and posterior, and between 5–7 pairs of smaller cell bodies, all of which have axons that ascend in VIT to the next anterior ganglion where they form terminal arrays of branches. 5-HT-immunoreactive neurons also occur in the abdominal ganglia of embryos, and in adults each neuromere of A1-3 and A7-10 (and probably the intervening ganglia) contains two pairs of such neurons. Some of these neurons may be the progeny of neuroblast 7-3 which forms three sibling neurons S1-3 (Taghert and Goodman, 1984). In embryos, S1 shows 5-HT-like immunoreactivity in all thoracic and abdominal ganglia; S2 in all except T3 and A1; and S3 only in the prothoracic ganglion.

5.10.2. Sensory neurons

Many of the nerves of the segmental ganglia are reported to contain small bundles of axons of sensory neurons that show 5-HT immunoreactivity (Lutz and Tyrer, 1988). In N1 of a thoracic ganglion these axons belong to sensory neurons from the chordotonal organ at the wing hinge (*see Chapter 11*), and in N5 from some of the neurons of the femoral chordotonal organ (*see Chapter 8*). However, more specific antibodies fail to stain these neurons, suggesting that the original descriptions may have resulted from cross-reactivity of the antibody. The conclusion is that 5-HT is probably not present in sensory neurons.

5.10.3. Efferent neurons

A pair of neurons (salivary neurons, SN2) with cell bodies in the labial neuromeres of the suboesophageal ganglion contains 5-HT and innervates the salivary glands (Tyrer *et al.*, 1984). These neurons also show immunoreactivity to GABA (Watkins and Burrows, 1989). A second pair of neurons (SN1) with cell bodies in the mandibular neuromere that contains dopamine (Ali *et al.*, 1993) also innervates the salivary glands. The axons of both pairs of neurons run in the same nerve (N7B), but the axon of an SN1 neuron is contralateral to its cell body whereas that of an SN2 neuron is ipsilateral. Both the SN1 and SN2 neurons supply the acinar cells of the salivary gland, whereas the meshwork of terminals of the satellite neurons (*see section 5.10.4*) is restricted to the surface of the salivary nerve as it passes along the salivary duct and its branches. These different patterns of supply may indicate a different role in regulating the secretion of saliva during feeding. It is assumed that SN1 and SN2 primarily control salivation by releasing dopamine and 5-HT, respectively. Their role may be further subdivided, with the SN1 neurons controlling the parietal cells that are mostly responsible for the water and ion content of the saliva, and the SN2 cells controlling the central cells that probably produce and release the enzymes in the saliva. The satellite neurons may regulate the reabsorption activity of the duct cells to which they are closely apposed. Confirmation of these proposals will hinge on the distribution of the different amine receptors in the salivary gland. The role of GABA and whether it is coreleased with 5-HT by the SN2 neurons remains unknown, as is the role of the octopamine and FMRFamide-like peptides released by neurons from the thoracic ganglion that also supply the salivary glands.

Several of the muscles in the head also receive terminals of 5-HT-containing neurons. Some efferent DUM neurons in the fused thoracic and abdominal ganglion of *Rhodnius prolixus* may contain 5-HT (Orchard *et al.*, 1989). Neurons in the last abdominal ganglion of the cricket supply the genitalia and the hind gut (Hustert and Topel, 1986) and the Malpighian tubules.

5.10.4. Neurons of the satellite nervous system

Seven pairs of neurons in the suboesophageal ganglion which have 5-HT-like immunoreactivity are the sole contributors to a complex peripheral plexus (Fig. 5.4) (Bräunig, 1987). They have 35–40 μm diameter cell bodies at the anterior of the ganglion medial to the emergence of the connectives to the brain, many fine neurites in the dorsal neuropil of the mandibular neuromere, and axons in N1, the mandibular nerve. Three pairs of these neurons contribute to a fine nerve that emerges from N1 and then divides, so that each branch follows closely the paths of all the other suboesophageal nerves with the exception of N3, N8, the single median nerve, and the tritocerebral nerves to the labrum. Along each nerve they form a dense meshwork of varicose terminals, and even in N1 short branches repeatedly emerge from the axons that are deep within it to form varicose endings on the surface. The other four pairs of neurons also have axons in N1, but their meshwork of terminals is limited to the branches of N1 that supply the mandibular muscles. The spikes in these neurons are generally longer than 5 ms and are conducted away from the suboesophageal ganglion. The function of these neurons is unknown, but it must be presumed that they release 5-HT locally around the structures innervated by the suboesophageal ganglion, perhaps during feeding when they normally spike (Schachtner and Bräunig, 1995). Neurons that form branches around N7B and supply the salivary gland, thus provide a further 5-HT supply in addition to the SN2 neurons. The advantage of this locally distributed design for neurohaemal release, as opposed to the concentration of terminals in neurohaemal structures such as the corpora cardiaca or the median nerve perivisceral organs, is that release can be precisely targetted without loading the whole of the haemolymph. Moreover, it overcomes the problem that the released substance might be degraded before it could even reach distant target sites. In this way, the innervation of the salivary glands and muscles moving the mouthparts can be modulated specifically during feeding. These neurons and the SN1, SN2 to the salivary glands generally spike only during feeding (Schachtner and Bräunig, 1993). The spikes begin a few seconds before the mandibular closer muscles begin to contract, continue throughout the period of feeding, and spikes in SN1 alone may persist for a few seconds after the muscles have stopped contracting. The effect will be to modulate the contractions of the mandibular muscles, the secretions of the salivary glands, and possibly the performance of the chemoreceptors on the mouthparts. The salivary glands are also innervated by DUM neurons (DUM1b) (Bräunig *et al.*, 1994), but their action during feeding is not known.

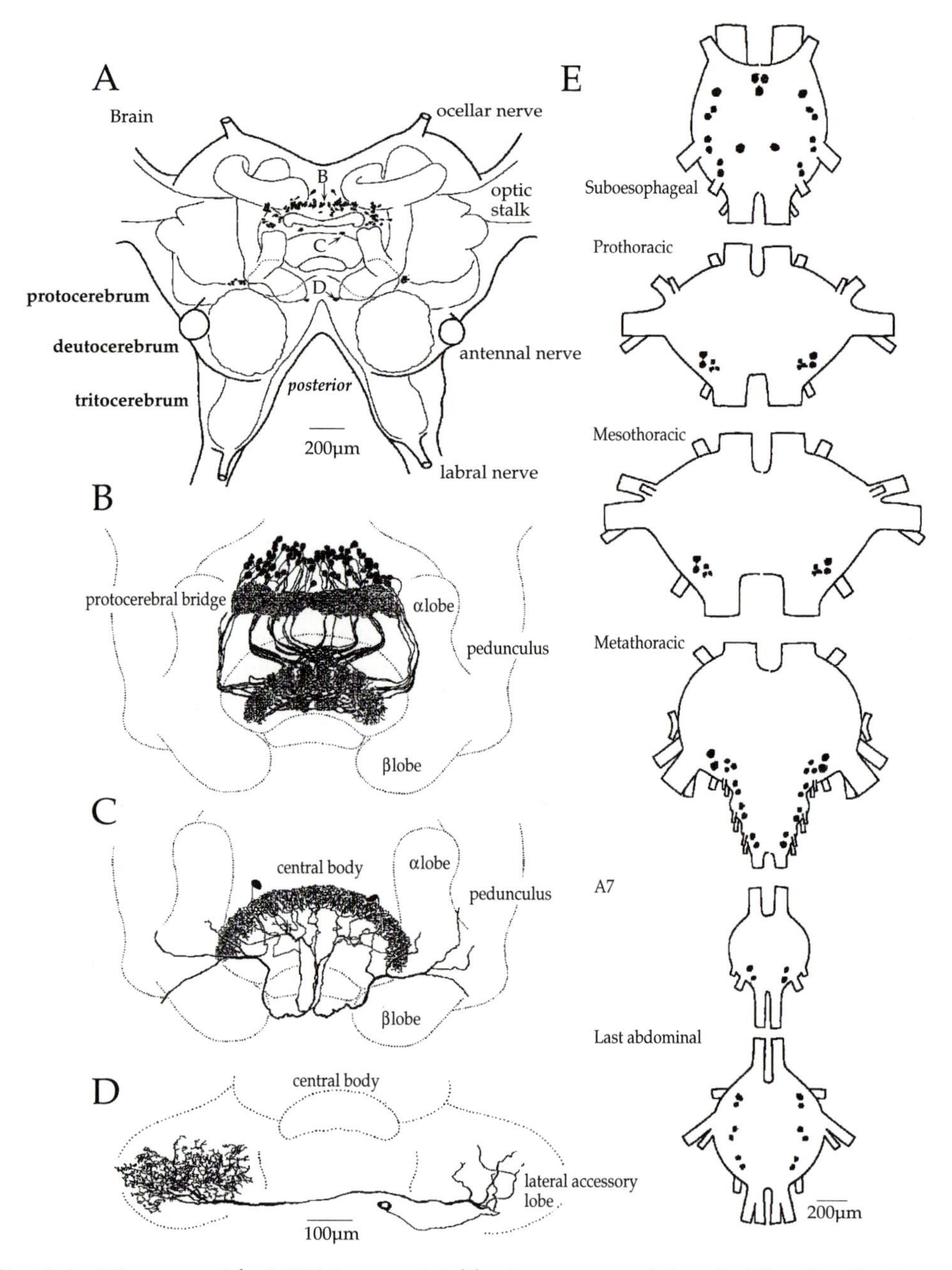

Fig. 5.4 Neurons with 5-HT (serotonin)-like immunoreactivity. A. The distribution of neurons in the brain. B. Reconstruction of the morphology of a group of neurons with branches in the protocerebral bridge and central body. C. A pair of neurons with branches in the central body. D. A neuron with branches in the lateral accessory lobes on both sides of the brain. E. Distribution of cell bodies in the suboesophageal, thoracic and abdominal ganglia. A-D are based on Homberg (1991), E on Tyrer *et al.* (1984).

5.10.5. Neurons in the stomatogastric nervous system

About 80 of the 200 neurons in the frontal ganglion show 5-HT-like immunoreactivity and have axons in the frontal nerve and in the recurrent nerve to the hypocerebral (occipital) ganglion, where they form branches before forming a plexus along the pharynx (Klemm *et al.*, 1986). Many immunoreactive fibres form a dense meshwork on the nerves and ganglia of the entire stomatogastric nervous system. The hypocerebral ganglion contains some 2–4 stained cell bodies. The extrinsic muscles of the fore gut are supplied by 5-HT-immunoreactive fibres and the midgut has a plexus of fibres on its outer longitudinal layer of muscles.

5.10.6. Receptors for 5-HT

There are at least three types of receptor for 5-HT, associated with the cell bodies of neurons (Table 5.1) that differ in their pharmacology both from one another and from the 5-HT_{1-3} receptors of vertebrates (Bermudez *et al.*, 1992). Activation of the first evokes an inward current that is maximal at –60 to –70 mV and results from a decrease in K^+ conductance. The current is blocked by Cs^+ and Rb^+, only partially blocked by 4-AP and TEA and not blocked by Mn^{2+}. Activation of the second produces an inward current from –30 to –80 mV that is carried by Na^+ and which is abolished in Na^+-free saline, but is unaffected by Cs^+ and Mn^{2+}. Activation of the third evokes a fast current at potentials more positive than –50 mV that is probably carried by K^+ as it is blocked by 4-AP.

There is almost nothing known about the distribution of 5-HT receptors and the actions they might mediate in the neuropils of the central nervous system, against which to compare these responses. 5-HT-sensitive adenylate cyclases occur in the central nervous system, and a binding site representing a G protein-coupled receptor has the overall features of a 5-HT receptor (Wedemeyer *et al.*, 1992). Three different 5-HT receptors have been cloned from *Drosophila* that have significant homologies with some of the receptors in vertebrates (Saudou *et al.*, 1992).

5.11. TYRAMINE

The presence of tyramine in the central nervous system may reflect its role as a precursor of octopamine, but its distribution does not parallel that of octopamine so that it may be more than simply a metabolic intermediary (Downer *et al.*, 1993). Moreover, tyramine can be released in a Ca^{2+}-dependent manner within the brain, and there are specific uptake mechanisms and perhaps even specific receptors for it. Together, this evidence indicates that tyramine may be a neurotransmitter or neuromodulator in its own right, and that the neurons containing it should be identified and their actions analysed.

5.12. OCTOPAMINE

Octopamine has many actions in controlling behaviour, and may act in the following ways.

1. As a **neurotransmitter.** It controls the light emitted from the firefly light organ. It may also act as a neurotransmitter when released from the paired octopaminergic interneurons in the brain and segmental ganglia.
2. As a peripheral **neuromodulator** acting on skeletal and visceral muscle, on the terminals of motor neurons, and on sensory neurons.
3. As a central **neuromodulator** changing the performance of motor patterns, influencing the integration of sensory signals, and altering sensitisation, habituation and memory.
4. As a **neurohormone** mobilising lipids and carbohydrates.
5. It may also alter the action of the haemocytes so that they isolate foreign material in nodules and is present in high concentrations when the formation of these nodules occurs in response to bacterial infection (Dunphy and Downer, 1994). Octopamine may also bind to certain invading bacteria, so facilitating their removal from the haemolymph by the haemocytes.

5.12.1. Paired octopaminergic neurons

Only a small number of paired neurons that have cell bodies scattered in different parts of the nervous system show octopamine-like immunoreactivity (Fig. 5.5). From what little is known of their morphology some, at least, may be interneurons with presumptive release sites in the central nervous system.

The **brain** has a number of paired neurons that contain octopamine. In the tritocerebrum is a pair of neurons with large diameter cell bodies, the axons of which enter the tritocerebral commissures and then loop back to the contralateral side of the brain. Their targets and action are unknown. At the border of the trito- and deutocerebrum is a group of 6–8 smaller cell bodies that have octopamine-like immunoreactivity. The tritocerebral giant neuron (TCG) (*see Chapter 11*) also has its cell body in this cluster but does not itself show octopamine-like immunoreactivity. The rest of these neurons have extensive arborisations arranged in layers throughout the lobula and medulla, but not the lamina, of an optic lobe (Fig. 5.6). One pair has branches only in the ipsilateral optic lobe, and another only contralateral branches. In keeping with this extensive supply by octopaminergic neurons, the density of octopamine receptors here is about three times higher than elsewhere in the nervous system (Roeder and Nathanson, 1993). This reflects this distribution of only one receptor subtype (*see section 5.12.3*), and because there may be other subtypes a different distribution may result if they were considered. In the lateral protocerebrum there is one large pair of cell bodies and a number of smaller ones. The pars intercerebralis also contains neurons with octopamine-like immunoreactivity that have axons in NCCI that are said to modify secretions from the corpora cardiaca, and there are also many scattered octopaminergic neurons in the medulla. Neurons

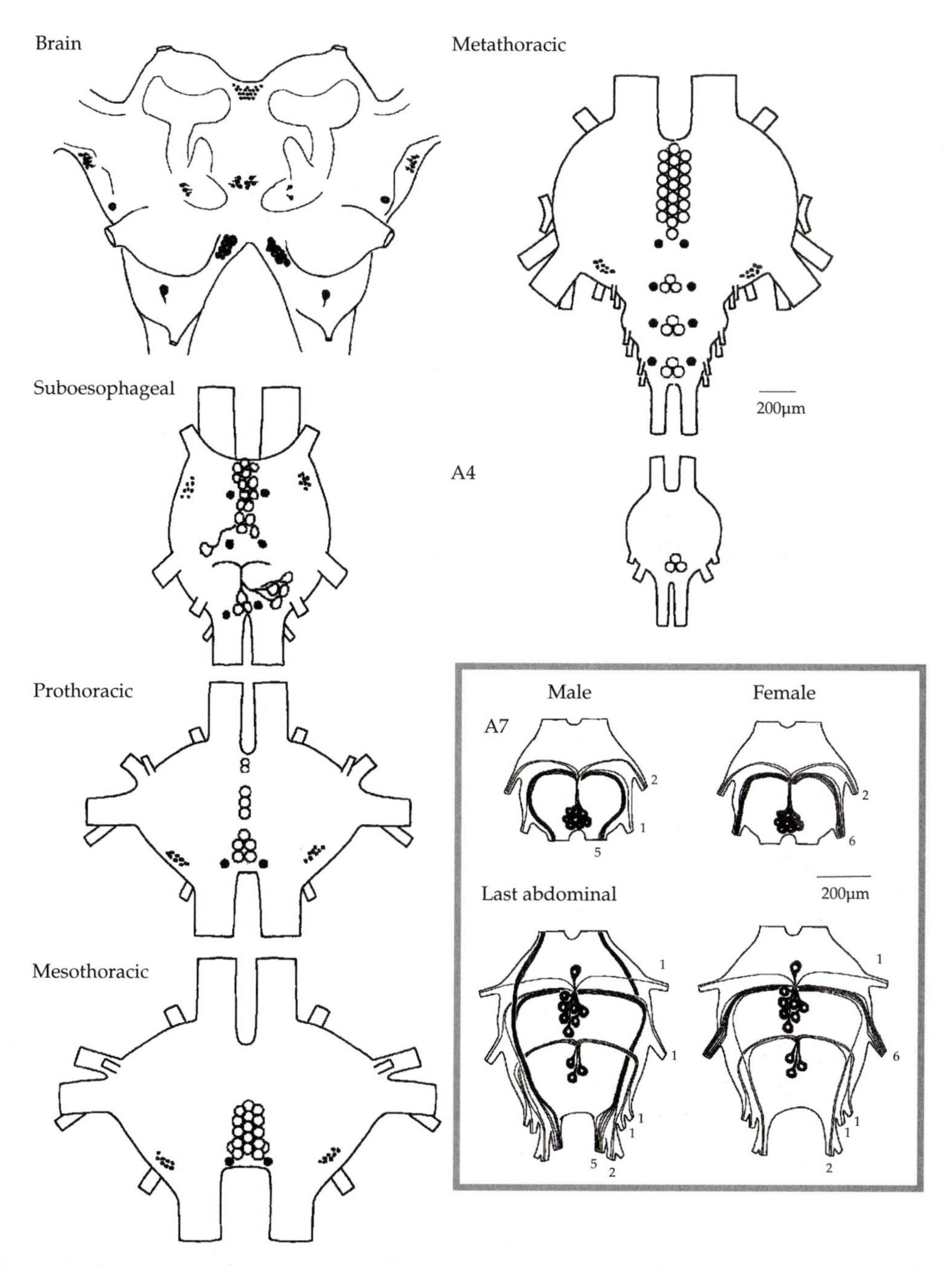

Fig. 5.5 Neurons with octopamine-like immunoreactivity. In the brain, suboesophageal, thoracic and A4 the paired neurons are shown in black and the dorsal unpaired median (DUM) as open circles. In A7 and the last abdominal ganglia only the DUM neurons are drawn, to show their different arrangement in males and females. Based on data courtesy of P.A. Stevenson and Stevenson *et al.* (1994).

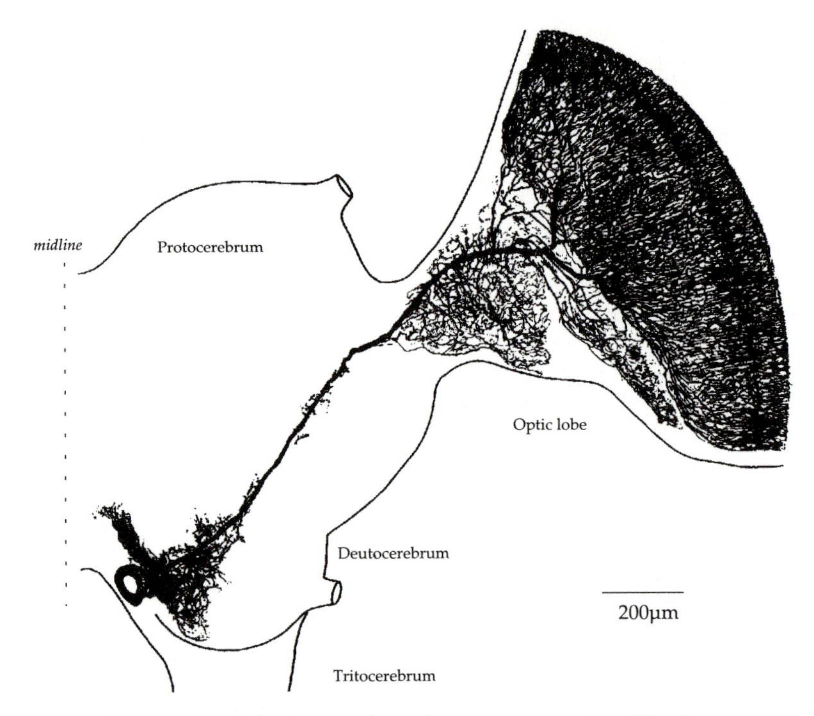

Fig. 5.6 Paired neuron in the brain that shows octopamine-like immunoreactivity. It has extensive branches in the lobula of one optic lobe. Based on data courtesy of P.A. Stevenson.

with similar projections may also contain dopamine and 5-HT (*see sections 5.9 and 5.10*).

In the **suboesophageal ganglion** there are four types of neurons that show octopamine-like immunoreactivity in addition to those that have axons that emerge in nerves from the brain (*see Chapter 3*). Most of these are probably interneurons, but with actions that have yet to be elucidated.

1. Five interneurons send axons to the brain, but do not have axons that leave the brain. One of these octopaminergic neurons (called VUMmx1), with its cell body in the mandibular neuromere of the suboesophageal ganglion and projections to the brain, has received particular attention in the bee for its actions during associative olfactory learning (Hammer, 1993). It responds to the unconditioned stimulus signal (sucrose) and normally spikes before the proboscis is extended. Stimulation of this neuron can substitute for the sucrose reward in olfactory conditioning.
2. Neurons with cell bodies in the posterior part of the suboesophageal ganglion have axons that project in the cervical connectives to unknown destinations. These cell bodies are variable in position and are not always at the midline.

3. A single neuron has an axon in each of the anterior and posterior connectives and is thus similar in appearance to the H neuron in thoracic ganglia (*see Chapter 4 and section 5.24*).
4. Local interneurons with branches restricted to the suboesophageal ganglion (*see Fig 3.18C*).

In the **prothoracic ganglion** there are two anterior cell bodies that have axons in the posterior connectives, and in the other thoracic ganglia and abdominal ganglia A1-6 there are two more posterior and ventral cell bodies. These may be intersegmental interneurons with projections into the ventral neuropil that contains the terminals of many mechanosensory neurons.

Some of the neurons that contribute to the dense meshwork of varicose endings that surrounds sympathetic nerves, some other nerves and the connectives all show octopamine-like immunoreactivity (Spörhase-Eichmann *et al.*, 1992).

Thus, virtually the complete set of paired neurons that contain octopamine is known. The variety of different projections of their axons and their neuropilar branches suggests no single or common action on other neurons or effectors during behaviour. Therefore, it would seem likely that octopamine has a range of specific actions during the processing of particular sets of signals, or in altering the responses of specific effectors.

5.12.2. Unpaired octopaminergic neurons (DUM neurons)

In each of the segmental ganglia are a group of neurons with cell bodies at the dorsal midline and with axons that emerge in the lateral nerves to supply structures on both sides of the body. These are the efferent dorsal unpaired median (DUM) neurons. Their cell bodies are intermingled with unpaired intersegmental interneurons and unpaired local interneurons (*see sections 3.6.2.4 and 3.8.1.1*).

The initial indications that efferent DUM neurons contain octopamine (Hoyle, 1975) were confirmed when cell bodies of individual DUM neurons in the metathoracic ganglion were extracted and shown to contain 0.1 pmol of octopamine at a concentration of $9{\times}10^{-4}$ mol.l^{-1} (Evans and O'Shea, 1978). This has now been extended to other DUM neurons, in particular those in the seventh abdominal ganglion (A7) that supply the oviducts and which contain octopamine at a concentration of about $1{\times}10^{-3}$ mol.l^{-1} (Orchard and Lange, 1985). Finally, efferent DUM neurons throughout the nervous system show immunoreactivity to an antibody that appears specific for octopamine (Stevenson *et al.*, 1992).

Staining with other antibodies indicates that taurine may be co-localised with octopamine in many efferent DUM neurons (Nürnberger *et al.*, 1993), and an FMRFamide-like peptide may also be present in the DUMheart neuron in each of the A2–7 ganglia (Stevenson and Pflüger, 1994). It may well be, therefore, that all efferent DUM neurons are octopaminergic and that some, at least, of their peripheral effects are mediated by the release of octopamine.

5.12.2.1. *Distribution of efferent DUM neurons*

The number of efferent DUM neurons is constant from locust to locust in one species (Table 5.3) indicating that they have the potential to be treated as identified individuals, but despite their size and accessibility only a few have been identified. Much of what we know has to be extrapolated from detailed studies of one neuron backed by information from a few others in various insects. The largest number of efferent DUM neurons occurs in the meso- and metathoracic ganglia, which each have 18–20 efferent DUM neurons, compared to only 8–10 in the prothoracic ganglion (Fig. 5.5), suggesting an involvement in the control of flight. The argument in favour of a role in flight is supported by the pattern of muscles that they supply, but is weakened because in cockroaches and crickets there are only half this number. In the abdomen, the largest number occur in A7 and A8, suggesting a role in the control of the reproductive organs.

5.12.2.2. *Thoracic DUM neurons*

Each efferent DUM neuron in the meso- and metathoracic ganglia has a cell body of 30–80 μm in diameter and a single primary neurite that enters the neuropil in

Table 5.3. Number of neurons with octopamine-like immunoreactivity in the central nervous system

Region of nervous system	Number of octopaminergic DUM neurons	Number of other octopaminergic neurons
Brain	0	many
Suboesophageal ganglion		
mandibular neuromere, S1	7–8	2
maxillary neuromere, S2	7–8	2
labial neuromere, S3	7–8	2
Prothoracic ganglion, T1	8–10	4
Mesothoracic ganglion, T2	18–20	2
Metathoracic ganglion		
metathoracic neuromere, T1	18–20	2
abdominal neuromere, A1	3	2
abdominal neuromere, A2	3	2
abdominal neuromere, A3	3	2
Abdominal ganglia 4-6, A4-6	3	2
Abdominal ganglion 7, A7	8	0
Last abdominal ganglion		
neuromere 8, A8	8	0
neuromere 9, A9	3	0
neuromeres 10,11, A10-11	0	4

either the superficial (SDT) or deep (DDT) DUM tracts before dividing into processes that form symmetrical axons in the lateral nerves of the right and left sides of the ganglion (see Fig. 3.19). All of these neurons therefore have at least two axons, but in the majority the lateral processes divide further to form axons in several nerves. The two lateral processes also form symmetrical arrays of fine neurites in the more dorsal regions of the neuropil. Most neurons have projections restricted to a segmental ganglion and have no processes projecting to adjacent ganglia. The axons project to targets within the segment of origin, with the exception of the metathoracic DUM1b neuron which supplies the mesothoracic spiracle, and some neurons in the prothoracic ganglion which have axons only in nerves of the suboesophageal ganglion (Fig. 5.7).

Some of the DUM neurons in the thoracic and abdominal ganglia are given names according to the nerves in which they have axons, or according to the muscle in which their terminals are distributed. This does not necessarily imply that the neurons are identified, as more than one neuron can have the same pattern of axonal branches, nor does it imply that the muscle after which a neuron is named is its only target. In the metathoracic ganglion, most of the efferent DUM neurons have been characterised in this way (Campbell *et al.*, 1995), but only three are identified as individuals: DUMETi (DUM neuron to the extensor tibiae muscles), DUMDL (DUM neuron to the dorsal longitudinal muscles), and DUM1b (DUM neuron to thoracic ventral longitudinal muscles, spiracle closer muscle and the salivary gland) (Table 5.4). These three neurons have been extensively studied, but it is still not certain that the targets found so far are the only ones.

There is a greater dearth of information about the other thoracic DUM neurons, with the exception of an unusual neuron in the prothoracic ganglion. This has axons that project to the suboesophageal ganglion in the MDT of the connectives, but none in the peripheral nerves of the prothoracic ganglion (Fig. 5.7) (Bräunig, 1988). Within the suboesophageal ganglion, each axon forms branches in lateral neuropil on one side of the maxillary and labial neuromeres, but not the mandibular neuromere, and then sends axons into the maxillary, labial and occipital nerves that supply the mouthparts and the neck muscles.

5.12.2.3. *Brain*

The brain itself apparently has no DUM neurons, but it does have paired neurons that contain octopamine (Bräunig, 1991a; Stevenson and Spörhase-Eichmann, 1995) (*see section 5.12.1*).

5.12.2.4. *Suboesophageal DUM neurons*

All the DUM neurons in the suboesophageal ganglion have bilaterally paired axons that project either to the brain or to more posterior ganglia, so that there are no segment-specific efferent neurons with axons in any of the lateral nerves (Bräunig, 1991a). Only those with axons that pass through the brain to peripheral targets fit at all closely to the definition of an efferent DUM neuron. Seven neurons in three sets with cell bodies in the anterior of the suboesophageal ganglion, extensive bilateral

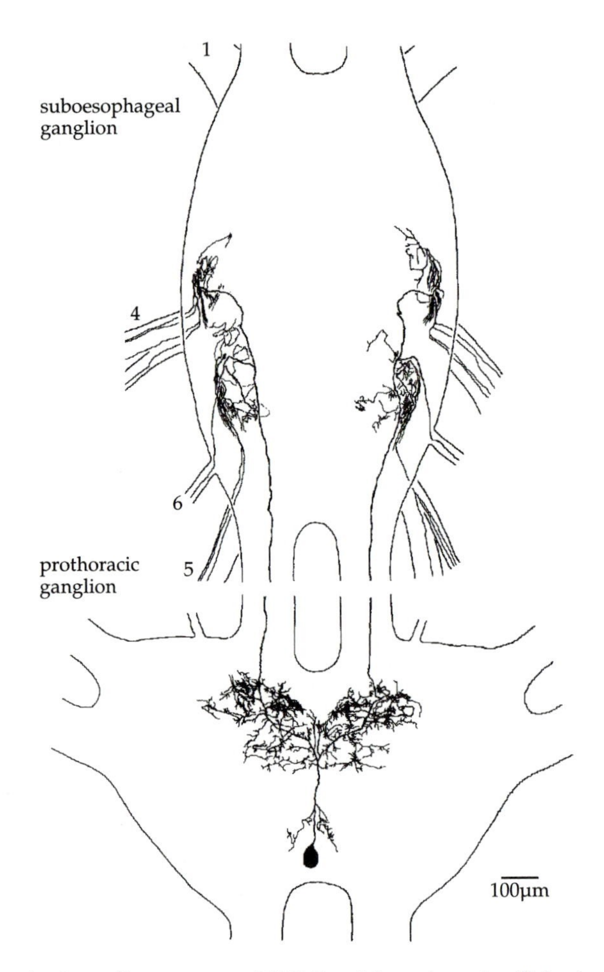

Fig. 5.7 An unpaired median neuron (DUM) with a dorsal cell body in the prothoracic ganglion that has octopamine-like immunoreactivity. It has no axons in the lateral nerves of the prothoracic ganglion, but instead its two axons project to the suboesophageal ganglion where they branch to emerge from particular lateral nerves. There are branches in the neuropil of both the thoracic and suboesophageal ganglion. Based on Bräunig (1988).

arrays of neurites in this ganglion and with axons in brain nerves have been stained (Bräunig, 1990a, 1991a).

The two neurons in the first set have axons in the nervus corporis cardiaci III (NCCIII) of the brain to supply the retrocerebral complex and the pharyngeal dilator muscles, in the frontal connectives to supply the pharyngeal dilator muscles, and in the antennal nerves to supply antennal muscles and the antennal heart that pumps blood into an antenna.

The four neurons in the second set have no branches in the brain and axons only in NCCIII, but their targets are unknown.

Table 5.4. Efferent DUM neurons with octopamine-like immunoreactivity in the metathoracic ganglion

Name	Number of neurons	Tract	Nerve	Known muscle and other targets
DUM1a (DUMDL)	1	DDT	1	Dorsal longitudinal (112)
DUM1b	1	DDT	1	Mesothoracic spiracle closer, thoracic ventral longitudinal muscles, salivary gland, median nerve neurohaemal organ
DUM 3	5	SDT	3	?
DUM 3,4	6	SDT	3,4	Subalar (129), posterior tergocoxal (119), tergosternal (113)
DUM 3,4,5	3	SDT or DDT	3,4,5	Pleuroaxillary (114), second posterior tergocoxal–remotor 2 (120): flexor tibiae (136)
DUM5a (DUMETi)	1	DDT	5	Extensor tibiae (135)
DUM5b	2	DDT	5	?

Numbers of neurons are based on Campbell et al. (1995)

The two neurons in the third set have axons only in the frontal connectives but again their targets are unknown. It must be assumed that they can exert neuromodulatory effects on some end organs, or that they release octopamine directly into the haemolymph through the retrocerebral complex.

5.12.2.5. *Abdominal DUM neurons*

The three neurons in each of the A1-6 neuromeres are characterised morphologically and two are named DUM1 and DUM2 according to the nerves containing their axons (Fig. 5.8) (Pflüger and Watson, 1988). The targets of these two neurons are the skeletal muscles innervated by the motor neurons with axons in the same nerve. The third neuron has an axon in N1, but is called DUMheart as its terminals are close to the muscles that generate the movements of the heart (Ferber and Pflüger, 1992). This neuron contains both octopamine and a FMRFamide-like peptide (Stevenson and Pflüger, 1994). Any of these three neurons can have their cell bodies in either the dorsal or ventral cortex, so that use of this criterion alone for their characterisation results in them being called either DUM (dorsal unpaired median) or VUM (ventral unpaired median) neurons. They are nevertheless the same neurons, and are not to be confused with VUM neurons in *Drosophila* that are the progeny of a different neuroblast.

Abdominal ganglia A7 and A8 each contain five extra DUM neurons (Stevenson *et al.*, 1994). In males, the axons of these extra neurons leave the last abdominal

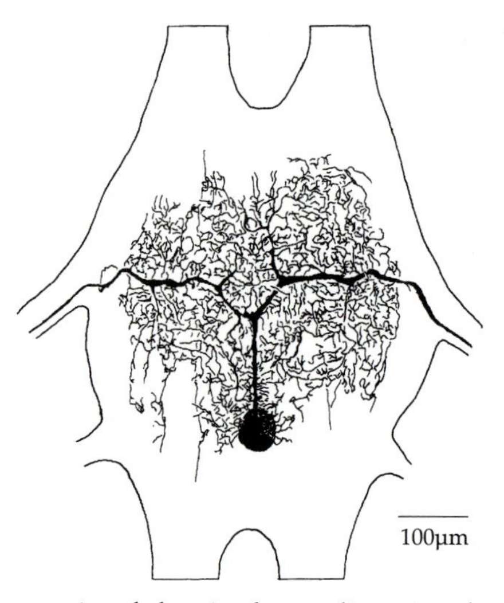

Fig. 5.8 A DUM neuron in abdominal ganglion A5 that shows octopamine-like immunoreactivity. It has a cell body at the dorsal midline and axons in lateral nerves 1 to supply the heart. Based on Ferber and Pflüger (1992).

ganglion in the genital nerve, but in females, where this nerve is absent, their axons are in N2 of their respective ganglion (Fig. 5.5). Some (perhaps five or six) of these DUM neurons in A7 supply the oviducts (Lange and Orchard, 1984; Orchard and Lange, 1985). DUM neurons in the last abdominal ganglion innervate the genitalia and accessory glands.

5.12.2.6. *Structure and distribution of synapses*

Input synapses are made onto the fine neurites and the spines and not onto the main lateral neurites or their larger branches. They are generally arranged with a spacing of about 7 µm, which is a much lower density than on other neurons. Some synapses from the same or from different neurons can, however, sometimes occur within 1 µm. The different input synapses contain round or pleomorphic agranular vesicles, or larger granular vesicles. Usually, the DUM neuron neurite is one of two postsynaptic processes at a synapse and sometimes the second neurite belongs to another DUM neuron. GABA immunoreactivity is present in 39% of the presynaptic processes and immunoreactivity to glutamate in a further 21% of synapses (Pflüger and Watson, 1995).

Only a few sites that might represent output synapses have been seen on the central neurites of thoracic efferent DUM neurons (Watson, 1984). Some spines contain agranular vesicles of 40–60 nm in diameter close to an apparent presynaptic density in a structure that may represent an output synapse. The primary and lateral neurites contain scattered granular vesicles 100–180 nm in diameter, but these are not grouped together and are not found in the fine neurites. It is not known what sort of vesicles might contain octopamine, or whether it is released from conventional synaptic structures. The ultrastructure suggests either that few output synapses are made from these neurons in the central nervous system, or that release is from sites at which a few vesicles accumulate and that are not normally recognised as synapses. However, numerous processes with octopamine-like immunoreactivity do occur in dorsal neuropil of the thoracic ganglia, but the neurons to which they belong are not identified.

At the muscles, the axon terminals of these DUM neurons contain scattered granular vesicles with diameters of 60–230 nm that are similar to those associated with Golgi complexes in the cell body, and small agranular vesicles with diameters of 28–66 nm clustered around membrane densities that appear similar to presynaptic dense bars (Hoyle *et al.*, 1980). It is uncertain which vesicles may be associated with the release of octopamine, but the smaller ones seem the more probable candidates. The efferent DUM neurons do not make conventional synapses on their target muscles, but instead end in swellings in the amorphous matrix of mucopolysaccharides and strands of collagen, so that the membranes of the pre- and postsynaptic cells are not closely apposed. They must exert their effects on a muscle through release sites other than conventional ones. Similar peripheral structures are associated with the axon terminals of DUM neurons that affect the light organs of fireflies (Oertel *et al.*, 1975) and the oviducts of locusts (Kiss *et al.*, 1984).

Release in the periphery The efferent DUM neurons release octopamine at peripheral targets such as skeletal, heart and oviducal muscle, and perhaps also into the haemolymph. The release sites are close to receptors on the terminals of the motor neurons and on the muscle fibres, but do not involve conventional synapses. The effect of octopamine is an alteration in the performance of the muscles and their response to particular features of their motor innervation (*see Chapter* 6). The release can be demonstrated directly by measuring the amount of octopamine in the perfusate of a muscle challenged with a saline containing a high concentration of K^+ or when a particular DUM neuron that innervates the muscle is stimulated electrically (Morton and Evans, 1984). The release is Ca^{2+}-sensitive and is dependent on the frequency of spikes in the particular DUM neuron, so that it is likely that release occurs from the terminals of the DUM neuron at the muscle. Of course, there might well be release of other neuroactive substances at the same time.

Release in the central nervous system As yet, there is no evidence that the efferent DUM neurons release octopamine in the central nervous system. Electron microscopy has not demonstrated the expected density and frequency of release sites, and physiology has not demonstrated an effect of spikes in an efferent DUM neuron on other central neurons. It is unlikely that clear relationships will be found if the primary action of these neurons is neuromodulatory and release is nonsynaptic. Receptors for octopamine are, however, present on many central neurons (Konings *et al.*, 1989b) and octopamine can alter the actions of central neurons and the processing that they perform (*see Chapter* 6). As the efferent DUM neurons are the most common neurons showing octopamine-like immunoreactivity, it is tempting to suppose that they do actually release in the central nervous system. If they do not, then central effects of octopamine will have to be exerted through the action of the few octopaminergic interneurons, or by its presence in the haemolymph. The latter would seem unlikely, for not only would it have to penetrate the sheath of the ganglia to reach the synaptic sites, but it would also preclude a local action. This would relegate octopamine to a role in general arousal rather than as a modulator of specific actions and specific pathways.

5.12.2.7. *Membrane properties*

Spike-initiating zones Spikes are generated in each of the two symmetrically arranged neurites which form the bilaterally symmetrical axons of an efferent DUM neuron. In addition, the cell body itself can support spikes (Crossman *et al.*, 1971b, 1972; Heitler and Goodman, 1978), and thus differs from those of motor neurons and interneurons. The axon spikes are actively propagated to the periphery but spread passively into the cell body, where they appear with a variable amplitude of 8–15 mV. The cell body spikes are themselves overshooting, with an amplitude of 70–90 mV, a long duration of 3–7 ms at half-height, and with a pronounced after-potential of about 15 mV (Goodman and Heitler, 1979). Very rarely, spikes of a size intermediate between the cell body spikes and axon spikes occur in the cell body and these may originate in the primary neurite, indicating the presence of voltage-dependent channels throughout much of the membrane.

Currents The axon spikes are generated primarily by an inward Na^+ current, as they are blocked by tetrodotoxin (TTX) and by the removal of external Na^+, but the cell body spikes are carried by a mixed Ca^{2+} and Na^+ current, as they can be blocked by Co^{2+} (which leaves the axon spikes intact), or by TTX. Na^+ can carry all the current of the cell body spikes following axotomy, or if the Ca^{2+} and delayed rectifying K^+ channels are blocked. Ca^{2+} can carry all the current when there is no external Na^+ and the same K^+ channels are blocked. This suggests that the membrane of the cell body, and perhaps other parts of the neuron, has voltage-dependent Ca^{2+} and Na^+ channels, both of which normally carry currents underlying spikes. This might explain why the spikes of isolated cell bodies from a cockroach abdominal ganglion (which are thought to be from DUM neurons) have spikes that are blocked by saxitoxin, indicating that the current is carried by voltage-sensitive Na^+ channels (Lapied *et al.*, 1990).

Two Ca^{2+} currents have been identified, one of which is activated at about –50 mV and the other at –30 mV and which differ in their pharmacological profile (Wicher and Penzlin, 1994). For example, while both are blocked by Cd^{2+}, the low voltage threshold current is enhanced by neurohormone D (= myotropin I) released by the corpora cardiaca, whereas the high voltage threshold current is depressed. The low voltage threshold current is independent of the membrane voltage and normally flows at the resting potential (about –52 mV) of isolated cell bodies. When this current is enhanced, the regular frequency of spontaneous spikes of isolated cell bodies is increased (Wicher *et al.*, 1994), but these DUM neurons do not spike in this regular way when enmeshed in their circuits in an intact nervous system, or when their resting potential is higher.

The membrane of the cell body also shows marked outward rectification at depolarised potentials that reduces the membrane resistance of isolated cells by more than half. This delayed current is carried by K^+ and is probably activated by Ca^{2+} (Lapied *et al.*, 1989). Two further outward K^+ currents that can be distinguished by their dependence and sensitivity to 4-AP are also present in these neurons that may contribute to their spontaneous spiking, particularly when they are isolated in culture (Grolleau and Lapied, 1994, 1995).

Integrative properties The integrative consequences of this arrangement of spike initiation and the participation of the active membrane of the cell body and neurite are largely unexplored. A synaptic depolarisation of the membrane at the synaptic sites in the neuropil is assumed to spread equally to the two axonal spike-initiating sites so that the same sequence of spikes is delivered to the two sides of the body. While the ability to produce different spike patterns in the two axons may be present, it has not been demonstrated during normal behaviour. At the same time, the axon spikes are passively conducted to the cell body, where they may trigger a cell body spike, although the reliability with which these spikes occur is reduced as the frequency of the axon spikes increases and as the temperature approaches the preferred range of the locust. Synaptic inputs from any particular presynaptic neuron are, however, likely to be restricted to one side of the bilateral neurites because most do not have branches on both sides of the midline. It may be possible, depending on the space constant of the intervening membrane, for these inputs to initiate spikes at only one site so that they are differentially channelled to just one side of the animal, but as yet there is no evidence for this.

The ability of the cell body to generate spikes may simply reflect the presence of a smaller K^+ current than in the cell bodies of other neurons, allowing the voltage-gated channels in the membrane to be more effective. Nevertheless, the importance of these channels may lie in the boost that they give to the synaptic input signals to other parts of the neuron rather than in generating spikes, because there is no evidence that the cell body is involved in releasing neurotransmitters or neuromodulators. It is also unlikely that the spikes control metabolism in the cell body as they may not be a reliable indicator of axonal activity. The only correlation that can be made is that the longer duration spikes that characterise the central processes of these neurons are common in other neurosecretory neurons (Gosbee *et al.*, 1968; Cook and Milligan, 1972), and may thus be necessary for release. The anatomical feature that all neurosecretory neurons have in common is that their axons have neurohaemal-like endings, suggesting that the longer duration spikes ensure invasion of these fine varicose branches and thus the reliable release of their secretions. The temporal spacing of spikes, determined by their associated currents, may be important to ensure that all the terminals in the varicose network of endings on the peripheral targets are invaded. This assumes, of course, that the axon spikes will be of long duration like those recorded in the central processes (usually the cell body), even though the balance of currents carrying these spikes may not be the same. Alternatively, the greater influx of Ca^{2+} into the cytoplasm could be important in controlling nonsynaptic release.

5.12.3. Octopamine receptors

In keeping with the widespread actions of octopamine, it is not surprising that receptors activated by octopamine, released either as a neurotransmitter or a neuromodulator, are found widely on central neurons, on the terminals of motor neurons in the periphery and on muscle fibres. The following types of receptors have been characterised, chiefly from experiments on the extensor tibiae muscle of a hind leg (Evans, 1981).

The first receptor on muscles is called the octopamine$_1$ receptor and its activation leads to an increase in intracellular Ca^{2+} levels. It inhibits the myogenic rhythm of contractions in a bundle of fibres in the extensor tibiae muscle of a hind leg. The effects of DUMETi and octopamine applied to the muscle on this receptor are blocked by phentolamine.

Receptors whose activation leads to an increase in the level of cAMP are also present on the terminals of the excitatory motor neurons (type 2a), or on the muscle fibres themselves (type 2b). A third type of receptor (unnamed) is also present on the muscle fibres and has a high affinity for 5-HT with a threshold at 10^{-9} mol.l^{-1}, but a low affinity for octopamine. Blocking phosphodiesterase that normally breaks down cyclic nucleotides, or the addition of forskolin, leads to an increase in cAMP. In the extensor tibiae muscle, both of these changes lead to an enhancement of the myogenic rhythm of contractions shown by this muscle, and can alter the force produced by the slow excitatory motor neuron.

Octopamine$_1$ and octopamine$_2$ receptors can be distinguished pharmacologically, even though both are stereospecific for the D(–) isomer of octopamine. Octopamine$_1$ receptors are blocked more effectively by yohimbine than by metoclopramide, whereas the reverse is true for the octopamine$_2$ receptors. Similarly, clonidine is a more effective agonist at octopamine$_1$ receptors than is naphazoline, while the reverse is true at the octopamine$_2$ receptors.

It is still uncertain whether all of these receptor types, or additional ones, are present on neurons and glial cells in the central nervous system. Currently, there is a tendency to treat the actions of octopamine on the central nervous system as activating one type of receptor, and therefore to point to differences with those receptors already characterised in the periphery and to classify them as a third type (Roeder, 1992, 1994). More likely is that a mixed population of receptor types is being activated. On the cell bodies of neurons, the application of octopamine leads to a slow depolarisation accompanied by a small decrease in membrane resistance. The underlying current reaches a peak at –70 mV and is blocked by Cs^+, and by Na^+-free saline.

A better resolution of the different receptor subtypes would appear to be around the corner with an increased attention on agonist-specific coupling, and with the prospect that they can be cloned and then expressed in particular tissues.

An octopamine/tyramine receptor has been cloned from *Drosophila* (Arakawa *et al.*, 1990) and when expressed in a Chinese hamster cell line inhibits adenylate cyclase activity and leads to the elevation of Ca^{2+} levels by separate G protein-coupled pathways (Robb *et al.*, 1994). It is therefore similar to the $octopamine_1$ type of receptor in locusts. The study of this receptor indicates that members of the G protein-coupled family may be linked to many effector pathways through different G proteins, thus increasing the complexity of effects that must be considered for these sorts of neuromodulators. A single receptor may therefore have a different pharmacological profile, depending on the tests used to determine its efficacy and different agonists may couple it to different second messenger pathways.

5.13. PEPTIDES AS NEUROMODULATORS AND NEUROHORMONES

The recognition that peptides may play a diverse role in the nervous system has come largely from the demonstration of their presence in both neurosecretory neurons and other neurons intrinsic to the central nervous system by immunocytochemistry and radioimmunoassays. Roles for peptides that have been envisaged range from signalling at synaptic and nonsynaptic sites through neuromodulation to trophic actions. The impetus for the burgeoning number of studies has been the expectation that a peptide could be localised to particular identified neurons so that the task of understanding its action would be eased. This remains a largely unfulfilled hope because the neurons stained generally have small cell bodies which are not readily distinguishable from many other surrounding cell bodies. It is nevertheless an objective worth pursuing and which should be achievable once the subject moves beyond its present descriptive phase and into an analytical one.

Our knowledge of peptides in the nervous system has accrued along two fronts. First, we know something of their actions as hormones that are released into the haemolymph and control behaviour such as moulting, reproduction, contractions of the heart, visceral and skeletal muscle, and diverse responses such as colour change, osmoregulation and metabolism. We know relatively little of the action of peptides on neurons within the central nervous system. Second, many maps have been produced of the locations of neurons immunoreactive to particular peptides, often mammalian peptides, or to partial sequences of these peptides extracted from a number of species of insects (see Nässel, 1993).

The problem of the specificity of the antibodies, particularly those raised against peptides that have not been sequenced from the animal to which they are then applied, is now widely recognised, but still makes interpretation of much of the literature difficult. The frequent use of the phrase '...-like immunoreactivity' is testament enough to the uncertainty of what is being described. This problem should eventually diminish as more peptides are sequenced and *in situ* hybridisation techniques become more widely used. A problem that is likely to remain, however, is variability of staining and how to interpret it. It is common to find neurons within one animal that stain intensely with a particular antibody, those that stain less strongly, and those that are difficult to distinguish from background. This could represent true differences in expression of the antigen or simply differences in methodology, such as the penetration of the antibody. Moreover, in any one group of locusts a set of neurons may stain, whereas in another group the same neurons may not stain. This could again represent true differences in expression, but could also be caused by different levels of behavioural activity in the two groups leading to different levels of depletion, or again could be due to the methodology. Differences in the expression of peptides can occur at different stages in the moulting cycle and at different stages of development, so that neurons that stain intensely in larvae may stain only weakly in adults. Sometimes, however, under conditions where two groups of locusts are closely matched, the variability still occurs and cannot be caused by the methodology alone, but the true cause remains unanswered.

The mapping of immunoreactive neurons indicates at one extreme that a peptide may be restricted to a very few neurons, thus suggesting a limited function, while at the other that it may be widely distributed in many neurons, suggesting that it may have widespread actions. It also indicates that some of the peptides may be co-localised in certain neurons with other neuroactive substances such as amines. This mapping approach has largely outstripped the functional understanding of what the peptides might be doing, and has clearly outpaced our knowledge of their receptors. While at present it remains unclear how to advance functional studies in the face of the avalanche of immunocytochemical information about the possible occurrence of particular peptides, what we can look for are the few peaks of insight that no other approaches would have revealed. These are enticing enough to encourage a deeper analysis of more peptides and more comprehensive descriptions of the neurons in which they are located.

5.14. PROCTOLIN

Proctolin was first extracted from the hind gut (proctodeum, hence its name) of the cockroach and was the first insect peptide to be sequenced (Starrat and Brown, 1975). Radioimmunoassays show that it is present in the nervous system with the highest concentrations in the last abdominal ganglion, and immunocytochemistry indicates that it is distributed in only a small number of neurons in different parts of the nervous system (see Fig. 5.13). It was once thought to be a putative transmitter at the hind gut but evidence for this action has not been forthcoming. It may be a neurohormone because terminals showing proctolin-like immunoreactivity occur at release sites into the haemolymph. Its presence in the haemolymph is, however, likely to be short lived. Proctolin is now regarded as a neuromodulator at many peripheral targets where it may be co-released with a fast transmitter, and at central targets, although the evidence for these actions in the locust is still limited.

5.14.1. Distribution of neurons with proctolin-like immunoreactivity

In the brain, there are no descriptions of the neurons that contain proctolin, although it is possible that some antennal motor neurons show proctolin-like immunoreactivity (Bauer, 1991). In the cricket, many of the antennal motor neurons, including ones with slow and fast actions at the muscles, show proctolin-like immunoreactivity (Bartos *et al.*, 1994). In the fly *Calliphora erythrocephala*, by contrast, about 100 immunoreactive neurons occur in each optic lobe and about 40–50 pairs in the rest of the brain and suboesophageal ganglion, but none in the deutocerebrum (Nässel and O'Shea, 1987).

Only about 35 pairs of neurons in the ventral nerve cord of the locust show proctolin-like immunoreactivity (Keshishian and O'Shea, 1985b). In the suboesophageal ganglion are an anterior group of cell bodies at the ventral midline, a lateral group, and bilateral posterior groups. Some of the neurons in the anterior lateral group may be motor neurons that innervate the closer muscles of the mandibles (Baines *et al.*, 1990). In the thoracic ganglia, three clusters of immunoreactive cell bodies are recognised. First, in the pro- and mesothoracic ganglia are one pair, and in the metathoracic ganglion two pairs at the level of N5. These may be interneurons with axons that project in the connective contralateral to the cell body at least as far anteriorly as the suboesophageal and posteriorly as far as the metathoracic ganglia. Second, are 1–5 pairs, depending on the ganglion, of more anterior neurons in the region that contains most of the leg motor neurons and which may themselves be motor neurons. Third, are 1–3 pairs of more posterior and dorsal cell bodies. The seventh abdominal ganglion contains three pairs of neurons that are probably motor neurons innervating the oviducts (Lange *et al.*, 1986). The last abdominal ganglion has anterior clusters of three pairs of neurons, and a posterior cluster of smaller diameter neurons. The anterior neurons innervate the midgut and not the hind gut as in some other insects, so that they represent only a small subset of the 30 that innervate this region of the gut. They probably act as neuromodulatory neurons because of the long-lasting changes in excitability that experimentally applied proctolin evokes on the gut. At the opener muscle of the ovipositor valves, proctolin acts in a concentration-dependent manner, increasing the frequency of miniature end plate potentials (mepps) from motor neurons, enhancing evoked twitch contractions and then increasing the level of basal force (Belanger and Orchard, 1993).

The cockroach has a similarly limited number of neurons with proctolin-like immunoreactivity, and one of these has been identified. In the metathoracic ganglion, one pair of cell bodies on the dorsal surface is identified as the slow motor neuron (called Ds) to the coxal depressor muscle (a muscle that actually extends the femur and is located in the thorax and coxa and consists of five parts, numbered 177A-D), one of five motor neurons to this muscle group (Bishop and O'Shea, 1982). In the unfused abdominal ganglia there are again several groups of stained cell bodies, one pair of which, called the lateral white neurons, was originally identified as proctolinergic (O'Shea and Adams, 1981). While these neurons are undoubtedly neurosecretory neurons, it now seems unlikely that they contain proctolin. Other somata may belong to motor neurons that provide the proctolinergic innervation of the intersegmental and ventral diaphragm muscles (Witten and O'Shea, 1985). The last abdominal ganglion has many stained somata that supply the hind gut and the oviducts, both of which generate tonic contractions to the application of 10^{-11}mol.l^{-1} proctolin (Sobek *et al.*, 1986; Puiroux *et al.*, 1993).

5.14.2. Is proctolin co-released from motor neurons?

The evidence for co-release comes from experiments on the slow coxal depressor motor neuron (Ds) of the cockroach that is immunoreactive for proctolin. Electrical stimulation of the nerve containing the axon of this motor neuron, or bathing the muscle that it supplies in high K^+ saline, causes a Ca^{2+}-dependent release of a substance that co-elutes with proctolin, and whose action on the locust extensor tibiae muscle can be mimicked by proctolin (Adams and O'Shea, 1983). Each spike in this motor neuron causes a synaptic potential in a depressor muscle fibre accompanied by a conductance change, and a twitch contraction of the muscle. Repetitive spikes (evoked by electrical stimulation of the nerve containing the axon of this and other neurons) cause repetitive synaptic potentials and repetitive twitches, but these are then superimposed on a more slowly developing and longer lasting contraction that is not accompanied by a depolarisation of at least some of the muscle fibres, or a change in their conductance. Only one record of these effects has been published, although it has been reprinted many times. The rapid contractions caused by the EPSPs can be mimicked by glutamate and the slow contractions by proctolin, and are therefore thought to be caused by the co-release of glutamate and proctolin. The interpretation is that proctolin is acting here as a conventional transmitter, albeit one that does not act on receptors directly linked to ion channels, and that it has an additional modulatory role in prolonging the twitch contractions caused by the fast transmitter when they occur sequentially. A similar pattern of contractions, interpreted as being of similar origin, can be produced in cricket antennal muscles by stimulation of certain motor neurons (Allgauer and Honegger, 1993).

An alternative explanation of the twitch and slow contractions is that different populations of muscle fibres contribute to the force of the whole muscle, with some developing force rapidly to generate the twitch contractions, and others developing force more slowly to produce the slow contractions. Such an arrangement is well known in arthropods and has been the source of many paradoxical results where attempts are made to correlate electrical events in one fibre with the force produced by the whole muscle (Atwood and Hoyle, 1965). The fast and slow contractions might therefore be explained by the release of just one transmitter, without the need to invoke the co-release of two transmitters. Similarly, the prolongation of the twitch contractions could be due the heterogeneous contractile properties of the muscle fibres.

5.14.3. Actions of proctolin

Clear demonstrations of the effects of neurally released proctolin are few, with some earlier claims failing to stand the test of time. For example, spikes in a pair of neurons of a free abdominal ganglion of the cockroach called the lateral white neurons (these neurons have not been found in locusts) produce a long-lasting increase in the force and frequency of the heartbeat without causing synaptic potentials in the heart muscle (Adams *et al.*, 1989). These neurons have peripheral axons that may project to the heart (O'Shea *et al.*, 1982) and are said to contain proctolin and adipokinetic hormone (AKH), an amine, and a bursicon-like substance. The evidence for these claims has never been published and it now appears that they do not contain proctolin, so that any effects that they may have on the heart must be mediated by the other substances they contain. The suggestion is that the release of these cardioacceleratory substances increases blood flow and thus allows a rapid change in the distribution of hormones throughout the body. Implicit in this is the idea that proctolin also acts as a neurohormone (O'Shea and Adams, 1986), although there is no extant evidence for this surprising action, given that proctolin is rapidly inactivated in the haemolymph.

Proctolin has an excitatory effect on the myogenic contractions of the extensor tibiae muscle, in contrast to the inhibitory effects of octopamine. These opposing effects are also repeated for the myogenic contractions of the antennal hearts in cockroaches which act as auxiliary pumps and neurohaemal sites (Hertel *et al.*, 1985). These organs consist of chambers with a vessel leading into each antenna that are linked by an innervated dilator muscle, but with no demonstrated proctolinergic innervation. Proctolin at threshold concentrations of 5×10^{-9} mol.l^{-1} causes a steeper rise in the pacemaker potentials in the muscle fibres, while octopamine inhibits the rhythm by hyperpolarising the membrane.

In all, there are more examples of the effects of applied proctolin on various effectors than can be explained by the direct patterns of proctolinergic innervation. While there is no doubting the potency of proctolin in these examples, there remains a question as to whether it is normally released by neurons in the vicinity of the effector, or how it could be effective if borne in the haemolymph. This can only be resolved by demonstrating release from identified proctolinergic neurons and by the demonstration of proctolin receptors on the target. This is a stringent test that has not been fully met for any putative target of proctolin, but has of course been achieved for a small number of putative insect neurotransmitters such as glutamate, GABA and octopamine that control the contractions of the skeletal muscles.

5.14.4. Are any locust motor neurons proctolinergic?

In locusts, the slow motor neuron (SETi) to the extensor tibiae muscle of a hind leg has often been said during the last 10 years (Witten *et al.*, 1984; Worden *et al.*, 1985; O'Shea *et al.*, 1988; Adams *et al.*, 1989) to show proctolin-like immunoreactivity and to act in the same way as the slow coxal depressor motor neuron of the cockroach by co-releasing glutamate and proctolin. However, the evidence to substantiate these claims has never been published. All that is known is that some cell bodies in the region of the ganglion containing those of leg motor neurons show proctolin-like immunoreactivity (Keshishian and O'Shea, 1985b), which is hardly a convincing identification of a motor neuron that has been well characterised anatomically and physiologically. Application of proctolin in low concentrations (10^{-9} mol.l^{-1}) to the extensor tibiae muscle does, however, accelerate the frequency of the myogenic rhythm of contractions, and this effect has become a sensitive bioassay for proctolin. The explanation of this effect presumes a proctolinergic innervation of this muscle, but, as yet, proctolin receptors are known only from the ability of low concentrations of proctolin to cause a tonic contraction (May *et al.*, 1979).

A common theme running through all these varied observations is that proctolin may be present in some slow, but not fast, motor neurons [some fast motor neurons of cricket antennae are an exception (Bartos *et al.*, 1994)] and that it may be co-released with a fast-acting transmitter at some skeletal and visceral muscles. The fast transmitter causes conductance changes and twitch contractions, but proctolin itself acts *via* second messengers either to potentiate the contractions caused by the fast transmitter or to cause slower and more prolonged changes in force. The extrapolation has been made that the 'catch-like' forces generated by many muscles may be attributable to the co-release of a slower acting transmitter (O'Shea *et al.*, 1988). Such a proposition must, however, allow many of the slow motor neurons to release substances other than proctolin, because the restricted distribution of this peptide is insufficient to explain all the known examples of 'catch-like' forces. The contractions of many muscles can be altered by the application of proctolin, but for most there is little evidence to suggest that this is their normal mode of activation.

Release of proctolin in the central nervous system is to be expected from the demonstration that some of the neurons that contain it are interneurons, but there are no reports of unequivocal actions.

5.15. ADIPOKINETIC HORMONES (AKH)

The adipokinetic hormones are a large family of neuropeptides, which in addition to AKH-I and AKH-II are known by names such as red pigment concentrating hormone and neurohormone D (myotropin I). About 20 members of this family have been sequenced so far. In the locust, AKH-I and -II and at least three other peptides (adipokinetic precursor-related peptides, APRPs) are synthesised in the glandular cells of the corpora cardiaca from co-localised but different prohormones that are themselves formed from different mRNAs (O'Shea and Rayne, 1992). During postembryonic development, the relative and absolute levels of these peptides changes, so that in adults the level of AKH-I is five times higher than in the larval instars (Siegert and Mordue, 1986). The relative proportion of AKH-I to AKH-II changes from 1: 1 in the first instar larvae to 5: 1 in adults (Hekimi *et al.*, 1991). This correlates with a huge increase in the number of cells in the gland from about 50 in first instar larvae to about 6000 some 4 weeks later in adults (Kirschenbaum and O'Shea, 1993), and with the development of wings and the ability to fly. The new cells are formed throughout larval life and into early adulthood from undifferentiated precursor cells, and proceed to express AKH peptides.

When released into the haemolymph, most notably about 15 min after the start of flight, the AKH hormones affect protein synthesis, heart rate, and the mobilisation of lipids from fat bodies and carbohydrates from flight muscles. A primary action is to raise the level of internal Ca^{2+} in their target cells. Neurohormone D, for example, when applied to the isolated cell bodies of efferent DUM neurons enhances a transient Ca^{2+} current that is active at the resting potential of these isolated cells (Wicher and Penzlin, 1994) (*see Chapter 3*). Neurons with AKH-like immunoreactivity also occur in the brain but it is uncertain whether the same antigen is being recognised as that in the corpora cardiaca, and there are few clues as to what their function might be.

5.16. FMRFAMIDE-LIKE PEPTIDES

The FMRFamide family of peptides consists of a number of closely related peptides that are cleaved from a common precursor protein. In insects, there appear to be three classes of FMRFamide-like peptides which are N-terminally extended – those with the FMRFamide sequence (not found in locusts), sulphakinins (drosulphakinins), and those related to SchistoFLRFamide. In *Drosophila*, sequencing of the genes indicates that FMRFamide itself is not used as a signalling molecule (Schneider and Taghert, 1988), although it does not follow that all insects will conform to this pattern. In locusts, six chromatographic peaks of FMRFamide-like immunoreactivity occur in extracts of the nervous system with high concentrations in the brain, corpora cardiaca and thoracic ganglia (Robb and Evans, 1990). The major peak is caused by SchistoFLRFamide. Antibodies against three related molecules, FMRFamide, bovine pancreatic polypeptide (BPP) and SchistoFLRFamide stain sets of neurons throughout the nervous system with considerable overlap (Fig. 5.9) (Myers and Evans 1985a,b, 1987; Swales and Evans, 1995a). The differences presumably represent the uncertainty over just what is being stained and encapsulate well the problems of interpreting immunoreactivity.

In the **brain**, about 170 neurons stain with the BPP antibody and, of these, some 50 also stain with the FMRFamide antibody. Most of these neurons occur in the pars intercerebralis and in proximal regions of the optic lobes. The storage lobes of the corpora cardiaca also stain with the BPP antibody.

The **suboesophageal ganglion** has two groups of cell bodies at the midline that are stained by the BPP and FMRFamide antibodies but not by the SchistoFLRFamide antibody. These neurons have branches within the suboesophageal ganglion and axons in NCAII which bypass the copora allata to project to the corpora cardiaca (Bräunig, unpublished). Some of these same neurons also show immunoreactivity to locustamyotropin and to myomodulin, and the four larger cells in the anterior group may also show immunoreactivity to crustacean cardioactive peptide (*see section 5.18*). They would thus seem to be neurosecretory cells containing a number of different peptides.

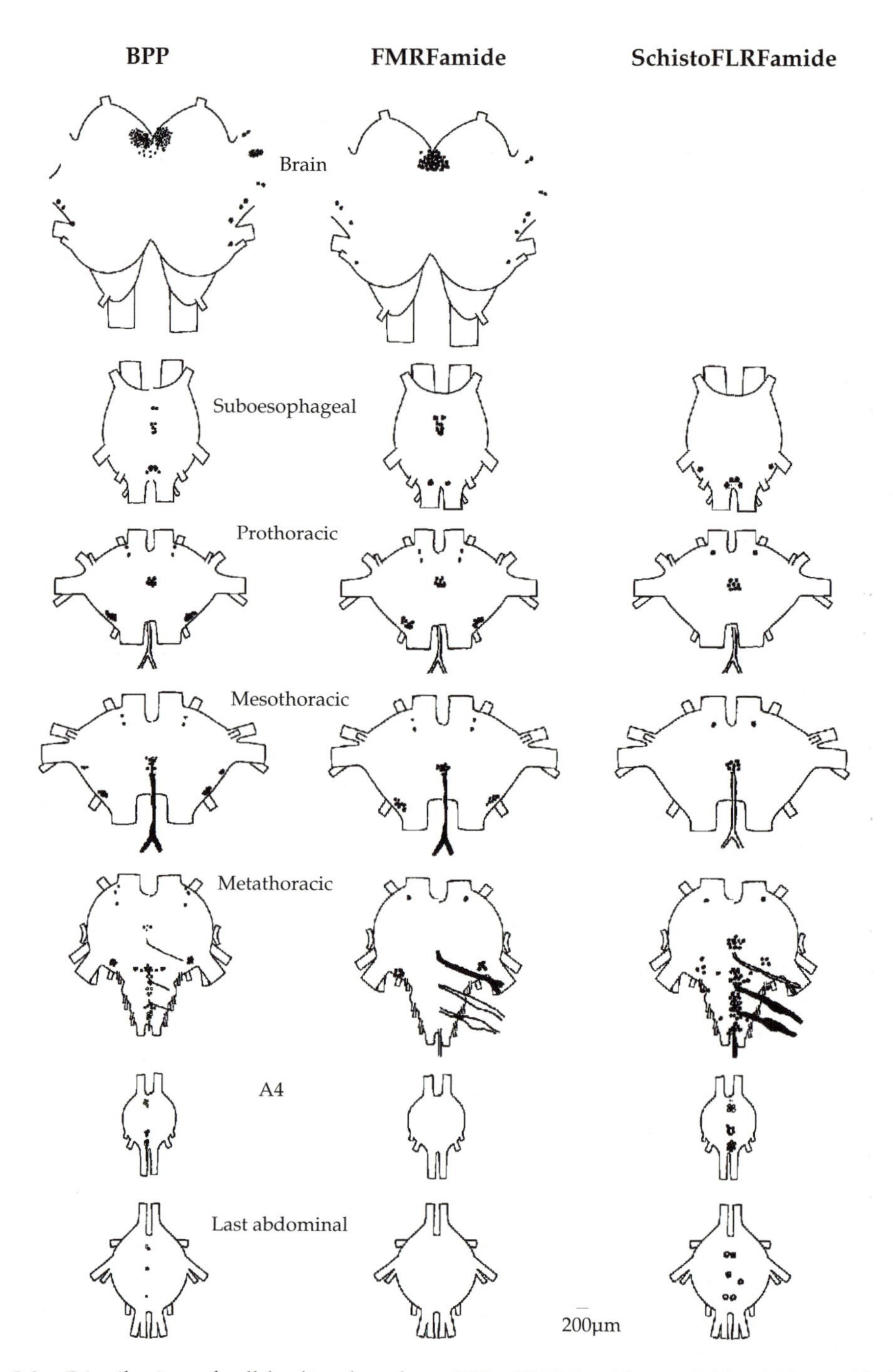

Fig. 5.9 Distribution of cell bodies that show BPP-, FMRFamide- or SchistoFLRFamide-like immunoreactivity. The median nerves that show strong immunoreactivity are drawn in black. Based on Swales and Evans (1995a) and Myers and Evans (1987).

In each of the **thoracic ganglia** there are three groups of neurons; first, are two (three in the metathoracic ganglion) pairs of neurons with anterior cell bodies, but nothing further is known of their anatomy, or of their action. Second, is a single group consisting of 8–10 (four in the metathoracic ganglion) cell bodies at the ventral midline, but again nothing further is known of these neurons. Third, is a bilaterally paired group, each consisting of 8–10 posterior and lateral cell bodies with diameters of 15–20 μm and axons in the median nerve that end in release sites at the median neurohaemal organs. At least some of these cells also show myodulin-like immunoreactivity (*see section 5.24*), but are not stained by the SchistoFLRFamide antibody. The median nerves, and their transverse branches, of the meso-, metathoracic and abdominal ganglia show immunoreactivity to some of the three antibodies, particularly in neurohaemal regions where they are enlarged.

Abdominal ganglia 1–7 have about 16 cell bodies in three groups at the midline, but the last abdominal ganglion has fewer, that stain with the BPP and SchistoFLRFamide antibody. The FMRFamide antibody does not stain any of these cells. The efferent DUM neuron to the heart from the abdominal ganglia may be one of these neurons, as it alone of the three abdominal DUM neurons shows both octopaminergic and FMRFamide-like immunoreactivity (using a different antibody) (Stevenson and Pflüger, 1994). The other neurons are probably the bilaterally projecting neurons (BPN) that seem to supply the heart and the genitalia.

The staining of some of these neurons with an FMRFamide antibody changes with the cycle of egg production and laying in females (Sevala *et al.*, 1993). Similarly, the level of FMRFamide-like activity in the haemolymph increases while the eggs are being made but falls once they have been laid. The same FMRFamide-like substance alters the contractions of the oviducts, suggesting that its release may control part of the reproductive cycle.

5.16.1. A neuron to the heart and retrocerebral complex

A pair of neurons in the suboesophageal ganglion that shows bovine pancreatic polypeptide-like and SchistoFLRFamide-like immunoreactivity has an extraordinary structure. The cell bodies are at the ventral midline of the labial neuromere, the neurites are restricted to a medial area of neuropil of the suboesophageal ganglion, but the axons project extensively (Bräunig, 1990a, 1991b). They are sometimes called the VPM (ventral posterior median) neurons. An anterior axon projects into NCCIII of the brain without forming any branches in the brain itself to end probably in the corpora cardiaca. The posterior axon projects at least as far as abdominal ganglion 5 (A5). In each of the ganglia A1-5, both neurons have an axon in N1, then in the segmental heart nerve and finally in a cardiac nerve cord of the heart. The function of these neurons is unknown as stimulation of them does not affect the frequency or the amplitude of the heartbeat, but it is assumed that they have some neurosecretory action. Release at the heart may not necessarily indicate an action on the heart, but the optimal placement for effective dispersal throughout the body. Many FMRFamide analogues can alter the heartbeat, particularly when it is irregular (Cuthbert and Evans, 1989), so that a direct action on the heart cannot yet be ruled out. It is possible that these neurons are the segmental homologues of the two maxillary neurons that have vasopressin-like immunoreactivity (*see section 5.17*).

5.16.2. Neurons with sulphakinin-like activity

Only four pairs of neurons in the brain show sulphakinin-like activity (Agricola and Bräunig, 1995). Three pairs have processes that stay within the brain and are thus interneurons. The fourth pair have a remarkable structure akin to that of the vasopressin neurons. They have branches in the brain and an axon that descends in the connective contralateral to the cell body as far as the last abdominal ganglion. In each ganglion they form extensive branches characterised by terminal varicosities. They do not have axons that emerge from any of these ganglia, so they appear to be neurons with a widespread effect on the processing performed by these ganglia, but there are no clues as to what this action might be.

In the thoracic ganglia, the groups of 8–10 small diameter lateral cells that project into the median nerve and to the neurohaemal organ also show immunoreactivity to sulphakinin. These cells are in the same group as those that show immunoreactivity to bovine pancreatic polypeptide (BPP), to myodulin (*see section 5.24*) and to locustamyotropin. It is unclear whether the same cells contain all these peptides or whether they are contained in different cells of the same population.

5.16.3. Effects of FMRFamide-like peptides on muscle and the heart

FMRFamide and some other FMRFamide-related peptides increase the amplitude, contraction rates and relaxation rates of twitch contractions of the extensor tibiae muscle caused by the slow but not the fast motor neuron (Evans and Myers, 1986). The effects depend on the frequency of the motor spikes and there may be regional differences in the responses of the muscle, reflecting different proportions of muscle fibres that are innervated by the slow and not the fast motor neurons. The effects are in addition to those caused by octopamine and proctolin and are mediated by a receptor that does not raise cyclic nucleotide levels. There is no supply to the extensor tibiae muscle by neurons that contain FMRFamide-like peptides.

The occurrence of FMRFamide-like peptides in neurohaemal release sites and the corpora cardiaca suggests that they can act as neurohormones and cardioregulators, in addition to the effects that their presence in the haemolymph will have on skeletal muscle. FMRFamide has an excitatory action on the heart, increasing both the amplitude and the frequency of the heartbeats, with effects persisting for several minutes after application (Cuthbert and Evans, 1989). Other FMRFamide-like peptides may either excite or inhibit the heart, with their effects often dependent on the concentration. All the effects are slow to be expressed and then persistent, indicating that the pathway involves a second messenger which is probably not cAMP, cGMP or arachidonic acid.

SchistoFLRFamide appears to be present in the oviducts, to reduce the amplitude and frequency of their rhythmic myogenic contractions and to reduce the amplitude of neurally evoked contractions (Lange *et al.*, 1991). It also reduces, in a dose-dependent way, the amplitude and frequency of contractions of the heart with a threshold at about 10^{-8} mol.l^{-1} and produces effects that last for a few minutes (Robb and Evans, 1994). It can also cause a small (10–20%) increase in the amplitude of twitch contractions of the extensor tibiae muscle evoked by spikes in the slow motor neuron, but depending on the concentration the enhancement can be preceded by a diminution of the contractions, perhaps indicating the presence of two receptors on the muscle membrane. Again, there is no demonstrated supply to the extensor tibiae muscle by neurons that contain SchistoFLRFamide. In the cockroach *Periplaneta americana*, however, some 6–8 fine axons run to a leg from cell bodies at the ventral midline and at a posterior and lateral region of a thoracic ganglion that are immunoreactive to an antibody which recognises FMRFamide and FLRFamide peptides (Elia and Orchard, 1995).

5.17. NEURONS WITH VASOPRESSIN-LIKE ACTIVITY

Arginine vasopressin and a peptide immunoreactive to an antisera raised against FLRFamide are co-localised in a single pair of neurons that have cell bodies some 55 µm in diameter in the maxillary neuromere of the suboesophageal ganglion (Fig. 5.10) (Evans and Cournil, 1990). These neurons are unique in that no others in the entire nervous system contain both these substances, at least as revealed by immunocytochemistry. They have been given various names, the most common of which is the VPLI neurons (vasopressin-like immunoreactive neurons). In fact, they contain three arginine vasopressin-like peptides, the F1 monomer, and the parallel (PDm) and antiparallel (F2) dimers (Baines *et al.*, 1995). They have branches in many parts of the nervous system, from an axon in LDT that extends to the brain and to the last abdominal ganglion. Although these two neurons occur in many different species of locusts, grasshoppers and other insects, they have striking differences in their patterns of branches in *Schistocerca gregaria* and *Locusta migratoria* that seem to reflect phylogeny (only members of the subfamily Oedipodinae are different) and not habitat (Tyrer *et al.*, 1993). In all ganglia of both species there are varicose branches in the dorsal and lateral regions of the ipsilateral neuropil, but in the brain of *Schistocerca gregaria* there are extensive branches in the trito- and protocerebrum but none in the mushroom bodies and optic lobes, whereas in *Locusta migratoria* there are extensive branches throughout the optic neuropils. In *Locusta migratoria* but not *Schistocerca gregaria*, bundles of fine processes enter most of the lateral nerves of the suboesophageal and more posterior ganglia to form a varicose plexus that ends before any obvious target is reached (Thompson *et al.*, 1991). These branches are in the dorsal parts of the nerves close to the motor neurons. It seems unlikely that they represent peripheral release sites as there is no access to the haemolymph at these points and the fibres are not close to the surface of the nerves. More probably, therefore, they could be acting on the axons or glia of certain sets of neurons in these regions of the nerves. The only action they could exert on the axons would be to shunt their spikes, so perhaps their actions are indeed on the glial cells. In *Schistocerca gregaria* but not *Locusta migratoria* there can be sexual dimorphism in the patterns of branches, with males having a branch in the genital nerve of the last abdominal ganglion which is lacking in females. It is possible that these neurons are the segmental homologues of the two labial neurons that have bovine pancreatic polypeptide-like immunoreactivity (*see section 5.16*).

Both neurons receive many synaptic inputs, amongst which is a prominent excitatory input that originates from an extraocular light detector somewhere in the brain with a sensitivity to wavelengths of light that suggest a rhodopsin-based pigment (Thompson and Bacon, 1991; Baines and Bacon, 1994). The location of this extra-ocular receptor is not known, but as yet no staining for S antigen (arrestin) has been carried out even though this is considered to be a marker for both retinal and extra-retinal receptors in many animals. In the fly *Calliphora erythrocephala*, for example, S antigen immunoreactivity occurs in small groups of neurons distributed in the optic lobes and in the proto-, deuto- and tritocerebrum, suggesting that extra-ocular light sensitivity may be scattered (Cymborowski and Korf, 1995). The synaptic input gradually stops in the light and resumes in the dark and is responsible for the greater spike activity of these neurons in the dark than in the light. The visual input is not dependent on either the compound eyes or the ocelli, and the time course of the changes does not indicate direct control by visual interneurons from these photoreceptors. The same presynaptic neurons providing the light input are also excited by mechanosensory stimuli of various parts of the body and probably mediate their effects by the release of acetylcholine.

The function of these vasopressin-containing neurons still remains obscure. The immunoreactivity of the neurons coupled with the extraction from the suboesophageal ganglion of a neuroactive substance (F2) with diuretic effects thought to be present in these neurons (Proux *et al.*, 1987), has led to the suggestion that they are involved in osmoregulation. There is, however, no change in the secretion of these neurons in locusts challenged with hygrometric variations (Remy and Girardie, 1980), nor are there changes in their synaptic inputs when the osmolarity of the bathing Ringer solution is changed. Synthetic F2 has no effect on the secretion of fluid from the Malpighian tubules (Coast *et al.*, 1993). Moreover, an osmoregulatory action would involve release into the haemolymph and in *Schistocerca gregaria* there are no such release sites; even the branches in the nerve roots of *Locusta migratoria* may have no access to the haemolymph.

The occurrence of these neurons in so many species (Davis and Hildebrand, 1992) suggests a conserved function, and the presence of vasopressin and the extra-ocular input suggests a possible parallel with neurons in vertebrates that control circadian rhythms of behaviour. Perhaps they act on the glial cells to alter the level of K^+ and thus decrease the overall level of motor activity during the night. In keeping with this would be the possible modulatory effects on spikes of efferent neurons. The reasons for the different patterns of brain projections in *Locusta migratoria* and *Schistocerca gregaria*, however, remain obscure.

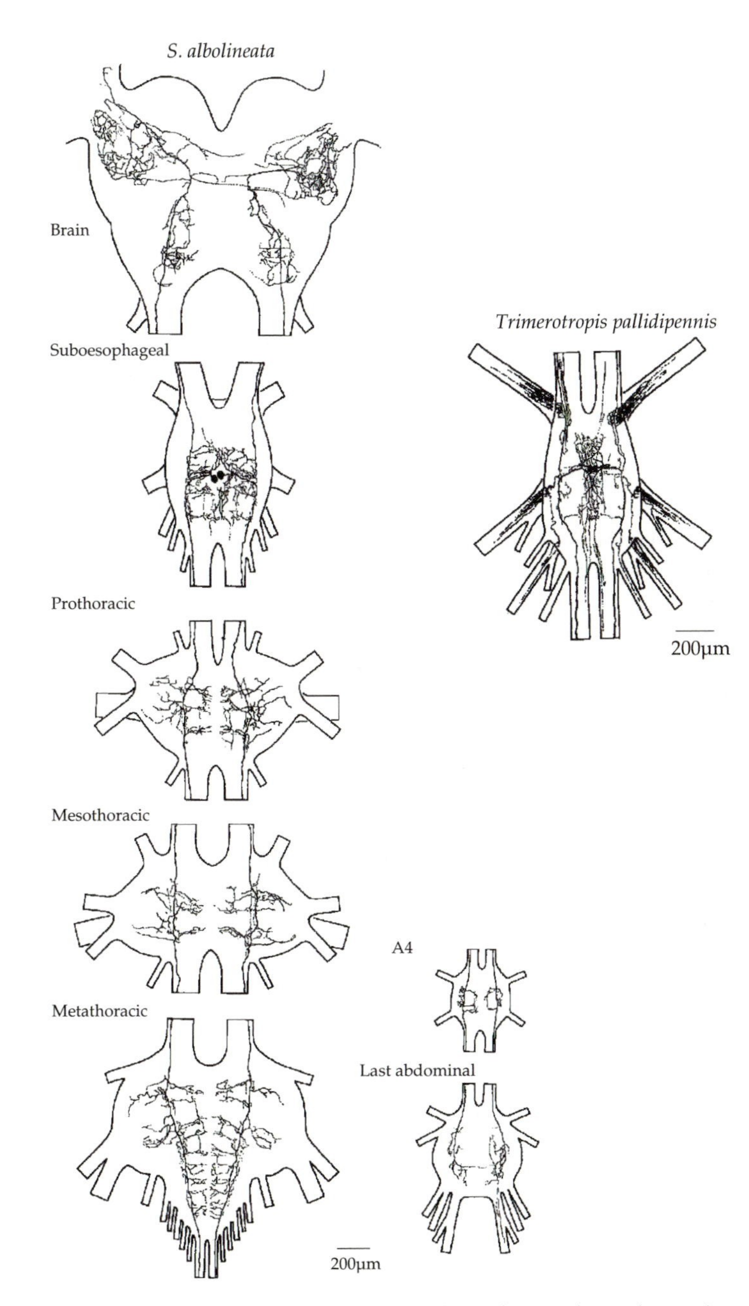

Fig. 5.10 A pair of neurons with cell bodies in the suboesophageal ganglion that show vasopressin-like immunoreactivity. These neurons in *Schistocerca albolineata* have extensive branches throughout the central nervous system. In *Trimerotropis pallidipennis* (Oedipodinae) the branches extend into the proximal parts of the lateral nerves as illustrated in the suboesophageal ganglion. These neurons are also present in *S. gregaria* and in *Locusta*. Based on Tyrer *et al.* (1993).

5.18. NEURONS WITH CRUSTACEAN CARDIOACTIVE PEPTIDE (CCAP)-LIKE ACTIVITY

The cyclic nonapeptide CCAP was first found in the pericardial organs of crustacea where it has a potent excitatory effect on the heart, but it also occurs in the nervous system of locusts where its actions are largely unknown, save for inducing contractions of the hind gut (Stangier *et al.*, 1989) and modulating contractions of skeletal muscles. Its distribution suggests that it may act as a neuromodulator released by interneurons or as a neurohormone released into the haemolymph.

Each half of the **brain** contains more than 250 cell bodies that show CCAP-like immunoreactivity, with most occurring in the optic lobes (Dircksen and Homberg, 1995). The protocerebrum has some 12 groups of cell bodies and the extensive arborisations of many of these neurons have been reconstructed so that they can almost be treated as identified individuals. By contrast, the deutocerebrum has only three local interneurons, and the tritocerebrum a single neuron with an axon that projects to the contralateral frontal ganglion.

Two neurons in the **suboesophageal ganglion** project to the corpora cardiaca and to the pharyngeal dilator muscles.

Neurons that are immunoreactive to this peptide also occur throughout the **ventral ganglia**, often as apparently homologous repeating sets (Fig. 5.11) (Dircksen *et al.*, 1991). Neurons with either dorsal or ventral cell bodies in the anterior and lateral regions of each ganglion have axons that descend to more posterior ganglia in the connective contralateral to their cell bodies. Many seem to be interneurons, but some of those in the thoracic ganglia may have axons in N5 to unknown destinations and an axon that descends in the connective contralateral to the cell body. Similarly, one neuron in both the pro- and mesothoracic ganglia has an axon in N6 and forms varicosities that may indicate release sites at the point where N6 joins N1 of the next posterior ganglion. In the suboesophageal ganglion, four anterior somata at the midline belong to the same neurons that show BPP-like and myomodulin-like immunoreactivity, and have axons in NCAII to the corpora cardiaca where they form putative neurosecretory endings (Bräunig, unpublished). One neuron in both the meso- and metathoracic ganglia has axons in N1,3 and 6 contralateral to its cell body. The branches of these axons in N1 and N6 end near the spiracle muscles or near the heart, and the branch in N3A runs between the vertical flight muscles to a place where larval muscles have degenerated in adults.

In the abdominal ganglia are a pair of neurons with 30–40 μm diameter cell bodies that have axons in the ipsilateral sternal nerve that can be traced to the alary and diaphragm muscles and to the neurosecretory structures associated with the median nerves.

In addition to these repeating neurons there are a few CCAP-containing neurons that are found only in specific parts of the nervous system. A prothoracic neuron has an axon that ascends ispilaterally, forming branches in the suboesophageal ganglion and the brain, and some neurons in the last abdominal ganglion have axons that project anteriorly. Despite these descriptions of these neurons, we have little idea of

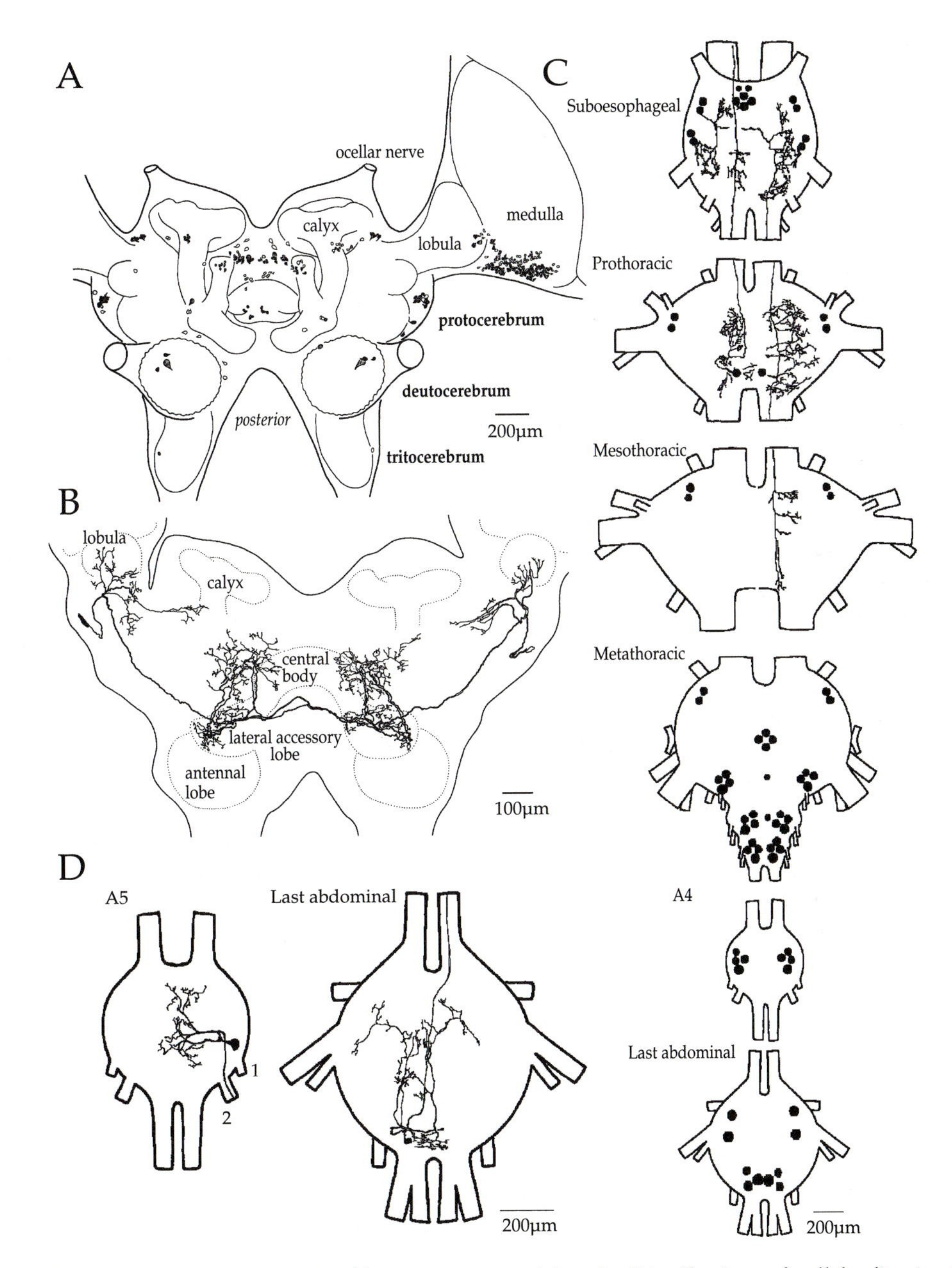

Fig. 5.11 Neurons with CCAP-like immunoreactivity. A. Distribution of cell bodies in the brain. B. Reconstruction of a neuron with bilateral branches in the lobula of the optic lobes, and in the lateral accessory lobes of the protocerebrum and the central body. C. Distribution of cell bodies in the suboesophageal, thoracic and abdominal ganglia. The branches of two neurons with cell bodies in the suboesophageal ganglion are also drawn. D. Reconstruction of neurons in abdominal ganglion A5 and in the last abdominal ganglion. A, B are based on Dircksen and Homberg (1995), C, D on Dircksen *et al.* (1991).

what their actions might be, whether CCAP is actually released, or whether it is co-localised with other peptides or with other neuroactive substances.

5.19. NEURONS WITH PIGMENT-DISPERSING HORMONE (PDH)-LIKE ACTIVITY

These peptides are named according to their ability to disperse pigments in crustaceans but they also occur in many insects, including grasshoppers (Homberg *et al.*, 1991). Particular groups of neurons with PDH-like immunoreactivity occur in the tritocerebrum but the most prominent ones have cell bodies at the base of the medulla in the optic lobes and widespread branches in the lamina and medulla both ipsilaterally and contralaterally (Fig. 5.12). This anatomy has led to the suggestion that they may be involved in some regulatory function associated with the visual system, or in controlling circadian rhythms.

5.20. NEURONS WITH MELANOTROPIN (MSH)-LIKE ACTIVITY

A small number of neurons for which no function can be ascribed show αMSH (melanophore stimulating hormone or melanotropin)-like immunoreactivity in *Locusta migratoria* (Schoofs *et al.*, 1987) (Fig. 5.13). Two groups of these cells occur in the optic lobes dorsal to the lamina, one pair in the pars lateralis of the brain and one pair in the suboesophageal ganglion. Five groups of about 15 cells in total occur

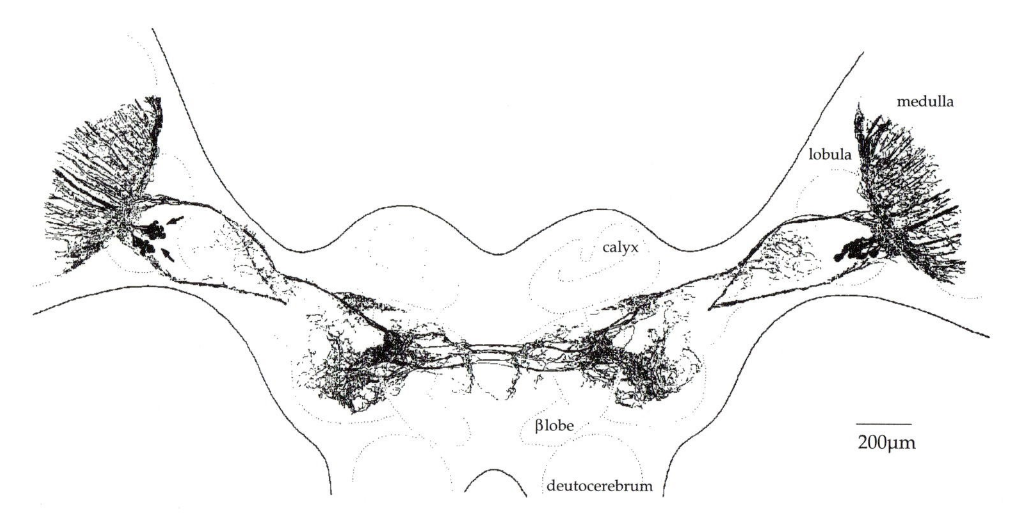

Fig. 5.12 Reconstruction of neurons in the brain that show PDH-like immunoreactivity. The neurons have branches in the medulla and lobula of the optic lobes and in the protocerebrum. Based on Homberg *et al.* (1991).

in the metathoracic and A1-7 neuromeres. Only a few cells at the base of each of the three ocelli show βMSH-like immunoreactivity, but again there is no hint as to their likely function, or whether the antigen is like its vertebrate counterpart to which the antibody was raised. The association of these neurons with visual neuropils might

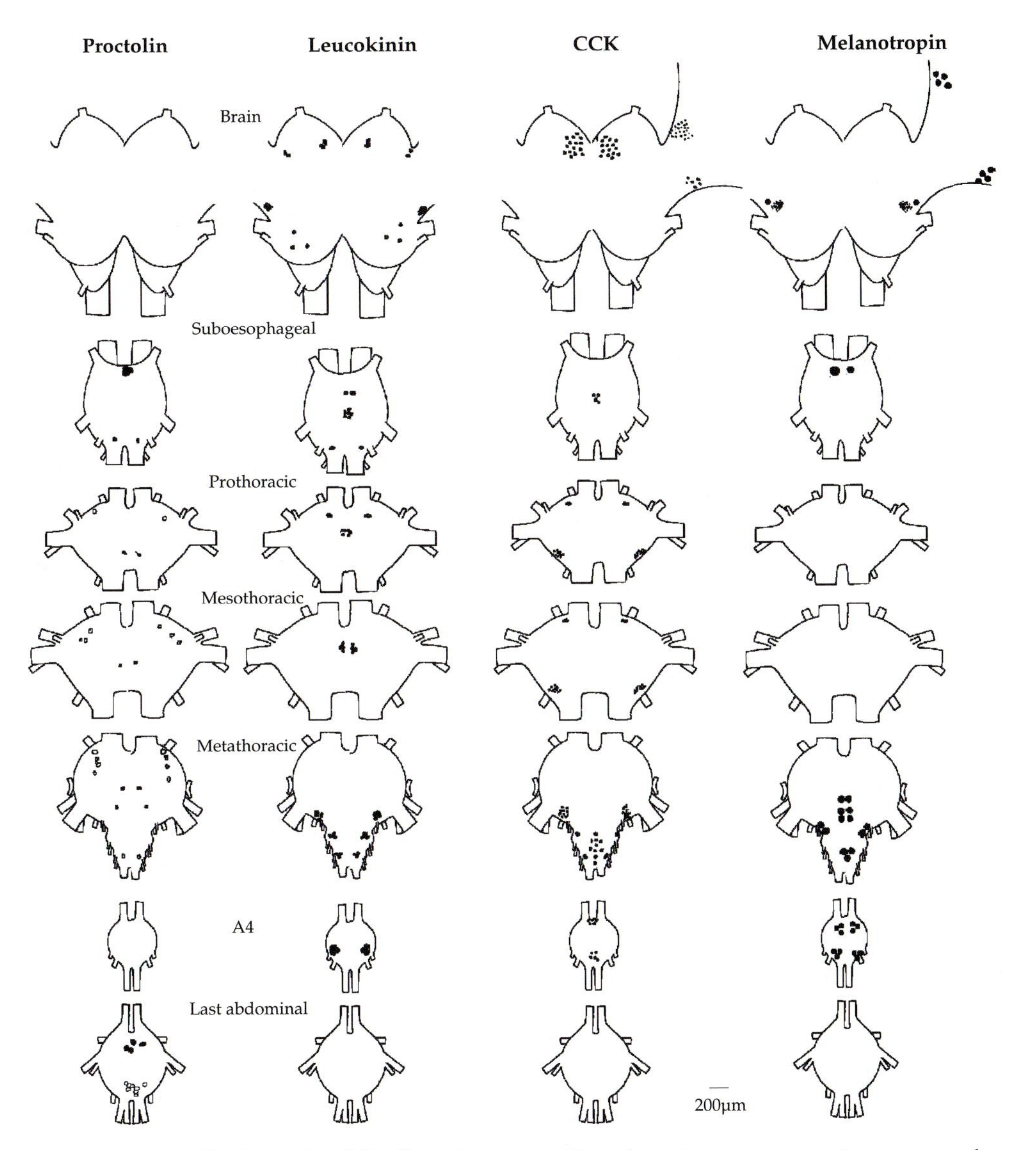

Fig. 5.13 Distribution of cell bodies of neurons that show immunoreactivity to proctolin, leucokinin, CCK and melanotropin. The drawing of the proctolin neurons is based on Keshishian and O'Shea (1985b), the leucokinin neurons on Chen *et al.* (1994), the CCK neurons on Tamarelle *et al.* (1988), and the melanotropin neurons on Schoofs *et al.* (1987).

suggest that they are in the pathways that link visual information to processes such as cuticular melanisation, phase polymorphism, or, by analogy with vertebrates, to circadian rhythms.

5.21. NEURONS WITH CHOLECYSTOKININ (CCK)-LIKE ACTIVITY

In the brain and in the thoracic and abdominal ganglia are neurons with gastrin cholecystokinin (CCK)-like immunoreactivity (Tamarelle *et al.*, 1988) (Fig. 5.13). In the brain some are intrinsic neurons while others project to the corpora cardiaca, and in the ganglia some have branches in neuropil while others project just below the sheath. This suggests a possible role for this peptide as both a neuromodulator and neurohormone, but its actions are unknown and largely untested. In the protocerebrum there are about 25 immunoreactive neurons, some of which may give rise to the stained axons in NCCII to the corpora cardiaca, 6–8 cell bodies near the calyx, and 4–6 near the corpus pedunculata. About 40 stained neurons also occur between the lamina and medulla of an optic lobe. By contrast, the deutocerebrum contains only four stained neurons and the tritocerebrum, one. The suboesophageal ganglion has two pairs of neurons, while each of the thoracic ganglia has one pair with ventral and anterior cell bodies and a bilateral posterior and lateral group of 10 cell bodies. Each of the A1-7 neuromeres has a pair of large lateral cell bodies, while A4-7 have 10 cell bodies at the anterior midline and eight at the posterior midline. The last abdominal ganglion has no stained cell bodies. This distribution gives no clues as to the likely actions of these neurons, nor does it provide much information that could be used in the future to identify these neurons.

5.22. NEURONS WITH NEUROTENSIN-LIKE ACTIVITY

A substance similar to bovine neurotensin can be demonstrated biochemically in the brain and immunocytochemically in particular neurons (Ammermüller *et al.*, 1994). Three pairs of neurons with cell bodies ventral to the calyces form a diffuse network of varicose processes in the protocerebrum, particularly around the pedunculi, and a fourth pair have somata close to the antennal lobes. A group of 8–10 neurons are also intrinsic to the medulla, with projections in those layers that also contain dopamine. There is no indication of their possible function, but analogies with vertebrates might suggest a modulatory effect on the release of dopamine, even though many of their branches occur in regions of neuropil that do not contain dopamine.

5.23. NEURONS WITH MYOTROPIN-LIKE ACTIVITY

A myotropin has been isolated from the accessory gland of males that has no sequence resemblance to other peptides and no known function (Paemen *et al.*, 1992). Three other myotropins, called locustamyotropins have been isolated from the retrocerebral complex and immunoreactivity to the terminus of these molecules is shown by about 100 neurons distributed in the tritocerebrum of the brain and in the nerves to the corpora allata, and in the suboesophageal and abdominal neuromeres (Schoofs *et al.*, 1992). Many of these cells may be the same as those which show myomodulin-like activity (*see section 5.24*).

5.24. NEURONS WITH MYOMODULIN-LIKE ACTIVITY

Myomodulin-like immunoreactivity is present in all ganglia of the ventral nerve cord with some sexual dimorphism in the cell bodies that are labelled, particularly in the abdominal ganglia (Swales and Evans, 1994) (Fig. 5.14). Its distribution suggests that it may act in the central nervous system as a neuromodulator or neurotransmitter and be released into the haemolymph to act as a neurohormone where it may affect muscles, and influence the secretions of the corpora allata and neurohaemal organs.

The **brain** contains up to 100 cell bodies that show myomodulin-like immunoreactivity (Swales and Evans, 1995b). These cell bodies occur in groups in the medulla and between the medulla and lobula of the optic lobes, in the deutocerebrum, and two groups in the tritocerebrum, one at the margin with the deutocerebrum and the other more posterior and lateral. There are also a few cell bodies scattered in the protocerebrum and the pars intercerebralis. The cells in the posterior group of the tritocerebrum have axons in NCCI that extend to the corpora cardiaca and the corpora allata.

The **suboesophageal ganglion** contains five groups of cell bodies, mostly at the ventral or dorsal midline. Some of the ventral midline cells have processes in NCAII (N3) to the corpora allata and appear to be the same cells that have locustamyotropin-like activity, while some of the anterior ventral midline cells appear to be the same as those that show FLRFamide- and vasopressin-like immunoreactivity, and others that show CCAP-like immunoreactivity.

The thoracic neuromeres of the three **thoracic ganglia** all have the same distribution of stained cell bodies. A single unpaired neuron is stained that is probably the H neuron (*see Chapter 4 and section 5.12.1*). Lateral and posterior cell bodies have axons that project into the median nerve and are the same neurons that show FMRFamide- and BPP-like immunoreactivity (*see section 5.16*).

In the **abdominal neuromeres**, all the stained cell bodies are at the ventral or dorsal midline and many processes in the median nerve are stained. It is hard to know how much of the staining of the same groups of neurons with the different antisera is due to cross-reactivity of the antisera, but more probably it indicates the presence of several peptides within the same neurons.

5.25. NEURONS WITH ALLATOSTATIN-LIKE IMMUNOREACTIVITY

A peptide called allatostatin was so named because of the ability of peptides contained in neurons with axons in NCAI to inhibit the secretion of juvenile

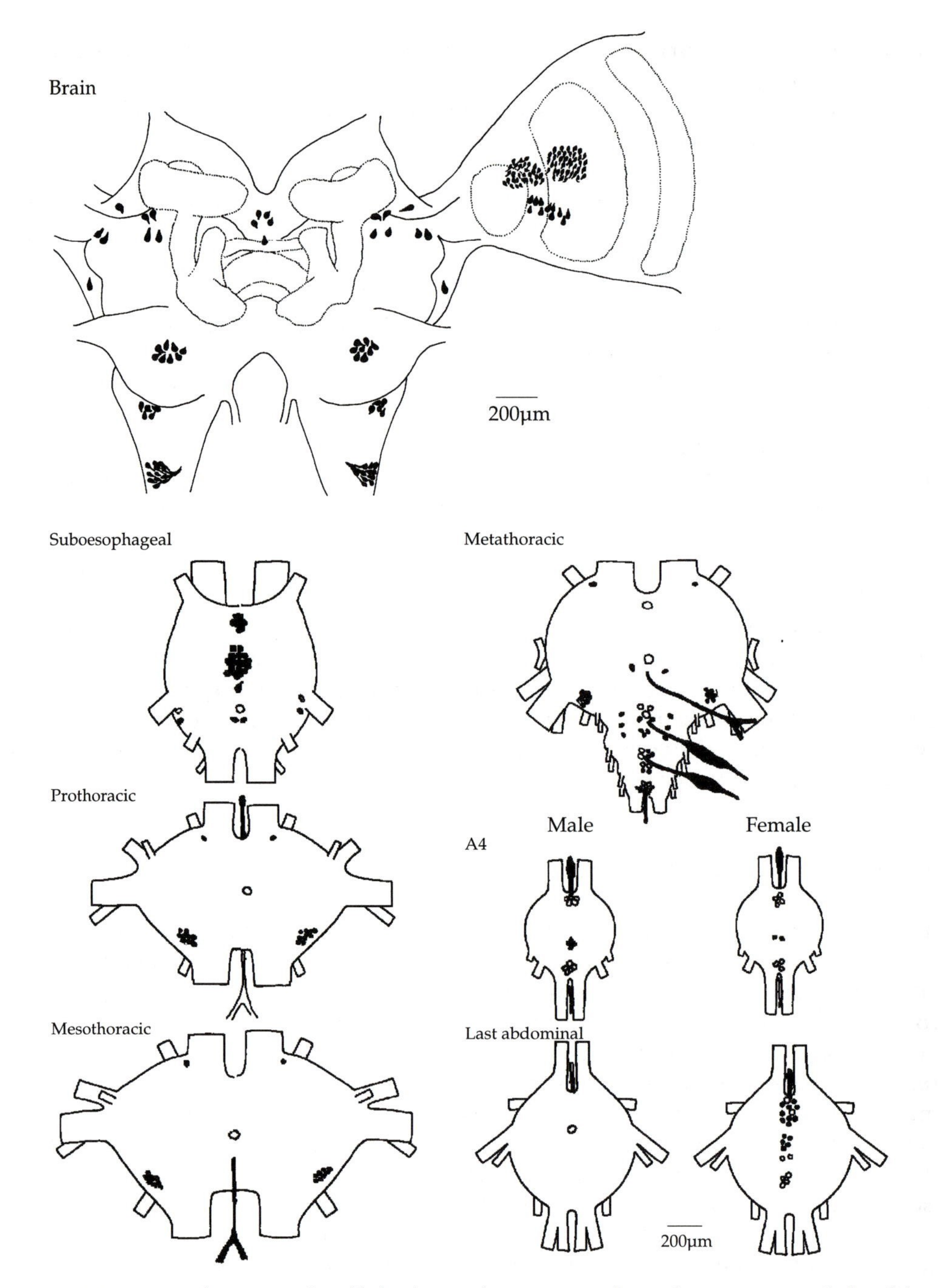

Fig. 5.14 Distribution of cell bodies of neurons that show myomodulin-like immunoreactivity. There are differences between males and females in their distribution in abdominal ganglion A7 and in the last abdominal ganglion. Based on Swales and Evans (1994, 1995b).

hormone from the corpora allata. This role is still recognised, but the more widespread occurrence of neurons that show immunoreactivity to antibodies raised against an allatostatin, and the identification of at least 13 different allatostatins, suggest more diverse roles.

In the locust (Veelaert *et al.*, 1995) and the cockroach *Diploptera punctata* (Stay *et al.*, 1992), some of the neurons with cell bodies in the pars lateralis of the brain and axons projecting in NCAI to the ipsilateral corpora allata show allatostatin-like immunoreactivity. This would be in keeping with the suggested role of allatostatin in regulating the production of juvenile hormone. Other cell bodies in the brain also show allatostatin-like immunoreactivity, such as the two pairs of large cell bodies of interneurons in the protocerebrum. There are also other immunoreactive cell bodies close to the junction of the protocerebrum with the optic lobes and with the deuterocerebrum, and processes in the glomeruli of the deutocerebrum. In the suboesophageal ganglion, neurons that innervate the antennal hearts are immunoreactive, but allatostatin applied to the antennal hearts in the cockroach *Periplaneta americana* does not appear to alter their contractions (Hertel and Penzlin, 1992). Immunoreactive neurons are also present in the segmental ganglia, most notably in a pair of neurons in the metathoracic ganglion of cockroaches and which have projections to all the abdominal ganglia.

Only a few effects of allatostatins have so far been demonstrated in insects, although more can be expected. One of these is their ability to decrease the frequency and amplitude of contractions of the hind gut (Lange *et al.*, 1993). There are no indications of their effects on the processing by neurons in the central nervous system, but such are to be expected given their actions on neurons in the stomatogastric ganglion of crustacea (Skiebe and Schneider, 1994).

5.26. OTHER PEPTIDES

It seems clear that these descriptions of peptides with neuroactive properties are just the tip of an iceberg. For example, glucagon is present in some neurohaemal areas and in peripheral neurosecretory neurons (Raabe, 1986), but has not been localised to individual neurons in the central nervous system. Corazonin, galanins, neuroparsins, somastostatin, neuropeptide Y, substance P, and substances closely related to Leu- or Met-enkephalin have also been found in the nervous system of many insects using antibodies raised against vertebrate peptides. Eleven new neuropeptides have been isolated from the retrocerebral complex alone that affect the motility of the gut. Four members of the tachykinin family of peptides have been isolated from the brains and corpora cardiaca and called locustatachykinins (Schoofs *et al.*, 1990) and about 200 neurons immunoreactive to these peptides occur in the central nervous system of the blowfly (Lundquist *et al.*, 1994). Leucokinins, first found in the cockroach *Leucophaea maderae*, also occur in locusts and are distributed in a small number of neurons in the brain and segmental ganglia, with three pairs in each abdominal neuromere that have apparent release sites in neurohaemal organs (Chen *et al.*, 1994) (Fig. 5.13). At least eight different leucokinins are now recognised. They may affect the contractions of visceral muscle, and ion transport and fluid secretion in the Malpighian tubules (Holman *et al.*, 1990).

5.27. BRINGING ORDER TO THE DESCRIPTIONS OF PEPTIDERGIC NEURONS

And so continues the seemingly ever-expanding descriptions of peptides. Nearly all of them have, as yet, undefined actions, and the identity and actions of the neurons in which the immunoreactivity is located are also largely unknown. In 1975, the structure of the first insect peptide, proctolin, was described (Starrat and Brown, 1975). By 1990, the number of peptides known had grown to 40 (Holman *et al.*, 1990) and at present exceeds more than 100. These fall into about 20 families based on homologies, only some of which are the same in all species. Many of the descriptions pose the problem of how the neurons containing the antigen can be identified, and how the actions of the peptide(s), should it be released, can be determined. Examples are rare, but rewarding, where the distinctive anatomy of a neuron offers real hope that more can be learnt about it and the peptides that it might contain. Generally, the maps of the immunoreactive neurons give, by their very nature, insufficient information for the neurons to be recognised when they are studied with other techniques. For most of these peptides, therefore, the daunting task to be faced is to probe with microelectrodes in regions where the immunoreactive neurons are located, work out their actions and connections, stain them with dyes to reveal their anatomy, and then at the same time stain alternate sections with an antibody to see if the identified neurons contain the antigen being sought. Given the large number of neurons in the areas designated by the immunocytochemistry this must represent an enormous undertaking, and yet it still leaves unanswered whether the antigen has anything to do with the action of the neurons and whether it is ever released. The lesson from the abundance of papers on the fluorescence of amines, and from the descriptions of neurosecretory cells revealed with a large number of stains (Rowell, 1976) and the relatively small increment of understanding that accrued, is there for us all to ponder before more descriptive papers are written.

Actions of neuromodulators

The vast number of substances identified in the preceding chapter as potential neuromodulators means that any assessment of their actions will be complex. What role do the neurons containing these substances play in the expression of behaviour? Are these substances released locally at specific sites in the central and peripheral nervous system? Is their release more widespread so that they will affect many neurons at the same time, assuming, that is, that the neurons have the appropriate receptors? Can we predict the likely effect of a particular substance when it is present in a cocktail of many others? Clues to the answers of at least some of these questions can be gained by restricting attention to a few modulatory substances and to the actions of particular neurons that contain these substances. Much of the following description thus relates to the action of the efferent DUM neurons and the effects of octopamine.

6.1. WHAT LINKS THE ACTIONS OF NEUROMODULATORY NEURONS TO THE MOVEMENTS THEY MIGHT INFLUENCE?

Central to any understanding of the possible effects of neuromodulators on behaviour is a knowledge of when the neurons that release these substances are active. Information with the required detail is difficult to obtain because the axons of these neurons are small so that their spikes are hard to discern in intact animals amongst the plethora of motor and sensory signals. Moreover, the time scale of action of these neurons may be quite different from that of the sensory neurons, interneurons and motor neurons that are involved in generating a movement appropriate to the current circumstances of the animal. Motor spikes typically produce effects on movements within milliseconds, whereas the effects of a neuromodulator can take many tens of seconds to be observable, may persist for minutes, and even then may be so subtle that they are resolvable only by the measurement of many physiological variables.

Virtually all that we know about this question is derived from the octopaminergic, efferent DUM neurons, largely because recordings can be made from their large cell bodies. The common observation in recordings from these neurons is that they receive a continuous synaptic input, but spike only sporadically and never at frequencies as high as those that are common in most interneurons and motor neurons. Stimuli that arouse a locust lead to an increase in the number and frequency of spikes, but most of the sources of these inputs do not produce consistent and reliable responses, leading to a frustration that soon becomes overt in the writing of those who have attempted to analyse this problem. This may explain why no sensory neuron or interneuron has been identified that causes synaptic inputs to any of the known DUM neurons. The apparent lability in responses suggests that the synaptic inputs are not caused directly by spikes in sensory neurons, but instead result from many layers of processing between the stimulus and what is observed in the DUM neuron itself. Until the presynaptic neurons are identified, the role of the DUM neurons in behaviour is likely to remain perplexing. The accepted wisdom on the action of the DUM neurons is that they spike in advance of the motor neurons to the muscle(s) that they also supply, but most of the reports tend to be rather anecdotal and the correlations are, at best, weak. Despite this surprising deficit in our knowledge, some real clues as to their possible actions are starting to emerge that may give broader clues to the actions of neuromodulatory neurons in general.

6.1.1. Common synaptic inputs

Paired recordings from DUM neurons of the same thoracic segment reveal that, like most motor neurons, none are electrically or synaptically coupled to each other. In early embryos, the firstborn DUM neurons are strongly electrically coupled to each other, but during embryogenesis the coupling gradually disappears, as it does for other cells (Goodman and Spitzer, 1981) (*see Chapter 4*). Some of the neurons with the same peripheral distribution of axons may be driven by common patterns of synaptic inputs to produce similar patterns of spikes (Hoyle and Dagan, 1978). These same pairs of neurons can, however, readily show independent actions to other inputs that are not in common. The DUM neurons with common inputs are, however, not identified, so that the relationship of these inputs to a common behavioural action is not established. It would seem likely that those with many inputs in common are those that supply the same muscles, such as the DUM3,4 neurons that supply the subalar and remotor flight muscles. This type of control would then place them in the same category as motor neurons innervating the same muscles (*see Chapter 3*).

In the abdominal ganglia of moths of all stages are two unpaired median neurons with ventral cell bodies (VUM neurons) and bilateral axons that have octopamine-like immunoreactivity (Pflüger *et al.*, 1993). They are probably equivalent to efferent DUM neurons, or are, at least, the progeny of the median neuroblast, the somata of which are merely displaced ventrally. Neurons in the same or adjacent ganglia receive many inputs in common that seem to derive from neurons in the suboesophageal ganglion. This pattern of inputs suggests that the neurons in different ganglia can be activated as a unit by descending signals, or independently by inputs that originate in their own segment, but the context in which they may be activated is not known.

It seems likely that each efferent DUM neuron, or a small subset of DUM neurons, will be active at the same time as the muscles that they supply and will therefore be driven by some of the same neurons that evoke the specific motor patterns. A working assumption must be that the different DUM neurons with their different patterns of innervation will be used in different ways. This makes it essential to identify the DUM neuron when making correlations with particular movements. The failure to do so may explain much of the extant confusion over the role of these neurons. Nevertheless, the temporal correlation between the spikes in a DUM neuron and those in the other motor neurons may not need to be exact given the slow onset and the long duration of their effects. What is not known, however, is the effectiveness on a motor pattern of spike patterns and frequencies seen in DUM neurons during natural movements. Are a few spikes sufficient to effect or maintain a changed state of the muscles and what is the time scale of their action? Must we assume that the time-to-onset of their action will be the many seconds that are taken to exert their experimentally observable effects? Do the experimental manipulations necessary to make these observations mask the effects that would be caused by the natural patterns of DUM spikes?

6.1.2. DUM neurons to the leg muscles

Bursts of a few (3–8) spikes in any metathoracic DUM neuron are said to precede any walking, jumping or kicking movement of the hind legs (Hoyle and Dagan, 1978; Sombati and Hoyle, 1984b). If this unlikely result is true, it implies a general nonspecific action of these neurons, because there can be little point in altering the properties of, for example, the dorsal longitudinal flight muscles by spikes in DUMDL during walking. During tethered walking, some prothoracic DUM neurons in the cricket spike several times at the start of the leg movements but their subsequent and occasional spikes are not linked to the stepping cycle or the forward speed (Gras *et al.*, 1990). Similarly, when a cricket is running away from a wind stimulus delivered to its cerci, some DUM neurons spike only when the movement is initiated and not during its execution, while others spike sporadically during running but not during walking.

These results are hard to interpret because many of the DUM neurons from which the recordings have been made may not affect leg muscles. What is needed, therefore, is specific information about the role of identified DUM neurons during leg movements and the identification of their targets.

This has now been provided by an analysis of the actions of identified DUM neurons during the specific and repeatable motor pattern that underlies jumping and kicking (Fig. 6.1) (Burrows and Pflüger, 1995) (*see Chapter 9* for detailed description of the motor pattern). These movements are powered by contractions of the extensor and flexor tibiae muscles of the hind legs which co-contract for several hundred milliseconds before the stored force that they produce is suddenly released when the flexor is inhibited. This results in both a rapid and powerful extension of the tibia(e). During this motor pattern, only three of the 20 DUM neurons in the metathoracic ganglion spike and these are the neurons that supply the leg muscles. The remaining DUM neurons are either silent or spike only sporadically. The three active DUM

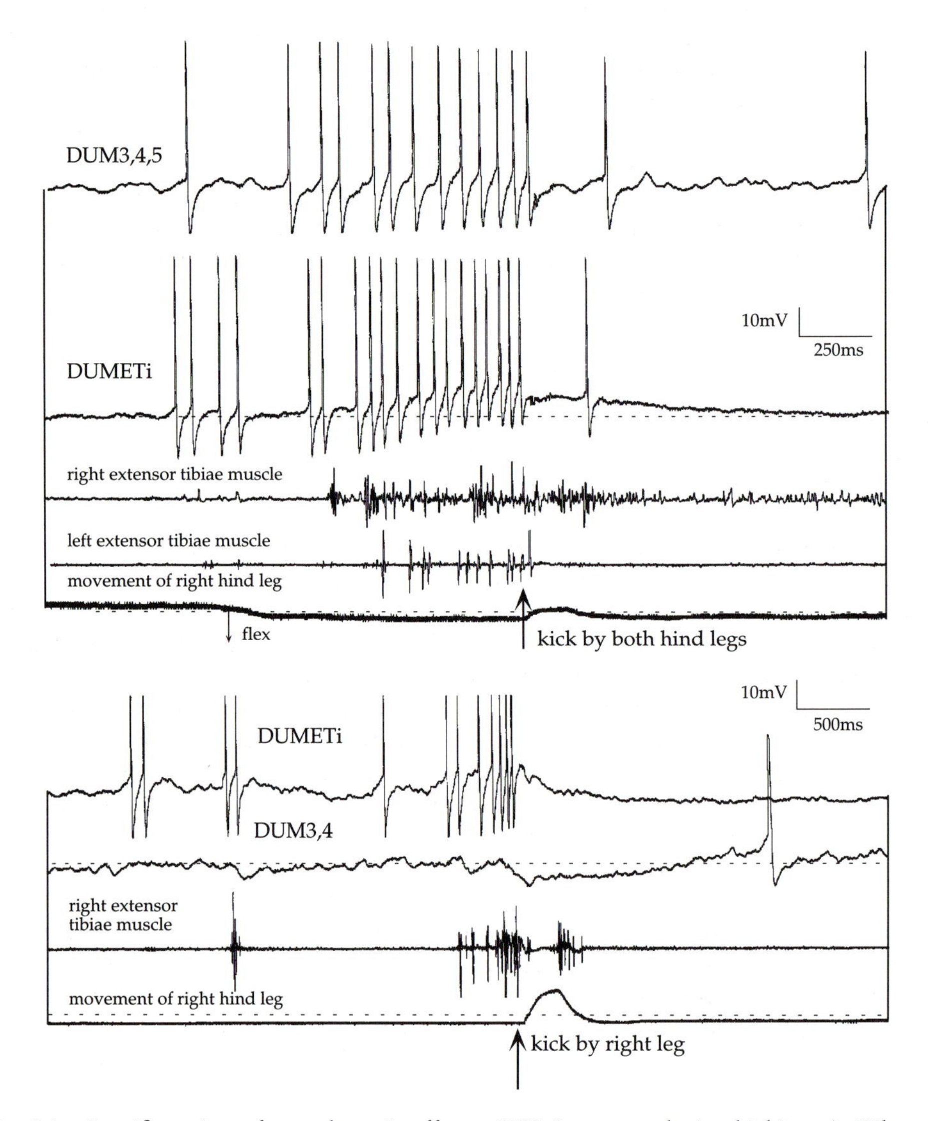

Fig. 6.1 Specific action of metathoracic efferent DUM neurons during kicking. A. When the tibia of a hind leg is flexed in preparation for a kick both DUMETi and a DUM3,4,5 neuron start to spike. The frequency of their spikes accelerates during the co-contraction phase but the spikes stop as soon as the kick occurs. Both of these neurons supply muscles involved in moving the tibia. B. A DUM3,4 neuron, which does not innervate leg muscles, does not spike during the motor pattern for kicking. Based on Burrows and Pflüger (1995).

neurons produce an accelerating sequence of spikes during the co-contraction phase of the pattern that parallels the actions of the extensor and flexor tibiae motor neurons. Their spikes stop when the tibia is extended. In this specific motor act, therefore, only a subset of DUM neurons is active which supplies the muscles that are actively involved in the movements. It is assumed that the release of octopamine at these muscles will in some way modulate their performance either during the performance of the kick or jump in which they are active, or during movements subsequent upon these movements. This makes sense because it is hard to imagine why DUM neurons that supply muscles not used in a particular movement should release their contents, when the only effect this could have would be to raise the level of octopamine in the blood. The specific supply of muscles and effectors by DUM neurons suggests that their release is targeted and local. The readily testable hypothesis generated by these results is that different subsets of DUM neurons will spike during different movements.

6.1.3. DUM neurons to the wing muscles

The stimuli that lead to the initiation of flight also excite some of the thoracic efferent DUM neurons (Ramirez and Orchard, 1990). DUMDL, DUM3, DUM3,4 and a DUM5 in the mesothoracic ganglion all begin to spike before a fictive flight pattern is generated, and then spike at a lower and gradually declining frequency when the flight rhythm is expressed. They may stop spiking even before the end of these fictive flight sequences, which only last for a few seconds, so that we do not know what their action is during more sustained flight, and therefore whether they are driven by the same neurons that generate the flight pattern. It is also unclear whether it is only a subset of DUM neurons that are active during the start of flight, because there is no indication whether certain members of the population are silent. Nevertheless, the unspecified DUM5 neuron must supply leg muscles and not those actively generating the wing movements. What is now needed is for more certain identification of the DUM neurons during this experiment and longer term recordings during more sustained flight to see if individual DUM neurons are indeed used specifically.

It is also uncertain whether this initial flurry of spikes is sufficient to set in motion the effects that octopamine has on the flight machinery, and whether a consistent pattern in time with the wing movements, or just occasional spikes, would be necessary to maintain those effects during more sustained flight. At the start of flight, the level of octopamine in the haemolymph also rises but the source of this is unknown. Its presence in the haemolymph suggests a role as a neurohormone, perhaps involved in the mobilisation of fats, that shows a similar time course at the start of flight.

6.1.4. DUM neurons to the oviducts

The spikes in some DUM neurons that supply the oviducts are linked to the motor pattern that controls the movements of the oviducts (Kalogianni and Theophilidis, 1993). The muscle fibres of the oviducts may be electrically coupled and capable of generating myogenic contractions (Orchard and Lange, 1986). They are also innervated by motor neurons that can cause neurogenic, rhythmic contractions. In otherwise intact female locusts, cutting the connectives anterior to abdominal ganglion A7 releases a rhythmic pattern of spikes in the oviducal motor neurons from this ganglion that then causes rhythmic neurogenic contractions of the lower parts of the oviducts with a periodicity of 0.1–0.5 Hz. These neurogenic rhythms normally occur during the first stage of egg-laying, when a hole is made in the ground by digging movements of the ovipositor and when the premature ejection of the eggs must be inhibited (Lange *et al.*, 1984). The contractions retain the eggs in the oviducts and tend to propel them back towards the ovaries. During these neurogenic contractions, at least some of the five efferent DUM neurons that supply the oviducts produce a few spikes at frequencies of about 0.5 Hz towards the end of each cycle of motor spikes. Stimulation of an individual DUM neuron reduces the amplitude of the rhythmic contractions (Kalogianni and Pflüger, 1992), and may reduce the frequency of the motor spikes and the amplitude of the EPSPs that they evoke in muscle fibres. Octopamine applied directly to the oviducts inhibits the myogenic contractions and reduces the amplitude of neurally evoked contractions (Lange and Tsang, 1993). These actions of the DUM neurons would seem to oppose the actions of the motor neurons, because they should reduce the neurogenic rhythms. During the myogenic contractions of the upper parts of the oviducts that occur at frequencies of 0.03–0.08 Hz and that lead to the expulsion of the eggs during the second phase of oviposition behaviour, both the motor and the DUM neurons spike tonically at low frequencies. Application of octopamine reduces the amplitude of these myogenic contractions so that again spikes in any of the DUM neurons could potentially regulate the contractions. The probable action of the DUM neurons is to enhance the rates of relaxation of the neurogenic contractions of the oviducal muscles and to inhibit the expression of the myogenic contractions during the neurogenic rhythm.

6.1.5. DUM neurons to the antennae

The seven muscles that move each antenna of a cricket are innervated by two DUM neurons with cell bodies in the suboesophageal ganglion, by 17 excitatory motor neurons, and all but one by a common inhibitor (Allgäuer and Honegger, 1993). Stimulating both these DUM neurons at 10 Hz for at least 2min before stimulating motor neurons with a limited range of frequencies, reduces the tetanic contraction produced by a slow motor neuron, but enhances that produced by a fast motor neuron in about 10% of the crickets tested. Bath-applied octopamine can mimic these effects more reliably. The DUM neurons do not, however, spike in advance of normal antennal movements, although they may spike during the movements, so it is uncertain whether they would normally have a modulatory effect.

6.1.6. DUM neurons as motor neurons

In the last abdominal ganglion of the firefly *Photuris versicolor*, four DUM neurons that contain and apparently release octopamine cause the emission of light from the paired lanterns (Christensen *et al.*, 1983). The terminals of the DUM neurons are not directly associated with the photocytes but with tracheolar cells, so it is unclear how light production is stimulated. Nevertheless, the amount of light produced is linearly related to the frequency of spikes induced by intracellular stimulation of a single DUM neuron (Christensen and Carlson, 1982). Each spike in any of the four DUM neurons is followed, after a delay of about 1.5 s, by a detectable and simultaneous output of light from both lanterns, and as the frequency of spikes is increased the pulses of light eventually sum to give a smooth emission. Octopamine stimulates the production of cAMP, and an octopamine-sensitive adenlyate cyclase is present in the lantern (Nathanson, 1979). Light emission is associated with the release of octopamine, and both it and the release of octopamine are abolished if the lanterns are either denervated or bathed in Ca^{2+}-free Ringer (Carlson and Jalenak, 1986). The evidence thus points to octopamine being released as a neurotransmitter, but instead of causing the contraction of a muscle, it causes light to be emitted.

6.2. MODULATION OF MUSCULAR ACTIONS

6.2.1. Efferent DUM neurons

All the efferent DUM neurons have octopamine-like immunoreactivity, a few have been shown to contain octopamine, and some have been shown to mediate their effects by the release of octopamine (Evans and O'Shea, 1978; Morton and Evans, 1984) (*see Chapters 3 and 5*). This still leaves open the possibility that they both contain and release other neuroactive substances, and this has been shown for those in the abdominal ganglia that supply the heart (Stevenson and Pflüger, 1994). The effects of many DUM neurons can nevertheless be mimicked by the application of octopamine, with threshold effects at concentrations of 10^{-10} to 10^{-9} mol.l^{-1}, and reduced or abolished by substances that block octopamine receptors.

6.2.1.1. *Effects on skeletal muscle*

Efferent DUM neurons alter the level of force that a muscle can produce, the rate at which it is attained and the rate at which it declines (Fig. 6.2). These effects are caused by direct actions on the muscle fibres themselves and on the terminals of the excitatory motor neurons. They are mediated by receptors on the muscle membrane and on the terminals of the motor neurons that may be of different types (Evans, 1981) (*see Chapter 5*). All the effects greatly outlast the period of spikes in a DUM neuron. The history of spikes in the DUM neurons can thus set the responsiveness of a muscle to its current pattern of motor spikes (Evans and Siegler, 1982).

The **direct** effects on the muscle are on the contractile mechanisms, although they may be accompanied by a small (3–5 mV) hyperpolarisation of the membrane of a muscle fibre (Evans and O'Shea, 1978). Spikes in a specific DUM neuron or the application of octopamine, with the threshold concentrations lying between 10^{-9} and 10^{-8} mol.l^{-1}, reduce the basal tonus in a particular muscle, alter the 'catch' property,

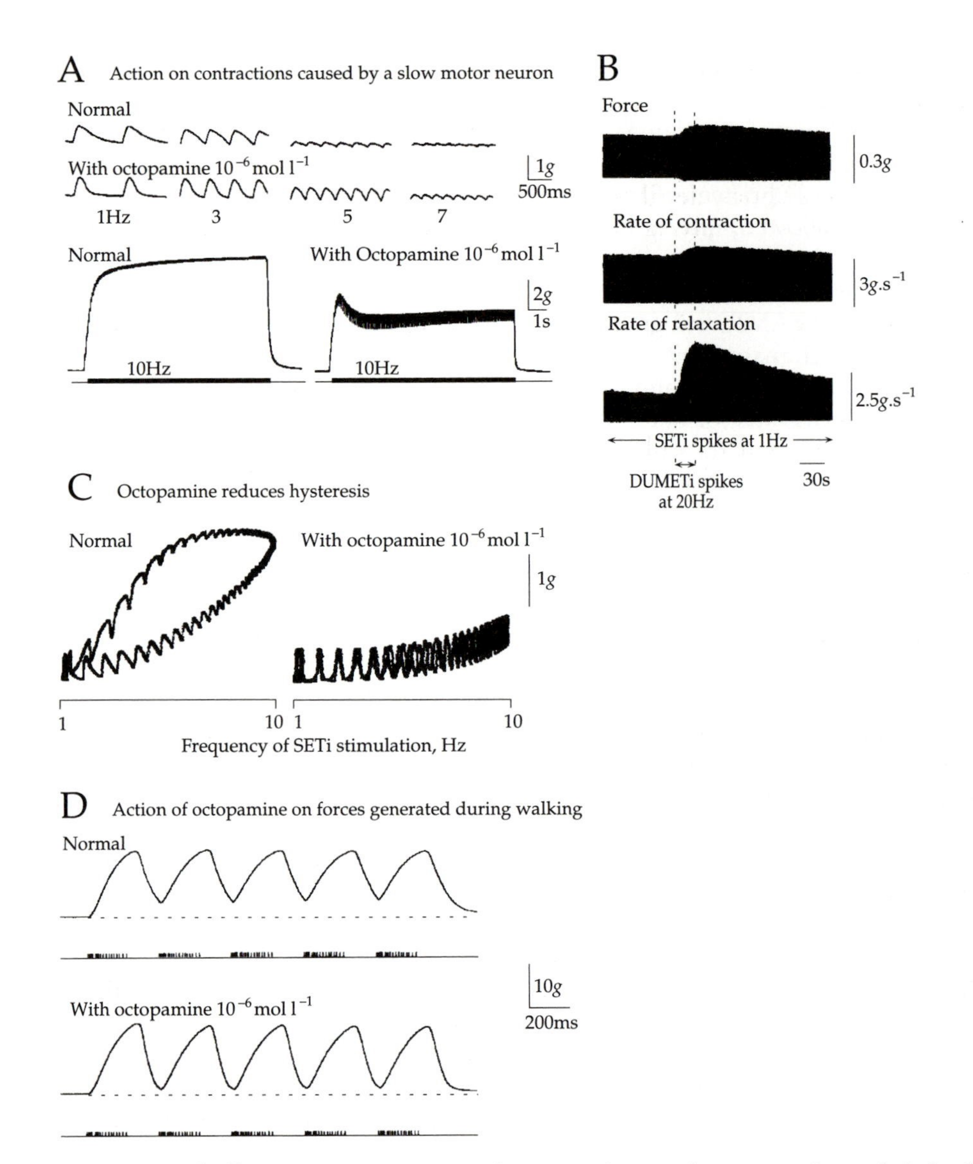

Fig. 6.2 Action of efferent DUM neurons and octopamine on the contractions of skeletal muscles. A. Twitch contractions of the extensor tibiae muscle in a hind leg caused by spikes at low frequency in its slow motor neuron (SETi) are enhanced in amplitude when octopamine is present, but tetanic contractions are reduced. B. Spikes in DUMETi enhance the force, the rate of contraction and the rate of relaxation of twitch contractions caused by SETi. C. Octopamine reduces the hysteresis caused by an increasing and decreasing frequency of spikes in SETi. D. Simulation of the action of SETi during walking. The pattern of spikes during normal walking is recorded and then played back to an isolated leg as stimuli to the axon of SETi. The relaxation of force at each simulated step is enhanced in the presence of octopamine so that between each contraction it returns almost to baseline. Based on Evans and Siegler (1982) and Evans (1981).

and reduce the frequency of rhythmic myogenic contractions in part of the extensor tibiae muscle.

The **indirect** effects involve the alteration of force produced in the muscles by the action of slow motor neurons, the amplitude of their synaptic (junctional) potentials in the muscle fibres, and the frequency of their miniature end plate potentials (mepps), implying both pre- and postsynaptic actions (O'Shea and Evans, 1979).

These effects appear to be similar on all skeletal muscles, including those that move the legs and the wings, but are best illustrated by the control of the extensor tibiae muscle of a hind leg.

Effects on the extensor tibiae muscle This muscle appears to be innervated by just one DUM neuron, called DUMETi (Hoyle, 1974), although it is not known whether branches of the DUM3,4,5 neurons may also be able to exert an effect. Octopamine or DUMETi have four effects on contractions of this muscle.

First, stimulation of DUMETi or the application of octopamine potentiates the twitch force caused by the slow motor neuron, SETi, by as much as 30%, and reduces the duration of the twitch (Fig. 6.2A,B). The effect of the spikes in DUMETi can persist for up to 2 min, but the relationship between number or frequency of these spikes and the duration of their effect is not known. The effect on the rate of relaxation of the muscle contraction has a lower threshold to applied octopamine and an earlier onset than the effect on amplitude, suggesting that different receptors or mechanisms may be involved. Both effects are dependent on the frequency of spikes in the slow motor neuron. At 1 Hz or less, both the amplitude and rate of relaxation are increased, between 1 and 20 Hz individual twitch amplitudes are still potentiated but maintained force is reduced, and above 20 Hz the greatest effect is on relaxation. The mechanisms that lead to changes in the rate of relaxation are not known, and it is possible that some of the differences are due to the contribution from the different types of muscle fibres that are present in virtually all muscles. Thus, the faster relaxation could be caused by the greater contribution of more phasic fibres which produce more force in the presence of octopamine. In the muscles that move the wings, the large effects of octopamine on the rate of relaxation are temperature dependent and disappear at the temperatures that occur in the thorax during flying (Whim and Evans, 1988; Malamud *et al.*, 1988).

Second, the amplitude of the synaptic potential evoked by the slow motor neuron SETi in a muscle fibre is also increased, but the effect is variable and not comparable to the consistent effect on force, suggesting that the two effects are not causally linked.

Third, the frequency of mepps produced by the slow motor neuron is also increased but the effect that this may have on the force produced by the muscle is not known. All three of the above effects are blocked by the α-adrenergic blocker phentolamine.

Fourth, the extensor tibiae muscle, like many other muscles, shows 'catch-like properties' (Blaschko *et al.*, 1931; Wilson and Larimer, 1968; Wilson *et al.*, 1970).

This is seen with three patterns of motor spikes; a period of high frequency interposed between periods of low frequency, the interjection of a few (sometimes just one) extra spikes in a continuous sequence, or a steadily increasing and then decreasing frequency of spikes. Greater force is generated by a particular frequency of spikes if it is preceded by a high spike frequency, so that there is a difference from the expected amount of force produced, and hence hysteresis. The force needed to maintain posture can thus be maintained by a low frequency of spikes, but slow changes in force would be an impediment during the rapid movements that are required during locomotion. These effects are reduced or abolished by spikes in DUMETi, which acts on the contractile mechanisms in the muscle and not on the facilitatory or other mechanisms at the terminals of the slow motor neuron (Fig. 6.2C).

The effects of DUMETi and octopamine thus change the response of the extensor muscle from one favouring posture to one favouring rapid change so that the force generated by the muscle is no longer so dependent on the history of the motor spikes it has received. The effectiveness of this change has been elegantly shown by delivering to the muscle the sequence of slow motor impulses that it normally receives during walking, in the presence or absence of DUMETi spikes or octopamine (Fig. 6.2D). When octopamine is present, the fluctuations in force at each simulated step are much greater even though the total force remains the same, so that the force generated for extension during the stance phase will not overlap and impede that which must be produced by the antagonistic flexor during the swing phase. The rapidity of the limb movements will therefore be enhanced, if indeed octopamine is released by DUMETi during walking, but as yet it is not known how the different DUM neurons are activated in this behaviour.

6.2.1.2. *Effects on visceral muscle*

Octopamine acts on the visceral muscle that controls the movements of the oviducts in a way similar to its action on skeletal muscle in that it reduces basal tension, inhibits any myogenic contractions and relaxes the muscle, but differs in that it reduces the amplitude of neurally evoked twitch contractions instead of potentiating them (Orchard and Lange, 1985). The difference in action on twitch tension could result from an overriding effect on the increased rate of relaxation of the twitch. These effects are blocked by phentolamine and are probably mediated by receptors on the muscles. Proctolin increases the amplitude of neurally evoked contractions, the basal tonus and the frequency of myogenic contractions but has no consistent effect on synaptic potentials (EJPs), although it may hyperpolarise the membrane by a few millivolts (Orchard and Lange, 1986).

6.2.1.3. *Effects on myogenic rhythms*

Some visceral muscles, such as those controlling the heart and the movements of the oviducts, are able to generate myogenic rhythms of contraction, but so too are some skeletal muscles, the exemplar being the extensor tibiae muscle of a hind leg. These movements can be altered by octopamine and by other putative neuromodulators.

Rhythmic contractions of the extensor tibiae muscle occur in a small bundle at its proximal end that consists of 8–12 fibres, each 150 μm in diameter, innervated by SETi and CI1, but not by FETi (May *et al.*, 1979). The bundle is also supplied by nearby endings of DUMETi. The fibres insert onto a short apodeme that joins the main apodeme and can generate only about 2% of the force of the whole muscle. These muscle fibres are electrically coupled to each other and are depolarised slowly in time with the rhythm (Burns and Usherwood, 1978), even when their motor nerve is cut. The contractions occur with a variable periodicity of up to 4 cycles.min^{-1} with each contraction lasting a few seconds, but all these parameters change in a single animal over time. In a quiescent animal, they are sufficient to extend the tibia through an angle of about 30° and may initiate resistance reflexes as a result of the movement. The rhythmic contractions are independent in the two hind legs (Hoyle, 1978a). They may aid the flow of haemolymph, or possibly air in the tracheae, along the long hind legs because they do not occur in the smaller front or middle legs. Similar pulsatile contractions of other structures aid haemolymph flow into the antennae and wings.

The frequency of the contractions is reduced by the application of octopamine, with the threshold lying between 10^{-10} and 10^{-9} mol.l^{-1}, and by spikes in DUMETi. Both act directly on the muscle membrane (Evans and O'Shea, 1978; Hoyle and O'Shea, 1974). The reliability of this rhythm has enabled an analysis of the receptors by which octopamine and other potential neuromodulators might mediate their effects on muscle.

6.2.2. Control of the heart

The heart is a muscular tube that runs dorsally along the length of the body and is constricted into a series of thoracic and abdominal chambers by ostial valves and suspended in the abdomen by alary muscles. It contracts rhythmically at 30–40 beats.min^{-1} to pump the haemolymph forward from the abdomen to the head and laterally though paired vessels that branch from some of the chambers in the middle of the body. These contractions are generated myogenically but are modulated by neural and hormonal secretions.

In cockroaches, the heart is innervated by neurons in the paired lateral cardiac nerve cords that run alongside it, and by neurons from the central nervous system. Two types of neurons have cell bodies in the cardiac nerves; neurosecretory cells, and cardiac ganglion cells that are thought to act as motor neurons and which may show some inherent rhythmicity. Each abdominal chamber is innervated by three pairs of these neurons that deliver rhythmic bursts of spikes to the cardiac muscles when they are contracting. Three types of neurons from the central nervous system also innervate the heart (Johnson, 1966; Miller and Usherwood, 1971). First, are presumed neurosecretory neurons that do not make endings on either the myocardial muscle or the cardiac neurons, but contain large dense core vesicles. Second, are also presumed neurosecretory neurons that synapse directly on the muscle and cardiac neurons, and contain small granular vesicles. Third, are neurons that synapse on the cardiac neurons, but not the muscle, and contain no granular vesicles. It is unclear which of these neurons can be equated with the DUM or bilaterally projecting neurons. Thus, the myogenic rhythm of the heart is modulated by intrinsic cardiac neurons and by neurons from the central nervous system that act directly on the muscle and on the cardiac neurons. There has been extensive pharmacological study of the heart but little knowledge of the nature of these effects or whether the substances tested normally control the heart (Collins and Miller, 1977). ACh increases the heart rate, perhaps by activating the cardiac ganglion cells and the neurosecretory cells, and may result from a sensory input signalling stretch. 5-HT increases the heart rate but decreases the amplitude of the beats. Other amines act with decreasing effectiveness in the sequence synephrine, octopamine, tryptamine, dopamine, tyramine.

6.2.3. Control of the gut muscles

Many of the same problems apply to the control of the gut as to the control of the heart; we know many of the pharmacological effects of applied substances on the intrinsic rhythmicity of the muscles, but have little information on the normal control mechanisms. The motor innervation can override the intrinsic contractions and also operates the various valves that allow food to pass from one region to another.

The fore gut is innervated by the stomatogastric nervous system and by neurons from the tritocerebrum that pass through the frontal ganglion (Aubele and Klemm, 1977). The hind gut in cockroaches is innervated by paired proctodeal nerves from the last abdominal ganglion, but may also be affected by nerve cells on its outer surface. Glutamate and proctolin seem the most likely candidates for exerting control. The effects of applied substances are hard to interpret and often have different effects in different species or on the different parts of the gut of one species. Dopamine has an inhibitory effect on locust fore gut and may be a naturally occurring neuromodulator. 5-HT, which is present in the muscles of the gut, increases the frequency and amplitude of contractions.

6.3. MODULATION OF SENSORY RESPONSES

The peripheral cell bodies of most sensory neurons are exposed to the substances carried in the haemolymph and to those released in their vicinity by neurosecretory neurons that are either targetted at them directly or at the nearby muscles. Any neuromodulators could thus have access to the sensory neurons and change the way they code and transmit the signals detected by their receptor organs. The effect of these neuromodulators is to stretch the time course of changes that occur intrinsically in sensory neurons during use from milliseconds and seconds to minutes or even hours. They can change the sensitivity of the receptors to keep them in tune with the daily fluctuations of environmental stimuli, and match their responses to the performance of specific motor acts such as flying or walking. Such changes in sensitivity indicate that sensory neurons are not the providers of a carefully calibrated and immutable signal, but instead shift their responses in the context of the release of certain chemicals. The single cell stretch receptor at a wing hinge (*see Chapter 11*) and the multicellular chordotonal organ at the femoro-tibial joint of a hind leg (*see Chapters 7* and *8*) are receptors that have been examined in detail and thus give the best indications of the sorts of effects that occur. It seems that we have only just begun to appreciate the possibility that specific neuromodulatory neurons may supply particular sense organs and to digest the consequences that this may have for the integration of sensory signals.

6.3.1. Wing stretch receptor

The number of spikes produced by a stretch receptor in response to an imposed movement of a wing is increased by raising octopamine levels in the haemolymph to those normally found at the start of flight (about 2×10^{-7} mol.l^{-1}) (Ramirez and Orchard, 1990). 5-HT is as effective as octopamine, but it is not known whether it activates the same receptors, while other amines are less effective and dopamine is ineffective. An obvious source of the octopamine is from the efferent DUM neurons (DUMDL) that supply the nearby dorsal longitudinal muscles. Stimulating the axon of this neuron in N1 on the opposite side of the body to the stretch receptor being tested leads to an increased response to wing movements, but the enhancement is greater if the other lateral nerves containing the axons of other DUM neurons are also stimulated. Neurons other than DUMDL could only affect a stretch receptor by releasing their octopamine into the haemolymph as their endings are not close to the receptor.

The physiological significance of these changes has not been explored, but a stretch receptor of an isolated wing hinge can generate the same frequency of spikes as is recorded in the intact flying locust in the absence of octopamine. Assuming that octopamine does increase the frequency of stretch receptor spikes during normal flight, then this would imply a shift in the coding of wing position based on frequency that would have to be balanced by the decoder of these signals, unless the amplitude of the wing movements is to be curtailed by the effects of the connections made by the stretch receptor with central neurons (*see Chapter 11*).

6.3.2. Femoral chordotonal organ

Of the many sensory neurons in this organ that code the various parameters of joint position and movement, only those that signal position are affected by octopamine at a threshold concentration of 5×10^{-7} mol.l^{-1} (Ramirez *et al.*, 1993). The frequency of spikes in the position-sensitive neurons reaches a maximum within 2 min of application and persists for about 10 min before declining to previous levels despite the continuing presence of octopamine. The response is then desensitised to subsequent applications of octopamine. The effect is thought to be directly on the sensory neurons and not on the basal tension in the surrounding muscles, although the type of octopamine receptor involved is not known. These effects imply that the coding of position based on the frequency of spikes in the sensory neurons would not alone be sufficient to indicate absolute position. Instead, it suggests that the frequency of spikes in a sensory neuron can only indicate a particular position of the joint if it is read against the effects of this and probably other neuromodulators, perhaps by concomitant changes in the neurons performing the decoding. It is also curious that an effect restricted to the neurons that provide static information should be transient, for it implies again that the frequency code for position will be constantly shifting as the release of octopamine first increases, then decreases and finally desensitises their responses.

For neither of these sensory receptors is the mechanism of action of octopamine understood, but in the femoral tactile spine of the cockroach it seems to act by reducing the adaptation of the spikes (Zhang *et al.*, 1992). The spikes in the single sensory neuron of this receptor adapt rapidly to an imposed deflection of the spine, following the membrane potential with a delay that has two exponential components; a slow component with a time constant of 1000 ms that seems to involve a Na^+ pump, and a fast component with a time constant of 100 ms that involves inactivation of Na^+ channels. Octopamine reversibly raises the resting threshold by increasing the fast but decreasing the slow component. This could explain the increased spiking of the wing stretch receptor if the primary effect were on adaptation.

Perhaps the most important message from the effects of octopamine on these sensory receptors is that the signals in their sensory neurons cannot be viewed as invariant for a particular stimulus. The sensory neurons clearly possess receptors for various substances that could act as neuromodulators and their coding of the same sensory stimulus will be altered when these substances are present.

6.3.3. Auditory receptors

Some of the DUM neurons in the prothoracic ganglion (to which the auditory sensory neurons project) of the cricket respond to sound (Gras *et al.*, 1990) and have some branches in the auditory neuropil. High concentrations of octopamine increase the responsiveness of certain interneurons to sound. All of this is circumstantial evidence for a role of octopamine in modulating the auditory pathway, but whether this is carried out by the efferent DUM neurons is yet to be established. In the locust, many of the local DUM neurons that contain GABA also respond to sound, but their contribution to the processing of these stimuli is not known (Thompson and Siegler, 1991).

6.4. MODULATION OF CENTRAL SYNAPSES

Determining the effects of putative neuromodulatory substances on central synapses is fraught with difficulties that, for the present, have to be accepted, until synaptic connections can be established in culture between identified neurons that retain their inherent characteristics. The problems are of three main types.

First, establishing that the putative neuromodulators are actually released in the central nervous system. For many of these substances, immunocytochemistry shows them to be present in the terminals of neurons at probable release sites. For others, such as octopamine, the evidence is less clear. For example, the efferent DUM neurons that contain octopamine have many branches within the central nervous system but convincing release sites have not been demonstrated, let alone release itself. Its release and potential effects are still anticipated, however, because an adenylate cyclase that can be activated by low concentrations of octopamine is present in the central nervous system (Nathanson and Greengard, 1973), and octopamine receptors are locally abundant.

Second, gaining access to the synapses is restricted by a neural sheath and the tight packing of the neurons. The sheath around the nervous system restricts entry, and even the exchange of ions can take up to an hour, with little certainty about the completeness of the exchange. Furthermore, the tight packing of neuronal processes within the neuropil severely restricts the passage of larger molecules. It must be assumed that many neuromodulators will be released locally in the neuropil, close to the neurons and synapses that are to be affected, so that bath application of drugs, or their injection into specific regions of neuropil will only, at best, be a weak approximation of the real events.

Third, determining whether effects observed on a particular neuron or synapse are direct. All the central neurons are embedded in networks of synaptic connections with other neurons, so that an applied substance is likely to affect many neurons that can then have secondary effects on the neuron being studied. It takes the most sophisticated of experiments and the most careful of interpretations to learn about the sites of actions of these substances and the mechanism by which they may be exerting their effects. Often it is necessary to be satisfied with a description of the effects without being able to attribute them to causal mechanisms.

6.4.1. Synapses between leg motor neurons

Only one motor neuron has been found to make a direct connection with another motor neuron in the central nervous system, but fortunately these neurons are large and allow recordings to be made simultaneously pre- and postsynaptically. This means that there is a good chance to dissect the effects of potential neuromodulators on this synaptic connection. The fast motor neuron (FETi) to the extensor tibiae muscle of a hind leg makes direct excitatory connections with flexor tibiae motor neurons (Burrows *et al.*, 1989; *see also Chapter 9*). The connection does not occur between the comparable neurons of the front and middle legs.

6.4.1.1. *5-HT*

In these leg motor neurons of the locust, 5-HT causes an increase in membrane resistance and a broadening of the spikes and reduction of their after-hyperpolarisations (Parker, 1995a). These effects are mimicked by imipramine, which inhibits the uptake of 5-HT, and are blocked by the 5-HT_2 receptor antagonist ketanserin and probably involve the modulation of a K^+ channel. This conclusion is supported by the fact that if the spike is first prolonged by the application of TEA then the addition of 5-HT has no further effect. The modulatory effect can also be mimicked by the intracellular injection of cAMP, and by the bath application of dibutryl cAMP and the phosphodiesterase inhibitor 3-isobutyl-1-methyl xanthine, and blocked by a protein kinase inhibitor. The receptor is thus similar in its properties to the first type of receptor found on cell bodies (see Table 5.1). The precise K^+ current that is modulated in locusts has yet to be elucidated but in neurons cultured from the antennal lobe of *Manduca sexta*, 5-HT reduces a rapidly activating A-type K^+ current, a relatively slowly activating K^+ current resembling the delayed rectifier type, and a voltage-dependent Ca^{2+} current (Mercer *et al.*, 1995). Not all the neurons were affected in this way, suggesting that there are differences in the distribution of 5-HT receptors on different neurons.

The result of these effects is to increase the excitability of motor neurons to synaptic inputs from other neurons and to increase the efficacy of their spike-mediated output synaptic connections. Thus, the spike of FETi is broadened and the amplitude of the EPSP generated in a flexor motor neuron by its monosynaptic connection is increased (Parker, 1995a). Moreover, the increased membrane resistance of the flexor motor neurons means that synaptic inputs from other sources are more likely to make them spike. The result will be that more motor spikes and at higher frequencies will be generated, which should result in changes in the forces produced by the leg muscles during such movements as walking and jumping. It seems likely that endogenous 5-HT normally has access to these neurons and their synapses because blocking its uptake enhances the transmission at the synapses between them.

5-HT produces similar effects in interneurons of the antennal lobes of *Manduca sexta* (Kloppenburg and Hildebrand, 1995). In the retina of locusts, the photoreceptors show diurnal changes in the properties of certain K^+ channels and these effects can be mimicked by 5-HT (Cuttle *et al.*, 1995). During the day, depolarisation caused by light activates a delayed rectifier K^+ current, but during the night this current is reduced and instead depolarisation activates a transient current. 5-HT thus matches the response properties of the photoreceptors to the differing behavioural requirements of day and night and promotes the efficient coding of visual signals over a range of light intensities. The common thread through the various effects of 5-HT in insects, and many other animals, therefore seems to be its modulation of particular K^+ channels.

6.4.1.2 *Octopamine*

The various effects that octopamine produces on both the pre- and the postsynaptic neurons (Parker, 1995b), following bath application or injection into the neuropil, reflects its extensive arousal effects on many neurons. In general, it reduces the amplitude of the synaptic potential in a flexor evoked by a spike in FETi, but the concomitant increase in the excitability of the postsynaptic neuron means that the reduced potential is still able to evoke more spikes. The contributory causes of this change are many. First, octopamine may have a direct depolarising effect on the postsynaptic flexor motor neurons. Second, the duration of the spike in the presynaptic neuron is increased and its after-hyperpolarisation reduced, but this is an indirect effect that results from an increased sensory input, in turn caused by the enhanced contraction of the extensor muscle in the presence of octopamine caused by spikes in FETi. Third, in both the FETi and flexor motor neurons there is an increase in synaptic inputs resulting from the increased spike activity in other neurons. This causes a substantial depolarisation of the postsynaptic flexor motor neurons.

6.4.1.3. *Functional consequences*

The potency of 5-HT and octopamine suggests that they can substantially alter the contribution of these neurons to behaviour. For example, during jumping (*see Chapter 9*), the increase in excitability caused by octopamine should lower the threshold for initiation of the motor pattern. The initial cocking phase of the jump, in which the tibia of a hind leg is flexed fully about the femur, will be shortened because the increase in membrane resistance of the flexors caused by 5-HT will increase the number of spikes produced in response to the same synaptic inputs. The co-contraction phase, in which contraction of the flexor muscle restrains the contraction of the more powerful extensor muscle, will be enhanced because the broadening of the FETi spikes will release more transmitter onto the flexors, and the reduction in their after-potential should allow spikes to be generated at a higher frequency. The flexors will thus receive an enhanced synaptic input to which they should produce a greater spike response, thus reinforcing the co-contraction by allowing greater forces to be produced, and perhaps reducing the duration of this phase. The resulting jump should thus be produced more quickly and be more powerful. FETi of a hind leg is rarely, if ever, used in walking so these modulatory effects will not come into play during these movements, but similar effects can be envisaged at the numerous other synapses that could increase the participation of certain other motor neurons during walking.

6.4.2. Altering the properties of interneurons

The excitability and the membrane properties of some interneurons, particularly those that are involved in the generation of flight (*see Chapter 11*), are altered when ganglia are bathed in octopamine (Ramirez and Pearson, 1991a,b). Many of these intrinsic changes are due to the expression of voltage-dependent plateau potentials (*see section 3.3.4.3*). The result of activating these currents is that a synaptic potential, or a brief pulse of depolarising current injected into a flight interneuron, produces a depolarisation and spikes that outlast the input. This effect is dependent on the voltage of the interneuron and can be terminated by a brief hyperpolarising pulse. In some interneurons, a prolonged depolarisation that would normally only produce a tonic sequence of spikes will, in the presence of octopamine, generate bursts of spikes that are caused by the intrinsic properties of the neurons and do not result from feedback from other neurons. During sequences of fictive flight activity induced by high external concentrations of octopamine, the interneurons produce rhythmic sequences of spikes. Short hyperpolarising pulses injected into an interneuron during the spiking phase of the cycle reset its membrane potential, suggesting that the plateau potentials contribute to the overall pattern. The ability of some interneurons to express voltage-dependent nonlinearities is thus established, but what is harder to assess is whether these properties are expressed during normal behaviour and therefore contribute to the motor pattern. The ability of at least two interneurons involved in flight to produce bursts of spikes appears to be intrinsic and can be triggered by injecting current in the absence of octopamine, or by octopamine itself (Ramirez and Pearson, 1993). During flight, the inputs that these interneurons receive from particular sets of sensory neurons are also able to activate these currents so that they produce bursts of spikes. The consequence is that the sensory input is amplified in a way that is potentially able to alter the movements of the wings during flight. Presumably, the presence of octopamine during flight can lower the threshold that must be exceeded by the synaptic input to induce these properties.

The concentrations of octopamine in the bathing haemolymph that are necessary to expose these properties are nevertheless much higher than those which occur naturally during flight, but just what sort of release occurs within the central nervous system itself is unknown. The efferent DUM neurons that are active at the start of flight are known only to release octopamine in the periphery, and it is not known whether the octopaminergic interneurons (*see Chapter 3*) are activated during flight. It can be argued that the high concentrations used experimentally are necessary simply to allow enough octopamine to cross the sheath of the ganglion, but it will then affect many neurons at the same time, whereas release by normally activated interneurons may result in high local concentrations and thus the selective activation of a limited number of target neurons. The resolution of this issue awaits the selective activation of interneurons that release octopamine or other neuromodulators during the expression of a normal motor pattern. Meanwhile, it is necessary to be mindful of the sorts of changes in central neurons that neuromodulators have the power to affect and the drastic consequences this may have on integrative mechanisms. The powerful message about the effects of neuromodulators on the motor patterns produced by the stomatogastric nervous system of crustaceans must be firmly incorporated into our thinking about the functioning of interneurons and the networks that they form in other nervous systems.

6.4.3. Altering the properties of efferent DUM neurons

The neurons that release octopamine are themselves affected by octopamine and have octopamine receptors (Morton and Evans, 1984). Their membrane resistance is increased in a concentration-dependent manner and, probably as a consequence, the frequency of spontaneous spikes caused by synaptic inputs also increases (Washio and Tanaka, 1992). This means that if the central neurites of the efferent DUM neurons are exposed to the octopamine that is released from their own terminals, a positive feedback will be set up.

6.5. MODULATION OF MOTOR PATTERNS AND BEHAVIOUR

If neuromodulators can alter the performance of both the individual components that make up a neural circuit and the muscular effectors of those circuits, then it is reasonable to expect that these effects will be expressed as a change in behaviour. This, of course, depends on the demonstration that these substances are normally present, that there are neurons containing them, that there are release sites and that there are specific receptors for them on other neurons. The presence of putative neuromodulators can be demonstrated by immunocytochemistry, often at the level of individual neurons, while the demonstration of binding sites (= receptors or uptake sites?) has to rely on the lower resolution of autoradiography or biochemistry applied to whole regions. Nevertheless, discrepancies can be detected in particular regions of neuropil where binding sites are present but where neurons containing the putative activator of this receptor are not present, or *vice versa*. For example, in the calyces of the mushroom bodies of the brain of bees there is no 5-HT immunoreactivity but high densities of 5-HT binding sites, and in the β lobe there is no octopamine immunoreactivity but a high density of octopamine binding sites (Erber *et al.*, 1993). Thus, there can be receptors for a particular substance in regions where there are no neurons that release the substance, perhaps implying a role for these receptors in neurohormonal rather than local neuromodulatory responses.

The usual method adopted to test for the effects of putative neuromodulators is to inject them into ganglia, topically apply them to parts of the nervous system, or simply inject them into the whole animal. The resulting effects are then ascribed to the neuromodulatory action of the substance, but in reality are much more difficult to interpret. Injections into a ganglion will necessarily indiscriminately activate receptors on any, and many, nearby neurons, whereas we imagine that release from the neuromodulatory neurons themselves will be more closely directed at specific neurons that may not all be in the same anatomical location. Furthermore, each receptor generally has a high affinity for one chemical but lower affinities for some others, so that the high concentrations of putative neuromodulators that are used experimentally may activate receptors other than those for the substance being tested. Injections into the body typically require higher concentrations than are ever measured in the haemolymph of a normal animal to produce an observable change in the action of central neurons. Changes in muscle performance and responses of sensory neurons can, however, be produced by physiological concentrations. The rationale for the use of such high concentrations is that the barrier around the central nervous system limits access, so that the concentration reaching the central neurons will be much lower. This logic inevitably, therefore, ignores the effect that octopamine may have on cells forming the barrier itself (Schofield and Treherne, 1986) and hence indirectly on the performance of cells within the central nervous system. The experiments are trying to simulate changes that would be caused by release within the central nervous system, but of course they do this only in the most crude of ways, for their effects will be much more widespread that those caused by local injections. To overcome some of these problems, it is thus always necessary, but not always practicable, to test a range of putative substances and show that any effects can be blocked by specific antagonists, or reproduced by specific agonists, before they can be attributed to a particular substance. It is also necessary to eliminate possible spurious effects such as those caused by changes in pH. Finally, it must be recognised that these methods represent only an interim step in the analysis and must eventually be replaced by the activation of specific neurons or groups of neurons so that the site of action, concentrations and perhaps the mix of substances released will be governed by the connections and distribution of terminals of these neurons.

These caveats therefore invite extreme caution in the interpretation of the effects of putative neuromodulators on behaviour. Nevertheless, the positive side of these experiments is that they emphasise the widespread distribution and action of receptors for various neuromodulators, and that the behaviour generated by neural circuits can be moulded to show much plasticity.

6.5.1. Changes in reflex pathways

Octopamine can affect all the neuronal elements in a reflex pathway and also the effector muscles themselves, so that it is impossible to predict what its effect might be on the resulting movements. The effects have been measured in the stick insect for the reflexes that are evoked in leg muscles by imposed movements of the apodeme of the femoral chordotonal organ (Büschges *et al.*, 1993). In a quiescent animal, the imposed movements evoke a resistance reflex that is reversed when the animal is active (*see Chapter 8*). Injection of 50–100 µl of 5×10^{-2} mol.l^{-1} octopamine into the haemolymph initially causes the animal to become active so that the evoked reflexes are similar to those in a normally active animal, but this is then followed by a prolonged suppression of any reflex response. When the octopamine is applied directly to the surface of the mesothoracic ganglion, the resistance reflexes of the middle legs are blocked, but when the animal is stimulated to be active the reversed reflex can still be evoked. These results are interpreted to indicate that the two reflexes involve different pathways and that the effect of octopamine is to reduce resistance reflexes that would impede locomotion when the animal is active. The levels of octopamine that are required to produce these effects, however, result in the animal being unable to support the weight of its own body, which is an essential requirement for locomotion. Again, therefore, the possible ways that neuromodulators might alter the processing in these sorts of pathways are indicated but fail to show whether they might normally do so.

6.5.2. Flight motor pattern

The motor pattern for flight can be induced in larval locusts that have no wings, and would not therefore normally express a flight motor pattern, by perfusing the thoracic ganglion with octopamine. In adult locusts flight patterns can also be induced by injection of octopamine into specific areas of neuropil (Sombati and Hoyle, 1984c; Stevenson and Kutsch, 1987, 1988) (*see section 6.8.1 and Chapter 11*). Similarly, in the moth *Manduca sexta*, injection of dopamine into specific regions of the mesothoracic ganglion evokes a flight motor pattern with a latency of a few seconds, even in the absence of sensory feedback (Claasen and Kammer, 1986). Octopamine injected into these same regions of the moth either evokes unpatterned activity in flight motor neurons, or the flight motor pattern in the presence of sensory feedback. The effective site is different to that at which dopamine is most effective, and produces effects only after delays of 2–20 min. 5-HT suppresses the patterns induced by either dopamine or octopamine. There are no explanations for the different sites of actions of dopamine and octopamine, or for the long delays in the effects of octopamine. All of the descriptions of these sorts of experiments emphasise that the injections must be precisely located in specific regions of the neuropil to be effective. It remains uncertain what is specific in these regions and upon what neurons the injected substances may be acting. There is no evidence that the neurons which contain these substances have release sites in these areas and therefore that this is a mechanism by which flight is normally initiated. Indeed the generally slow time course of action of modulators would make it seem unlikely that they are used to initiate a movement that needs to be initiated rapidly in many life threatening situations.

6.5.3. Ventilatory motor patterns

Possible effects on the ventilatory movements of locusts have not been tested, but the frequency of ventilation in the aquatic larvae of the dobson fly *Corydalus cornutus* is increased by about 16% by the application of 10^{-5} mol.l^{-1} octopamine and is decreased by a similar amount by the same concentration of dopamine (Bellah *et al.*, 1984). It is not known whether these effects are specific to the motor pattern of ventilation or reflect wider changes in excitability.

6.5.4. Motor pathways for escape

Escape running in cockroaches can result from the activation of ascending interneurons by wind-sensitive hairs on the abdominal cerci (*see Chapter 10*). The response of the motor neurons to wind directed at the cerci, or to electrical stimulation of the ascending interneurons, is reduced when 10^{-3} mol.l^{-1} 5-HT is applied topically to the metathoracic ganglion, and enhanced when 10^{-3} mol.l^{-1} dopamine is applied (Goldstein and Camhi, 1991). When these substances are applied to the last abdominal ganglia they have no effect on the summed activity of the ascending interneurons, whereas octopamine, which is without effect on the metathoracic ganglion, enhances their response. The synaptic potentials generated in some thoracic interneurons by ventral giant interneurons are increased in amplitude when the thoracic ganglia are bathed in 10^{-4}–10^{-2} mol.l^{-1} octopamine, and decreased when bathed in 10^{-4}–10^{-3} mol.l^{-1} dopamine (Casagrand and Ritzmann, 1992). The specificity or primacy of these effects on the thoracic interneurons is unknown, but they could contribute to the changes in the responses of the motor neurons when these high concentrations of amines are applied. If this is to be a mechanism used to change the responsiveness or normal behaviour, then the neurons that trigger the movement must activate the neurons that are to release the neuromodulators in advance of activating the thoracic motor circuitry. Given the rapidity of the transmission in the pathways that activate the motor neurons and the long time course of the action of neuromodulators, this would seem unlikely. The activation of the neuromodulatory neurons could thus only be a mechanism to alter responsiveness to subsequent stimuli, but this is clearly a valuable effect.

6.5.5. Antennal and mouthpart movements

In bees, the antenna and proboscis are involved in three motor responses that can be modified by learning (Erber *et al.*, 1993). First, a bee will extend its proboscis when small drops of sugar water are applied to an antenna. Second, the antennae are moved towards a source of olfactory or gustatory stimuli and then scan the stimulus with rapid movements that contact the source. Third, the antennae point in the opposite direction to a vertically moving pattern of black and white stripes; they point down when the stripes move up and *vice versa*. All three responses are enhanced by the injection of octopamine and reduced by the injection of 5-HT, and the storage and retrieval of signals learnt during these responses is enhanced by octopamine and reduced by 5-HT.

6.5.6. Solitary and gregarious phases

One of the biggest behavioural and morphological transformations that occurs in the life of a locust results from its living in crowded conditions with many other locusts. Differences between these solitary and gregarious phase locusts have been correlated with higher levels of octopamine in the heads of solitary *Locusta migratoria* (Fuzeau-Braesch and David, 1978). In *Schistocerca gregaria*, however, no differences are found in the levels of octopamine in the nervous system or in the muscles in the two phases (Morton and Evans, 1983). The levels found in *S. gregaria* are more in keeping with reasonable expectations, so that these results are not likely to reflect a species difference and seem to rule out a causal role for octopamine in determining phase differences.

6.6. MODULATION OF EPITHELIA

Epithelia, in general, may receive a direct aminergic innervation. The plasticisation of the cuticle following feeding of the blood-sucking bug *Rhodnius prolixus* may be under direct neural control and may be mediated by 5-HT receptors (Maddrell, 1966). Similarly, a direct aminergic innervation may control the erection of hairs on mosquito antennae (Nijhout, 1977). The salivary glands receive a complex innervation from neurons containing GABA, dopamine, 5-HT and octopamine which regulate the rate of secretion and the relative concentrations of the constituents in the saliva. Some of these neurons may be active in association with feeding movements so that release is regulated in accordance with the behavioural demand (Schactner and Bräunig, 1995).

6.7. MODULATION OF METABOLISM

The metabolism of a locust must change to meet diverse behavioural needs, and at no time is there a greater and more dramatic need for a rapid change than when the locust starts to fly. So great are the energy demands of flight that they cannot be met from the small reservoirs in the muscle, so other stores must be mobilised. During the first 20min of flight, the energy requirements are met by an increase in trehalose levels in the haemolymph, but this then declines and is replaced by a steady-state usage of lipids from the fat bodies that accounts for 75% of the energy needs. It is not known what triggers the rise in trehalose concentration or whether its fall sets in motion the sequence of events that leads to the mobilisation of lipid stores.

The mobilisation of lipid from fat bodies is controlled mainly by the adipokinetic hormones (AKH- I and -II) that are released from the corpora cardiaca into the haemolymph during flight. The level of these hormones in the haemolymph begins to rise about 10 min into a flight, after an initial peak in the levels of lipids and cAMP, but preceding a sustained rise in their levels. The release of these hormones is, in turn, controlled by neurons in the brain with axons in nervus corporis cardiaci II (NCCII) that make synapses directly onto the cells secreting the hormones. Release of the hormones can be caused by stimulation of NCCII and mimicked by the application of high concentrations of octopamine in the presence of a phosphodiesterase inhibitor, thereby implicating cAMP in mediating the effect. Whether neurons with axons in NCCII actually release octopamine to bring about these effects is uncertain.

In addition to its putative effect on the release of AKH, octopamine may act directly on the fat bodies to mobilise their stores, and in keeping with this proposition the level of octopamine in the haemolymph increases to 2×10^{-7}mol.l^{-1} during the first 10 min of flight when there is also an initial peak in the level of lipids (Goosey and Candy, 1980). It could thus provide an initial boost to the mobilisation of the lipid stores before the secretion of AKH takes over, and its presence may be essential for the action of AKH to be fully expressed. It may also act directly on the flight muscles themselves to increase the metabolism of, for example, glucose and trehalose (Candy, 1978).

In cockroaches, octopamine may also affect the release of other hormones (Thompson *et al.*, 1990). The production of juvenile hormone, which regulates metamorphosis and reproduction, by the corpora allata is reduced by octopamine which probably acts to elevate levels of cAMP, to hyperpolarise the neurosecretory cells and alter the electrical coupling between them, and to promote the production of an inhibitory neuropeptide allostatin.

6.8. INTEGRATED ACTIONS OF NEUROMODULATORS

Many of the actions of the neuromodulators when they are combined are often envisaged as shifting the animal from one behavioural state to another or, in other words, altering the state of arousal. In flight, for example, just one neuromodulator, octopamine, has the following effects on the performance of the muscles, the central neurons, and the sensory neurons.

1. Increases the mobilisation of lipids from fat bodies.
2. Initiates, or modulates the release of AKH from the corpora cardiaca.
3. Increases metabolism of flight muscles.
4. Increases the force and accelerates the rate of relaxation of flight muscles.
5. Reduces the 'catch-like' property of muscle.
6. Increases the responsiveness of sensory neurons to movements of the wings.
7. Changes the membrane properties of flight interneurons so that they generate plateau potentials and bursts of spikes.
8. Facilitates or initiates the production of a flight motor pattern.

Some of the octopamine must originate from the increased release from some of the efferent DUM neurons when they spike during flight, but it is unlikely that this activity can account for all the effects, most notably those on the central neurons. The summed effect of all these changes is that the locust is now in a more responsive state for flight. Octopamine cannot be the sole cause of the changes that occur and its actions must be seen in the context of other neuromodulators and other neurons that are activated.

6.8.1. Orchestration of motor responses

The role of neuromodulatory neurons has been elevated to that of initiating particular movements by selecting from the ensemble of neurons which should be activated at a particular time. This role has even been given the status of an hypothesis called the 'Orchestration hypothesis' (Sombati and Hoyle, 1984c). There are two basic propositions. First, that by a general release of their modulator(s), neuromodulatory neurons can raise the excitability of neuronal networks. Second, that the action of particular neuromodulatory neurons causes particular movements to be initiated. These ideas were developed from experiments on efferent DUM neurons and from the application of the modulator that they release, octopamine, though it is easy to see that they may have more general applicability. The efferent DUM neurons were suggested to act in one of three ways. First, the neural commands to execute a motor pattern activate the DUM neurons which then in turn activate selected parts of the motor circuitry. Second, the neural commands activate in parallel the DUM neurons and the appropriate motor circuitry. Third, the neural commands independently activate the DUM neurons and the motor circuitry. It is envisaged that different sets of DUM neurons are activated during different movements and by virtue of their projections to different parts of the neuropil would be able to influence different sets of neurons. Support for this idea may be taken from the observation that injection of octopamine into specific locations within the neuropil can release the motor pattern for flying (Sombati and Hoyle, 1984c; Stevenson and Kutsch, 1987, 1988) (*see Chapter 11*). Moreover, in kicking movements at least (*see Chapter 9*) only certain DUM neurons are active and these are the ones that supply the muscles used in these movements. It would seem probable that the other DUM neurons are also used selectively in other movements. None of these observations imply a causal role for the neuromodulatory neurons in initiating movements. Furthermore, it is not known whether efferent DUM neurons release octopamine in the central nervous system, whether the distribution of the different DUM neurons is sufficiently different to activate different subsets of neurons, and more importantly whether this is the actual route by which movements are initiated. No experiments have sought to disrupt the actions of particular DUM neurons and then observe possible effects on movements. The slow time course of most modulatory effects would suggest that this is not an appropriate initiating route. It seems more likely that the modulatory neurons act in parallel with other components of the motor circuits, influencing the properties of the neurons that are activated and perhaps which ones are recruited, and certainly altering the performance of the peripheral motor and muscular machinery. The hypothesis certainly encourages the need to think of the neuromodulatory neurons as integral parts of the circuits that generate and control movement, and it suggests clear experiments by which it could be tested. In the 10 years since its formulation, however, not much progress has been made.

There is much that is attractive in these ideas and much supportive evidence in other animals (particularly crustaceans) for neuromodulators being able to alter the participation of certain neurons in different movements. In locusts, however, the evidence is almost totally lacking because it is not known whether any of the identified neuromodulatory neurons actually influence neurons in the central nervous system.

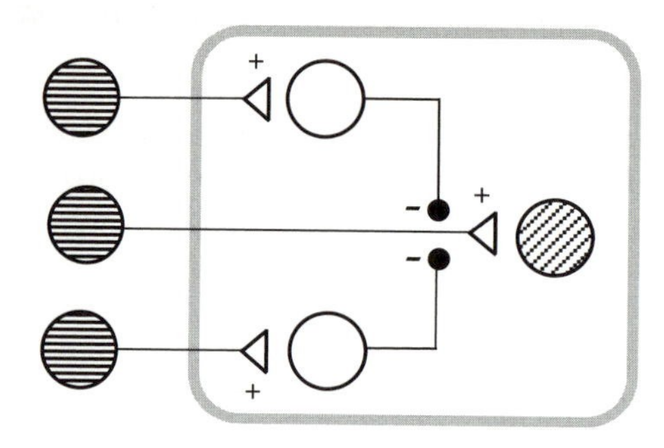

7

Controlling local movements of the legs

7.1. LOCAL MOVEMENTS AND LOCAL PROCESSING

Local movements are defined as those caused by the contractions of the muscles in a single segment. These are most often expressed as a movement of a single appendage such as a leg, wing or antenna in response to a mechanical stimulus detected by receptors on that appendage; a withdrawal movement of a leg when it touches an object, a twisting of a wing when its loading changes, and the movement of an antenna to a change in air currents. The neural processing underlying these movements is carried out within a restricted part of the central nervous system. For the movements of a leg, for example, the processing is performed in a single segmental ganglion so that the movements can still be produced when that ganglion is isolated from the rest of the central nervous system. These local movements of a single leg are therefore the result of local processing within a single ganglion.

The movements of a leg rely for their execution on a combination of signals generated by the central nervous system, from sensory receptors on or in the appendage itself, and on signals from receptors on other parts of the body, most notably the head. To perform its devolved functions, a segmental ganglion processes the sensory signals from its appendages and makes the necessary adjustments to its motor output. Much of this corrective role can be performed without reference to events elsewhere in the body and without requiring signals to be sent to those other parts. During normal behaviour, however, it must place these local movements in the context of the movements of the whole animal and thus frequently refers to the actions of the other ganglia. This means that the information processed by an individual ganglion is conveyed to other parts of the nervous system, and similarly its own action is placed in the context of the operation of the other ganglia. The many local actions of one leg are dependent upon, and in turn influence, the movements of the other legs through the reciprocal exchange of signals between segmental ganglia.

Analysis of neuronal mechanisms underlying the control of local movements is thus simplified because each ganglion contains the complete complement of neurons necessary for their production, and this also means that a restricted number of participating neurons must be considered. The analysis of local movements is best illustrated by reference to the movements of a leg.

7.2. SENSORY NEURONS OF A LEG

The sensory receptors associated with a leg can be divided into two types, **exteroceptors** and **proprioceptors**, although for some the distinction may not be clear cut.

Exteroceptors provide the locust with specific information about the environment through which a leg is moving, and tell it whether it is in contact with an object. This frees the eyes from the task of observing what the legs are doing and where they must be placed (an almost impossible task with six legs), and allows them to concentrate on surveying the route ahead and scanning for other hazards. Most of the exteroceptors on a leg respond only to touch and are therefore of no predictive value, but merely allow a movement to be made when contact has been detected. A few are sensitive to the air currents that might be generated by an approaching predator and thus can initiate an escape response (*see Chapter 10*). Other exteroceptors give information about contact with chemicals, and the design of some, at least, of these chemoreceptors allows them to detect strong airborne odours.

The proprioceptors give information about the movement and position of joints and about the strains that are created in the exoskeleton and muscles as a leg is moved. Essentially, these receptors monitor the performance of the muscles, joints and exoskeleton against the changing loads and forces that the leg experiences. They are also essential in dealing with the nonlinear properties of much of the motor machinery.

The exteroceptors are far more numerous than the proprioceptors and they provide a fine grain map of the surface of the body and limbs so that contact can always be detected and located spatially. They are spaced over a leg with the highest density on the most exposed parts that are most likely to be stimulated. The proprioceptors, by contrast, are concentrated at the joints of a leg and at particular regions on the cuticle where the greatest strains are generated. Consequently, of the 10 000 sensory axons from receptors associated with a hind leg perhaps only a few hundred are proprioceptors, whereas the rest are mechano- and chemoreceptive exteroceptors.

7.3. EXTEROCEPTORS

The surface of the body and its appendages are sparsely covered with hairs of different types that occur in different densities and proportions in different regions.

There are basically two types of hairs, **trichoid** and **basiconic** sensilla, that are distinguishable on a number of criteria other than the length and thickness of their shafts, which show a more continuous distribution.

7.3.1. Trichoid sensilla

The trichoid sensilla act as mechanoreceptors signalling tactile stimulation, although on other parts of the body, such as the cerci (*see Chapter 10*) and head and neck (*see Chapter 11*), they also respond to air currents. Each hair is separately innervated by a single sensory neuron with its cell body just below the socket in which the hair shaft is articulated. The dendrite of this sensory neuron extends into the shaft and transduces movements into sequences of spikes that are conducted along the axon to the central nervous system. The thickness of the shaft varies for hairs in different regions of the body so that at one extreme the hairs are called filiform and at the other, bristle hairs. Correlated with these differences in the shafts are variations in the stiffness of the socket. Together, these factors set limits on the mechanical responses of the hairs and on the spike coding of their sensory neurons, although this is also influenced by the membrane properties of the sensory neurons themselves (*see Chapter 3*). For example, the long filiform hairs on the underside of the neck (prosternum) that may be involved in the control of steering during flight (*see Chapter 11*) generate a maximal spike response to a 1° displacement (Klee and Thurm, 1986), which translates to a sensitivity to air currents with a velocity as low as 0.03 $m.s^{-1}$ (Pflüger and Tautz, 1982). By contrast, the air speed of a freely flying locust may be 4.6 $m.s^{-1}$. The hairs on the cerci (*see Chapter 10*) are also filiform and can be excited by even slower air movements, or even by sounds of high intensity. The hair receptors on the cerci of cockroaches, for example, respond to air currents with velocities as low as 0.012 $m.s^{-1}$ and with an acceleration of 0.6 $m.s^{-2}$. In crickets, the long hairs are velocity sensitive and respond to low-frequency movements, while the shorter hairs are acceleration sensitive and are insensitive to low frequency movements (Shimozawa and Kanou, 1984).

7.3.1.1. *Tactile hairs on the legs*

The hairs on a leg are sparsely but not uniformly distributed in a pattern that is consistent from locust to locust, so that many hairs can be recognised as individuals, particularly when set against other landmarks such as the tibial spines, or the cuticular ridges on the femur (see Fig. 9.2). The hairs are most abundant on the distal tibia and proximal femur and coxa, but are sparse on the proximal tibia near the femoro-tibial joint. Although there is some variability, certain hairs can always be found, and the number of hairs along the femoral ridges differs only within narrow limits in different adult locusts of a particular species. The shafts of leg hairs are stiff so they respond only to tactile stimulation and not to air currents, and range in length on a hind leg from 60 to 780 μm (Newland, 1991a). Long hairs (450–780 μm) are present only on the tibia at the base of each tibial spine.

Each hair moves back and forth most readily in the plane parallel to the long axis of a leg, due to the eccentric placement of its shaft in its socket (Nicklaus, 1965), and the shafts are all aligned so that their tips point along the long axis of a leg towards the distal tarsus. Partly as a result of this anatomical arrangement, the sensory neuron from each hair shows a directional sensitivity by producing its greatest spike discharge to a mechanical stimulus that bends the shaft towards the proximal end of a leg (Fig. 7.1). A movement in the opposite direction (distally) evokes fewer spikes, but the directional tuning is such that a stimulus over an arc of some 100° still evokes spikes. The hair sensory neurons can thus give some indication of the direction of a stimulus, but their signals will be ambiguous because a weak stimulus in the preferred direction will give the same signal as a strong stimulus in the opposite (nonpreferred) direction. The ambiguity about direction could be resolved by sequential stimulation of the hairs as a stimulus moves in a particular direction.

The hairs are of two physiological types as defined by the angle through which their shaft must be moved to evoke a spike, but both are directionally sensitive (Fig. 7.1). **High threshold hairs** must be deflected by a 0.1 Hz stimulus through an angle of 37–47° before a spike is evoked, have velocity thresholds of about $20°.s^{-1}$, and adapt completely to a repetitive stimulus so that their response is typically phasic (Newland, 1991a). **Low threshold hairs** are more sensitive as they need to be deflected by only 10–28°, have velocity thresholds of $3°.s^{-1}$, and continue to respond to a repetitive stimulus, so that their response is phaso-tonic. They will also respond to a maintained deflection with a less rapid adaptation of their response than the high threshold hairs. The high threshold hairs occur only on the tibia, particularly at the base of the spines, and although these are among the longest hairs on a leg, their physiological properties correlate more with position than size, because hairs of comparable length on the femur have low threshold properties.

7.3.2. Basiconic sensilla

Basiconic sensilla are widely distributed over the body, including the legs, but are particularly concentrated on the antennae and on the tips of the mouthparts. They are peg-like structures with a shaft that is typically much shorter (30–45 μm) than that of the trichoid sensilla and which has a pore at its tip. The pore provides access for contact of chemicals or odourants with the dendrites inside the shaft, so that typically these receptors act as contact chemoreceptors or odour receptors. Each sensillum is innervated by a number of sensory neurons (range 5–12 depending on their location and the species of insect; locusts typically have five) of which one is a mechanosensory neuron and the others are chemosensory, responding to different, but overlapping, ranges of chemicals. In flies, the four chemosensitive neurons of basiconic receptors on a leg are thought to comprise one that responds to sugar, one to salt, one to water and one for which no stimulus has yet been identified. Their presence on the legs means that locusts can taste and possibly smell with their legs, and although this may initially seem to be a bizarre attribute, it could enable them to

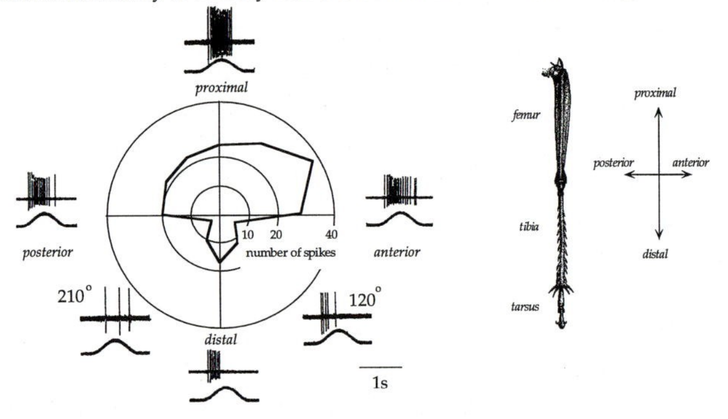

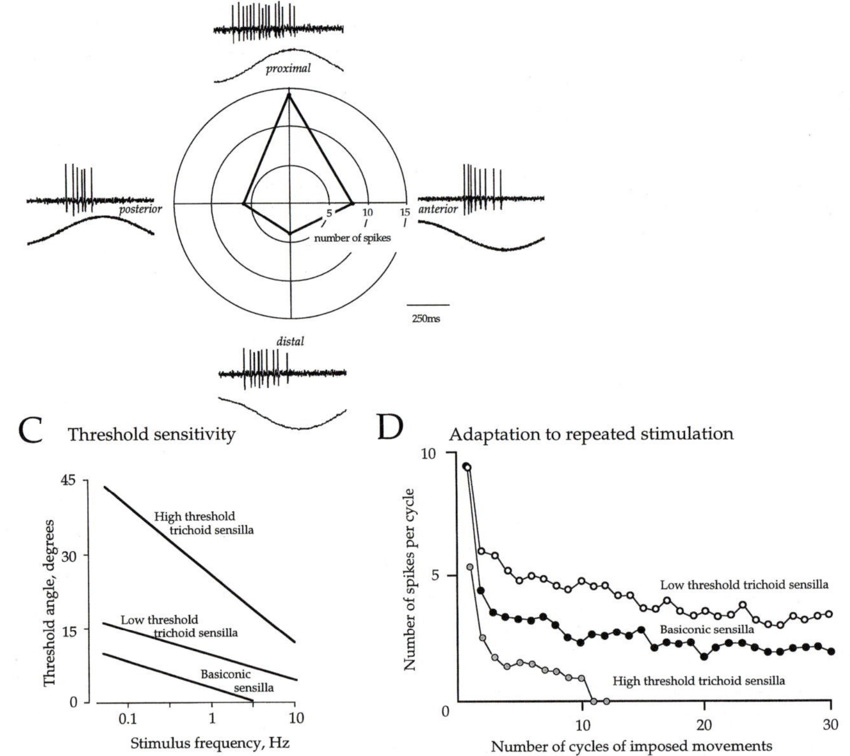

Fig. 7.1 Properties of mechanosensory neurons of trichoid and basiconic sensilla on a leg. A. Directional sensitivity of a sensory neuron from a trichoid sensillum. B. Directional sensitivity of a sensory neuron from a basiconic sensillum. C. Threshold sensitivity. The mechanoreceptor of a basiconic sensillum has the lowest threshold to an imposed movement. D. Responses to repetitive movements of a sensillum. Sensory neurons from high threshold trichoid sensilla adapt rapidly, but those from low-threshold trichoid sensilla and from basiconic sensilla continue to respond to repeated movements. Based on Newland (1991a) and Newland and Burrows (1994).

sample the vegetation through which they may be walking, and may give additional information on where the feet should be placed.

On the legs, these receptors are intermingled with the larger trichoid sensilla, and like them have a sparse but nonuniform distribution. For example, along a ventral ridge of the femur there are 15 trichoid and 15 basiconic sensilla, whereas on a dorsal ridge there are 19 trichoid and 26 basiconic sensilla. They are more sensitive to mechanical displacements than are either of the two types of surrounding trichoid sensilla (Fig. 7.1) (Newland and Burrows, 1994). They have a much lower velocity threshold, so that to evoke a spike the shaft needs only be deflected by 4–11° at 0.1 Hz, falling to 3–7° at 1 Hz. They have a velocity threshold of less than $1°.s^{-1}$ and are able to code the velocity of a stimulus in the frequency of their spikes. They initially adapt rapidly to repetitive stimuli but continue to spike at a lower level to each stimulus, so that their best response is phasic and provides information about intermittent rather that sustained stimuli. Like the trichoid sensilla they are directionally sensitive, responding best to a deflection that bends the shaft towards the proximal end of a leg, with sensilla on different parts of a leg oriented in the same way.

An approaching object will therefore first stimulate the long, high threshold trichoid sensilla before finally stimulating the short, low threshold basiconic sensilla. Some stimuli may therefore deflect only the trichoid sensilla whereas other stimuli that contact the cuticle will deflect both the trichoid and basiconic sensilla so that the combined responses will ensure that a larger sensory signal is delivered to the central nervous system.

7.4. LOCAL MOVEMENTS OF A LEG

7.4.1. Tactile stimuli

The hairs of all different types and sizes provide a mechanism for detecting when a leg has bumped into an object during walking, or when something touches a leg when the locust is standing still. Many stimuli will simultaneously stimulate an array of receptors that includes both the long trichoid sensilla and the shorter basiconic sensilla. Generally, many hairs must be stimulated to evoke a movement of the leg, but occasionally, touching a single hair is sufficient to produce a coordinated movement (Pflüger, 1980). Strong stimuli that cause deformations of the cuticle will also excite campaniform sensilla and perhaps may move a joint and thus excite internal proprioceptors. It is thus often difficult to say that a movement has been evoked only by stimulation of the exteroceptors. The hairs are not usually stimulated by the movements that a locust makes with its own legs, so that they cannot be used to provide reliable information during the cycle of movements used in walking. Some specialised and shorter hairs are grouped together in patches (hair plates) and are used as proprioceptors to monitor the position of one joint relative to another (*see Chapter 8*).

Touching hairs on one hind leg of a locust causes the leg to move away from the stimulus, but there need be no accompanying movement of the contralateral or other legs; the movement can be entirely local. For example, touching hairs on the tarsus often causes a leg to be lifted and put down again in a new position (Laurent, 1986). Such movements result in appropriate changes in posture, and during the swing phase of walking lead to the avoidance of the object that the leg has contacted. The joints that are moved depend on the site of stimulation, and indicate that the sensory information must be sufficient for the central nervous system to detect at least four areas on, for example, the femur; dorsal, ventral, anterior and posterior (Fig. 7.2 and Table 7.1) (Siegler and Burrows, 1986).

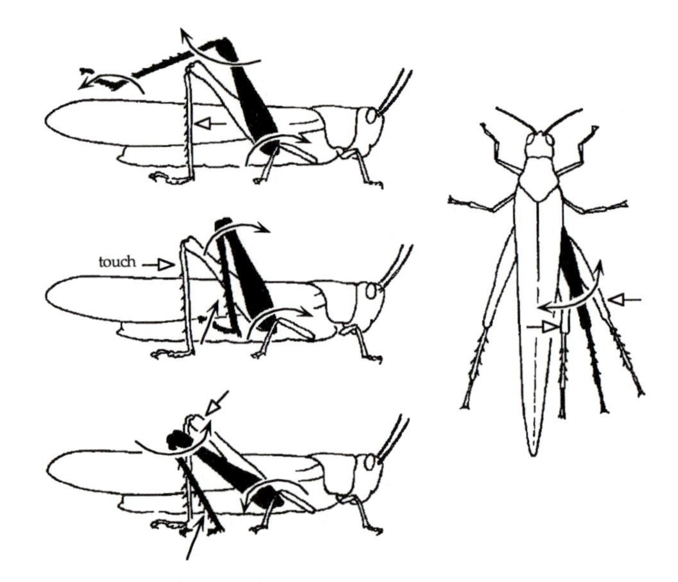

Fig. 7.2 Movements of a hind leg elicited by mechanical stimulation of different arrays of hairs. The open arrows indicate the hairs stimulated, the curved arrows show the direction of movement about the joints, and the silhouettes indicate the final position of the leg. Each movement takes the leg away from the source of stimulation. Based on Siegler and Burrows (1986).

Table 7.1. Movements of a hind leg elicited by tactile stimulation of its receptors

Site of tactile stimulation	Joint movements			
	Coxa	Trochanter/femur	Tibia	Tarsus
Dorsal femur		Depress	Flex	Levate
Ventral femur		Levate	Extend	Depress
Dorsal tibia		Levate	Flex	Levate
Ventral tibia		Levate	Extend	Depress
Dorsal tarsus		Levate	Flex	Levate
Anterior femur, tibia	Rotate posteriorly			
Posterior femur, tibia	Rotate anteriorly			

Touching hairs on the **dorsal** surface of the femur of a hind leg causes the trochanter and the femur, to which it is fused, to depress, the tibia to flex and the tarsus to levate, but touching hairs only a short distance away on the ventral femur causes the opposite sequence of movements. Depending on the antero-posterior location of the femoral hairs, a posterior or an anterior rotation of the coxa is added to the movements.

Touching hairs on the **ventral** tibia causes the same movements as touching the ventral femoral hairs, but touching the dorsal hairs on the tibia causes a different movement, in which the trochanter/femur is levated, the tibia flexed and the tarsus levated. If flexion is prevented then the tactile stimulation leads to depression of the tarsus, but this is overridden when the tibia moves by an interjoint reflex mediated by the femoral chordotonal organ, emphasising that the responses depend heavily on the behavioural context in which they are elicited.

Touching hairs on the **anterior** distal femur or proximal tibia causes the coxa to be rotated (retracted) posteriorly, while stimulating **posterior** hairs on these parts of a leg cause the coxa to be rotated (protracted) anteriorly.

From these observations of movements, the motor neurons can be pictured as having receptive fields defined by the arrays of hairs that either excite or inhibit them in these local movements.

7.4.2. Odour stimuli

A leg will also move away from an odour placed near it, in coordinated movements (Slifer, 1954, 1956) that are similar in form to those elicited by mechanical stimuli. The most effective odours are those from certain acids. The movements elicited are adaptive in that they move the leg away from the source of the stimulus, and there is some suggestion that spatial information is available to the central nervous system. The receptors responsible for the detection of these odours are on a leg and are assumed to be the basiconic sensilla, which must therefore be acting as true olfactory receptors as they do in flies (Dethier, 1972) and certain moths (Städler and Hanson, 1975). Any spatial information must be extracted from the signals in the array of receptors that are excited by the odour, and will necessarily be rather diffuse. The basiconic sensilla also act as contact chemoreceptors. Certain noxious substances that discourage feeding elicit waving movements of a leg when they come into contact with it (White and Chapman, 1990), and in flies, potential sources of food detected by the tarsi lead to extension of the proboscis (Dethier, 1976). Mechanical contact with a chemical may also excite the mechanoreceptor in a basiconic sensillum, so that a signal about the location of contact and the nature of the chemical signal itself can be sent to the central nervous system.

7.4.3. The circuitry for the local movements is contained within a segmental ganglion

All the neural circuitry for these movements of a hind leg, or for the grooming movements of the front legs is contained within the corresponding segmental ganglion. If the appropriate ganglion is isolated from the rest of the nervous system the movements are unchanged, but the reliability with which they can be elicited is enhanced (Pflüger, 1980; Rowell, 1961). The local movements of one leg can, however, take place only if the posture or movements of the other legs are appropriate. This implies that information which places the action of one leg in the context of the movements of the other legs must be transferred between ganglia. For example, a leg can only be lifted if the balance is maintained by the other legs.

7.4.4. What these movements reveal about the neural organisation

The local movements of a leg suggest the following features of the organisation of the underlying neural pathways.

1. The input from the large number of exteroceptors must be reduced to activate only a small number of output elements; several thousand input (sensory) channels must be reduced to about 70 output (motor) channels (see Table 8.3).
2. The spatial information delivered to the central nervous system in the spikes of particular arrays of sensory neurons must be preserved throughout the central processing because stimulation of hairs on particular regions of a leg leads to specific movements of a leg. This also implies that the synaptic connections of the sensory neurons with central neurons are specific and reliable from animal to animal.
3. The neuronal circuitry within a segmental ganglion must be sufficient to organise all aspects of the local movements, although normally it would be subject to influences from signals generated elsewhere in the central nervous system so that the movements of one leg are related to the prevailing behaviour.
4. The sensory neurons from the exteroceptors must be sorted onto the neurons of the local circuits to construct functional maps for the performance of specific movements and not simply to preserve spatial information. The spatial information must be assembled according to functional constraints with specific regard to the behaviour of the locust.

7.5. SENSORY PROJECTIONS AND MAPS

If the central nervous system is to integrate the signals conveyed by the large number of sensory neurons, there must be some coherent presentation of this information to the neurons performing the first integrative steps. Sensory maps are used by many animals and for many different modalities of sensory signals. The central projections of neurons from the eyes of insects are arranged retinotopically (Strausfeld, 1976), from the ear tonotopically (Oldfield, 1982; Römer *et al.*, 1988), and from limbs and cerci somatotopically (Murphey, 1981; Kent and Levine, 1988; Peterson and Weeks, 1988).

7.5.1. Collecting the axons together

The solution adopted for the mechanosensory neurons from exteroceptors on a leg is that their axons, along with those of other receptors on their segment, are initially collected together in different nerve branches. For the metathoracic segment, dorsal hairs on the thorax have their axons in N1, ventral ones in N2 and lateral ones in N3 and N4, whereas hairs on the distal parts of a leg have axons in N5. Along the surface of the leg itself, different branches of N5 supply specific patches of cuticle (Mücke, 1991). For example, N5B2 divides into two branches just as it enters the tibia. The anterior branch (N5B2a) innervates all the hairs anterior to the midline of the dorso-ventral axis, the two anterior moveable spurs, the tibio-tarsal chordotonal organ, receptors on the anterior half of the tarsus. The posterior branch (N5B2b) innervates all the hairs posterior to this axis and the posterior moveable spurs, and receptors on the posterior half of the tarsus.

7.5.2. Mapping the mechanosensory terminals in neuropil

Once the axons of the sensory neurons enter a segmental ganglion, they branch repeatedly so that their terminals form a somatotopic map of the surface of a leg in the ipsilateral ventral neuropil (Fig. 7.3). The branches of all the sensory neurons from the trichoid sensilla on a leg project to the ipsilateral neuropil of the ganglion of their segment. All the terminals are restricted to the most ventral regions of neuropil (a,v,lVAC, see Fig. 2.5) that are characterised by the presence of numerous fine fibres and the absence of branches from leg motor neurons and the majority of branches from nonspiking local interneurons. None of the sensory neurons from hairs on a leg itself have branches to the contralateral neuropil, but sensory neurons from a few hairs close to the midline on the ventral thorax of a cricket do cross the midline (Hustert, 1985). Similarly, only sensory neurons from a few of the long hairs on the thorax, but none from the leg hairs, project to the next anterior ganglion. The implication is that the sensory signals are processed locally to effect changes in the movements of the leg from which they originate and that they must then be passed both to intersegmental interneurons and neurons with contralateral branches if they are to influence circuits controlling the movements of the other legs.

The map formed by the terminals of the sensory neurons within the ventral neuropil represents the three axes of a leg (Fig. 7.3) (Newland, 1991b). The proximo-distal axis is represented antero-posteriorly, the dorso-ventral axis is represented dorso-ventrally and the antero-posterior axis is represented medio-laterally. This means that the tarsus is represented in lateral and posterior neuropil and more proximal segments progressively more medially and anteriorly, with the hairs on the sternum represented most medially (Pflüger *et al.*, 1988). Along this axis, ventral hairs are represented more ventrally than dorsal hairs, and anterior hairs are represented more medially than posterior hairs. The location of a hair on a leg can thus be read directly from the position of its terminals in this restricted region of neuropil although there is some overlap between the branches from hairs in different locations.

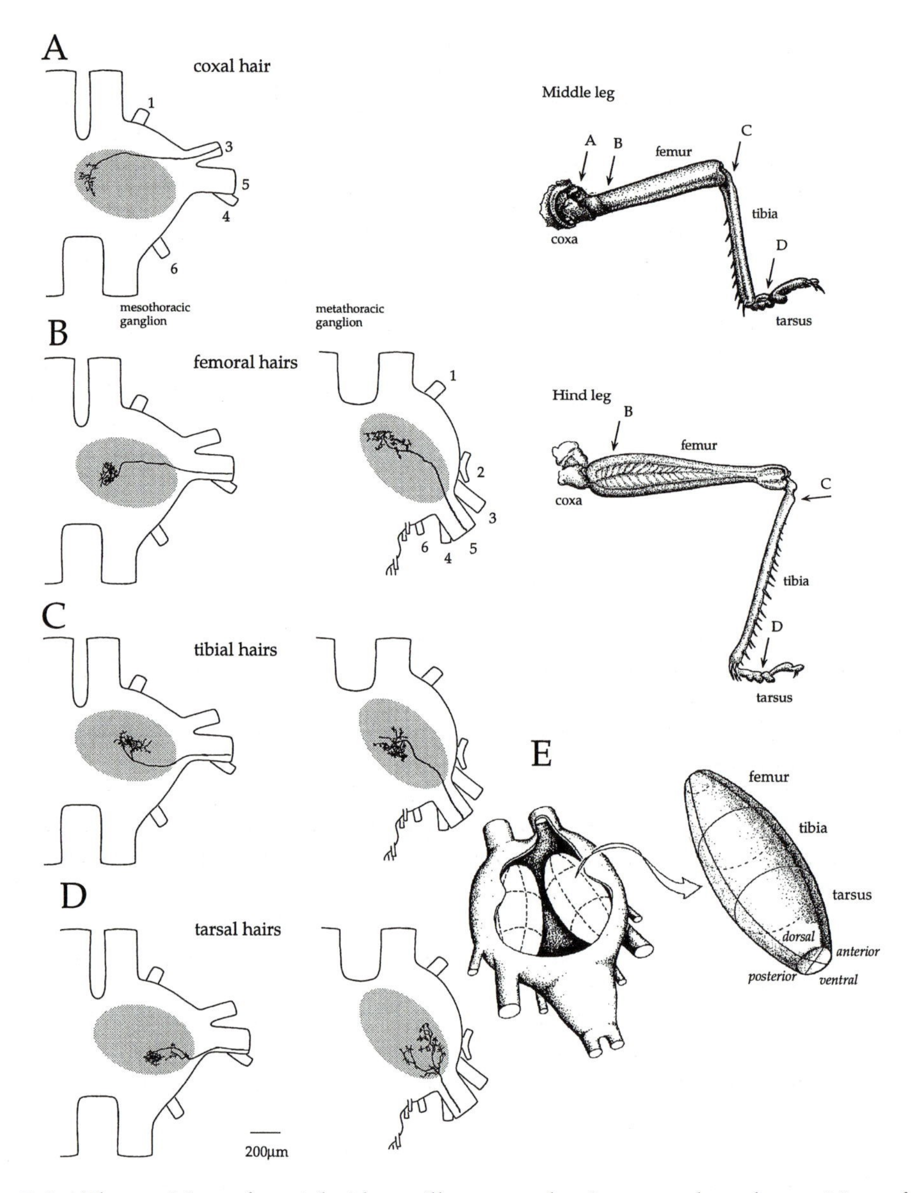

Fig. 7.3 The position of a trichoid sensillum on a leg is mapped as the position of the branches of its sensory neuron in the ventral neuropil of a thoracic ganglion. The left-hand column shows the mesothoracic ganglion with projections from hairs on a middle leg and the right column shows the metathoracic ganglion with projections from hairs on a hind leg. A. Coxal hair. B. Femoral hair. C. Tibial hair. D. Tarsal hair. The ventral area of neuropil to which the sensory neurons project is shaded. The positions of the hairs on the legs are marked on the drawings on the right. E. Diagram of the metathoracic ganglion to show the map of the surface of a hind leg in the ventral neuropil. The metathoracic stains are based on Newland (1991b). E is based on Burrows and Newland (1993).

The projections from the mechanosensory neurons of the basiconic sensilla also seem to conform to this map, but the chemosensory neurons project to different neuropil regions that are slightly more dorsal. A similar arrangement occurs in the fly *Phormia regina*, where one sensory neuron from a basiconic sensillum ends in VAC and is assumed to be mechanosensory, while the other four neurons project more dorsally and are assumed to be chemosensory (Murphey *et al.*, 1989).

Projections from the proprioceptors, such as the campaniform sensilla, hair plates, chordotonal organs, strand receptors and joint receptors are all absent from the ventral neuropil that contains the terminals of the mechanosensory neurons of the exteroceptors. Instead their branches end in more lateral and dorsal areas of neuropil, where there are branches of motor neurons and nonspiking local interneurons, suggesting a segregation of the initial processing of the signals from the different types of sense organs on a leg.

7.5.3. Organisational features of the mechanosensory map

The terminals of the sensory neurons from exteroceptors on a leg are thus organised anatomically in the central nervous system in the following ways.

1. The terminals of the mechanosensory neurons from the trichoid sensilla provide a three-dimensional map of the surface of a leg that represents the spatial location of each receptor.
2. The terminals of mechanosensory neurons from the two types of exteroceptors (trichoid and basiconic sensilla) are brought together in the same region of ventral neuropil so that the spatial arrangement of receptors on a leg is preserved.
3. The terminals of exteroceptive and proprioceptive sensory neurons are segregated, implying that much of the initial processing of the signals from these receptors is performed in separate channels.
4. The terminals of mechanosensory and chemosensory neurons, even from the same basiconic sensillum, are also segregated, again implying that initial processing of their signals occurs in separate channels.
5. Signals in these different channels must eventually converge within the local circuits because similar movements of the legs are generated by exteroceptive, proprioceptive and chemosensory stimuli.

7.6. SYNAPTIC CONNECTIONS UNDERLYING LOCAL PROCESSING

The signals provided by the sensory neurons associated with a leg are processed by the local networks in a segmental ganglion that contains both local and intersegmental interneurons, and all the motor neurons that control the movements of a leg. Virtually all of the motor neurons and many of the interneurons can be identified. The local interneurons belong to several different populations which each show common anatomical features and common features of the processing that they perform. Some are spiking and others are nonspiking neurons. Many appear to be unique individuals in that both their anatomy and the main features of the processing that they perform are the same in all locusts of one species. Nevertheless, the usual caveats about identification apply to these as to other local interneurons (*see Chapter 3*). The intersegmental interneurons convey the processed sensory signals from one leg to other parts of the central nervous system. The actions of all these neurons and the processing they perform may be modulated by the local, or more widespread, secretions of neurosecretory or neuromodulatory neurons (*see Chapter 5*).

7.6.1. Strategies in analysing the synaptic connections

One of the first steps in understanding how the sensory signals can elicit movements of a leg is to unravel the connections between the sensory neurons and other neurons in the central circuitry, and then follow the signals through the various stages in their processing until they emerge as patterns of spikes in identified motor neurons. A knowledge of all the connections, were that ever to be possible, would not in itself constitute an explanation of how the movement is generated. This is because the operation of a network depends on influences at many different levels and over many different time scales, from the molecular and cellular properties of the neurons to the emergent properties of the network. Moreover, as a large number of neurons is likely to be involved, it is not a practical strategy to attempt to discover how they are all connected. A more rewarding and feasible strategy is to define the types of component neurons in the networks (Fig. 7.4) and then to analyse selected examples and see if they give an insight into the general principles underlying the processing. This has been the guiding strategy in the analyses of these local networks.

7.6.2. Synaptic connections of mechanosensory exteroceptive neurons

The mapping of the terminals from the trichoid sensilla presents information about the spatial order and distribution of the sensory receptors to neurons in the central nervous system in a highly ordered and coherent fashion. Which neurons read this map and how do they handle the signals? The general rule that emerges is that the signals from one region of the leg are processed in parallel, both by several neurons of the same type and by neurons of different types. It appears to be a general principle that all the connections between mechanosensory neurons and central neurons are excitatory. Inhibitory effects are always caused by an interposed interneuron that reverses the sign of the excitatory connections from the primary sensory neurons.

7.6.2.1 *Connections with spiking local interneurons*

Spiking local interneurons appear to be the primary integrators of the signals from the trichoid and basiconic sensilla. About 100 of these interneurons, belonging to at least two distinct populations (named midline and antero-medial according to the position of their somata in the ventral cortex of a segmental ganglion, *see Chapter 3*), receive direct, chemically mediated, excitatory synaptic potentials from the mechanosensory

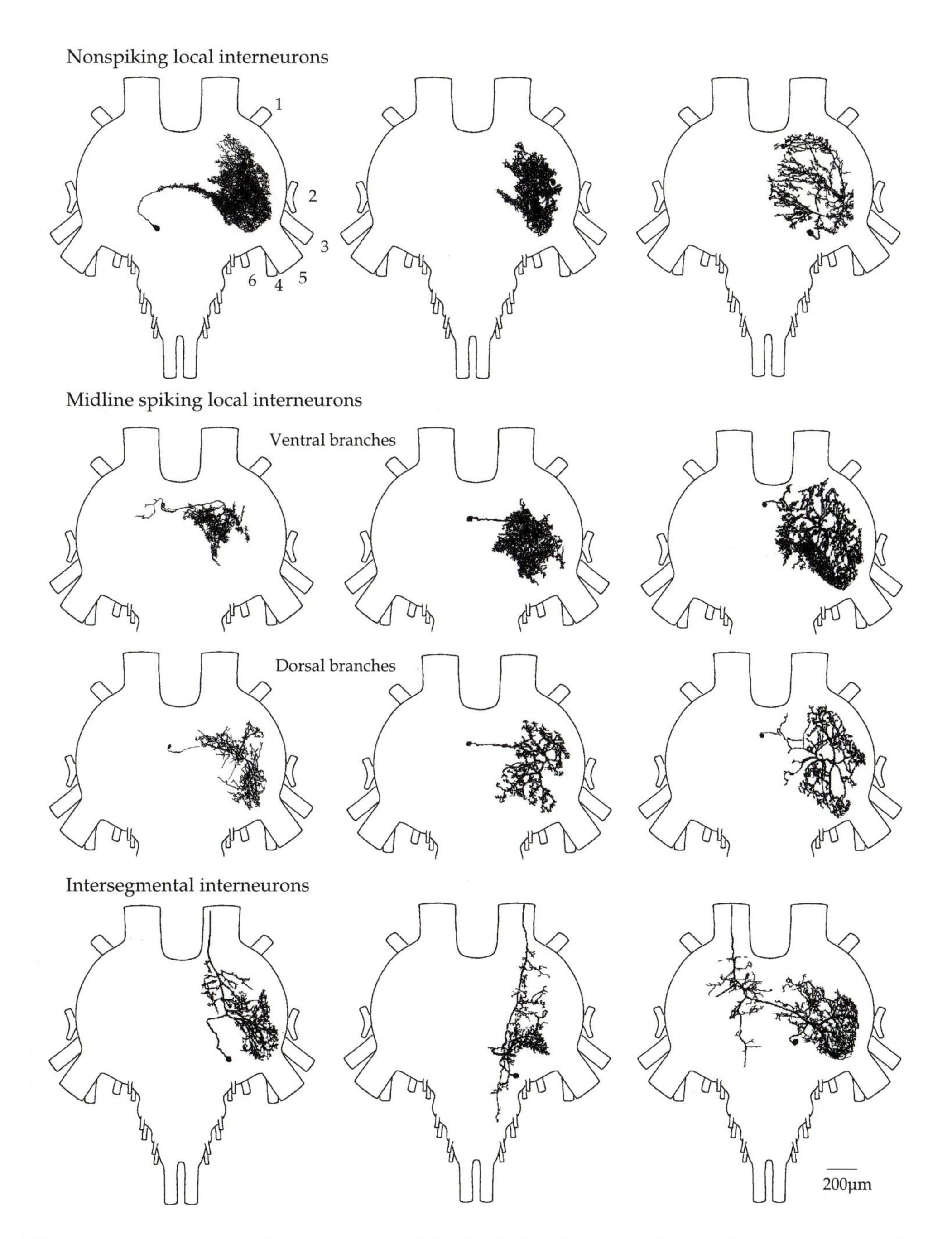

Fig. 7.4 Major neuronal components of the local circuits processing mechanosensory signals. Three examples of nonspiking local interneurons, spiking local interneurons of the ventral midline group, and three intersegmental interneurons with ascending axons are shown. Based on Watkins *et al.* (1985), Burrows and Siegler (1984) and Newland (1990).

neurons of these receptors (Siegler and Burrows, 1983; Nagayama and Burrows, 1990). The evidence for these properties is as follows (Fig. 7.5).

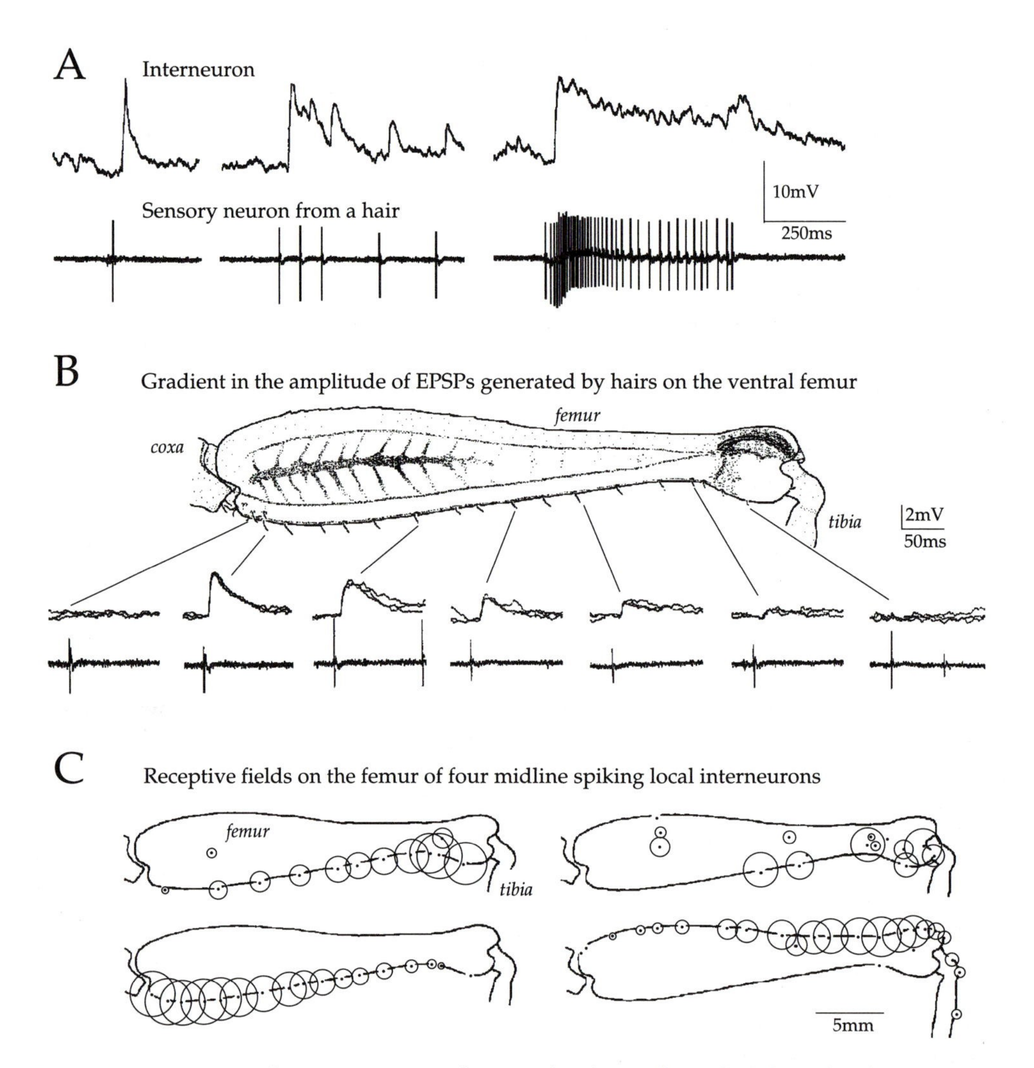

Fig. 7.5 Connections of sensory neurons from trichoid sensilla with spiking local interneurons of the ventral midline group. A. Each sensory spike is followed by an EPSP in an interneuron, but the depolarisation resulting from a burst of spikes is of no greater amplitude than that caused by a single spike. B. The receptive field of an interneuron is defined by the array of hairs whose sensory neurons make direct synaptic contact. Within the field, the amplitude of the EPSPs evoked by the different hairs varies. In this example, the most proximal hairs on the ventral femur cause the largest amplitude EPSPs, with a gradient of amplitude stretching distally. C. Receptive fields of the four interneurons that map the surface of the femur. The amplitudes of the EPSPs from the different hairs are represented by the diameter of the circles surrounding them. Based on Burrows (1992).

First, the physiological evidence suggests that the connections are direct because each spike in a mechanosensory neuron from a trichoid sensillum is followed consistently and without any failures by a depolarising synaptic potential in a particular spiking local interneuron. The latency between the entry of the sensory spike into the segmental ganglion and the appearance of a synaptic potential as recorded in the soma of an interneuron is 1.5 ms. This would allow time for only one chemical synapse and not the participation of another interposed interneuron.

Second, the potentials evoked are depolarising at the normal resting potential of the interneuron and involve an increase in the conductance of the membrane. They can be increased in amplitude if the membrane is hyperpolarised and decreased when it is depolarised, so that the reversal potential must be more positive than resting potential. Some single synaptic potentials and many summed synaptic potentials lead to spikes. The transmitter used by these sensory neurons has not been determined, but a pervasive assumption is that mechanosensory neurons use acetylcholine by analogy with the sensory neurons of cercal hairs in the cockroach (*see Chapter 5*). Some caution over these extrapolations is necessary because in *Drosophila*, but not in some other flies, the sensory neurons of the hair sensilla are immunoreactive to histamine (Buchner *et al.*, 1993).

Third, the branches of the sensory neurons and the interneurons overlap in the ventral neuropil (VAC) so that direct connections are at least possible. The branches of the spiking local interneurons in this region of neuropil have predominantly input synapses (Watson and Burrows, 1985).

Similar lines of evidence are available for all the connections between the components of the local circuits that are thought to be direct. Moreover, where pairs of neurons that are thought to be interconnected on physiological criteria have been labelled for electron microscopy, direct synapses between them can be observed (Burrows *et al.*, 1989).

The synaptic potentials evoked by a sensory neuron of either a basiconic or trichoid sensillum initially decrement rapidly, even when individual spikes are spaced at intervals of 2 s. A deflection of a single hair in its preferred direction evokes a burst of spikes at high frequency, but the depolarisation they cause in an interneuron is longer lasting although of no greater amplitude than that caused by an isolated single spike (Fig. 7.5). The properties of both the sensory neurons themselves and the connections with the interneurons are therefore designed to accentuate novel events and to decrement rapidly during the continuous stimuli that are likely to be encountered by the legs as a locust walks through dense vegetation. The connections show none of the facilitation and long-term potentiation that characterise the connections between sensory neurons from hairs on a proleg of a moth and certain motor neurons (Trimmer and Weeks, 1991).

Convergence from different classes of receptors The mechanosensory neurons from both trichoid and basiconic sensilla on a particular region of a leg converge onto the same interneurons (Fig. 7.6) (Burrows and Newland, 1994), and for some interneurons the campaniform sensilla may also converge (Siegler and Burrows,

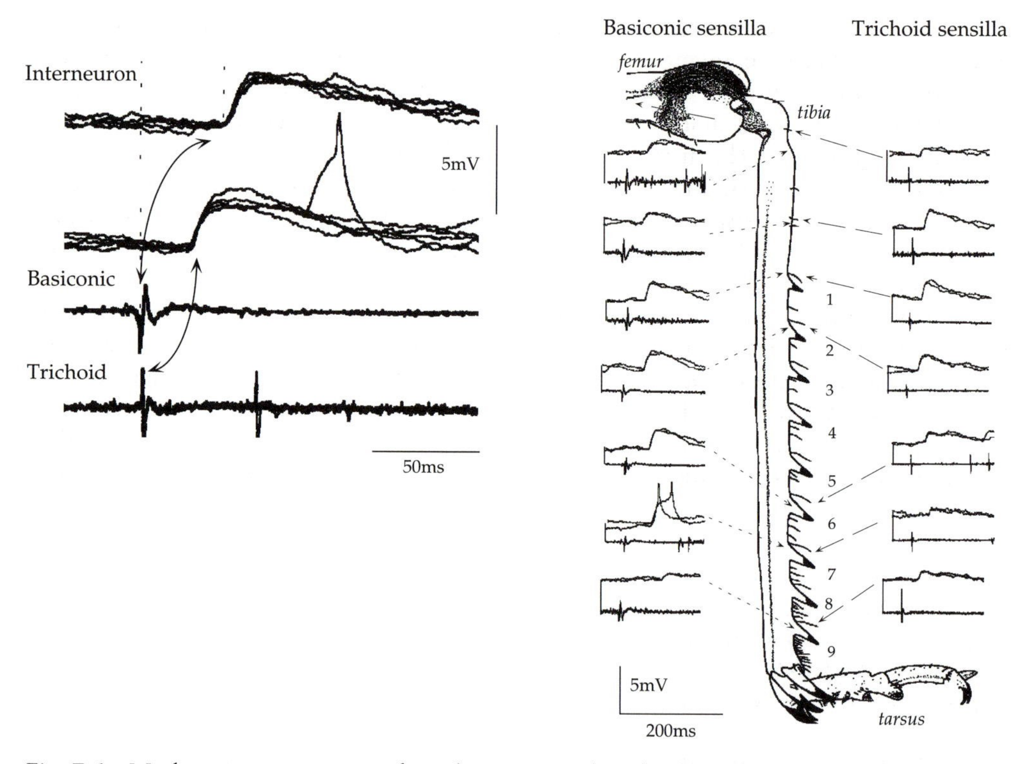

Fig. 7.6 Mechanosensory neurons from basiconic and trichoid sensilla converge on the same interneuron. The conduction velocity of the sensory neuron from a basiconic sensillum is slower than that from a trichoid sensillum, so that the EPSP evoked by its spike occurs later. The arrays of the two types of receptors that define the receptive field of an interneuron have a similar distribution and the gradients in the amplitudes of the EPSPs they evoke are similar. The interneuron shown has a receptive field on the dorsal tibia. Based on Burrows and Newland (1994).

1983). Each interneuron is excited by a contiguous and overlapping array of intermingled trichoid and basiconic receptors that form its receptive field. Different spiking local interneurons thus have receptive fields which consist of different arrays of these receptors.

The connections from both types of receptors are excitatory and apparently direct, with the anatomy of the sensory projections suggesting that the synapses from them both are made onto the same region of branches of an interneuron. The spikes of the mechanosensory neurons from the basiconic sensilla are, however, conducted more slowly than those from the trichoid sensilla so that the response of an interneuron to a mechanical stimulus will be an initial depolarisation caused by the trichoid sensory neurons, followed up to 20 ms later by a depolarisation caused by the basiconic sensory neurons. The synaptic potentials evoked by the different types of receptor and by receptors of the same type located on the same region of a leg sum to increase the depolarisation of an interneuron. This convergence and summation between the signals from the different sensory receptors will increase sensitivity and the reliability with which a stimulus can activate a functionally appropriate movement of a leg. It preserves spatial information and could also ensure that appropriately directed movements are made in response to the information from the chemosensory neurons of the basiconic sensilla that occurs at the same time. A high frequency of spikes in one sensory neuron can, however, depress the response of a postsynaptic interneuron to an input from an adjacent hair, either through a presynaptic or a postsynaptic action (Burrows, 1992).

The spiking local interneurons may also have inhibitory regions to their receptive fields which are formed by inhibitory connections from other interneurons and not by direct sensory inputs. The inhibitory regions of the receptive fields of the antero-medial spiking local interneurons are caused by connections from particular midline spiking interneurons. It is assumed that the inhibitory regions in the receptive fields of the midline spiking interneurons result from connections between members of this population.

Receptive fields The convergence onto a spiking local interneuron of many mechanosensory neurons from trichoid and basiconic sensilla on a particular part of a leg defines the excitatory region of the receptive field of the interneuron (Burrows and Siegler, 1985). The receptive fields of the different interneurons are organised according to the following rules.

1. In the midline group of spiking local interneurons, some interneurons have receptive fields with only excitatory regions from receptors on one area of cuticle, others have both excitatory and inhibitory regions, but none have exclusively inhibitory regions. The hairs providing the excitatory receptive field are always contiguous. The excitatory and inhibitory regions may also be contiguous, or they can be on separate parts of a leg.
2. The receptive fields of some interneurons are complementary, so that the excitatory region of one is the inhibitory region of another and *vice versa*.
3. The direct synaptic connections from the sensory neurons define the excitatory region of a receptive field, whereas the indirect connections that are made through other interneurons define inhibitory regions. The intracellular recordings used to define the receptive fields record the synaptic inputs and the spike outputs, so that the complete field is defined without any of the subliminal fringes that plague receptive fields defined only by the spikes of a neuron.
4. The boundaries of the receptive fields are set by the major antero-posterior and dorso-ventral axes of the leg, and along the proximo-distal axis most boundaries are set by the joints, although some that occur along the femur or tibia do not correspond to obvious anatomical boundaries. Some fields may be extensive and others restricted to small regions of a leg.
5. The receptive fields of the interneurons overlap so that one region of a leg will be represented by several interneurons, and the mechanosensory neuron from a single hair may synapse directly with several interneurons. Nevertheless, a

particular sensory neuron connects with only a small subset of the total population of spiking local interneurons, implying that the connections are highly specific. The signals from one patch of receptors are thus processed in a parallel and distributed fashion by a number of interneurons of the same type.
6. Within the confines of a receptive field, most, if not all, of the exteroceptors synapse on a particular interneuron. A notable exception is that the high threshold but not the adjacent low threshold trichoid sensilla may contribute to the receptive fields of particular local interneurons (Burrows, 1992).
7. The receptive fields preserve spatial information about a leg in a way that is functionally appropriate for the production of local movements.

Different interneurons have receptive fields that differ in overall size and in the complexity of their excitatory and inhibitory regions. Three examples illustrate the general features. One interneuron is excited by hairs on the posterior surfaces of the entire leg and is inhibited by contiguous regions on the anterior surfaces. A second interneuron is excited by hairs on the ventral coxa, femur and proximal region of the tibia, and is inhibited by hairs on the dorsal tarsus and a second patch of hairs on the dorsal surface of the distal tibia and proximal tibia. By contrast, a third interneuron is excited only by hairs on the distal tarsus and is unaffected by inputs from hairs elsewhere on a leg.

Synaptic gain of the sensory connections Within a receptive field, the contribution of individual receptors to the excitation of an interneuron can be markedly different (Burrows, 1992). Typically, a few hairs clustered together and not necessarily at the centre of a field generate the largest amplitude EPSPs in an interneuron and are the most effective in making the interneuron spike. For the sensory neurons from the trichoid sensilla, and probably also for those from the basiconic sensilla, the amplitude of the EPSPs equates with the gain of the synaptic connection defined by the probability with which a single sensory spike will evoke a spike in the interneuron. The gain of the connection for some sensory neurons is 1, so that a single sensory spike can cause a spike in an interneuron. For most of the hairs in a receptive field the gain is less, so that the input from one hair must sum with inputs from other hairs or other sources to evoke a spike. Most stimuli to a hair will result in a burst of spikes in its sensory neuron. The high gain of the synapses ensures that transmission is reliable, with the consequent preservation of much of the sensory information flowing into the central nervous system from a leg.

The mechanosensory neuron from each hair connects with several interneurons but the amplitude of its synaptic connection is different for each interneuron. Thus, where the receptive fields of interneurons overlap, one hair can provide a large EPSP to one interneuron and a small EPSP to another interneuron. The gain of the connection made by a sensory neuron with an interneuron is thus determined by the position of its receptor within the receptive field of an individual interneuron and not by any particular anatomical feature of the receptor itself.

Gradients of decreasing effectiveness radiate from the most effective hairs to the boundaries of the fields (Fig. 7.5). These gradients are aligned with the axes of the legs and mean that a receptive field has just one area of receptors of a particular type with the highest synaptic gain. The gradients can be steep at the edges of a field, so that the boundaries are sharp. For example, for an interneuron that has its receptive field on the femur, proximal hairs on the ventral surface provide the most powerful inputs while the more distal hairs along the proximo-distal axis of the ventral surface provide progressively weaker inputs. None of the hairs on the other surfaces of the femur or on the tibia make connections with this interneuron, so that its field is sharply limited to the ventral surface of the femur. For other interneurons, the boundaries may be less steep, with the gradients of effectiveness spreading along all three axes of a leg from the most effective hairs.

The most effective basiconic and the most effective trichoid sensilla within the receptive field of most interneurons are close together, but in some cases they can be distant from one another (Fig. 7.6) (Burrows and Newland, 1994). The gradients in the amplitudes of the EPSPs in a particular interneuron are also the same for the inputs from both the trichoid and the basiconic sensilla. In some interneurons, the most effective inputs come from trichoid sensilla, but in others they come from the basiconic sensilla. This mapping of the different receptors onto specific interneurons and the effectiveness of their respective inputs is remarkably constant from animal to animal, so that the same microstructure occurs in the receptive fields of some interneurons. The same developmental and organisational constraints may thus determine the establishment of the connections of both types of receptor.

This different contribution of sensory neurons from receptors within a receptive field seems to be a general characteristic of the processing of mechanosensory signals because it also occurs for the neck hairs of a locust (Pflüger and Burrows, 1990) (*see Chapter 11*), for the cercal hairs of a cricket (Shepherd *et al.*, 1988) (*see Chapter 10*), and for the hairs on the proleg of a moth larva (Weeks and Jacobs, 1987).

Correlating interneuron shapes and receptive fields The surface of a leg is represented in the central nervous system by an anatomical map formed by the terminals of the mechanosensory neurons, and by a physiological map formed by the overlapping receptive fields of the local spiking interneurons. Both maps are functional representations of the information used to adjust posture and locomotion to changes in the environment. There is a strong correlation between the anatomical projections of the sensory neurons, the receptive field of a local or intersegmental interneuron and the structure of an interneuron (Fig. 7.7). This is well exemplified by midline interneurons that have one field of branches in the ventral neuropil, which also contains the terminals of the sensory neurons, and a second in more dorsal neuropil, which contains the processes of motor neurons and other interneurons. The ventral branches form predominantly input synapses and the dorsal branches predominantly output synapses. The arrays of ventral branches of the different interneurons match the array of branches of mechanosensory neurons which define their receptive fields. For example, mechanosensory neurons from receptors on the tarsus project only to the posterior region of the ventral neuropil and an interneuron with a receptive field on the

Projections of sensory neurons from trichoid sensilla on a hind leg

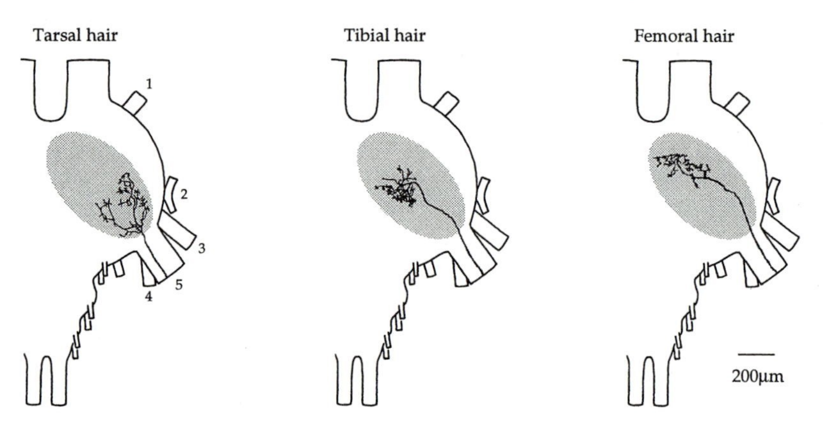

Projections of ventral branches of midline spiking local interneurons

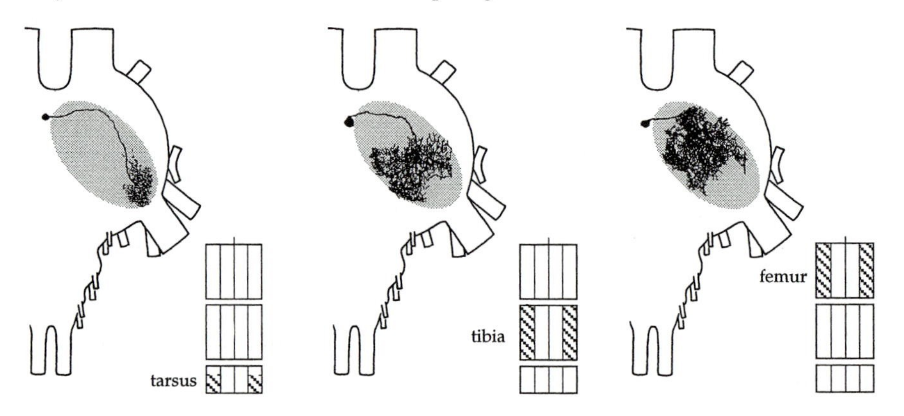

Fig. 7.7 Correlation between the sensory map and the morphology and receptive field of an interneuron. The top row shows the projections of individual sensory neurons from trichoid sensilla on the tarsus, tibia and femur, and the lower row the ventral branches of particular spiking local interneurons of the midline group with receptive fields on the tarsus, tibia and femur. The receptive fields are shown diagramatically by the rectangular boxes that represent the surface of a hind leg with all the joints extended, opened along the ventral midline and then laid flat. The excitatory region of a receptive field is hatched. For more receptive fields *see Fig. 7.19*. Based on Burrows and Newland (1993) and Newland (1991b).

tarsus only has ventral branches here. Similarly, mechanosensory neurons from receptors on the femur project medially and anteriorly and an interneuron with its receptive field on the femur has branches in this region of neuropil and none in the region to which the tarsal sensory neurons project. Tibial sensory neurons project to neuropil in between that occupied by the tarsal and femoral sensory neurons and here there are the branches of interneurons with tibial receptive fields. Based on these arrangements, it is possible to predict the receptive field of an interneuron from its array of ventral branches, with the only caveat being that overlap of branches does not necessarily imply that connections are made. The absence of branches from particular regions of the neuropil implies that the corresponding regions of a leg are not part of its excitatory receptive field. The organisation of the sensory projections into maps makes the vast amount of sensory information available to the central neurons in an easily accessible way. It also defines simple rules for the formation of synaptic connections between sensory neurons and interneurons that lead to the construction of functionally important receptive fields. These statements, however, say nothing about the use to which these interneurons put this collated sensory information.

Coarse and fine coding The two populations of spiking local interneurons with which the sensory neurons from the exteroceptors connect provide different representations of the surface of a leg. The ventral midline interneurons essentially provide a detailed and fine coding of the surface of one leg and give no indication of stimuli to other legs. By contrast, interneurons in the antero-medial group provide a coarser coding, because they have more extensive fields on one leg, and are also excited or inhibited by inputs from combinations of the other legs (Nagayama, 1990). This may be of greater use in the coordination of the movements of one leg with those of the other legs during posture and locomotion. For these neurons, the excitatory region of their receptive field from the ipsilateral leg of their own segment is caused by direct synaptic inputs from the sensory neurons (Nagayama and Burrows, 1990), whereas the inhibitory region from this leg, and both the excitatory and inhibitory regions on other legs are caused by inputs from other interneurons. The fields are of three basic types; first, those with only excitatory regions that are always contiguous on a particular leg; second, those with only inhibitory regions; third, those with both excitatory and inhibitory regions. While the field on the ipsilateral leg of its own segment may contain both excitatory and inhibitory regions, the other legs contribute either excitation or inhibition, but not both. Most antero-medial spiking local interneurons in the metathoracic ganglion have fields on the ipsilateral middle leg that have the same general characteristics.

The organisation of these receptive fields conforms to the same rules as for the midline interneurons. This means that only a subset of hairs on the leg contributes to the receptive field of one interneuron, but that each receptor will, in different combinations with others, be represented in the receptive fields of a few members of the population. This, in turn, means that the input from each hair will be processed in a parallel and distributed fashion by spiking local interneurons belonging to different populations. In the antero-medial interneurons, the local processing will be carried out against a background of the inputs from the other legs that are also part of their receptive fields.

7.6.2.2. *Divergence of signals to other components of the networks*

The spiking local interneurons seem to provide the major route through which the mechanoreceptive signals are first channelled and collated before integrated signals derived from them are passed to other neurons in the networks. They are, however, not the only route because the sensory neurons make divergent connections in parallel with other types of neurons.

Intersegmental interneurons Signals about the actions of the legs controlled by one segmental ganglion and the stimuli detected by their receptors are sent to the ganglia controlling the other legs. At the same time, a segmental ganglion receives signals about the performance of the other legs. This information is largely conveyed by intersegmental interneurons because only a few sensory neurons from receptors close to the midline of the thorax (Hustert, 1985), and none from receptors on a leg, project intersegmentally. It is information that is essential for the legs to be moved in coordinated gaits, and for the adjustment of one leg to be placed in the proper context of the movements or posture of the other legs. If this information were not available, the legs might become tangled, and inappropriate postural adjustments might be made if the leg to be moved were load bearing at the time.

In the metathoracic ganglion there are about 150 pairs of intersegmental interneurons with cell bodies in groups in the dorsal and ventral cortices and with axons that either remain ipsilateral or cross to ascend in the contralateral connective and form branches in all three thoracic ganglia (Laurent and Burrows, 1988; Newland, 1990). In the mesothoracic ganglion there are about 30 symmetrically arranged pairs of interneurons with cell bodies ventral and posterior, and with axons that descend in the ipsilateral connective to the metathoracic ganglion (Laurent, 1987) (*see Fig. 3.16*). Perhaps other groups of intersegmental interneurons with descending axons also exist. We know that there are ascending interneurons with axons that link all the thoracic ganglia controlling the three pairs of legs, but we do not know whether there are interneurons in the prothoracic ganglion with axons that descend to the meso- and metathoracic ganglia. The prothoracic legs will need information only from more posterior legs, the metathoracic legs only from more anterior legs, but the mesothoracic legs will need information from both anterior and posterior. There are occasions when the pro- and mesothoracic legs may move in a tetrapod gait without the participation of the metathoracic legs (*see Chapter 8*).

Sensory neurons from exteroceptors make direct synaptic connections with at least some of these intersegmental interneurons and these determine the excitatory regions of their receptive fields. In addition, an interneuron may have an inhibitory region of its receptive field that results from inputs from local interneurons. Most of these interneurons respond only to stimuli on the leg that is ipsilateral to its branches in the ventral neuropil of the same ganglion as its cell body, but a few metathoracic interneurons have bilateral branches and receive inputs from both the hind legs. The rules that determine the organisation of their receptive fields are the same as those for the spiking local interneurons, and this means that interneurons with small receptive fields have small areas of branches in regions of neuropil that match the sensory map, and those with extensive fields have correspondingly larger areas of branches.

Little is known about the output connections of these interneurons in general except for one of the populations (*see below section 7.6.3.3*). It is assumed that the branches from the axon in the distant ganglion are the sites of synaptic outputs, but it is probable that these are also modulated by input synapses. These branches are usually in the dorsal regions of neuropil that contains the branches of motor neurons and local interneurons, but some are in ventral neuropil that contains the branches of sensory neurons. The output synapses of the intersegmental interneurons could thus regulate both the inputs and outputs of another local circuit. In the ganglion that contains their cell bodies, and where they may receive direct synaptic inputs from sensory neurons, their axons also give rise to sparse branches in dorsal neuropil. A reasonable interpretation of this anatomy would be to suggest that they make output connections onto other elements of the local circuitry in the ganglion containing their cell bodies, so that their spike outputs are fed back, perhaps even into the pathways that shape their own receptive fields.

The presence of these interneurons indicates that sensory signals from a leg can bypass the local circuitry and be supplied to the other ganglia as the spikes of the intersegmental interneurons, altered only by the processing they perform. The integrated signals of the local circuitry are also passed to intersegmental interneurons for wider distribution, emphasising yet again the parallel and distributed nature of all the processing.

There are, of course, many additional interneurons that originate chiefly in the brain and suboesophageal ganglion that have branches in all three thoracic ganglia, and interneurons originating in abdominal ganglia (particularly the terminal one) that have ascending axons with branches in all thoracic ganglia. Finally, there are interneurons with cell bodies in the metathoracic ganglia that have axons descending to abdominal ganglia, implying that these neurons influence the contractions of the abdomen in accordance with the movements of locomotion. Most notably, the abdomen may need to be raised at the start of walking and when stepping over obstacles. They may also play a role in coordinating the movements of the abdomen and hind legs during steering manoeuvres in flight (*see Chapter 11*).

Nonspiking interneurons Some nonspiking interneurons are directly excited by trichoid sensilla on a leg (Laurent and Burrows, 1988), but most are inhibited. The branches of the nonspiking interneurons and the sensory neurons from the mechanosensory hairs can overlap in the more lateral parts of the ventral neuropil (lVAC), but generally do not overlap in the more medial parts (mVAC, vVAC). This again means that the nonspiking interneurons have receptive fields consisting of excitatory regions formed by direct connections of sensory neurons from a contiguous array of receptors, and inhibitory regions resulting from the sensory signals that have been processed first by other interneurons such as the spiking local interneurons. The receptive fields are therefore organised according to the same set of rules as for the spiking local interneurons and the intersegmental neurons, with the boundaries defined by the major axes of the legs, so that the fields are functional representations for use in controlling appropriate local movements of a leg.

The contribution of these interneurons means that the sensory signals are processed in parallel by local neurons with distinct physiological properties, some of which use spikes and some of which use analogue signals to effect intercellular communication.

Motor neurons In general, sensory neurons from exteroceptors do not connect directly with any of the leg motor neurons. The only exceptions to this rule that have been found in locusts are the sensory neurons from a small number of stiff trichoid sensilla on the tarsus whose sensory neurons connect with one of the motor neurons that innervates the retractor unguis muscle moving the tarsus (Laurent and Hustert, 1988). The hair receptors providing this direct input, however, have a specialised function, as they point downwards from the ventral surface of the tarsus and are thus stimulated each time that the tarsus is placed on the ground. They thus serve more of a proprioceptive role and reinforce the contraction of the retractor unguis muscle to improve the grip on the ground. In moth larvae, sensory neurons from hairs on the prolegs also connect directly with particular motor neurons (Weeks and Jacobs, 1987). The failure of all other mechanosensory neurons from exteroceptive hair sensilla on locust legs to connect directly with motor neurons is consistent with the anatomy of the motor neurons, which have no branches in the ventral neuropil to which these sensory neurons project.

7.6.3. Synaptic connections between the interneurons

Most of the local circuitry that processes the mechanosensory signals and shapes the motor output is formed by an extensive web of interconnections amongst the local interneurons, and with outputs to the motor neurons and the intersegmental interneurons. Unravelling these connections is greatly aided by a knowledge of the behaviour and thus of the objective underlying the processing, because it is then possible to interpret the connections in terms of possible roles in controlling certain leg movements. Not all of the connections can be understood in these terms and it would be surprising if they could, because a single neuron has many actions that depend on the interplay between the other neurons that are active at the same time. A single interneuron does not have a single action.

7.6.3.1 *Spiking local interneurons*

Two of the groups of spiking local interneurons so far identified make different connections with the other members of the local circuits. The midline interneurons make inhibitory connections and the antero-medial interneurons make excitatory connections.

Midline spiking local interneurons Spiking local interneurons of the midline group make inhibitory output connections with particular nonspiking interneurons, with some members of the antero-medial group of spiking interneurons, with some intersegmental interneurons and with some motor neurons (Fig. 7.8) (Burrows, 1987; Laurent, 1988; Nagayama and Burrows, 1990). This means that once the

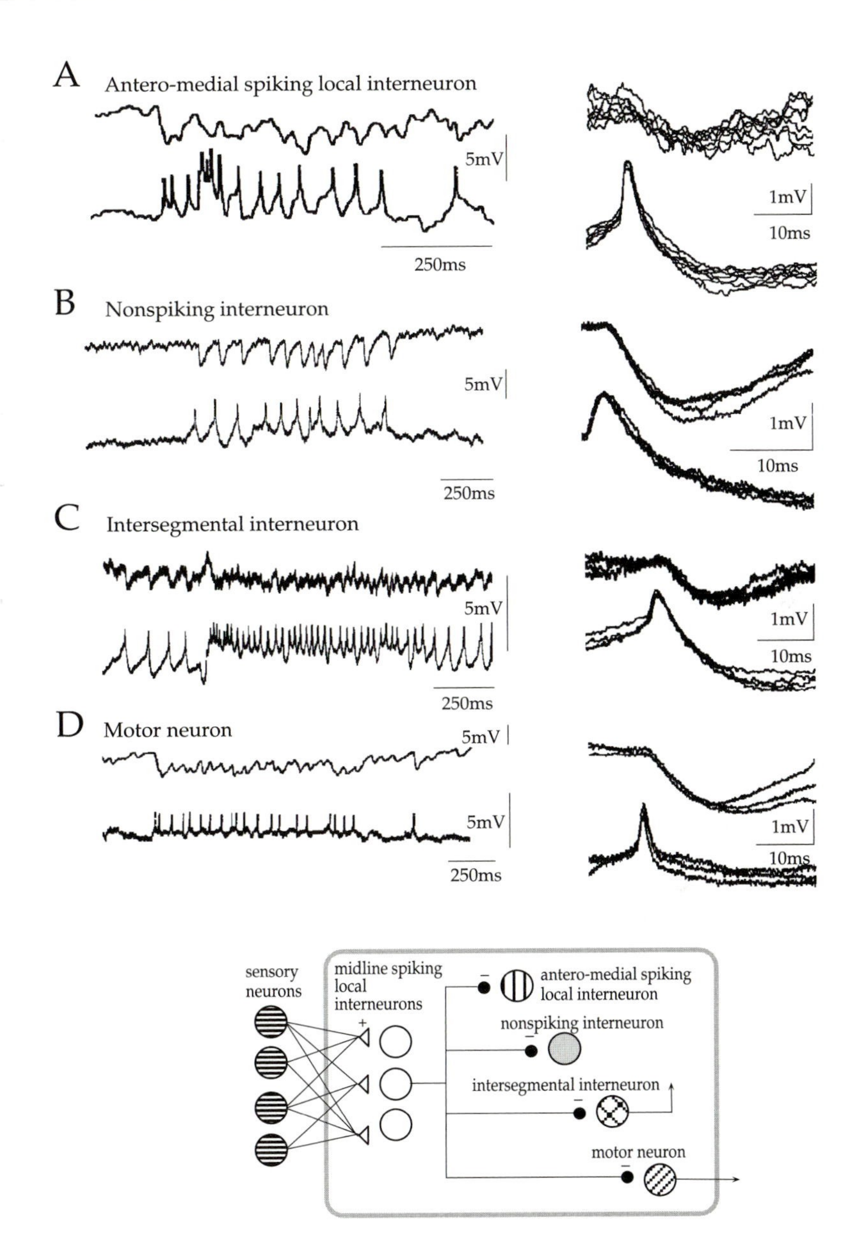

Fig. 7.8 Divergent output connections of spiking local interneurons of the ventral midline group with (A) an antero-medial spiking local interneuron, (B) a nonspiking interneuron, (C) an intersegmental interneuron, (D) a motor neuron. The left side of each panel shows the effect of mechanical stimulation of an array of hairs, and the right side shows the IPSPs caused by the spikes in the midline interneurons. The diagram below these shows the connections established. Excitatory connections are represented by open triangles and inhibitory ones by filled circles. Each type of neuron in the network is represented by a circle with a particular fill pattern. These conventions are also used in subsequent diagrams.

signals from a particular array of sensory receptors have been collated, they are then distributed again more widely by these divergent connections within the circuit, and inverted from excitatory to inhibitory. The connections are unidirectional; no reciprocal connections back to the midline spiking local interneurons are known. Each of the midline local spiking interneurons makes a specific set of connections with only a limited number of the other neurons. The connections that each spiking local interneuron makes with different postsynaptic neurons differ in their effectiveness because the quantal content of the different output synapses can range from 2 to 10 (Laurent and Sivaramakrishnan, 1992). These differences in the probability of release imply that the effectiveness of connections may depend on the frequency and patterning of presynaptic spikes and perhaps on the presence of neuromodulators that could selectively alter the probability of release at the different synapses (*see Chapter 3*).

Some of these interneurons connect directly with motor neurons (Burrows and Siegler, 1982) so that they provide the shortest pathway between certain sensory neurons and the inhibition of specific motor neurons. A single spiking local interneuron is presynaptic to several nonspiking interneurons and to several intersegmental interneurons. These connections with local and intersegmental interneurons are thus largely responsible for shaping the inhibitory regions of the receptive fields of postsynaptic neurons, and for sharpening the boundaries of contiguous excitatory regions. Moreover, the inhibitory regions of the receptive fields of the midline spiking local interneurons themselves could also be formed by inputs from other members of the same population. These connections have not been demonstrated, but the complementary receptive fields of some of these interneurons are appropriate for such interactions. The inhibitory output connections also exclude, or reduce, the contribution of postsynaptic neurons during certain movements, in such a way that the formation of these connections may be dependent on the output connections made by the postsynaptic neurons. An understanding of these pathways is as dependent on a knowledge of the outputs of the interneurons as it is on their inputs.

The connections imply that the receptive fields of the interconnected interneurons must overlap and that the inhibitory regions of the receptive fields of the postsynaptic interneurons are determined by the excitatory receptive fields of the presynaptic spiking local interneurons. The connections between the interneurons that shape the receptive fields involve both divergence and convergence of signals, but with the dominant constraint being that spatial information is preserved (Fig. 7.9). Thus, a stimulus to hairs on the dorsal femur excites several spiking local interneurons which, in turn, make divergent connections with several nonspiking and intersegmental interneurons. The connections are specific because not all of the spiking local interneurons excited by this stimulus converge on the same nonspiking interneurons.

The inhibitory regions of the receptive fields of some interneurons are caused by inputs from a single spiking local interneuron, but in others are caused by the convergence of several. This implies that the receptive fields of the latter interneurons will be larger than those of any of the presynaptic spiking local

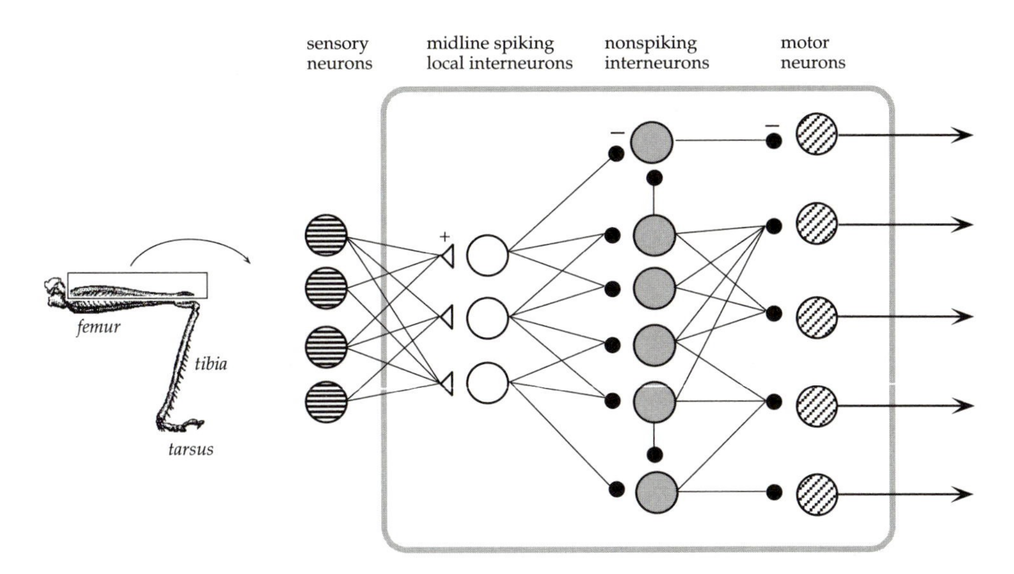

Fig. 7.9 Diagram of the convergent connections made by sensory neurons, by midline spiking local interneurons and by nonspiking interneurons that control the leg motor neurons.

interneurons, as they themselves have overlapping but not identical fields. An overlap in the receptive field of a spiking local interneuron and another interneuron does not, however, necessarily imply that there will be connections between these interneurons. For example, a nonspiking interneuron does not receive inputs from all the spiking local interneurons that have parts of their receptive fields in common.

The inhibitory effects of at least some of the spiking local interneurons on postsynaptic interneurons are probably caused by the release of GABA. Many cell bodies in the ventral midline region of the ganglion, where those of midline spiking interneurons are located, stain with an antibody raised against GABA, and some, but not all, individually identified spiking local interneurons show GABA-like immunoreactivity (Watson and Burrows, 1987). The inhibition that they cause in postsynaptic neurons is abolished by picrotoxin, which blocks chloride channels that are usually opened by GABA, and can be mimicked by the application of GABA. The inhibitory potentials evoked by these interneurons in nonspiking interneurons reverse at about –80 mV (Laurent and Sivaramakrishnan, 1992). The implication is that some of these interneurons contain, and probably release, GABA.

Antero-medial spiking local interneurons Spiking local interneurons of the antero-medial group make excitatory output connections with motor neurons (Fig. 7.10) (Nagayama and Burrows, 1990). It is not known whether they also make connections with other elements of the circuits, but it is presumed that they do. Each interneuron makes divergent connections with some, perhaps all, members of the

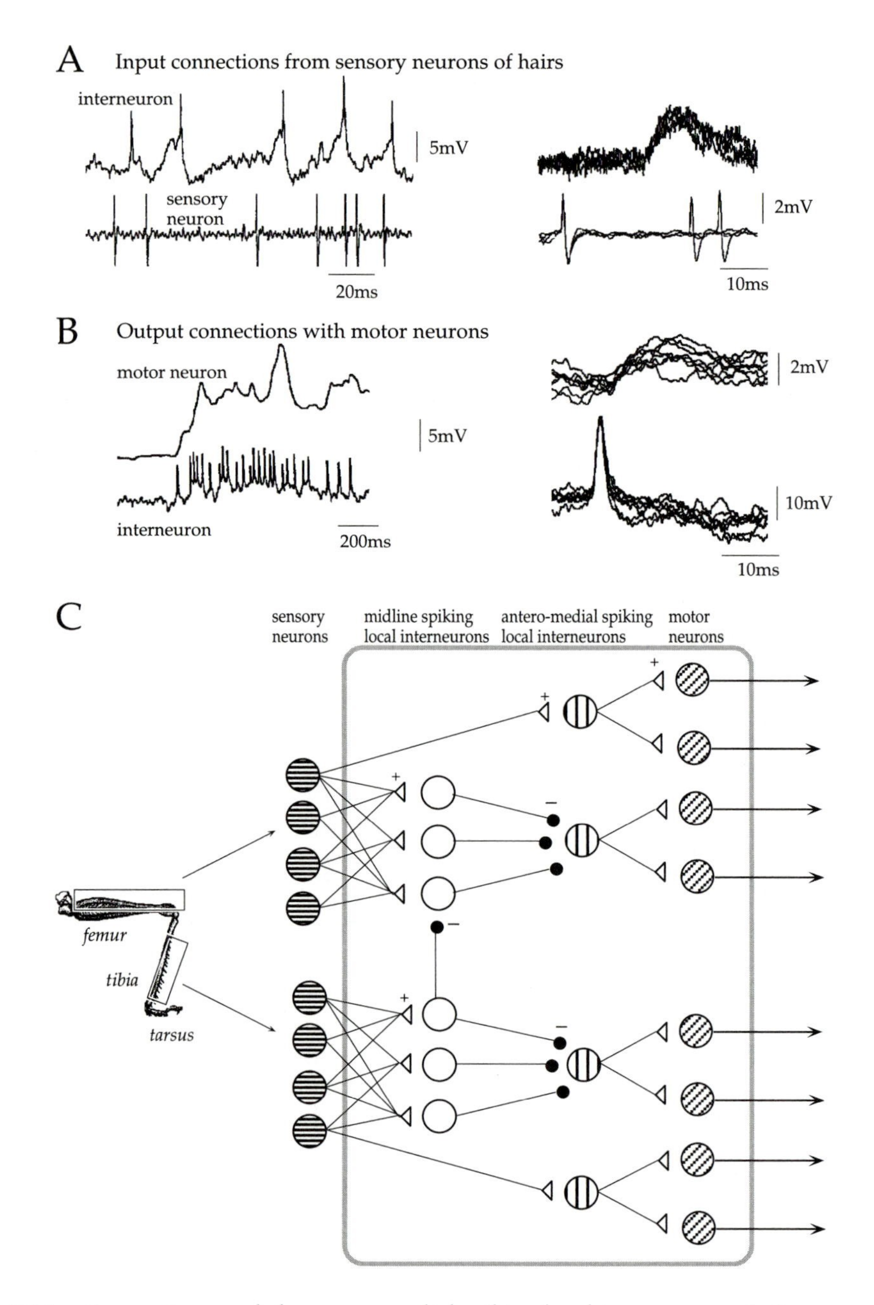

Fig. 7.10 Connections made by antero-medial spiking local interneurons. A. Sensory neurons from trichoid sensilla make direct excitatory connections. B. An interneuron makes an excitatory output connection with a leg motor neuron (fast depressor tarsi). C. Diagram of the connections made by an array of receptors on the dorsal femur and another on the dorsal tibia with midline and antero-medial spiking local interneurons. A and B are based on Nagayama and Burrows (1990).

motor pool supplying one muscle, and, in turn, each motor neuron receives convergent inputs from a number of interneurons that have overlapping receptive fields. The connections are again highly specific because not all the interneurons with overlapping receptive fields connect with the same motor neurons. Moreover, because these interneurons can have receptive fields on several legs, their control of the motor neurons of one leg will be placed in the context of the movements of the other legs and the stimuli they receive. The transmitter used by these interneurons in mediating their excitatory effects on other neurons is not known.

7.6.3.2. *Nonspiking interneurons*

The nonspiking neurons make a complex web of inhibitory synaptic connections with each other in addition to the extensive connections that they make with motor neurons (Burrows, 1979b, 1980a). All the connections are one way so that no direct reciprocal connections seem to occur. The action of an individual interneuron can be defined in terms of the movement that it produces when a depolarising current is injected into it, so that, for example, one interneuron might cause flexion of the tibia of a hind leg and another may cause extension. The abundant connections between the nonspiking interneurons can also be understood in terms of these sorts of actions. These interneurons are, however, not normally activated in this separate fashion and their actions will be much more complex when other parallel channels are also active. This means that the action of an interneuron must be related to the context in which it is observed, rendering a simple and singular definition of its action at best confusing and at worst misleading.

Inhibitory connections occur between interneurons that drive the same set of motor neurons, implying that these interneurons are almost certainly used during different movements. This is confirmed by the fact that their receptive fields are different, so that they are likely to be activated by different sensory inputs. When interneurons such as these are experimentally activated together their effects sum in a nonlinear way at the motor neurons, but normally their different responses to sensory signals from movements of the leg ensure that they would not be activated at the same time.

Other nonspiking interneurons that normally bring about antagonistic movements may be linked by inhibitory synaptic connections. For example, an interneuron that causes flexion of the tibia may inhibit another interneuron that causes extension of the tibia. Similarly, an interneuron causing extension may inhibit one causing flexion so that there are reciprocal pathways between the interneurons controlling the antagonistic pools of motor neurons. Thus, the reciprocity of action by the extensor and flexor motor neurons is organised by the nonspiking interneurons. There are times, however, when the two motor pools must act synergistically, such as when the flexor and extensor tibiae motor neurons must cause a co-contraction during the preparation for a jump or a kick (*see Chapter* 9). They must then be driven by different sets of interneurons that either prevent the action of the interneurons that would cause reciprocal actions, or reduce the efficacy of their connections.

Some inhibitory connections involve interneurons that affect the movements of different joints of a leg. An interneuron that causes extension of the tibia, as would occur during the stance phase of walking, inhibits an interneuron that causes movements of the coxa that are likely to be used during the swing phase.

Other connections are less easy to fit into a logical framework, but this may simply be because current concepts are defined too rigidly by the necessary constraints of the experiments. The muscles of the legs are used in many different combinations when walking horizontally, vertically and upside-down (*see Chapter 8*), and when walking forwards, sideways and backwards. It is the flexibility imparted by the connections between the nonspiking interneurons and the summation of their effects on different sets of motor neurons controlling different muscles that makes this variety of action attainable. It is, however, almost impossible to determine all the connections made by an interneuron and thus to predict all its effects and actions. It thus becomes necessary to rely on the examples that can be revealed, and to accumulate evidence that indicates whether these are representative of the whole pattern of connections. What can be concluded so far is that the inhibitory one way connections made with other nonspiking interneurons and the connections made with motor neurons are characteristic of the nonspiking interneurons in the circuits controlling leg movements. Even this statement, however, does not allow the extrapolation that all the nonspiking interneurons in a segmental ganglion make this pattern of connections. There is no reason to suppose that the nonspiking interneurons have a uniform set of connections and effects just because they share one feature of their physiology in common, namely the ability to communicate with other neurons without the intervention of spikes. The connections also emphasise still further that the action of an individual interneuron can only be viewed in the context of the actions of the many other interneurons to which it is either connected directly or indirectly.

The effectiveness of these connections between the nonspiking interneurons depends both on the duration and the frequency of the inputs to the presynaptic interneuron. The change in the postsynaptic voltage gradually increases as the duration of a presynaptic pulse is increased, reaching a plateau with durations of 200–300 ms that can be sustained for pulses lasting many seconds or even minutes. Conversely, the postsynaptic voltage change starts to decline with frequencies of only a few cycles per second, but effects can still be produced by frequencies in excess of 20 Hz. These durations and frequencies are well within the range of durations of the swing and stance phases of walking, and of the frequency of the step cycle, so that the interactions can be expected to operate during walking and to be essential mechanisms in the patterning of the motor signals.

7.6.3.3. *Intersegmental interneurons*

The only intersegmental interneurons whose output connections are known in any detail are those with cell bodies in the ventral and posterior region of the mesothoracic ganglion and with axons that descend in the ipsilateral connective to the metathoracic ganglion (Laurent, 1987). These interneurons make parallel connections with nonspiking interneurons and motor neurons in the local circuitry controlling the ipsilateral hind leg (Laurent and Burrows, 1989b). Connections with midline spiking local interneurons have not been found, but connections with antero-medial interneurons from intersegmental interneurons such as these must be postulated to explain their receptive fields on a middle leg. About 80% of these intersegmental interneurons make excitatory connections with metathoracic neurons and the remainder make inhibitory connections; this is in keeping with the proportion of these interneurons that show GABA-like immunoreactivity (Watson and Laurent, 1990).

When a particular region of a middle leg is touched, a specific set of intersegmental interneurons is excited which, in turn, activates a particular set of metathoracic nonspiking interneurons and motor neurons (Fig. 7.11). Stimuli to different parts of a middle leg excite different interneurons whose outputs may then also converge on the same metathoracic neurons. Most stimuli to a particular region of one of the middle legs excite several intersegmental interneurons whose outputs may converge onto a metathoracic neuron. Some of the intersegmental interneurons make excitatory and some make inhibitory connections with the same nonspiking interneuron. The apparently conflicting information delivered to a nonspiking interneuron could have three possible effects. First, the inputs could create a gradient of excitation and inhibition that could blur the boundaries between the excitatory and inhibitory receptive fields of the nonspiking interneuron. Second, the inputs could be distributed to distinct branches of the nonspiking interneuron where they might act with some degree of independence. Third, the combination of excitation and inhibition could provide information about the direction of the tactile stimulus sweeping across the receptive field.

A single intersegmental interneuron can make divergent connections with members of a motor pool supplying one muscle and with nonspiking interneurons that are presynaptic to some members of that pool. The result is that the signals from the middle leg are amplified by the nonspiking interneurons and distributed widely to the motor pool, where they sum on certain motor neurons with the direct inputs they receive from local interneurons. The inhibitory connections between the nonspiking interneurons will also spread the intersegmental signals further through the network and could allow an intersegmental interneuron to exert opposite effects on two pools of motor neurons that may have antagonistic actions.

7.6.4. Central drive to the interneurons

Such analyses of connectivity as described above and summarised in Fig. 7.12 are performed in quiescent locusts that are not walking so that the number of active pathways can be more under the control of the experimenter. When the locust is walking, these pathways can be modified by the actions of the neurons that generate the motor patterns. The spiking local interneurons of the midline group receive a synaptic input in time with the motor patterns that are produced in isolated ganglia and during normal walking (Wolf and Laurent, 1994). This means that the effectiveness of the sensory signals in making the interneurons spike is related to the

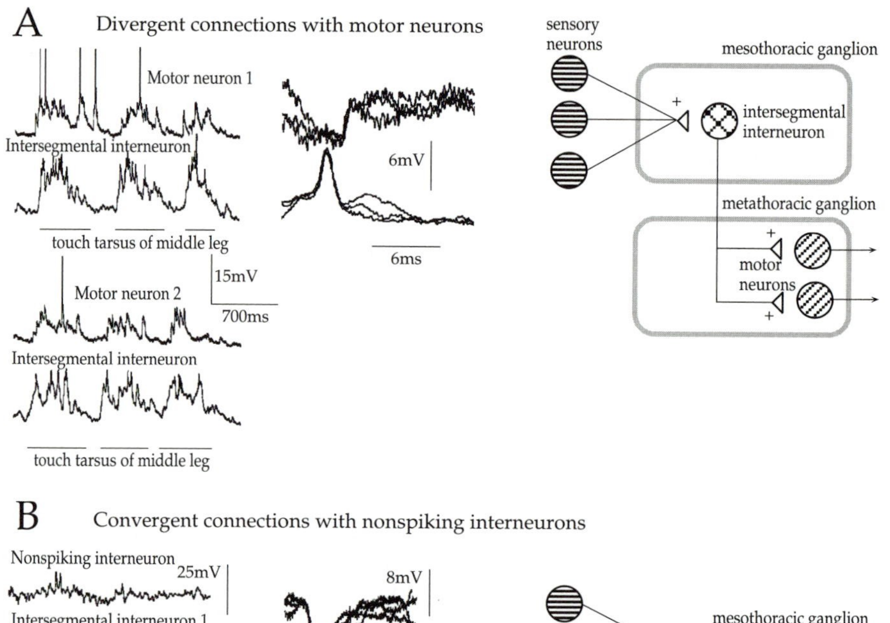

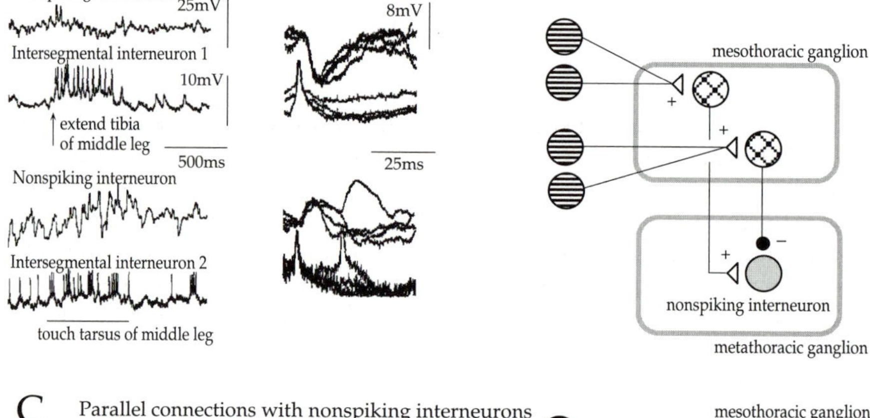

C Parallel connections with nonspiking interneurons and motor neurons

mesothoracic ganglion

metathoracic ganglion

Fig. 7.11 Connections made by mesothoracic interneurons with (A) motor neurons and (B) nonspiking interneurons in the metathoracic ganglion. C. Some interneurons make parallel connections with both motor neurons and nonspiking interneurons. The diagrams on the right show the input connections from sensory neurons of trichoid sensilla in the mesothoracic ganglion and the output connections in the metathoracic ganglion. Based on Laurent and Burrows (1989b).

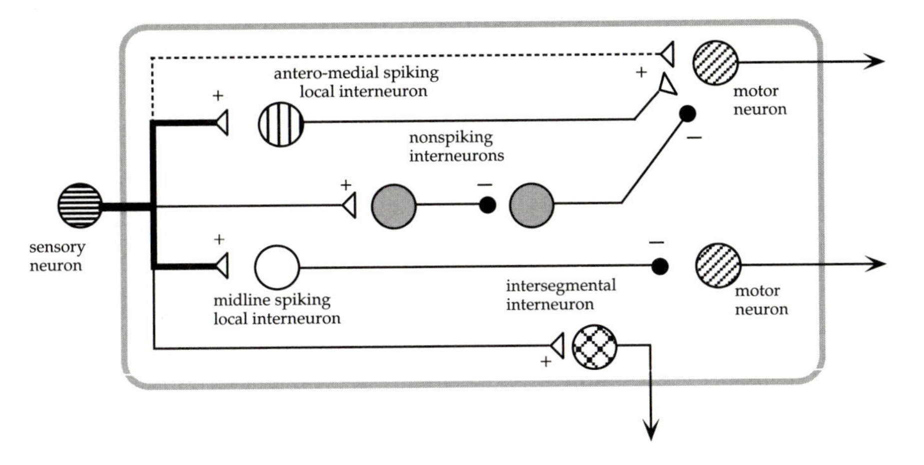

Fig. 7.12 Diagram of the connections made by mechanosensory neurons from exteroceptors (trichoid and basiconic sensilla). Each synaptic connection has been established by paired intracellular recordings from the component neurons. The diagrams indicate only the general pathways and give no flavour of the numbers of the different types of neurons involved or of the convergence and divergence of connections. Compare with Fig. 7.13.

phase of the leg movements in the step cycle. If the sensory signals occur during the depolarising phase of the central input their contribution will be enhanced, whereas if they occur during the hyperpolarising phase it will be decreased. This alteration in the response to the sensory signals will thus operate in parallel to that exerted at the sensory terminals themselves (*see section 7.14*). The central input might even mean that the receptive field of an interneuron could be altered by the central drive. A further consequence will be that the central signals will be distributed widely in the networks through the inhibitory output connections that the spiking local interneurons make with other neurons.

7.7. PROPRIOCEPTORS

Proprioceptors signal the movements and positions of the joints, and monitor the force generated by the muscles and the strains that are consequently set up in the cuticular exoskeleton. The receptors are of several different types. Joint movements and positions are monitored by chordotonal organs, by individual or small groups of strand receptors, by hair plates, and by a small number of multipolar joint receptors. Muscle force is measured by multipolar tension receptors, which are known to be present as individual cells in only a few muscles. Strains in the cuticle are largely monitored by strategically placed individual or small groups of campaniform sensilla (*see Chapter 8* for details of these different receptors).

Each movement of a joint excites many sensory neurons from several different types of receptors in parallel, so that it is difficult to separate the contribution to a motor response of individual receptors, let alone individual sensory neurons. The use of many parallel channels points to the importance of accurate information about the joint, but why it is necessary to devolve this to several different types of receptors that appear to provide parallel information is unknown. The paradox is at its most acute in the femoro-tibial joint where a chordotonal organ with almost 50 neurons signalling joint movements acts in parallel with the single neuron of a strand receptor, a single tension receptor, a number of campaniform sensilla, and five multipolar joint receptors. Just what the small number of other receptors can add to the signals of the chordotonal organ remains obscure because both the strand receptor and three of the joint receptors signal extension movements of the tibia over the same range that is covered by many neurons in the chordotonal organ. This arrangement points to the operation of the receptors in ways that are different from those used when experimental measurements are made on the joint.

Two main approaches have been used to study these proprioceptors. First, the feedback that they provide during a normal movement is perturbed, to investigate whether any changes in the motor pattern result. The disruption may, however, affect receptors other than the one under investigation, so that interpretation of the results can be ambiguous. Second, individual receptors, or even individual sensory neurons within a receptor, are stimulated in a quiescent animal and effects are sought in the actions of muscles, motor neurons or interneurons. The effects produced may be subthreshold in many of the neurons in the central pathways, so that complete pathways may not be activated and there may be no output in the motor neurons and in the muscles. Moreover, the effects that are produced may be different from those exerted in a normally behaving animal, when many pathways are active in parallel and when the effects of neuromodulators may change the effectiveness of signal transmission and the membrane properties of many neurons.

7.8. THE BEHAVIOUR ELICITED BY THE MECHANORECEPTORS

Stimulation of this wide range of receptors, not surprisingly, leads to a wide range of motor effects. The campaniform sensilla can increase the force generated by the muscles to compensate for the increased loading or forces to which they respond. Joint movements lead to stabilising reflex responses that normally resist an imposed deviation but under different conditions can assist it (*see Chapter 8*). When the animal is quiescent, an imposed flexion of the femoro-tibial joint excites extensor tibiae motor neurons and inhibits flexor tibiae motor neurons so that the joint is stabilised. The effects of moving one joint may radiate within the central nervous system to motor neurons innervating muscles of neighbouring joints to provide further stability and allow posture to be maintained. Under constant conditions these resistant reflexes are reliable, but they are not rigid immutable responses and, depending on what the animal is doing, can change in gain and sign. An imposed flexion of the tibia can, under appropriate conditions, lead to excitation of flexor tibiae motor neurons and inhibition of the extensors so that the imposed movement is now assisted by the motor response. This high degree of flexibility allows voluntary movement in the face of connections that would otherwise oppose these movements.

7.8.1. What the behaviour reveals about the neural organisation

These sorts of behavioural response point to the following design constraints on the circuitry that must process the proprioceptive signals.

1. The processing must be reliable, to permit the expression of a consistent motor response to repeated movements of a joint. This implies reliable and perhaps short pathways linking sensory neurons and motor neurons.
2. The motor responses generated by proprioceptive inputs must be sustainable over long periods if they are to contribute to the maintenance of posture.
3. The gain in the pathways must be variable, to permit the expression of different levels of muscular force to the same stimulus in different circumstances. This probably implies that there must be flexibility in the responses of the local interneurons that receive converging inputs from different sources.
4. The pathways must be modifiable, to permit the expression of voluntary movements. This could imply modification at various different stages in the processing, but probably indicates changes at the level of the sensory neurons and local interneurons.
5. The proprioceptive and exteroceptive signals must be brought together at some stage in the circuitry so that an input of one modality can alter the expression of a motor response caused by the other modality.

7.9. SENSORY PROJECTIONS

The axons of sensory neurons from proprioceptors are usually of larger diameter than those of the tactile hairs and thus conduct their spikes at a faster velocity. The branches within a ganglion are also of larger diameter and some are more widely distributed than the branches from exteroceptive sensory neurons, but the number of branches from each sensory neuron is small and terminal varicosities are common. The terminals of the sensory neurons of proprioceptors are spatially separate from those of the exteroceptors on the leg because they do not project to the most ventral regions of the neuropil that contains the terminals of the exteroceptive sensory neurons, but instead project to slightly more dorsal and lateral regions, the aLAC and pLAC. The pattern of branching from most individual sensory neurons from campaniform sensilla and chordotonal organs is very similar so that, at present, only a few can be distinguished on anatomical grounds alone. Some of the different sensory neurons from complex chordotonal organs such as that at the femoro-tibial joint have different patterns of branches (Matheson, 1990).

7.10. SYNAPTIC CONNECTIONS UNDERLYING THE PROCESSING

The proprioceptive sensory neurons connect with the same types of neurons as the exteroceptive signals so that many of the same organisational principles apply to the processing of their signals (Fig. 7.13). A distinct difference, however, is the direct connections that the proprioceptive sensory neurons from both campaniform sensilla and chordotonal organs make with the leg motor neurons (Fig. 7.14) (Burrows, 1987; Burrows and Pflüger, 1988). In addition to this direct route to the motor neurons, the sensory neurons also make parallel connections directly with midline spiking local interneurons, intersegmental interneurons, and nonspiking interneurons. Some of the processing is carried out by parallel sets of the local interneurons that process the exteroceptive signals, while some interneurons process convergent inputs from both proprioceptive and exteroceptive sensory neurons.

The connections made by the sensory neurons of proprioceptors are again always excitatory, thus reinforcing the rule that the central synaptic connections of mechanosensory neurons are invariably excitatory. Any inhibitory effects must be mediated by interneurons that reverse the sign of the incoming sensory signals.

7.10.1. Spiking local interneurons

Individual spiking local interneurons in the midline group may receive inputs exclusively from exteroceptors, exclusively from proprioceptors, or convergent inputs from both types of receptor (Burrows, 1985). The morphology of these individual interneurons also correlates with their patterns of sensory inputs; thus, those with inputs only from proprioceptors have few branches in their ventral field and none in the most ventral regions of VAC to which the exteroceptive sensory

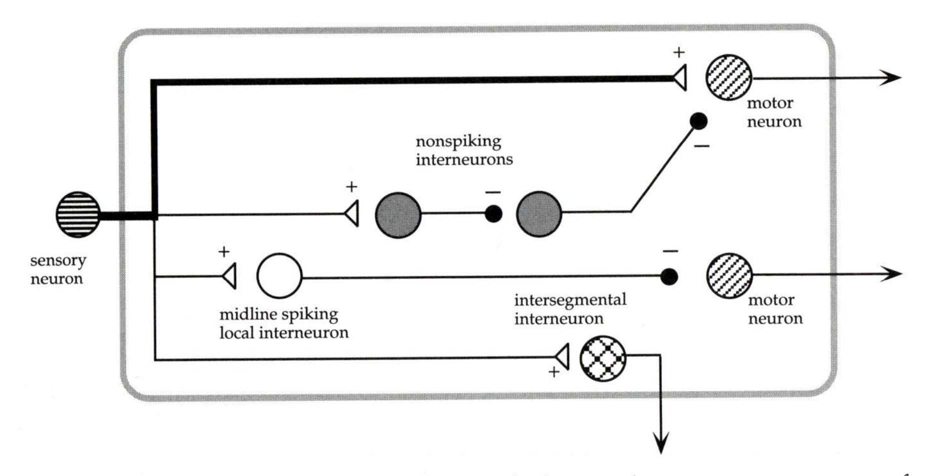

Fig. 7.13 Diagram of the connections made by mechanosensory neurons from proprioceptors (chordotonal organs and campaniform sensilla). Compare with Fig. 7.12.

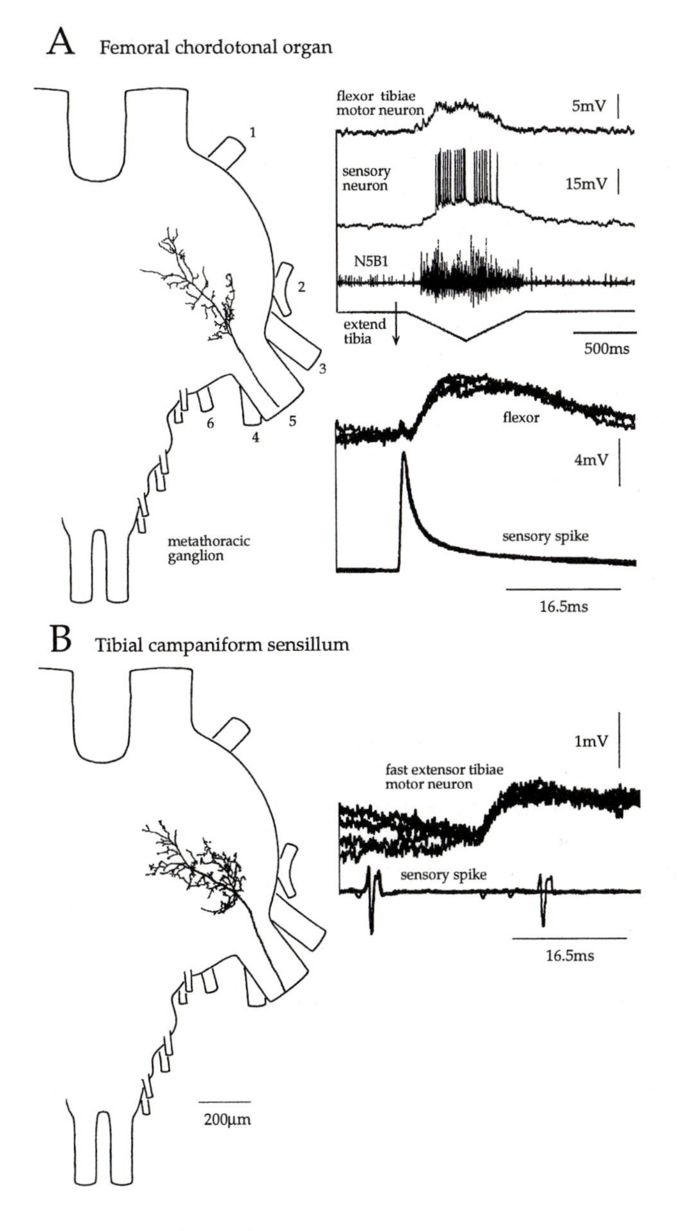

Fig. 7.14 Connections made with motor neurons by a sensory neuron from (A) a chordotonal organ and (B) a campaniform sensillum. A. Extension of the tibia evokes spikes in many sensory neurons of the femoral chordotonal organ and in the one impaled by a microelectrode. At the same time, a flexor tibiae motor neuron is depolarised. Each spike in the impaled sensory neuron is followed by an EPSP in the motor neuron. B. Each spike in a sensory neuron from a campaniform sensillum on the proximal tibia is followed by an EPSP in the fast extensor tibiae motor neuron. The drawings on the left show the projections of the sensory neurons in the metathoracic ganglion. B is based on Burrows and Pflüger (1988).

neurons project. Their branches are instead in the more dorsal and lateral regions of neuropil, aLAC and pLAC where they overlap with the projections of the proprioceptive sensory neurons. The connections that the proprioceptive sensory neurons make with the midline spiking interneurons appear to be direct (Burrows, 1987), but while the antero-medial interneurons also respond to joint movements and cuticular strains, direct connections have not been demonstrated.

The direction, amplitude, velocity and probably acceleration of the movements, and the positions of a leg joint, such as the femoro-tibial joint are coded by different spiking local interneurons (Burrows, 1988). Some respond to extension, some to flexion and some to both directions of movement. Those that respond to movement in one direction may be inhibited or unaffected by movement in the opposite direction. All respond to repetitive movements within the range normally used during walking (*see Chapter 8*). It is possible that only one interneuron of each type exists so that there is little parallel processing of a particular feature of a stimulus within this midline population. Other interneurons in this population receive direct inputs from campaniform sensilla that are placed to detect particular strains in the cuticle, or to detect deflections of the tibial spurs as the tarsus is placed on the ground (Burrows and Pflüger, 1986).

Some interneurons that code the position of the joint show less hysteresis in their responses than the sensory neurons that provide their input. This reduction in hysteresis may result from the feedback control of the terminals of the sensory neurons (Hatsopoulos *et al.*, 1995) (*see section 7.14*). Other interneurons respond only phasically to a movement, with the velocity of the movement determining the sign of their responses; slow movements might cause excitation and faster movements inhibition. This implies that the responses of an interneuron depend both on the direct connections of the sensory neurons and the parallel connections through other interneurons.

All the salient features of the movement of a joint are therefore coded in a number of parallel channels. Most of the coding is fractionated among several interneurons so that, for example, no interneuron codes position over the entire range of joint movement although some sensory neurons of the femoral chordotonal do so. This organisation means that each movement will excite a number of interneurons and inhibit a number of others, resulting in a representation of joint movement that is very similar to the representation of the surface of a leg in the receptive fields of other members of this population. Those interneurons that receive inputs from both proprioceptors and exteroceptors have responses to joint movement that are dependent on simultaneous exteroceptive inputs and, conversely, responses to tactile inputs that are dependent on the position and movements of particular joints. These combinations of responses often represent the pattern of inputs that are likely to be generated during normal walking movements.

7.10.2. Intersegmental interneurons

Proprioceptive sensory neurons synapse directly on some intersegmental interneurons belonging to several different populations. In some of these interneurons, the proprioceptive and the exteroceptive signals converge, but others receive inputs only from proprioceptors, and others only from exteroceptors. The morphology of these interneurons again correlates with these features of their receptive fields so that those which receive inputs only from proprioceptors have either no branches, or very few, in the most ventral regions of neuropil (Burrows and Newland, 1993).

7.10.3. Nonspiking interneurons

The nonspiking interneurons play an important part in the production of local reflexes initiated by proprioceptors by virtue of the direct excitatory connections that they receive from certain proprioceptive sensory neurons, the inhibitory connections they receive from the midline spiking local interneurons, and the output connections that they make with other nonspiking interneurons and with pools of motor neurons. Any movement of a joint or strain in the cuticle affects several nonspiking interneurons and their divergent connections mean that many other neurons in the circuits controlling the movements of the legs will also be affected. Despite this parallel and distributed processing, experimental manipulation of the membrane potential of a single nonspiking interneuron can alter the expression of a local motor response (*see section 7.13*). This implies either that the number of nonspiking interneurons in particular pathways is limited or that the effects of certain individual interneurons are powerful.

Individual nonspiking interneurons respond in different ways to the same movement of a particular joint; some may be depolarised by direct synaptic inputs from sensory neurons, whereas others may be hyperpolarised by the action of spiking local interneurons that themselves receive direct input from the sensory neurons (Burrows *et al.*, 1988). For an individual interneuron, movement of the joint in one direction may cause a depolarisation, and movement in the other direction a hyperpolarisation. For other interneurons, only one direction of movement may have an effect, and this can be either depolarising or hyperpolarising, others may be depolarised by both directions of movements, while still others may be inhibited by both directions. The depolarisation in the different nonspiking interneurons can result either from direct excitatory synaptic inputs from the sensory neurons, or from the withdrawal of inhibition caused by the spiking local interneurons. This finding emphasises the widespread occurrence of disinhibition in the processing of sensory signals.

7.10.3.1. *Hysteresis and joint position*

When a joint moves to a new angle and is held there, as when the posture of a leg is altered, some nonspiking interneurons show a sustained shift in their membrane potential that can be maintained for minutes (Fig. 7.15) (Siegler, 1981a). At least part of this shift is due to the change in the maintained pattern of spikes in the sensory neurons of particular chordotonal organs and hence their direct action on the interneurons, although part may be due to changes in the other interconnected interneurons and to the inherent properties of the nonspiking interneurons themselves. These sustained shifts show considerable hysteresis that depends on the preceding direction of movement and would represent a substantial error if joint

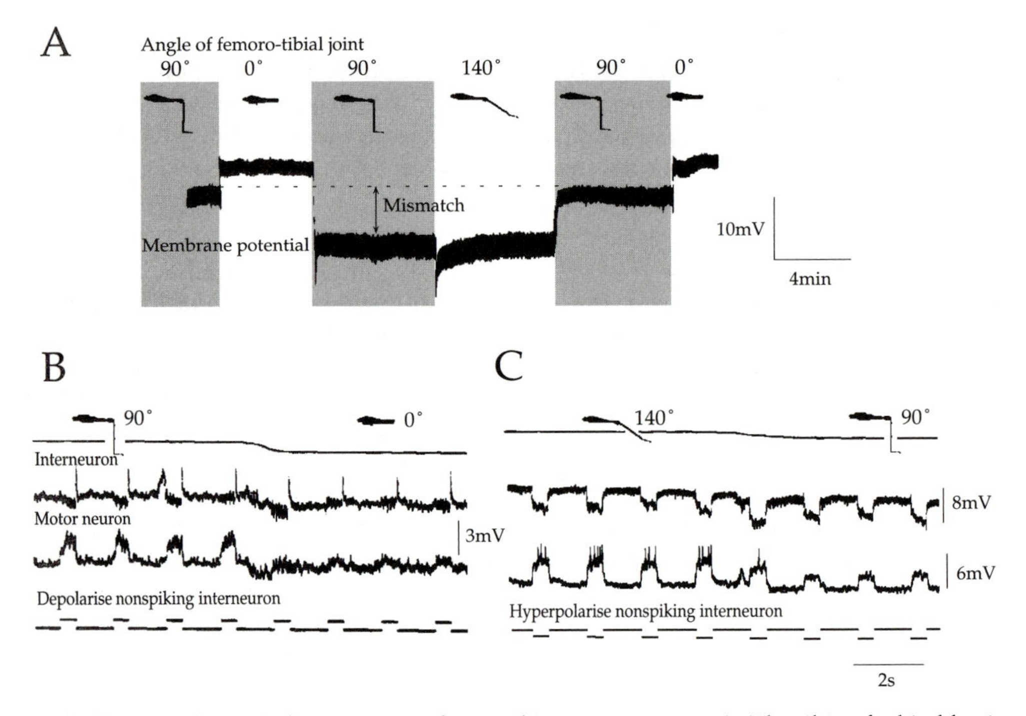

Fig. 7.15 Hysteresis in the responses of nonspiking interneurons. A. The tibia of a hind leg is moved forcibly to different positions and then held at those positions. The membrane potential of a nonspiking interneuron at a femoro-tibial angle is different depending upon whether this position was approached from a preceding flexed or extended position. B, C. The effectiveness of the output connections with motor neurons depends on the position of the femoro-tibial joint. B. At 90°, each pulse of depolarising current injected into the interneuron causes a large depolarisation in a motor neuron, but at 140° this effect is substantially reduced. C. At 140°, each pulse of hyperpolarising current injected into another nonspiking interneuron causes a depolarisation and spikes in a motor neuron, but at 90° no spikes are evoked. Based on Siegler (1981a,b).

position were coded as membrane potential in these interneurons. The net result is that the sustained membrane potential of a nonspiking interneuron at the same joint position can differ by as much as 5 mV, depending on the direction from which this position was approached. Long-lasting changes of these magnitudes in the nonspiking interneurons are sufficient to change the release of transmitter onto postsynaptic neurons (*see Chapter 3*) and may partly explain the hysteresis in the motor neurons that occurs during sustained alterations in posture. They also mean that the effectiveness of the output synapses of a nonspiking interneuron are dependent on the position of the leg (Siegler, 1981b). Both the history of movement and the current position of a joint will therefore alter the transmission of signals from nonspiking interneurons and such mechanisms help to explain how the same stimulus can lead to different responses in different behavioural contexts.

7.10.3.2. *Strategies for flexibility in the patterns of connections*

A common pattern in the connections made by a proprioceptive sensory neuron is a direct connection with a motor neuron and a parallel direct connection with a nonspiking interneuron that affects the same motor neuron (Burrows and Pflüger, 1988). This is a potent way of amplifying the effects of the sensory signals on a motor neuron, and is a mechanism also used by some intersegmental interneurons. To ensure the flexibility of action by the members of a motor pool, a proprioceptive sensory neuron may connect directly with only some motor neurons, but may make parallel connections with nonspiking interneurons that either connect with all members, or with a particular subpopulation of the pool. For example, sensory neurons from campaniform sensilla that monitor strain in the tibia when it encounters resistance during walking synapse directly on the fast extensor tibiae motor neuron but not the slow (*see Fig. 9.12*). They also synapse on a nonspiking interneuron that causes excitation of both motor neurons by disinhibition. This design means that the sensory effects on the slow motor neuron can be regulated by inputs to the nonspiking interneuron, so that sometimes they can be expressed and at other times gated, whereas the effects of the nonspiking interneuron on the fast motor neuron can only be added to the direct inputs from the sensory neurons.

7.10.4. Motor neurons

Direct connections of proprioceptive sensory neurons with particular motor neurons is the rule rather than the exception, and this sets them apart from the exteroceptive sensory neurons where direct connections with motor neurons are the rare exception. The sensory neurons from campaniform sensilla, from chordotonal organs and from muscle tension receptors synapse directly on particular motor neurons. Sensory neurons from a chordotonal organ synapse directly on the motor neurons innervating muscles that move the same joint; those from the femoral chordotonal organ synapse directly on flexor and extensor tibiae motor neurons, and those from the tibio-tarsal chordotonal organ connect directly with levator and depressor tarsi motor neurons (Laurent and Hustert, 1988). We do not, however, know whether these sensory neurons also connect directly with motor neurons of adjacent joints that participate in the interjoint reflex adjustments (Field and Rind, 1981).

The direct connections that the sensory neurons make with the motor neurons are such that single sensory spikes do not lead directly to spikes in a motor neuron. Instead, the sensory spikes must occur at high frequency and sum with inputs from other sources, including the outputs of nonspiking neurons that are processing the same sensory signals.

7.11. CONTROLLING THE MOTOR OUTPUT

The motor neurons are the points for convergence of the processing carried out in the local circuits (Fig. 7.16). Integration also occurs at the level of the muscle fibres through the interactions between excitatory and inhibitory motor neurons and with the efferent DUM neurons. The motor neurons receive inputs from virtually all the different classes of neuron, including proprioceptive sensory neurons, spiking local interneurons, nonspiking local interneurons, and from intersegmental interneurons. The convergence of inputs from the different classes of interneurons emphasises the different routes that can be taken by the sensory signals to influence the motor output. The spike output of a motor neuron is thus determined by a balance between the following influences.

1. The different pathways involved in the processing of local sensory signals.
2. The different pathways involved in the processing of sensory signals from the other legs.
3. The pathways from neurons that place the action of a motor neuron in the context of sensory inputs to other parts of the body that occur at the same time.
4. The pathways that signal what other behaviour may be being performed at that time, so that movements are behaviourally appropriate.
5. The intrinsic membrane properties of the motor neuron itself that determine thresholds for spiking and the spike response to a given synaptic current, the time course of action of the synaptic input and whether the changes affect voltage-sensitive channels that might lead to plateau potentials.

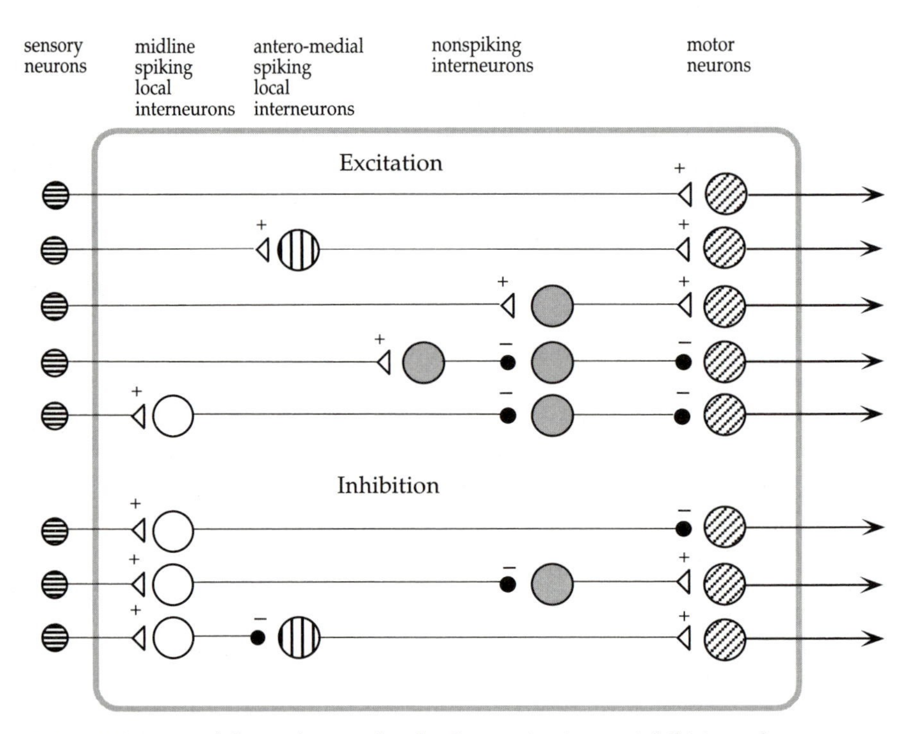

Fig. 7.16 Diagram of the pathways that lead to excitation or inhibition of motor neurons.

6. The feedback pathways that might be activated by the action of the motor spikes in the central nervous system or in the periphery.
7. The neuromodulatory neurons that might be activated by the sensory signals and the effects they may have on motor performance.
8. The ambient levels of ions, neuromodulators, neurohormones and other substances in the haemolymph.
9. The history of its own activity and of its presynaptic neurons.

7.11.1. Nonspiking interneurons

In all these converging pathways onto the motor neurons, the role of the nonspiking interneurons stands out as crucial in shaping the sequence of spikes in a single motor neuron, in recruiting the members of a motor pool innervating one muscle, and in coordinating the output in the motor neurons to muscles moving different joints so that a coherent movement of a leg results (Burrows, 1980a).

7.11.1.1. *Control of a single motor neuron*

Each nonspiking interneuron connects directly with several motor neurons and, in turn, each motor neuron receives either inhibitory or excitatory inputs from several nonspiking interneurons. The control exerted is graded (*see Chapter 3*) so that an individual interneuron can alter the frequency of spikes in a motor neuron in a continuously graded fashion. Often this can lead to graded changes in the force of a particular muscle that parallel the fluctuations in the membrane potential of an interneuron, with phase changes caused by the transmission times in the pathway. This type of organisation is important, for the locust has only a small number of motor neurons that innervate individual muscles, with the consequence that it must rely more on changes in the frequency of motor spikes than on the recruitment of more motor neurons to cause changes in force. Most of the skeletal muscle fibres are multiterminally and polyneuronally innervated and do not generate propagated spikes.

The control of an individual motor neuron is fractionated among a number of nonspiking interneurons and no individual interneuron can elicit the full dynamic range of motor spikes. For example, a slow motor neuron can spike at frequencies up to 200 Hz during normal locomotion, but the excitation from a single interneuron may be able to change this frequency only over a small part of this range. Different interneurons can change the frequency of spikes in a motor neuron only over narrow but different ranges, whereas others can cause larger changes. The full range of spike frequencies must thus result from the concerted actions of many nonspiking interneurons (at least 12 control the slow extensor tibiae motor neuron) and from many other types of interneurons, with the contribution of each depending on the prevailing behavioural context. In certain conditions, an individual nonspiking interneuron can release transmitter tonically so that a motor neuron is held continuously depolarised or hyperpolarised. Additional inputs to such an interneuron can thus increase or decrease the rate of release of transmitter and alter the membrane potential of a motor neuron in either direction.

Nonspiking interneurons that converge on the same motor neurons with effects of the same sign are often themselves driven by inputs originating from common presynaptic neurons. This means that they are likely to be activated at the same time, thereby increasing the probability that their effects will sum at the motor neurons. This summation can be nonlinear, so that the combined effects of two interneurons is greater than the linear sum of their individual effects (Burrows, 1979b).

7.11.1.2. *Control of a motor pool*

A motor pool consists rarely of just a single motor neuron, more commonly of two motor neurons which cause slow and fast contractions, respectively, of their muscle, and frequently of several (up to nine) motor neurons that evoke a range of different contractions in their muscle. The different motor neurons in a pool can be recruited in an orderly sequence, as when the power requirements of a movement gradually increase, or some of the individuals can be used separately in different movements. For example, postural adjustments are usually carried out by the slow motor neurons, but when fast movements are used both the slow and the fast motor neurons may be recruited together, with the greater force generated by the fast ones dominating the movement. All of these different patterns of control and recruitment can be seen in the actions of nonspiking interneurons on the motor neurons.

The same nonspiking interneuron can be presynaptic to the fast and the slow motor neurons of the same muscle (Fig. 7.17). The relationship between the current injected into the interneuron and the voltage change in the motor neurons is steeper for the slow than for the fast motor neuron, and, consequently, less current is needed to make the slow motor neuron spike. The different responses of the two motor neurons can be explained by their different membrane properties in response to the common synaptic drive without the need to invoke different efficacies of the synapses, or the number and distribution of synaptic contacts, although these may indeed also differ. This organisation means that depolarisation of such an

Fig. 7.17 Control of motor neurons by nonspiking interneurons. A. Control of the only two members of a motor pool innervating a particular muscle. Pulses of increasing current injected into an interneuron cause progressively larger depolarisations in both motor neurons and spikes in one. B. Control of antagonists. Current injected into an interneuron causes a depolarisation of an extensor motor neuron and a hyperpolarisation of a flexor. The connection with the flexor appears to be direct, but the effect on the extensor is most probably explained by disinhibition of another nonspiking interneuron. C. Recruitment of members of a motor pool by increasing sinusoidal current. The force produced by the muscle, its myogram and the intracellular membrane potential of one motor neuron are monitored. D. Convergence of many nonspiking interneurons onto one motor neuron. The graph shows the change in voltage of the motor neuron induced by current injected into the different interneurons in one animal. E. Diagram of the connections made by nonspiking interneuron with motor neurons. Based on Burrows (1980a).

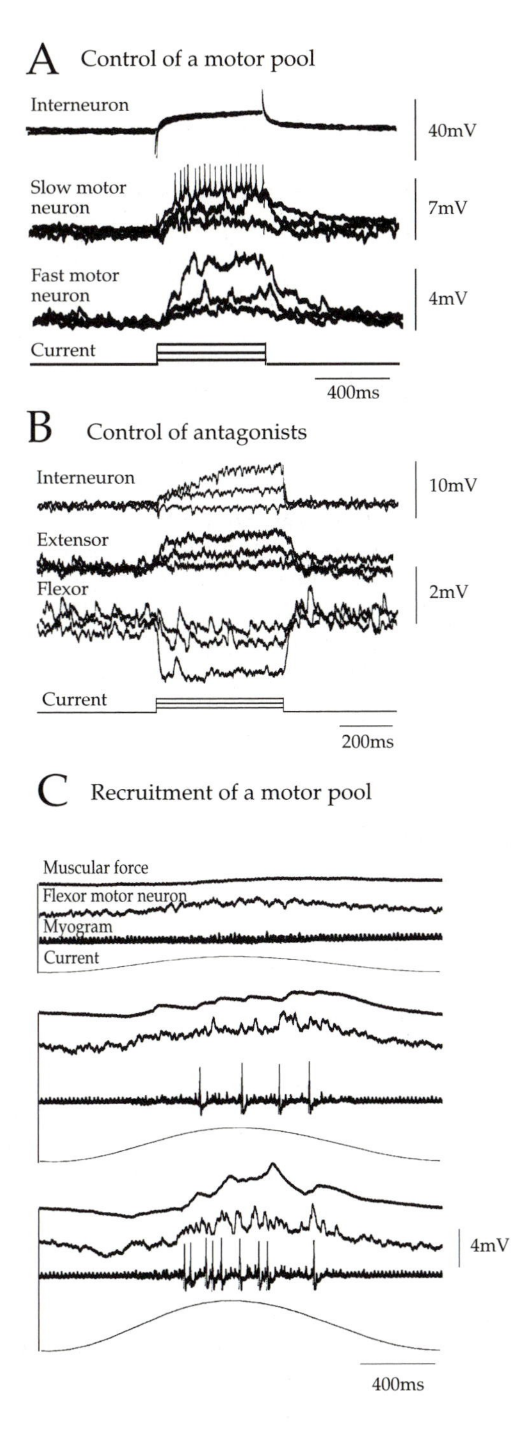
A Control of a motor pool
Interneuron
40mV
Slow motor neuron
7mV
Fast motor neuron
4mV
Current
400ms
B Control of antagonists
Interneuron
10mV
Extensor
Flexor
2mV
Current
200ms
C Recruitment of a motor pool
Muscular force
Flexor motor neuron
Myogram
Current
4mV
400ms

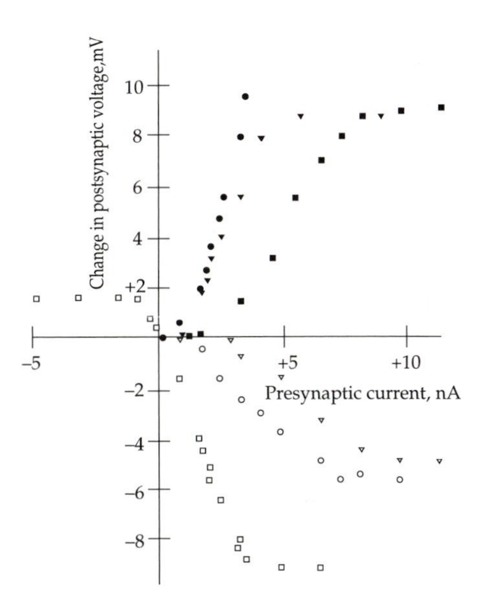
D Convergence onto a motor neuron
Change in postsynaptic voltage,mV
10
8
6
4
+2
–2
–4
–6
–8
–5
+5
+10
Presynaptic current, nA

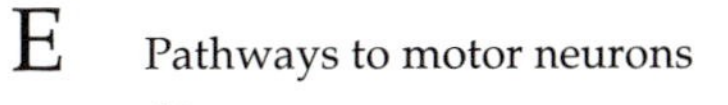
E Pathways to motor neurons

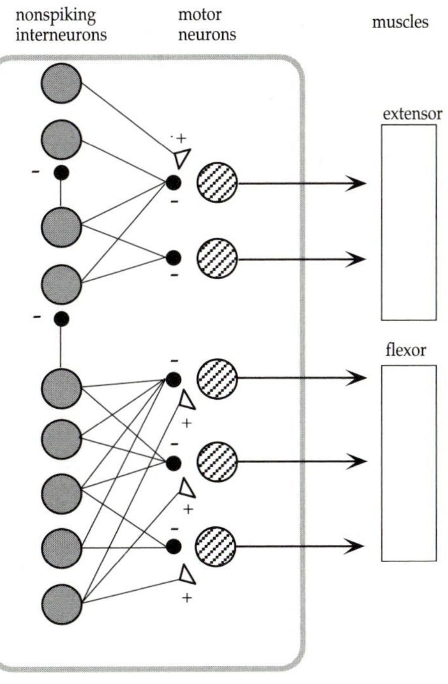
nonspiking interneurons
motor neurons
muscles
extensor
flexor

interneuron will recruit first the slow and then the fast motor neuron in a way that reflects their normal action during leg movements.

Other nonspiking interneurons can excite the slow motor neuron to one muscle but have no effect on the fast motor neuron, while still others can inhibit the fast but have no effect on the slow. These actions allow postural changes to occur without the recruitment of the fast motor neuron which would cause rapid twitch contractions of the muscle and hence more jerky movements of a joint. Similarly, they allow activation of the fast without the slow when the force must increase or decrease quickly in certain rapid movements.

The same organisation appears to operate for motor pools containing larger numbers of motor neurons, but the combinations in which they can be activated obviously increase. Some interneurons recruit a large proportion of the pool in an orderly sequence. Current injected into such an interneuron first recruits the slow motor neurons, so that the force generated by the muscle follows the waveform of the injected current. Increasing the current increases the frequency of the slow motor neuron spikes so that the force increases smoothly until the spike threshold of faster motor neurons is exceeded and they add more discontinuous contractions to the force. Even with high currents, a single interneuron may be unable to depolarise the fastest motor neurons sufficiently to make them spike. These fast motor neurons are thus recruited by the concerted action of several nonspiking interneurons, or by a single interneuron operating against an increased pattern of inputs from other neurons. Other nonspiking interneurons control only a small subset of the motor pool.

7.11.1.3. *Control of several motor pools*

Motor pools of the different muscles must be activated in an appropriate sequence during the leg movements of walking and other locomotory movements. Some must be activated at the same time and others with different phase relationships relative to the step cycle, and these patterns may need to change during different gaits. Many muscles must act antagonistically and this generally means that their motor neurons must be activated reciprocally. This action can be achieved by many of the nonspiking interneurons that inhibit one set of motor neurons directly and excite another set through the inhibitory connections that they make with other nonspiking interneurons (Fig. 7.18). Any tonic release of transmitter from these interneurons sets the balance of action between the two sets of motor neurons and this can be altered by synaptic inputs of either polarity to the nonspiking interneurons. Under the action of such interneurons, the two sets of motor neurons maintain a strict reciprocal action, but in other movements they may be required to act together or with different phase relationships. These actions are then controlled by different nonspiking interneurons that make either excitatory or inhibitory connections with both sets of motor neurons. Other nonspiking interneurons affect just one pool of motor neurons, so that they can be activated with some degree of independence from the other pools.

The nonspiking interneurons are also capable of organising the complex sequences of action of the different motor pools of the various leg muscles that are

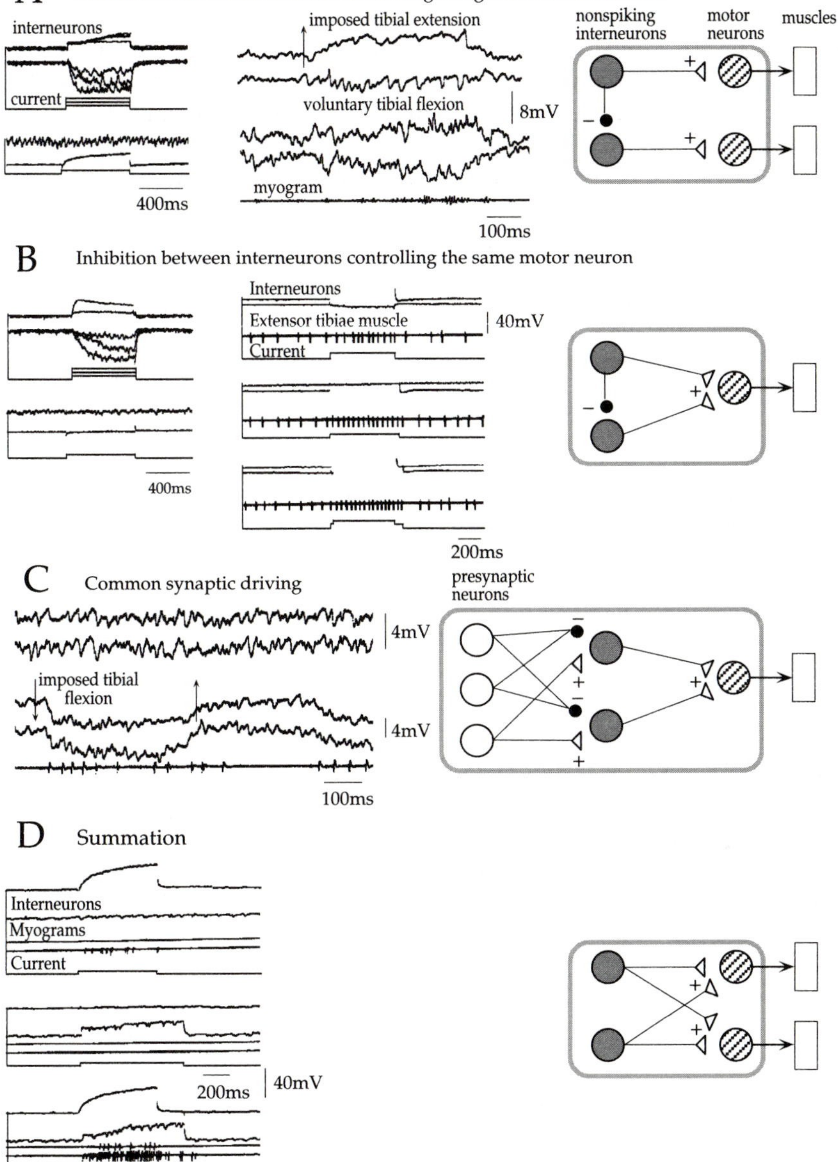

Fig. 7.18 Interconnections between nonspiking interneurons. A. An inhibitory connection between two interneurons that cause excitation of antagonistic motor neurons. B. An inhibitory connection between two interneurons that cause excitation of the same motor neuron. C. Interneurons with convergent connections on the same motor pools have many of their synaptic inputs in common. D. The effects of two nonspiking interneurons on motor pools sum in a nonlinear way. The diagrams on the right show the connections that are made. Based on Burrows (1979b).

needed to produce a coordinated movement. They do this by virtue of their connections with the different motor neurons and by connections with each other that allow their influence to spread widely through the local circuit (Fig. 7.18). Thus, current injected into a single nonspiking interneuron can cause movements of the joints in a way that is characteristic of their use in walking. These effects are independent of the feedback that results from the movement of any one joint, but normally would be supplemented and modified by such feedback.

7.11.1.4. *Control of inhibitory motor neurons and DUM neurons*

The skeletal muscles of a locust are also innervated by inhibitory motor neurons and supplied by neuromodulatory (DUM) neurons. Unlike the excitatory motor neurons, these neurons, in general, supply more than one muscle. The contribution of the inhibitory motor neurons is important in controlling the force of the slow muscle fibres (*see Chapter 9* for details of innervation and properties of two skeletal muscles) and especially in controlling their rate of relaxation. The three common inhibitory neurons innervate two sets of muscles in a leg: CI1 has a wide distribution to muscles in the more proximal segments, while CI2 and CI3 appear to innervate the same muscles in more distal segments (*see Fig. 8.6*). Despite this difference in distribution, synaptic inputs to these three neurons are common and they produce similar patterns of spikes during many movements of the legs. The sensory effects that influence these neurons are mediated by the same types of pathways as for the excitatory motor neurons; exteroceptors do not make direct connections, and of the proprioceptors only the tension receptor is known to make direct connections (Matheson and Field, 1995). They also receive inputs from midline spiking local interneurons (Schmidt and Rathmayer, 1993) and from nonspiking interneurons, although these connections have not been explored in any detail.

Particular DUM neurons appear to participate in the movements controlled by the muscles whose action they are able to modulate (*see Chapter 5*). Many of their synaptic inputs and spike outputs follow the same general pattern as those of the excitatory motor neurons of the muscles that they supply, but the neurons responsible for controlling their membrane potentials have not been identified. Given the important role that nonspiking neurons play in the organisation of motor patterns, it would seem worth exploring their role in controlling the DUM neurons at the same time.

7.11.1.5. *Organisational principles of the control by nonspiking interneurons*

From these experimental observations it is possible to assemble a picture of how the nonspiking interneurons shape the motor output.

Each of the small number of motor neurons is driven by several nonspiking interneurons that, in turn, interact with other nonspiking interneurons through a web of inhibitory connections. The interneurons must be organised so that each contributes its share of the driving of the motor neurons according to the context of the actions of the other interneurons to which it is connected. This is achieved by common synaptic inputs to interneurons with similar actions, the exclusion of the contribution of certain interneurons through inhibitory connections, and the summation of the effects of interneurons with both excitatory and inhibitory actions in a nonlinear fashion at the motor neurons.

The nonspiking interneurons marshal the motor neurons into overlapping sets that are appropriate for particular leg movements. This means that an individual motor neuron will be driven together with a certain array of other motor neurons by one particular nonspiking interneuron and with a different array of motor neurons by a different nonspiking interneuron. A single interneuron cannot call forth the full dynamic range of actions of an individual motor neuron, nor can it produce a complete coordinated movement of a leg. Instead, an individual interneuron can produce only part of a movement, so that it is the concerted action of a number of interneurons which produces each coordinated movement. The actions of the motor neurons and the combinations in which they are used are thus largely determined by the sets of nonspiking interneurons that are active at any one time. This implies that during active movements there may not be a simple relationship between the membrane potential of an individual nonspiking interneuron and that of one of its postsynaptic motor neurons. Indeed, there are often occasions when the membrane potentials of the two neurons move in opposite directions from that predicted from the connections between them. This again emphasises that movements result from the cooperative actions of many interneurons and their connections with different combinations of motor neurons. Shifting the balance of excitation and inhibition between the sets of nonspiking interneurons results in the motor neurons being activated in different combinations and thus allows them to produce movements with different characteristics.

7.12. MAKING FUNCTIONAL SENSE OF THE SYNAPTIC CONNECTIONS

The complex array of connections between the neurons processing the sensory information from the legs seems to be organised according to simple design principles for the functional control of movements. Functional sense can be made of many of the connections in the following way.

7.12.1. Excitatory and inhibitory receptive fields of local interneurons place stimuli in the context of normal leg movements

The arrangement of receptive fields (Fig. 7.19) with inhibitory and excitatory regions, and the convergence of signals from exteroceptors and proprioceptors can be understood in terms of the way a leg is moved during walking. Consider the combination of inputs to a particular midline spiking local interneuron that is excited by touching hairs on the distal, ventral tibia and by an extension of the tibia, but is inhibited by touching the ventral surface of the tarsus (Burrows, 1988). The interneuron makes a direct connection with the levator tarsi motor neuron. An extension of the tibia is thus accompanied by a depression of the tarsus in the absence of other sensory input. However, when the tarsus is placed on the ground

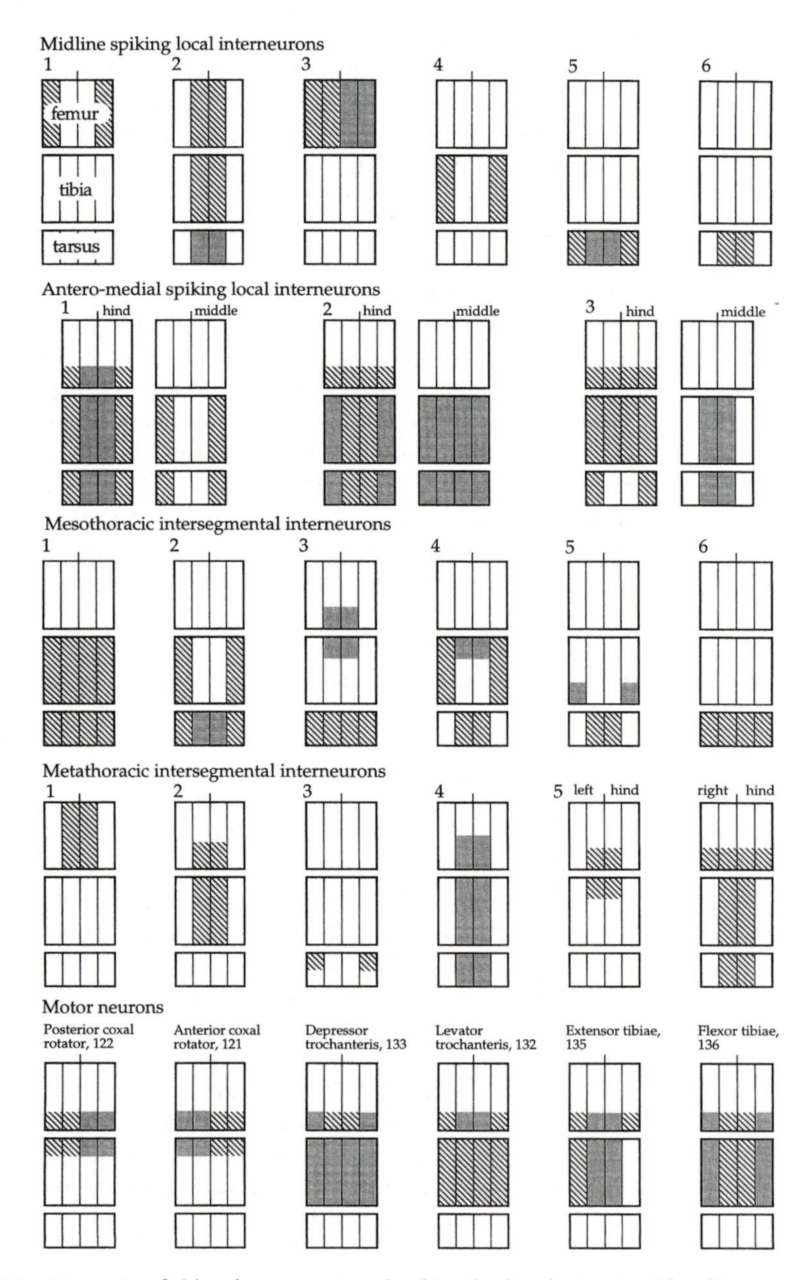

Fig. 7.19 Receptive fields of neurons involved in the local circuits. The diagrams are drawn as described in Fig. 7.7. Excitatory regions of a receptive field are hatched, inhibitory regions are shaded.

during the stance phase of walking, the interneuron is inhibited even though the tibia may now be extended. This means that the inhibition of the levator will be delayed until the end of the stance phase of locomotion when the tarsus is lifted from the ground.

Similar functionally important interactions between the different types of interneurons can result from the overlap of their inhibitory and excitatory receptive fields. Consider an intersegmental interneuron with a receptive field consisting of an excitatory region caused by direct sensory input, and an inhibitory region caused by at least one spiking local interneuron with an overlapping receptive field consisting again of excitatory and inhibitory regions (Laurent, 1988). Stimulation of the unguis leads to inhibition of the local interneuron, and to excitation of the intersegmental interneuron through direct sensory input and release from inhibition by the local interneuron. Stimulation of the region around the femoro-tibial joint excites the intersegmental interneuron and makes it spike but has no effect on the local interneuron as this region is outside its receptive field. If both the unguis and the femoro-tibial region are stimulated simultaneously, both interneurons are depolarised but the intersegmental interneuron now fails to spike because the inhibition from the local interneuron counteracts the direct excitation. Thus, the effectiveness of an excitatory region of a receptive field depends on the prevailing movements of the leg and the stimuli that are received. In this way, the strains and stresses on a leg that vary during a step cycle can influence the response to a tactile stimulus and *vice versa* and could lead to phase-dependent expression of particular reflex responses.

7.12.2. Interactions leading to postural adjustments

Consider the processing that occurs when hairs on the ventral femur or tibia are stimulated. This stimulation excites particular sets of spiking local interneurons and leads to a movement in which the trochanter/femur is levated, the tibia is extended and the tarsus is depressed (Fig. 7.2). Some of the antero-medial interneurons directly excite the depressor tarsi motor neurons, and probably motor neurons of the other joints. Some interneurons in the midline group make inhibitory connections with specific nonspiking interneurons that excite the flexor tibiae motor neurons, thus suppressing their drive to these motor neurons and allowing the tibia to extend. Some also make divergent inhibitory connections with antero-medial interneurons to suppress their excitation of particular motor neurons. Others inhibit a nonspiking interneuron that excites both the flexor tibiae and the levator tarsi motor neurons, while still others directly inhibit the levator tarsi motor neuron. The tarsus can thus depress because the depressor motor neurons have been excited, because the excitatory drive from certain nonspiking interneurons to the antagonistic levator has been withdrawn, and because the levator has been directly inhibited.

7.12.3. Interactions that modify walking movements

Much of the processing of the exteroceptive and proprioceptive signals is at the very heart of the mechanisms that lead to the appropriate changes in locomotion that are required in an unpredictable world. One example illustrates the type of contributions that these pathways can make.

On the posterior surface of a hind leg are two rows of spines, at the base of each of which is a single campaniform sensillum. On a hind leg, only the two most distal pairs are moveable and are called spurs, but on a middle leg all the spines on the ventral surface are moveable. None of the sensory cells of these spurs or spines are excited by contractions of the tibial muscles but instead respond to strains or displacements imposed on the spurs by outside forces. Only the distal spurs act as true proprioceptors, but when walking on a flat horizontal surface they are not normally displaced when a tarsus is placed on the ground. When, however, a tarsus is placed on rough terrain they may be deflected and their sensory neurons directly excite particular spiking local interneurons (Burrows and Pflüger, 1986), and depressor tarsi motor neurons (Laurent and Hustert, 1988). These pathways thus ensure a more powerful depression of the tarsus and thus increase the traction of the tarsus on uneven ground during the stance phase of walking.

7.12.4. Interactions of intersegmental interneurons place movements of one leg in the context of movements of the other legs

The connections of the intersegmental interneurons place the operation of a local circuit controlling the movements of one leg in the broader context of the actions of the other legs. Through their connections with the nonspiking interneurons, a local reflex can be altered by the intersegmental inputs, and, conversely, the efficacy of an incoming intersegmental input can be altered by the local sensory inputs. Some of their connections also result in antero-medial spiking interneurons and some nonspiking interneurons in the metathoracic ganglion having receptive fields on the middle legs. In contrast to the simplicity of the pathways that underlie the receptive fields on a hind leg, these pathways are more complex and, as a consequence, the fields are even more dependent on the actions of other neurons in different behavioural contexts. The convergent signals from different intersegmental interneurons can mean that a part of the receptive field is detectable only when some of these interneurons are inhibited, either by stimuli to the inhibitory region of their own receptive fields or by the actions of other neurons during certain movements.

7.12.5. Resistance and assistance reflexes

Both nonspiking and spiking local interneurons are involved in processing signals from proprioceptors, such as the femoral chordotonal organ, that lead to resistance or assistance responses in different behavioural contexts. In both locusts and stick insects, different nonspiking interneurons that influence the output of the extensor and flexor tibiae motor neurons respond differently to movements of the femoro-tibial joint. Often their responses would appear to be in conflict if the objective of the processing is seen to be the production of one type of motor response. Thus, some interneurons that excite the extensor motor neurons when depolarised with injected current are depolarised phasically during an imposed flexion, others are depolarised tonically during a maintained flexion, and still others are depolarised during both flexion and extension movements (Burrows *et al.*, 1988; Büschges, 1990). Similarly, interneurons that inhibit the extensors when depolarised with injected current are themselves hyperpolarised during flexion and depolarised during extension movements of the joint. A similar arrangement of nonspiking interneurons is involved in the responses of coxal retractor and protractor motor neurons that result from stimulation of a hair plate on the coxa (Büschges and Schmitz, 1991). Thus, there are interneurons that may be directly involved in exciting or inhibiting the motor neurons during resistance reflexes, and some whose connections would apparently lead to an assistance response. This organisation must therefore be seen as only one of a number of parallel pathways that lead to a particular motor response in any one context, and that a change in the weighting given to these different pathways could lead to the expression of one movement or the other.

7.13. GAIN CONTROL

It is essential that the operation of these pathways and connections should be capable of change to meet the fluctuating demands of different forms of locomotion. At its simplest, this may mean a change in the force that must be produced by the muscles to meet an increased load, and at its most complex may require a change in the gait, particularly if one leg is disabled. There are mechanisms intrinsic to the circuitry controlling one leg that result in flexible motor outputs, and mechanisms that allow changes in the patterns of coordination between the different legs. Nonspiking interneurons are important elements in adjusting the output of both local and intersegmental responses, although other elements throughout the networks could also contribute. The nonspiking interneurons are particularly well placed for this action because they are the summing points for signals transmitted directly by proprioceptive sensory neurons, the signals from the spiking local interneurons, and the signals from other legs delivered by intersegmental interneurons, and because they make direct and graded synaptic connections with sets of motor neurons.

7.13.1. Local responses

A single nonspiking interneuron can alter the efficacy of a local reflex as gauged by the number and frequency of spikes in the participating motor neurons (Fig. 7.20) (Burrows *et al.*, 1988). Movements of the femoro-tibial joint can, for example, cause resistance reflexes in the muscles that move this joint and effects that spread to alter the motor output to the muscles of neighbouring joints. Several nonspiking interneurons are either depolarised or hyperpolarised during such movements. If one that is depolarised by synaptic inputs is artificially hyperpolarised by progressively larger applied currents, then the excitation of one set of motor neurons by the

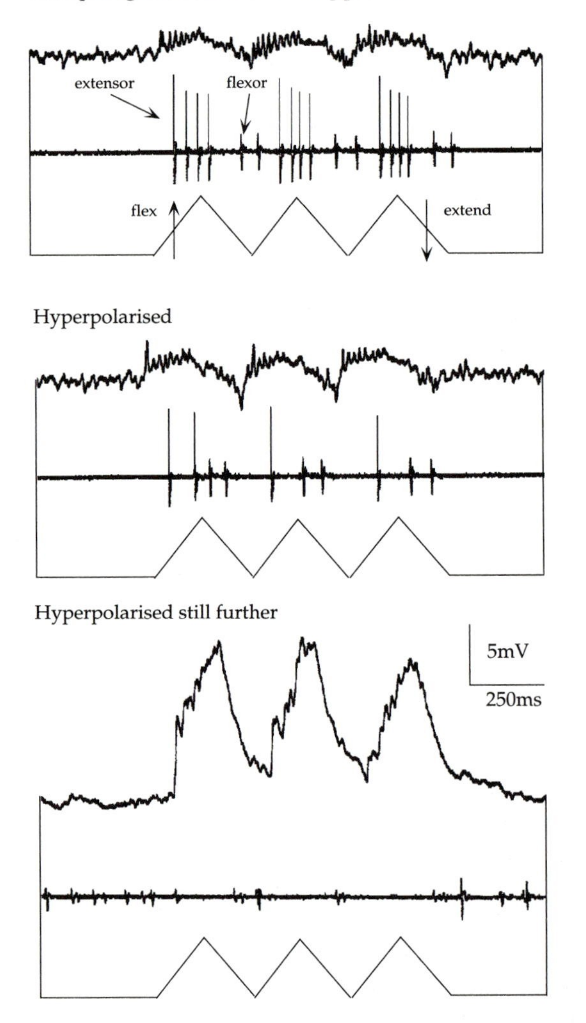

Fig. 7.20 The expression of a local reflex can be controlled by a nonspiking interneuron. With the nonspiking interneuron at its resting potential, imposed flexion and extension movements of the tibia of a hind leg cause spikes in extensor and flexor tibiae motor neurons, respectively. If the nonspiking interneuron is hyperpolarised, the number and frequency of spikes is reduced. If the hyperpolarising current is increased further then spikes in the extensor motor neuron are suppressed.

imposed movement is reduced in a graded fashion. Similarly, if the depolarisation of an interneuron is increased with applied currents then the excitation of motor neurons to adjacent joints is also increased in a graded fashion that depends on the current injected. These observations imply that the synaptic inputs that a nonspiking interneuron receives from many different sources are capable of altering the reflex response to inputs from one particular sensory receptor. This is, therefore, a mechanism for matching a reflex response to the prevailing behaviour of the animal as expressed by the inputs impinging on certain nonspiking interneurons in the appropriate local networks. The same mechanism is also used to adjust and match the signals from intersegmental interneurons.

7.13.2. Intersegmental responses

A nonspiking interneuron can change the effectiveness of the output connections of an intersegmental interneuron and thus alter the expression of its effects on the local circuitry. This effect is a consequence of the parallel connections that an intersegmental interneuron makes with nonspiking neurons and with motor neurons, coupled with the fact that a nonspiking interneuron may have a receptive field on both a hind leg and the ipsilateral middle leg. Conversely, an input from an intersegmental interneuron can alter the expression of a local response in which a nonspiking interneuron participates (Laurent and Burrows, 1989a). These effects are caused by the convergence at the nonspiking interneurons of sensory signals from a leg belonging to the same segment, and the inputs from the intersegmental interneurons that signal movements of a distant leg. The intersegmental inputs act by altering the expression of local responses, so that, for example, touching the tarsus of a middle leg enhances the response of certain hind leg motor neurons to an imposed movement of the tibia of the hind leg (Fig. 7.21). The mesothoracic stimulus excites several intersegmental interneurons that make parallel connections with the participating motor neurons and with nonspiking interneurons. Part of the enhancement could thus result from summation of inputs at the level of the motor neurons, but the more effective enhancement results from the connections with the nonspiking interneurons. It is the state of the nonspiking interneurons that largely determines the expression of the local response, so that their contribution can be either increased or decreased by an intersegmental input. The effects can be produced in several ways, as illustrated by the following three examples.

First, the inhibition in a motor neuron caused by a sensory input from a hind leg is continuously modulated by inputs from intersegmental interneurons responding to stimulation of a middle leg (Fig. 7.21). For example, the sensory neurons on a hind leg evoke EPSPs in a particular nonspiking interneuron, and each EPSP controls the release of transmitter evoking corresponding IPSPs in a motor neuron. Inputs from the intersegmental interneurons hyperpolarise the nonspiking interneuron but have no direct effect on the motor neuron. The effect of this inhibitory intersegmental input is to reduce, in a graded fashion, the transmission caused by the EPSPs in the local interneuron and thus reduce the inhibition of the motor neuron.

Second, sensory signals from a hind leg evoke common EPSPs in both a motor neuron and a nonspiking interneuron that also makes a parallel excitatory connection with the same motor neuron. When certain stimuli are given to a middle leg, the nonspiking interneuron is hyperpolarised but the motor neuron is unaffected, because of the pattern of connections made by the intersegmental

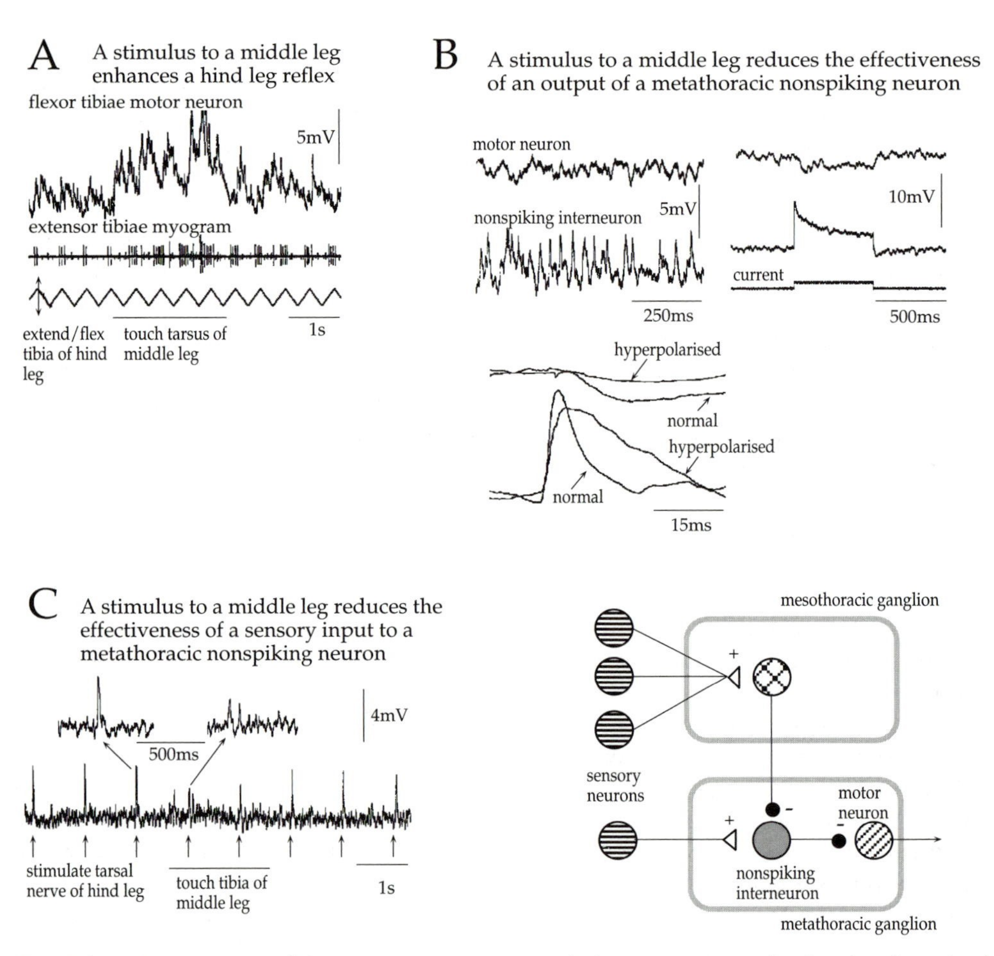

Fig. 7.21 Intersegmental interneurons can control the expression of a local reflex. A. A mechanical stimulus to a middle leg enhances the response of a hind leg flexor tibiae motor neuron to imposed movements of the femoro-tibial joint of a hind leg. B. A mechanical stimulus to a middle leg reduces the effects of a metathoracic nonspiking interneuron on a metathoracic motor neuron. C. A mechanical stimulus to a middle leg reduces the depolarisation caused by a metathoracic sensory neuron in a metathoracic nonspiking interneuron. The diagram shows the connections established. Based on Laurent and Burrows (1989a).

interneurons. The result is that the amplitude of the depolarisation in the motor neuron caused by the metathoracic stimulus is reduced because of the reduction in the contribution from the nonspiking interneuron. The intersegmental interneurons have access only to one part of the metathoracic parallel pathway to the motor neuron.

Third, the inhibition of a motor neuron in response to a metathoracic stimulus can be overridden by a simultaneous stimulus to a middle leg. Some intersegmental interneurons synapse on a nonspiking interneuron that releases transmitter tonically.

This, in turn, excites a metathoracic motor neuron, so that depolarising or hyperpolarising inputs to the interneuron cause a shift in the inputs to the motor neuron of either polarity. The local inputs evoke a hyperpolarisation and this results in an inhibition of the motor neuron, while intersegmental inputs evoke a depolarisation and this overrides the local inhibition and effectively gates the local response.

Most of the synaptic inputs caused by the intersegmental interneurons produce detectable conductance changes in postsynaptic nonspiking interneurons. In some nonspiking interneurons no conductance change can be seen, indicating that it must be restricted to sites that are not accessible to the recording electrode. The relative spatial arrangement of synapses on the neurites of nonspiking interneurons from the intersegmental and local pathways thus becomes important if the two are to interact. Indeed, only some of the intersegmental inputs can alter the local inputs to nonspiking interneurons, suggesting that their respective synapses are segregated and that some of the processing in the nonspiking interneurons is compartmentalised.

These types of pathways ensure that the expression of a local movement of one leg is related to the movements and posture of the other legs, and rely heavily on the participation of the nonspiking interneurons. The nonspiking interneurons themselves can act as a unit in which all inputs are capable of affecting any other input, but often they seem to act as compartmentalised units in which only some of the intersegmental inputs can alter the local processing. The implication is that the inputs from the different intersegmental interneurons are segregated on the different branches to which their conductance changes are limited. Only certain inputs can therefore interact with certain others under certain conditions. These conditions can nevertheless be changed, perhaps simply as a result of still further inputs from other sources. All these effects have been analysed in locusts that are not walking, so that during walking and other voluntary movements, the possible contributions of the many neurons that must then be activated in parallel offer considerable scope for the further modulation of these pathways and the communication between the various compartments of a particular nonspiking interneuron.

7.14. FEEDBACK CONTROL OF THE SENSORY NEURONS

The spike signals that a mechanosensory neuron delivers to the central nervous system are influenced by many factors, the most important of which are the following.

1. The mechanics of the receptor. For the exteroceptive hairs, this means the length and stiffness of the hair shaft and the flexibility of its mounting in its socket. For a campaniform sensillum, this means the suspension of the receptor from the cuticular dome and the rigidity of the surrounding cuticle. For a sensory neuron in a chordotonal organ, it means the way that its dendrites are stretched or

relaxed by the apodeme and suspensory ligaments. The mechanics will also change with the phase of the moult cycle.

2. The membrane properties of the sensory neuron as determined by the presence of different channels, their arrangement and density. This means that the same mechanical stimulus may be transduced by receptors with identical mechanical properties but different membrane properties as either a brief burst of spikes or, at the other extreme, as a sustained sequence of spikes.
3. The action of neuromodulators. These may alter the performance of a sensory neuron in a way that depends on the cocktail of substances that are either released locally in the vicinity of a receptor or that may have access to it through the haemolymph. If such effects are substantial then some recalibration will be necessary in the central nervous system if accurate information of, for example, joint position is to be extracted. More probably, it emphasises the importance of changed rather than sustained sensory signals being delivered to the central networks.

Some mechanosensory neurons from the legs of *Drosophila* have synapses between their axons in a peripheral nerve before they reach the central nervous system (Shanbhag *et al.*, 1992), and in crustacea electrical coupling occurs between mechanosensory neurons at sites within the central nervous system (Wildman and Canone, 1991; El Manira *et al.*, 1993). Neither of these types of interactions have been demonstrated in locusts. Synaptic potentials can, however, be recorded in the central terminals of many locust sensory neurons, including those from chordotonal organs, strand receptors and hairs. Presynaptic inhibition of the terminals of sensory neurons caused by such synaptic inputs is a ubiquitous phenomenon throughout the animal kingdom and is a mechanism for changing the effectiveness of a spike in signalling to postsynaptic neurons. The modulation of the sensory spikes in this way is thought to have several actions in vertebrates including enhancing the spatial discrimination in cutaneous sensory neurons (Schmidt, 1971) and, in muscle sensory neurons, of increasing the input to one set of motor neurons and decreasing it to another (Rudomin, 1990). During fictive walking in both mammals and crustacea, the terminals of sensory neurons receive a synaptic input that can modulate reflex responsiveness during the step cycle. Analysis of the inputs to the terminals of locust mechanosensory neurons from a particular proprioceptor show that the spikes are also inhibited presynaptically but reveal a new facet to the action of this inhibition.

7.14.1. Synaptic potentials in the terminals of sensory neurons from a chordotonal organ.

The terminals of a sensory neuron from the femoral chordotonal organ receive a barrage of synaptic inputs when the femoro-tibial joint moves actively, and when it is moved passively (Burrows and Matheson, 1994; Wolf and Burrows, 1995). Many of these synaptic inputs result from spikes in other sensory neurons from the same chordotonal organ, because they occur when it alone is stimulated by movements of its apodeme. This implies that sensory neurons from the same sense organ, and activated by the same movement, interact in some way that could potentially regulate the efficacy of transmission from their fellow sensory neurons to postsynaptic target neurons in the central nervous system. Many of the potentials are discrete events that would appear to be caused by spikes in presynaptic neurons. They do not arise directly from spikes in the other sensory neurons and must therefore be caused by spiking interneurons. While this is the most likely explanation, the contribution of nonspiking interneurons cannot be excluded, particularly given their ability to release transmitter in response to single synaptic potentials.

7.14.1.1. *What is the nature of the synaptic potentials?*

The potentials recorded in the terminals of sensory neurons of the femoral chordotonal organ are depolarising inhibitory potentials probably caused by the release of GABA (Fig. 7.22) (Burrows and Laurent, 1993). At the normal membrane potential of the sensory neurons these potentials are depolarising, but the reversal potential is only a few millivolts more positive than normal membrane potential. This, and the strong outward rectification of the membrane at potentials around resting level, means that the synaptic inputs do not sum to cause much change in the membrane potential and consequently they are unable to evoke spikes in the terminals. In crustaceans and mammals, the synaptic inputs have a reversal potential that is more distant from resting level and they may induce antidromic spikes. The injection of GABA into the neuropil of locusts has an apparently direct action on the terminals of the sensory neurons, evoking potentials that reverse at the same membrane potential as the normal synaptic inputs. GABA, or its agonist muscimol, mimic all the usual actions of the synaptic input. Moreover, the synaptic input is blocked by picrotoxin but not reliably by bicuculline, as is usual for most suspected GABAergic synapses in insects. The pharmacology suggests a receptor with properties more in keeping with a vertebrate $GABA_A$ than a $GABA_B$ type of receptor. Neurons that are immunoreactive for GABA make direct synapses onto the fine branches of the sensory neurons close to the output synapses that the sensory neurons make with other central neurons (Watson *et al.*, 1993). There are also other presynaptic neurons which do not show GABA-like immunoreactivity, suggesting that the inputs to the sensory neurons may be more complex than has so far been revealed by simple pharmacology.

7.14.1.2. *Pathways that generate the synaptic inputs*

If GABA is the transmitter used by the majority of neurons causing the synaptic inputs in the terminals of the sensory neurons, then direct interactions between the sensory neurons themselves cannot be the cause. The sensory neurons show no GABA-like immunoreactivity but instead probably contain acetylcholine. Moreover, recordings from pairs of sensory neurons reveal no direct connections between them. In occasional pairs, however, spikes evoked in one sensory neuron are followed

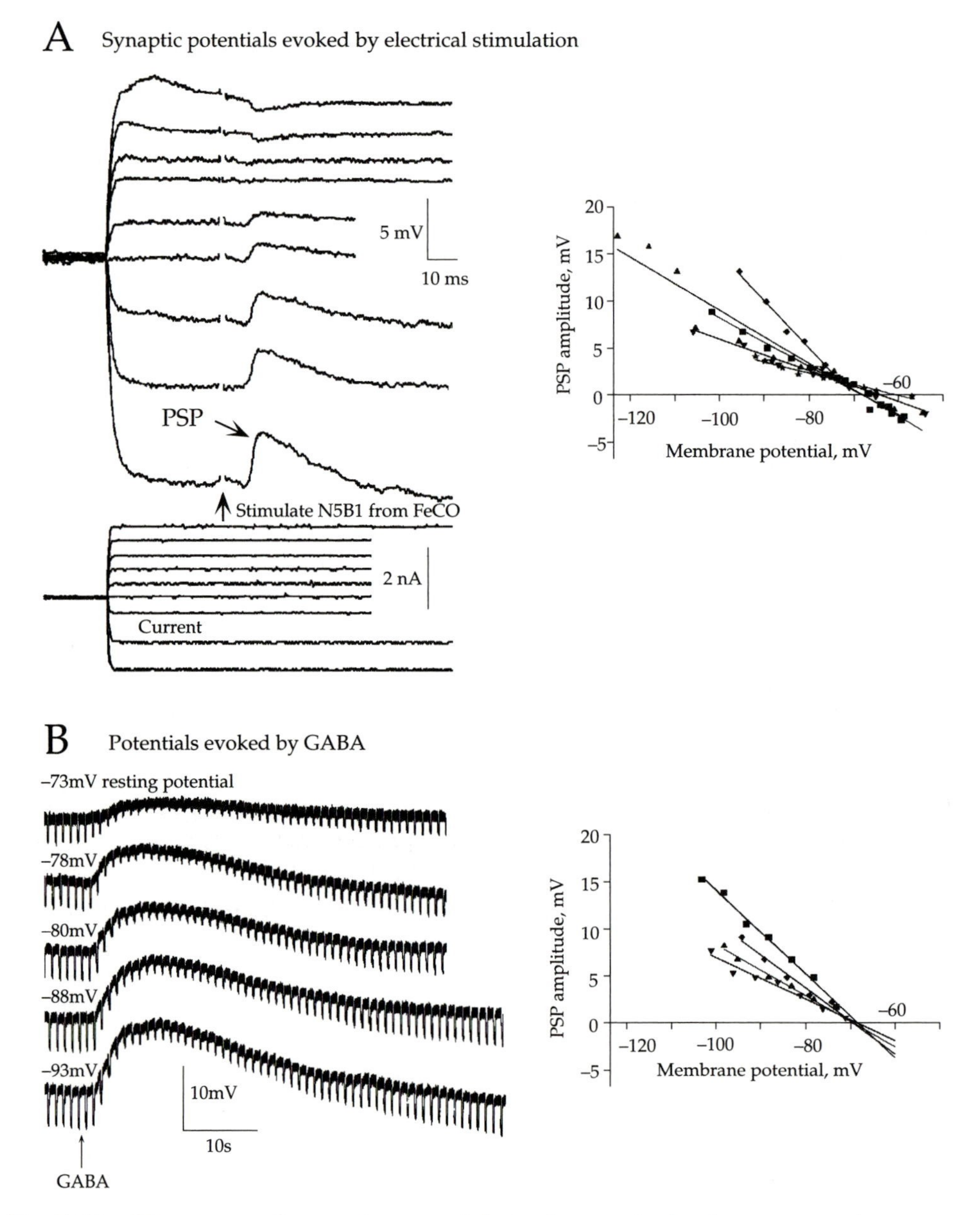

Fig. 7.22 The nature of synaptic potentials in the terminals of proprioceptive sensory neurons. A. A sensory neuron from the femoral chordotonal organ is shifted to various potentials under current clamp while a synaptic potential is evoked by electrical stimulation of other sensory neurons from the same receptor. The potential reverses a few millivolts more positive to resting potential. B. Injection of GABA into the metathoracic ganglion close to the terminals evokes a depolarisation accompanied by an increase in the conductance of the membrane. These evoked potentials reverse at the same potentials as the naturally evoked synaptic potentials. Based on Burrows and Laurent (1993).

consistently by synaptic potentials in the other but with a latency (2.5 ms) that is too long for a direct connection, and which suggests an interposed interneuron. The inputs must therefore be generated by interneurons. Particular sensory neurons must therefore make excitatory synaptic connections with a group of GABAergic interneurons which then make direct inhibitory synapses with a certain population of other sensory neurons (Fig. 7.23). The gain of the connection between the sensory neurons and the interneurons must be high, to ensure the consistency with which sensory spikes in some sensory neurons are followed by synaptic potentials in others. Although the interneurons are not yet identified, they are not the GABAergic midline spiking interneurons that process many of the mechanosensory signals from a leg.

7.14.1.3. *When are the synaptic potentials generated?*

The general picture to emerge from imposing movements on the femoro-tibial joint in a locust that is not walking is that the synaptic inputs to a particular sensory neuron of the chordotonal organ are greatest for the stimuli that also elicit its best spike response (Fig. 7.24). The input is therefore related to the response properties of a particular sensory neuron and must be caused by a specific set of other sensory neurons and not by the whole population. The input can be both phasic and tonic so that it depends on the characteristics of the movement itself, and on the position of the joint over which the movements occur. The different patterns of synaptic inputs and their relationship to the response properties of the different sensory neurons is complicated by the enormous range of response types among the population, but the following examples illustrate the general principles.

Phasic synaptic inputs A sensory neuron that responds to particular velocities of flexion movements receives a phasic synaptic input only during flexion movements, so that its spikes are always superimposed on this input. Similarly, sensory neurons that respond to the velocity of extension receive a phasic input only during extension movements. The amplitude of the input depends on the position of the joint and

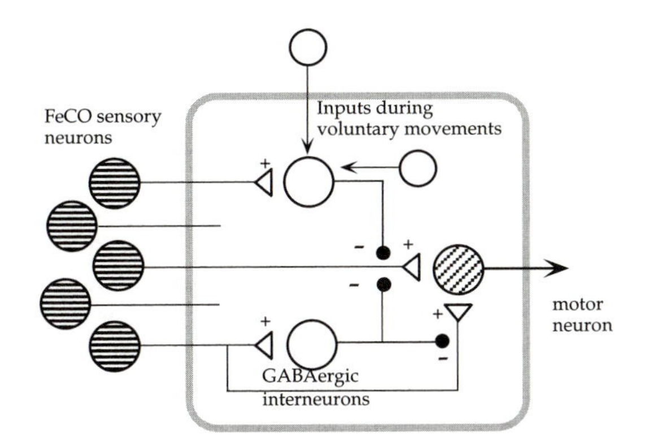

Fig. 7.23 Diagram of the connections that explain the synaptic potentials in the terminals of the proprioceptive sensory neurons.

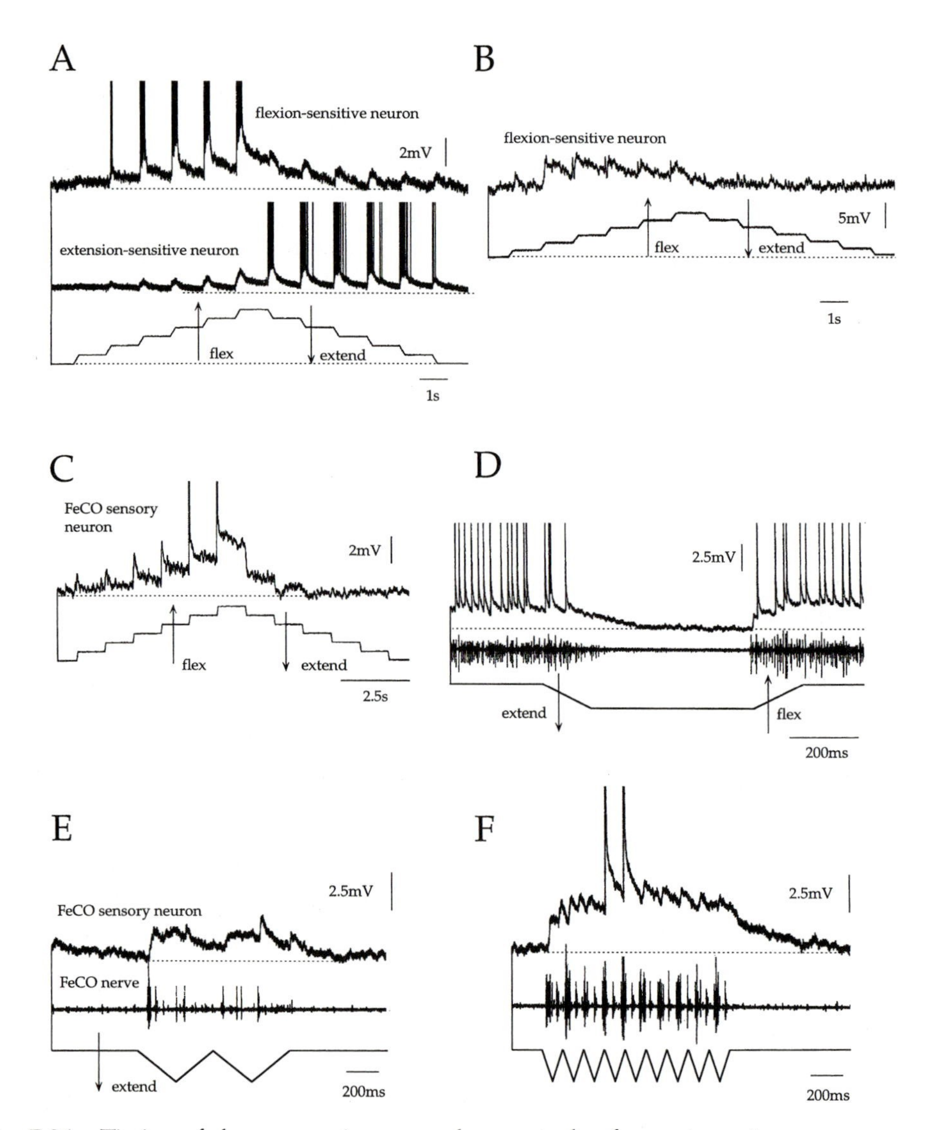

Fig. 7.24 Timing of the sensory inputs to the terminals of proprioceptive sensory neurons from the femoral chordotonal organ of a hind leg. A. A flexion and an extension-sensitive sensory neuron are recorded simultaneously while the apodeme of the chordotonal organ is moved to simulate movements of the femoro-tibial joint from the fully flexed to the fully extended position and back. The synaptic inputs are greatest for the movements and positions that evoke the greatest spike response. B. A flexion-sensitive neuron receives synaptic inputs only during flexion movements. C. A sensory neuron responding phasically to flexion movements receives both phasic and tonic synaptic inputs. D. A sensory neuron responding to position receives a tonic synaptic input upon which its spikes are superimposed. E, F. A sensory neuron responding to the velocity of extension movements receives a synaptic input at all frequencies of movement but this is larger for movements to which it responds with spikes. Based on Burrows and Matheson (1994).

corresponds to the range of joint positions at which the sensory neuron will spike in response to a particular velocity of movement. Range fractionation of the joint position is distributed among the sensory neurons. The amplitude of the input also depends on the velocity of the movement and is largest for those velocities to which the sensory neuron responds best. This means that there are inputs to a sensory neuron over ranges of joint movement and velocities to which it does not normally respond with spikes. For some sensory neurons that spike in response to only one direction of movement, synaptic inputs may occur during movements in both directions, but they are always greater during movements in the preferred direction. The inputs that occur when there are no spikes obviously cannot modulate the spike signals of a sensory neuron, so that their action is unclear. Perhaps they simply reflect the inevitable consequences of interactions set up by a limited population of sensory neurons from one receptor.

Tonic synaptic inputs The synaptic input to a sensory neuron may have a tonic component that is related to the position of the joint (Fig. 7.24). This applies both to sensory neurons that respond to the velocity of a movement with a phasic burst of spikes, and to those that code the position of a joint in their tonic sequence of spikes. A sensory neuron responding to the velocity of flexion movements not only receives a phasic input at each movement but also a sustained input that is related to the position of the joint over which these movements are made. This input is again largest for the range over which it gives its best spike response. Sensory neurons that signal the position of the joint in their tonic sequence of spikes also have a synaptic input that is related to joint position, which means that their tonic spikes are superimposed on a tonic synaptic input; the faster a sensory neuron spikes to signal a new position of the joint, the greater will be the synaptic input that it receives. At each shift to a new joint position, the synaptic input may precede the change in spike frequency and must be caused by sensory neurons responding phasically to the movement. This means that sensory neurons responding either phasically or tonically can receive both a phasic and a tonic synaptic input, so that their spike signals are directly related to the underlying movements and positions of the joint.

The design principle behind the delivery of synaptic inputs to the terminals of these sensory neurons appears to ensure that the best spike signals in any of the channels will be met by the strongest inhibitory synaptic inputs generated by other sensory neurons from the same sense organ responding to the same stimulus. The net result is to place the spike discharge of an individual sensory neuron in the broader context of what the other sensory neurons are signalling about the same movement.

7.14.1.4. *What is the effect of the synaptic potentials?*

The effect of the synaptic input to the sensory neurons is to reduce the excitability of the presynaptic terminals in two ways (Fig. 7.25). First, the occurrence of a synaptic potential reduces the ability of the membrane to generate an action potential for about 100 ms, an action that far outlasts the conductance change associated with the

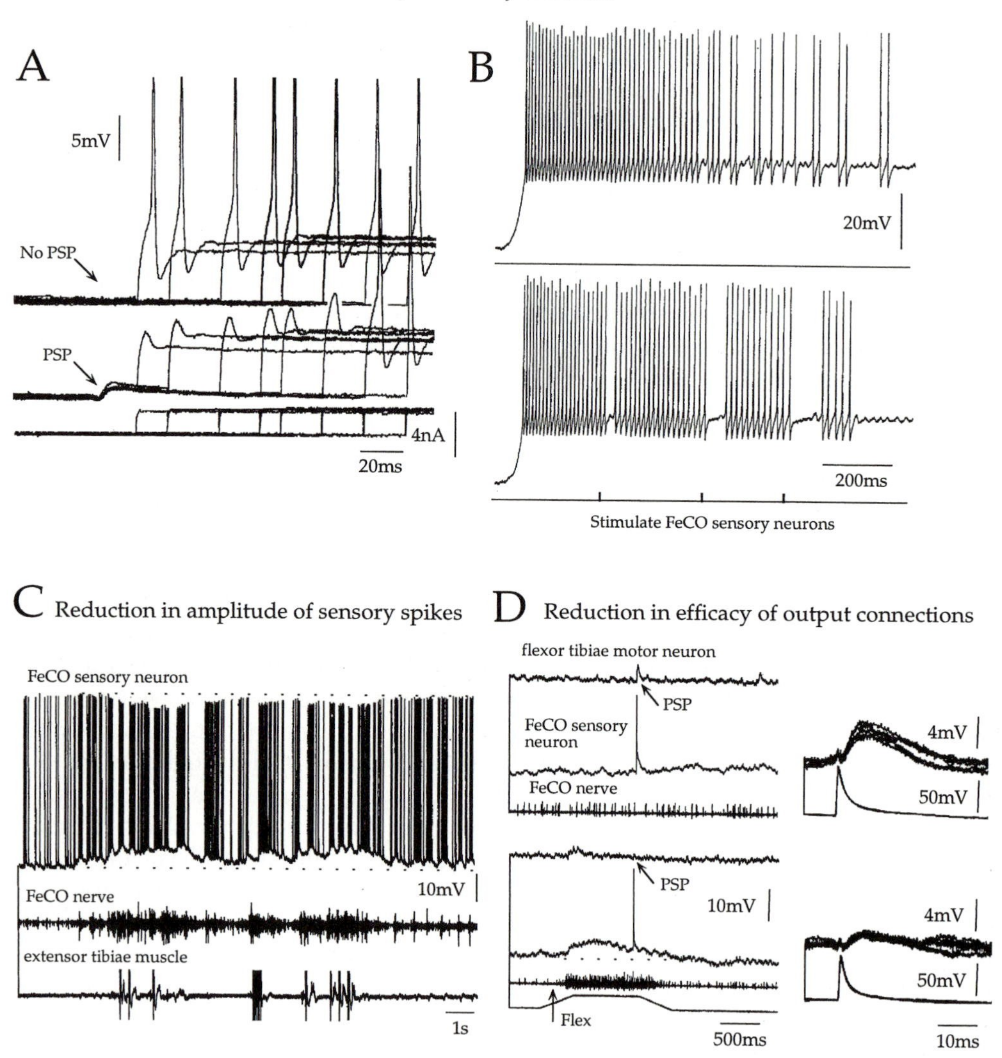

Fig. 7.25 Effects of synaptic inputs to the terminals of proprioceptive sensory neurons. A. The effectiveness of a pulse of depolarising current injected into the axon of a sensory neuron is reduced if it is preceded by a synaptic input to the terminals. B. A sequence of spikes evoked on release of hyperpolarising current injected into the axon of a sensory neuron is interrupted by a synaptic input to its terminals. C. Synaptic inputs reduce the amplitude of spikes in the terminals. D. Synaptic inputs reduce the amplitude of an EPSP in a motor neuron evoked by a sensory neuron. The top panel shows the EPSP evoked when there are few spikes in other sensory neurons from the femoral chordotonal organ, and the lower panel shows the reduction that occurs when many other sensory neurons spike. A, B are from Burrows and Laurent (1993), C, D are from Burrows and Matheson (1994).

synaptic potential itself. The effect would be to reduce the ability of the sensory spikes to propagate into the branches and should therefore result in a spike of smaller amplitude at the output synapses. Second, a synaptic potential may be able to block the conduction of spikes into particular terminal branches so that the output synapses on these branches are not activated.

The synaptic inputs also reduce the amplitude of spikes that invade the sensory terminals, with the reduction proportional to the amplitude of these inputs. It is difficult to estimate how great will be the reduction in the amplitude of the spikes at the terminals where there are output synapses because recordings can only be made some distance away. Even at these distant sites, however, the maximum reduction can reach almost 30%, and follows a time course that matches the synaptic input. In the branches leading to and containing the terminals, the spike appears to propagate passively so that the effects at the output synapses can be expected to be much larger.

All these effects on the presynaptic spikes are then manifested as a reduction in the amplitude of the EPSPs that the sensory neurons evoke in some of their postsynaptic neurons, and a reduction in the gain of the synaptic transfer (Fig. 7.25). Thus, when a spike occurs in a sensory neuron and there is little synaptic input, it evokes an EPSP of large amplitude in a postsynaptic motor neuron. The spike is able to reach the terminals at its normal amplitude and the excitability of the terminals is not reduced, so the output synapses work at their maximum levels. By contrast, when a sensory spike is superimposed on a synaptic input, the amplitude of the EPSP in a postsynaptic neuron is reduced in amplitude. The reduction is continuously graded with respect to the synaptic input and presumably depends on the changes in the spike and changes in the excitability of the membrane. The result is that the response of the motor neurons to the input from this sensory neuron will be reduced.

7.14.2. Automatic gain control

The synaptic input to the terminals of a chordotonal sensory neuron is caused by interneurons that are activated by spikes in other sensory neurons from the same sense organ which respond to the same movements. This places the signals of one sensory neuron in the context of the signals from a particular population of other sensory neurons. This design means that if a sensory neuron were able to spike on its own, then the effectiveness of its output synapses with motor neurons and other interneurons would be maximal, but when it spikes together with certain other sensory neurons its effectiveness is reduced in proportion to the number of other active neurons. The occurrence of spikes in one neuron alone is, however, extremely unlikely because the movements and positions of the joint are signalled by many sensory neurons acting in parallel. Thus, when a sensory neuron spikes at the same time as other sensory neurons, the effectiveness, or gain, of its output synapses will be reduced by the synaptic input it receives from these other sensory neurons. The more spikes in a greater number of sensory neurons responding to the same movement, the greater will be the reduction in the gain of the output synapses. In this way, the whole system of interactions could act as an automatic gain control mechanism, limiting the dynamic range of the output in the context of the input. The pattern of synaptic inputs to a sensory neuron fits well with this idea because the strongest synaptic inputs occur for stimuli that elicit the best spike response from a sensory neuron. They do not occur at the extremes of the operating range of a sensory neuron as would be expected if the input were a form of lateral inhibition to sharpen its responses. It will, however, be necessary to test whether the synaptic input and the spike responses saturate at the same stimulus parameters.

When a joint is moved, the following sequence of events can be envisaged. The first spikes in sensory neurons with the largest diameter axons (likely to be those responding to acceleration) and arriving first in the central nervous system will generate large amplitude EPSPs on their postsynaptic neurons because none of the feedback pathways will be activated. The gain of their synapses will be maximal and cannot be influenced by the other sensory neurons because their spikes have yet to arrive. This is similar to the 'winner-take-all' design in electronic circuits. These initial spikes will be the first to activate the automatic gain control circuits, so that subsequent spikes in these fast-conducting neurons and spikes in the other more slowly conducting neurons will be subjected to presynaptic inhibition. As more spikes arrive in more neurons the feedback inhibition will increase, but not all the active sensory neurons will contribute to the inhibition of the others. The interactions are selective and depend on the response properties of the neurons and, hence, on the properties of the stimulus, such as the velocity of the movement and the range of joint angles through which the movement is imposed. The result of these interactions will be that the amplitude of the EPSPs evoked by these neurons in their postsynaptic neurons will be reduced. In other words, the effectiveness of the synaptic connection of a particular sensory neuron is related to the action of the other sensory neurons that are also active. Were such a mechanism not to operate, the summation of inputs expressed with maximal effectiveness onto a postsynaptic target would soon drive the membrane potential beyond the level where it could elicit a greater response. The feedback thus prevents saturation of the responses of the postsynaptic neurons and extends their dynamic range.

It seems likely that these interactions may represent a general mechanism for the operation of joint proprioceptors, for there is nothing special about the femoral chordotonal organ that would limit its usefulness to this organ alone. Indeed, similar events can be seen in different types of receptors at other joints of the legs of a locust. The explanation of these feedback interactions between the sensory neurons of one joint may provide a better understanding of at least some of the ubiquitous presynaptic inhibition of sensory neurons in many other animals.

7.14.3. Inhibition occurs during movements of the other legs

Movements of the middle or front legs cause synaptic inputs to chordotonal sensory neurons of a hind leg. These potentials have the same properties as those generated by other sensory neurons from the chordotonal organ itself and could therefore be caused by the same interneurons or by parallel sets of similar interneurons. These inputs mean that the signalling by a proprioceptor of one leg is placed in the context of the movements of the other legs. These inputs presumably modulate the signalling by the sensory neurons of the other legs at particular phases of the movements during walking. It demonstrates the interdependence of the processing that might, at first, seem to be the province of a local circuit. To produce a coordinated movement with all the legs clearly requires the exchange of information between the local controllers, but it would not be readily predicted that these interactions should extend to the alteration of the sensory signals themselves. During walking, the sensory neurons are depolarised rhythmically in time with movements of the legs. Again, the input is greatest during those movements which the particular sensory neuron signals with its best spike response (*see Chapter 8*).

7.15. DESIGN OF REFLEX PATHWAYS

From what has been unravelled so far, the networks (Fig. 7.26) that process the sensory signals from both exteroceptors and proprioceptors can be seen to have the following eight design features.

7.15.1. The sensory to motor pathways are short

The information from a sensory neuron passes through only a few serially arranged synapses before it reaches the motor neurons (see Fig. 7.16). For the exteroceptive signals, direct connections with the motor neurons are rare so that almost all of the signals are processed by networks of spiking and nonspiking interneurons before being delivered to the motor neurons. Nevertheless, even in these networks only two or three synapses are interposed between the sensory neurons and the motor neurons. The networks for the processing of exteroceptive and proprioceptive signals are essentially the same, with an obvious difference being that it is the rule for the proprioceptive sensory neurons to synapse directly onto motor neurons, whereas for the exteroceptive sensory neurons it is the exception. Many of the interneurons process signals from both types of receptors, so that their receptive fields can be complex. The processing of one modality can thus determine the processing of the other modality in ways that are functionally relevant for locomotion.

7.15.2. There is massive convergence of sensory neurons onto interneurons

The sensory neurons show massive convergence onto the local interneurons, but in a way that preserves spatial information so that it can be used in particular local responses for the adjustment of posture and locomotion. The terminals of the sensory neurons from exteroceptors form a spatial map of the leg in a particular region of ventral neuropil, so that information about the spatial location of a stimulus can be readily accessed by the local interneurons. The terminals of the sensory neurons from proprioceptors are segregated in a different region of neuropil. The connections of the sensory neurons are specific, so that one interneuron receives

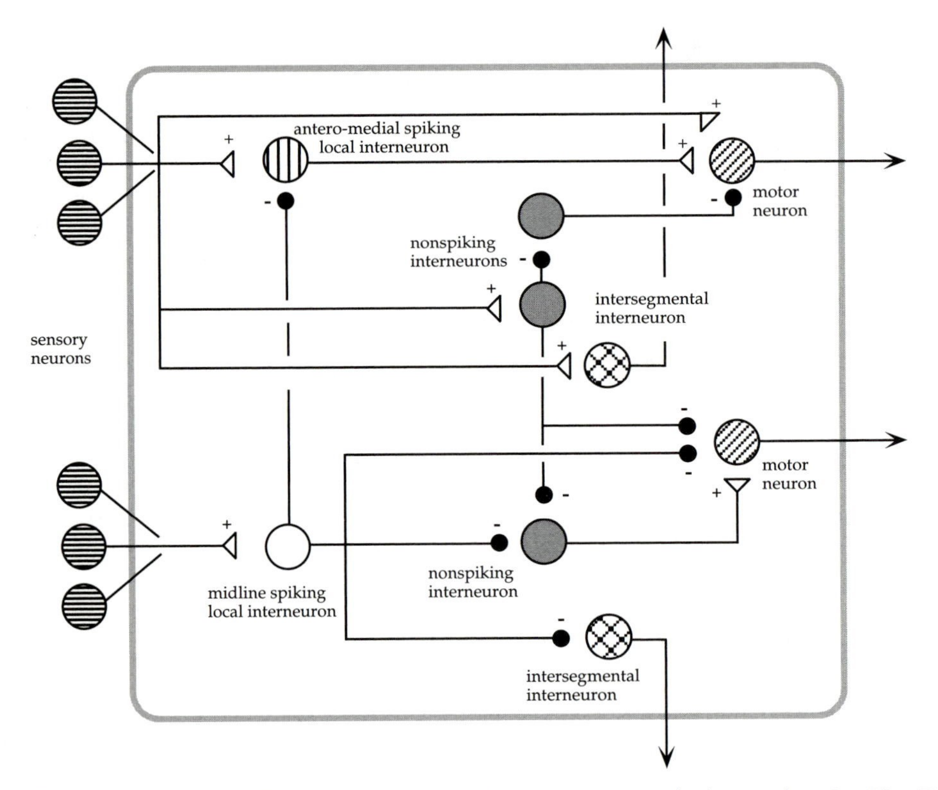

Fig. 7.26 Pathways for the processing of mechanosensory signals from a leg. See Figs 7.12 and 7.13 for the contributions made by exteroceptors and proprioceptors.

inputs only from sensory neurons of particular arrays of spatially arranged receptors. The gain of the connections between the exteroceptive sensory neurons and some of the local interneurons is high, so that if the sensory spikes occur at low frequency the probability with which they will evoke a spike in an interneuron is also high. In this way, each interneuron has a discrete and characteristic receptive field on a leg, and, in turn, the whole surface of a leg is mapped onto the interneurons as a series of overlapping receptive fields. Some of the interneurons provide a fine representation of a single leg while others provide a coarser coding into which information from other legs is also incorporated.

7.15.3. The processing occurs in a parallel and distributed fashion

Each sensory neuron makes connections with several interneurons of the same class so that, for example, the receptive fields of the midline spiking local interneurons overlap. They also make connections with interneurons of different types so that there is a high degree of divergence in the connections of a sensory neuron. A single sensory neuron from a hair may synapse with at least two types of spiking local interneurons, with nonspiking interneurons and with intersegmental interneurons. The sensory signals arising from a stimulus to one region of a leg are thus processed in a parallel and distributed fashion by many neurons of the network. In this way, a stimulus to one part of the leg is represented by the actions of many interneurons and not by a single interneuron, with the result that the interpretation of spatial location emerges only from the interactions of interneurons with overlapping receptive fields.

7.15.4. Connections in the networks are specific

The connections made by the sensory neurons and by the interneurons are specific. Each sensory neuron connects with only a small subset of the interneurons and the interneurons connect only with certain other interneurons, or with a particular subset of the motor neurons. Specificity of action thus results from the specificity of the connections, and from the processing performed by specific interneurons. As a result of this design, the receptive field of a particular interneuron can be defined and is always the same in different animals.

7.15.5. Analogue and digital processing operate in parallel

Two distinct classes of local interneuron, nonspiking and spiking, dominate the processing in the networks and have distinct mechanisms of intracellular and intercellular communication. The nonspiking interneurons use the graded signals of their synaptic inputs to cause the graded release of their transmitter onto postsynaptic neurons, whereas the spiking local interneurons use spike signals to effect communication with postsynaptic neurons. The reason for the distinction between analogue and digital coding in these networks remains as intriguing as it is obscure. Both types of neuron have only local branches, so the reason that one generates spikes and the other does not cannot reside in the need to conduct signals over long distances. The morphological design of the neuron may, however, have some bearing. The spiking local interneurons have largely segregated input and output regions linked by a single narrow process, and perhaps only a spike signal is able to transmit the summed signals from the input region to the output synapses through this bottleneck. Alternatively, it may be desirable to activate all the output synapses that are distributed on many fine branches, and this could only be achieved by flooding these branches with a spike. By contrast, the design of the nonspiking neurons appears to be different in that the input and output synapses are intermingled on all of the branches. This design suggests the possibility that these neurons act as compartmentalised units with each neuron capable of performing several independent computations at any one time. Strategic inputs or the actions of neuromodulators could alter the relative independence of these computations. Spikes propagating throughout the neuron would largely preclude this powerful form of integration.

The spiking interneurons could also act as a filter, by generating an output only when their summed input is sufficient to exceed spike threshold. The very nature of the nonspiking interneurons means that they cannot operate in this fashion because each synaptic input is potentially capable of altering the release of transmitter. The graded outputs of the nonspiking interneurons could thus be seen as a mechanism to provide a more precise control of the responses of a postsynaptic neuron than could be provided by the limited range of spike coding in a limited number of spiking interneurons. This may be particularly important in networks where the number of neurons is small and where precise movements must still be produced by only a few motor neurons. To achieve the same degree of control it is probable that many spiking local interneurons would need to replace a single nonspiking interneuron.

7.15.6. Intersegmental interneurons regulate the action of the local circuits

The processing performed by the different networks controlling the movements of each of the legs is linked by signals in intersegmental interneurons. These signals place the processing of one local circuit in the context of the processing by the local circuits in other ganglia, so that a movement of one leg is always appropriate to the movements of the other legs. These local circuits can thus be considered as the local controllers for a particular limb. Coordination of these local controllers is achieved largely by the summation of the intersegmental signals with the local signals, particularly at the level of the nonspiking interneurons. In this way, the intersegmental signals do not initiate a new movement but modify the performance of a local response. The design allows devolvement of control of the limb movements largely to the local controllers, upon which other signals from the brain can be superimposed to effect any desired changes in those movements.

7.15.7. Nonspiking interneurons act as gain controllers

The nonspiking interneurons can control the expression of a local response in a graded fashion by acting as the summing points for local and intersegmental signals. Despite the parallel and distributed processing of all the signals, alterations in the membrane potential of a single nonspiking interneuron can alter the contribution of sets of motor neurons to a particular movement and hence alter the movement, so that the same input now generates a different output in terms of the number of motor spikes and the force produced by the muscles.

7.15.8. The effectiveness of sensory signals is regulated by presynaptic inhibition

The effectiveness of the signals delivered to the central nervous system by a particular proprioceptive sensory neuron is dependent on the network of actions of the other sensory neurons that are active at the same time. The result is that the effectiveness of the signals in a sensory neuron is reduced in a graded fashion, depending on the type and number of other sensory neurons that signal the same stimulus. This is achieved by presynaptic inhibition of the terminals of the sensory neurons through the action of interneurons that are activated by other sensory neurons from the same sense organ. This will prevent saturation in the responses of the postsynaptic neurons when many sensory neurons are active at the same time. It may also extend the dynamic range of the postsynaptic neurons and reduce the hysteresis of their responses in comparison to that of the sensory neurons themselves.

The effectiveness of the sensory signals is also altered during movements of the other legs and at a particular phase of the step cycle during coordinated walking movements. This presynaptic inhibition may be caused by the same interneurons that operate the gain control circuit and presumably matches the performance of the sensory signals in the context of the movements of the other legs (*see Chapter 8*).

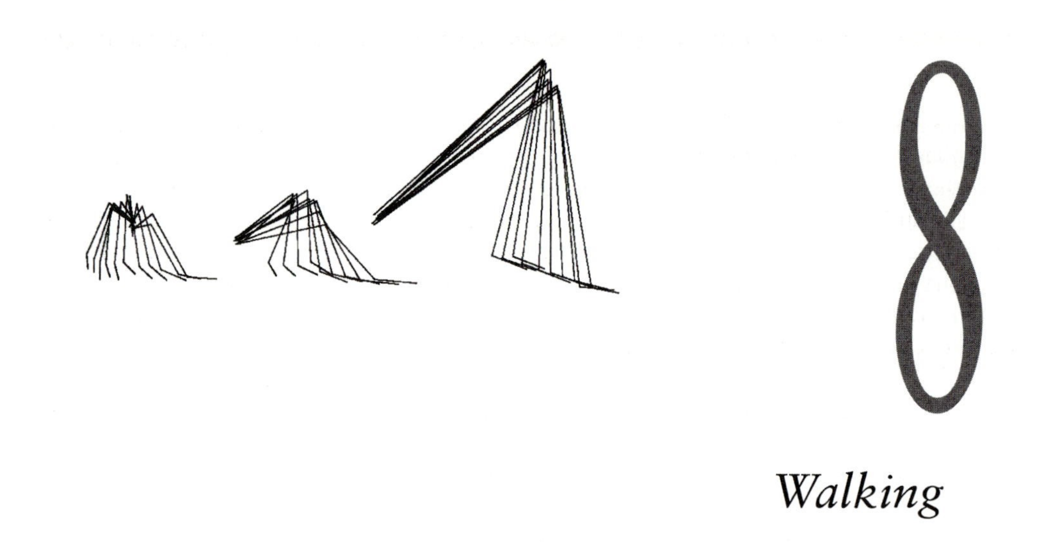

Walking

8.1. WALKING MOVEMENTS

The disparate sizes of their three pairs of legs would seem to make locusts more rationally designed for moving forward in a series of hops or jumps rather than in a steady progression by walking; the hind legs are greatly enlarged and specialised for jumping and even the middle legs are some 1.2 times larger than the front ones. Locusts, nevertheless, can still walk well at stepping rates of 1–4 Hz with a tripod gait typical of that used by most other insects. There is no transition of the gait that would indicate walking changing into running, but instead forward speed, particularly of the larvae, is increased by a change to hopping propelled by the simultaneous extension of the large hind legs. Locusts are renowned for the awesome invasions of vegetation by marching bands of gregarious larvae in which the individuals can walk and hop for many kilometres.

8.1.1. Gaits used in walking

When walking, the legs must be able to support the weight of the body by raising it above the ground, provide stability while the different legs are lifted and moved forwards, and of course provide forward thrust when they are in contact with the ground. The legs are at the sides of the body and protrude laterally from it so that they provide a stable support. The need for stability when an insect is walking slowly becomes clear from the following illustration. If a dog is walking at a stride frequency of 1 Hz and with its body 0.5 m above the ground, it will fall within 300ms if there is no support (Ting *et al.*, 1994). By contrast, a cockroach walking slowly at 2 Hz and with its body 0.02 m above the ground will take only 50ms to fall. Large mammals may be able to make dynamic adjustments to avoid falling, but the solution adopted by insects during slow walking is always to have three legs on the ground. This is the minimum requirement for static stability and works as long as the centre of mass of the body falls inside this tripod of support.

Locusts use essentially the same gait, with a considerable amount of variation, for virtually all speeds of walking, but the movements of the legs are more variable than those of some other insects; the hind legs can often become uncoupled from the pattern and the front legs can be moved about in apparent searching movements before they are placed on the ground (Burns, 1973).

The gait is an **alternating tripod**, in which the front and hind legs on one side and the middle leg of the other side move together and alternate with the remaining three legs, so that the body is always supported by a stable tripod of legs placed on the ground (Fig. 8.1). This gait gives good stability while still allowing the locust to walk over complex terrain. The tarsi are also able to grasp, so that a locust can walk along branches of trees and up the stems of plants. The alternating tripod is widely used by other insects for most speeds of walking, and insect walking mechanisms provide good designs for legged robotic vehicles. Nevertheless, although the tripod gait provides good stability, the body is accelerated to its maximum velocity at the midpoint of the contact of the legs with the ground and is decelerated almost to zero at the end. These periodically applied accelerations can therefore result in jerky movements of the body. As each tripod of legs makes a step there is a peak in the vertical force and as the body comes down on a tripod of support it decelerates in the forward direction and its vertical force increases above body mass. The body is then lifted upwards and accelerated so that the vertical force decreases, and the cycle then repeated at the next step. In quadrupeds, a more constant velocity is maintained by each leg contributing as it first touches the ground.

Some insects, however, may adopt slightly different gaits when walking slowly. For example, large grasshoppers such as *Romalea microptera,* which walk at stepping rates of no more than 2 Hz, have gaits in which the forward movements of the front and hind legs of the two tripods overlap (Burns, 1973). The hind legs of a tettigoniid grasshopper may sometimes take steps that are 2–3 times larger than normal, with the consequence that they miss steps relative to those of the other legs, but then rejoin the stepping pattern with the correct timing (Graham, 1978). During fast walking, this may mean that the hind legs step at only half the rate of the other legs. Cockroaches (*Periplaneta americana*) may walk slowly with a wave of leg movements that sweep forwards, first on one side and then on the other. At stepping frequencies from 2 to 24 Hz, the alternating tripod gait is maintained (corresponding to forward velocities of 0.05–1.0 $m.s^{-1}$) (Delcomyn, 1971a). At the still higher speeds of running (1.0–1.5 $m.s^{-1}$ or 50 body $lengths.s^{-1}$), the front of the body may be raised to give an angle of attack of some 30° so that then only the middle and hind pairs of legs contact the ground and provide the thrust (quadrupedal gait), or even more remarkably the hind legs alone (bipedal gait) may contact the ground (Full and Tu, 1991). This means that, like rapidly running ghost crabs, they can have an aerial phase to their movements. Speed is increased by increasing the stride length of the longer hind legs while the frequency of movements stays the same. First instar stick insects use either the tripod gait or a tetrapod gait in which four legs remain on the ground at one time, while a leg on each side of the body is lifted and moved forwards (Fig. 8.2). Adult stick insects also use a regular tetrapod gait when walking slowly, but as they walk faster (the highest stepping frequency of an adult is only

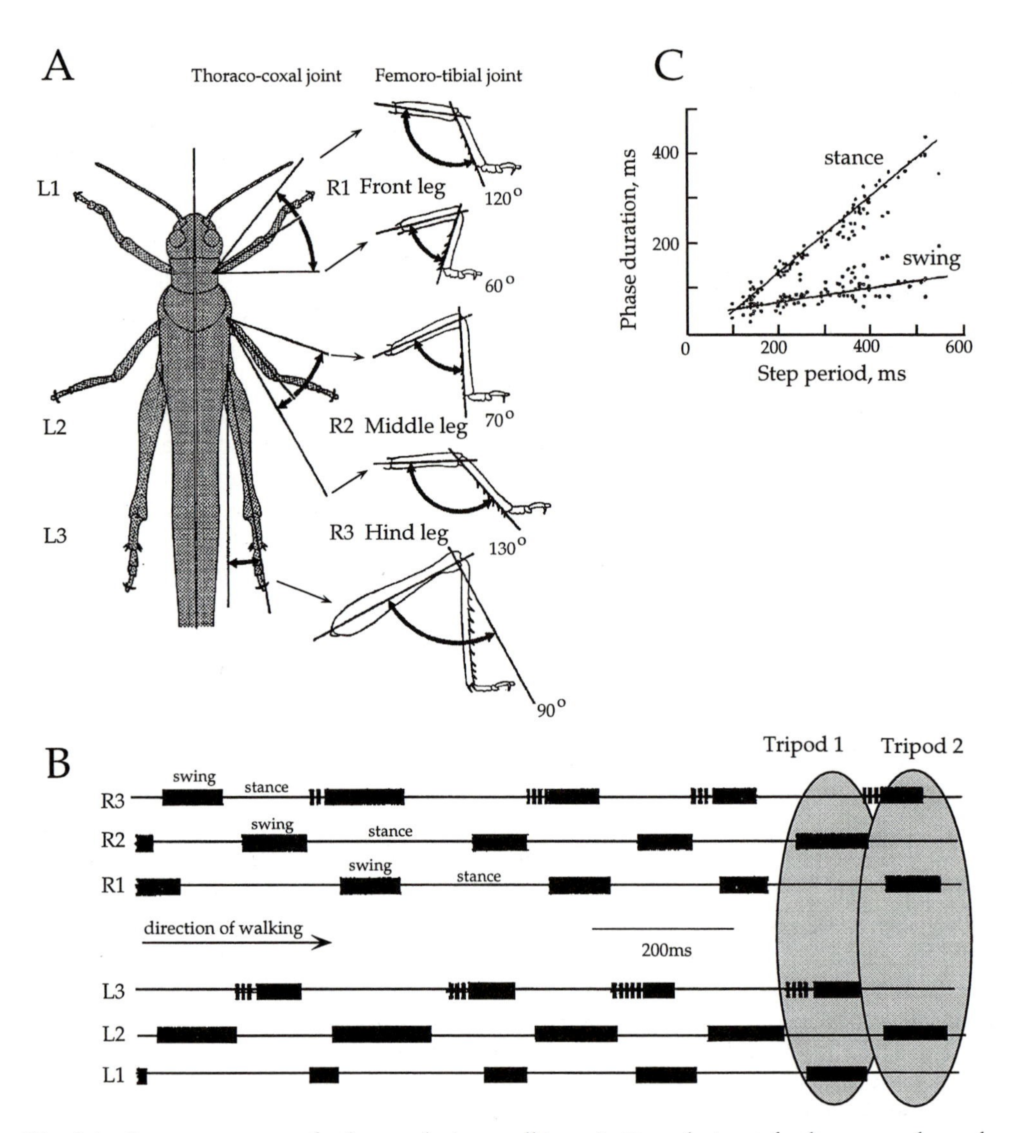

Fig. 8.1 Leg movements of a locust during walking. A. Dorsal view of a locust to show the angles through which a front, a middle and a hind leg move during walking. B. Diagram of the tripod gait used when walking. The thick black bars indicate the time when a leg is off the ground during the swing phase, the thin lines show when it is in contact with the ground during the stance phase, and the broken bars when the tarsus is dragged on the ground. The two alternating tripods of support are outlined. C. The duration of the swing phase changes less than the duration of the stance phase as the speed of walking changes. Based on Burns (1973).

3 Hz) the sequence of movements of the legs starts to overlap. Their gait also depends upon the load against which they are required to work, the tripod operating when the load is low and the tetrapod gait when it is high.

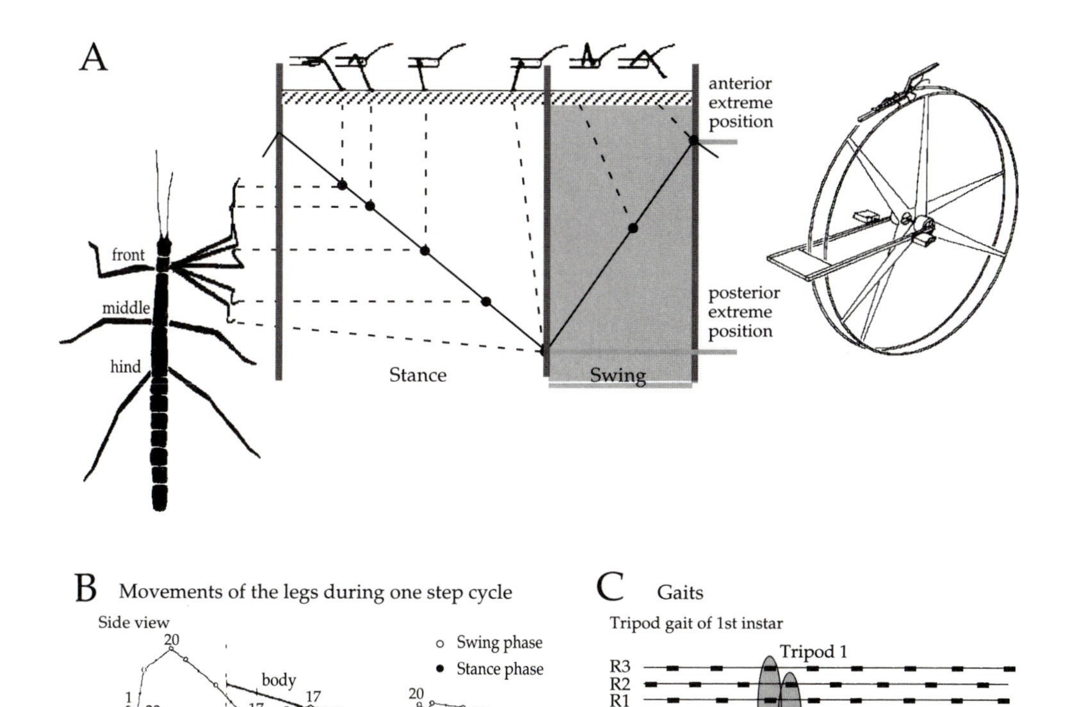

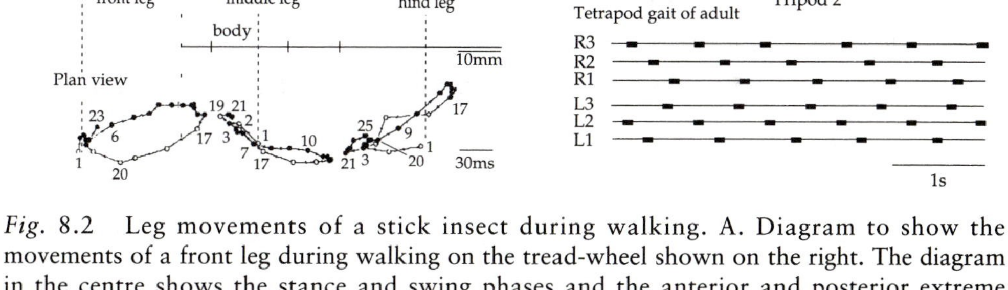

Fig. 8.2 Leg movements of a stick insect during walking. A. Diagram to show the movements of a front leg during walking on the tread-wheel shown on the right. The diagram in the centre shows the stance and swing phases and the anterior and posterior extreme positions of the leg. B. Movements of the three legs on one side of the body as viewed from the side and from above. The numbers indicate the sequential positions of a leg during one cycle of movement. C. Gaits during walking, plotted as in Fig. 8.1. First instar stick insects using an alternating tripod gait, but adults use a tetrapod gait. A is based on Cruse (1990), B on Cruse (1976b), and C on Graham (1972).

8.1.2. The step cycle of a leg

Each of the legs moves through a repetitive cycle of movements during walking that are of the same general form but differ in important details. Each cycle of movement can be divided into two phases (Fig. 8.3).

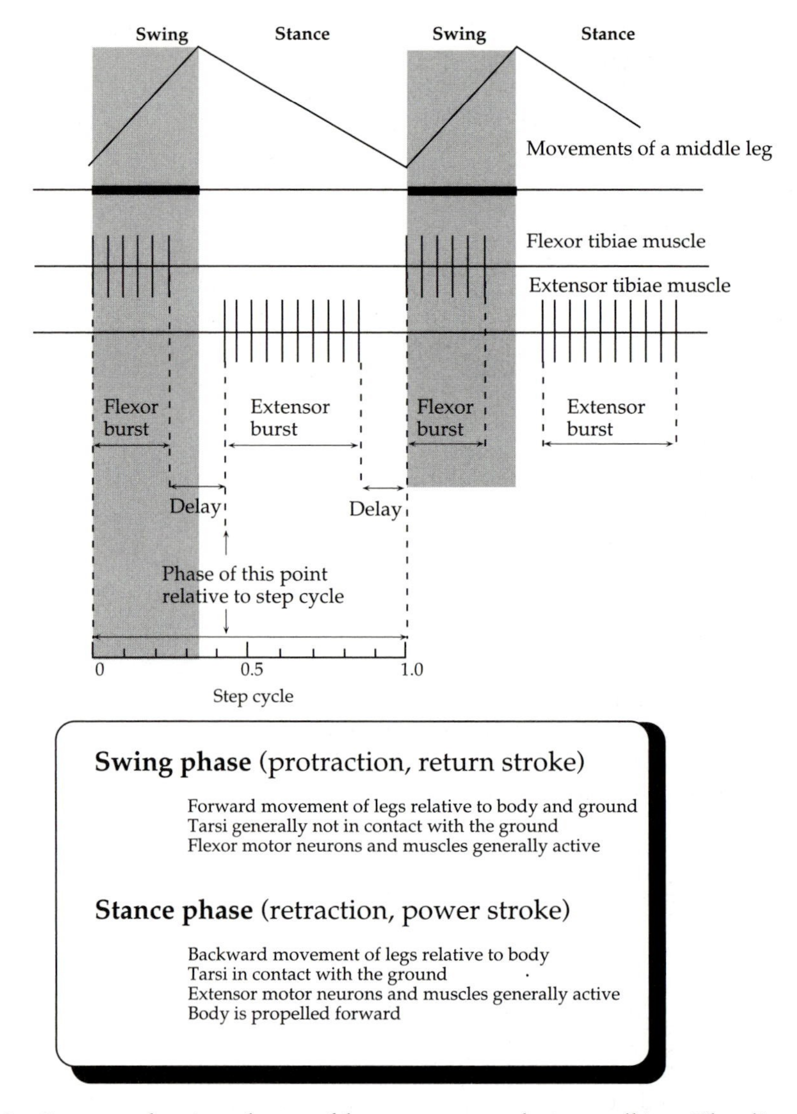

Fig. 8.3 Stance and swing phases of leg movements during walking. The diagram is of a stylised movement of a middle leg during two step cycles. The action of two antagonistic muscles that move the tibia are shown, with the timing differences between them exaggerated so that the terms general to walking can be more easily defined.

The **stance phase**, (often called retraction, the support phase, or the power stroke), in which the tarsus is in contact with the ground and the body is propelled forwards.

The **swing phase**, (protraction, the recovery phase, or the return stroke), during which the leg is swung forwards and the tarsus is off the ground.

Both phases change linearly in duration as the frequency of stepping varies, but the stance phase changes at a faster rate, so that the ratio of the swing to stance phases (usually called the protraction/retraction ratio, P/R) can be greater than 1.0 for the middle legs. This also means that the duration of the swing phase is more constant than that of the stance phase, a feature common to walking in vertebrates. It also parallels the relationship between the two phases of breathing (*see Chapter 12*) and flying (*see Chapter 11*) motor patterns.

The movements of one leg relative to another can be described by their phase relationships, derived by dividing the delay between the movements of the two legs that are being compared by the cycle period of one of those legs. Thus, in a strict alternating tripod, the front and hind legs on one side move together and thus have a phase of 0 or 1.0, while the middle leg alternates with these two and thus has a phase of 0.5. Legs on the opposite side of the same segment that move alternately also have a phase of 0.5. In real walking, the phase relationships between the three legs moving together as a tripod can show some drift, which implies that correcting influences are needed to maintain the coordination, but the legs of one segment that alternate show less variation in their phase relationships. Nevertheless, all the phase measurements for walking by locusts are more variable than those in stick insects or cockroaches (Delcomyn, 1971a; Graham, 1985). This all points to a neural control mechanism that must allow for much variation and that must generate movements of great variety. It is, however, important to bear in mind that much of the analysis of locust walking has been carried out on brief bursts of activity encouraged by a particular sensory stimulus. These movements may have the urgency of escape responses and may therefore differ from the slower sequences of more sustained walking analysed in some other insects.

8.2. THE MACHINERY FOR WALKING

8.2.1. The legs

Each of the six legs consists of five parts; a short, proximal **coxa**, a small **trochanter** that in the front legs can move relative to the femur, but in the middle and hind legs is fused to a long **femur**, a **tibia** the same length as the femur, and a distal **tarsus** that is divided into a series of moveable parts, with a terminal claw or **unguis**. The disparity in the sizes of the legs is such that the mass of the femur of a hind leg is about ten times that of a femur of a front leg; this is associated with the specialisations of the hind legs for jumping (Table 8.1 and *see Chapter 9*). The tibia of a hind leg has two rows of fixed and pigmented spines on its dorsal surface which are used to push away adversaries. The most distal two pairs of spines can be moved passively and probably serve a proprioceptive function when walking over rough terrain. By contrast, the spines are on the ventral surfaces of the front and middle tibiae, and all can be moved passively but not by muscular action. Both the fixed and moveable spines have a single large campaniform sensillum at their base.

Table 8.1. Characteristics of the three pairs of legs

	Front leg	Middle leg	Hind leg
Tibial length (mm)	7	9	24
Tibial mass (mg)	3.7	9	23
Femoral length (mm)	7	8.5	26
Femoral mass (mg)	8	9.5	88
Mass of extensor muscle (mg)	–	–	70
Mass of flexor muscle (mg)	–	–	15
Ratio of mass of extensor to flexor	–	–	4.7:1

Measurements are from mature adult *S. gregaria*

8.2.2. The muscles

Most of the mass of muscles that control the movements are at the base of a leg so that its moment of inertia is kept to a minimum (Fig. 8.4). The joint between the thorax and coxa allows a universal movement and is controlled by three sets of muscles, so that any particular movement will depend on the balance of action by them all. The three distal joints are hinge joints that are each controlled by a pair of muscles which cause either flexion or extension.

The muscles in the three pairs of legs are homologous, but there are small differences between the different legs, particularly in the arrangement of the muscles that move the coxa and the femur (Table 8.2). Some of these differences appear to be real but others probably represent quirks of the descriptions. Details of the innervation and actions of the muscles that move a hind leg are given in Table 8.3.

8.2.2.1. *Coxa*

The coxa is moved by six muscles and, like all the muscles in the legs, they are given functional names based only on anatomical analyses. These names are particularly inadequate when applied to the coxal muscles as they imply specific antagonistic actions, when, in reality, the six muscles must act as a group with each movement of the joint involving different combinations of actions. The general action of these muscles can be summarised by stating that the promotor lifts the anterior edge of the coxa upwards, whereas the remotors lift the posterior edge upwards. The anterior rotators rotate the anterior of the coxa forwards and closer to the body, whereas the posterior rotators rotate the posterior of the coxa backwards and closer to the body. The abductors pull the coxa forwards and away from the body, whereas the adductors pull the coxa towards the body. The action of each individual muscle will, however, depend on the action of the other muscles in the group, so that it is difficult to attribute a particular function to one muscle.

8.2.2.2. *Trochanter*

In the middle and hind legs where the trochanter is fused to the femur, the action of the levator trochanteris is to flex the trochanter/femur about the coxa whereas the

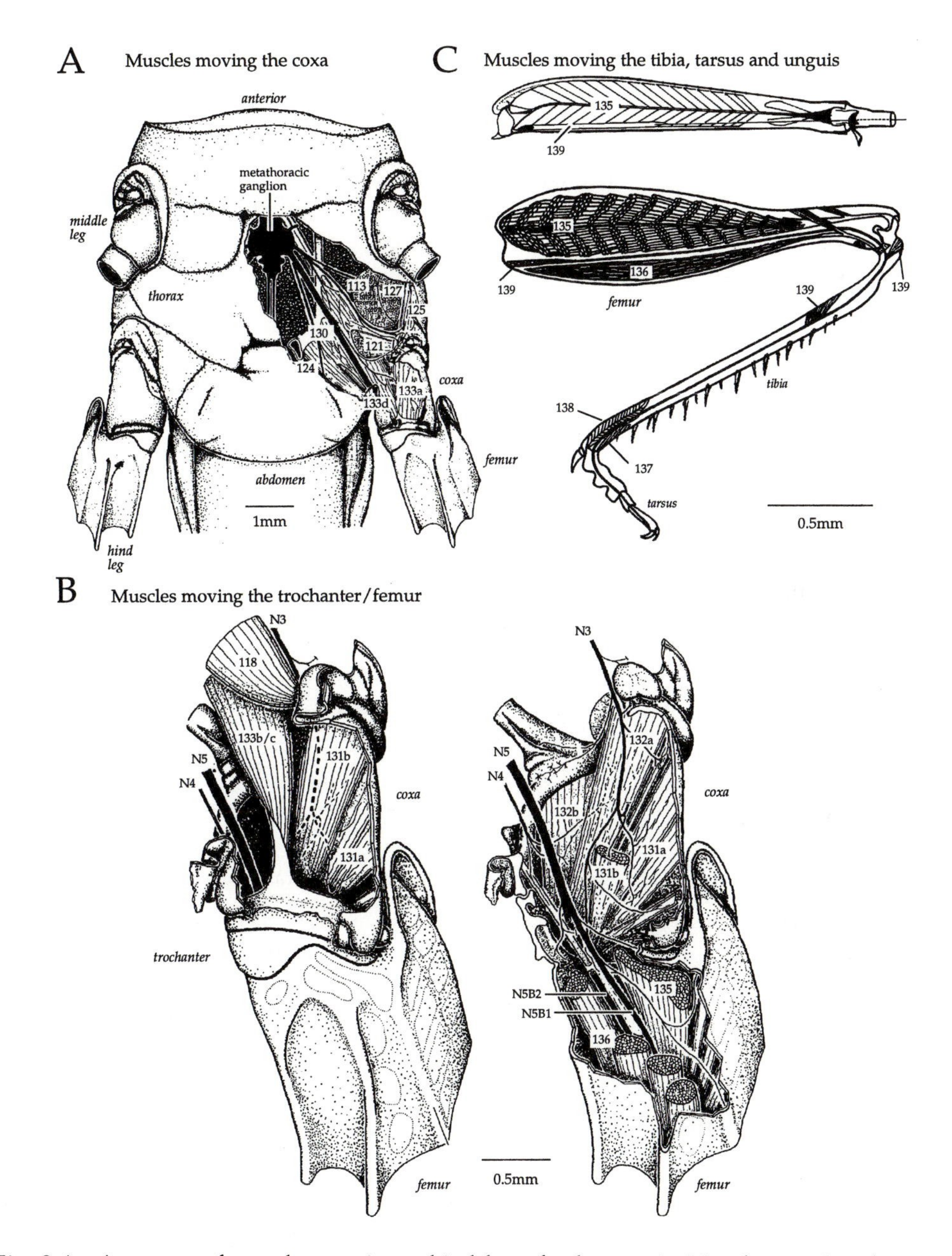

Fig. 8.4 Anatomy of muscles moving a hind leg of a locust. A. Muscles moving the coxa. The thorax is opened ventrally to reveal the metathoracic ganglion and the muscles inserting in the coxa. B. Muscles in the thorax and coxa that move the fused trochanter/femur. The leg is in the same orientation as in A. C. Muscles that move the tibia and tarsus. The femoral muscles are shown as viewed dorsally and laterally. The muscles are numbered according to Snodgrass (1929); details are given in Tables 8.2 and 8.3. A and B are based on Bräunig (1982), C on Hoyle (1978b).

Table 8.2. Muscles in the front, middle and hind legs

Joint	Muscle name		Origin	Insertion	Muscle number Front	Middle	Hind
Thoraco-coxal	Promotor		Tergum	*F.* Trochantin	62		
				M. Apodeme between trochanter and coxa		89	
				H. Anterior coxa			118
	Remotors	(1)	Tergum	Posterior coxa	63	90	119
		(2)			64	91	120
		(3)			65	–	–
	Anterior rotator		Mesosternum	Anterior coxa	–	92	121
	Posterior rotators	(1)	*F.* Sternal apophysis	Posterior coxa	66		
			M. Spina			93	
			H. Sternal apophysis				122
		(2)	*F.* Spina	Posterior coxa	67		
			H. Sternal apophysis			–	123
		(3)	Sternal apophysis	Posterior coxa	–	–	124
	Abductors	(1)	Episternum	Outer rim of coxa	68a,b	94	125
		(2)			–	95	–
		(3)			–	96	126
	Adductor		Sternal apophysis	Inner rim of coxa	69	100	130
						101	
Coxo-trochanteral	Levator trochanteris		Dorsal, anterior coxa	Dorsal rim of trochanter	70	102	131a,b
			Dorsal coxa	Dorsal rim of trochanter	–	–	132a,b
	Depressor trochanteris		Ventral coxa	All converge on an apodeme that inserts	71 71a	103 103a	133 133a
			F. Ant episternum *M, H.* Lateral scutum	on ventral trochanter	71b	103b	133b

Table 8.2. Continued

Joint	Muscle name	Origin	Insertion	Muscle number		
				Front	Middle	Hind
		F. Ventral pleural arm		71c		
		M, H. Lateral scutum			103c	133c
		F. Tergum		71d		
		M, H. Sternal apophysis			103d	133d
Trochantero-femoral	Reductor femora	Ventral trochanter	Femur	72	104	–
Femoro-tibial	Extensor tibiae	Anterior and dorsal femur	Tibia	74	106	135
	Flexor tibiae	Ventral femur	Tibia	75	107	136
Tibio-tarsal	Levator tarsi	Ventral distal tibia	Tarsus	76	108	137
	Depressor tarsi	Anterior and posterior wall of distal tibia	Tarsus	77	109	138
Tarso-ungual	Retractor unguis	Proximal femur	Apodeme in femur that inserts on pretars us (claw)	78a	110a	139a
		Dorsal, proximal tibia	Apodeme in proximal tibia	78b	110b	139b
		Ventral, tibia approx. 1/3 from proximal end	Apodeme in proximal tibia	78c	110c	139c

Data are from *S. gregaria*
F = Front leg
M = Middle leg
H = Hind leg

depressor trochanteris extends it. In the front legs, the trochanter may be able to move by a small amount about the femur under the control of a single reductor trochanteris muscle.

Table 8.3. Muscles of a hind leg: their action and innervation

Joint	Muscle name	Muscle number	Nerve	Action	Number of efferent neurons			Motor neuron group
					Excitors	Inhibitors	DUM neurons	
Thoraco-coxal	Promotor	118*	3A4	Not used in walking?	3	–	–	1/2
	Remotors (1)	119*	4D1	Not used in walking?	3	–	–	7
	(2)	120	4D2	Lifts posterior edge of coxa upwards	2 large 4 small	CI1	–	7
	Anterior rotator	121	3C1	Rotates anterior of coxa forwards and closer to body	2	CI1	?	2
	Posterior rotators	122	4B1	Rotate posterior of coxa	4	CI1	?	7/8
		123	4C	backwards and	3	CI1	?	7/8
		124	4C	closer to body	3	CI1	?	7/8
	Abductors	125	3A3	Pulls coxa forwards	3	CI1	?	1/2
		126	3A3	Pulls coxa away from body and forwards	3	CI1	?	1/3
	Adductor	130	3C2	Pulls coxa towards body	1	CI1	?	3
Coxo-trochanteral	Levator trochanteris	131a,b 132a,b	3B6a,b 4A3,4	Flexes trochanter/femur	7 6	CI1 CI1	3,4 3,4 + 3,4,5	6/7
	Depressor trochanteris	133a 133b/c 133d	5A2a 3C3 3C2a	Extends trochanter/femur	2 6	CI1 CI1	?	2–5

Table 8.3. Continued

Joint	Muscle name	Muscle number	Nerve	Action	Number of efferent neurons			Motor neuron group
					Excitors	Inhibitors	DUM neurons	
	CxTrMRO		4A2	Sets length/ tension of the receptor	1	–	?	7
Femoro-tibial	Extensor tibiae	135	5B1	Extends tibia	2	CI1	DUMETi	4/6
	Flexor tibiae	136	5B2	Flexes tibia	~9	CI2, CI3	3,4,5	5
Tibio-tarsal	Levator tarsi	137	5B2a	Flexes tarsus	1	CI2, CI3	?	4/5
	Depressor tarsi	138	5B2a	Extends tarsus	~3	CI2, CI3	?	4/5
Tarso-ungual	Retractor unguis	139 a,b,c	5B2	Curls the unguis	2	CI2, CI3	?	4/5

8.2.2.3. *Femur*

The femur of a hind leg is sturdy because of the requirements for jumping, and contains the large extensor muscle and the smaller flexor tibiae muscle that can move the tibia through approximately 150°. The extensor muscle of a hind leg has approximately five times the mass of the flexor and is specialised to generate large forces rather than fast contractions (*see Chapter* 9). Despite this disparity in mass, the flexor tibiae is innervated by some nine excitors and two inhibitors while the extensor tibiae is innervated by only two excitors and one inhibitor. In the front and middle legs, the flexor muscles are larger than the extensors (Burns and Usherwood, 1978), suggesting that these muscles must act in different ways in the different legs. During jumping, the flexor in the hind leg must lock the femoro-tibial joint to restrain the co-contraction of the extensor muscle (*see Chapter* 9). In walking, flexors in the different legs may be active at different times during the step cycle of their particular legs.

8.2.2.4. *Tibia and tarsus*

The tibia, even of a hind leg, is light and the muscles that it contains are small and are located at its distal end. The levator tarsi flexes the tarsus and the depressor extends it. The depressor tarsi is innervated by about three excitatory motor neurons while the levator tarsi is innervated by just one, or sometimes by this and by an electrically coupled supernumerary neuron (Siegler, 1982). In both the femur and the tibia, therefore, one muscle is innervated by a far greater number of motor neurons than its antagonist, but the design implications behind this disparity have never been explained. If these patterns of innervation were the result of an evolutionary fusion of muscles then it might be expected that the motor neurons would innervate distinct parts, but the limited evidence available suggests that the terminals of the motor neurons are instead distributed widely throughout the muscles to the extent that an individual muscle fibre can be innervated by many motor neurons. This organisation suggests a functional explanation related to the use of the different types of muscle fibres in diverse movements, but it still leaves unexplained the difference between the antagonists of one joint.

8.2.2.5. *Unguis*

The most bizarre muscular arrangement in the leg is that of the retractor unguis muscle, which curls the distal part of the tarsus and thus enables the locust to grasp an object over which it is walking or climbing. The muscle consists of three parts, one in the proximal femur and two in the tibia, all attached to a common apodeme (tendon) that runs from the proximal femur to the tip of the tarsus. There is no antagonist to this muscle so that it must therefore work against elastic forces. Why is there such an elaborate arrangement just to curl up the distal end of the tarsus, and how does the movement of the unguis remain independent of the movement of the femoro-tibial and tibio-tarsal joints through which its apodeme must pass?

Some answers to these questions are available for stick insects, where the organisation of the muscle is similar. The apodeme passes through the femoro-tibial joint ventral to the axis of rotation, but remains close to the axis of rotation through the tibio-tarsal joint (Radnikow and Bässler, 1991). If the apodeme is moved within the femur then it produces movements of the unguis that are independent of the position of the femoro-tibial joint. If, however, the apodeme is clamped in one position in the femur and the angle of the femoro-tibial joint is changed, then movements of the unguis result. The distance between the origin of the muscle bundle in the proximal femur and the insertion point of the apodeme in the unguis therefore depends on the position of the femoro-tibial joint, and during a full rotation of this joint would be sufficient to move the unguis through its complete range were the proximal muscle to be stiff. The force that this muscle can produce is, however, relatively independent of its length. The probable solution lies in using the tibial parts of the muscle for most movements of the unguis, as their actions will be independent of the position of the femoro-tibial joint provided that the femoral part of the muscle is not stiff. Their small size means that they will not produce a large moment of inertia. The larger femoral part can then be reserved for movements that require more force, and because it is proximal, it also will not produce a large moment of inertia. This solution requires no neural compensatory mechanism, but to work it would require the differential action of the three parts of the muscle. This could be accomplished by different patterns of motor innervation, or by different contractile properties of parts of the muscles to the same motor commands.

8.2.3. The motor neurons

Each of the leg muscles, like those elsewhere in the body, is innervated by only a small number of neurons ranging from one to about nine (Table 8.3). The total set of motor neurons controlling the movements of all the joints of a leg is thus only 70, of which the majority are involved in controlling the complex joint between the thorax and the coxa, and the movements of the trochanter about the coxa. The success of this design, which uses such a small number of output elements, is nevertheless exemplified by the range, delicacy and precision of the movements performed. Locusts can balance on a twig, walk on horizontal and vertical surfaces, or even upside-down, march for large distances in bands as young larvae, jump from one grass stem to another to escape predators or to launch into flight, and kick accurately to fend off an adversary or remove an unwanted object.

The majority of the motor neurons have an excitatory action on the leg muscles (Fig. 8.5). Each innervates a subset of muscle fibres within one muscle, with the proportion of fibres innervated by any one motor neuron varying considerably between the different muscles. In the extensor tibiae muscle, for example, the single fast motor neuron provides the sole excitatory innervation to 76% of the fibres, the single slow motor neuron to 8.5%, whereas both innervate the remaining 15.5% (Hoyle, 1978b) (*see Chapter 9*). These motor neurons can cause contractions that range from fast to slow, enabling at one extreme the production of fast movements for use in running or walking, and at the other extreme the maintenance of posture or the adjustment of slow movements. They therefore contrast with the power producing flight muscles, where the motor neurons produce only twitch contractions (*see Chapter 11*). Generally, the fast motor neurons innervate the muscle fibres that are able to contract the fastest, and the slow motor neurons those fibres capable of more sustained contractions. There is, however, often considerable overlap and in large motor pools the occurrence of motor neurons with intermediate properties further confuses any clear separation in the patterns of innervation.

Apparently correlated with this diversity of motor neuron actions and muscle properties is the occurrence of motor neurons that have an inhibitory action on the muscle fibres themselves. These inhibitory neurons do not innervate only a single muscle, as do the excitatory motor neurons, but instead innervate several, or even many, muscles with distinct and often opposing actions during walking. Each leg is innervated by three of these inhibitory neurons, which are called common inhibitory motor neurons (CI1, CI2 and CI3) because they supply several muscles. One (CI1) innervates 12 muscles that move the proximal parts of a leg and the pleuroaxillary muscle of a wing, while the other two (CI2, CI3) innervate the same three muscles in the more distal parts of the leg (Fig. 8.6).

All the motor neurons have cell bodies and neuropilar branches on the same side of the ganglion as the muscle that they innervate (Fig. 8.5). Most of the cell bodies are in the ventral cortex and are amongst the largest cell bodies of any in the thoracic ganglia. The cell body of the fast extensor tibiae motor neuron in the metathoracic ganglion is the largest in the whole central nervous system. The cell bodies are

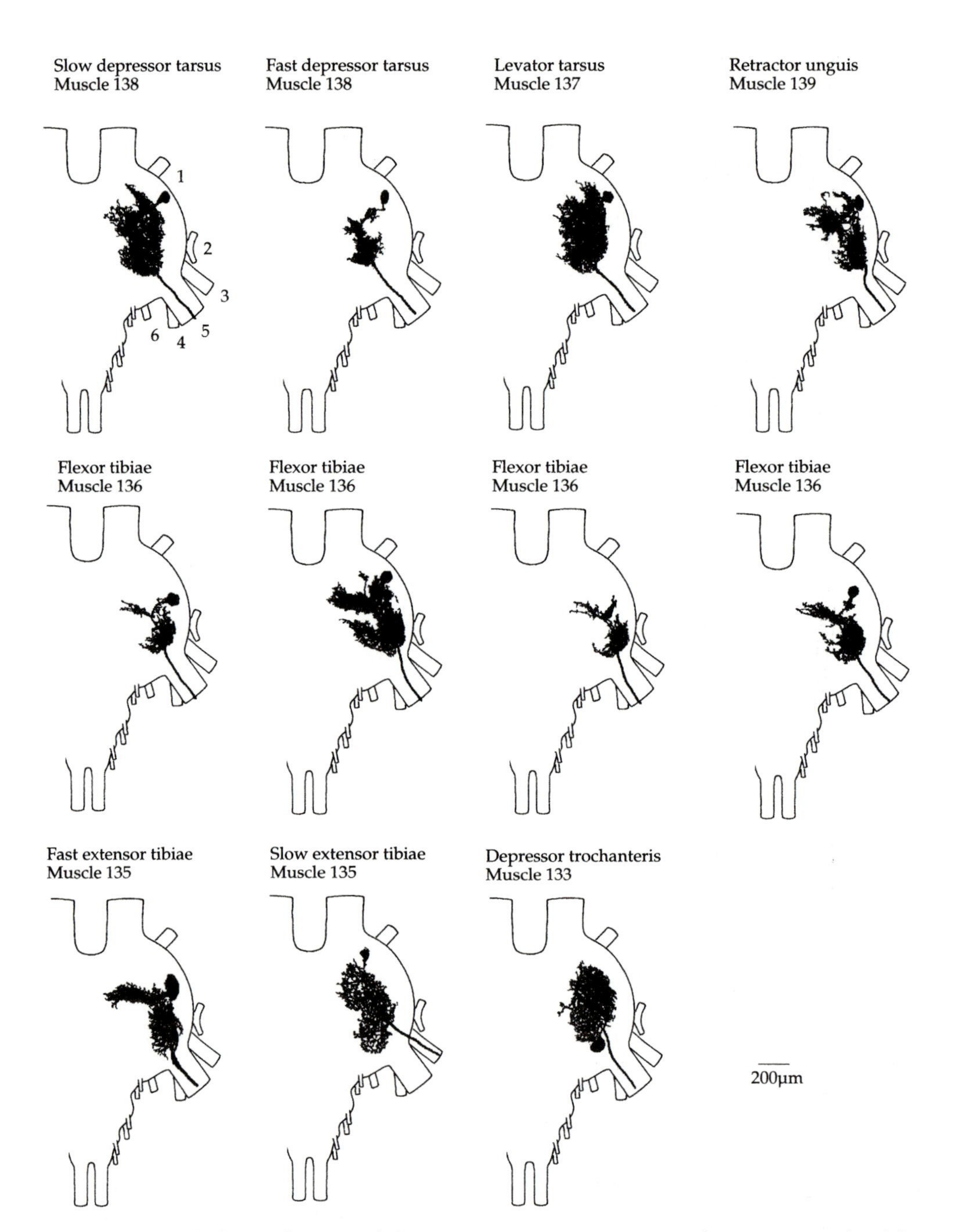

Fig. 8.5 Morphology of some of the excitatory motor neurons that innervate hind leg muscles. Each neuron was identified and then stained intracellularly by the injection of cobalt. The drawings show one half of the metathoracic ganglion with the lateral nerves to the thorax and a hind leg numbered.

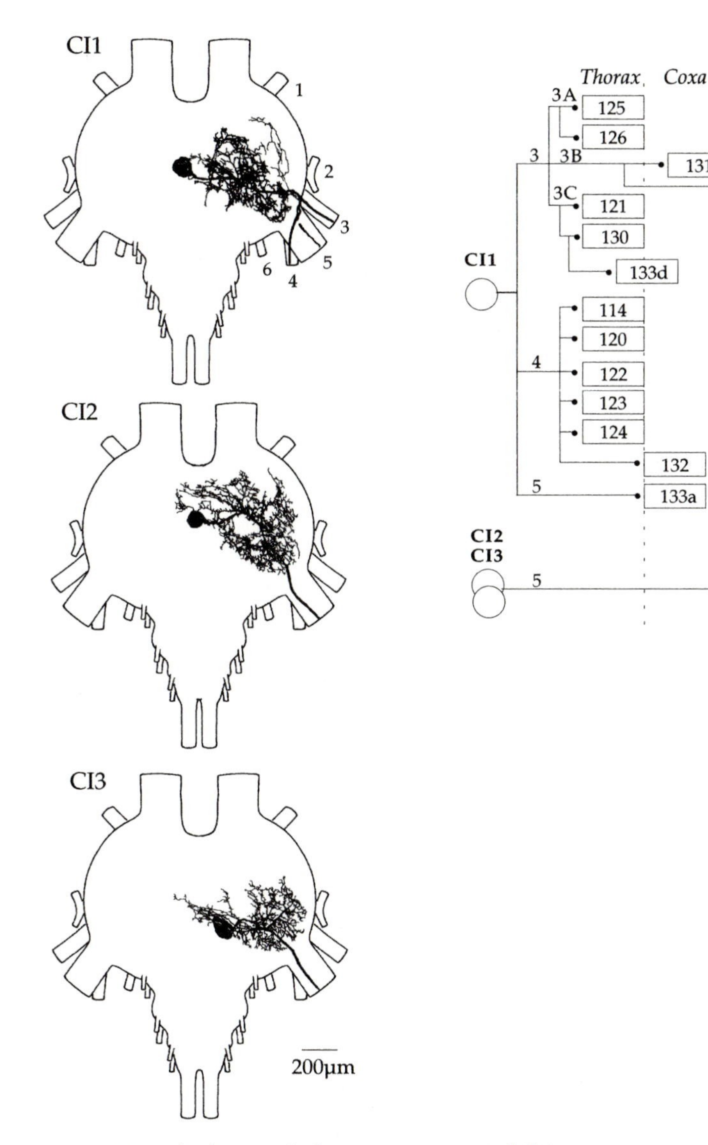

Fig. 8.6 Morphology of three common inhibitory motor neurons (CI1–3) in the metathoracic ganglion and their patterns of innervation of thoracic and leg muscles. The nerves in which their branched axons run are labelled in the diagram.

packed in groups surrounded by glial cells, and the primary neurites from members of a particular group may follow the same path through the neuropil so that the neurons have similar overall shapes (Siegler and Pousman, 1990a,b). Each group may represent the progeny of an individual neuroblast (*see Chapter 4*) but do not represent a common function, or the nerve containing the axons of their neurons. This arrangement means that motor neurons innervating the same muscle may belong to different groups, a notable example being the slow and fast extensor tibiae motor neurons (*see Chapter 3*).

The small size of the motor pools to the various muscles of a leg means that most of the leg motor neurons can be treated as identified individuals that are characterised morphologically and physiologically.

8.3. MOTOR ACTIVITY DURING WALKING

The actions of the muscles during walking can be determined by recordings made with fine wire electrodes that do not impede the leg movements, so that the locust is able to walk horizontally, vertically or even upside-down (Fig. 8.7). This means that the loading of a leg is relatively unaffected so that the sensory signals generated should reflect those that would occur normally. The wire electrodes can either record the motor spikes in the nerves (neurograms) or the potentials that they evoke in the muscle fibres (myograms). The success of neurograms is limited by the accessibility of small nerves in which the spikes of known neurons can be identified. Most accessible nerves contain the axons of many neurons whose individual actions cannot be distinguished. Myograms have been used more successfully, and some of the most favourable recordings allow the analysis of the patterns of spikes in individual motor neurons, but only when the number innervating a particular muscle is small and when the currents generated in the muscle are large enough.

8.3.1. Front (prothoracic) legs

The movements of the front legs during walking can sometimes be quite variable because they often perform searching movements to test the terrain. The front tarsi have a larger number of tactile and chemosensory receptors than the hind tarsi, many of which are on the underneath, so fitting them for exploring the way ahead (Table 8.4) (Kendall, 1970).

Table 8.4. Tarsi and their receptors during postembryonic growth

Stage	Front leg		Middle leg		Hind leg	
	length of tarsus (mm)	number of receptors	length of tarsus (mm)	number of receptors	length of tarsus (mm)	number of receptors
1st instar	1.5	170	1.6		2.0	
2nd instar	1.8		1.9		2.5	
3rd instar	3.5	230	3.7		5.0	
4th instar	3.7		3.9		5.7	
5th instar	4.7		4.8		6.5	
Adult	5.6	700	5.7	640	8.1	450

Each value is the mean from 10 males. Data from Kendall (1970).

Most of the movements of the front and middle legs during walking are produced about the coxal joints; rotation lifts a leg from the ground at the start of the swing

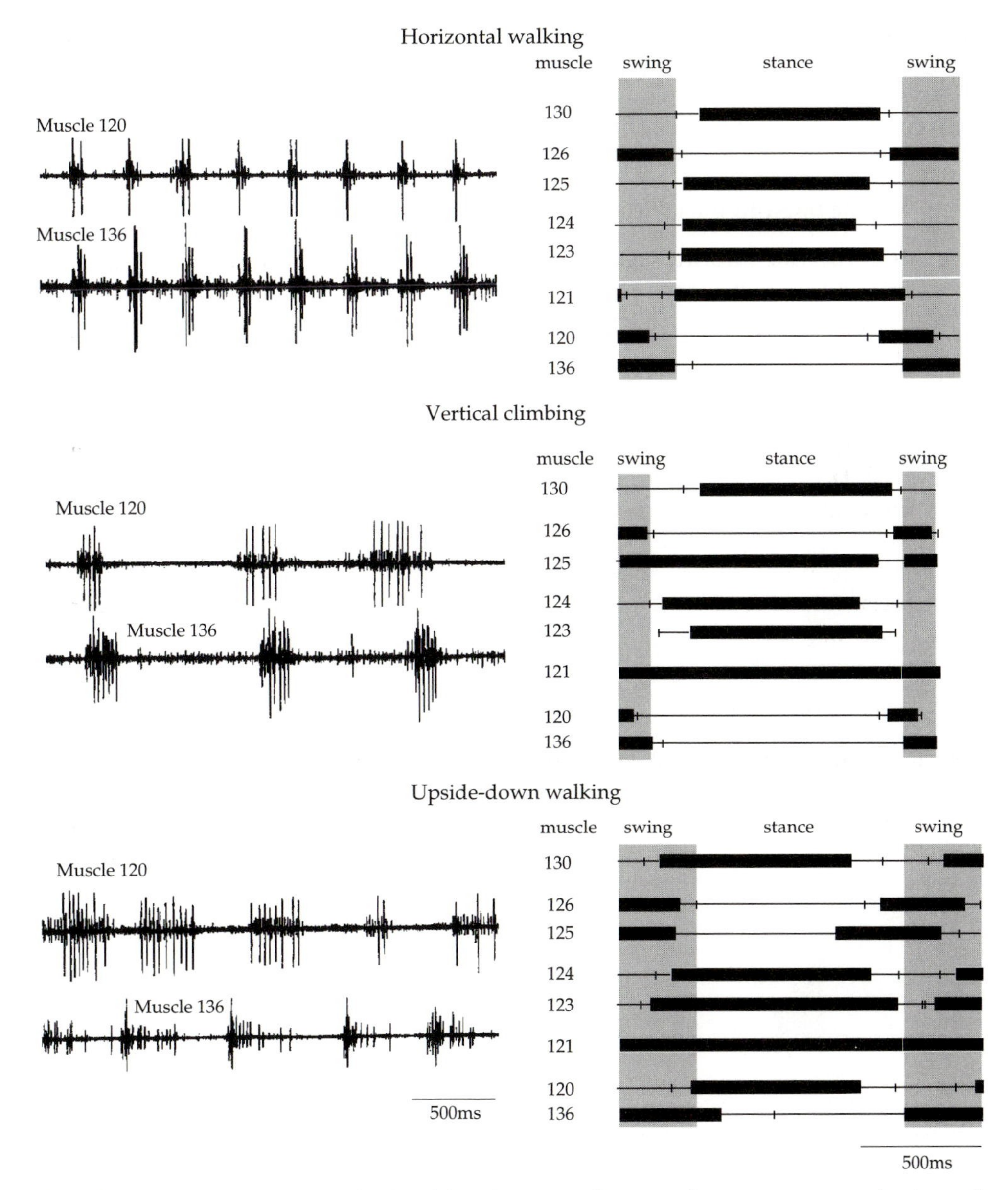

Fig. 8.7 Actions of muscles of a hind leg during walking in three orientations; horizontal, vertical and upside-down. The left-hand column shows simultaneous recordings from muscle 120, a coxal remotor, and from muscle 136, the flexor tibiae. The right-hand column shows a summary of the actions of the coxal muscles and the flexor tibiae. The thick bars indicate when a muscle is active. Based on Duch and Pflüger (1995).

phase, moves it forwards during the swing phase and then backwards during the stance phase while twisting it around its long axis. The tibia of a front leg extends during the swing phase (protraction) to place it in front of and lateral to the head, and flexes during the stance phase (retraction) so that its contribution is to pull the body along by the action of its flexors (Fig. 8.8 and Table 8.5). The slow extensor tibiae motor neuron (SETi) is active to about the same extent during both phases of the movement, but the fast (FETi) is most active during the swing phase (Burns and Usherwood, 1979).

8.3.2. Middle (mesothoracic) legs

The tibiae of the middle legs flex during the swing phase as the legs move forwards, and extend during the stance phase when they are in contact with the

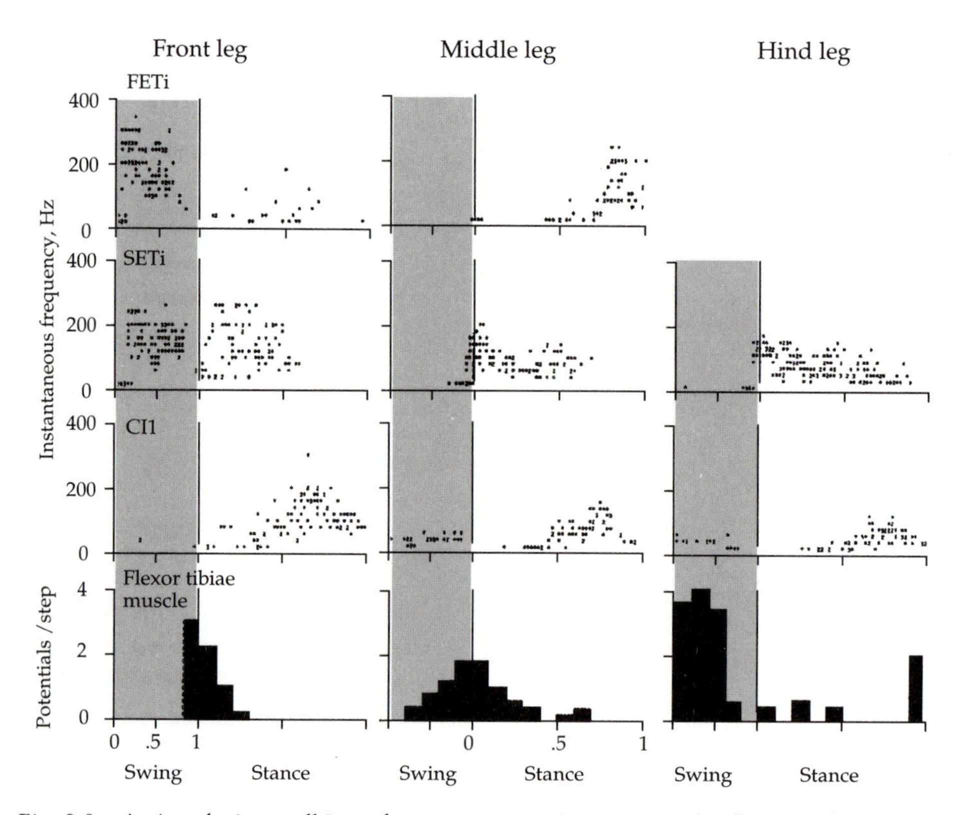

Fig. 8.8 Action during walking of motor neurons innervating the flexor and extensor tibiae muscles of a front, middle and a hind leg. The graphs show the instantaneous frequency of spikes of extensor motor neurons during the swing and stance phases. The action of flexor motor neurons is represented as histograms of summed activity because the large number of motor neurons means that the spikes of an individual neuron cannot be recognised in recordings from the muscle. The activities during 10 steps are superimposed for the front and hind legs, and for nine steps for the middle leg. Based on Burns and Usherwood (1979).

Table 8.5. Phases of action of tibial motor neurons during walking

Leg	Swing Protraction Promotion	Stance Retraction Remotion
Front (prothoracic)	Extensors, SETi + FETi	Flexors and extensors, SETi only
Middle (mesothoracic)	Flexors; SETi starts to spike towards the end	Extensors, SETi + FETi
Hind (metathoracic)	Flexors	Extensors, SETi only

ground and thus provide propulsive force. The slow extensor tibiae motor neuron (SETi) starts to spike before the end of the swing phase, reaching a peak before the tarsus is placed on the ground, and continues to spike during the stance phase, whereas FETi spikes only towards the end of the stance phase (Fig. 8.8). The flexor motor neurons are activated half way through the swing phase. These actions mean that the legs push the body forwards by the contraction of the extensors. During the stance phase, the depressor tarsi and retractor unguis muscles are active, presumably to maintain traction with the ground (Laurent and Hustert, 1988), but during the swing phase they are silent and the levator tarsus muscle is active.

8.3.3. Hind (metathoracic) legs

The femora of the hind legs are held almost parallel to the body so that the swing and stance phases consist mostly of flexion and then extension of the tibiae, with a smaller rotation of the coxae than in the front and middle legs lifting the legs from the ground at the start of the swing phase. During the swing phase, contractions of muscles moving the coxa rotate the leg forwards and lift it upwards, and rotate it backwards and downwards at the start of the stance phase. The sequence of movements of the hind legs is thus the same as for the middle legs, but flexion of the tibia may precede lifting of the leg with the result that the tarsus is dragged along the ground at the start of the swing phase. A notable difference, however, is in the contribution of the two extensor tibiae motor neurons; only SETi contributes to the propulsive force generated during the stance phase, with FETi remaining silent, at least when walking along a horizontal surface (Fig. 8.8). These legs thus again push the body forwards by the action of the extensors.

The different actions of the two extensor motor neurons during walking are correlated with a reversal in the positions of their somata and the nerves in which their axons run (Wilson and Hoyle, 1978). In the pro- and mesothoracic ganglia, the soma of SETi is the larger, is located posteriorly, and its axon runs in N5, whereas that of FETi is smaller and anterior, and its axon runs in N3B. In the metathoracic ganglion, SETi has a smaller anterior soma and an axon in N3B, while FETi has a large posterior soma and an axon in N5. In the metathoracic but not in the pro- and mesothoracic ganglia, FETi makes direct excitatory connections with flexor tibiae motor neurons in the central nervous system, suggesting that these connections are correlated with the specialised role of the hind legs in jumping (*see Chapter* 9).

8.3.4. Analysis of intracellular recordings

Further analysis requires that intracellular recordings are made from individual motor neurons in the central nervous system. The advantage then is that the synaptic input to the motor neurons as well as their spike outputs can be measured, but the disadvantage is that the preparation necessary to make the recordings alters the loading on the legs and hence the sensory feedback. The most successful approach so far has the locust fixed by its thorax and walking upside-down with the left and right sets of legs stepping on the inside circumference of linked tread-wheels (Wolf, 1990a). A small window is made in the ventral thoracic cuticle to give access to the thoracic ganglia, so that the dissection is minimal, but because the animal is not supporting its own weight, even though the inertial force it must exert on the tread-wheels is adjusted for body mass, the sensory feedback may not be normal. Locusts do, however, normally walk upside-down, during which the loading must be the opposite to that experienced during upright walking. Similar recordings can also be made from upright but restrained stick insects (Godden and Graham, 1984) or locusts (Ramirez and Pearson, 1988), but the dissection is more extensive and the data so far collected are limited.

In locusts, this approach has shown when the common inhibitory motor neurons are active and what their contribution is towards the movements of a leg (Wolf, 1990a). When the locust starts walking, the normally silent inhibitors begin to spike and are depolarised rhythmically in time with the stepping movements so that they spike most rapidly as the tarsus is lifted from the ground at the start of the swing phase and continue until it is once again placed on the ground at the start of the next stance phase. Similar patterns of spikes occur even when the leg that they innervate is removed, suggesting that most of the patterning is derived from central pathways rather than from feedback from sensory neurons monitoring the leg movements. All three inhibitors have similar patterns of spikes but the peak frequency in CI1 is reached earlier and the durations of the CI2 bursts of spikes are shorter. If the number of spikes in CI1 is reduced by the injection of hyperpolarising current, the velocity of the leg movement during the swing phase is reduced.

Inhibitors innervate those muscle fibres that are innervated by slow and not by fast motor neurons. They can act both presynaptically on the terminals of the motor neurons and postsynaptically on the muscle fibres to reduce or prevent their activation and to accelerate their relaxation after a contraction. The increased number of spikes in the inhibitors during walking should therefore decrease the maintained force in these tonic muscle fibres; force that is necessary for a sustained posture, but which would impede walking movements. Thus, the high frequency of CI spikes enables the swing phase movements to proceed rapidly and without the stiffness that would result from slowly declining contractions in various muscles.

This explains the apparent anomaly that the inhibitors innervate muscles that can have antagonistic actions during locomotion.

8.4. MOTOR PATTERNS FOR WALKING

8.4.1. Can the central nervous system alone generate walking patterns?

Walking movements are generated by a patterned motor output to the muscles of one leg so that its joints are moved in the correct sequence, and to the different legs so that they are coordinated in the appropriate gait. Moreover, the maintenance of the same alternating tripod gait over a wide range of walking speeds involves the active regulation of the durations of the swing and stance movements of one leg, the delays between these phases, and the timing of the movements of the legs relative to each other. As with other repetitive movements (*see Chapters 11 and 12*), a prominent question has been: what parts of the underlying neuronal activity are generated by the central nervous system and what aspects are regulated by the sensory signals generated by the movements and forces of the legs? Under certain experimental conditions, a pattern can be induced in the motor neurons innervating the leg muscles that does not depend on sensory feedback. The problem is then to assess whether these induced motor patterns bear any resemblance to the patterns that produce walking, but this is made harder to unravel because the legs are often used rhythmically in many other movements. The experimental strategies used to analyse this question have been to progressively reduce the sensory feedback to the central nervous system until it is finally abolished, and then to reduce the number of participating ganglia with the hope of inducing rhythmic motor patterns from the stumps of the nerves to the legs. Such experiments have been performed largely on the fast moving cockroach, for which it is concluded that a basic pattern can be produced by the central nervous system alone, and on the slowly moving stick insect, for which it is concluded that the sensory feedback plays a dominant role.

8.4.1.1. *Motor patterns in cockroaches*

In cockroaches, the main muscles active during the stance phase are the femoral depressors (extensors) (which also support the weight of the body when standing) and the tibial extensors, whereas the femoral levators (flexors) and tibial flexors are active during the swing phase (Fig. 8.9). The flexor and extensor motor neurons innervating these muscles are reciprocally active during most speeds of walking, with a clear overlap between the two tibial muscles present only at slow walking speeds (Krauthamer and Fourtner, 1978). As the frequency of stepping increases there is a decrease in the duration of the bursts and in the number of spikes per burst, but an increase in the instantaneous frequency of those spikes. There is no indication of bursts of spikes from one acting against more continuous spikes from the other, as occurs when a cockroach grasps a plastic ball and does not support the weight of its body (Delcomyn, 1973).

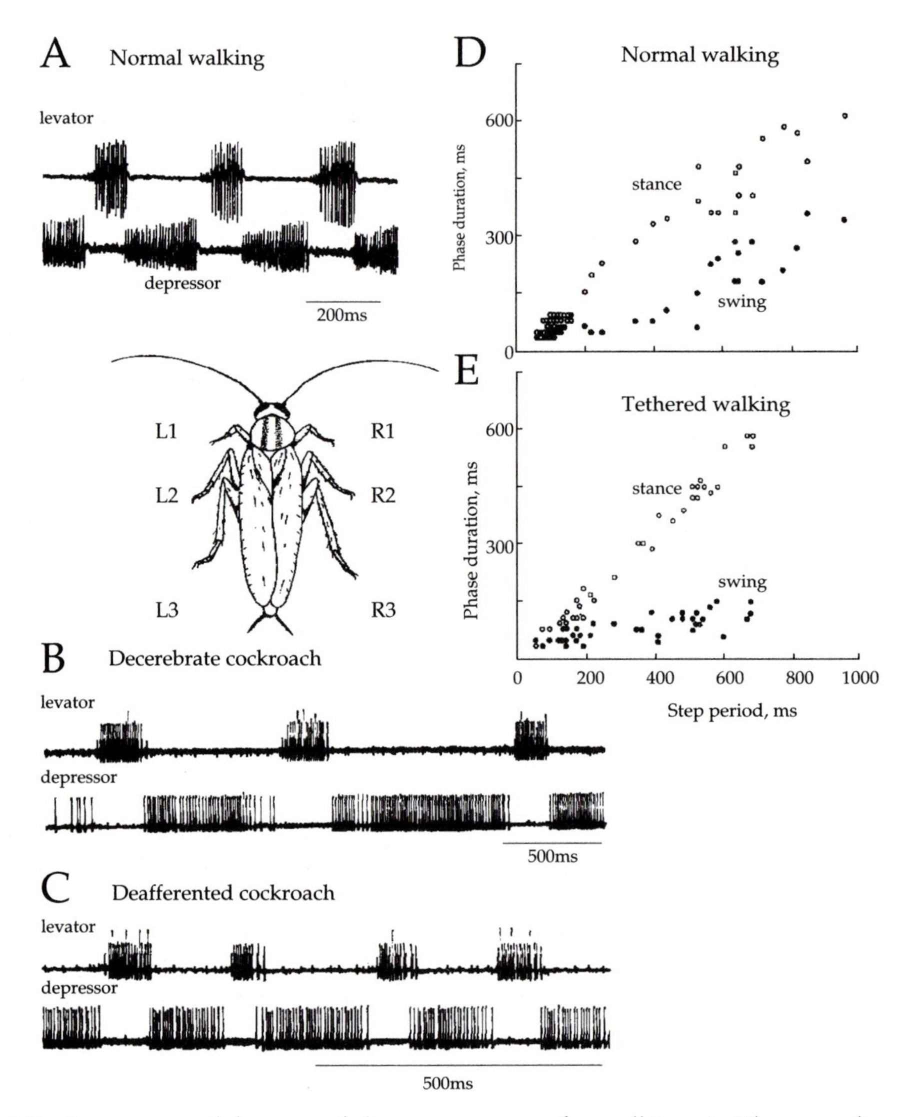

Fig. 8.9 Intrasegmental features of the motor pattern for walking. A. The normal motor pattern expressed in coxal levator and depressor muscles in a dissected cockroach. B. The action of the same two muscles after decerebration. C. The motor pattern after removal of all sensory receptors in the legs. D. The relationship between the duration of the stance and swing phases as the speed of walking changes in a freely moving cockroach and (E) in one that is tethered. A is based on Pearson (1972), B and C on Pearson and Iles (1970), D on Delcomyn and Usherwood (1973) and E on Delcomyn (1973).

With descending signals from the brain removed by cutting the anterior connectives, an otherwise intact cockroach lying on its back can produce rhythmic leg movements that last some seconds (Fig. 8.9). Are these movements an expression of the motor pattern normally used during walking? In these movements, the femoral levator and depressor muscles are alternately activated by rhythmic bursts of spikes, but these typically last for only a few cycles (Pearson and Iles, 1970; Pearson, 1972). With sensory information from all the legs eliminated by cutting nerves, stimulation of a cercus still produces rhythmic bursts of spikes in levator and depressor motor neurons recorded from the stumps of the nerves to one leg. Sometimes, the levator bursts can occur without the depressors, but not the other way around. In both experimental situations, the duration of the levator bursts varies less with cycle period than that of the depressor bursts, and the bursts in both are separated by clear delays. These features are thus similar to the actions of these muscles and motor neurons during normal walking, when the femur is levated during the relatively constant swing (protraction) phase and depressed during the more variable stance (retraction) phase (Delcomyn, 1971a). Finally, the cycle period can be similar to that in slow walking, especially when sensory feedback from the legs is present. There are, however, differences from the activity seen in normal walking; the number of spikes produced by individual motor neurons at each cycle may not be the same, slow motor neurons may be active throughout the whole cycle, and the durations of the bursts may not be directly related to cycle period. It is not surprising, therefore, that the pattern is variously interpreted, but it seems prudent to await more information before concluding that it represents walking.

8.4.1.2. *Motor patterns in stick insects*

In stick insects whose body weight is supported experimentally, a walking rhythm persists in the leg muscles and nerves as long as one leg is intact and can move and send its sensory signals to the central nervous system, but when all sensory information is removed, no walking pattern is produced. A denervated preparation can, however, produce motor rhythms with two patterns. First, touching the abdomen can sometimes elicit a rhythm in the motor neurons innervating the coxal protractor and retractor muscles, but many features of this rhythm are different to normal walking. Many of the motor neurons have an incorrect pattern of activity, to the extent that some spike tonically instead of in bursts, and the phase relationship between the activity to the middle and hind legs is incorrect. Second, a rhythm with a period of 1–5 Hz can be produced that would result in all the legs on one side of the body moving together and out of phase with the legs on the opposite side. Neither rhythm is thought to represent walking, the first being an expression of struggling movements and the second rocking movements which are usually performed by a quiescent animal stimulated by a variety of stimuli (Pflüger, 1977).

8.4.1.3. *Rhythms in an isolated ganglion*

The elimination of sensory effects on potential walking motor rhythms can be taken still further by completely isolating thoracic ganglia in a dish and then treating them with particular drugs. If a thoracic ganglion (especially the metathoracic ganglion) of a locust isolated in this way is treated with the cholinergic (muscarinic) agonist pilocarpine, then a rhythmic motor pattern is produced with a frequency of up to 0.2 Hz that is dependent on the applied concentration of the drug (Fig. 8.10) (Ryckebusch and Laurent, 1993). The mechanism by which the rhythm is released is unknown, but muscarinic receptors can affect the release of transmitter at cholinergic synapses and can alter postsynaptic excitability and regulate some inward currents (*see Chapter 5*).

Motor output to one leg The rhythm in the motor discharge involves all the motor neurons that normally move the joints of a leg and consists of two phases that are at least suggestive of the swing and stance phases of walking; a levator phase, which occupies only 20% of the period and whose duration is independent of the cycle period, and a depressor phase, whose duration varies with cycle period. The motor

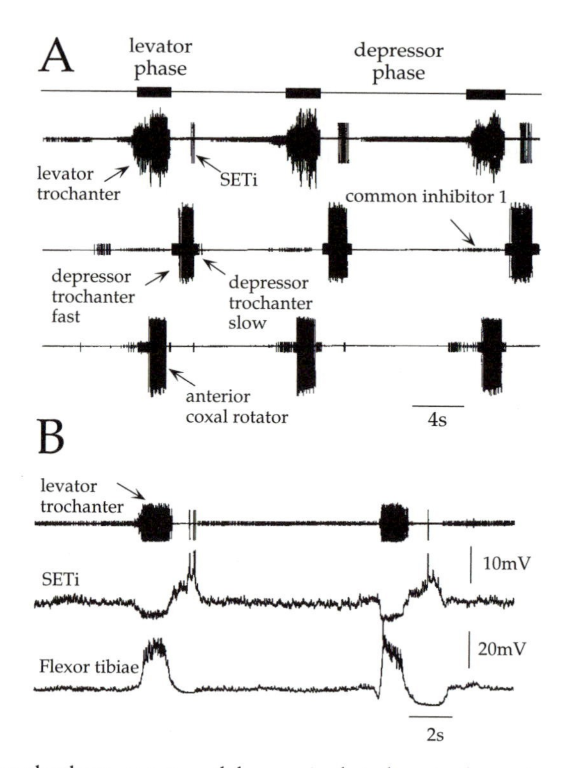

Fig. 8.10 Motor rhythms expressed by an isolated metathoracic ganglion of the locust when treated with the cholinergic (muscarinic) agonist pilocarpine. A. Extracellular recordings are made from particular nerves on one side of the ganglion in which the spikes of known motor neurons can be recognised. The rhythm is characterised by two phases defined by the action of trochanteral levator and depressor motor neurons. B. An intracellular recording from the slow extensor tibiae motor neuron (SETi) and from a flexor tibiae motor neuron reveal the synaptic drive underlying the rhythm. Based on Ryckebusch and Laurent (1993).

neurons to muscles that move the different joints are coupled in the rhythm, so that a constant phase relationship between them is maintained and could be imagined to produce a coordinated movement of the leg were it to be present. For example, trochanteral levator and flexor tibiae motor neurons are excited during the levator phase, and trochanteral depressor, extensor tibiae and depressor tarsi motor neurons are excited during the depressor phase. The phase relationships between the activity of particular groups of motor neurons can be quite stable, even when the frequency of the rhythm changes. In stick insects, pilocarpine also releases rhythmic alternating bursts of spikes in motor neurons to the leg muscles, but here the pattern appears to be further subdivided, although again there is no strict coupling between the motor output to the legs on opposite sides of the same segment. A slow rhythmic output in levator and depressor coxae motor neurons may be repeated in other leg motor neurons, but these can also show several weaker bursts of spikes during one cycle of the slower rhythm (Büschges *et al.*, 1995). This may give the appearance that the bursts of spikes in antagonists of one joint of a leg are not coupled to those innervating other joints of the same leg, so that the different joints would move with different rhythms. These observations suggest that the interneurons that generate coordinated movements of a leg are not rhythmically active and indicate that the effects of the pilocarpine may be limited to a few premotor interneurons and the motor neurons themselves. Inevitably, they also raise doubts about whether a complex pattern generator is really being allowed to be expressed by the application of this drug.

Motor output to the two legs of a segment The rhythm emerging from the two sides of a locust metathoracic ganglion to the two legs occurs at different frequencies, but with a strong probability that spikes in the trochanteral levators on one side accompany spikes in trochanteral depressors on the other.

Motor output to all the legs If all three thoracic ganglia remain linked together when they are isolated from the rest of the nervous system each can produce a rhythm but they rarely do so at the same time and more usually their output is irregular and even arhythmic (Ryckebusch and Laurent, 1994). When they do produce rhythms, they are of different frequencies in each ganglion and correlations between the bursts of spikes in particular sets of motor neurons are not maintained with time either in one locust, or in different locusts. There are tendencies, however, for spikes in trochanteral levator motor neurons on one side of the mesothoracic ganglion to follow those in levators on the same side of both the pro- and metathoracic ganglia, and to occur at the same time as increases in the frequency of trochanteral depressor spikes. These weak linkages between the ganglia are interpreted to indicate the involvement of coordinating pathways activated by pilocarpine, but give no information about the nature of the neurons involved and the connections that they might make.

It seems strange that pilocarpine induces just one rhythm that is similar to that used in walking and not any of the other rhythms, such as breathing and flying, that can be generated by the thoracic nervous system. The rhythm is, however, coupled to the ventilatory rhythm in such a way that spikes in trochanteral levator motor neurons do not occur during inspiration. Perhaps the fact that only one rhythm is produced is telling us something that we do not yet understand about the organisation of the neurons and the circuits responsible for the different patterns. Alternatively, the pilocarpine-induced rhythms may be more complex than is so far apparent and have little relationship to the motor patterns that are normally expressed.

8.4.1.4. *Fictive and real locomotion*

These experiments indicate that the central nervous system can, without receiving feedback from the consequences of its actions, generate a pattern of motor spikes to muscles that seem appropriate for moving the legs in walking. Do the rhythmic motor bursts recorded from an insect lying on its back, or otherwise prepared to gain access to its neurons, and from isolated ganglia, represent walking or some other motor pattern? As with all fictive motor patterns, it is necessary to be cautious in concluding that a pattern is an expression of a particular behaviour, and essential to compare the fictive with real patterns to determine the similarities and differences. The rhythmic patterns recorded from leg motor neurons in insects unable to move properly could represent walking, righting responses (when the animal tries to turn itself the right way up), grooming, scratching, a sequence of struggling movements or, in stick insects, rocking movements. Pearson (1972) clearly believed that the patterns he recorded from cockroaches represented walking, but often the bursting activity is of short duration and the pattern variable so that both Reingold and Camhi (1977) and Zill (1986) have suggested that it represents struggling and attempts to right the body, and not walking. The data are inadequate to support either contention, but it should be remembered that some of the original patterns said to represent walking involved the movements of a single leg in response to stimulation of a cercus. It is equally clear that Bässler (1993) believes that the rhythms recorded from the isolated central nervous system of stick insects represent struggling movements and not walking. Despite this, models of how walking is produced still rely on controllers for each leg in which some element of a rhythmic central drive appears essential.

Is the rhythm recorded from isolated locust ganglia related to walking? The phase relationships between the motor neurons are, in general, appropriate for the levator phase to correspond with the swing (protraction) phase, and the depressor phase with the stance (retraction) phase. The coupling between the two legs of one segment is nevertheless stronger than that seen in recordings from levators in more intact animals (Pearson and Iles, 1970). There is, however, a dearth of quantitative data on the actions of the muscles and motor neurons during normal walking with which the now more detailed information from the isolated ganglia can be compared. The much lower frequency and greater variability of the isolated rhythm could be explained by the absence of signals from sensory neurons that either raise the level of excitability in a tonic fashion, or initiate the next phase of a cycle and correct the irregularities. Given, however, that the legs are capable of many sorts of rhythmic movements, and that they can be moved in different patterns of coordination, caution is necessary in deciding what the fictive rhythms might represent. Where longer sequences of rhythmic activity are recorded, from so-called 'semi-intact preparations' such as those developed for crustacea (Chrachri and Clarac, 1990; Elson *et al.*, 1992), and for rhythms in many other animals (from leeches to lampreys), the phase relationships between the motor neurons seem appropriate for normal locomotion, even though the repetition rate of the bursts can often be lower than seen in real behaviour. In these examples, there seems little doubt that the central nervous system can produce the framework for the motor patterns. Such preparations now need to be developed for insect walking.

While all these experiments leave the question of whether the observed rhythms represent walking largely unresolved, they do start to emphasise the importance of sensory feedback, particularly at low rates of stepping or when walking in complex environments. It is an apparent balance between these two mechanisms that produces successful walking, and in this may lie some of the differences between the results from cockroaches and stick insects. When a cockroach is walking rapidly, the central motor commands seem to dominate because there is little time for the sensory signals generated by each cycle of movement to have much immediate effect, whereas in a stick insect walking slowly, sensory influences appear to dominate. If the central elements are thought of as a **predictor** (*see Chapter 11*), then during running the leg movements will be more predictable, thus explaining the dominance of the centrally generated pattern, whereas during irregular walking they will be less predictable and thus the sensory feedback will exert considerably more influence. This scheme still leaves a role for essential feedback of the unpredictable during rapid movements and for the central components to form the framework of slow movements. What is clear is that sensory control of each cycle of leg movements is essential when walking in an unpredictable environment, to the extent that argument about the 'central' or 'peripheral' nature of the rhythm becomes sterile.

8.5. INTERNEURONS AND THE GENERATION OF WALKING MOTOR PATTERNS

Little is known of the neuronal network interactions that might be responsible for generating a walking motor pattern. This contrasts with the considerable knowledge of the way that interneurons control the motor output to the leg muscles, process the sensory signals from proprioceptors and exteroceptors associated with the legs, and organise the many local reflexes in which all of these neurons participate (*see Chapter 7*). The analysis of the networks has proceeded by seeking to solve specific problems about the processing that they perform during leg movements, rather than how the interneurons might be connected to generate a repetitive pattern. This strategy seems eminently sensible in view of the complexity of the walking movements and the large number of neurons that must be involved in its generation and control. Future analyses of the networks will undoubtedly take advantage of the development of preparations of isolated ganglia that generate rhythmic activity in leg motor neurons, and preparations that can move their legs rhythmically in walking sequences.

8.5.1. Nonspiking local interneurons

Some nonspiking interneurons that can lead to excitation of cockroach flexor (levator) motor neurons are rhythmically active during episodes of rhythmic motor activity (Pearson and Fourtner, 1975). They are inferred to be part of the network generating the rhythm, as manipulation of their membrane potential at particular phases of the cycle can reset the rhythm. It would, however, be surprising if such neurons were not rhythmically active in time with the movements, given their involvement in processing proprioceptive signals from one leg and integrating them with signals from other legs and supplying a drive to the motor neurons (*see Chapter 7*). They undoubtedly play some part in shaping the motor activity, along with many other nonspiking and spiking interneurons yet to be identified. It seems unlikely that a single nonspiking interneuron is responsible for controlling the bursts of spikes in the levator motor neurons as was originally proposed (Pearson and Fourtner, 1975). All subsequent investigations point to the concerted action of many local interneurons in generating a particular motor output. In both stick insects and locusts, recordings can be made from nonspiking interneurons during walking, but as the legs do not support the weight of the body the sensory signals generated by the leg movements will not be the same as during normal walking. The membrane potential of these interneurons is modulated in time with that of the leg motor neurons which they either excite or inhibit (Büschges *et al.*, 1994). In some, the peaks of the rhythmic depolarisations occur at the transition from the stance to swing phases, suggesting that they may be involved in controlling the switch between these phases. Alternatively, their membrane potential may simply be following the drive to one set of neurons or the pattern of sensory inflow from the movements. Experimentally altering the membrane potential of individual interneurons can cause changes in the motor activity, indicating the importance of particular interneurons despite the many parallel pathways that must be operating in controlling these complex movements.

8.5.2. Spiking local interneurons

Spiking local interneurons receive direct inputs from arrays of sensory neurons on one leg so that some process signals only from exteroceptors, others only from proprioceptors, and others from both types of receptors (*see Chapter 7*). The majority of these different types of interneurons show patterns of depolarising and hyperpolarising synaptic inputs that are linked to the slow motor rhythms expressed in isolated ganglia (Wolf and Laurent, 1994). These inputs can alter the response of an interneuron to its sensory signals, depending on the phase of the cycle of the rhythm. If the rhythm does represent some form of locomotion, it implies that sensory processing will be directly related to the movements. This is confirmed by recordings during walking that show a rhythmic synaptic input to these interneurons in time with the step cycle of a leg. Most of this input comes from the rhythmic activation of the sensory neurons, but part is still present when the sensory input is abolished. The interneurons are thus driven in time with the locomotory rhythm by a central input that has the potential to modify their processing of the sensory signals in keeping with the phase of the leg movements. Thus, the sensory inputs delivered to the central nervous system are modified at two different stages in the central processing; first, by presynaptic inputs to the terminals of the sensory neurons themselves (*see Chapter* 7), and second by a central drive to the interneurons that play a prominent early role in the sensory processing.

8.5.3. Intersegmental interneurons

Many interneurons are known that receive sensory inputs from one leg and have axons that project to the ganglia controlling the movements of the other legs (Fig. 8.11). Some are even known to make connections with the nonspiking interneurons that control the output of the motor neurons in an adjacent ganglion and that can change the gain of the local reflexes expressed in the distant ganglion (Laurent and Burrows, 1989a,b) (*see Chapter* 7). Nothing, however, is known of what they do during walking and thus whether they are involved in generating or merely coordinating the motor patterns of different legs. Different sets of interneurons carry signals anteriorly and posteriorly, either linking several ganglia, or having their output synapses restricted to the adjacent ganglion. Some spikes can be recorded from the connectives of a cockroach that are linked to the motor bursts in levator muscles, but the neurons responsible are not identified. Moreover, no neurons that could ensure the correct coordination of the limbs on opposite sides of the same segment have yet been found. Many of the intersegmental interneurons have extensive neurites on one side of the ganglion and axonal branches on the other, so that they could potentially play a role in bilateral coordination. The ability of an isolated ganglion to produce rhythms of different frequencies to the legs on either side, and the apparent weakness of reflexes across the body, has, however, led to the assumption that much of the bilateral coordination results from mechanical coupling and the sensory inflow that consequently results. Central mechanisms must exist, however, and are implicated from the fact that rhythms from the two sides of an isolated ganglion are coupled, even if their frequency is not the same. Similarly, stick insects that are suspended and allowed to walk on mercury so that the mechanical coupling between the legs is minimal nevertheless maintain a coordinated gait, although it is not as precise as in the normal insect (Graham and Cruse, 1981).

8.6. COORDINATING THE MOTOR OUTPUT TO DIFFERENT LEGS

Experiments to unravel the mechanisms responsible for coordinating the movements of the legs have relied largely on defining the sequence of movements of the legs and the actions of some of their muscles, and then on disrupting the sensory feedback. The latter is achieved by immobilising legs, by using animals that have lost parts or

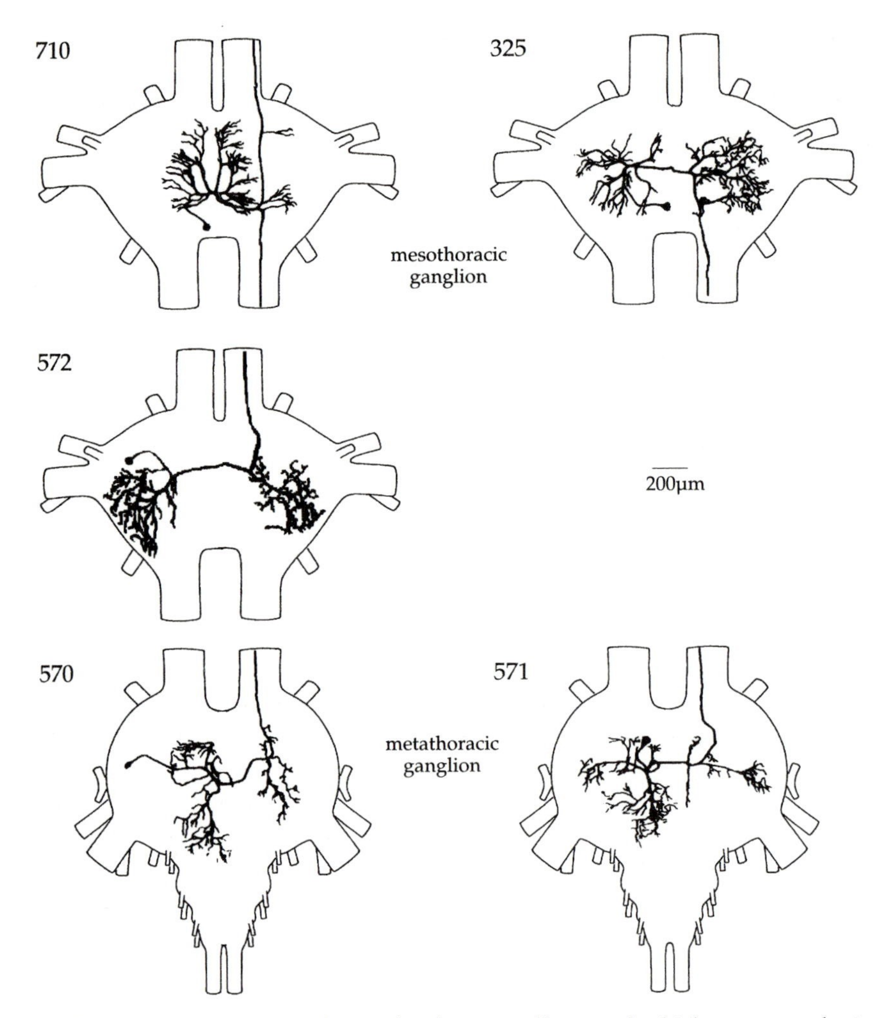

Fig. 8.11 Thoracic interneurons that spike during walking and which appear to be involved with generating and controlling the motor pattern. Based on Ramirez and Pearson (1988).

even the whole of a leg (many insects will autotomise a leg at a preformed breakage plane to escape from a predator), by using prostheses, and by manipulating the signalling of particular sense organs. The results are then modelled and predictions are made about possible neuronal mechanisms, but few recordings have been made from the neurons to test the validity of the predictions.

8.6.1. Gaits at different stepping frequencies

Changes in the stepping patterns are possible when the following patterns of coordination between the muscles of the different legs are altered. First, the durations of both the stance and the swing phases change with the speed of locomotion, but the stance phase changes the most. In locusts, the stance phase can be considerably longer than the swing phase, but it is always the stance phase that changes with speed when walking horizontally, vertically or upside-down. As a consequence, the ratio between the two phases increases linearly with the frequency of leg movements and the gait remains constant for all but the very slowest speeds of walking. In cockroaches walking slowly, the stance phase can be three times longer than the swing phase, but at high speeds it is shorter (Fig. 8.9) (Delcomyn, 1971a). These changes in the leg movements are paralleled by decreases in the duration of the bursts in the femoral levator motor neurons during the swing phase, but they decrease less than the duration of the femoral depressor bursts during the stance phase. In some experimental examples of tethered walking, where the load of the body is not carried by the legs, the duration of the swing phase stays more constant as walking speed varies. Second, the bursts of spikes in the levator trochanter of, for example, a middle leg, always occur half way through the stepping cycle (phase 0.5) of the ipsilateral hind leg. As speed increases, the delay between the end of the levator bursts of the hind leg and the start of them in the middle leg gets shorter (Pearson and Iles, 1973). This coordination is disturbed when sensory feedback from the legs is removed, but the rudiments remain so that generally the levator bursts to the middle leg still follow those to the hind leg, but without the normal delays. This suggests that neurons generating the output to the different legs are linked by interneurons carrying information between the respective segmental ganglia. Bursts of spikes coupled to the rhythm can be recorded from the connectives linking the ganglia and may be from such postulated coordinating interneurons, although they have not been characterised further. Behavioural observations indicate that signals must pass both from a posterior leg to a more anterior one and also in the opposite direction, but not all models have incorporated this information (see Fig. 8.13).

8.6.2. Gaits in insects that have lost legs

The gait stays constant when distal parts of a limb are lost, provided that the remaining proximal part can touch the ground and support the weight of the body. A prosthesis can replace the missing part of a middle leg and result in an almost normal gait in locusts (ten Cate, 1936). If the whole of a leg is lost, so that there is no longer any possible support, then the alternating tripod gait is replaced by one that varies with the speed of walking and which provides stable support for the body (Fig. 8.12) (Delcomyn, 1971b). The form of the new gait depends on which leg (or legs) is lost, and differs from that in an insect where the legs are all intact but one is prevented from moving. All these observations on disabled insects point to the importance of sensory feedback in coordinating the movements of the different legs. The changes result, in large part, from the altered loading on the remaining legs, rather than from the loss of sensory signals from the missing legs, and thus indicate the influence of sensory signals monitoring load. If the loading is maintained constant by supporting the weight of the body, the gait is still altered,

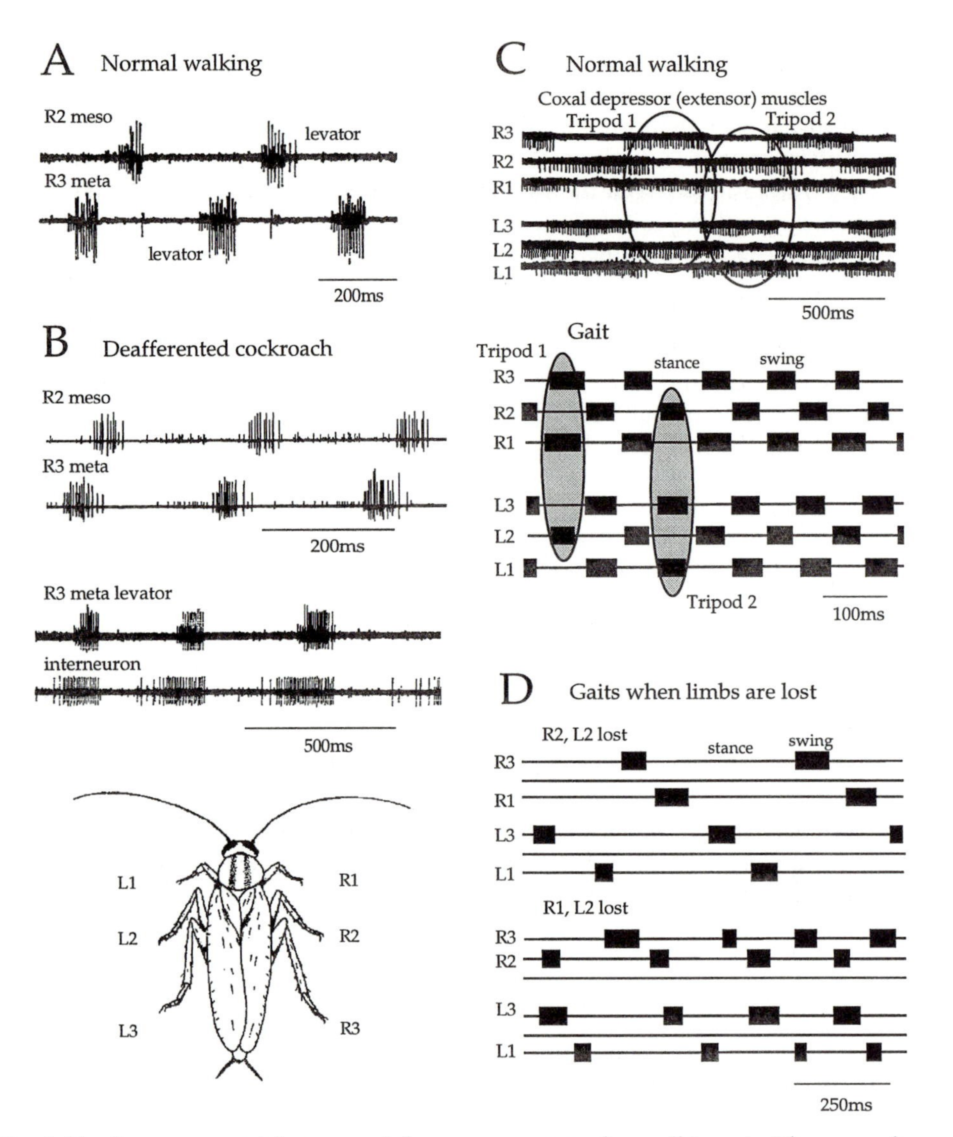

Fig. 8.12 Intersegmental features of the motor pattern for walking. A. The normal motor pattern expressed in coxal levator muscles of ipsilateral middle and hind legs of a dissected cockroach. B. The same muscles after removal of all sensory receptors in the legs. The lower panel shows that the spikes in the metathoracic levator motor neurons are preceded by spikes in an interneuron recorded in the connectives between the meso- and metathoracic ganglia. C. The alternating tripod gait during walking in an unrestrained cockroach. The upper panel shows the action of coxal depressor muscles in each of the legs, and the lower panel the times when the tarsi are in contact with the ground (stance phase, represented as the thin horizontal line) and when being moved forwards and off the ground (swing phase, represented as thick bars). D. Alteration of the gait when particular legs are lost. A and B are based on Pearson and Iles (1973), C (upper panel) on Delcomyn (1993), C (lower panel) on Delcomyn (1971a), D on Delcomyn (1971b).

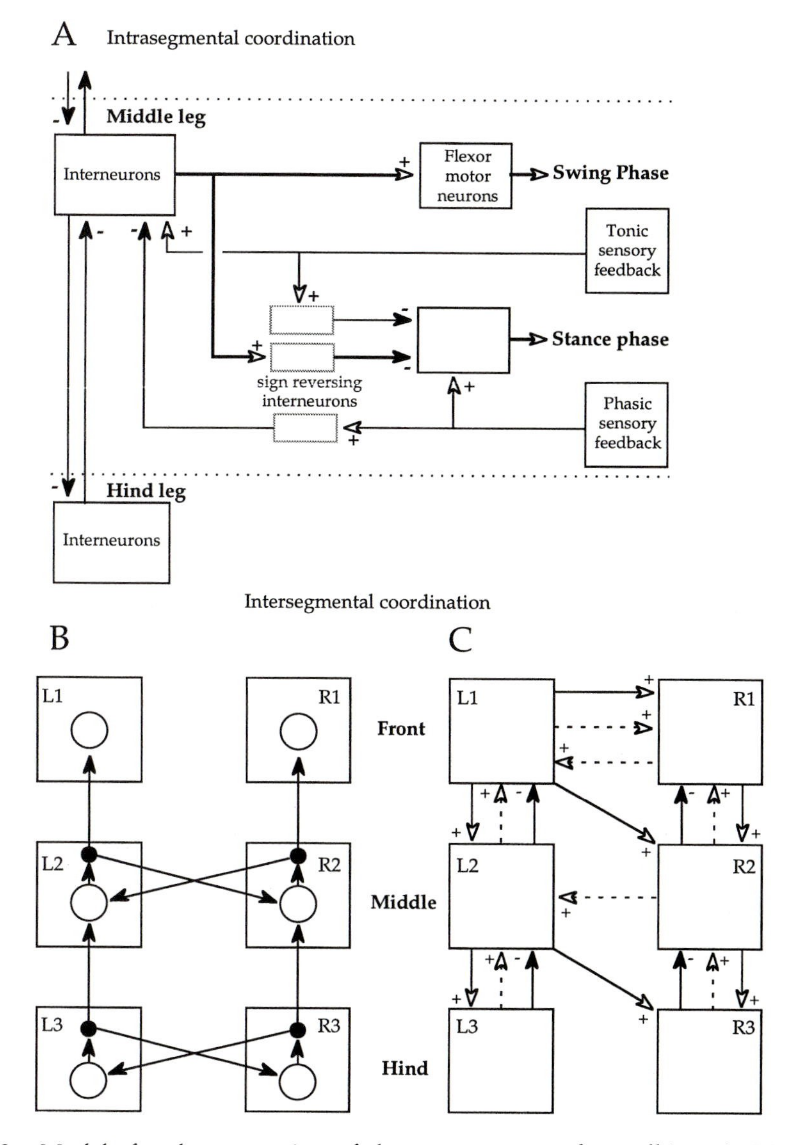

Fig. 8.13 Models for the generation of the motor pattern for walking. A. Model of the asymmetric pattern generator controlling the movements of an individual leg. The interneurons and the sensory neurons are represented by rectangles as they are not identified individually, and are shown converging onto flexor and extensor motor pools that are active in the swing and stance phases of walking, respectively. B. Graham's model for the production of coordinated gaits by the three pairs of legs. The open circles represent relaxation oscillators in each half of a thoracic ganglion (box) and the filled circles represent delay oscillators. C. Cruse's model for intersegmental coordination. The arrows indicate the direction of flow of both excitatory (open arrowheads) and inhibitory (filled arrowheads) signals between the controllers for each leg, whose components are not defined. A is based on Pearson and Iles (1970) and Pearson (1972), B on Graham (1977), and C on Cruse (1980a,b, 1990).

indicating this time that the loss of the sensory signals from the missing leg can influence the motor output to at least the more posterior legs (Delcomyn, 1991a). Moreover, recordings from the stump of a leg that has been lost show that its muscles can sometimes produce more than one burst of spikes during one cycle of movement of the intact legs so that the stump moves more frequently. Stick insects will walk on mercury, the viscosity of which is so low that the mechanical coupling between all the legs is abolished, but the surface tension of which is high enough to prevent the tarsi sinking (Graham and Cruse, 1981). Under such conditions, the stick insect steps at a higher frequency in a coordinated gait but the pauses that normally accompany the switch from one phase of a leg movement to another are absent.

8.7. MODELS FOR GENERATING THE WALKING PATTERN

8.7.1. Generating the motor output to one leg

The asymmetry in the motor pattern to one leg, in terms of the relative durations of swing and stance phases, that is used in normal walking and that is also generated by an isolated nervous system leads to the suggestion that the flexors (active during the swing phase) are driven directly and rhythmically by a set of interneurons which, at the same time, indirectly inhibit the extensors (active during the stance phase). This has led to a model for cockroaches (Pearson and Iles, 1970; Pearson, 1972) (Fig. 8.13) which has dominated thinking about the control of insect locomotion, but for which there is little evidence about the neural components, or in favour of the pathways that are included rather than those that are excluded. The network of interneurons envisaged to generate this rhythmic pattern receives different levels of general excitation from intersegmental interneurons that set the period of the walking rhythm and the duration of the motor bursts in the flexors. The output of the network will, in turn, be converted into spike frequencies in different sets of motor neurons that are a function of the stepping rate. The excitatory level of the network is also thought to be set by tonically active sensory neurons which also excite the extensors. Specific phasic sensory feedback influences certain extensor motor neurons and modifies the action of the interneurons generating the rhythm. The model fits the available data for the movements of one leg, but it suffers from the conceptual separation of interneurons generating the rhythm from those performing all the other essential integrative tasks. It emphasises certain sensory pathways, but excludes those that must exist, for example, to the levator motor neurons. As yet, it has not had great predictive value in helping to elucidate the way the rhythm is controlled and only a few of the pathways or elements in this model have been found in reality. In cockroaches, for example, campaniform sensilla on the trochanter are thought to excite the slow extensor motor neuron of the femur and inhibit the bursting output of the motor neurons through presumed connections to interneurons, but the evidence for this supposition is weak.

8.7.2. Generating the coordinated motor output to all the legs

To explain the movements of all the legs it is assumed that the individual neuronal controllers of each leg, modelled as relaxation oscillators, are coupled to each other. The different nonhierarchical ways in which this coupling could be achieved have led to four models of different complexities which are able to explain particular features of the walking pattern, but none of which can explain all the variations that are seen. The first two models rely on coupling signals running from the rear forwards, the third on the signals running posteriorly, and the fourth on information flowing in both directions.

First, Wendler's model (1978a) has the two legs of one segment influencing each other to the same extent, and the hind leg of one side more strongly influencing the middle leg than it does the front leg on the same side of the body. Walking speed is governed by a tonic input to all the oscillators.

Second, Graham's model (1977) gives the individual oscillators variable thresholds which, when exceeded, trigger the swing phase and are then reset to zero (Fig. 8.13). The natural frequency of the oscillators also decreases from the front to the rear, which would mean that on their own the front legs would walk faster than the hind legs. The legs of one segment are reciprocally coupled, and on one side of the body the hind leg influences the middle leg, and the middle leg influences the front leg. This arrangement means that, when a leg is in the swing phase, it prevents the leg in front of it and the leg on the opposite side of the body from entering the swing phase for a period that depends on the frequency of stepping. The posterior legs thus constrain the movements of the more anterior legs so that all move at the same stepping frequency. Walking speed is again governed by a tonic input to all the oscillators and the delays between the movements of the legs are determined by delay oscillators, each driven by the relaxation oscillator of each leg. A delay oscillator is turned on when the relaxation oscillator is reset and while it is running raises the threshold of the relaxation oscillator in the next anterior segment. Changing the level of tonic input to these oscillators allows either a tripod or tetrapod gait to be expressed.

Third, Cruse's model (1979a) has information flowing posteriorly to legs on both the same and on the opposite sides of the body. The output of the oscillator of an anterior leg excites the next posterior one so that it reaches its threshold earlier.

Fourth, the more complex model of Cruse (1980a,b) has information flowing anteriorly and posteriorly both in and out of phase with the bursts produced by one leg controller (Fig. 8.13). The controller itself consists of elements that can produce the stance and the swing phases, with sensory signals determining transitions, and the controllers for each leg have the same natural frequency. Three signals influence the transition from stance to swing phase (Cruse, 1990). First, an inhibitory ascending signal hinders an anterior leg from starting its swing phase while the adjacent posterior leg is performing its swing phase. This influence can prolong the stance phase of the anterior leg by shifting the most posterior position it reaches. Second, an ascending signal can elicit a swing phase in an anterior leg when the posterior leg starts its stance phase. Third, a descending signal causes the posterior leg to start its swing phase before the anterior leg has begun its swing phase. The further the anterior leg moves backwards in its stance phase, the earlier is the start of the swing phase of the posterior leg.

Such ideas have now been incorporated into a neural model that can generate walking movements in six legs, each with three joints, and that is able to respond to disturbances in sensory inflow (Cruse *et al.*, 1995). It points a clear way to the development of six-legged robots with their inherent stability and to the greater understanding of the real neuronal networks that will be forced by the initial inadequacies of these models.

8.7.2.1. *Mathematical models*

Mathematical models of the networks underlying the generation of walking can explain most of the gaits that an insect performs (Collins and Stewart, 1993, 1994). Such models envisage identical oscillators for each limb, each coupled to its two nearest neighbours around a hexagonal ring. The different oscillators could be grouped in three simple ways.

If alternate oscillators are clumped together to form two interlaced sets of three, then two patterns emerge. First, all oscillators are synchronised so that all legs move together, as when the locust jumps. Second, one set of oscillators moves together and the other set also moves together but half a cycle out of phase with the first. This pattern gives rise to the alternating tripod.

If the six oscillators are clumped into three sets of two then two prominent patterns emerge. First, the pairs move together and successive pairs are one third of a cycle out of phase, so producing a rotating wave of activity around the ring. Locusts may use this gait when walking slowly on a flat surface. Second, one pair of oscillators may cycle at half the rate of the other four with the result that one pair of legs takes one step for every two of the other legs. This gait is often adopted by the hind legs of a locust.

If the oscillators are not clumped, then successive oscillators around the ring are one sixth of a cycle out of phase with each other. This pattern results in a metachronal wave gait as used by cockroaches when walking slowly.

Gaits with different symmetries can, according to this model, be generated by varying the coupling between the oscillators in ways that can be readily envisaged as having a neurobiological basis.

8.7.2.2. *Matching models and real neurons*

None of the models seek to incorporate real neurons or the few central pathways that have been identified, or to take account of the signals sent by the many sensory pathways. They define the patterns of coordination seen from largely behavioural experiments, and highlight the numbers of pathways and elements that still need to be identified. Despite the success of these models in simulating walking and in predicting the sorts of pathways that must now be sought, a series of such networks coupled by central pathways cannot explain the multiple bursts of spikes that are seen emerging from the central nervous system to the stumps of amputated legs, or when the apodeme of the femoral chordotonal organ has been reversed. This implies a strong sensory influence on the central mechanisms.

8.8. SENSORY EFFECTS ON WALKING

If the motor patterns generated in the absence of any sensory signals represent walking, then clearly the central nervous system can, by itself, set the timing of the swing and stance phases and determine when each is initiated. Sensory signals can, however, alter all the features of the pattern and can determine the time at which a particular leg begins its phase of the movement. It takes time, however, for the sensory signals to be generated, conducted to the central nervous system, processed, to influence the motor output, and then for the motor commands to produce a movement of a joint. This processing time becomes a significant proportion of the step period during fast walking, so that the potential effects of a sensory signal must depend on the speed of walking. Thus, in a stick insect stepping at only 1–2 Hz there is ample time for the sensory signals to influence the motor output at each step, but in a cockroach stepping at 24 Hz influences may have to be spread over several steps.

Most of the detailed information about the effectiveness of sensory signals during walking again comes from cockroaches and stick insects, because all that is known of locust walking is that it becomes uncoordinated if the signals provided by the femoral chordotonal organ or tarsal receptors are disturbed or eliminated (Usherwood *et al.*, 1968; Usherwood and Runion, 1970). In all these insects, however, there still remains the difficulty of making sense of the enormous barrage of sensory signals during walking, so that attempts to assign a function to one particular receptor in these circumstances are almost doomed to failure because there are simply too many pathways acting in parallel. There have, nevertheless, been three experimental approaches to circumvent this problem.

First, to analyse the effects of the different receptors individually in animals that are not walking, and which are by necessity often dissected and in highly unnatural postures. This strategy allows the response properties of the sensory neurons to be defined accurately and their connections with the central neurons determined, so that the pathways integrating their signals can be analysed. It may also allow the motor effects of these controlled and selected inputs to be defined. The problem that arises is in predicting whether these effects remain the same during normal behaviour, because the processing in one pathway is likely to be highly dependent on the other pathways that are active at the same time; the actions of the neurons are dependent on the prevailing behavioural context. Moreover, the effectiveness of sensory signals is altered actively through presynaptic inhibition of the sensory terminals in the central nervous system at particular times during locomotion (Wolf and Burrows, 1995).

Second, to ablate a particular receptor and look for changes in the movements and phasing of the legs. This approach highlights the dilemma of the previous approach, because such ablations frequently fail to produce a measurable effect on the behaviour, even though connections with, or effects on, particular motor neurons may be known. What such experiments reveal is the parallel nature of the processing and the flexible response of the nervous system to the loss of some of its normal inputs.

Third, to apply an unusual stimulus to a sense organ, or a stimulus at an inappropriate time, and seek effects on the movements. While these can produce some bizarre results, it is not always clear that they point to the normal effects of these sense organs during walking.

The gulf between these different experimental approaches cannot easily be bridged and, at present, any understanding only emerges from the patient analysis of pathways combined with behavioural and physiological observations of the real behaviour. It has always to be recognised that the match between the different approaches will not always be perfect.

8.8.1. Phase-dependence of sensory action

Sensory effects on leg movements observed in highly dissected preparations may not be expressed in the same way when the locust is walking. Moreover, sensory effects that are seen when the locust is standing still may also not be expressed when the locust is walking. When the body is supported by all six legs, as when standing, a movement of any one leg will not disrupt the posture, but when only three legs are in contact with the ground during the tripod gait used in walking, an unexpected movement of one of these legs could have serious consequences. It is thus imperative to examine the responses evoked by the sensory neurons during walking and even to relate their actions to the particular phase of the step cycle of an individual leg.

Consider the signals generated by receptors on a tarsus when it is placed on the ground and supports the weight of the body (Wolf, 1992). When the locust is standing still, electrical stimulation of the tarsal receptors causes the leg to be lifted from the ground by a series of coordinated movements of its various joints in an apparent avoidance response. The effect of the same stimulus during walking depends on the phase of the step cycle the leg that is stimulated is in. During the stance phase, when the tarsus is on the ground, the stimuli have no effect, but when applied towards the end of the swing phase they cause an additional small swing (protraction) movement, as if the leg were moving to avoid an obstacle in its path, or initiate an early swing phase movement before the leg has reached its most posterior position. These changes are expressed as excitatory synaptic inputs to the motor neurons active during the swing phase and inhibitory inputs to those active during the stance phase.

8.8.2. The switch from one phase of the movement to the next

A limb normally moves through a particular arc delimited by the posterior extreme position, at which the stance phase of the movement switches to the swing phase, and by the anterior extreme position, at which the swing phase switches to the stance phase. Are there sense organs that signal when these positions have been reached and which can therefore determine the switch to the next phase of movement? In stick insects, positional information is important in signalling the end-point of one phase of a movement, whereas velocity information determines the force, and hence speed, of the leg movements (Dean and Cruse, 1986).

8.8.2.1. *Stance to swing phase*

If, during walking, a locust lifted a leg that was supporting the weight of its body before the other legs adopted their correct positions then it is obvious that its balance would be disrupted. Both positional and load signals can determine when the stance will give way to the swing, provided that the leg is in the correct phase of its cycle, and will thus ensure the correct maintenance of balance during locomotion.

Positional signals A leg of a stick insect can be persuaded to grasp a rod while the other legs continue to walk. The rod is then moved backwards until the leg reaches its posterior extreme position, at which point it lets go and moves forwards in a swing phase movement (Wendler, 1964). The posterior position of the leg is signalled by the hair rows and one of the hair plates on the coxa, for when these are immobilised the posterior extreme position is now more posterior, and the leg may have to be pulled from its attachment to the ground by the action of the other legs. This information is thus akin to that provided by the hip joint receptors of mammals in being able to delay the onset of the next swing phase (Grillner and Rossignol, 1978).

Force signals When walking up a slope or working against an increased load, force may need to be exerted for a longer period and hence the duration of the stance phase must be increased. The tarsi are then often extended further forward than normal during the swing phase, so that they remain in contact with the ground for a longer period during the stance phase, and the gait may also change (Spirito and Mushrush, 1979). A stick insect will walk with the legs of one side on one tread-wheel and those on the other side on another tread-wheel so that the load on the two sets of legs can be varied independently (Foth and Graham, 1983). Increasing the load on both sides causes the tarsi to be placed more anteriorly and to be picked up more posteriorly. Reducing the loading on the legs of one side causes them to produce double steps, with alternate steps in the correct timing relative to the more heavily loaded legs. The increased load is also met by an increased force output from the muscles that may be monitored by the campaniform sensilla on the proximal part of the leg. An increased motor output to one leg leads to an increased output to neighbouring legs (Cruse, 1985).

During the stance phase, the distribution of the loading of a leg will change once it has passed a certain point as it moves posteriorly relative to the body. This means that changing the load could provide a signal that the leg no longer has to support the body and can therefore be moved forward in the next swing phase. If a small clamp is applied to the trochanter of a stick insect so that sensory signals, probably from campaniform sensilla, signal that loading of the leg is continuing then the stance phase is prolonged, the next swing phase is delayed, and the tarsus is dragged along the ground by the movements of the other legs (Bässler, 1977). Velocity and acceleration information from receptors at the joints, probably supplied by chordotonal organs, will also be necessary to control the muscular force.

8.8.2.2. *Swing to stance phase*

Sensory signals can also trigger the switch from the swing to the stance phase, because the swing phase can be prolonged by altering proprioceptive signals from the femoro-tibial joint (Graham and Bässler, 1981). The apodeme of the chordotonal organ in a stick insect can be switched from its insertion close to the extensor tibiae muscle to a position near the insertion of the flexor tibiae muscle, so that it is now stretched by extension of the joint instead of by flexion. During slow walking the operated leg moves quite normally, but during faster walking it is suddenly extended and held up and away from the body in a 'salute' posture that can be maintained while the other legs perform several steps in a gait that compensates for the lack of contribution from the saluting leg (Fig. 8.14). Only if the tarsus contacts an object will the prolonged swing phase be ended and a stance phase produced. The explanation for this response is that the chordotonal organ would normally signal that the leg has **flexed** by a certain amount during the swing phase and that an extension would be appropriate. When the insertion of the apodeme is reversed, however, the chordotonal organ now signals that the leg has **extended** a certain amount and thus reinforces the actions of the extensor tibiae muscle; a negative feedback loop has been converted into a positive one. Continuously stretching the chordotonal apodeme also causes the leg to be held in its saluting posture.

8.8.3. Where the feet are placed on the ground

The six legs do not become entangled during walking because the tarsi of the middle and hind legs are placed relative to the secure footholds found by the front tarsi. On rough terrain or when climbing, a locust uses three strategies to place its tarsi (Pearson and Franklin, 1984). First, the front legs search for a secure footing towards the end of their swing phase, often with rhythmic up and down movements at frequencies of up to 8 Hz. Second, if a leg strikes an object when it is moving during its swing phase, it is lifted so that it avoids the object. Third, small and rapid searching movements of individual tarsi are used to locate a suitable foothold, before weight is placed onto them. The result is that the tarsi are placed at different angles relative to the body compared with their placements on flatter terrain, and the gait may no longer be the alternating tripod, but instead may become irregular, with the two legs of one segment moving together and not alternately. Such in-phase movements of the legs of one segment may also occur at the start of walking on flat surfaces. Stick insects, but not locusts, probably use additional tactile information from the antennae to guide the placement of the front tarsi. The position of the anterior tarsi then guide the placement of the more posterior tarsi (Fig. 8.14) (Cruse, 1976a, 1979b) in much the same way as quadrupeds guide the placement of their feet. A hind leg is normally set down just outside and behind the preceding footfall of the middle leg no matter how variable this placement might have been. Ablating hair

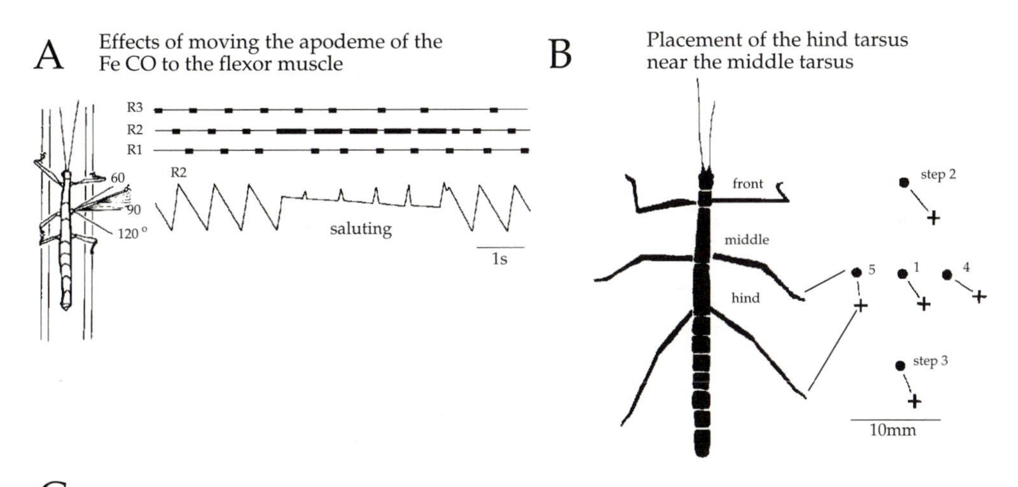

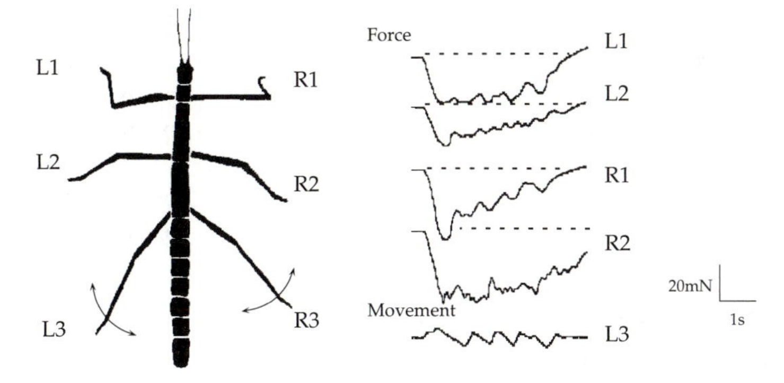

Fig. 8.14 Sensory influences on walking in a stick insect. A. The apodeme of the femoral chordotonal organ is moved from the tendon of the extensor tibiae muscle to that of the flexor muscle. Instead of executing a normal swing phase the operated leg is often raised in a 'saluting' posture. B. The placement of the tarsus of a middle leg influences the placement of the tarsus of the ipsilateral hind leg. C. Force is generated by the muscles of all the legs even when only two are allowed to walk on a tread-wheel. A is based on Graham and Bässler (1981), B on Cruse (1979b), C on Cruse and Saxler (1980).

plates results in the hind leg being placed more posteriorly, and ablating coxal hair rows results in a more anterior placement (Cruse *et al.*, 1984). This implies that proprioceptive signals about the placement of an anterior leg are available to the networks controlling a more posterior leg.

8.8.4. Avoiding an obstruction

It is inevitable that the legs also encounter obstructions as they are moved. Such encounters with small obstructions are met by an increased elevation and extension of the limb so that it steps over the object, or onto it, in responses that closely resemble the foot placing reactions of mammals (Forssberg, 1979). Stick insects reverse the direction of movement of the leg before swinging it forward again, and then repeat this sequence at 3–4 Hz until the object is surmounted. If the obstruction prevents movement completely then the response depends on the time in the step cycle that it is encountered. The earlier in a swing phase that an object is encountered, the more prolonged is the attempt to move the leg forward and the more likely is the leg to stop its stepping movements while those of other legs continue, but with appropriate adjustments. The later in a swing phase that an object is encountered, the more likely is the leg to resume its stepping with the correct timing relative to the other legs (Dean and Wendler, 1982). The timing of the leg in front is also reset and its posterior extreme position is altered (Cruse and Epstein, 1982).

The first clues that a large object such as a step or a ditch is likely to be encountered are visual, with the result that walking is stopped and the gait is adjusted either to move around the object or to surmount it by directing the front legs to a suitable foothold. If a front or middle leg steps into the ditch and has to search for a secure footing on the bank ahead, the hind leg is more likely to reach straight for the secure footing. This strategy can occasionally result in the hind leg stepping on the tarsus of the middle leg, but when this happens the hind leg is immediately lifted and placed to one side. Even if the hind leg is positioned correctly, it can be induced to reposition itself by touching the tarsus of the middle leg. This repositioning response only occurs when the hind leg is placed on the ground and not when it is in its swing phase, so that, like many other reflexes, it is thus phase dependent.

Some clues as to how the positioning of the legs might be controlled during walking are given by a local neuron in the metathoracic ganglion of stick insects which is depolarised when the hind leg moves forwards and when the middle leg moves backwards (Brunn and Dean, 1994). When depolarising current is injected into this interneuron it produces graded inhibitory effects on protractor and graded excitatory effects on retractor motor neurons, which would thus prevent the hind leg from moving far enough forward to bump into the ipsilateral middle leg. This interneuron could be coding the spacing between the movements of the two legs and therefore be part of the circuitry that ensures the correct placement of a leg at the start of the stance phase.

8.8.5. The height of the body from the ground

To maintain the correct height of the body relative to the ground when either standing or walking, the position and velocity of the six legs must be continuously adjusted. A stick insect keeps the height of its body constant relative to the ground by bending each of the body segments, even in the face of imposed loads (Wendler, 1964). Ablation of certain hair plates on the trochanter reduces the effectiveness of this response, indicating that height is controlled by a servo mechanism in which these hair plates are the sensors. When standing or walking over rough terrain each leg acts like a vertical spring that is independent of the other legs, so that mechanical coupling of all the legs to the body suffices to maintain the height of the body without the need for neuronal coupling between the different legs (Cruse *et al.*, 1989, 1993).

8.8.6. Sensory effects on gait

Strong sensory influences in determining the gait are also indicated by experiments in which a stick insect was encouraged to walk with its left legs on one side of a double tread-wheel and its right legs on the other side (Graham, 1981). Normally, the insects walk on such a wheel with a tripod gait, but the legs on the two sides can uncouple so that they walk at different frequencies. Changing the friction on one of the wheels also disrupts the coordination between the legs of the two sides. The potential for independence of action by individual legs is also demonstrated when five legs walk on a wheel and the sixth on a belt that is driven at different velocities. The leg on the belt will then step at different frequencies to the legs on the wheel (Foth and Bässler, 1985). Even more dramatic, however, are the oscillations of force that are produced by a fixed leg in phase with the movements of a leg that is able to walk (Fig. 8.14) (Cruse and Saxler, 1980). If four of the legs are fixed to force platforms, they each produce oscillations in force that are in the correct phase with each other and with the remaining two legs that produce walking movements. This suggests that the controllers for each of the fixed legs are driven by the sensory signals generated by the walking legs and not by the sensory signals from the small movements and changes in force generated by the fixed legs themselves.

8.9. WHAT ARE THE SENSE ORGANS AND WHAT DO THEY DO?

Each leg of a locust has a large array of diverse types of receptors that are divisible into those which act chiefly as exteroceptors and those that act as proprioceptors. This functional distinction may be blurred for such cuticular receptors as campaniform sensilla that can respond to externally imposed deformations of the cuticle and those caused by contractions of the muscles. Each joint, in turn, has several different types of these sense organs so that, for example, the thoraco-coxal joint is monitored by six internal and four external proprioceptors (Table 8.6). While the properties and actions of particular receptors and even their individual sensory neurons can be characterised, it is usually impossible to ascribe an effect on walking to just one, because many provide overlapping information and act in parallel during most movements.

8.10. CHORDOTONAL ORGANS

A prominent type of proprioceptor in the legs, and in other parts of the body, is the chordotonal organ, which can consist of many sensory cells that each encode

Table 8.6. Sensory neurons of the middle and hind legs

Type of sensory receptor	Location	Number in middle leg	Number in hind leg
Exteroceptors			
Trichoid sensilla	on all parts	1500–2000[k]	>2000
Basiconic sensilla	on all parts	?	?
	on tarsus	90[h]	60[h]
Canal sensilla	on tarsus	110[h]	70[h]
Proprioceptors			
Hair rows (RH)			
Coxa	CxRH1, dorsal posterior	5–7[m]	not present[m]
Coxa	CxRH2, ventral posterior	10–14[m]	6–9[m]
Coxa	CxRH3, ventral anterior	6–10[m]	10[m]
Tibia	TiRH1, proximal anterior	9–10[l]	not present[l]
Hair plates (HP)			
Coxa	CxHP1, ventral anterior	3[m]	40[m]
Coxa	CxHP2, dorsal anterior	30[m]	40[m]
Trochanter	TrHP1, dorsal anterior	30[m]	40[m]
Tibia	TiHP1, proximal anterior	8–13[l]	not present[l]
Campaniform sensilla (CS)			
Individual	on all parts of leg	30[l]	?
Fields			
Trochanter	TrCS1, anterior proximal	11[g]	15[g]
	TrCS2, anterior distal	10[g]	9[g]
	TrCS3, anterior proximal	12[g]	4[g]
	TrCS4, dorsal	3[g]	11[g]
	TrCS5, posterior	12[g]	not present[g]
Femur	FeCS1, anterior proximal	12[g]	not present[g]
	FeCS2, posterior proximal	14[g]	not present[g]
Tibia	TiCS1, proximal dorsal, anterior,	6[e]	not present[g]
	TiCS2, proximal dorsal, medial	10[e]	not present[g]
	TiCS3, proximal dorsal, posterior	6[e]	5–6
Multipolar receptors			
Coxa	anterior, associated with ajCO field close to dorsal articulation	?	1–2[c]
Femur	RDAL	1[o]	1[d]
	RDPL	2[o]	2[d]
	RVPL (lump receptor)	2[o]	2[d]
Tibia	distal in four groups	5+[l]	?
Tarsus	segment I	1[h,i,l]	?
Tarsus	segment II	1[h,i]	?
Tarsus	arolium	2[h,i]	?
Tension receptors	femur, distal	1[n]	1[k]
Strand receptors (SR)		20[a]	20[a]
Thorax	PlTnSR, spans thorax and trochantin	?	?
Coxa	CxTrSR1, spans coxa	?	8–10[a]

Table 8.6. Continued

Type of sensory receptor	Location	Number in middle leg	Number in hind leg
	CxTrSR2,		1
Femur	FeTiSR, strand of FeCO	1[b,c]	1[b,c]
Chordotonal organs (CO)			
Thorax,	CxCO, ventral coxal rim	10[c]	10[c]
thoraco-coxal joint	ajCO, anterior coxal rim	30[c]	30[c]
	pjCO, posterior coxal rim	5[c]	5[c]
Femur	FeCO, proximal cell group	few hundred[f]	50[j]
femoro-tibial joint	distal cell group	42[f]	50[j]
Tibia			
tibio-tarsal joint	distal	5[i]	?
Tarsus	segment III proximal	5[h]	?
Tarsus	segment III distal	5[h]	?
Myochordotonal organ	thorax, associated with anterior coxal rotator muscle 121	not present	30[c]
Muscle receptor organ (MRO)	CxTrMRO, coxo-trochanteral joint	?	1[c]
Brunner's organ	femur	not present	few CS and hairs
Subgenual organ	tibia, proximal	32[l]	32[l]

Comparable information is not available for the front legs, but it is assumed that an array of receptors similar to that of a middle leg is present.

Abbreviations

Sense organs

CO	chordotonal organ
ajCO	anterior joint chordotonal organ
pjCO	posterior joint chordotonal organ
CxCO	coxal chordotonal organ
FeCO	femoral chordotonal organ
CS	campaniform sensilla
HP	hair plate
MRO	muscle receptor organ
RH	hair row (row hair)
RDAL	récepteur dorso-antéro-latéral
RDPL	récepteur dorso-postéro-latéral
RVPL	récepteur ventro-postéro-latéral
SR	strand receptor

Parts of a leg

Cx	Coxa
Fe	Femur
Ta	Tarsus
Ti	Tibia
Tn	Trochantin
Tr	Trochanter
Pl	Pleuron

References

a Bräunig (1982)
b Bräunig (1985)
c Bräunig *et al.*, (1981)
d Coillot and Boistel (1968)
e Emptage (1991)
f Field and Pflüger (1989)
g Hustert *et al.*, (1981)
h Kendall (1970)
i Laurent and Hustert (1988)
j Matheson (1990)
k Matheson and Field (1995)
l Mücke (1991) and Newland *et al.* (1995)
m Pflüger *et al.* (1981)
n Theophilidis and Burns (1979)
o Williamson and Burns (1978

particular features of the movement, or position of a joint. A chordotonal organ is generally slung from the cuticular wall or from the muscles by a combination of stiff ligaments and elastic strands, and is linked to the joint articulation by an apodeme. This arrangement means that movement of the joint in one direction will pull on the apodeme and excite one subset of sensory neurons, whereas movement in the other direction releases the strain on the apodeme and excites a different subset of neurons.

A chordotonal organ has a distinctive structure in which 1–3 sensory neurons are associated with two accessory cells to form a structure called a scolopidium, so that a complete organ consists of several scolopidia with groups of these wrapped by additional sheathing cells. The distal part of the dendrites of the sensory neurons have the structure of a stereocilium while their proximal parts have basal bodies. One of the accessory cells, called the scolopale cell (probably equivalent to the trichogen cell of other sensilla), secretes the intracellular and electron-dense scolopale that forms a fenestrated cylinder (which may sometimes appear as a series of rods) around the dendrites and a cap around their distal ends. The second accessory cell, the attachment cell (= tormogen cell), anchors the whole structure to associated strands or directly to the cuticle. It is assumed that the sensory cells respond to stretch by the presence of stretch-activated channels in their membranes. If this is so, then the mechanics of the organ must be arranged so that the movement of a joint in one direction excites one population of sensory neurons and movements in the other excite a second population.

Seven chordotonal organs monitor the movements of a locust leg (Table 8.7), and others in the thorax [e.g. aCO (anterior chordotonal organ of the thorax) in Fig. 8.15] may also be stimulated during walking but more probably during flying. Three chordotonal organs monitor the universal movements of the joint between the coxa and the thorax so that each movement is signalled by a particular combination of actions of their sensory neurons (Fig. 8.16). Presumably, this information, along with that from other types of receptors at the joint, can be interpreted by the central nervous sytem to indicate a particular movement.

The chordotonal organ that has been examined in most detail is that at the femoro-tibial joint (usually called the femoral chordotonal organ, FeCO), and while it has its own characteristic features, it also typifies the way that these receptors work at the other joints and the effects that they can have on motor patterns. The location of this organ differs in different insects and in the different legs of one species. In the locust, the organ is in the proximal femur of the front and middle legs and in the distal femur of the hind legs, but in the tettigoniid *Decticus albifrons* it is in the proximal femur of all three pairs of legs. There is no functional explanation of these differences, although in *D. albifrons* it may simply reflect a packing problem, with space particularly restricted in the distal femur close to the femoro-tibial joint.

8.10.1. Femoral chordotonal organs of the front and middle legs

In the front and middle legs, the femoral chordotonal organ is clearly divided into two parts called proximal and distal even though they are both located near to each other in the proximal part of the femur (Fig. 8.17). Each part is suspended by separate strands of connective tissue from the anterior wall of the femur and from

Table 8.7. Receptors associated with the joints of a hind leg

Joint	Type of receptor	Name of receptor
Thoraco-coxal universal movement in three primary planes levation-depression protraction-retraction rotation		
	Hair rows (RH)	CxRH2 CxRH3
	Hair plates (HP)	CxHP1 CxHP2
	Campaniform sensilla (CS)	none
	Multipolar receptors	1–2 neurons close to ajCO several neurons close to dorsal articulation
	Strand receptors (SR)	PlTnSR
	Chordotonal organs (CO)	CxCO ajCO pjCO CO system in thorax (COS)
	Muscle receptor organ (MRO)	none
	Tension receptor	none
Coxo-trochanteral hinge movement long axis of leg levation–depression		
	Hair rows	none
	Hair plates (HP)	TrHP1
	Campaniform sensilla (CS)	TrCS1 TrCS2 TrCS3 TrCS4
	Multipolar receptors	none
	Strand receptors	CxTrSR1 CxTrSR2
	Chordotonal organs	none
	Muscle receptor organ	MRO
	Tension receptor	none
Femoro-tibial hinge movement long axis of leg flexion–extension		

Table 8.7. Continued

Joint	Type of receptor	Name of receptor
	Hair rows	none
	Hair plates	none
	Campaniform sensilla	none
	Multipolar receptors	two groups lump receptor
	Strand receptors (SR)	FeTiSR
	Chordotonal organs (CO)	FeCO
	Muscle receptor organ	none
	Tension receptor	yes
Tibio-tarsal hinge movement long axis of leg levation—depression		
	Hair rows	none
	Hair plates	none
	Campaniform sensilla	groups on proximal tibia
	Multipolar receptors	four groups
	Strand receptors	none
	Chordotonal organs (CO)	TiCO
	Muscle receptor organ	none
	Tension receptor	none
Tarsal joints curling of distal segments		
	Hair rows	none
	Hair plates	none
	Campaniform sensilla	several groups
	Multipolar receptors	three groups
	Strand receptors	none
	Chordotonal organs	two
	Muscle receptor organ	none
	Tension receptor	none

See Table 8.6 for list of abbreviations

the flexor tibiae muscle, but both join the same apodeme that runs the length of the femur to insert on the tibia just dorsal to the pivot point of the femoro-tibial joint. This arrangement means that flexion of the joint will stretch the apodeme and that extension will unload it. The anchorage to the flexor tibiae muscle means that the sensory neurons are also excited by isometric contractions of this muscle with their response depending on the position of the joint and hence on the extent to which the muscle is stretched (Burns, 1974). The distal part of the organ contains the cell bodies of about 40 sensory neurons that are 15–40 μm in diameter, whereas the

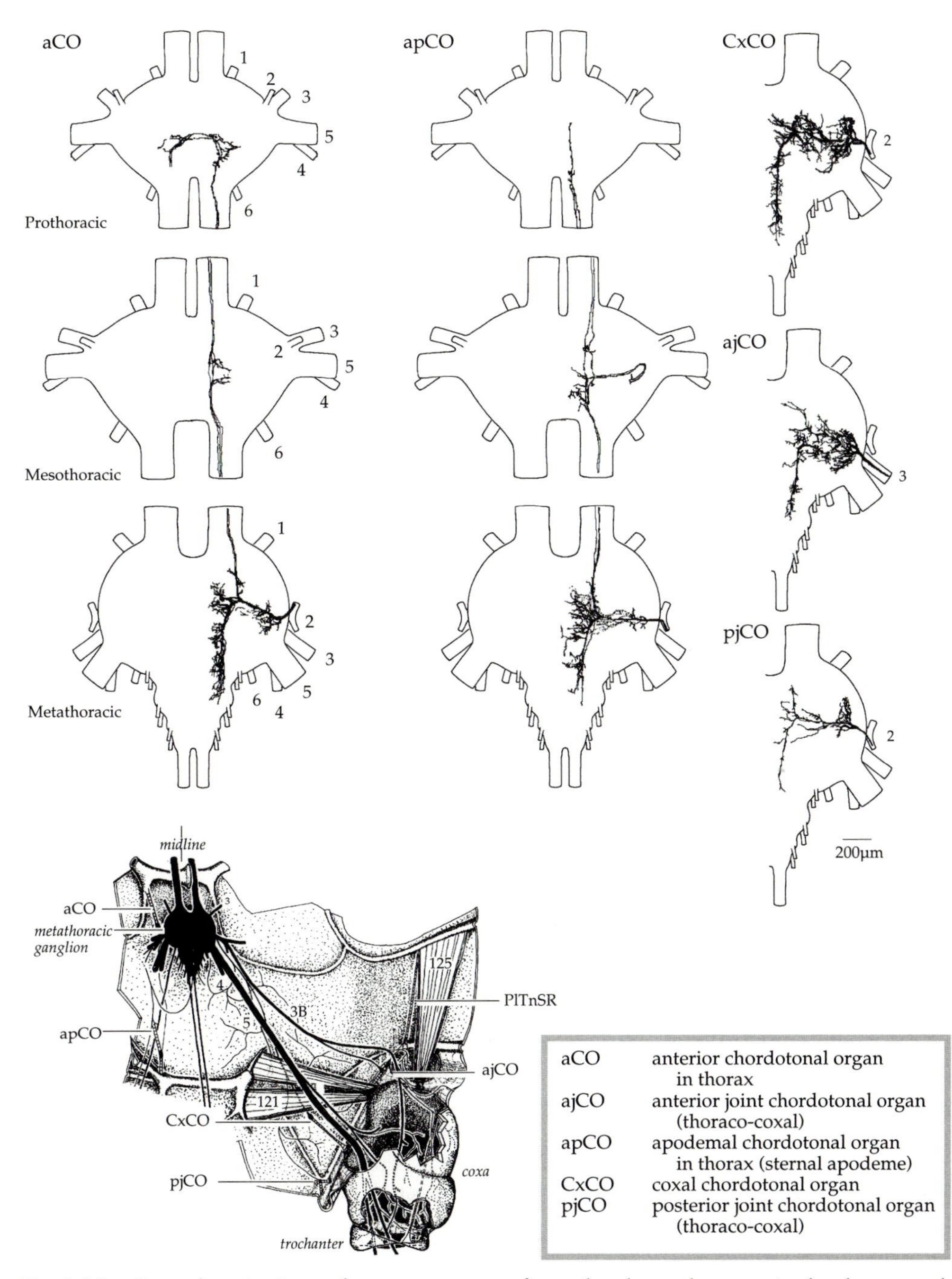

Fig. 8.15 Central projections of sensory neurons from chordotonal organs in the thorax and proximal part of the hind leg of a locust. Thorax receptors have projections to the three thoracic ganglia, but those associated with the thoraco-coxal joint project only to the metathoracic ganglion. Based on Bräunig *et al.* (1981).

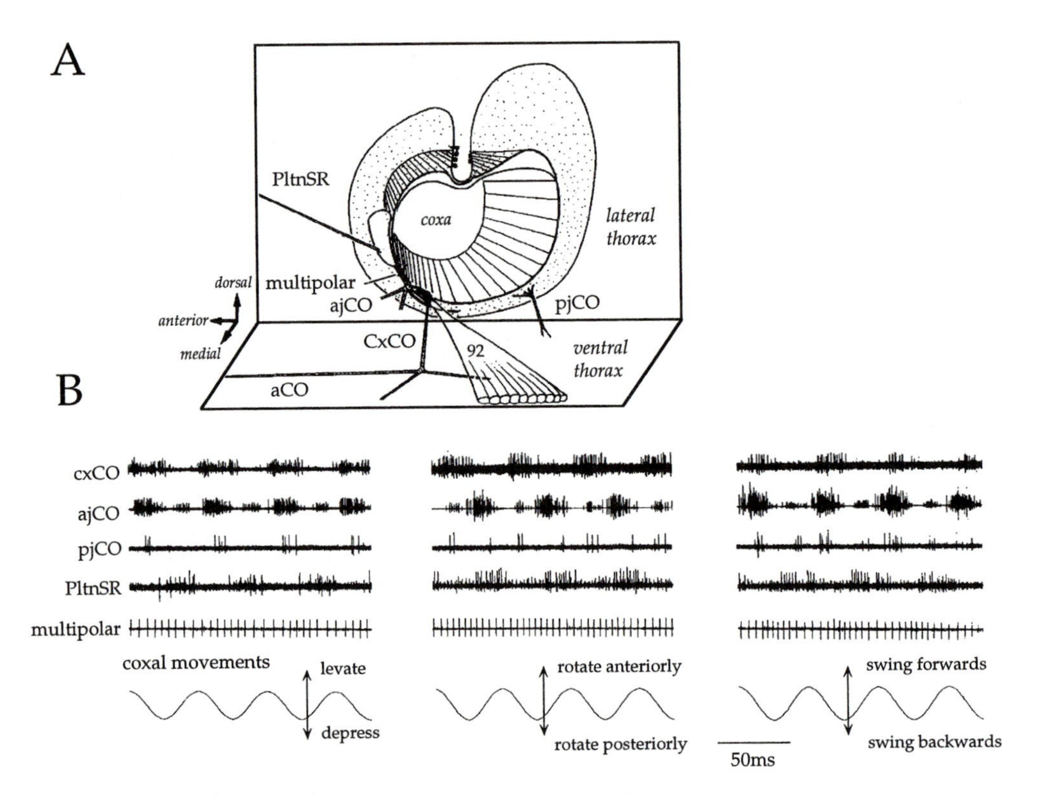

Fig. 8.16 Distribution of chordotonal organs at the thoraco-coxal joint and their action during coxal movements. A. The anatomy of various organs seen from the thorax and looking into the coxa of a right hind leg. B. Movements of the coxa in each of its three major planes is signalled by different combinations of spikes in the different chordotonal organs and in a strand receptor. See Tables 8.6 and 8.7 for the names of the various receptors. Based on Hustert (1983).

proximal part contains several hundred smaller cell bodies (3–11 μm in diameter) (Field and Pflüger, 1989). The larger, distal neurons signal movements of the femoro-tibial joint with different neurons providing information about position, velocity, acceleration and direction. A few of the smaller proximal neurons also provide this sort of information, but the majority respond to vibrations of 50–300 Hz, and thus act in parallel with the vibration-sensitive subgenual organ, campaniform sensilla and multipolar joint receptors, perhaps with a fractionation of frequencies among these different receptors. Only the distal part of the organ, however, controls the motor responses to imposed movements of the joint.

8.10.2. Femoral chordotonal organ of a hind leg

The femoral chordotonal organ in a hind leg consists of a single part in the distal femur, so that the apodeme is much shorter than those in the middle and front legs (Fig. 8.18).

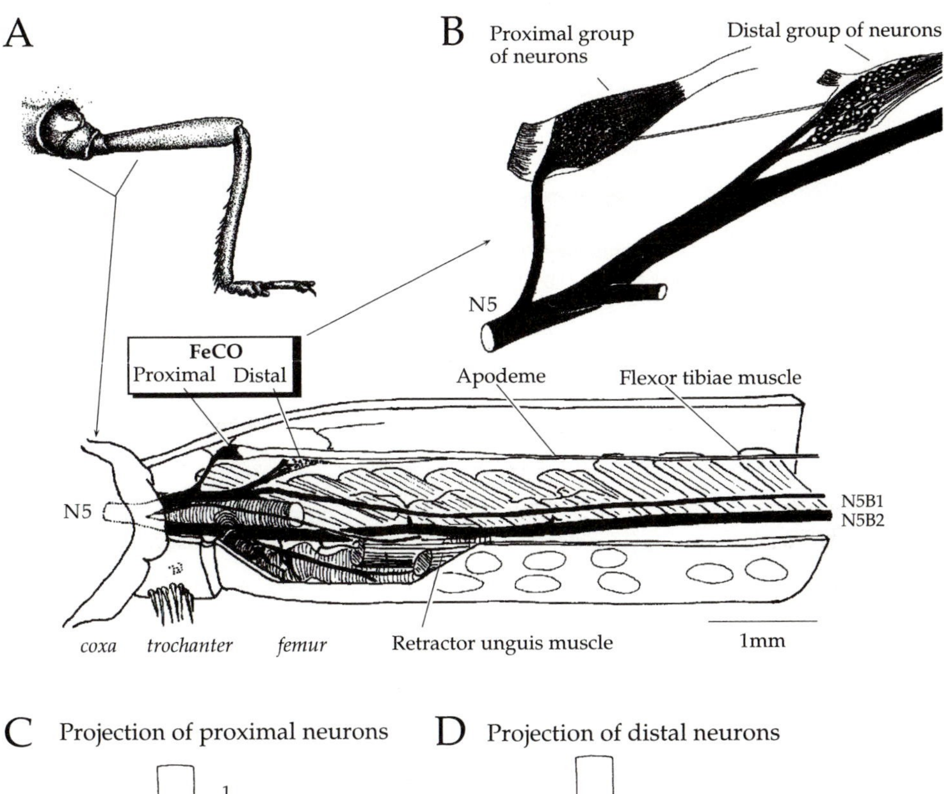

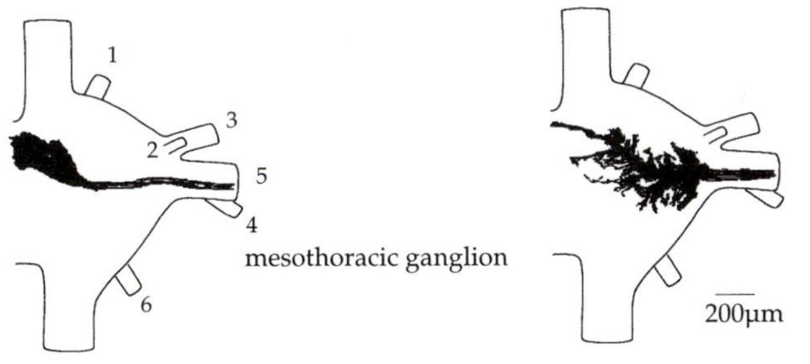

Fig. 8.17 Femoral chordotonal organ (FeCO) in a middle leg. A. Anatomy of the receptor in the proximal femur. The femur is dissected to expose the two groups of receptors and their apodeme in relation to the flexor tibiae muscle. B. The receptor cells occur in a proximal group with small diameter cell bodies and a distal group of neurons with larger diameter cell bodies. C, D. Projection of the axons of neurons from the proximal (C) and distal (D) groups of cell bodies into the mesothoracic ganglion. Based on data given by Field and Pflüger.

The distal placement of the receptor may have resulted from the need to improve the high frequency signalling by reducing the damping that occurs from the elasticity of a long apodeme in the other legs. One consequence of this design will be that the time taken for the sensory spikes to reach the central nervous system will increase in a hind leg, and this is apparently not compensated for by larger diameter axons.

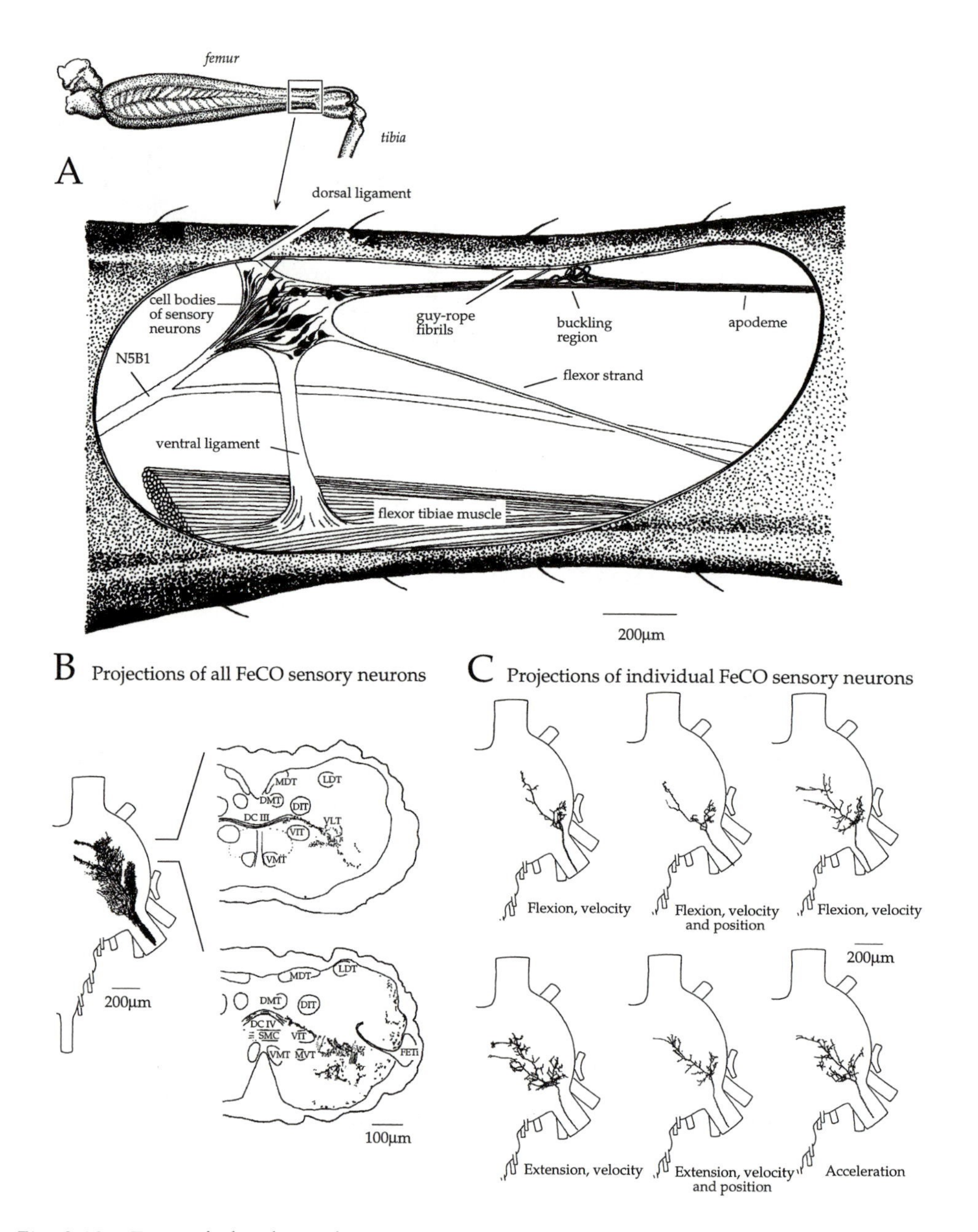

Fig. 8.18 Femoral chordotonal organ (FeCO) in a hind leg. A. Anatomy of the receptor in the distal femur. The sensory neurons all occur together and are anchored by dorsal and ventral ligaments and linked to the tibia by an apodeme with a buckling region and itself anchored to the cuticle by guy-rope filaments. B. Projections of sensory neurons into the metathoracic ganglion. The transverse sections show the relationship of the sensory terminals to known tracts and commissures. C. Projections of individual sensory neurons whose response properties are characterised. C is based on Matheson (1992).

Within the single part of the receptor are two groups of sensory neurons that almost certainly correspond to the two distinct parts of the organ in the other legs, even though their relative positions are reversed (Matheson and Field, 1990). The proximal group consists of 30–44 sensory neurons with cell bodies 10–50 μm in diameter and axons of 0.3–5.1 μm, whereas the distal group consists of approximately 50 neurons with cell bodies of 7–15 μm and axons of 0.3–2.6 μm in diameter. The majority of neurons providing the main source of information about movements and position of the femoro-tibial joint are in the proximal group and are thus reversed in position to those in a middle leg. It is presumed that the other neurons give information about vibrations, but this has not been adequately explored.

8.10.2.1. *Structure of the receptor*

The organ is attached anteriorly by a stout dorsal ligament to the anterior dorsal wall of the femur and by a ventral ligament that inserts on the apodeme of the flexor tibiae muscle some 5 mm from the articulation of the femoro-tibial joint (Fig. 8.18). A smaller strand, called the flexor strand, inserts more distally on the flexor apodeme close to the accessory flexor muscle. An extension of the dorsal ligament attaches distally to a narrower 3 mm long, rod-like apodeme that inserts on the tibia dorsal to the pivot of the femoro-tibial joint and close to the insertion of the apodeme of the extensor tibiae muscle. This distal ligament is also linked to the dorsal cuticle by fine elastic strands called either unloading strands (Field, 1991) or guy-rope fibrils (Shelton *et al.*, 1992). The complex structure of the dorsal ligament may largely determine the response of the sensory neurons to movements of the joint. It consists of many parallel fine strands of amorphous material (not collagen) 0.5–1.25 μm in diameter with the largest ones towards the outside. Proximally, some of these strands splay out over one side of the chordotonal organ so that each of the finest strands connects to one scolopidium, and in turn therefore to two sensory neurons. Distally, the strands are grouped into two bundles: a dorsal bundle that attaches directly to, and articulates with, the end of the apodeme, and a ventral bundle, the strands of which attach sequentially along a 900 μm length of the apodeme. The ligament therefore becomes thicker closer to the organ as shorter strands progressively join it. Each of the strands is probably an attachment cell for the sensory neurons in the receptor. The strands of the dorsal bundle are associated with the small cell bodies of neurons in the distal group and some of the larger neurons in a ventral region of the proximal group and remain slack over most of the range of movements; it buckles at extended positions and becomes taut only at very flexed angles. Strands of the ventral bundle seem only to be associated with neurons of the proximal group. There is much that is not understood about this complex anatomy and of the way the mechanics are likely to influence the signalling of the sensory neurons.

8.10.2.2. *Mechanics during joint movements*

When the femoro-tibial joint is flexed, the ligament of the apodeme is stretched, and when the joint is extended the apodeme is allowed to relax. This means that when the joint is in the middle of its range, the distal strands of the ventral bundle of the ligament are taut, while the proximal ones are slack. As the joint extends, the apodeme and ligament move proximally and the guy-rope fibrils (unloading strands) take the strain, allowing progressively more strands of the ligament to become slack. Conversely, when the joint flexes, the ligament moves distally, the guy-rope fibrils are unloaded and the ligament itself takes up the strain. Its strands, which are of different lengths, therefore become stretched in a sequential fashion according to their insertions along the apodeme, so that at full flexion all are fully stretched. The force associated with joint movements will thus not be transferred uniformly to the sensory neurons, and could therefore result in the sequential excitation of those linked to the different strands. This could therefore provide a mechanical mechanism for the range fractionation of the responses of the sensory neurons. It does not, as yet, explain why some neurons respond to flexion and others to extension, unless it is assumed that some neurons can be excited by stretch and others by relaxation.

Movements of the elastic flexor strand also influence the responses of the sensory neurons even though it is moved in a reciprocal way relative to the main apodeme during movements of the joint; the flexor strand relaxes during flexion and is stretched during extension (Field and Burrows, 1982). Stretching the flexor strand and stretching the apodeme nevertheless produce similar motor responses that correspond to extension and flexion of the joint, respectively. It is possible, therefore, that each activates different populations of sensory neurons, but perhaps more likely that the responses evoked from the flexor strand are an artefact of the way that it was moved. This leaves unexplained the role of this strand during joint movements.

Similarly unexplained is the anchorage of the receptor to the flexor muscle and whether the sensory neurons are sensitive to the isometric force generated by the flexor muscle, as is proposed for the femoral chordotonal organs in the other legs.

8.10.2.3. *The sensory neurons*

Only the responses of the neurons in the proximal group that regulate the action of leg motor neurons are known, and it is assumed that the smaller distal neurons, like their counterparts in the other legs, respond mostly to vibration. The proximal neurons give information on the position of the joint, and on its direction, velocity and acceleration of movement (Zill, 1985; Matheson, 1990; Hofmann *et al.*, 1985). These features of the joint movements may be coded separately, or in various combinations, by individual neurons, so that many different response types can be recognised in both the stick insect and the locust (Table 8.8). These types may simply result from the way the experimental tests are performed, as white noise analysis suggests a continuum of responses with individual sensory neurons often able to code several features of a stimulus and that these characteristics may be velocity dependent (Kondoh *et al.*, 1995).

The limited number of high sensitivity neurons code a large range of joint positions and movements by range fractionation; individual neurons respond over

Table 8.8. Response properties of sensory neurons in the femoral chordotonal organ of a hind leg

Response properties	Neuron type	Directionality	Joint angle at which response is maximal (°)
Position	1	–	0
	2	–	60
	3	–	120
Velocity	4	flexion	–
	5	extension	–
Acceleration	6	flexion	–
	7	extension	–
	8	flexion/extension	–
Position and velocity	9	flexion	0
	10	extension	0
	11	flexion	120
	12	extension	120
	13	flexion/extension	120
Velocity and acceleration	14	extension	–
	15	flexion/extension	–
Position, velocity and acceleration	16	extension	20–80
	17	extension	120

Data combined from locust (Zill, 1985; Matheson, 1990) and stick insect (Hofmann *et al.*, 1985).

restricted ranges of positions, velocities and accelerations, while the summed responses of the population as a whole cover the entire range. For example, different position-sensitive, and therefore relatively tonic, neurons have their highest frequencies at different positions; velocity-sensitive neurons code for ranges of velocities, or even ranges of velocities over particular ranges of joint positions. The response range of the neurons is still large, however, with half of the tonic neurons spiking at all positions of the joint and three quarters of the phasic neurons spiking to the fastest velocities at all leg positions. A few neurons give more precise information about the joint close to full flexion or full extension. This inevitably means that central neurons will need to look at information in many parallel lines to get accurate positional information. A further complication is that tonic neurons, in particular, can show a considerable (20%) hysteresis in the frequency of their spikes that depends on the direction from which a given position is approached, suggesting that the cause is the mechanical coupling rather than the intrinsic properties of the neuronal membrane. The feedback interactions to the terminals of the sensory neurons in the central nervous system could compensate for some of these nonlinearities (*see Chapter* 7), but hysteresis is still present in the responses of the nonspiking interneurons and in the motor neurons that process the inputs from the femoral chordotonal organ.

Location of sensory neurons in the organ Does the location of a sensory neuron within the organ determine its response properties, given that the strands of the ligament distribute forces unequally as the joint moves?

Position-sensitive neurons have cell bodies close to the proximal attachment of the dorsal ligament.

Velocity-sensitive neurons have cell bodies in two groups: one, near the insertion of the flexor strand, contains neurons sensitive to high velocity flexion movements at flexed positions, and the other, located centrally, contains both extension- and flexion-sensitive neurons that respond over a wide range of joint angles.

Combined *position and velocity* neurons, which spike tonically at a set position and give a burst of spikes to a movement, are also in dorsal and central groups.

Acceleration-sensitive neurons responding to either flexion or extension movements are dorsal, whereas those combining acceleration and velocity are ventral.

From this, it would seem that neurons with similar response properties occur in particular regions of the organ, so that if we knew more about the distribution of the strands of the ligament and the forces they exert during natural movements of the leg, the distribution might have a functional explanation. It would seem unlikely that the mechanics of the organ could explain all the different response properties of the sensory neurons. There is, moreover, no obvious correlation between the location of the cell bodies and directional sensitivity, and therefore whether the neuron is activated by stretch or relaxation of the apodeme; extension- and flexion-sensitive neurons are intermingled and are not arranged in two groups as proposed by Zill (1985), each with different orientations of their scolopidia. The correlation is made more complicated because the intrinsic membrane properties of the sensory neurons also contribute to their response properties. For example, pulses of depolarising current injected into the cell bodies of position-sensitive neurons give sustained sequences of spikes, whereas in velocity or acceleration neurons they give only a transitory burst of spikes.

Projections in the central nervous system The projections of the FeCO sensory neurons are restricted to the ipsilateral half of the metathoracic ganglion. The general rule is that chordotonal organs in the thorax (aCO and apCO in the metathoracic segment) have intersegmental projections, but that those associated with the coxa (CxCO, ajCO and pjCO) and more distal parts of the leg (femur, FeCO and tibia, TiCO) have projections only to their segmental ganglion (Figs 8.15 and 8.18) (Bräunig *et al.*, 1981). The axons of the larger neurons in the proximal part of the femoral chordotonal organ in a middle leg project to lateral neuropil (aLAC and pLAC) and medially between tracts VIT and DIT, whereas those of the smaller, distal neurons project more medially and ventrally in the mVAC between tracts MVT and VIT, and have no projections in the lateral neuropil. In the hind leg, only neurons in the distal group have been stained and their projections correspond to those of the proximal group in a middle leg. The projections of the neurons with a proprioceptive action are thus similar to those from the campaniform sensilla but are distinct from those of the mechanosensory exteroceptors that project to the most ventral regions of neuropil. The sensory neurons from the distal part of the organ that project to mVAC most closely resemble auditory and vibration receptors in the distribution of their central branches.

Is there a correlation between the response properties of a sensory neuron and the pattern of its projections in the central nervous system? Different sensory neurons certainly have different projection patterns and markedly different extents of branching that are variations on a common theme (Fig. 8.18). These differences are, however, related to function only in complex ways (Matheson, 1990). Part of the difficulty arises from trying to correlate the projection of a neuron with a particular parameter of the movement when an individual neuron may code for a combination. What can be said is that tonic or phasic extension-sensitive neurons that give maximal responses near full extension have longer medial branches than those with a maximal response near full flexion. The implication of these projections in terms of the connections that are made with central neurons has not been explored.

Altering the signals of the sensory neurons The organ receives no direct efferent innervation that could set the sensitivity of the sensory neurons, as could happen for the muscle receptor organ in the coxa and does happen for some proprioceptors in crustacea and vertebrates. Furthermore, there is no indication that there are synapses between the sensory axons in the peripheral nerves as occurs in *Drosophila* (Shanbhag *et al.*, 1992). There are also no peripheral synapses in the leg nerves between the sensory neurons and motor neurons that could create a local peripheral reflex similar to that invoked to explain the spontaneous closing movements of *Limulus polyphemus* claws when isolated from the body (Rane and Wyse, 1982).

Neuromodulators such as octopamine can, however, alter the responses of the sensory neurons, in particular those giving information about the position of the joint in stick insects (Büschges *et al.*, 1993) (*see Chapter 5*). This means that the quantitative nature of the signals supplied by the organ will be dependent on some measure of the state of the animal as determined by the concentration and types of modulators that are present. This would require recalibration elsewhere in the nervous system if the signals are to be interpreted correctly as indicating a particular position of the joint. Sensory neurons in the femoral chordotonal organ show immunoreactivity to choline acetyltransferase (Lutz and Tyrer, 1988), which is in accordance with the likely role of ACh as their excitatory transmitter in the central nervous system. The ability of octopamine to modulate the responses of the sensory neurons signalling position suggests that this effect is mediated by the efferent DUM neurons (DUMETi or DUM3,4,5) (*see Chapter 3*) that have terminals in the nearby extensor and flexor tibiae muscles, or that it is mediated by octopamine borne in the haemolymph from more distant release sites. Finally, the signalling of the sensory neurons to their postsynaptic targets in the central nervous system is regulated by presynaptic inputs to their terminals (*see Chapter 7 and below section 8.20*). These interactions place the actions of an individual sensory neuron in the context of the actions of the other sensory neurons signalling the same movement, and in the context of the behaviour of the locust. The result is that the gain of their signalling is reduced when other neurons respond to the same movement in an automatic gain control mechanism, and the overall effectiveness of their signalling is reduced during walking and then reduced further at particular phases of the step cycle.

8.10.2.4. *Motor effects of the femoral chordotonal organ*

Imposed movements of the tibia can potentially excite the whole array of sense organs at the joint, whereas experimentally applied movements of the apodeme of the chordotonal organ excite only the receptors within the chordotonal organ itself. The motor responses that result from stimulation of the chordotonal organ depend on what the animal is doing and on the position of the femoro-tibial joint when the imposed movements are applied. The responses may change in gain, so that the same input gives a different output; in mode, so that the response to opposite movements is much the same; and in sign, so that the response to a particular direction of movement is reversed. This means that, in controlled conditions, the responses evoked are reliable and consistent, but the pathways allow enormous flexibility as soon as conditions change. We know little about the signalling of the sensory neurons during walking and about the effectiveness of those signals because most analyses are performed under controlled experimental conditions. Then, there are three main categories of responses; resistance reflexes, so-called flexor mode responses, and assistance reflexes.

Resistance reflexes These occur when the locust is standing still, and in most experimental situations, so that an imposed movement is resisted by a response that opposes the disturbance. For example, an extension of the femoro-tibial joint excites flexor motor neurons and inhibits extensor motor neurons (Fig. 8.19) so that the position of the joint is stabilised and posture is maintained. These responses are reliable, in that they can be elicited readily, although they are dependent on the velocity, acceleration and amplitude of the imposed movement, and the joint angle at which the movement is applied, and may change in gain or decrement with repetition (Field and Burrows, 1982). For example, the slow extensor tibiae motor neuron (SETi) is excited to about the same extent by all velocities of movement, whereas the fast (FETi) responds better to faster movements. Similarly, the slow flexor motor neurons respond best to slow movements, and the fast motor neurons to fast movements. This suggests that the sensory neurons coding different velocities make different connections with members of a motor pool, with the fast motor neurons receiving greater inputs from neurons coding higher velocities. The excitation of a particular motor neuron also depends on the set position of the joint, so that, for example, both SETi or the antagonistic flexor motor neurons are excited more strongly by movements that

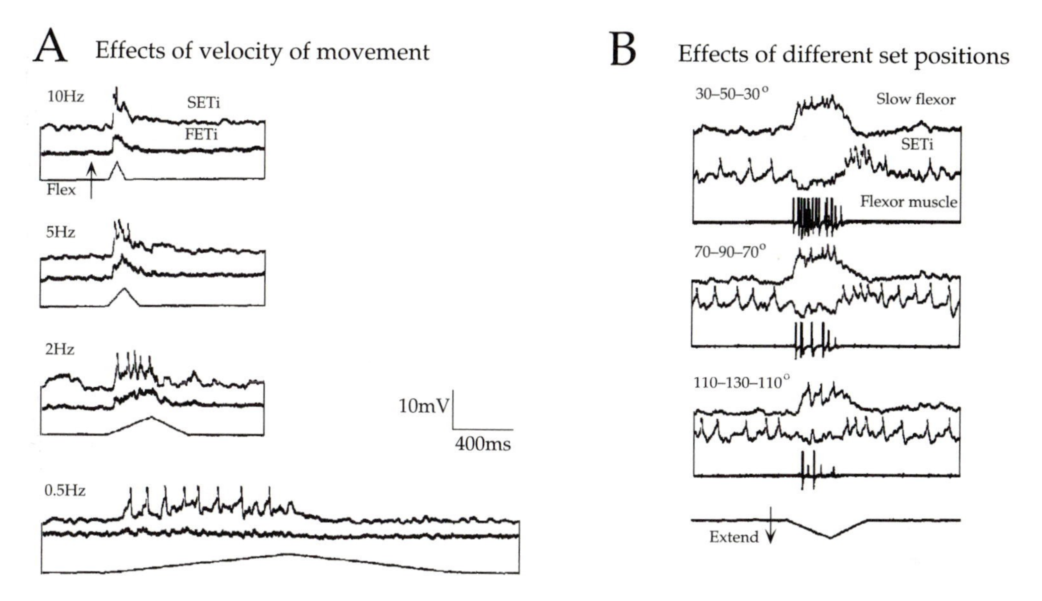

Fig. 8.19 Resistance reflexes evoked in leg motor neurons by stimulation of the femoral chordotonal organ. A. Imposed movements of the FeCO apodeme that simulate flexion movements of the femoro-tibial joint excite both of the excitatory motor neurons (SETi and FETi) innervating the extensor tibiae muscle. Their responses depend on the velocity of the imposed movements. Slow movements fail to excite FETi. B. Imposed movements that simulate extension excite a flexor tibiae motor neuron and inhibit SETi. The response of these motor neurons depends on the starting position from which the movement is made. Based on Field and Burrows (1982).

occur with the joint set close to full flexion (Field and Coles, 1994). Repetition of the movements usually leads to a gradual decline in the response of the fast motor neurons but a more durable response of the slow motor neurons. In both types of motor neurons, however, the response can suddenly change, depending on the other inputs the locust receives, or the other movements that it undertakes. The motor responses may show considerable hysteresis, so that there can be a mismatch between the frequency of their spikes before and after an imposed step movement, and depending on the previous direction of movement. This means that the hysteresis that is present in the responses of the sensory neurons and the local interneurons is also expressed by the motor neurons and must have a behavioural consequence.

The evoked motor response can also include the muscles of adjacent joints so that extension of the femoro-tibial joint is thus also accompanied by a depression of the tarsus and by an increase in activity of some trochanteral depressor and levator motor neurons so that the trochantero-femoral joint is stiffened (Field and Rind, 1981). In this way, a coordinated movement of the leg results from the signals generated at one joint.

The chordotonal organs, and other proprioceptors, thus maintain posture in the face of perturbations and allow the locust to compensate for changes in loading on its individual legs. These control mechanisms are readily expressed when a locust stands on a moving platform and maintains its posture, largely by contracting those muscles that decrease the angles of the joints (e.g. the flexor tibiae) (Zill and Frazier, 1990). If the displacement is large, or the initial angles of certain joints are approaching their extremes, then it may make additional compensatory steps. The human parallel is our ability to stand upright on a bus that is continually changing velocity and direction (Nashner, 1976). A locust must do this by sampling sensory signals from the individual legs because it apparently has no sense equivalent to that provided by our vestibular system. Qualitatively similar reactions result when sudden displacements occur during walking that are similar to the responses of a stick insect to sudden displacements of its femur.

From this, it is apparent that the signals from the femoral chordotonal organ are in large part responsible for setting the basal level of activity of the motor neurons that innervate the muscles of several joints which both maintain and adjust posture. It seems reasonable to suppose that the full complement of similar receptors at the other joints acts on particular groups of motor neurons and sets the level of tone for all the muscles in the leg, with overlap in their spheres of influence that is appropriate for the adopted posture.

These negative feedback reflexes also persist in a walking animal, as can be elegantly shown by allowing a stick insect to step with one of its legs upon a platform that can be suddenly raised or lowered, causing the femoro-tibial joint to be forcibly flexed or extended (Cruse and Pflüger, 1981). The leg then makes an appropriate compensatory movement, although the gain of the response is less than when it is standing.

Flexor mode responses These responses occur when a locust actively moves its legs and when an imposed movement is applied to the joint that is close to being fully flexed. In these conditions, an imposed movement of the femoro-tibial joint in either direction excites the flexor tibiae motor neurons. The active movements during which these responses can occur are described as searching movements so as not to imply that they represent locomotory movements. The implication, nevertheless, is that these are the sort of responses that can be expected to be caused during locomotion. The fact that these particular responses only occur when the leg is near full flexion suggests that they reinforce the flexion movements that are produced in response to a novel stimulus, so that the leg is prepared for a new movement.

Assistance reflexes These responses are primarily reported from stick insects in an active state. Although this state is not accurately defined, it is usually induced by stimulating the animal so that it moves its legs. The implication is that these responses are the ones that primarily occur during locomotion, because if resistance reflexes were always present, the desired voluntary movement would be opposed. It is, of course, essential that the signals generated by changes at the femoro-tibial joint be placed in the context of the insect's movements. Therefore, as the name implies, instead of resisting the imposed movements, the responses now assist the voluntary movement; an imposed extension of the joint excites extensor motor neurons and inhibits flexor motor neurons. A change in the responses of an insect from resistance to assistance mode is often referred to as reflex reversal, implying some sort of state change within the central nervous system, rather than an accommodation to the expression of a voluntary motor command. It probably represents no more than the context dependence of many sensory (or other) effects within the central nervous system. While a reflex can reverse completely, at other times the response of a motor neuron can show components of both resistance and assistance reflexes. With movements below a certain velocity, extension of the femoro-tibial joint in a stick insect evokes a resistance reflex, but the initial response of the flexor motor neurons to higher velocities can be an initial inhibition followed by the expected excitation, with the converse responses in the extensors (Bässler *et al.*, 1986).

Effects on locomotion The number of different receptors acting in parallel to monitor the movements and positions of a joint mean that it is almost impossible to ascribe an action during walking just to a chordotonal organ, although the large number of receptors that it contains makes it a likely candidate for some, at least, of the observed effects. Even allowing for this constraint, there is little evidence implicating the chordotonal organ in specific actions during walking. Removing the FeCO is said to abolish jumping, but the effect of removal on walking is less clear. The leg from which it is removed may extend more than normal, but its participation in the walking rhythm is otherwise normal and the timing of the movements of the other legs also seems to be unaltered. In stick insects, crossing the apodeme to the extensor muscle leads to saluting movements (*see section 8.8.2.2*). Attempts have been made to stimulate the chordotonal organ electrically to provide inputs at different times during the step cycle in order to see how this input might disrupt walking, but the experiments failed to control the spread of the stimulation to other structures, including the motor nerves, and cannot thus be interpreted (Macmillan and Kien, 1983).

Catalepsy A stick insect mimics the twigs on a tree by standing still with the legs seemingly held at one position for a long time and at a variety of angles. The ability of the tibiae to maintain these positions results from the high gain and velocity sensitivity of the feedback loop from the femoral chordotonal organ. By contrast, in a locust, which is unable to show such cataleptic mimicry, the gain of the feedback loop formed by the femoral chordotonal organ is lower. If a leg of a stick insect is moved slowly to a new position it moves only very slowly back to its original position because the evoked activity in the relevant motor neurons declines with a long time constant. Thus, to a casual observation, the joint may appear to remain in its new set position. If it is moved rapidly then the evoked motor responses decline more rapidly, indicating that the gain in the feedback loop is velocity sensitive, with gains as high as 5 for velocities between 0.1–0.5 Hz.

8.11. STRAND RECEPTORS

The characteristic feature of strand receptors is that the sensory neurons, unlike all other sensory neurons in the leg, have their cell bodies in the central nervous system. The axonal branches of these neurons innervate fine elastic strands that are strung across joints or attached to the apodemes of muscles, so that they signal the stretch caused by joint movements or muscle contractions.

8.11.1. Coxal strand receptors

Of the two strand receptors in the coxa, the first, CxTrSR1, attaches to the lateral coxal wall, spans the coxa and inserts on the apodeme of the trochanteral levator muscle (Bräunig, 1982). It is innervated by 8–10 sensory neurons. The second, CxTrSR2, is attached to the dorsal rim of the coxa and inserts more laterally, and is innervated by a single sensory neuron. Sensory neurons of both receptors are excited when the trochanter is depressed, coding in phasic bursts of spikes the position of the joint when a movement is made and the velocity of the depression movement (Fig. 8.20) (Bräunig and Hustert, 1985a). During a full excursion of the joint, the elastic strands of these organs will be stretched almost twofold, but despite this elasticity, no hysteresis in the response of the sensory neuron is reported.

Depression of the coxo-trochanteral joint excites levator trochanteris motor neurons and inhibits depressors in a typical resistance reflex that would stabilise the joint against changes in the external loading (Fig. 8.20). The depression is signalled by the strand receptors, which then make direct excitatory connections with at least some of the levator motor neurons (Skorupski and Hustert, 1991).

8.11.2. Flexor strand receptor

The flexor strand is closely associated with the femoral chordotonal organ, attaching to the outer lamella of the organ, but unlike the strands of the chordotonal ligament it has no connections with individual neurons within the FeCO (Bräunig, 1985). Instead, its single sensory neuron has an axon in N5B1c and a cell body in the metathoracic ganglion. The strand inserts onto the apodeme of the flexor tibiae muscle and is therefore stretched during extension, when the apodeme of the chordotonal organ is relaxed. The sensory neuron codes the velocity of extension movements over most of the range of joint movements, and codes the static position of the joint in the frequency of its spikes which are linearly related to joint position. It would thus appear to give information in parallel to many of the sensory neurons in the femoral chordotonal organ during movements of the femoro-tibial joint. It is possible that it can code for some feature of extension that is not so far revealed, or for active contractions of the flexor muscle, and that its connections in the central nervous system are different from those of the chordotonal sensory neurons. As yet, however, nothing is known of its connections within the central nervous system.

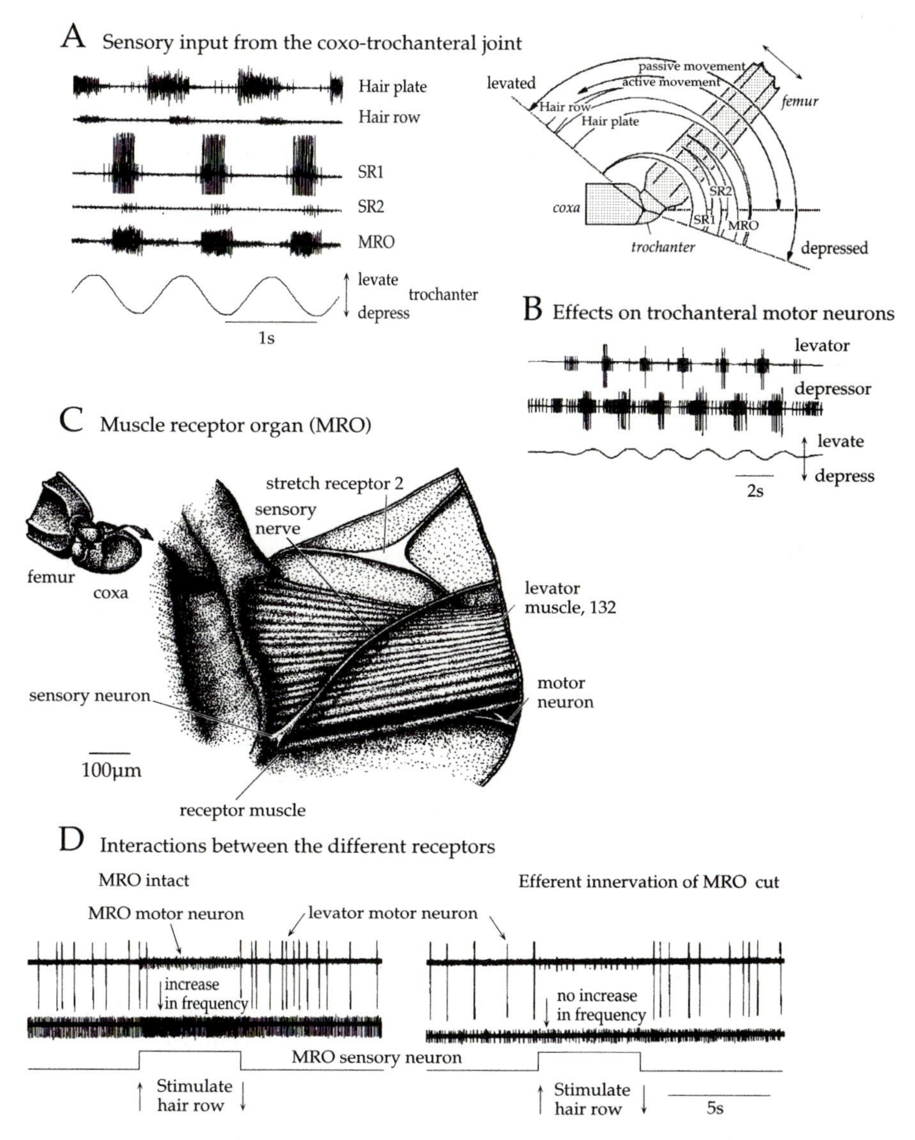

Fig. 8.20 Receptors that monitor the movements of the coxo-trochanteral joint. A. Movements of the joint evoke spikes in sensory neurons from a hair plate, a hair row, strand receptors and a muscle receptor organ. See Tables 8.6 and 8.7 for names of the receptors. The diagram shows the range of joint angles over which each of the receptors is active. B. The same movements of the joint evoke resistance reflexes in trochanteral motor neurons. C. Anatomy of the muscle receptor organ, its innervation and its relationship with trochanteral levator muscle 132. D. Interactions between the different joint receptors. Stimulation of the hair row when the efferent innervation of the MRO is cut no longer causes an increase in the frequency of spikes in the MRO sensory neuron. Based on Bräunig and Hustert (1985a,b).

8.12. MUSCLE RECEPTOR ORGANS (MRO)

At the coxo-trochanteral joint of each of the legs is a sense organ (the coxo-trochanteral muscle receptor organ, CxTrMRO) that is under direct efferent control (Bräunig, 1982), and because of this there is a complex interplay between the sensory neurons that monitor movements of this joint (Fig. 8.20). It is the only receptor in the leg for which efferent control has so far been demonstrated. The signals of this proprioceptor are thus controlled by the signals of the other proprioceptors monitoring the same joint. The only other receptor that can be considered to be under efferent control is the tension receptor in the flexor tibiae muscle, and the sensitivity of the single sensory neuron may also be set by the action of some flexor tibiae motor neurons (Matheson and Field, 1995).

In many respects, the muscle receptor organs at the coxo-trochanteral joint are similar to the crustacean muscle receptor organs that have been analysed extensively (Alexandrowicz, 1951). The muscle receptor organ of the locust consists of a single sensory cell associated with the distal attachment of a small muscle that consists of three fibres innervated by one motor neuron, and that runs parallel to part of the main levator muscle between the posterior wall of the coxa and the joint membrane at the proximal rim of the trochanter. The dendrites of the sensory neuron insert in series only on the collagenous insertion of the receptor muscle on the trochanteral membrane where tubular bodies, typical of sites of mechano transduction, occur in their tips (Bräunig *et al.*, 1986). This arrangement means that the sensory neuron should be affected by both force and length changes in the receptor muscle.

The spike frequency in the motor neuron is increased by inputs from other receptors of this joint and this makes the receptor muscle contract and, in turn, changes the output of the sensory neuron from the muscle receptor itself (Bräunig and Hustert, 1985a). The receptor is stretched by depression of the joint and relaxed by levation, but the spike frequency of the sensory neuron is high at both the fully depressed and fully levated positions of the joint and lowest in the middle of the range. The apparent anomaly of excitation at levated positions results from inputs to the motor neuron from the hair plate and hair row, and disappears when these inputs are abolished (Fig. 8.20). The strand receptors that are excited by joint depression also interact by causing a phasic increase in the discharge of the MRO motor neuron, and although the MRO sensory neuron can also excite its own motor neuron, these effects are too weak to contribute to its own tonic discharge. Positive feedback does not therefore lead to runaway excitation. The interactions between all the receptors probably serve to keep the muscle receptor taut at all joint positions so that its sensory neuron can immediately signal any new movement of the joint. The combined input from all the receptors at this joint can affect motor neurons that move the coxa and the trochanter, but they have no effect on the more distal joints. The hair plate and the hair row are excited by levation and excite depressor motor neurons and inhibit levators in typical resistance reflexes, while the strand receptors and muscle receptor organ are excited by depression and have the opposite motor effects (Bräunig and Hustert, 1985b).

Some coxal muscles are excited by the groups of receptors that respond to opposite movements of the joint and may therefore stiffen the joint as a support for movements of more distal joints.

8.13. MULTITERMINAL JOINT RECEPTORS

Five multiterminal receptor cells occur at the femoro-tibial joint of all the legs (Coillot and Boistel, 1968; Williamson and Burns, 1978). In the femur of a middle leg, two other groups of cells have been found that may also occur in the other legs but the action of these is not known (Mücke, 1991). Two of the five sensory neurons have their processes aligned with the transverse axis of the leg and are attached to the ventral arthrodial membrane, the distortions of which they signal. Two are close to the posterior articulation of the tibia, and the fifth is close to the anterior articulation with processes parallel to the long axis of the leg. These neurons signal the position of the tibia and extension movements beyond 80° by increases in their tonic spikes.

In the hind leg, the two ventral neurons are associated with the cuticular specialisations of the femur that are linked to jumping (*see Chapter* 9). These so-called lump receptors spike only when the tibia is fully flexed and it is assumed that they signal when the flexor apodeme lock is engaged ready for the start of a jump (Heitler and Burrows, 1977). During walking, the tibia is never fully flexed and it is assumed that they then do not provide any signals to the central nervous system. The three neurons close to the articulations of the joints code for extension of the tibia between 80 and 150° in the frequency of their spikes (Coillot, 1974), but again this range of tibial movements is largely outside that used when walking horizontally (see Fig. 8.1). Moreover, the velocity sensitivity of the receptors also saturates to the movements that normally occur when walking. Together, these facts suggest that they have no phasic involvement in normal walking, but instead monitor unusual strains in the joint articulation.

Almost nothing is known of the central processing of the signals from these afferents as no connections have been defined. They do not, for example, converge onto the spiking local interneurons that receive inputs from the femoral chordotonal organ and the femoral strand receptor (Pflüger and Burrows, 1987), nor do they synapse directly onto tibial motor neurons as do the chordotonal afferents. There is some indication that they may be able to exert reflex effects on some of the motor neurons innervating tibial muscles, but there are no adequate experiments that have eliminated the possibility that other receptors known to mediate such effects are acting in parallel.

8.14. TENSION RECEPTORS

Animals usually have receptors that monitor the force that is produced by their muscles. Mammals have Golgi tendon organs and crustaceans also have receptors on the apodemes of their muscles. Insects have receptors (campaniform sensilla) in their cuticle that can monitor the force produced by muscles when they distort the cuticle, and have individual receptor cells embedded in the muscles themselves. So far, these are only known in the flexor tibiae muscles but it would seem logical that equivalent neurons are present in at least some muscles that move each joint.

In the front and middle legs, the receptor cell is towards the distal end of the femur and its dendrites project within the muscle fibres of the second or third most distal and anterior bundles (Fig. 8.21) (Theophilidis and Burns, 1979). This location, rather than on the tendon itself, would suggest that it monitors the force produced by the local contractions of this specific part of the muscle rather than the whole muscle. There is no evidence, however, that this part of the muscle performs any individual function, or that it is separately innervated. By contrast, in the hind leg, the receptor is associated with a distinct accessory part of the flexor muscle at the distal end of the femur (Matheson and Field, 1995) whose fibres insert on the apodeme at a different angle to those of the main body of the flexor muscle. A separate function seems likely but again a separate innervation has not been demonstrated. The accessory muscle is only innervated by two of the total complement of about nine flexor motor neurons, and by CI2 and CI3, but all of these neurons also innervate the main body of the muscle.

The receptor responds best to the force generated by active, isometric contractions of the flexor muscle but only weakly to movements of the tibia caused by muscle contractions or when passively imposed. All of this indicates that it is signalling the force generated by the muscle. In a middle leg, the receptor signals the contractions of some of the fibres in the flexor muscle, but in a hind leg it responds only to the force generated by the accessory flexor muscle and not by the main body of the muscle. In the hind leg, the receptor is excited when individual flexor motor neurons that innervate the accessory flexor are made to spike and, in turn, the receptor excites these motor neurons and a limited number of others that innervate the main body of the flexor by pathways that involve interposed interneurons (Matheson and Field, 1995). The three common inhibitory neurons, by contrast, receive a direct excitatory input and the slow extensor tibiae motor neuron (SETi) is inhibited by a longer indirect pathway. This pattern of connections makes it difficult to infer the role of the receptor in influencing movements. The excitatory connections with the flexor motor neurons and the inhibitory effects on the slow extensor indicate that it forms a positive feedback loop that would reinforce the contractions caused by a particular subset of flexor motor neurons, but this would be offset by the increased relaxation caused by the inhibitors unless the timing of their actions were inappropriate. The excitation of the inhibitors indicates that resting tension is reduced in a negative feedback loop to allow a more rapid contraction caused by the excitatory motor neurons to the flexor muscle. For the hind leg at least, this suggests that there is a division of action within the pool of flexor motor neurons and the parts of the muscle that they innervate. The receptor spikes best when the tibia has to flex against a load and thus during the normal swing phase of walking should not be activated, unless the movement is impeded. During the preparation for a jump or a kick it should spike, but these movements can still occur

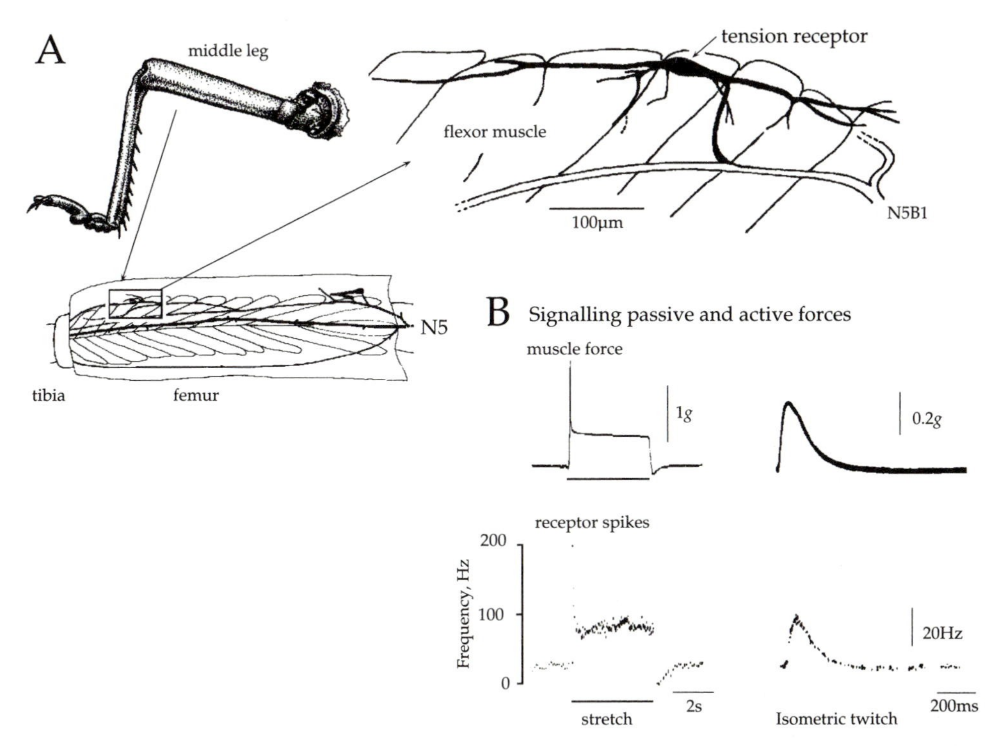

Fig. 8.21 Tension receptor in the flexor tibiae muscle of a middle leg. A. Anatomy of the receptor in the flexor tibiae muscle in the distal part of the femur. The single sensory neuron has an axon in a branch of N5B1. B. Passive tension in the flexor tibiae muscle caused by an imposed stretch is signalled by a transient and sustained sequence of spikes in the sensory neuron. An isometric twitch of the whole muscle is also signalled by spikes. Based on Theophilidis and Burns (1979).

when the receptor is destroyed. Perhaps the accessory muscle and its receptor act as a servo mechanism, with the motor innervation setting the level so that disturbances caused by increased force during movements could be signalled by deviations from this level. In this way, it might be responsible for setting the action of the main body of the flexor muscle in much the same way that muscle receptor organs might act. More indications of its likely actions would result if its signals during normal movements were known and the properties of the muscle fibres were analysed.

8.15. CAMPANIFORM SENSILLA

8.15.1. Structure of a campaniform sensillum

Each sensillum is an ovoid depression in the cuticle spanned by a flexible cuticular dome into which the dendrite of a single sensory neuron that contains a modified cilium inserts. Compression of the surrounding cuticle perpendicular to the long axis of the dome bends the flexible cuticle, thus deforming the tip of the dendrite and evoking spikes in the sensory neuron (Spinola and Chapman, 1975).

8.15.2. The sensory neurons and what they signal

The arrangement of the receptor means that its single sensory neuron is directionally sensitive, responding best to stresses in one particular direction, but this does not mean that its signals necessarily give unambiguous information about the movement of a joint. Passive movements in one direction and active movements in another can evoke the same sensory signals, but these could be interpreted differently by the central nervous system if it also has information about the actions of specific muscles (Delcomyn, 1991b).

8.15.3. Distribution of sensilla on a leg

Campaniform sensilla are widely distributed over a leg either individually or aggregated in fields where strains on the cuticle are likely to be highest during locomotion and perhaps other movements (Fig. 8.22). The highest concentration of receptors is on the proximal parts of a leg, where the strains are presumably highest, with the front and middle legs having five fields on the trochanter and two on the proximal femur, but with the hind leg lacking one of the trochanteral and both of the femoral fields (Table 8.6). Most of the femur of a hind leg has no campaniform sensilla, but on the proximal tibia there are three fields, and single sensilla are associated with each of the fixed spines and the moveable spurs. On the tarsus, the sensilla occur singly or in pairs associated with each of the segments. All the sensilla seem to be placed strategically to monitor strains in the cuticle caused either by loading of the limb, the force generated by particular muscles or the distortions of protuberances like the tibial spines. Typically, their effect during locomotion is to compensate for increases in load and force (*see Chapters 9* and *11*).

8.15.4. Trochanteral campaniform sensilla

Whenever there is an increasing likelihood of changes in load during a particular phase of locomotion, there is a general need for feedback. The challenge of increased forces, as when walking up an incline or when artificially made to carry an extra load, is thought to be detected by campaniform sensilla in the proximal cuticle of the leg. Stimulation of trochanteral campaniform sensilla in cockroaches excites particular coxal depressor motor neurons (Pringle, 1940), and they are also thought to be involved in inhibiting the rhythmic bursts of spikes that may represent a locomotory rhythm (Pearson, 1972). They are presumed to be excited during the stance phase when the leg is bearing the load of the body, and the signals they provide could therefore compensate for changes in load by increasing the output of the depressor motor neurons. Towards the end of the stance phase, the distribution of the loading on the leg will change and this, in turn, will reduce the positive feedback to the depressor motor neurons. If the leg is not unloaded then the signals of these sensory neurons inhibit the motor rhythm and thus ensure that the leg will not take its next step. The parallel here with the information provided by hip joint

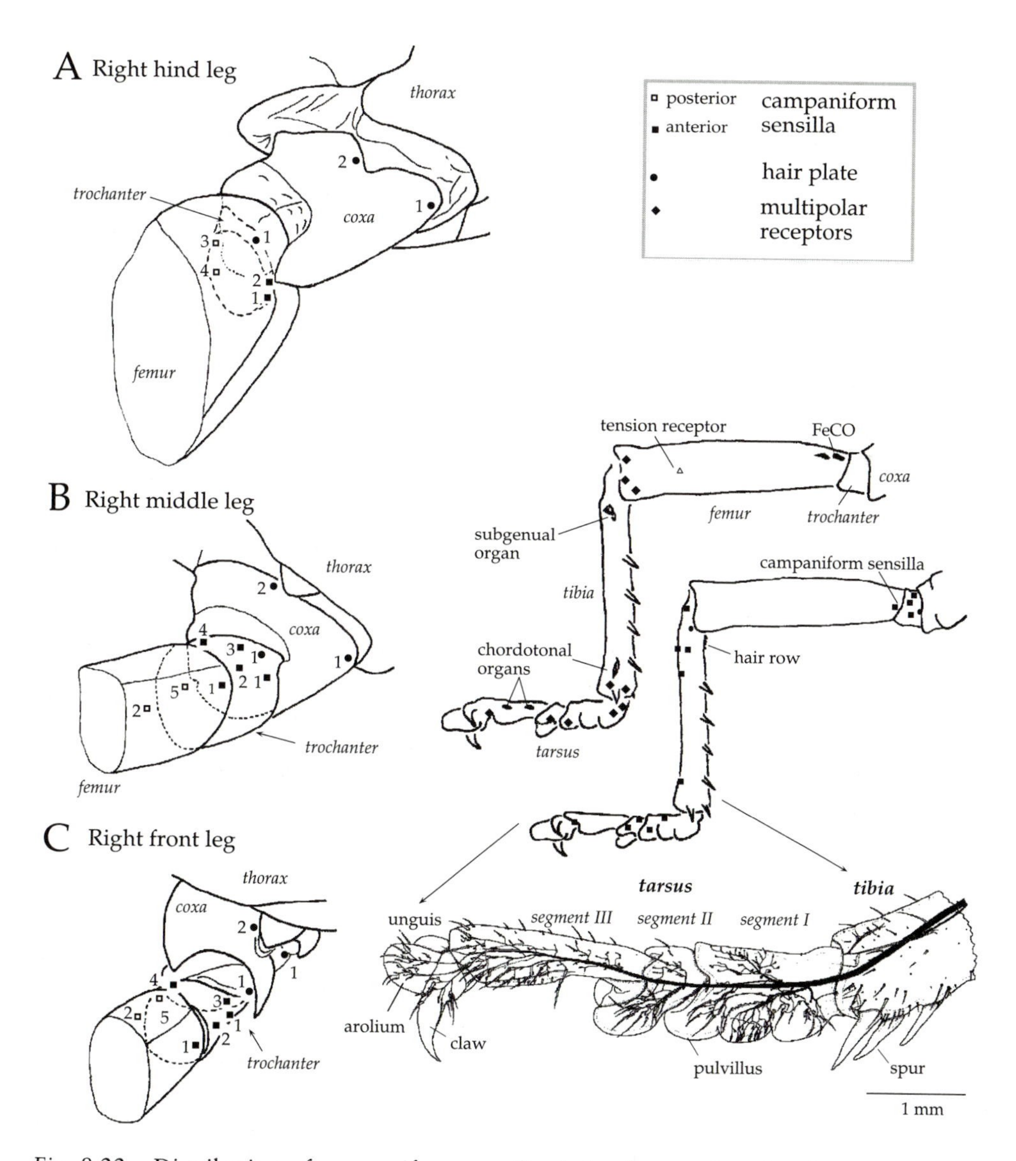

Fig. 8.22 Distribution of campaniform sensilla, hair plates and multipolar receptors in the legs. A. The proximal joints of a right hind leg. B. A right middle leg. The proximal joints are enlarged to show the numbers and distribution of receptors and the tarsus is also enlarged to show the innervation of its receptors. C. The proximal joints of a right front leg. The drawings of the proximal joints are based on Hustert *et al.* (1981), those of the whole middle leg on Mücke (1991), and that of the tarsus on Laurent and Hustert (1988).

receptors in mammals (Grillner and Rossignol, 1978) is thus strong. No recordings have, however, been made from sensory neurons of the campaniform sensilla in cockroaches and none of the pathways mediating their proposed effects have been elucidated. It is thus an assumption that the motor effects caused by pressing on the trochanter are in fact caused by these receptors, and an even greater extrapolation to suppose that the lack of input from them in an amputated leg is responsible for the changes in gait. In stick insects, however, the suggested role for these sensilla in delaying the onset of the next swing phase is more clearly indicated by experiments in which a clamp is placed on the trochanter to stimulate the nearby campaniform sensilla continuously without impeding movements (Bässler, 1977). Attempts to stimulate the campaniform sensilla electrically during locomotion (Macmillan and Kien, 1983) are not interpretable because of the spread of the applied current, as for the same experiments performed by these authors on the femoral chordotonal organ (*see section 8.10*).

8.15.5. Tibial campaniform sensilla

On the proximal tibia of the front and middle legs there are four groups of campaniform sensilla; a medial group of 10 sensilla, an anterior lateral group of six, a posterior lateral group of six, and a more distal group of some 2–3 (Emptage, 1991). The numbers of sensilla in these groups all increase during postembryonic development, so that in the first instar only half the final complement is present. The posterior group is innervated by N5B2 and the anterior group by N5B1, thus maintaining the spatial mapping of a leg that is apparent for the hairs (*see Chapter 7*), while both nerves supply the medial group. The caps of the sensilla in the lateral groups are oval with their long axes aligned with the proximo-distal axis of the tibia, while those of the medial group are more circular and thus have no obvious axis of orientation. The three proximal groups of sensilla give information about stresses that are caused by forces acting on the tibia in any plane. Receptors in the medial group signal tensile or compressive forces along the longitudinal axis, and those in the lateral groups signal forces in the transverse plane. The posterior group signal forces applied in an antero-posterior direction while those in the anterior group signal forces in the opposite direction. The individual sensilla respond either to compression or tension and in either a phasic or a phaso-tonic way, but all can code repetitive changes in force at the frequencies of leg movements that are used in walking. The direction of a force might be represented in central neurons by convergence of synaptic inputs from sensory neurons of the various groups.

The leg motor neurons can receive synaptic inputs from sensory neurons of the different groups of receptors. Stimulation of sensilla in the medial group excites flexor tibiae and depressor tarsi motor neurons, leads to inhibition of the slow extensor tibiae (SETi) and tarsal levator motor neurons, but has no effect on the fast extensor tibiae motor neuron or CI1. These motor responses are different from those elicited by touching the surrounding hairs, and the signals from the campaniform sensilla and the hairs are processed by different sets of spiking local interneurons of the midline group. The excitatory connections with the flexor motor neurons appear to be direct, but the inhibition of SETi is mediated by spiking local interneurons of the midline group (*see Chapter 7*) that receive a direct excitatory input from the sensory neurons and make a direct inhibitory connection with the motor neurons. Several sensory neurons from different sensilla within the group converge onto one flexor motor neuron, so that their effects sum, but not all the receptors within the group have sensory neurons that converge on the same motor neuron. This implies some selectivity in the connections that a particular sensory neuron makes with members of the motor pool. Similarly, the sensory neuron from one sensillum diverges to connect with several, but not all, of the motor neurons in the large flexor pool. Whether this selectivity of connections reflects a relationship between sensory neurons with different physiological properties and the different types of motor neurons has not been tested.

The pathways seem designed to produce a reliable and strong flexion of the leg when particular stresses are detected in the proximal tibia. The forces generated in the tibia when the leg is placed on the ground at the start of the stance phase are more than adequate to excite these receptors so that they can be expected to provide cyclical signals to the central nervous system at each step, and may be involved in regulating the transition from the stance to the swing phase. The forces can reverse when the locust makes certain turns, so that the role of the sensilla may be more important during such manoeuvres when they could amplify the action of the flexor tibiae and depressor tarsi motor neurons so that more force is generated to propel the body forward and to provide greater purchase on the ground.

In a hind leg, the lateral groups are each reduced to a single sensillum (*see Chapter 9*), the medial group is offset from the midline, and the distal group absent. While the actions of the lateral sensilla are known is some detail, almost nothing is known about the prominent medial group. Perhaps this is because its sensory neurons have no immediately obvious effects on the leg motor neurons.

8.15.5.1. *Tibial campaniform sensilla in the cockroach*

Groups of campaniform sensilla on the proximal tibia of a cockroach are positioned to detect the bending forces applied to the more distant tarsus and the larger forces generated by the nearby muscles of the femur (Fig. 8.23). All the sensilla signal strains that are applied in the plane of movement of the femoro-tibial joint. A proximal group contains some 6–10 sensilla that are oriented with their long axes at right angles to the long axis of the leg. They respond to axial compression caused either by a force that would bend the tibia dorsally, or by a contraction of the flexor tibiae muscle working against a tarsus that cannot move (Zill and Moran, 1981a). Stimulation of these receptors leads to excitation of the slow extensor tibiae and trochanteral depressor motor neurons, and inhibition of flexor tibiae and trochanteral levator motor neurons in negative feedback pathways that act as a load-compensating mechanism (Zill *et al.*, 1981); increased load on the legs causes dorsal bending which is met by increased force in the depressor and extensor muscles that straightens the legs, raises the body and decreases the stimulation of the receptors. None of the pathways leading to these effects are known.

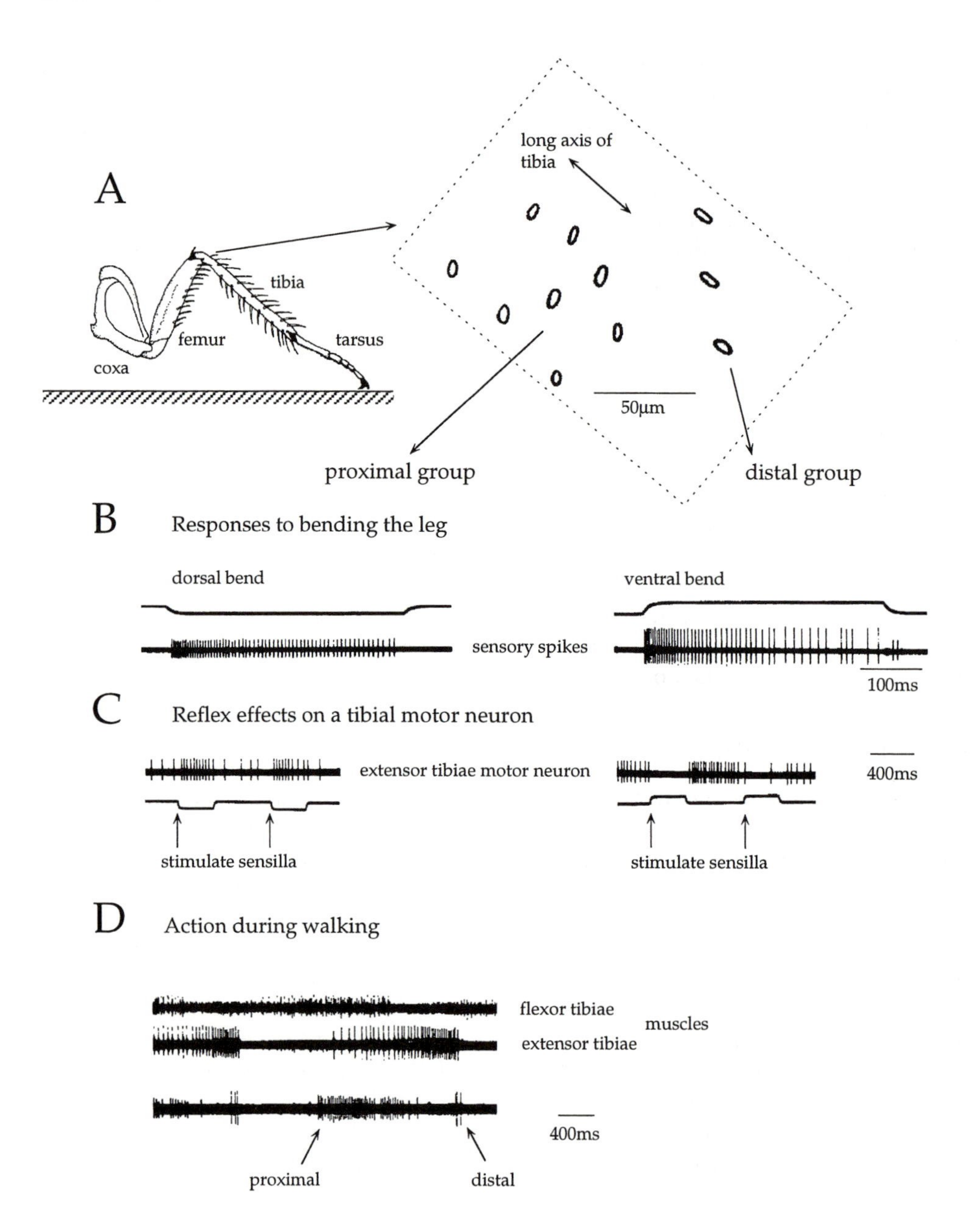

Fig. 8.23 Campaniform sensilla on the tibia of a cockroach. A. Position of two groups of receptors and the orientation of their caps. B. Sensory neurons from receptors in the more proximal group respond to a dorsal bending of the tibia whereas those in the more distal group respond to ventral bending. C. The different directions of bending lead to different responses in tibial motor neurons. D. During walking, the sensory neurons from the two groups of receptors spike at different times during the stepping cycle. A–C are based on Zill and Moran (1981a), Zill *et al.* (1981), and D on Zill and Moran (1981b).

A distal group approximately 50 µm away consists of 3–5 sensilla that are oriented with their long axes parallel to the long axis of the leg so that they respond to axial tension caused by a force that would bend the tibia ventrally, or by a contraction of the extensor tibiae muscle working against an immovable tarsus. Stimulation of these receptors mediates the opposite reflex response by exciting flexor tibiae and trochanteral levator motor neurons and inhibiting the extensor tibiae and trochanteral depressor motor neurons. They normally spike towards the end of the stance phase of walking, whereas the proximal sensilla spike earlier, and could therefore limit the force generated by the extensor muscles during this time (Zill and Moran, 1981b).

8.16. HAIR PLATES

Hair plates consist of tightly packed groups of stiff hairs positioned so that they will be deflected by the joint membrane or parts of the cuticle when a leg is moved (Fig. 8.22). Often there are two types of hairs associated with a plate, one of which is longer (more than 70 µm) than the other (less than 30 µm) and therefore more readily deflected during joint movements. These dense patches of hairs are thus distinct from the longer and thinner hairs that are more sparsely distributed over the surface of the legs and the body.

8.16.1. Locust hair plates

Each of the legs has two hair plates on the coxa and one on the trochanter, while the middle and front legs, but not the hind legs, have a fourth on the tibia. The longer hairs of these plates on the coxa and trochanter are deflected by movements of the adjacent joint, but the effective stimuli for the shorter hairs are unknown.

On the coxa of a middle leg, the ventral hair plate (called CxHP1 by Pflüger *et al.* 1981) consists of about 10 long and 25 short hairs that are stimulated when the coxa is rotated forwards during the swing phase (Kuenzi and Burrows, 1995). Stimulation of these hairs excites the motor neurons innervating the posterior rotator and adductor muscles which swing the leg backwards, and inhibit the anterior rotator muscle which would swing the leg forwards. The sensory neurons from the long hairs make monosynaptic excitatory connections with the posterior rotator and adductor motor neurons. This pattern of connections is thus appropriate for this receptor to signal the extent of the swing phase movement, and to initiate the start of the stance phase by causing a backward rotation of the leg.

Hairs of the tibial hair plate on the front and middle legs are longer and more sparsely arranged than in the other hair plate and their sensory neurons are deflected by an asymmetrical distal projection of the anterior femoral coverplate. They are therefore excited when the tibia extends from the fully flexed position to about 90° (Newland *et al.*, 1995), but their possible effects on the leg motor neurons that control this movement are not known. Whether they play a role in controlling walking movements is brought into some question because the hairs are not present until the fifth larval instar. Even in the fifth instar there are only three hairs, with the final complement of 11 hairs appearing only after the final moult to adulthood.

8.16.2. Cockroach trochanteral hair plate

The trochanteral hair plate consists of a patch of some 50–60 hairs located close to the articulation of the coxa with the trochanter (Wong and Pearson, 1976). The hairs are again of two anatomical types, one with shafts 30–70 μm in length that are deflected by the joint membrane as it flexes, the other with shafts 5–30 μm long that are not so deflected and the function of which remains unknown. The long hairs are, in turn, of two physiological types, one that responds to a displacement with a phasic response that adapts rapidly, the other with a more slowly adapting response that can even be maintained for large displacements. During walking, the long hairs are excited by flexion movements of the femur towards the end of the swing phase. The afferents from these receptors have two effects on central neurons. First, they excite the slow depressor (extensor) trochanter motor neuron directly, and inhibit, through longer pathways, the levator (flexor) trochanter motor neurons and the common inhibitor (Pearson *et al.*, 1976). These effects would thus tend to bring the swing phase to an end and initiate the stance phase, and would thereby prevent the leg from stepping too far. They may also help to stabilise the position of the joint when standing. Second, electrical stimulation of the afferents causes a slowing of the rhythmic bursting activity shown by the levator motor neurons in what may represent a locomotory rhythm. The result could be to advance the bursts in the depressors and limit the bursts in the levators. Removal of the hair plate in a walking animal leads to more intense and prolonged activity in the levators and to an exaggerated swing phase movement in which the leg may overstep and even collide with an adjacent leg. The initiation of activity in the extensors at the start of the stance phase is not delayed as predicted, indicating once again that many parallel channels of signals are used for any one part of the sequence of movements.

8.16.3. Stick insect hair plates

There are two hair plates on the coxa, one called BF3, near the pleural articulation, which has 15–20 hairs some 10–40 μm long, the other called BF2, near the trochantin, which has 20–30 hairs. There are also four hair rows on the dorsal surface. On the trochanter, there are also two hair plates, one consisting of 20–30 hairs at the anterior articulation of the coxo-trochanteral joint, the other with 25 hairs on the ventral surface and with an unknown function. Removal of the anterior trochanteral hair plate causes the operated leg to be held higher than normal, but does not alter the coordination of the other legs. Compensation in the height of the body when challenged with increased loading is less complete when the hair plates are removed from each leg, thus indicating that they are part of a control system that determines the height of the animal when standing (Wendler, 1972).

Shaving off the hair plates on the body of a stick insect that are bent by movements of the coxa results in a tendency for the legs to make exaggerated flexion movements and thus to take larger steps (Wendler, 1966). Waxing the hair plate on the coxa (BF2) of a stick insect so that its hairs are continuously deflected causes the leg to stay on the ground for longer and thus to move further back relative to the body, and delays the start of the next swing phase (Bässler, 1977). Removal of hair plate BF1 at the trochantero-femoral joint causes the operated leg to be held higher than normal, but does not alter the coordination of the other legs during walking.

8.17. HAIR ROWS

There are three hair rows on the coxa (Pflüger *et al.*, 1981) and a fourth on the tibia (Mücke, 1991). These rows are distinct from the plates, as their name suggests, and also from the surrounding and more sparsely distributed tactile hairs. The signals they provide are essentially unknown and their effects on central neurons have not been explored.

8.18. EXTEROCEPTORS

The exteroceptors consist of numerous tactile hairs (trichoid sensilla) and combined tactile and chemosensory hairs (basiconic sensilla) over the entire surface of a leg. The density of these receptors varies on different parts of a leg, being concentrated in exposed regions that are likely to be stimulated when the leg moves. Only a few are positioned so that they will be deflected by the movements of the joints, while those on the ventral surface of the tarsus will presumably be activated when the tarsus is placed on the ground. The mechanosensory receptors only give information about contact with external objects or with another part of the body, because their sensitivity is too low to be activated by wind stimuli. When these receptors are stimulated by contact with an obstacle, specific movements of the leg are generated that lead to appropriate compensatory movements. The nature of the movement depends on the spatial location of the array of receptors that are stimulated and on the phase of the walking movements. For details of the signals provided by these receptors and their processing in the central nervous system, *see Chapter 7*.

8.19. OTHER RECEPTORS

A number of other receptors also occur on a leg but either the stimuli to which they respond or their contribution to the generation and control of movement are unknown.

8.19.1. Brunner's organ

This is a small organ named after its discoverer (Brunner von Wattenwy, 1880) on the ventral surface of the femur of a hind leg (but not present on the other legs), consisting of a soft tubercle 150 μm long with typically three trichoid sensilla and two campaniform sensilla (Slifer and Uvarov, 1938; Heitler and Burrows, 1977). The tubercle is flattened onto the hairs when the tibia is fully flexed about the femur, as happens in preparation for a jump or a kick (*see Chapter 9*), but will not be distorted during normal walking movements. The function of this receptor has not been elucidated although its position on the hind leg and its deformation during full flexion suggest some role in jumping and kicking. The performance of these movements does not, however, appear to be altered by its removal.

8.19.2. Subgenual organ

The subgenual organ is a chordotonal organ that is not associated with a joint. It consists of some 32 sensory neurons that are slung from the proximal cuticle of the tibia and that may be associated with the trachea. The sensory neurons are exquisitely sensitive to vibration of the ground [in the cockroach *Periplaneta americana* they respond to a displacement of 2 nm (Shaw, 1994a)], but whether they also respond to sound as does the comparable organ in the cockroach (Shaw, 1994b) has not been explored. No effect of these sensory neurons on the actions of the muscles of one leg, or on the coordination of the legs, has been demonstrated during walking, but little is known about the connections that they make with central neurons and hence about their likely effect on behaviour. As vibration would seem to be a likely indicator of imminent danger, it would be worth exploring what access this sensory information has to the local circuits that control the movements of the legs.

8.20. ALTERATION OF THE SENSORY SIGNALS DURING WALKING

The sensory terminals of the proprioceptive sensory neurons are depolarised in a sustained fashion throughout walking and receive an additional rhythmical depolarisation in time with the movements of the leg whose actions they are signalling (Wolf and Burrows, 1995) (Fig. 8.24). These presynaptic inputs are probably mediated by GABAergic interneurons and have a reversal potential that is slightly more positive than resting potential (*see Chapter 7*). They have the same properties as those that are produced during passive and active movements of the femoro-tibial joint and could thus be caused by the same sets of interneurons. Thus, a sensory neuron that signals extension of the femoro-tibial joint in its pattern of spikes receives a rhythmic depolarising input during the stance phases of the step cycle when the leg is in contact with the ground and moving backwards relative to the body. Similarly, sensory neurons signalling flexion are depolarised rhythmically during the swing phases of stepping when the tibia is flexed, lifted from the ground and moved forwards. The maximal synaptic input is therefore delivered to particular sensory neurons at different times during the step cycle that correspond to the movements that they signal with their best spike response. The inputs could originate from three sources; first, from the actions of other sensory neurons from the same proprioceptors, or other receptors in the same leg, that can be elicited by passive or active movements of the femoro-tibial joint; second, from the movements of the other legs; third, from the networks of neurons that generate the basic motor pattern for walking. Probably, all three sources act together. A rhythmic synaptic input to a

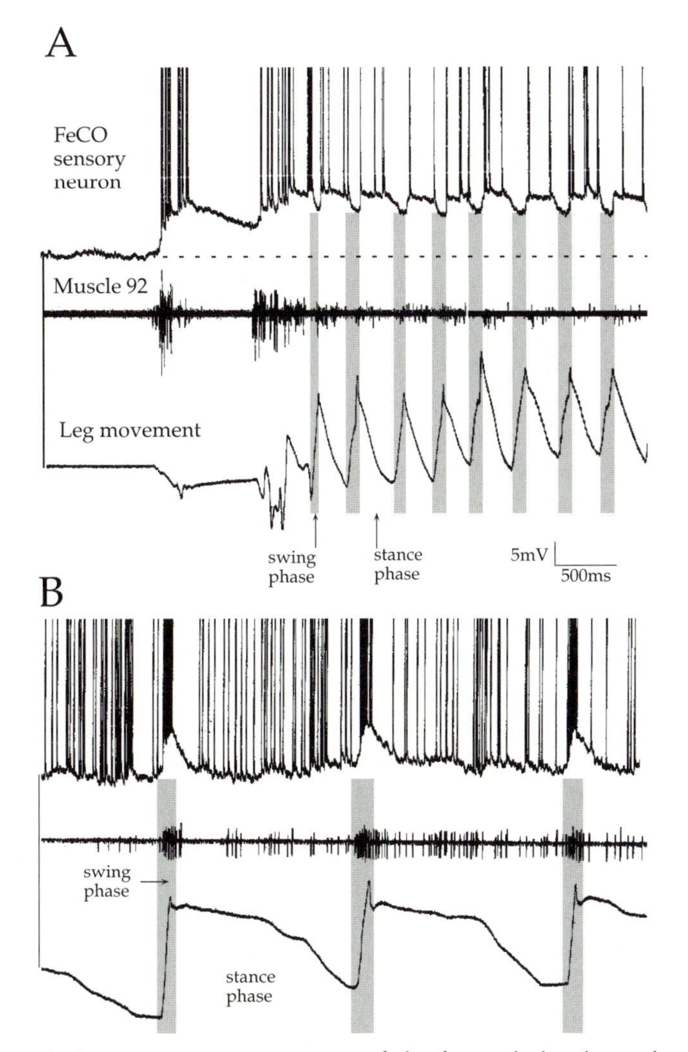

Fig. 8.24 Synaptic inputs to sensory neurons of the femoral chordotonal organ from the middle leg of a locust during walking. A. At the start of walking, the sensory neuron receives a synaptic input which is maintained throughout the walking sequence. During each stance phase, when the sensory neuron spikes it receives a further synaptic input so that its membrane potential is altered rhythmically in time with the stepping cycle. B. A different sensory neuron that produces a burst of spikes during the swing phase receives a depolarising synaptic input during this phase. Based on Wolf and Burrows (1995).

sensory neuron persists during walking even if the leg containing its receptor is removed. This indicates that part of the input must be caused by inputs from other receptors and other legs, or by the pattern generating network. If a metathoracic ganglion is isolated in a dish and treated with pilocarpine then it will express a rhythmic motor pattern in leg motor neurons that may have some relationship to the pattern used in walking (Ryckebusch and Laurent, 1993). During such a rhythm, and in the absence of sensory information from any of the legs, sensory neurons from the femoral chordotonal organ are depolarised rhythmically in time with the motor pattern. These inputs must thus be caused by the network of neurons that is responsible for generating the motor pattern. If this pattern is an expression of walking in some rudimentary form, then it indicates that at least some of the inputs seen during real walking result from the output of this central circuitry.

The consequence of these presynaptic potentials is that the efficacy of the spikes in sensory neurons is reduced during the movements that they signal with their best spike response. The synaptic input can thus be seen as a prediction expressed by the motor network that a particular movement is to be produced and that the resulting sensory signals should be accorded less importance. If, however, an unpredicted movement of a joint occurs, the sensory signals will occur without a synaptic input generated by the central circuitry and will thus exert a larger effect on their postsynaptic targets. The mechanism can thus be seen as a device for correcting the unexpected while allowing the expected to proceed in the face of possibly conflicting sensory information.

8.21. DESCENDING CONTROL

There are about 200 pairs of neurons in the brain that have axons descending to the thoracic ganglia, and a further unknown number in the suboesophageal ganglion. The assumption is that these neurons are in some way involved in controlling the actions of the body in relation to the sensory information that is received by receptors on the head, or in relation to the computations performed by the brain. Many may, of course, be neurons with neuromodulatory or neurosecretory actions, but many must also be involved in shaping the performance of motor acts over a short time scale. The best known are those that control the actions of the muscles moving the wings, to produce steering movements that are appropriate responses to combinations of sensory signals of several modalities (*see Chapter 11*). It is assumed that a similar controlling and regulating action must be exerted on the networks in the thorax that generate walking movements, but the identity and actions of most of these interneurons has yet to be elucidated.

8.21.1. Elucidating the role of descending interneurons

Little is known of the actions of individual interneurons during walking that is comparable to the information available for the descending control of flying (*see Chapter 11*). Instead, the analysis has relied on cruder methods of ablating parts of the nervous system, cutting nerves, or stimulating with extracellular electrodes. Inevitably, therefore, the interpretation of these experiments is difficult and few definitive results have emerged.

Cutting the connectives in the neck of mantids suggests that the descending control is largely inhibitory (Roeder, 1937), although because the operation affects so many pathways, a simple interpretation is not possible. Cutting the connectives of stick insects releases bouts of walking in some individuals (Graham, 1979) but, in others, no organised walking movements occur.

Many experiments have relied on stimulating regions of the brain with extracellular electrodes (Huber, 1960) or the axons of neurons in the connectives with electrodes that must excite many neurons at once (Kien, 1980, 1983). Stimulation at different sites within a connective from the brain or suboesophageal ganglion can sometimes evoke movements of individual legs, or coordinated movements of all the legs in patterns that resemble different forms of walking and struggling movements. Only rarely do these movements outlast the stimuli that are delivered, but most of the parameters of walking can be influenced. The interpretation of such experiments is complex, and the best that can be said is that motor activity can be produced when ensembles of neurons are activated in combinations and with activity patterns that are never normally used. The neurons whose axons are stimulated originate either in the brain or in the suboesophageal ganglion, but few have been characterised adequately by their input and output connections, by their normal modes of action during walking, or by their morphology. Almost no good data exist that show a causal role of any of these brain neurons in influencing walking, nor even a correlation between their spikes and particular features of the leg movements. What is known is that the processing performed by some of these neurons in crickets is dependent on motor activity, so that some respond better to moving visual patterns, or to patterns of sound when the animal is walking rather than standing still (Böhm and Schildberger, 1992). The assumption must remain that neurons such as these may also influence the generation of the walking movements, so that there is a two way interaction between the brain and the networks in the thorax that generate the motor patterns.

While the much trumpeted special role for the suboesophageal ganglion in controlling movement in general and walking in particular may prove to be correct, the data do not as yet define what role this might be. It is not surprising, therefore, that data of this sort should have spawned many models that attempt to explain the interactions between the brain, suboesophageal ganglion, and the thoracic motor controllers in generating movements. These models are little more than formal statements of the pathways that could theoretically exist and provide little insight into the possible functioning of the real networks. It is clear that better data are first needed before the models can become predictive and helpful in the admittedly difficult task of understanding the networks. It is nevertheless clear that physiological analyses by themselves will be inadequate in providing satisfying explanations, and that modelling will become an essential tool in helping to understand how complex assemblies of neurons control complex movements. To be useful, the models must have a reliable base and be of predictive value.

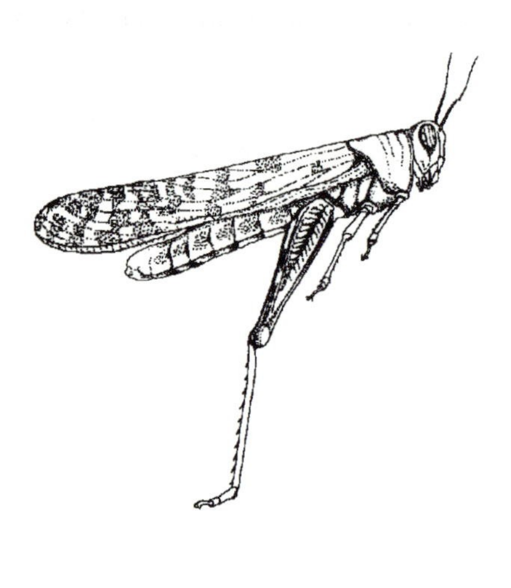

Jumping

9.1. JUMPING MOVEMENTS

Their large and powerful hind legs suggest that locusts are designed more for jumping than for walking, and indeed their performance when jumping approaches the maximum possible capabilities of a biological system. Adults jump by rapidly extending both hind legs at about the same time, to avoid predators and to initiate flight, while larvae boost their speed of locomotion by hopping. Locusts of all stages also repel adversaries by kicking rapidly with their hind legs, generally using them independently. The mechanisms underlying these jumping and kicking movements of the hind legs are basically similar, but are distinct from those used in walking. Both are ballistic movements that are only intermittently expressed at full power, but in kicking a hind leg is rotated at the coxa and the tarsus lifted from the ground, so that it can be aimed. The ability of a locust to leap from one grass stem to another is also testament to the accuracy with which jumping movements can be performed and implies that some aiming can take place at take-off.

The rapidity of the movements of the hind legs coupled with the need to generate considerable power imposes demanding constraints on the design of the cuticular and muscular machinery of these legs. The high velocity and high power output can only be produced by the leg muscles contracting slowly in advance of the movement and storing force in cuticular deformations. This requires mechanical and neural mechanisms to restrain the contractions and prevent movement of a leg until sufficient force is generated. The stored force must then be released suddenly to produce the rapid and powerful extension of the hind legs.

9.1.1. Mechanics of the jump

In preparation for a jump, the front two pairs of legs are first placed symmetrically and the coxae of the hind legs are depressed to bring the femora to an angle of 30–40° with the ground (Fig. 9.1A). The hind tibiae are then fully flexed against the femora and just before the jump the coxae depress still further to bring the femora and tibiae parallel with the ground. From this prejump crouching position, the energy for the jump can now be released by extending the metathoracic tibiae through about 120° in 20–30 ms (Brown, 1967) (Table 9.1).

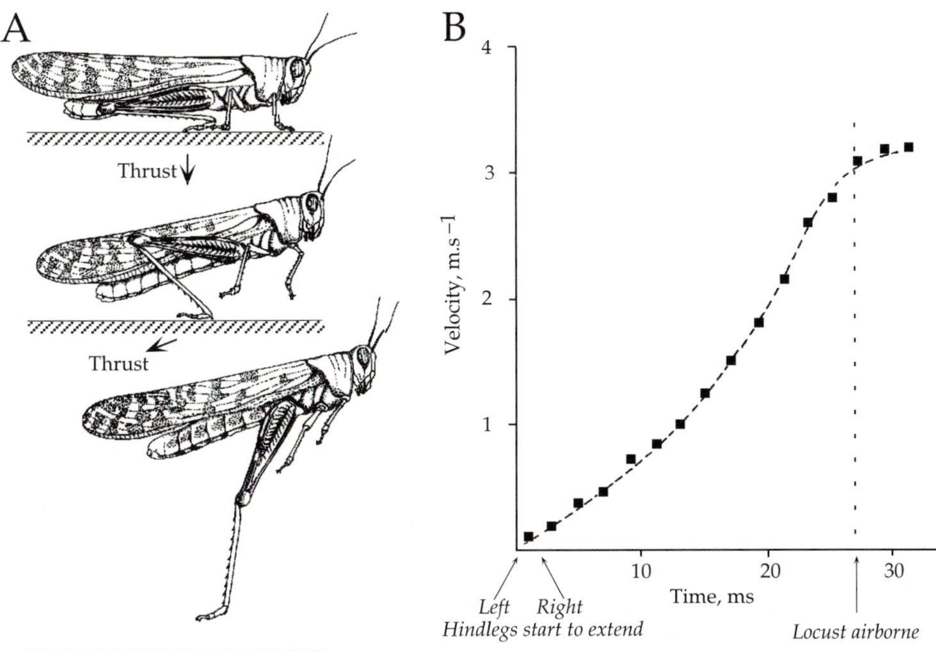

Fig. 9.1 Jumping behaviour. A. Three drawings of a locust in the act of jumping. The hind leg is initially rotated at the coxa so that the tibia, which is fully flexed about the femur, is parallel with the ground. The thrust is downwards. The tibia then begins to extend and lift the locust from the ground. The thrust is now angled backwards. Finally, all the legs leave the ground and the locust is propelled forwards at a take-off angle of about 45°. B. Graph (redrawn from Bennet-Clark, 1975) to show the velocity of the locust from the time that the tibiae start to extend. The locust is airborne in less than 30 ms.

Table 9.1. Features of a jump

Forward distance (m)	0.8–0.95 (20 body lengths) at 30°C
Take-off angle (°)	45
Take-off velocity (m.s^{-1})	3.1
Acceleration (m.s^{-2})	180
Final velocity (m.s^{-1})	2.8
Energy requirement (mJ)	7–9
Time to release energy store (ms)	20–30

The initial thrust of the extending tibiae is directed downwards, but as they continue to extend, the thrust develops a backward component which propels the locust forwards at a take-off angle of about 45°. This is the optimal angle to ensure that the greatest forward distance is covered (Fig. 9.1A) (Pond, 1972). The net result is that the locust can be airborne within 30 ms at a take-off velocity of 3.1 $m.s^{-1}$ (Fig. 9.1B). The two hind tibiae do not necessarily extend at precisely the same time, and the movement of one can be delayed relative to the other by a few milliseconds. If the time difference between the extension of the two legs could be varied systematically, then it suggests a mechanism by which the jump could be directed. Locusts will jump towards a target and can regulate the velocity of take-off, producing a faster take-off towards a more distant target (Sobel, 1990).

The jump requires that the hind legs be extended rapidly and with a power output that is several times greater than the usual capability of muscle, which can be specialised to produce either maximum force or maximum velocity, but not both. These conflicting requirements are met by a mechanism in which energy is stored before the jump and then released rapidly. The power is produced by the extensor tibiae muscles in each of the hind legs, and these muscles have a combined mass of about 140 mg. Each can produce a peak power output of 35 mW, or about 0.5 $kW.kg^{-1}$ muscle, so that their performance is similar to most other muscles, but each takes some 500 ms to generate its full force (Hoyle, 1955). During the jump, however, each leg produces about 375 mW, so that there must be a mechanism that provides a power amplification of 10 and at the same time allows the extensor muscles time to generate their maximum force. This implies that energy must be stored in advance of the movement.

The energy is stored in distortions of specialised cuticular structures in the femur which themselves weigh little and yet can store the energy produced by the large extensor tibiae muscle (Fig. 9.2). Most of the energy is stored in distortions of the semilunar processes of the femur near the joint with the tibia (4 mJ in one leg at peak stresses of 15N), in the apodeme (tendon) of the extensor tibiae muscle (3 mJ), and a small amount (0.3 mJ) in the walls of the femoral cuticle of each hind leg (Bennet-Clark, 1975). The energy that can be stored by the two legs is thus well in excess of the 7–9 mJ needed for a jump. These cuticular structures act like springs and can deliver considerable force at high velocity so that the power limitations of the muscles are surmounted.

This strategy for jumping and kicking means that the movement can be produced by a slowly contracting muscle that can generate considerable tetanic force. The price paid for avoiding the constraint imposed by the relationship between muscle force and velocity is the long time that is needed to prepare for a jump.

9.1.2. Development of jumping ability

Locusts increase enormously in mass from first instar hoppers to mature adults and this growth is achieved by a series of moults (Fig. 9.3A). This periodic softening and replacement of the cuticle so that the body can grow in size has a deleterious effect on locomotion and jumping in particular, because the ability to jump depends on the

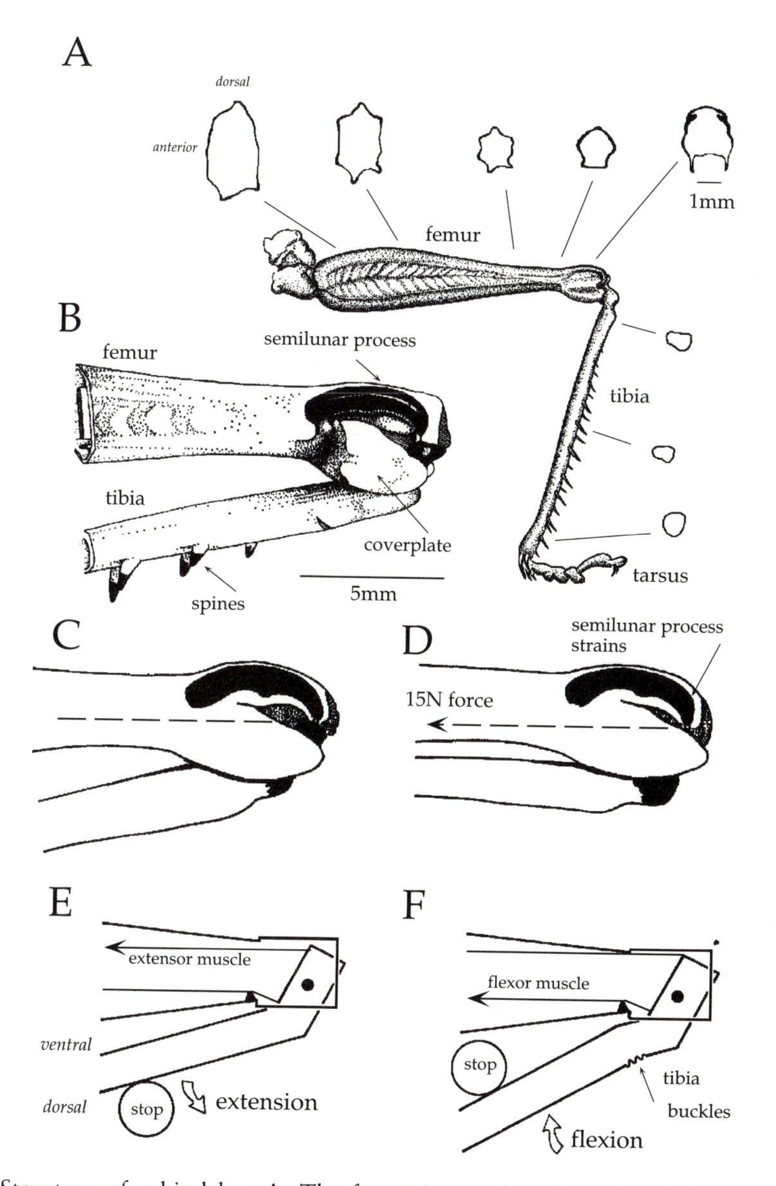

Fig. 9.2 Structure of a hind leg. A. The femur is greatly enlarged with herring-bone-like markings that indicate the insertion points of the extensor muscle, and the tibia is slender with prominent spines. In cross section, the femur is hexagonal. B. The femoro-tibial joint enlarged to show the black and heavily sclerotised semilunar processes, the thin coverplates, and the double row of tibial spines. C. The position of a semilunar process when the tibia is not fully flexed. D. The ventral distortion of the semilunar process when a 15N force is applied to simulate the forces generated during co-contraction. E. When the tibia extends and meets a resistance, the heavy scelerotisation on the ventral surface prevents distortion. F. When the tibia is prevented from flexing fully, it buckles at a predefined plane due to the weakness of the dorsal surface. B–D are from Bennet-Clark (1975), E and F from Heitler (1977).

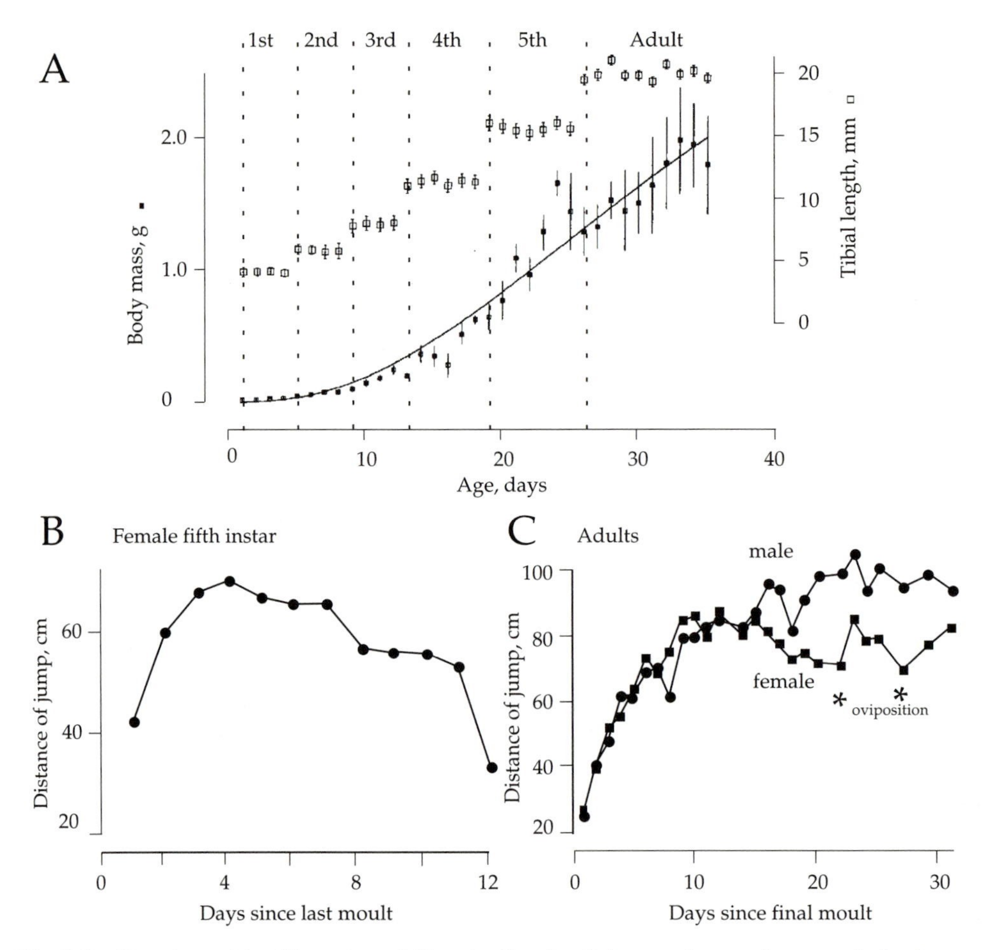

Fig. 9.3 Development of jumping ability. A. Graph of the continuous increase in body mass and the discontinuous increase in the length of a hind tibia during the five larval instars and as an adult. B. In a larval instar (the fifth is shown), the distance a locust can jump increases initially, reaches a plateau and then declines as a prelude to the next moult. C. As an adult, the distance jumped gradually increases, with males progressively outdistancing females. The distance a female can jump declines during periods when she is laying eggs. A is redrawn from Katz and Gosline (1992), B and C from Queathem (1991).

rigidity of the exoskeleton. The result is that jumping ability is not constant throughout life, but changes with the moulting cycle of the larvae and with the maturational changes that occur in adults.

First to fourth instar locusts can only jump about a third as far as adults, relative to their size, and even fifth instars can only perform about half as well as adults (Gabriel, 1985). All the instars can produce a take-off velocity that is only about half that of adults (Katz and Gosline, 1993), but as their wings are not functional this may not matter. The distance a locust can jump expressed per unit of body mass declines as it develops, so that the length of a jump scales as body mass to the 0.2 power (Queathem, 1991). By contrast, as the tibiae grow they become longer and thinner (Fig. 9.3A) yet increase in stiffness (Katz and Gosline, 1992). Stiffness therefore scales with body mass. In each instar, jumping ability gradually increases and then falls with the imminence of the next moult (Fig. 9.3B) (Queathem, 1991). Within one instar the body mass also increases, and this may explain why the energy that is produced for a jump peaks later than the peak of performance. The ability to jump further as the locust nears adulthood is associated with an increased length and mass of the femur, and with some thickening of the cuticular specialisations that are important in jumping. All this suggests that jumping has a different function in the flightless juveniles and in adults that can fly. In juveniles, repeated jumping (hopping) is used to increase the speed of locomotion by as much as 10 times above that possible during walking. In adults, by contrast, jumping is frequently used to achieve the minimum velocity for flight take-off (Weis-Fogh, 1956a; Katz and Gosline, 1993).

In adult males, jumping performance reaches a plateau in the first few weeks following the final moult, and in females it then fluctuates as they produce and lay eggs (Fig. 9.3C). Females are heavier (2.5–3.5 g) than the smaller males (1.5–2.0 g) and their jumping performance is correspondingly reduced (Fig. 9.3C). For most of the time, therefore, females will be operating below their maximum capabilities compared with males. The fluctuating jumping performance suggests that behavioural strategies must be adopted to avoid predation whenever this is low. Moreover, the risk of predation suggests a trade-off with the number of offspring that a female will produce, and with the number of steps in growth that must be undergone to reach adulthood.

9.1.3. Changes in jumping and kicking with the moulting cycle

If a locust jumps or kicks when the cuticle is soft then it is likely to damage the exoskeleton of a hind leg, with subsequent deleterious effects on its ability to compete for a mate and avoid predation. At all stages in the moulting cycle, contraction of the extensor tibiae muscle is potentially capable of damaging the exoskeleton (Norman, 1995), in much the same way that baseball pitchers and weightlifters can sometimes break bones in their arms as a result of their muscular contractions. When the cuticle is soft after a moult, contractions of the extensor muscle deform the walls of the femur so that force is not transmitted to the tibia, and if the force is not above a certain level then the deformations are reversible. This can be seen as a mechanism for protecting the other equally weak parts of the rest of the leg from irreparable damage. At the day of the moult all locusts show these deformations if the extensor contracts. A day after the moult, the most common damage is a rupture of the insertion of the extensor tendon on the proximal femur, or the snapping of the semilunar processes, because now the proximal femur does not collapse so the force is directly transmitted to the still weak distal structures. Five days after the moult, the incidence of damage falls, but now the most common effect is that the tendon of the extensor tibiae muscle breaks. Fourteen days after the moult, the incidence of damage declines still further with the only form of damage being the occasional break of the extensor tendon.

To lessen the probability of inflicting damage on its hind legs, the locust adopts the following strategies. First, the frequency with which kicks can be elicited by the same stimulus falls during the 10 days before a moult to reach zero on the day of the moult, recovering to its original level some 3 days after the moult. This means that the locust will only kick in the most exceptional circumstances. Second, newly moulted animals do not generate the motor pattern that characterises a kick or a jump. The extensor and flexor muscles do not co-contract, but instead are used separately with the result that all extension movements of the tibiae are much slower.

9.2. THE MACHINERY FOR JUMPING

The hind legs are clearly very different in design from the other two pairs of legs and this seems to be directly related to their role in jumping and kicking.

9.2.1. Femur

The femur is expanded dorso-ventrally, accommodates the large extensor tibiae muscle and has almost 10 times the mass of the femora of the front or middle legs. A series of longitudinal ridges gives a hind femur an approximately hexagonal cross section that should increase its ability to withstand longitudinal loads (Fig. 9.2A). Near the articulation with the tibia are two black semilunar processes (Fig. 9.2B) which are highly sclerotised, longitudinally laminated structures each weighing only about 1.3 mg. These are joined proximally by a ring of thick, but less brittle cuticle, which protrudes into the cavity of the femur to form a lump that is thickened on the inside by a layer of resilin (Fig. 9.4) (Heitler, 1974). At the distal end, the semilunar processes are joined ventrally by flexible cuticle to two flaps of cuticle called the coverplates. When the extensor muscle contracts isometrically, the semilunar processes are strained so that their distal ends move some 0.4 mm ventrally and at the same time the walls of the femur are strained (Fig. 9.2C,D) (Bennet-Clark, 1975).

9.2.2. Tibia

The tibia of a hind leg is long and thin with a strong tubular construction. In each instar it is twice the length of a tibia in a middle leg but has only the same diameter, even though it must withstand much greater forces during a jump (Katz and Gosline, 1992). On the dorsal side it is armed with a double row of immoveable spines that increase its stiffness and are useful in fending off adversaries when kicking (Fig. 9.2A,B). It contains only tracheae, nerves and the small muscles that move the tarsus, so that it is light and therefore minimises the energy lost in overcoming inertia during extension.

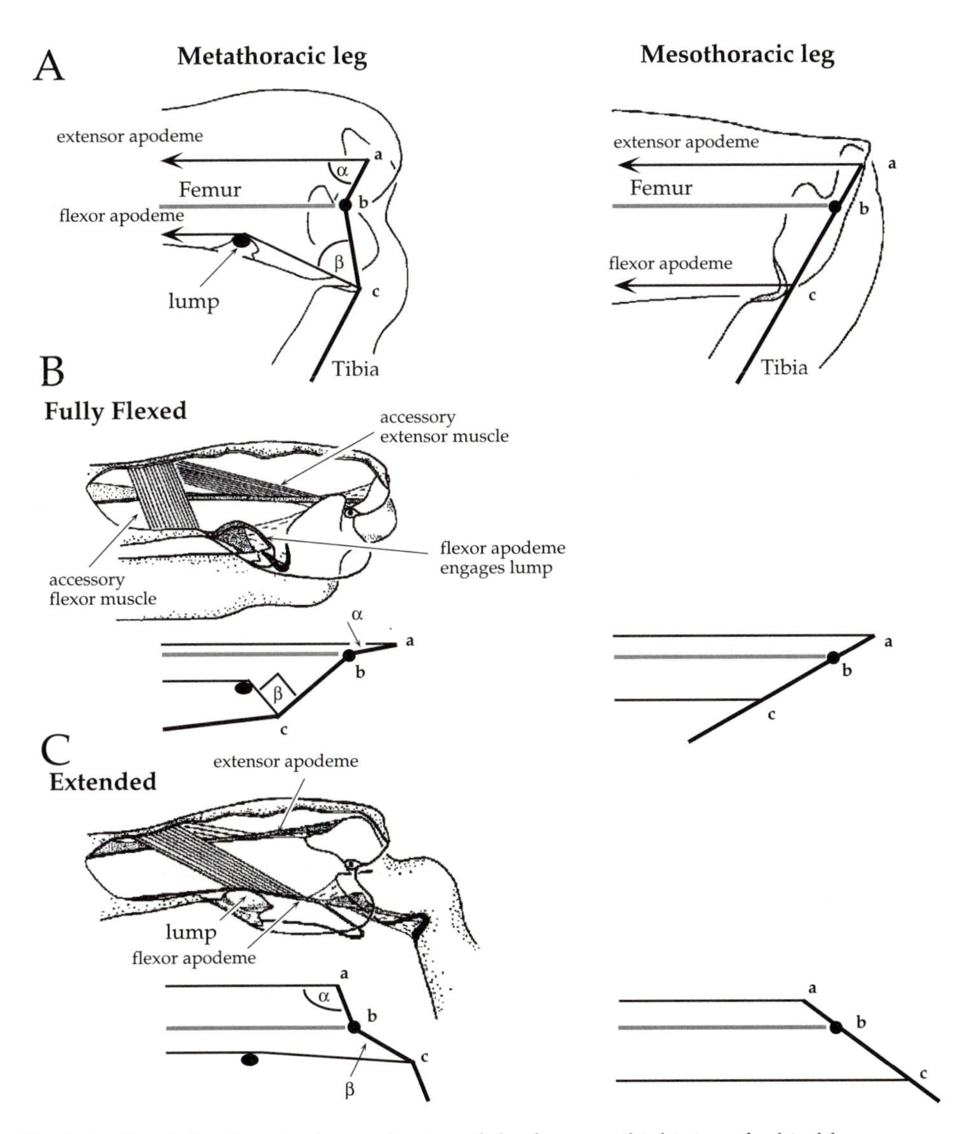

Fig. 9.4 Specialisations in the mechanics of the femoro-tibial joint of a hind leg as compared with a middle leg. A. Diagram of the main structural elements superimposed on outline drawings of the joints. The pivot (b), and the insertions of the extensor (a) and flexor (c) muscle apodemes are labelled. and indicate the angles formed by the lines of action of the extensor and flexor apodemes, respectively, with a line joining their insertion to the pivot. The lump on the ventral femur that provides the effective insertion for the flexor is indicated by a filled oval. B. The tibia is fully flexed and the pocket in the flexor apodeme engages the lump on the ventral femur. The small distal bundles of the extensor and flexor muscles are also shown. C. The tibia is extended and the flexor apodeme is disengaged from the lump. Based on Heitler (1974, 1977).

Contraction of the femoral muscles during the preparation for a jump or a kick puts the ventral surface of the proximal tibia under tension and the dorsal surface under compression. The ventral surface is heavily sclerotised so that it remains rigid under the stress of a jump, or when it meets an obstacle during extension (Fig. 9.2E). By contrast, the dorsal surface is lightly sclerotised and has little compressive strength (Bennet-Clark, 1975; Heitler, 1977). In an adult, the tibia can bend reversibly by 15–20° at a plane of weakness in this region (Fig. 9.2F), but in an immature locust the bending can be as great as 70–80° so that large forces can permanently damage the leg. During the full flexion of the femoro-tibial joint that precedes a normal jump, the tibia does not bend at the plane of weakness, but if a small obstruction prevents full flexion, the tibia does bend. The stiffness of the tibia means that it can withstand most of the bending forces exerted during jumping, but it may bend at the distal end during the preparation for a jump or kick, and therefore act as an additional elastic storage device contributing some 1.2 mJ of energy (Katz and Gosline, 1992).

9.2.3. Joint mechanics

The femoro-tibial joint of a hind leg has distinct specialisations in its mechanics that are associated with jumping (Fig. 9.4). The tibia of a middle or front leg can flex only to a minimum angle of about 25° about the femur, and the mechanical advantage of 2: 1 for the flexor tibiae muscle over the extensor remains constant throughout the entire range of movements. By contrast, the tibia of a hind leg can flex fully about the femur so that the segments become almost parallel, and can extend through 150°. It rotates about a fixed pivot made by two lateral pegs on the femur that fit into sockets on the head of the tibia. The arrangement of the joint is such that the lever ratio of the flexor and extensor muscles changes as the tibia extends. All this has two important consequences. First, it enables the femur and tibia to be nearly parallel with the ground just before take-off, and ensures the optimal application of thrust for a forward movement. Second, it enables the force necessary for jumping to be developed slowly, stored, and then released rapidly. Two features of the joint are important; its leverage and a flexor apodeme lock.

9.2.3.1. *Joint leverage*

The perpendicular distance between the lines of action of the apodemes of the extensor and flexor muscles and the pivot changes as the joint rotates (Fig. 9.4). The insertion of the extensor apodeme (a in Fig. 9.4) is distant from the pivot (b in Fig. 9.4) so that during rotation the apodeme always remains parallel to the long axis of the femur (Heitler, 1974). The extensor lever arm therefore varies with the sine of the joint angle, but always remains small so that the extensor muscle can develop high force with low torque. The flexor apodeme, by contrast, slides over a lump in the ventral wall of the femur that acts both as a pulley wheel and as the effective origin (real insertion is c in Fig. 9.4), and because this is near the pivot, the line of action of the flexor apodeme changes as the joint rotates. This, of course, assumes that the contractions of the muscles do not cause the lump to move. If this were to happen then clearly the lever ratios would change. When the joint is fully flexed, the angle between the line of action of the flexor and the line from the insertion to the pivot (flexor angle, β in Fig. 9.4B) is almost 90°, whereas the equivalent angle of the extensor is about 6° (α in Fig. 9.4B). The flexor lever arm is thus maximal and that of the extensor minimal. The flexor therefore exerts maximum leverage on the joint with a 21: 1 advantage over the extensor. As the joint extends to 90°, the flexor leverage drops while that of the extensor rises, but with further extension the leverage of both muscles decreases so that at 120° the mechanical advantage of the flexor over the extensor has fallen to 1.5: 1. All of this means that when the tibia is fully flexed, the small flexor muscle can hold the tibia in this position in opposition to the much greater force generated by the large extensor muscle. As the leg extends and the lever ratio between the flexor and extensor falls, any residual flexor tension will offer progressively less resistance to an extension.

9.2.3.2. *Flexor apodeme lock*

The flexor apodeme inserts on the tibia as two sclerotised strips joined by flexible connective tissue that forms a pocket. The two parts join about 1mm from the insertion to form a continuous apodeme through the body of the flexor muscle. At joint angles of more than 5°, the apodeme slides over the top of the inwardly protruding lump on the ventral wall of the femur (Fig. 9.4) (Heitler, 1974). As the joint approaches full flexion, however, the two arms of the apodeme slide down either side of the lump into grooves at its base and the lump slips into the soft pocket of connective tissue, effectively locking the tibia against the femur. Thus, when the flexor contracts and flexes the tibia fully, its effective lever ratio rises enormously and it engages the lock. If it continues to contract, the lock remains engaged and the tibia is unable to extend. The importance of the lock can be demonstrated quite simply by inserting a small piece of Plasticine between the tibia and the femur to prevent full flexion (Fig. 9.2F); the locust is then unable to jump or kick. It is not clear whether there is a special mechanism for disengaging the lock to allow extension to proceed, but at present it is assumed that a combination of the relaxation of the flexor and the continued development of force in the extensor suffices.

The combination of joint leverage and the flexor apodeme lock allows the weak flexor, which is able to generate only small forces, to resist the large forces produced by the more powerful extensor muscle. The extensor can therefore contract slowly and almost isometrically to produce the force required for a jump. The separation of power production by the extensor muscle and control of the expression of this power by the action of the flexor muscle has two functional consequences. First, it might be possible to set the threshold of the lock actively simply by changing the force in the flexor. Without this control, the setting would simply have to be lower than the maximum force that the extensor could produce, to allow a safety margin for circumstances, such as fatigue or differences in temperature, when the maximum force could not be produced. Second, the timing of the jump can be controlled precisely, rather than relying on the force of the extensor muscle gradually exceeding a threshold as it approaches its maximum.

9.3. THE MUSCLES

9.3.1. Apodemes of the extensor and flexor tibiae muscles

The apodeme of the extensor muscle is flattened dorso-ventrally at its insertion onto the end of the tibia. It twists through 90° in the distal part of the femur (Hoyle, 1955; Bennet-Clark, 1975) so that in the main part of the femur it is flattened antero-posteriorly to provide a large surface area for the attachment of muscle fibres. When strained by isometric contractions it can store some 3 mJ, or almost half the energy needed for a jump. Despite the high tensile strength of the materials from which it is made, the apodeme occasionally snaps in normal jumping some 3 mm from its insertion on the tibia. By contrast, the flexor apodeme is much stronger than its muscle and never breaks. The safety factors for these critical structures involved in the jump are therefore low, but increasing their strength would reduce the effectiveness of a jump because they would be able to store less energy (Alexander, 1981). This would seem to make it likely that the forces generated by the muscles during a jump must be monitored closely by proprioceptors to keep them within the capabilities of the load-bearing structures.

9.3.2. Extensor muscle fibres

The extensor muscle of a hind leg has approximately five times the mass of the flexor and can produce a maximum force of about 15N compared to 0.7N for the flexor. In the front and middle legs the arrangement is reversed, with the flexor muscles larger than the extensors and able to produce a greater power output (Burns and Usherwood, 1978). The hind leg extensor is specialised to generate large forces rather than fast contractions. Peak force is reached only slowly, in 350–500 ms, but the power output of 0.5 $mW.kg^{-1}$ is very high. There are about 3500 short (2.5–4.3 mm) fibres, with diameters of 40–150 μm (Hoyle, 1955, 1978b), that are arranged in chevrons so that the total cross-sectional area is about 16–17 mm^2 (Bennet-Clark, 1975). This arrangement is again a specialisation for the production of large forces rather than speed of contraction.

The muscle fibres have been classified into two types, based on the contractions they produce in high K^+ saline and on their ultrastructure. Tonic fibres give a sustained contraction, have well-aligned Z bands with mitochondria on either side and only 1% of their volume is occupied by sarcoplasmic reticulum. Phasic fibres contract only transiently, have poorly aligned Z bands but a more extensive sarcoplasmic reticulum that occupies 7% of their volume (Cochrane *et al.*, 1972). The length of the A bands of fibres throughout the muscle is about 5.5 μm.

The innervation is provided by two excitatory motor neurons, one slow (SETi, slow extensor tibiae) and one fast (FETi, fast extensor tibiae), by a branch of common inhibitory neuron 1 (CI1), and by a modulatory neuron (DUMETi, dorsal unpaired median neuron of the extensor tibiae). About 76% of the fibres are innervated by the fast but not the slow motor neuron, 15.5% by both, and only 8.5% by the slow but not the fast (Hoyle, 1978b). In general, the phasic muscle fibres are innervated exclusively by the fast, and the tonic fibres by the slow motor neuron. The tonic fibres occur in two distinct regions. One, at the distal end of the muscle, consists of two separate bundles each of 8–12 muscle fibres [called 135c and d (Snodgrass, 1929) and more commonly, the accessory extensors] which insert on the apodeme at a different angle to the main body of the muscle. They are innervated by SETi and CI1. The other bundle is at the proximal end of the muscle and consists of 8–12 fibres 150 μm in diameter that are innervated by SETi, CI1 and supplied by nearby endings of DUMETi (May *et al.*, 1979). This bundle of fibres generates myogenic rhythmical contractions (*see Chapter 5*).

Spikes in FETi evoke large, 30–50 mV synaptic potentials in muscle fibres that are initiated at closely spaced junctions along the length of a fibre, and which can give rise to graded, sometimes overshooting, active membrane responses. The potentials show little facilitation and each causes a powerful twitch contraction. By contrast, spikes in SETi evoke synaptic potentials (up to 20 mV in amplitude) that vary both in their initial amplitudes and in the amount that they facilitate. Graded active membrane responses may sometimes occur. The force produced by the muscle in response to the slow motor neuron is sensitive to the history of its patterns of spikes so that the rate of contraction can be doubled and the rate of relaxation prolonged 30-fold (Hoyle, 1978b).

9.3.3. Flexor muscle fibres

The flexor muscle has long (7 mm) fibres and a cross-sectional area of only about 2 mm^2, or 12% of that of the extensor. It can contract much faster than the extensor, reaching its peak force in 50 ms and relaxing again in about the same time. The force it produces, although smaller than that produced by the extensor, is sufficient to engage the locking mechanism and to restrain the contraction of the extensor muscle when the tibia is fully flexed. The fibres also insert onto the apodeme at low angles and this arrangement thus also contrasts with the extensor muscle. No values are available for the A band lengths of the fibres, but in *Calliptamus* the sarcomere lengths vary from 5 to 9 μm, with the longest in the proximal part of the muscle (Theophilidis and Dimitriadis, 1990).

The flexor is innervated by at least nine excitatory (Phillips, 1980) and two common inhibitory motor neurons (CI2 and CI3) (Hale and Burrows, 1985). In the front and middle legs there may even be more motor neurons (Theophilidis, 1983). This complex innervation of the flexor is paralleled in other insects; the equivalent muscle in cockroaches has at least eight excitatory motor neurons (Dresden and Nijenhuis, 1958) and in stick insects 14–15 (Debrodt and Bässler, 1989; Storrer *et al.*, 1986). The explanation for the large number of motor neurons must lie in a subdivision of function amongst members of this pool, or the subdivision of the muscle into units that are separately innervated. The motor neurons can be subdivided according to the contractions they produce in the muscle; individual motor neurons produce contractions ranging from slow to fast, with those at the extremes of this spectrum being used either for sustained contractions or for generating rapid increases in force. The evidence for the subdivision of the muscle into units is inconclusive, however, largely because the appropriate experiments have yet to be performed. For example, in *Schistocerca gregaria*, the terminals of all the motor neurons are said to be distributed uniformly throughout the muscle, although the evidence for this is weak (Phillips, 1980), whereas in the tettigonid *Decticus albifrons* there is some evidence that the proximal part of the muscle is innervated by some motor neurons that do not innervate the distal part (Theophilidis and Dimitriadis, 1990). The issue could be resolved by plotting the distribution of junctional potentials in the muscle that are evoked by intracellular stimulation of individual flexor motor neurons. The profuse innervation of the flexor contrasts with the restricted innervation of the extensor and has not been explained in any functional way.

The main body of the flexor muscle, like that of the extensor, does not extend into the distal part of the femur, but here there are a pair of muscle bundles that consist of thin sheets of fibres originating on the anterior and posterior surfaces of the femur and inserting on the flexor apodeme. These accessory muscles, like those of the extensor, are not present in the other pairs of legs (Myers and Ball, 1987), suggesting that they have a function related to the specialisation of the hind legs for jumping. The accessory flexor would seem to be appropriately placed to play some role in controlling the flexor apodeme lock (Fig. 9.4B,C), but when the leg is fully flexed and the lock engaged it is in its most relaxed state, although still capable of producing some force. A more likely role is in controlling the posture of the joint when the locust is standing (*see Chapter 8*). The accessory muscle contains a tension receptor (Matheson and Field, 1995), thus raising the possibility that it acts as a muscle receptor with the motor innervation setting the sensitivity of the receptor.

9.4. THE MOTOR PATTERN FOR JUMPING

The motor pattern that leads to the rapid extension of the hind legs in a jump or a kick consists of three phases: an initial **cocking** which is followed by **co-contraction** and finally by **triggering**. This pattern can be most clearly seen in recordings of the activity of the motor neurons that control the muscles moving the different joints of a hind leg (Figs 9.5 and 9.6). The main power is generated by the muscles that move the tibia, but the more proximal joints must be stiffened to allow this power to be delivered appropriately (Burrows, 1995).

9.4.1. Cocking

In this preparatory phase, the hind legs are rotated about the coxae to the appropriate position for making a jump or directing a kick. During kicking, the tarsus is lifted from the ground. The tibiae are then fully flexed when the flexor tibiae motor neurons spike (Fig. 9.5A,B). The locust thus assumes a preparatory crouching position. The flexor tibiae muscles are the dominant ones to be active, and at this stage the extensors have not started to contract.

9.4.2. Co-contraction

When a tibia of a hind leg is fully flexed, the activity of the flexor tibiae motor neurons increases still further and the extensor motor neurons begin to spike, so that the flexor and extensor tibiae muscles now co-contract (Figs 9.5 and 9.6). This phase is characterised by spikes in the FETi, which, in the hind legs, is not used in normal walking movements. It spikes for 150–500 ms at instantaneous frequencies of up to 100 Hz, while the SETi motor neuron produces more spikes and at frequencies up to 250 Hz. Many flexor motor neurons are also active, so that the flexor muscle is able to prevent the force generated by the extensor from extending the tibia. The mechanical advantage conferred by its lever ratio and the lump that engages the flexor apodeme enables it to perform this action. The duration of the co-contraction determines the power that will be available in the final movement.

9.4.3. Triggering

At the end of the co-contraction, the excitatory flexor motor neurons stop spiking abruptly and, at the same time, the inhibitory motor neurons that innervate the flexor muscles are excited rapidly (Fig. 9.5A,B). This combination of the cessation of excitation and active inhibition, most probably caused by interneurons, causes the flexor force to drop rapidly thereby releasing the lock engaged by the flexor apodeme. The spikes in the extensor motor neurons continue so that the extensor muscle now begins to shorten. As the tibia extends, FETi is inhibited and stops spiking, suggesting that at least part of the inhibition of FETi is caused by a changed sensory feedback as a result of the tibial movement. The end result of this complex motor pattern is that the energy produced by the extensor muscle, and stored in the distortions of the cuticle and apodemes is released suddenly and causes the tibia to extend with high force and velocity.

9.5. HOW IS THE MOTOR PATTERN GENERATED?

To explain how this motor pattern is produced requires an understanding of any interconnections between the motor neurons, the identification of the interneurons that control the motor neurons and a determination of their properties and connections, and an analysis of the effects of sensory feedback, particularly from a hind leg. The analyses so far have tended to concentrate on the connections of the neurons, largely to the exclusion of their cellular properties and the way that the synaptic connections may be modified. This approach is only an essential first step

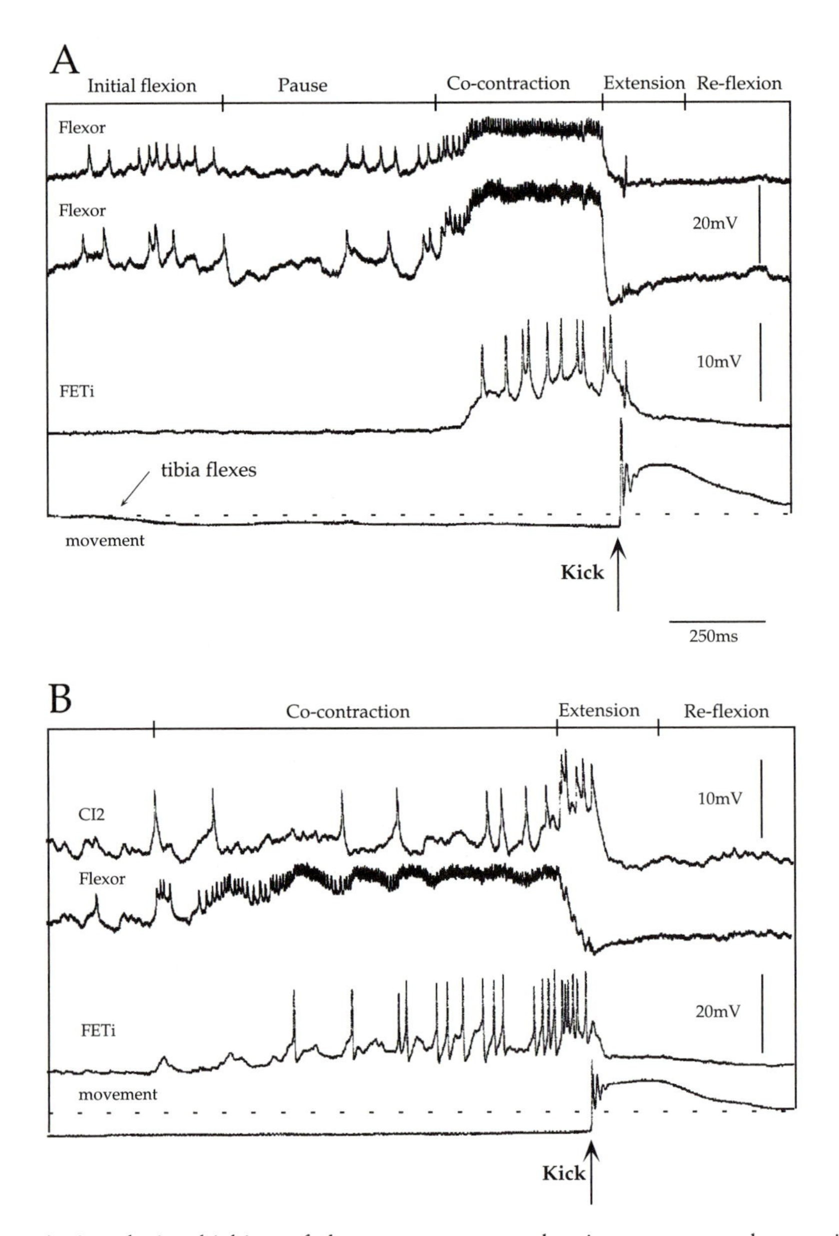

Fig. 9.5 Action during kicking of the motor neurons that innervate muscles moving the tibia of a hind leg. A. Simultaneous intracellular recording from two flexor tibiae motor neurons and the fast extensor tibiae (FETi). The tibia initially flexes when the flexor motor neurons spike. There is then a period of co-contraction. The flexor motor neurons are then inhibited and the tibia extends. B. Simultaneous recording from a flexor, FETi and common inhibitor 2 (CI2) that innervates the flexor tibiae muscle. The inhibitor produces a burst of spikes when the flexor is inhibited and just before the kick is produced. Based on Burrows (1995).

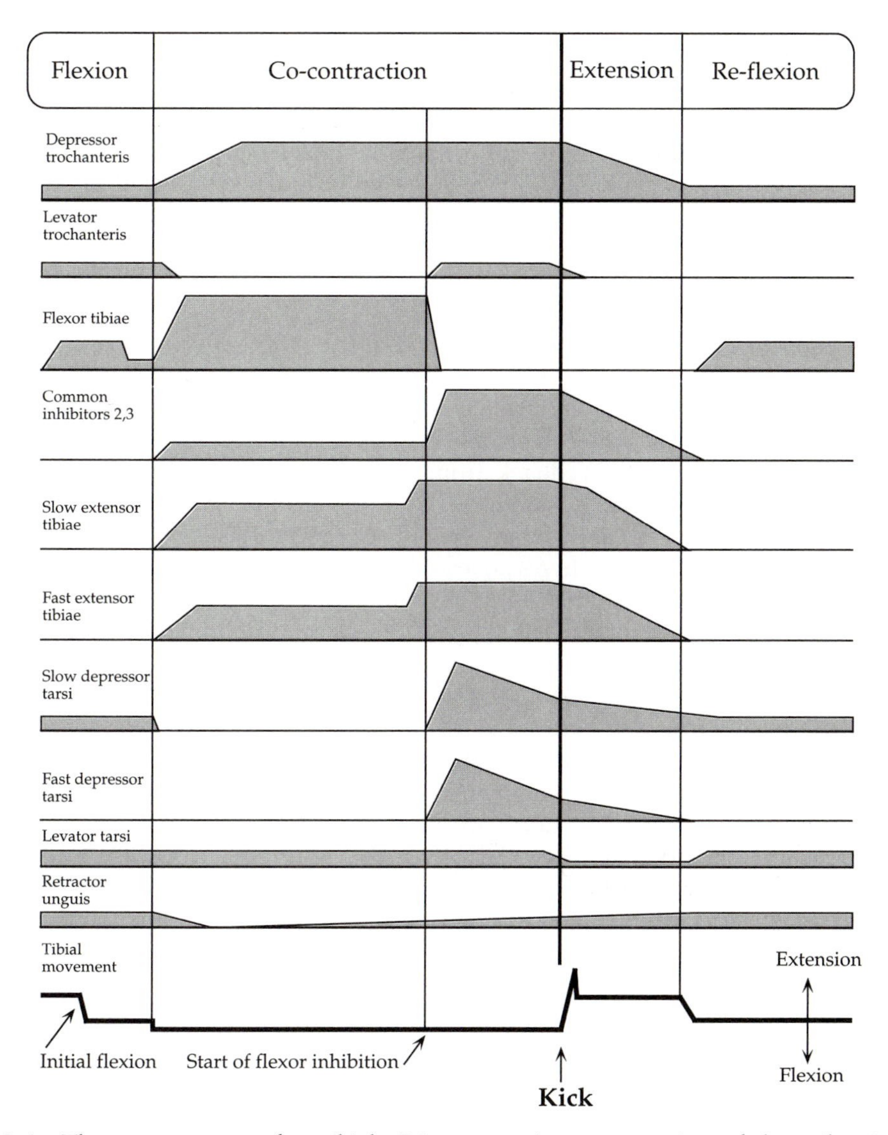

Fig. 9.6 The motor pattern for a kick. Diagrammatic representation of the spikes in the major motor neurons that innervate the various joints of a hind leg. The movement of the femoro-tibial joint is shown below. Based on Burrows (1995).

and cannot be expected to produce completely satisfying explanations of the way the circuitry works. An isolated metathoracic ganglion can generate the whole sequence of motor activity for a kick, so that the necessary set of interneurons must be contained within this ganglion. These interneurons are, however, influenced by interneurons carrying signals from other parts of the nervous system.

9.5.1. Central interactions between motor neurons

A central synaptic connection exists between FETi and many, perhaps all, of the flexor tibiae motor neurons innervating a hind leg (Fig. 9.7) (Hoyle and Burrows, 1973). The connection does not exist between the homologous neurons of the other pairs of legs (Wilson and Hoyle, 1978), suggesting a function specific to the role of the hind legs in jumping and kicking. Both physiological and ultrastructural evidence indicates that the major component of the synaptic connection is monosynaptic although polysynaptic pathways may also exist in parallel (Burrows *et al.*, 1989).

A single FETi spike can evoke a depolarisation and spikes in several flexor motor neurons (Fig. 9.7B). The EPSP, as recorded in the soma of a flexor, can be 20 mV in amplitude, reaches its peak in 20 ms and declines to half its maximum amplitude in 30 ms. The EPSP depends entirely on connections within the central nervous system and not on feedback from peripheral sense organs. Feedback can, however, add later synaptic potentials. The EPSP is chemically mediated as it is associated with an increase in the conductance of the flexor, is increased in amplitude when the flexor membrane is hyperpolarised, and will generate spikes, so that it is clearly excitatory. The transmitter released by FETi at these central synapses, as at its synapses with the extensor muscle, appears to be glutamate (Parker, 1994) (*see Chapter* 5).

At temperatures of about 20°C, the latency from the FETi spike to the EPSP in a flexor is about 2.0 ms, as measured close to the synapses in the neuropil, but can appear to be double this if the recordings are made from the somata. The first recordings from somata at these temperatures were therefore interpreted to indicate that the connection involved interneurons. Increasing the temperature to 35°C, however, causes the delay to fall to 0.5 ms with a slope of 0.75 ms.°C^{-1} (Burrows, 1989). In high Mg^{2+} and zero Ca^{2+} saline the amplitude of the PSP declines gradually. This physiological evidence therefore indicates that the connection is monosynaptic.

Fig. 9.7 A central connection between two motor neurons involved in kicking; FETi makes direct excitatory connections with flexor tibiae motor neurons. A. The structure of FETi and a fast flexor to show the overlap of their branches. B (left panel). A simultaneous intracellular recording from the soma of FETi, and from the soma and axon of the same flexor motor neuron. Two sweeps are superimposed and triggered from single FETi spikes; the first spike evokes only an EPSP in the flexor, whereas the second evokes an EPSP and two spikes. The attenuation of these spikes in the soma of the flexor is so great that they add little to the depolarisation. B (right panel). A simultaneous recording from the soma of FETi and the somata of three postsynaptic flexor motor neurons. FETi connects with each and consistently evokes an EPSP with a constant latency. C,D. Effects of manipulating FETi spikes during a kick. C. FETi is hyperpolarised to prevent its initial spikes. The flexor is still depolarised and spikes at high frequency. When the FETi spikes are allowed to resume, the frequency of flexor spikes rises and the kick occurs. D. The later FETi spikes are prevented by hyperpolarising current. The flexor is still inhibited in a way similar to a normal kick and the tibia extends. A and B are from Burrows *et al.* (1989). C and D are from Heitler and Bräunig (1988).

However, the EPSP consistently follows each FETi spike at frequencies of only up to 20 Hz and at higher frequencies its amplitude declines and failures may occur. This type of evidence is usually taken to indicate that a connection is not monosynaptic, but may simply be the result of rapid decrement.

Electron microscopy reveals direct but complex synaptic contacts between the two motor neurons when they are both labelled with horseradish peroxidase (HRP) (Burrows *et al.*, 1989). The physiological connection is estimated to be represented by several hundred contacts between the two neurons in three different regions of the neuropil. The flexor may be the only postsynaptic neuron, or one of a pair. FETi is also presynaptic to unlabelled neuronal processes which, in turn, are presynaptic to the flexor and to FETi. Spikes in FETi could therefore also affect flexors through

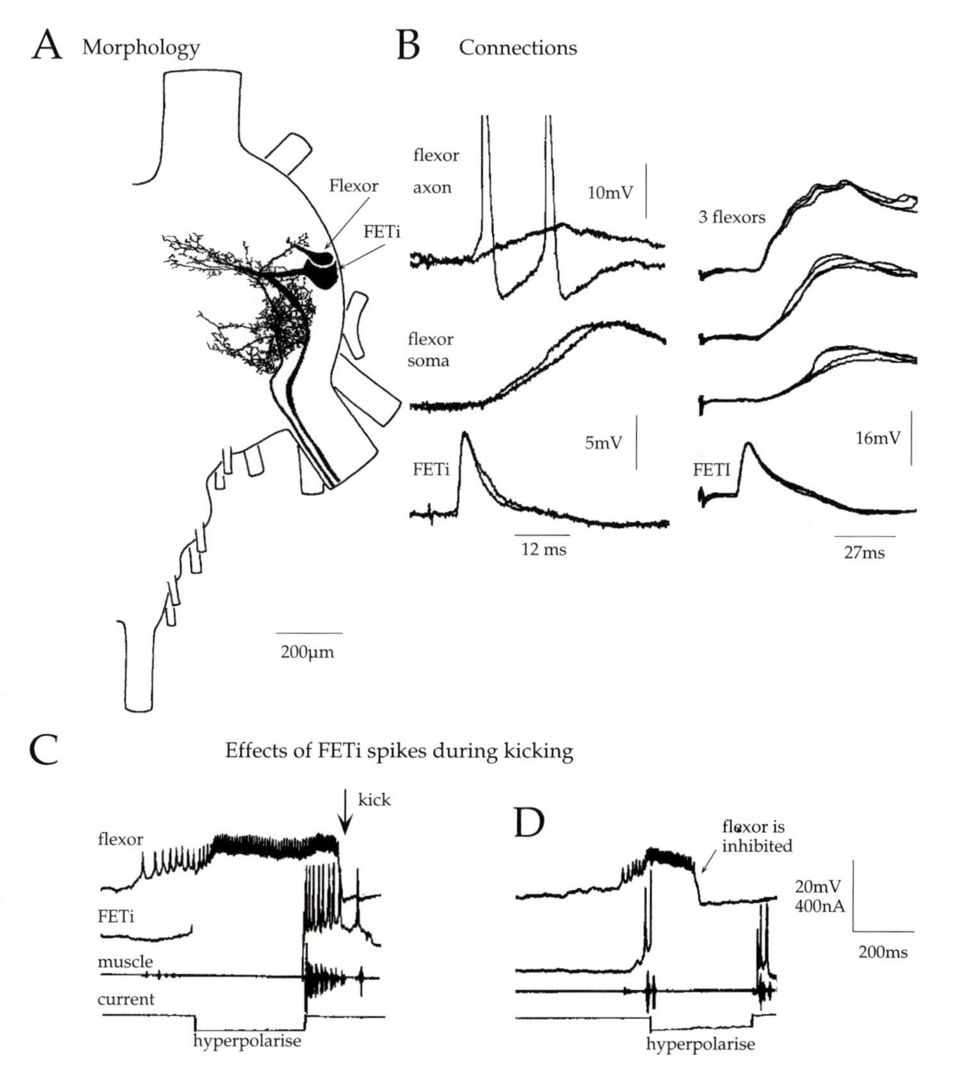

these serial synapses and limit their own effectiveness through the reciprocal synapses. The possibility of modulation of the connection is also raised by the synapses that are made by unlabelled processes onto both FETi and the flexor close to the synapses between these two motor neurons.

9.5.1.1. *Functional role of the synaptic connection*

What is the function of this monosynaptic connection between these motor neurons during a jump or a kick? Does it play a role in initiating the spikes in the flexor motor neurons? It clearly cannot initiate the first flexor spikes that move the tibia to a flexed position. Similarly, it cannot initiate the start of the co-contraction, because the depolarisation of the flexors often precedes the first spike in FETi. Other neurons must therefore cause the initial depolarisation of both the extensors and flexors. Could it help to maintain the co-contraction of the flexors? When FETi is made to spike repetitively by electrical stimulation, the amplitude of the EPSP in the flexor declines. For example, when the interval between pairs of pulses is more than 500 ms, the second EPSP shows little decrement, but at intervals of 50 ms it is reduced to 10% of the amplitude of the first EPSP (Burrows *et al.*, 1989). With sequences of FETi spikes that occur during the co-contraction phase, the decline of the EPSP is more rapid but again is frequency dependent. Most of the decline can be attributed to events in the presynaptic FETi motor neuron, and the failures of some EPSPs could be explained by the failure of the presynaptic spikes to invade all terminals. On this basis, therefore, the initial FETi spikes might be expected to ensure a high level of force in the flexor muscle, but later spikes should contribute progressively less to the excitation of the flexors as co-contraction proceeds and their frequency increases. Sensory feedback from receptors in a hind leg, however, gradually adds to the excitation of the flexors and will compensate for the declining effectiveness of the FETi spikes. For example, repetitive FETi stimulation (20 Hz for 2 s) still evokes spikes and an enhanced depolarisation in some flexors if the tibia is flexed but not if it is extended. Recordings during a kick show that the initial FETi spikes are indeed followed by a rapid depolarisation of the flexors and an increase in the frequency of their spikes that can be ascribed to the monosynaptic connection. The later FETi spikes occur when the flexors are already spiking at high frequency, so that they can contribute little more to the overall excitation.

To test directly the contribution of this connection, the FETi spikes can be abolished by injecting hyperpolarising current (Heitler and Bräunig, 1988), or by functionally severing the axon of FETi with a laser after intracellular injection of Lucifer Yellow (Heitler, 1995). If the spikes are prevented until the later stages of the co-contraction then the pattern of flexor spikes remains the same, although the frequency may not be as high (Fig. 9.7C). When the FETi spikes resume, the frequency of the flexor spikes immediately rises. If the FETi spikes are suppressed entirely, then the flexor shows a typical increase in the frequency of its spikes, followed by an inhibition that would coincide with triggering of the movement (Fig. 9.7D). The depolarisation and the frequency of spikes in the flexors may, however, be reduced. This experiment also implies that feedback from sense organs monitoring muscle force is not necessary for the inhibition that triggers the movement, although it may still have an effect (*see section 9.5.2.3*).

It appears that neither the increase in the frequency of the flexor spikes at the start of the co-contraction, nor the timing of the inhibition at the end of the co-contraction, are dependent on the spikes in FETi. The basic pattern of flexor spikes must be caused by other neurons because it can be expressed in the total absence of FETi spikes. Nevertheless, the frequency of flexor spikes during the co-contraction can be increased by FETi spikes and this must directly increase the force that can be produced in the movement. What has so far not been considered is the possibility that the input from FETi might lift the membrane potential of the flexors to a new stable state (plateau potential) at which the inputs from other sources could more effectively evoke flexor spikes. The central synaptic effects of FETi spikes on the flexor motor neurons are thus a contributory factor to the motor pattern of the jump, but do not by themselves initiate any particular aspect.

9.5.2. Interneurons

In a movement as complex as jumping or kicking, numerous interneurons can be expected to be involved with processing the sensory input from the environment and from the movements and forces generated in the hind legs. Similarly, many can be expected to shape the motor output. Many, perhaps all, will share a dual role either receiving a direct sensory input and making direct connections with the motor neurons, or being separated by only a few synapses from either the input or the output. There are many problems in seeking to establish that any of these interneurons play a causal role, because to establish causality implies that the result of experimentally manipulating the activity of one interneuron will have a measurable effect on the motor pattern. Initial experiments were undertaken with the expectation that a few key interneurons would be involved, and as a result much effort was devoted to the analysis of just three identified interneurons; the descending contralateral movement detector (DCMD), a multimodal (M) interneuron, and a cocking (C) interneuron. The initial optimism that these neurons contributed to significant features of the motor pattern has largely dissipated, leading to the conclusion that either the idea of a few key interneurons was wrong, or that the correct interneurons have yet to be identified. The more likely explanation is that the properties of the networks to which each of these and many other interneurons contribute have been overlooked. It seems unlikely that a particular interneuron is responsible for the control of each of the different phases of the movements, not least because these phases are in themselves essentially an abstraction of the experimental analyses.

9.5.2.1. *Descending contralateral movement detector (DCMD)*

This pair of bilaterally symmetrical interneurons have axons that descend to the metathoracic ganglion in the connective contralateral to their cell bodies in the brain. (Fig. 9.8A). A role for these interneurons in a jump or a kick was inferred from their response properties and the connections they make with some of the neurons closely implicated in controlling the movements of the hind legs. However, the initial idea that these interneurons were involved in either initiating or triggering the movement has not been substantiated by subsequent experiments. The search for a behavioural role for these neurons merely illustrates the almost impossible task of ascribing causality to a few neurons in a complex network of neurons controlling an intricate motor pattern.

Within the metathoracic ganglion, each of these interneurons makes an excitatory connection with both the left and the right FETi motor neurons (Fig. 9.8B) (Burrows and Rowell, 1973), and with the bilaterally paired M interneurons (*see section 9.5.2.3*) (Pearson *et al.*, 1980). They also connect with the C (=314) interneurons in the mesothoracic ganglion (Pearson and Robertson, 1981) and with some flight motor neurons in both the meso- and metathoracic ganglia (Simmons, 1980) (Fig. 9.8C). The EPSPs they evoke in FETi are only 0.5 mV in amplitude, whereas those in the interneurons are much larger at 4–6 mV. The connection with FETi has therefore been considered to be functionally weak (Pearson and Goodman, 1979), or unimportant (Pearson, 1983), because the EPSPs cannot themselves evoke motor spikes. Visual stimuli that may excite additional interneurons can, however, evoke motor spikes at temperatures above 30°C (Heitler *et al.*, 1977). The effectiveness of a synaptic input is always dependent on the action of many other neurons, so that the input to the FETi motor neurons could therefore be crucial in certain circumstances. The connection needs to be explained and not dismissed, but further doubt on its importance is cast by the fact that it is variable in different laboratory locusts, which often have been inbred for many generations. For example, only 61% of FETi motor neurons receive an input from both DCMDs, and as many as 28% receive no input at all (Pearson and Goodman, 1979). The lack of a connection correlates with the absence of a particular branch of the DCMD in the metathoracic ganglion. By contrast, the connections with the C (=314) and M interneurons occur reliably in all animals and are perceived as stronger because they generate larger EPSPs that can evoke spikes by themselves.

The effects of the DCMD synapses can also be variable because of probable presynaptic inhibition in their terminals (Pearson and Goodman, 1981). Depolarising synaptic potentials that can reduce the amplitude of the spikes can be

---

Fig. 9.8 The structure and action of a DCMD interneuron in kicking. A. Structure. It has a cell body in the brain, and an axon that descends in the contralateral connective beyond the metathoracic ganglion. It branches in the thoracic ganglia, but the most extensive branches are in the metathoracic ganglion. B. Output connections with the FETi motor neuron. C. Diagram of the connections made by DCMD with motor and interneurons (C and M) in the meso- and metathoracic ganglia. D, E. Responses of DCMD depend on the current behaviour. D. Spikes in a DCMD are evoked by a movement of a small object in its visual field when there is no activity in leg motor muscles. E. The response to the same visual stimulus is suppressed during the performance of a kick. A is from O'Shea *et al.* (1974), B is from Burrows and Rowell (1973), D and E are from Heitler (1983).

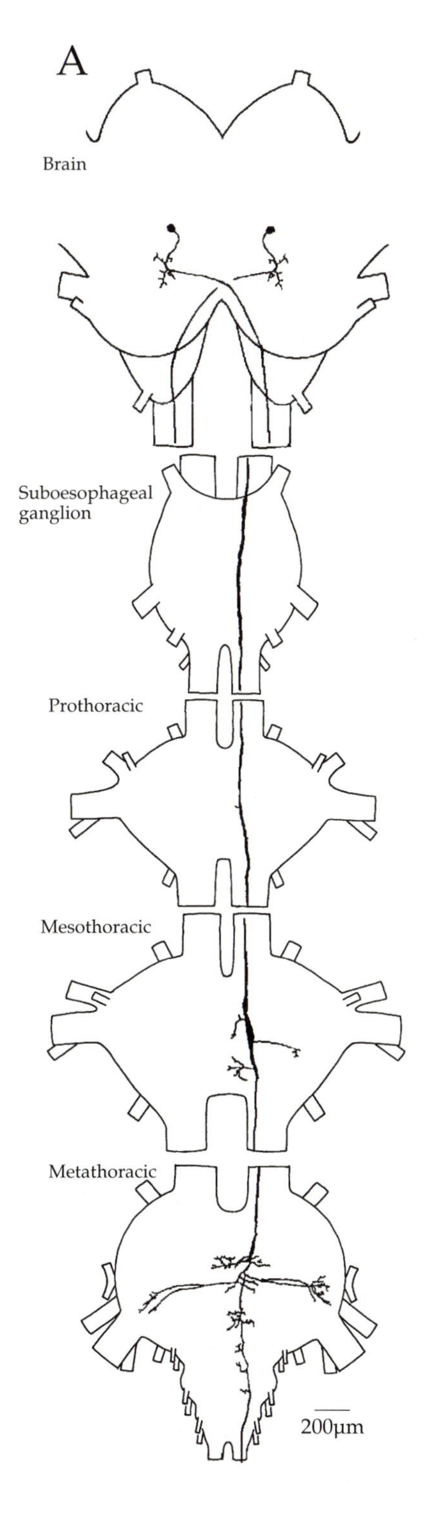
A
Brain
Suboesophageal
ganglion
Prothoracic
Mesothoracic
Metathoracic
200μm

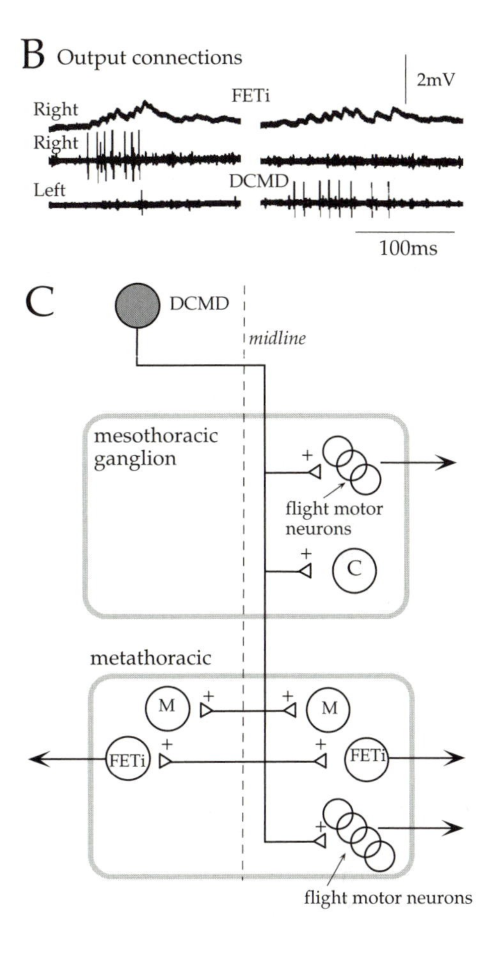
B Output connections
2mV
FETi
Right
Right
Left
DCMD
100ms
C
DCMD
midline
mesothoracic
ganglion
+
flight motor
neurons
+
C
metathoracic
M
+
+
M
FETi
+
+
FETi
+
flight motor neurons

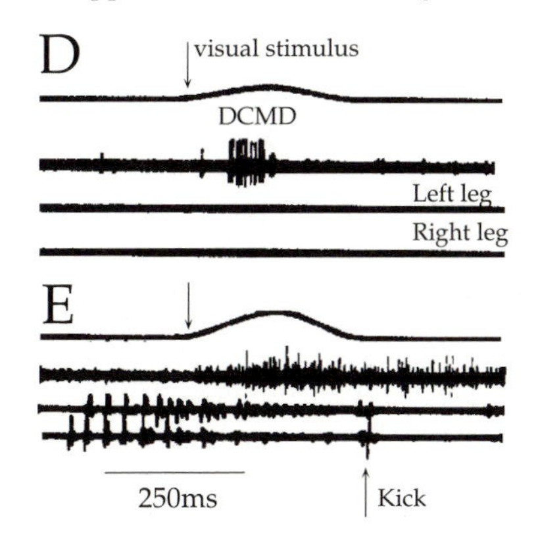
Suppression of DCMD during a kick
D
visual stimulus
DCMD
Left leg
Right leg
E
250ms
Kick

evoked in the terminals of a DCMD by its own spikes, by spikes in the other DCMD, and by auditory inputs. These inputs are associated with a conductance change that may involve chloride ions and that reverses a few millivolts more positive than the resting potential, suggesting that they are depolarising IPSPs. The latency from the spike in one DCMD to the appearance of a reliable synaptic potential in the other is 4 ms, indicating that the connection is probably mediated by an interposed interneuron in a pathway with a high synaptic gain. The same stimuli that evoke the IPSPs also cause a reduction in the amplitude of the EPSPs that a DCMD evokes in a postsynaptic M interneuron, but no simultaneous recordings have been made from pre- and postsynaptic neurons. The inference is that this reduction is, in part, due to the pre- rather than postsynaptic action of the stimulation. The action of this presynaptic inhibitory mechanism could be to limit the effectiveness of transmission to postsynaptic neurons, but when this would be appropriate has not been explored.

What role might the DCMD connections play in a jump or a kick? The preparation for these movements takes at least 500 ms, so that only a persistent visual stimulus could both initiate and terminate the motor activity. The DCMD, however, responds best to novel stimuli, such as small moving objects in a large visual field (Fig. 9.8D) (Rowell, 1971), or to objects, such as an approaching predator, that loom towards the locust (Rind and Simmons, 1992). Could the DCMDs trigger a jump by adding their inputs to those present during the later part of the co-contraction phase? The DCMDs do not, however, respond to a visual stimulus presented during the preparation for a kick, during the co-contraction (Fig. 9.8E), and for 1s after the kick itself because feedback from the motor circuitry in the thorax suppresses the generation of their spikes in the brain (Heitler, 1983). This implies that the DCMDs could not participate in the final triggering of the movement, and explains why there is no immediate correlation between its spikes and the movement (Rowell, 1971). The DCMDs could, however, still have a role in the initial part of the motor sequence, with the movement then following up to 500 ms later.

9.5.2.2. *C interneuron*

Paired C (= cocking) interneurons (given the cell number 314) with cell bodies in the mesothoracic ganglion and axons that descend to the contralateral side of the metathoracic ganglion (Fig. 9.9A) have been implicated in initiating the cocking phase of the jump by virtue of their input and output connections (Pearson and Robertson, 1981). The neurons were named according to this supposed functional role, but again this has not been substantiated by subsequent experiments.

Each interneuron receives excitatory input from the DCMDs, from neurons that respond to high frequency sounds, and from neurons that respond to tactile stimulation of the abdomen. None of these inputs alone make the C neurons spike. In turn, each C interneuron makes a direct excitatory connection with FETi and with some flexor tibiae motor neurons in the metathoracic ganglion contralateral to its cell body (Fig. 9.9B). The connection can be powerful enough that a single spike

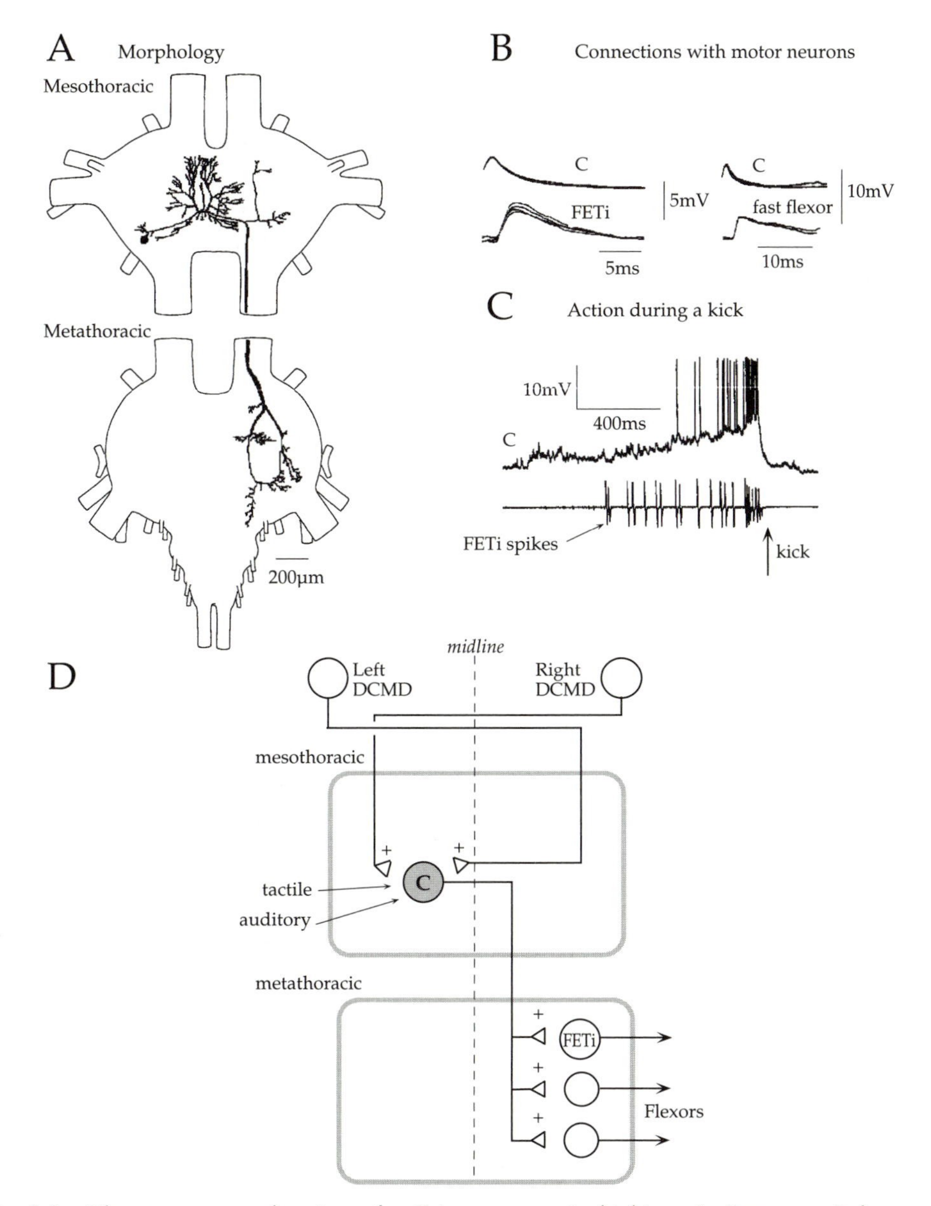

Fig. 9.9 The structure and action of a C interneuron in kicking. A. Structure. It has a cell body and many branches in the mesothoracic ganglion. An axon descends in the contralateral connective to form branches on one side of the metathoracic ganglion. B. It makes direct excitatory connections with FETi and flexor motor neurons. The traces are triggered by C interneuron spikes which are followed at a constant latency by EPSPs in the motor neurons. C. Action during a kick. During the co-contraction phase it is gradually depolarised and produces a burst of spikes when the tibia extends. D. Diagram of the input and output connections of a C interneuron. A and B are from Pearson and Robertson (1981), C is from Gynther and Pearson (1986).

in a C interneuron can evoke a spike in FETi and in some of the flexor motor neurons. Another effect adduced as evidence for their proposed functional role results from stimulating the axon of a C interneuron in a meso-metathoracic connective with extracellular electrodes. Such stimulation produces a larger depolarisation of the motor neurons if the hind tibia is flexed than if it is extended. This effect could, however, result from the different synaptic inputs to FETi from other neurons at different positions of the femoro-tibial joint, and from the action of the other neurons that are inevitably activated by stimulation of a whole connective.

These properties of the C interneurons lead to the suggestion that they are responsible for the initial flexion of the tibia in the cocking phase, and perhaps the initial co-activation of flexor and extensor tibiae motor neurons in the co-contraction phase (Pearson and Robertson, 1981). Four observations indicate that this is not their role. First, spikes in the flexor motor neurons often precede those in FETi, even in an intact freely moving locust (Gynther and Pearson, 1986), suggesting they are not caused by the same presynaptic neuron(s). Second, a C interneuron is depolarised only after the co-contraction phase has started, and spikes only towards the end of this phase (Fig. 9.9C). It cannot, therefore, be involved in the cocking phase. Third, if the mesothoracic ganglion is cut in half by a longitudinal incision at the midline, normal jumping can still occur even though the C interneurons will be severely damaged (Ronacher *et al.*, 1988). Fourth, cutting the connectives between the meso- and metathoracic ganglia does not prevent kicking or jumping. This leaves the possibility of a contributory role for the C interneurons during the co-contraction, but they cannot be involved in initiating the first 'cocking' phase of the jump.

9.5.2.3. *M interneuron*

The bilaterally paired M (= multimodal) interneurons were suggested to be responsible for triggering jumping or kicking (Steeves and Pearson, 1982). They were thought to sum visual and auditory inputs with the gradually increasing feedback from leg proprioceptors during the co-contraction phase to produce spikes that would then inhibit the flexor motor neurons, thus allowing a tibia to be extended. While they are active at the appropriate time, they do not act alone in triggering the movement.

Each of these interneurons has a cell body near the dorsal midline of the metathoracic ganglion and an axon that ascends in the anterior contralateral connective (Fig. 9.10A) (Pearson *et al.*, 1980). A distinctive branch arises from the axon in both the meso- and metathoracic ganglia to form lateral projections of fine branches that overlap branches of leg motor neurons. Each interneuron receives an excitatory input in response to several modalities of stimuli, but no modality on its own can evoke spikes. For example, the interneurons receive excitatory input from the DCMDs, from a mesothoracic G (=714) interneuron that is excited by sound (Fig. 9.10B), and from proprioceptors at the femoro-tibial joint, the identity of which remains elusive. For example, the femoral chordotonal organ has been suggested, as

EPSPs are evoked when single FETi spikes cause twitch movements of the tibia and are abolished if the nerve to the femoral chordotonal organ (amongst other receptors) is cut. Rapid but passive extension of the tibia that would also activate the chordotonal organ does not, however, evoke the EPSPs as would be expected. Similarly, joint receptors (*see Chapter 8*) might be involved, as electrically stimulating the lateral nerve in the femur that contains their axons also evokes EPSPs. However, if these receptors are the source, then the EPSPs should not have been abolished when the nerve to the chordotonal organ was cut. Tactile stimuli to a hind leg evoke no EPSPs. Perhaps the input is mediated by indirect pathways involving other interneurons with a low gain and therefore unreliable input from several types of proprioceptors.

An M interneuron makes direct inhibitory connections with flexor tibiae motor neurons of one hind leg (Fig. 9.10C) but no connection with FETi (Pearson *et al.*, 1980). Output connections to neurons in other ganglia have not been explored, but are to be expected from the anterior projection of its axon.

Support for a role in triggering jumping and kicking movements stems from the observation that the combined sensory inputs that excite an M interneuron (and many other interneurons) can evoke IPSPs in flexors, but that inputs of one modality cannot. The inputs were assumed to be summed by an M interneuron which then caused inhibition of the flexor motor neurons. Furthermore, the first recordings from an M interneuron showed a slow depolarisation during co-contraction that was attributed to proprioceptive input (Steeves and Pearson, 1982). Crucial to this idea of the M interneuron being a trigger, therefore, is that proprioceptive inputs depolarise it sufficiently during the co-contraction phase for additional inputs from visual or auditory stimuli to make it spike.

Subsequent recordings from M interneurons, however, show that they are hyperpolarised, not depolarised during co-contraction (Fig. 9.10D) (Gynther and Pearson, 1986). Moreover, the same pattern of activity occurs in an M interneuron even if the ipsilateral leg is unable to kick, provided that the contralateral leg does kick. This must indicate that proprioceptive input from the ipsilateral leg is not necessary for its usual pattern of activity. Finally, the rapid depolarisation and burst of spikes that the interneuron shows just before a movement is not dependent on an additional visual or auditory input. The idea is no longer tenable, therefore, that a gradual increase of proprioceptive inputs pushes the M interneuron to a point where another sensory input will elicit a burst of spikes.

Simultaneous recordings from the M interneuron and a flexor tibiae motor neuron show that the burst of interneuron spikes either precedes or follows (Fig. 9.10E) the rapid repolarisation of the flexor (Gynther and Pearson, 1989). This indicates that the repolarisation of the flexor is due either to the reduction of an excitatory drive, or to an active inhibition to which the M interneuron may contribute. If the M neuron is prevented from producing its normal burst of spikes, the end of the co-contraction still occurs and kicks are produced with the same latency, even though the time taken for repolarisation of a flexor is lengthened. Conversely, making the M interneuron spike at high frequency with a pulse of current during the co-contraction abolishes the flexor spikes and allows the tibia of

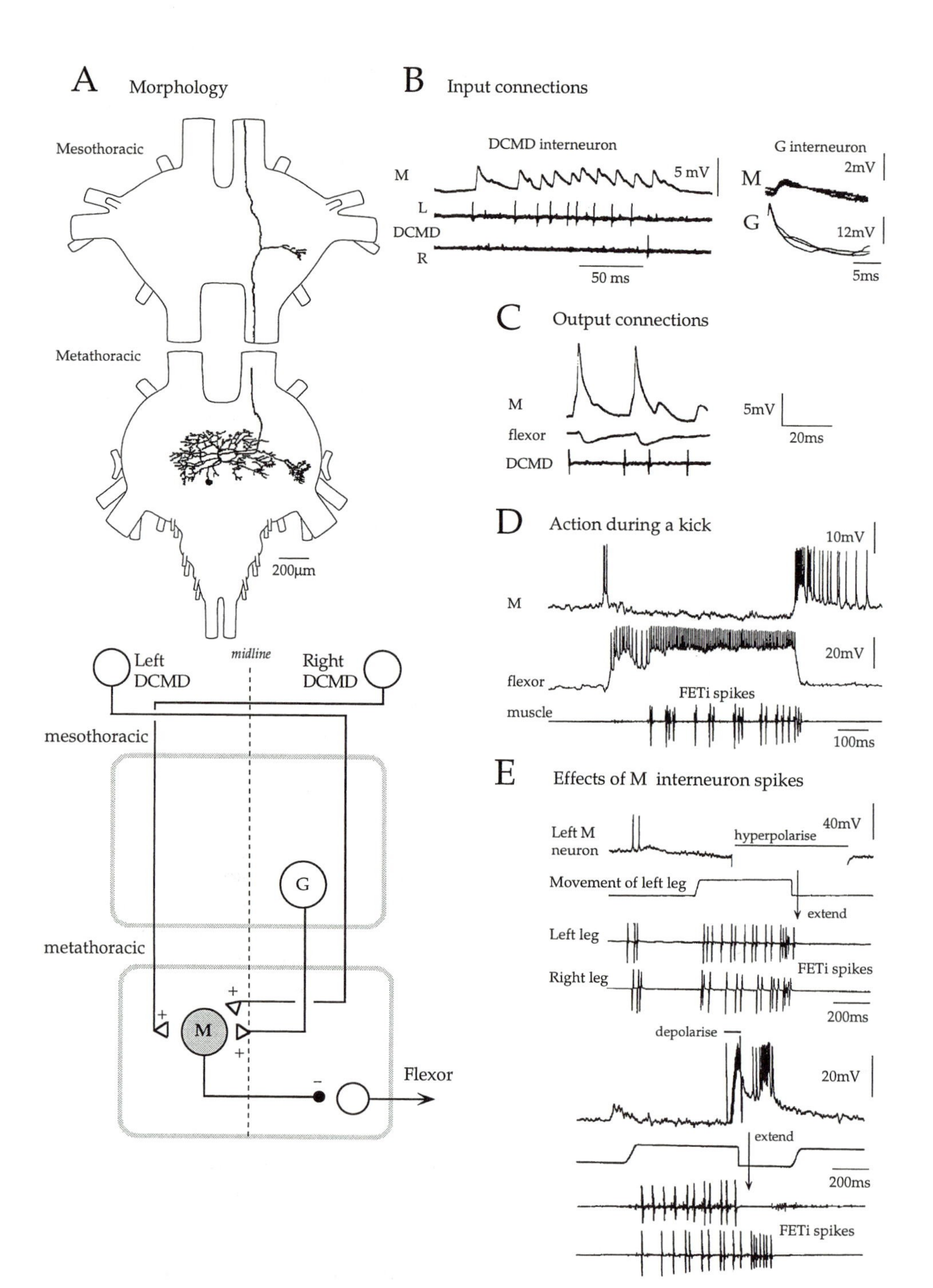
A Morphology
Mesothoracic
Metathoracic
200μm
Left DCMD
midline
Right DCMD
mesothoracic
metathoracic
G
M
Flexor
B Input connections
DCMD interneuron
M
L
DCMD
R
5 mV
50 ms
G interneuron
M
G
2mV
12mV
5ms
C Output connections
M
flexor
DCMD
5mV
20ms
D Action during a kick
10mV
M
20mV
flexor
FETi spikes
muscle
100ms
E Effects of M interneuron spikes
Left M neuron
hyperpolarise
40mV
Movement of left leg
extend
Left leg
Right leg
FETi spikes
200ms
depolarise
20mV
extend
200ms
FETi spikes

one leg to extend under the force already generated by the extensor muscle. The interneuron may then produce a later burst of spikes which is correlated with the movement of the contralateral leg. It does not, of course, connect with the contralateral flexor motor neurons. The end of the co-contraction, however, shows three differences from normal. First, the tibia extends with a delay that is twice as long as usual, suggesting either that not all the flexors are inhibited at the same time, or that the common inhibitory motor neurons that innervate the flexor are not excited. Second, the repolarisation of a flexor is neither as rapid nor as complete as normal. Both these observations suggest that, in a normal kick, other neurons must be involved in triggering the movement in addition to the M interneuron. Third, the frequency of spikes in the M interneuron needed to stop flexor spikes is 15% greater than that during normal behaviour; when the M interneuron is experimentally made to spike at the frequencies recorded during normal kicks the flexor spikes are not abolished, again indicating that another inhibitory input must normally occur.

All these observations indicate that the M interneuron does not normally act by itself and is not sufficient to trigger a kick. Other interneurons have been suggested to fill this role by virtue of their apparently appropriate response properties (Gynther and Pearson, 1989). For example, an interneuron called 707 produces a burst of spikes at the end of the co-contraction, but manipulating its pattern of spikes does not alter the expression of the normal activity in the flexors. From what is known of the response properties of interneurons processing the proprioceptive and exteroceptive signals from a leg (*see Chapter 7*), it would be surprising if many interneurons were not found that spike during a jump or a kick. Before the triggering can be better understood, we need to identify the neurons that excite the common inhibitory motor neurons to the flexor muscle. Perhaps the same putative neurons excite both the inhibitors and the M interneurons. Similarly, the

---

Fig. 9.10 The structure and action of an M interneuron in kicking. A. Structure. Its cell body is in the metathoracic ganglion where it also has a series of medial branches and a prominent lateral branch. Its axon ascends in the contralateral connective and gives rise to a similar lateral branch in the mesothoracic ganglion. B. Input connections. Both the left and the right DCMD make direct excitatory connections. Spikes in a G interneuron are followed at a constant latency by EPSPs in an M interneuron. C. Output connections. Each spike in an M interneuron, caused by spikes in a DCMD interneuron, evokes IPSPs in a flexor tibiae motor neuron. D. Action during a kick. Simultaneous intracellular recordings from an M interneuron and a flexor motor neuron, show that M spikes at the start of co-contraction, is then hyperpolarized for the remainder of this phase, before producing a burst of spikes as the tibia extends. E. Effects on the motor pattern of manipulating the spikes in an M interneuron. If an M interneuron is hyperpolarised in the middle of a co-contraction so that it can no longer produce spikes, a kick still occurs as judged by the movement of the tibia and the pattern of FETi spikes. If an M interneuron is depolarised by a pulse of current so that it produces a burst of spikes, extension of the tibia follows and FETi spikes stop. The interneuron also produces a later burst of spikes when the contralateral tibia extends. A–C are based on Pearson *et al.* (1980), D and E on Gynther and Pearson (1989).

interneurons responsible for the removal of excitation of the flexor motor neurons await identification. Further light would also be shed on the action of the M interneuron if the connections made by its branches in other ganglia could be explained.

The M interneuron, therefore, cannot be regarded as a trigger neuron, since causality is not established: it does not sum exteroceptive signals with proprioceptive signals to produce spikes that allow the jump to occur. Instead, it must be seen as one of a group of neurons whose combined activity may be part of the mechanism that leads to a jump or a kick. The M interneurons may, indeed, contribute to the inhibition of the flexors, but triggering the jump results from a complex series of events. There is little to be gained from assigning the depolarisation of the M interneuron at the end of the co-contraction phase to a 'trigger system' of interneurons, as causality is merely pushed back one further stage.

9.5.3. Sensory feedback

The many sense organs on the leg and at its joints provide considerable information to the central nervous system as the tibia is initially flexed, as force is developed by the co-contracting muscles, and as the tibia extends rapidly. For example, the movements of the femoro-tibial joint excite a chordotonal organ and various joint receptors (*see Chapter 8*), and the force produced by the muscles strains the cuticle and this, in turn, excites campaniform sensilla. Single multipolar sensory neurons in the main body of the flexor muscle in the front and middle legs, and in the accessory flexor in a hind leg, monitor the active force generated by their respective muscles (Theophilidis and Burns, 1979; Matheson and Field, 1995) (*see Chapter 8* for full description).

9.5.3.1. *Campaniform sensilla*

Two campaniform sensilla (CS) (Fig. 9.11A) on the proximal tibia spike only during extension and flexion movements when the leg is load bearing or meets a resistance (Burrows and Pflüger, 1988). During the co-contraction phase of a jump, each spikes continuously at about 200 Hz, with some instantaneous frequencies of 650 Hz (Fig. 9.11B). These actions can be readily seen if tibial movements are caused by spikes in FETi. If the tibia moves, then the sensory neurons from these campaniform sensilla do not spike, and instead the spikes of chordotonal sensory neurons signalling the movement cause a hyperpolarisation of FETi and a depolarisation of the flexors in a negative resistance reflex that opposes the movement (*see Chapter 8*). If the movement is restrained, however, both campaniform sensilla spike rapidly and a spike in FETi is now followed by a wave of excitatory synaptic potentials (Fig. 9.11C,D). The positive feedback is caused by the campaniform sensilla, because sequentially cutting the two nerves that innervate them, distal to the femoral chordotonal organ, progressively abolishes the input (Fig. 9.11D), as does cauterisation of the individual receptors themselves. The input from one of these receptors, induced by mechanical stimulation to spike at the frequencies seen during co-contraction, is sufficient to evoke a spike in FETi (Fig. 9.11E,F). The force generated in the muscle by each spike adds more force and hence more feedback so that a sequence of spikes results. The runaway excitation continues until the mechanical restraint is removed, or a wave of synaptic feedback fails to exceed the spike threshold in FETi.

The two sensory neurons from these sensilla synapse directly on the fast but not the slow extensor motor neurons, and on the fast but not the slow flexor motor neurons (Fig. 9.12). This powerful synaptic input therefore adds to the excitation that causes the co-contraction. The contribution becomes clear when the sensory neuron from a single campaniform sensillum is made to spike; both extensor and flexor motor neurons are depolarised, often to the extent that FETi is made to spike. The result is that the FETi spike then causes an additional depolarisation of the flexor through the central connection between these motor neurons. Longer pathways through nonspiking interneurons are also involved in exciting the motor neurons. Particular nonspiking interneurons are excited directly by inputs from the campaniform sensilla and their outputs lead to further excitation, probably through disinhibition, of both SETi and FETi (Fig. 9.12D).

The feedback provided by these campaniform sensilla therefore has two behavioural consequences. First, during the co-contraction phase of jumping and kicking, the excitation of the extensor and flexor motor neurons is reinforced. Second, as the tibia extends at the end of a kick, any resistance is met by an increased motor output causing the tibia to thrust more firmly against the obstacle.

9.5.3.2. *Chordotonal organ*

Cutting the apodeme of the femoral chordotonal organ prevents jumping (Bässler, 1967). A possible explanation for this is that the force on the apodeme of the receptor is now relaxed so that the receptors will signal that the tibia is held in an extended position. The suppression of jumping or kicking is thus an entirely appropriate response as it ensures that a co-contraction is not attempted at perceived joint angles where the flexor cannot resist the force of the extensor.

In inactive animals, moving the apodeme to simulate joint movements evokes resistance reflexes in which flexor motor neurons are excited by an imposed extension, and inhibited by a flexion, with the extensor motor neurons showing the opposite effects (*see Chapter 8*). In an animal actively moving its legs, these reflexes can reverse. Conceivably, therefore, the input from the chordotonal organ during the co-contraction phase could add to the excitation of the flexors. Whether the input from this organ is also necessary for the transition between leg movements that require the alternating action of the extensors and flexors (e.g. walking), and the jumping movements that require co-contraction remain to be explored. An established role is to allow the motor patterns for jumping and kicking to be expressed only when the tibia is fully flexed. In addition, it and possibly the joint receptors, may be responsible for the inhibition of the extensors as the tibia extends rapidly in the kick or jump movement.

9.5.3.3. *Joint receptors*

One of the three groups of joint receptors (*see Chapter 8*) (Coillot and Boistel, 1968, 1969) called the 'lump receptor' lies in a groove between the posterior wall of the femur and the ventral invagination that forms the lump (Fig. 9.13A) (Heitler and Burrows, 1977). The posterior arm of the flexor apodeme slides into this groove when the joint is fully flexed during a co-contraction, and thus comes to rest directly on the receptors. When the tibia is in this fully flexed position, the sensory neurons from these receptors spike as the flexor force rises (Fig. 9.13B) (Heitler and Burrows, 1977). They do not spike, however, unless the tibia is fully flexed, even though the flexor muscle may be producing an equivalent force (Fig. 9.13C). They thus signal that the flexor apodeme lock is engaged, which is a prerequisite for the production of extensor force during the co-contraction. Cutting the nerve to the receptors does not prevent a co-contraction but does greatly reduce the occurrence of jumping or kicking, again emphasising that many pathways act in parallel. The central effects of these sensory neurons are not known, but it would seem likely that they contribute to the excitation of the motor neurons during the co-contraction.

9.5.3.4. *Can the motor pattern be produced without sensory feedback?*

The forces generated during jumping are high, and during kicking are particularly unpredictable because the time at which an object is encountered by the extending tibia, and its rigidity, are not known. Logic would seem to indicate the wisdom of monitoring the force that the extensor muscle actually produces. Despite this, the basic motor pattern in the extensor and flexor tibiae motor neurons can be

Fig. 9.11 The role of tibial campaniform sensilla in kicking. A. The sensilla (CS) are on the dorsal surface of the proximal tibia. The posterior one is innervated by a branch of N5B2, which also innervates some nearby hairs, basiconic sensilla and a group of campaniform sensilla. The anterior one is innervated by N5B1, which again innervates nearby receptors, including the subgenual organ. B. Action of the sensilla during a kick. During the co-contraction phase, the single sensory neuron from a sensillum spikes continuously at high frequency. C–F. Excitation of FETi by the CS sensory neurons. C,D. A single antidromic spike is evoked in FETi by electrical stimulation of the extensor muscle and the movement of the tibia is prevented. C. The resulting strain in the tibia is signalled by spikes in the sensory neuron from the anterior CS and is followed by a depolarisation of FETi. D. Cutting N5B1 reduces the excitation, and then cutting N5B2 so that the input from both receptors is prevented, abolishes the depolarisation of FETi. E. Pressing on the anterior CS evokes high frequency spikes in its sensory neuron that depolarise FETi sufficiently to make it spike. Each motor spike causes a muscle contraction that increases the strain on the cuticle and is thus signalled by more spikes in the CS sensory neuron. FETi therefore spikes repetitively until the positive feedback fails to depolarise it sufficiently. F. Interaction between the sensory feedback from the CS sensory neurons and the central connection from FETi to the flexor motor neurons. The CS spikes depolarise both motor neurons. When FETi spikes, its central synapse adds to the depolarisation of the flexor. Based on Burrows and Pflüger (1988).

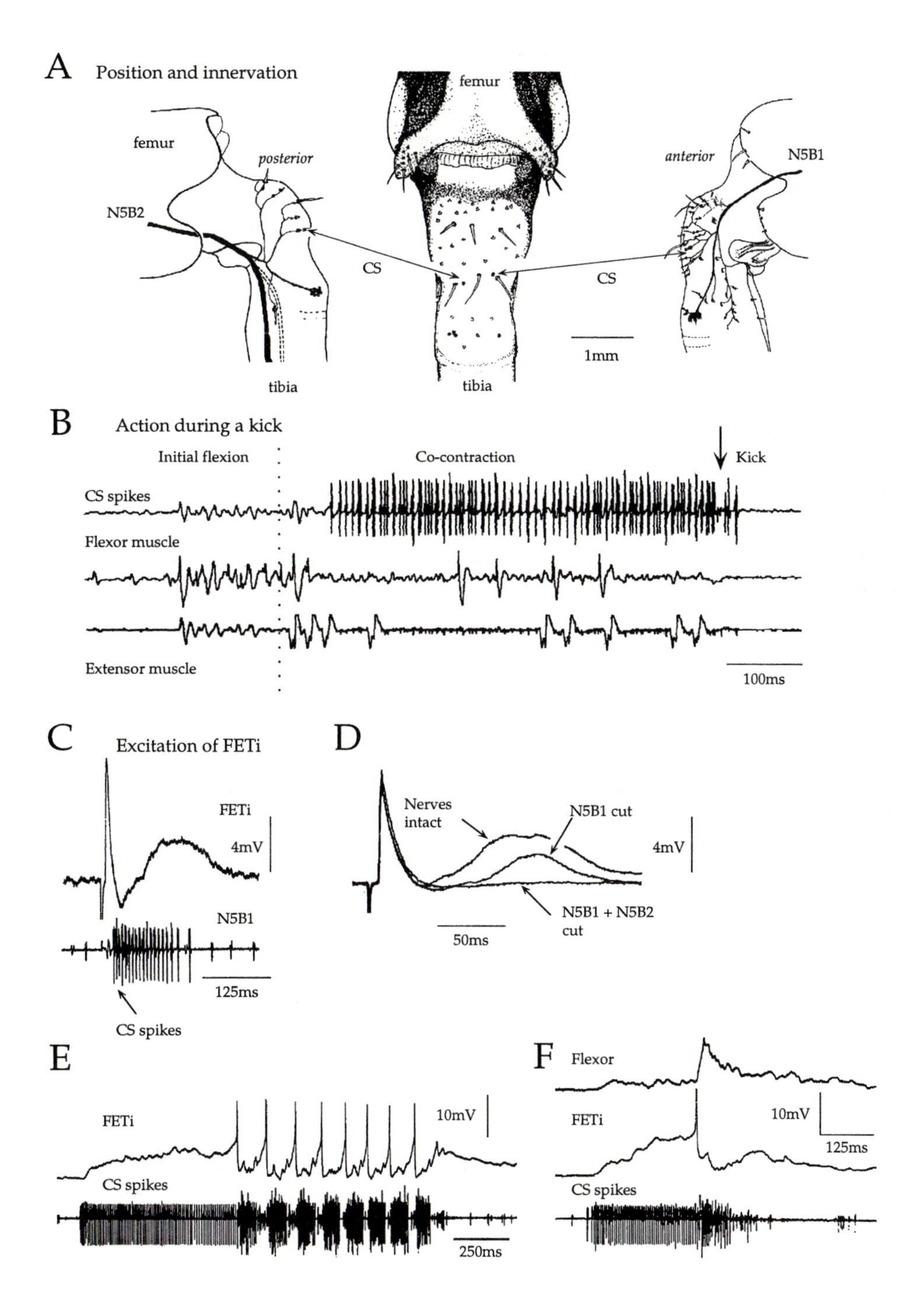
A Position and innervation
femur
femur
posterior
anterior
N5B1
N5B2
CS
CS
1mm
tibia
tibia
B Action during a kick
Initial flexion
Co-contraction
Kick
CS spikes
Flexor muscle
Extensor muscle
100ms
C Excitation of FETi
FETi
4mV
N5B1
125ms
CS spikes
D
Nerves intact
N5B1 cut
4mV
N5B1 + N5B2 cut
50ms
E
FETi
10mV
CS spikes
250ms
F Flexor
FETi
10mV
125ms
CS spikes

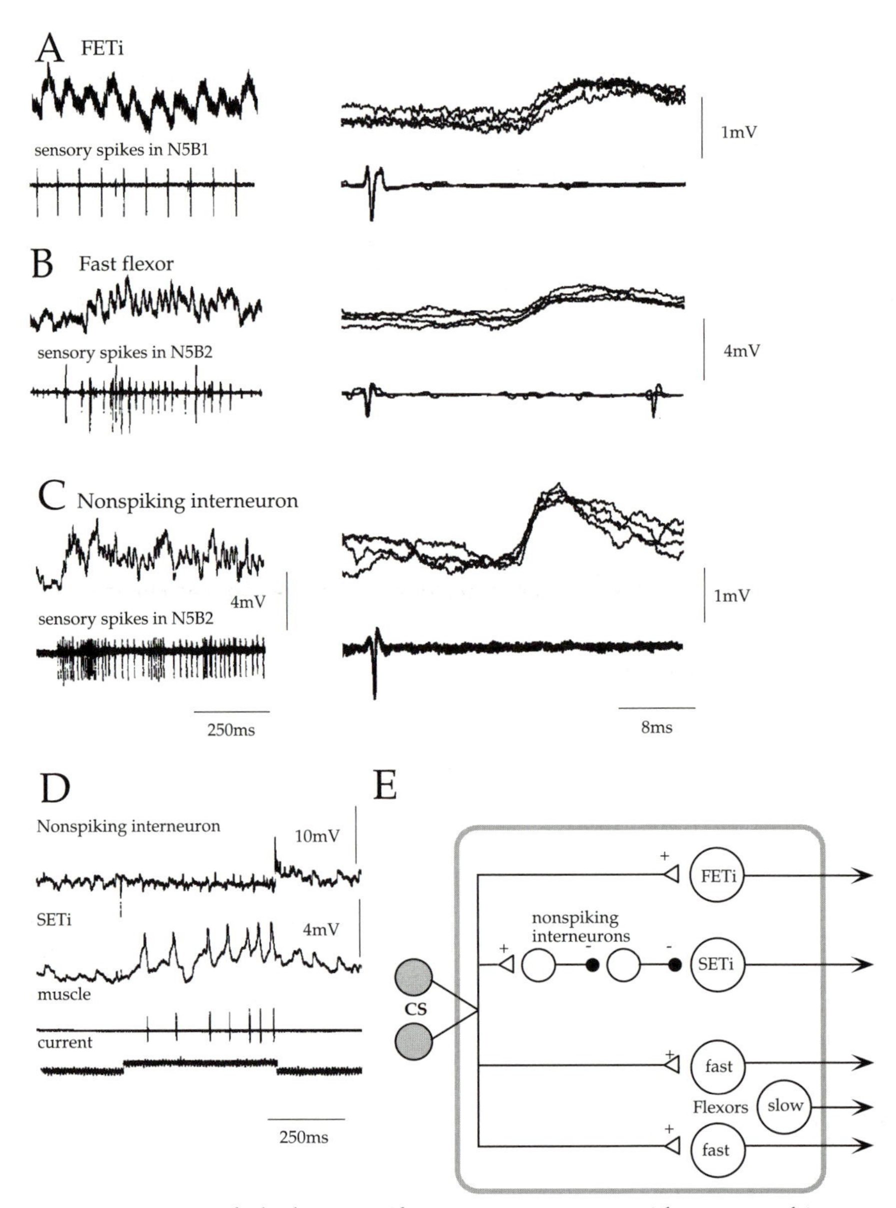

Fig. 9.12 Connections of tibial campaniform sensory neurons with motor and interneurons. A. FETi. B. A fast flexor. C. A nonspiking interneuron. The left panels show the sensory neuron spikes consistently evoking EPSPs in the central neurons. The right panels show sweeps triggered by the sensory neuron spikes recorded in the distal femur. Each spike is followed after a constant latency by an EPSP. D. The nonspiking interneuron that receives a direct input from the CS sensory neurons is depolarised by a pulse of current injected intracellularly. It causes the SETi motor neuron to spike, probably by disinhibiting another nonspiking interneuron that makes a direct inhibitory connection with SETi. E. The connections made by the CS sensory neurons. Based on Burrows and Pflüger (1988).

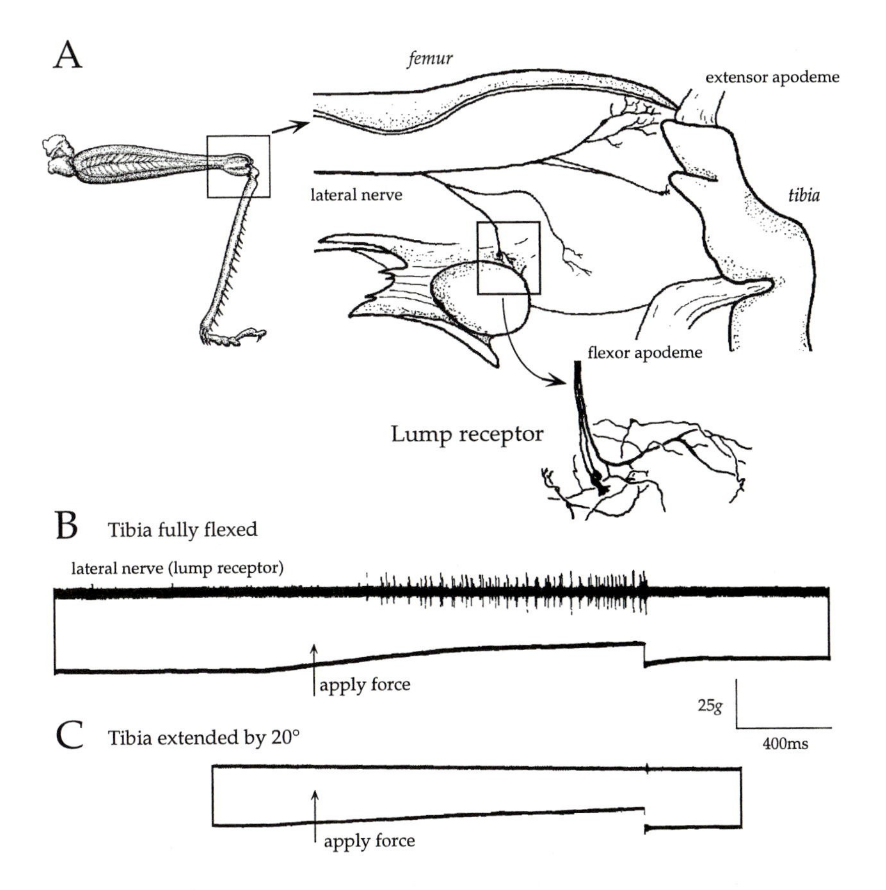

Fig. 9.13 The lump receptor. A. Location of the multipolar receptors associated with the lump in the ventral femur. The distal femur is exposed, the muscles removed and their apodemes deflected. The lateral nerve which innervates these and other joint receptors was stained with cobalt. B, C. Action of the lump receptor during tibial movements. B. The tibia is moved to a fully flexed position in which the flexor apodeme engages the femoral lump. Force applied to the flexor apodeme now evokes a series of spikes in the axons of the lump receptors. C. The same force applied when the tibia is extended, and hence the lock is not engaged, evokes no spikes in the axons of the lump receptors. Based on Heitler and Burrows (1977).

produced when the extensor muscle is prevented from producing any force (Heitler, 1995). This can be shown in an ingenious experiment in which the FETi motor neuron is filled with a fluorescent dye and spike transmission in its axon is blocked by illumination with a laser beam. The spikes generated in the undamaged central parts of FETi will then not reach the muscle, but by monitoring the spikes that are produced and using an electronic bypass, the same pattern of stimuli can then be delivered to the muscle as desired. All the other nerves except N5 containing the axon of FETi to the hind leg are also cut, so that SETi spikes cannot contribute to the extensor force. Even when no force is generated in the extensor muscle, the same basic pattern of spikes is generated in the flexor and extensor motor neurons, with the only differences being that the acceleration of the FETi spikes at the end of the co-contraction phase is absent and the inhibition of the flexor motor neurons is less rapid. This is thus another example of the parallel nature of neuronal processing that allows normal motor patterns to be produced when parts of the usual signals are removed. Presumably, the sensory neurons that signal the force contribute to a normal jump or kick by modifying the pattern and regulating the force, so that the mechanical limits of the muscles and skeletal elements are not exceeded. Alternatively, in the absence of feedback the central networks may produce only a weak approximation of the normal pattern, but when feedback is present the level of excitation is higher so that the pattern produced is closer to the mechanical limits.

9.5.4. Action of neuromodulatory neurons

The extensor muscles of both hind legs are supplied by a single efferent DUM neuron called DUMETi (*see Chapters 3 and 5*) and the flexor tibiae muscles by two other DUM neurons called DUM3,4,5 neurons. These neurons receive a synaptic input that is similar to that of the excitatory extensor and flexor motor neurons so that they produce a burst of spikes during the co-contraction phase and stop spiking when the tibia is rapidly extended (Burrows and Pflüger, 1988). None of the other efferent DUM neurons in the metathoracic ganglion that supply muscles that are not involved in jumping are activated in this manner, indicating that the release of octopamine from these neurons is timed to influence either the performance of the current kick or jump, or to influence leg movements consequent upon these rapid movements. The effects of the pattern of spikes in these neuromodulatory neurons and their close mirroring of the action of the motor neurons is yet to be appreciated in functional terms. The knowledge of their patterns of activity does, however, make it essential to consider them more fully as part of the circuits that generate and control motor patterns.

9.6. THE SAME MOTOR PATTERN IS USED FOR DIFFERENT MOVEMENTS

9.6.1. Jumping and kicking

Essentially the same motor pattern is used in both kicking and jumping, particularly in the actions of the motor neurons innervating the muscles that move the tibia. The differences lie in the actions of the motor neurons controlling the movements of the more proximal joints, and in the compensatory and balancing support provided by the other legs. In kicking, the leg must be lifted from the ground by a rotation about the thoraco-coxal joint that is more extensive than that needed to rotate the legs into their appropriate position for jumping. Other differences lie in the timing of the movements of the two legs. In jumping they are normally moved at about the same time but in kicking they are normally moved independently. Both movements are nevertheless directed: a locust jumps away from a threatening stimulus, and a kick is accurately directed at the source of the stimulation.

9.6.2. Swimming

It may seem odd that all stages of locusts, which are essentially desert living animals, are able to swim. Nevertheless, marching bands of hoppers can cross water successfully by swimming (e.g. Knechtel, 1938). First and second instars generally make only limited forward progress but adults can swim at about 130 mm.s^{-1}. Swimming is produced by rapid, synchronous extensions of the hind tibiae that resemble the movements used in jumping and kicking, followed by slower flexions that can be repeated many times at frequencies of 1–3 Hz (Pflüger and Burrows, 1978). Only the movements of the hind legs provide power, the other legs merely adopting a characteristic posture in which the front legs point forwards and the middle legs backwards, similar to that used when the locust is about to land after flight. Usually both hind legs are moved together and then the swimming is strong. The synchrony is not always maintained, however, and in some strokes the movements of the two legs may alternate, while in others only one leg may be used.

The sequence of movements starts when the flexor tibiae motor neurons spike and flex the tibia. They are then joined by the two extensor motor neurons, with FETi spiking 4–6 times at instantaneous frequencies of up to 30 Hz, so that there is a short period of co-contraction. The flexor motor neurons then stop spiking while the extensors continue, with the result that the tibia extends rapidly in 50 ms, slowed by the resistance of the water. The tibia is fully flexed against the femur, although some weak swimming strokes can be performed even if it is not fully flexed.

The same basic motor pattern therefore appears to be used to produce a spectrum of movements ranging from powerful jumping to weak swimming (Fig. 9.14). First, all movements are characterised by the participation of FETi, which is not used in normal walking. Second, the sequence of action of the motor neurons and the muscles that move a tibia are the same. Third, the movements of the two hind legs are not necessarily synchronised: in kicking and weak swimming, one leg is more likely to be used independently, whereas in powerful jumping and powerful swimming the two legs move at almost the same time. All these movements can vary both in rapidity and in the power generated. The greater force needed in jumping as compared to swimming is achieved by lengthening the period of co-contraction (from 80 ms to more than 500 ms), increasing the number of FETi spikes (from 4 to 20) and increasing their instantaneous frequency (up to 100 Hz). If the tibia is not fully flexed, the flexor apodeme lock cannot be engaged and hence little extensor force can be stored. The rapidity and force of the extension movements are then limited, as in

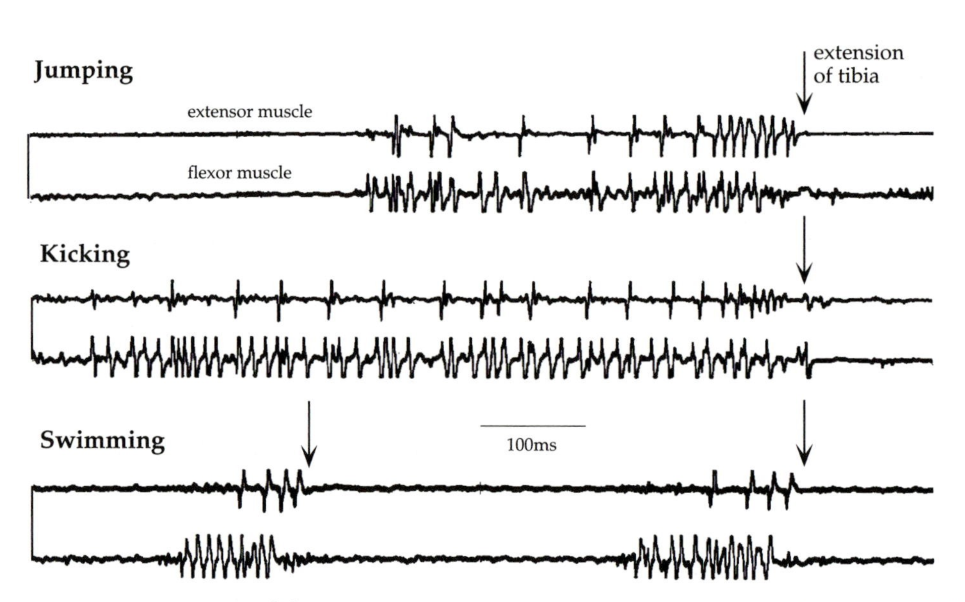

Fig. 9.14 Comparison of the motor patterns used in jumping, kicking, and swimming. The extracellular recordings were made from the extensor and flexor tibiae muscles of freely moving animals. Based on Pflüger and Burrows (1978).

walking or scratching, to what can be produced by the instantaneous action of the motor neurons.

9.7. JUMPING AND FLYING

A jump often launches an adult locust into flight (Fig. 9.15A). When this happens, the wings generally open 15–35 ms after the tarsi leave the ground, but sometimes they open before this (Camhi, 1969a; Pond, 1972). This means that there is insufficient time for opening of the wings to be elicited by sensory signals such as those generated by wind on the head as the body is thrust forward, or those from the tarsi indicating that contact has been lost with the ground. This suggests that there must be a close link between the interneurons controlling both flying and jumping. If the interneurons that initiate jumping also initiate flying, then a close relationship between the two movements is to be expected, but few natural stimuli that elicit jumping or kicking have been used to test this linkage. When jumping or kicking movements are initiated by mechanical stimulation of the abdomen, motor neurons and interneurons controlling the flight muscles receive a depolarising synaptic input. This gradually increases during co-contraction of the femoral muscles, and may lead to spikes in flight motor neurons at variable times before or after a kick (Pearson *et al.*, 1986). In motor neurons innervating wing elevator muscles, and in some interneurons such as 401 and 504 (*see Chapter 11*) flight activity may even be

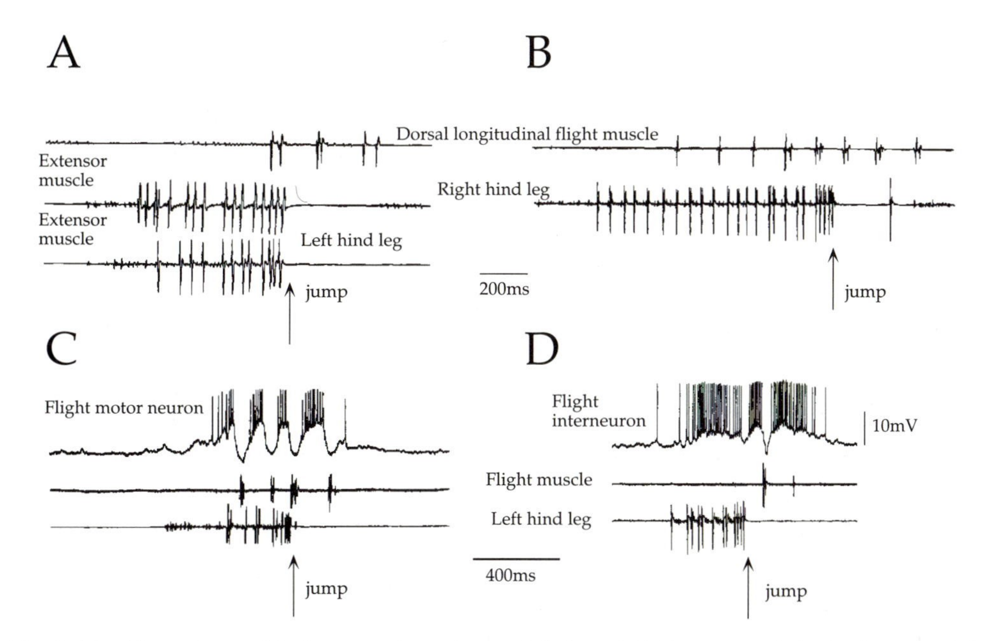

Fig. 9.15 The relationship between jumping and flying. A,B. Patterns of rhythmic potentials are established in a flight muscle before either of the hind legs leaves the ground in a jump. C. An intracellular recording from a flight motor neuron shows a series of rhythmic depolarisations before a kick, that are not subsequently maintained as a full flight pattern. D. Spikes occur in an interneuron involved in generating flight, before a kick occurs. No clear flight rhythm is established. Based on Pearson *et al.* (1986)

established before a jump is triggered (Fig. 9.15B-D). Only the 404 interneurons that can initiate flight receive a specific synaptic input at the time that the jump is triggered (Pearson *et al.*, 1985b). In an intact animal, this early expression of flight is presumably suppressed by sensory signals such as those indicating that the tarsi are still on the ground. The conclusion is that the interneurons which trigger jumping are not involved in initiating flying, but the interneurons which must maintain the close linkage between these two events remain to be identified.

9.8. COMPARABLE MECHANISMS IN OTHER ANIMALS

Is the mechanism used by a locust when jumping a unique solution to the particular problem faced, or do other insects use similar mechanisms? Unsurprisingly from the diversity of insects that exists, several different mechanisms for jumping appear to have arisen independently, not all of which involve rapid extension of the limbs. Nevertheless, an underlying design for the

Table 9.2. Jumping performance of six insects

Insect	Body mass (mg)	Height of jump (cm)	Distance of jump (cm)	Take-off velocity ($m.s^{-1}$)	Take-off angle (degrees)
Locust *Schistocerca gregaria*	1500–3500	30	80–95	3.1	45
Click beetle *Athous haemhorrhoidalis*	40	30	?	2	90
Flea beetle *Blepharida sacra*	7mm long	?	70	?	?
Ant *Harpegnathos saltator*	32	2	3–4	0.6–0.7	40
Fruit fly larva *Ceratitis capitata*	17	7	12	1	60
Flea *Spilopsyllus cuniculus*	0.45	3–5	–	1–1.2	50

production of rapid movement is a reliance on the slow increment of force in a joint that is locked in one position, the storage of this force in specialisations of the cuticle and a sudden release of the force so that acceleration is applied over only a very short period (Table 9.2).

Many **beetles** have powerful hind legs and enlarged extensor muscles whilst some, like the **flea beetles** (*Blepharida sacra*) (Alticinae, from the Greek meaning 'good at jumping'), enhance these features with a spring formed by the curling and chitinisation of the extensor tibiae tendon (Maulik, 1929; Furth *et al.*, 1983) to excel at jumping. Some are even reported to be able to jump 100 times their body length, and their springs to retain their properties after 1000 years in the soil. Powerful contractions of the extensor tibiae muscle dilate the spring which thereby stores the energy needed for the jump in its tension. The tibia is probably locked into the flexed position when a small triangular plate at the femoro-tibial joint is moved by contraction of the flexor muscle, so enabling the extensor to generate isometric force. The lock is probably released through changes in the force in the flexor and the stored energy is delivered rapidly as the spring snaps back to its original shape.

Only a few of the many species of **ants** jump to escape from predators or when they are themselves hunting prey. The jump, which takes some 15–25 ms from the first intentional raising of the head to the time that the legs leave the ground, propels the body forwards some 3–4 cm at an angle of about 40°, but with a low take-off velocity of only 0.6–0.7 $m.s^{-1}$ (Tautz *et al.*, 1994) The main problem that the ants, like locusts, must solve is to ensure that the forces at take-off act through the centre of their mass so that the body does not spin. This requires different strategies in ants of different shapes. In *Harpegnathos saltator* from India, the centre of mass is anterior to the mesothoracic legs, so the hind legs first push the body forwards before the rapid extension of both the middle and hind legs provides the main forwards and upwards thrust (Baroni Urbani *et al.*, 1994). In *Myrmecia piliventris* from Australia, the centre of mass is between the meso- and metathoracic legs and it is the simultaneous extension of both pairs of legs that provides the thrust. In *Gigantiops destructor* from Brazil, the centre of mass is behind the hind legs and the first action is that the gaster (swollen posterior part of the abdomen) is thrown forwards and upwards through 90° in 20 ms before the extension of the meso- and metathoracic legs thrusts the body forwards and upwards. In none of these species does the extension of the legs appear to require the preceding storage of energy, so that there is no co-contraction of the femoral muscles as in the locust.

The **flea** *Spilopsyllus cuniculus,* which weighs only 0.45 mg, can jump 3–5 cm into the air at a take-off velocity of about 1 $m.s^{-1}$. The energy required is produced by the rapid depression of the femora of the hind legs in about 1 ms (Bennet-Clark and Lucey, 1967; Rothschild *et al.*, 1972). These requirements for the jump can again not be met by direct muscular contraction and are produced by a prior contraction of the enlarged depressor muscle with energy stored in a pad of resilin. The stored force is released suddenly by the contraction of a small muscle that changes the point of action of the depressor muscle so that the femur can be depressed rapidly.

Some **eupelmine parasitic hymenoptera** (superfamily Chalcidodea, subfamily Eupelminae) jump by a rapid depression of the trochanter of a middle leg which is first retracted against the coxa. Curiously, the thoracic muscles that power the jump movements appear to differ anatomically between males and females of the same species (Gibson, 1986).

A **click beetle** *Athous haemorrhoidalis* uses the joints in its body instead of its legs for rapid propulsion. When it is lying on its back, it can propel itself vertically into the air for 0.3 m, the same height that a locust achieves in a jump, at take-off velocities of about 2 $m.s^{-1}$ by jack-knifing its 40 mg body at the junction between the pro- and mesothorax (Evans, 1972, 1973). The back is first arched so that a peg on the prosternum engages a socket on the mesosternum. With the joint locked in this position, a large dorsal intersegmental muscle, that accounts for as much as 10% of the mass of the beetle, contracts isometrically for between 0.4 and several seconds. The energy produced by this powerful contraction is stored in distortions of the cuticle, in the muscle apodeme and possibly also in resilin associated with the joint. The stored energy is then released suddenly, accompanied by an audible click, when the position of the peg is shifted by unknown mechanisms, thus catapulting the beetle into the air. Again, therefore, this rapid jumping movement is produced by a slow contraction followed by a rapid release of the stored energy.

The **Mediterranean fruit fly larva** *Ceratitis capitata* jumps by first arching its body to bring the tip of the abdomen in contact with the head and then contracting the body wall muscles while the mouth hooks anchor the head to the ground (Maitland, 1992). The body is held in this tensed and bulging condition with elastic energy stored in the cuticle until a wave of contraction passing forwards along the body releases the mouth hooks. The body is then straightened and the force exerted by the abdomen on the ground thrusts the body upwards and forwards in a somersaulting movement for a distance of some 12 cm, or about 14 body lengths, at a take-off velocity of 1 $m.s^{-1}$ to give an overall forward velocity that is some 200 times what it manages by crawling. The jumps can be repeated at a rate of about 8 min^{-1} and may represent a means of escaping predation by ants.

Adult *Drosophila* also jump to launch themselves into flight when confronted with a startling stimulus such as the sudden extinguishing of light. Most of the power is generated by a rapid extension of the middle pair of legs caused by contractions of the tergotrochanteral muscles, which also cause an initial elevation of the wings. The pathways responsible for initiating these movements so far seem to be simple, with the consequence that the muscles contract with a very short latency following the appropriate sensory stimulus (King and Wyman, 1980; Tanouye and Wyman, 1980). A pair of giant interneurons in the brain are excited by inputs from the eyes and send axons to the thoracic ganglion where they make electrical synapses directly with the motor neurons innervating the tergotrochanteral muscle on the same side of the body as their axon, and with a pair of interneurons. Each thoracic interneuron, in turn, makes a chemical synapse with the contralateral dorsal longitudinal motor neurons; strangely these are within the nerve containing the axons of these motor neurons. While a physiological analysis of the more detailed pathways that must be involved is likely to be difficult, a genetic analysis of the mechanisms involved now becomes a real possibility. For example, mutations which disrupt the *Passover* locus result in flies that cannot jump and in which the pathways to the muscles that would normally participate are disrupted by faults at the output synapses of the giant interneurons with the motor neurons and interneurons (Thomas and Wyman, 1984; Baird *et al.*, 1990, 1993). In some *Passover* genotypes, the branching of the tergotrochanteral motor neurons in the mesothoracic neuromere and contacts with the giant interneurons are reduced, but the mutations probably also affect cell recognition and synaptic function because the branches of the giants are still close to, even in contact with, the motor neurons. An enthusiasm for this powerful approach is thus tempered by the possibility of many functional consequences of a particular mutation.

Even in animals that do not jump, some of the same mechanisms found in the locust are used when a rapid and powerful movement must be produced. **Crickets** do not jump but they do perform well-directed kicks with a single hind leg to fend off adversaries or potential parasites such as wasps. To do this, the body is first tilted forward so that one hind leg is clear of the ground, and it is then aimed by rotation about the thoraco-coxal joint before the tibia is rapidly extended (Hustert and Gnatzy, 1995). The rapid extension is accomplished by a co-contraction of the flexor and extensor tibiae muscles but this lasts for no longer than 20 ms and the pattern of spikes in the two sets of motor neurons can be variable. The fast extensor motor neuron produces only a few spikes, often only one, and the flexor motor neurons are not inhibited before the extension can take place. Instead, the extension depends on the balance of forces produced by the two muscles, and not by a triggering as in the locust. The balance of forces between the large extensor and the small flexor is aided by the mechanics of the joint where, again, the flexor muscle has a greater mechanical advantage at more flexed positions. The flexor tendon does not have a pocket that engages a lump in the femur but it is expanded into a cushion that increases the mechanical advantage as it slides over a protuberance in the ventral femur. With this arrangement, the force and velocity of extension is determined by the differing forces developed in the flexor and extensor muscles. The design is driven by the need to respond rapidly once a tactile stimulus has indicated a threatening stimulus. Foregoing the long co-contraction phase means, however, that the force that can be produced is considerably reduced compared with that in either a kick or a jump by a locust and could not lift the body from the ground.

Mantid shrimps (various species of stomatopod crustaceans such as *Squilla empusa* and *Hemisquilla ensigera*) strike at prey or predators with highly specialised appendages (modified mouthparts) using mechanisms that are very similar to those of the locust when jumping (Burrows, 1969). The strike requires high force and high velocity and these features can only be met by a slow contraction of the power-producing muscles and the sudden release of stored energy. The joint of the striking appendage (meropodite-coxopodite) is locked in the flexed position by the contraction of two small flexor muscles operating two sclerites in the joint membrane. Once locked, it can be released only by a relaxation of the flexor muscles and thus allows the extensor muscles to slowly develop their maximum force. The flexor and extensor muscles then co-contract for periods of up to 1 s before the flexor motor neurons stop spiking and the full force produced by the almost isometric contractions of the larger extensor muscles is delivered suddenly. The extension movement takes only 4–5 ms before the prey is struck and has the power of a small calibre bullet; sufficient to stun a prey or crack open the shell of a crab.

A cocking and rapid release of the stored energy are also used by the **ant** *Odontomachus bauri* when closing its jaws (mandibles) on prey (Gronenberg *et al.*, 1993). The jaws can be held apart in a position that requires no muscular action. Contact with a potential prey is established by the antennae and this leads to contraction over periods as long as a few seconds of the powerful closer muscles, but it is unable to close the jaws from this position. As the ant continues to move forwards, two long hairs protruding anteriorly from the jaws touch the prey and within 8 ms the jaws then snap shut. The sensory neurons from these hairs have diameters of 20 μm and appear to directly excite two large motor neurons innervating a tiny muscle, the contractions of which alter the mechanics of the joint, allowing the stored force in the cuticle and tendon of the closer muscle to be delivered rapidly. The jaws snap shut in as little as 0.3 ms, making it one of the fastest known movements.

9.9. CONCLUSIONS

Jumping is often used to remove the locust from life-threatening stimuli. This behaviour therefore requires reliability from the ensemble of neurons generating the motor pattern, while allowing variability of action by individual locusts. It also demands that the mechanical components, the muscles, apodemes and cuticle, perform at close to their structural and functional limitations. The major requirement of the distinctive motor pattern for jumping and kicking is that the tibia can be locked in the fully flexed position so that the extensor and flexor muscles can co-contract slowly to build up the necessary force. The stored force is then released explosively by inhibition of the flexor motor neurons. The pattern results from the synaptic connections between motor neurons, the interplay between the actions of interneurons, and the effects of sensory feedback. In this complex parallel system of interacting elements, establishing a causal role for individual neurons has unsurprisingly proved elusive. Many different components contribute to this important behaviour, and the motor pattern can proceed without certain individual components because of the parallel and distributed nature of the processing. To complete the catalogue of components in the network, the interneurons initiating the cocking of the legs, those causing the excitation of the motor neurons during the co-contraction phase and those leading to triggering need to be identified. These interneurons might then provide clues as to how the jump motor pattern is selected over those responsible for other forms of leg movements.

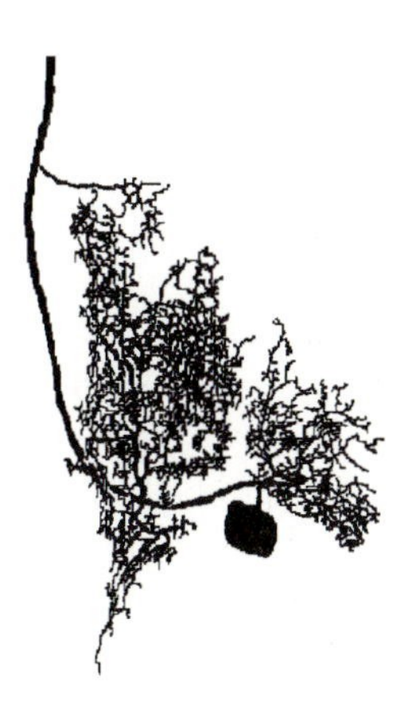

Escape movements

10.1. WHAT IS A GIANT INTERNEURON?

Giant interneurons in most animals have axons with large diameters so that their spikes can be conducted rapidly from one part of the nervous system to another. The largest neurons are therefore often involved in causing escape movements, either as motor neurons, as in squid, or as spinal interneurons, as in fish. In the locust, few neurons qualify to be called giants by virtue of their size. In crickets and cockroaches, however, a small group of neurons have large diameter axons in the abdominal connectives. These axons belong to interneurons with cell bodies in the terminal abdominal ganglion and may project to all ganglia in the body as far anterior as the brain. On the assumption that these interneurons are homologous in different insects, the name giant has been retained in locusts (see Tables 10.1–10.3). Even in those insects where some of the neurons can clearly be called giant interneurons, the line between these and other neurons is drawn somewhat arbitrarily, so that the number called giant interneurons has changed over the years. Moreover, the distinction of giant does not necessarily extend to all parts of these neurons such as their cell bodies, or even to their axons which can narrow within the abdominal ganglia and in the thoracic connectives. The attraction of these large interneurons has been the ease with which their spikes can be recorded and their excitation by air currents directed at the cerci which protrude from the rear of the abdomen.

10.2. WHAT BEHAVIOUR IS ASSOCIATED WITH THE ACTION OF THE GIANT INTERNEURONS?

In locusts, it is unclear what the giant interneurons may be doing. It is not known when they spike or what movements they might initiate, although one suggestion is that they elicit jumping, which is the locust's major response to a threatening stimulus.

Table 10.1. Giant interneurons in locusts

Interneuron name	Neuromere of terminal ganglion	Cell body diameter (μm)	Axon diameter (μm)	Conduction velocity (m.s^{-1})	Tract containing axon	Inputs from cercal hairs
GI1 (medial giant, MGI)	9	50	25	3.9	VIT	from both cerci
GI2	9	60	20	3.4	LDT then DIT	from both cerci
GI3	10	40	10–15	2.7	LDT then DIT	from both cerci
GI4	11	50	20	3.2	LDT then DIT	little or none from contralateral

DIT Dorsal intermediate tract
LDT Lateral dorsal tract
VIT Ventral intermediate tract

Table 10.2 Giant interneurons in crickets

Interneuron name	Neuromere of terminal ganglion	Tract containing axon
8-1a (Medial Giant)	8	VIT
9-1a (Lateral Giant)	9	VIT
9-2a	9	DIT
9-3a	9	DIT
10-2a	10	DIT
10-3a	10	DIT
11-1a	11	VIT

They may also be able to influence the performance of flight by the connections that they make with some thoracic interneurons. Moreover, in locusts, the cerci (paired appendages protruding posteriorly from the tip of the abdomen) are only small when compared with many other insects, and yet their mechanoreceptors provide the main excitatory input to the giant interneurons. To understand the possible role of these interneurons it is therefore necessary to turn to other insects, such as cockroaches and crickets, where the giant interneurons are readily recognisable, where the cercal receptor system is well developed, and where obvious behaviour can be associated with the actions of the giant interneurons. This approach is justified by the similarity in the organisation of the giant interneurons, their input connections from cercal receptors, and their output connections as far as these are known. There are also some similarities in the behaviour associated with the action of the interneurons in the different animals, although the main escape response shown by crickets and cockroaches is not shown by locusts.

Table 10.3. Features of giant interneurons in cockroaches

Properties and actions	Ventral giant interneurons GI1-4	Dorsal giant interneurons GI5-7
Spike conduction	Large diameter axons in ventral tracts; spikes arrive in thorax 4–5 ms ahead of those in dorsal giant interneurons	Smaller diameter fibres in dorsal tracts
Projections	Extend to the brain	Extend only to thorax
Connections	Thoracic intersegmental interneurons; possibly DUM and local interneurons	Interneurons 302, 404 involved with flying; interneuron 714; possibly other thoracic intersegmental interneurons
Walking and running	Silent	Some spike at each step
Response to air currents while walking	Continue to respond	Continue to respond
Effects of electrical stimulation on walking	Only GI1 gives weak and labile excitation of leg motor neurons	Evoke spikes in coxal depressor and levator motor neurons
Flying	Silent	Some spike intermittently at wingbeat frequency
Response to air currents while flying	Response is reduced	Response is maintained
Effects of electrical stimulation on flying	None	Can initiate flying

Air currents directed at the cerci cause a cockroach or a cricket to scuttle away rapidly, usually after first turning away from the stimulus, whereas locusts usually take little notice of such stimuli. The movements are preceded by a burst of spikes of large amplitude, as recorded extracellularly from the abdominal connectives, which are generated by the giant interneurons. The initial analyses of this escape behaviour (Roeder, 1948) suggested that the neural control was simple, with the signals from the air current sensors on the cerci evoking spikes in the giant interneurons that were then conveyed rapidly to the motor circuits in the thorax where appropriate movements of the legs were generated. At their simplest, the pathways were thought to consist only of sensory neurons from the hairs on the cerci that act as air current detectors, giant interneurons and leg motor neurons. This supposed simplicity of organisation, involving a small number of very large neurons, has been the attraction for many analyses at different levels over the last 50 years. The dominant assumption that has pervaded virtually all of these investigations is that the giant interneurons form a finely tuned mechanism for the detection of an approaching predator which can conduct signals rapidly to the motor machinery necessary to effect escape. This may well be one of their actions, but the dominance of this assumption has tended to obscure other possible actions of the cercal sensory neurons and associated interneurons, and has also led to the neglect of other mechanisms that might contribute to rapid escape movements. The cercal receptors and the giant fibres do provide a high sensitivity mechanism for detecting very low velocities of air movements and for the rapid conduction of information about the direction of this stimulus to the motor machinery in the thorax. Their function is not, however, solely to generate escape movements, because the cercal receptors can respond to air movements that are generated by conspecifics, and to sound. Receptors elsewhere on the body, such as those on the antennae, can also cause an escape response to air currents or to touch. Thus, the contribution of the giant interneurons to escape movements may be only one of their actions, and similarly, escape movements can be elicited by a variety of stimulus modalities detected by receptors on different parts of the body.

These observations emphasise that while analyses of the giant interneurons have concentrated on escape movements, the cerci may be detectors of stimuli involved in other behaviour, and this will have profound implications on the interpretation of the connections made by the component neurons. Not every connection found can be seen as part of the circuitry that causes escape movements. Similarly, it may not just be the giant interneurons that are involved in mediating the motor actions that result from the detection of these different stimuli. Other sensory receptors and other interneurons can therefore act in parallel.

What follows is an account of the cerci, their receptors and sensory neurons, the interneurons, including giant interneurons, nongiant interneurons and local interneurons, and the ways in which they may influence behaviour.

10.3. THE CERCI AND THEIR RECEPTORS

The cerci are paired appendages at the rear of many insects that can be moved independently by their own sets of muscles. They are covered in hairs sensitive to air currents and to touch. In locusts, the cerci are only short (1 mm long in females and 1.75 mm in males) representing only a small percentage of the body length and contrast with the longer and more elaborate appendages in some cockroaches and crickets that can be 46% of the body length. The reason for the difference in size between the cerci of male and female locusts is unknown, but may be related to copulatory behaviour in some way. Their small size relative to those in some other insects may be related to the ability of a locust to jump and then fly away from a threatening stimulus that may more often be detected by its eyes. On each cercus of a locust there are about 200 hairs that are of two main types: flexible, filiform hairs (Thomas, 1965) 20–500 μm long which respond to air currents and low frequency sound waves, and stiffer trichoid or bristle hairs 20–250 μm long which respond to touch (Boyan *et al.*, 1989a). Hairs of both types are each innervated by a single sensory neuron.

10.3.1. The cerci as air current detectors

The receptors can provide information about air movements or sound that result from activity in the animal's surroundings or from its own actions, while particular club-shaped receptors in some insects may give information about gravity. The receptors can respond to the air movements caused by an approaching predator and their signals can initiate turning movements when walking, running rapidly, or flying. Stimuli will be effective only when the potential predators are close, so that the dominant feature is the velocity of the air particles and not changes in air pressure.

The different lengths and stiffness of the hairs means that their sensory neurons code different features (velocity and acceleration) of an air current. In crickets, the long hairs are velocity sensitive and respond to low frequency movements, while the shorter hairs are acceleration sensitive and are insensitive to low frequency movements (Shimozawa and Kanou, 1984). The hair receptors on the cerci of cockroaches respond to air particle velocities as low as 12 $mm.s^{-1}$ and with an acceleration of 0.6 $m.s^{-2}$. Such air movements are created by a predatory toad as it lunges forward and flicks out its tongue in an effort to catch a cockroach, with the acceleration component being particularly important for the initiation of a reliable turn in the escape response (Plummer and Camhi, 1981). The sensory neurons from these hairs act as bandpass filters with peak frequencies of about 100 Hz that are largely determined by the mechanical properties of the hairs themselves (Kondoh *et al.*, 1991b). Many of the sensory neurons associated with these hairs may spike spontaneously, in large part because of the properties of the dendritic membrane rather than mechanical responses to movements of air particles (Hamon and Guillet, 1994). By contrast, the characteristics of their responses to air currents are largely determined by the mechanics of the receptor rather than by the properties of the spike-generating mechanism in the membrane of the dendrites. More slowly moving predators such as spiders, whose movements would not excite the air current receptors on the cerci, are probably detected only when they touch particular receptors such as those on the antennae (Comer *et al.*, 1994).

10.3.2. The cerci provide signals that may be used in different behaviour

The ability of the cerci to detect air particle movements of low velocity may be used in behaviour other than escape. Male cave crickets, for example, flick their wings forwards once through angles of about 180° during aggressive encounters, or 4–7 times in succession at frequencies of 8–12 Hz during courtship once they have taken up a preferred position relative to a female (Heinzel and Dambach, 1987). These movements set up vortex rings that travel up to 15 cm with an initial velocity of 400 $mm.s^{-1}$ and sweep over the female. The air particle movements in the vortices are detected by the cercal receptors of the female and evoke spikes in giant interneurons (Heidelbach and Dambach, 1991). In other species of crickets, the receptors on the cerci also respond to the low frequency (25–30 Hz) component of the sound that is generated by another cricket stridulating up to 15 cm away, or to the sound produced by its own song (Kämper and Dambach, 1981). The spikes in the sensory neurons of these receptors, in turn, evoke spikes in interneurons with axons projecting anteriorly that are in phase with the syllables of the different types of song. Inputs from the cerci can reset the stridulatory movements (Dambach *et al.*, 1983). When a locust is flying, some of the interneurons that receive inputs from the cercal hairs are excited rhythmically in time with the movements of the wings, perhaps reflecting the rhythmic movement of the air over the cercal receptors. During the brief periods when a cockroach will fly, the dorsal giant interneurons spike at the start and then intermittently during the flight with a preferred phase relationship in the wingbeat cycle (Libersat *et al.*, 1989). This suggests some role for these pathways in influencing the flight performance.

10.3.3. Cercal hairs act as directional air movement detectors

In locusts, the filiform hairs on a cercus all have the same preferred directional sensitivity, responding best to air moving parallel to the long axis of the body. The hairs are unevenly distributed on a cercus, in contrast to the precise ordering of hairs in rows and columns on the cerci of cockroaches (Fig. 10.1A), and the precise circumferential location of hairs with different directional sensitivities on the cerci of crickets (Palka *et al.*, 1977; Bacon and Murphey, 1984). The 19–21 annuli on a cercus of a cockroach have rows of long hairs, ordered in such a way that similarly placed hairs on each form columns along the long axis of the cercus. The hairs of a particular column are all oriented in the same direction, with this direction varying between the different columns along a row. Each hair moves back and forth most readily in one plane due to the eccentric placement of its shaft in the socket (Nicklaus, 1965), and its sensory neuron is excited by air moving through an arc of approximately 180° (Fig. 10.1B) (Westin, 1979). The combined responses of the sensory neurons from all the columns of receptors can therefore code an air current from any direction (Dagan and Camhi, 1979). Although adult cockroaches have more than 200 sensory receptors on each cercus that can code the direction of an air current, a first instar cockroach has only two and yet can still make appropriately directed turns away from the source of an air current (Dagan and Volman, 1982).

10.3.4. Central projections of the sensory neurons

The axons of the sensory neurons project into the terminal abdominal ganglion to form a compact area of branches in ipsilateral neuropil. In crickets, the sensory neurons from hairs with a particular directional sensitivity project to discrete areas within this region of neuropil so providing a topographical representation of the sensory surface of the cerci as a spatial map (Bacon and Murphey, 1984; Jacobs and Nevin, 1991). In both crickets and cockroaches (Thompson *et al.*, 1992), the position of a hair around the circumference of a cercus and its proximo-distal position determine the branching pattern of its sensory neuron in the central nervous system. In locusts, where the receptors have similar directional responses, it would be informative to know how the terminals of their sensory neurons are arranged in the central nervous system.

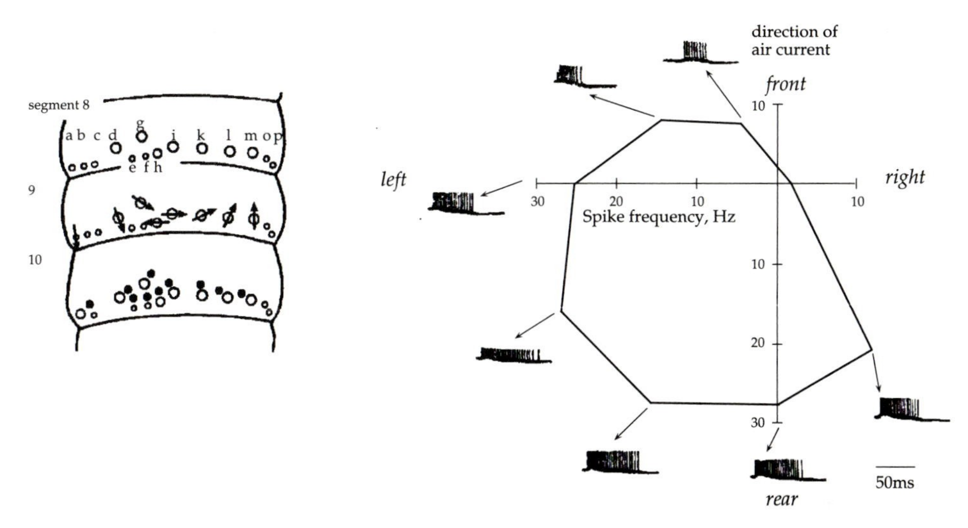

Fig. 10.1 Hair receptors on the cerci of a cockroach. A. The arrangement of hair receptors in rows and columns (lettered a-p) on the annuli of a cercus. The arrows indicate the direction of an air current that elicits the largest spike response of a hair. B. Directional response of a sensory neuron from a cercal hair to an air current. The graph is a polar plot of the frequency of spikes to air currents from different directions, with sample intracellular recordings from the axon of the sensory neuron. Based on Westin (1979).

In cockroaches and crickets, interneurons access this map by having branches in particular places, and the connections that are formed with specific populations of sensory neurons will thus determine their directional responses to air currents. The most obvious of the connections made by these sensory neurons are with the giant interneurons, but other connections are made with some intersegmental interneurons and with local interneurons.

10.4. PROPERTIES OF THE GIANT INTERNEURONS

10.4.1. Structure of the giant interneurons

In the locust, only four pairs of interneurons (GI1-4) (Fig. 10.2) qualify to be designated as giant interneurons (Cook, 1951), in contrast to 16 in some stick insects, nine in mantids and seven in cockroaches and crickets. The four giant interneurons in the locust have cell bodies in neuromeres 9–11 of the terminal abdominal ganglion and axons that ascend in the contralateral connective to the brain, forming branches in the ganglia through which they pass, with the most extensive ones in the thoracic ganglia. The spikes are conducted at 2.7–3.9 $m.s^{-1}$ (for comparison, a sensory neuron from a hair on a cercus conducts its spikes at about 1.0–1.5 $m.s^{-1}$), and apparently without the

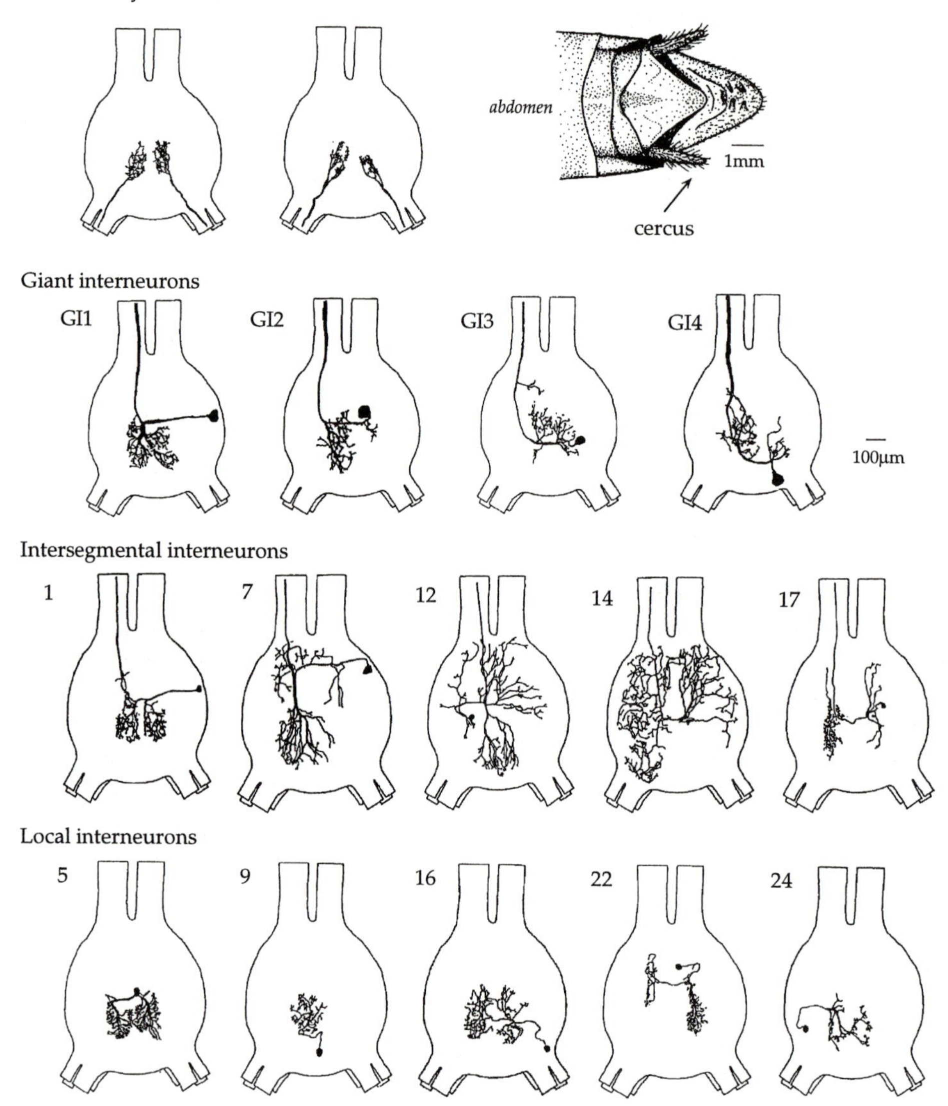

Fig. 10.2 Morphology of neurons processing signals generated by air currents directed at the cerci of a locust (*Locusta migratoria*). The axons of the sensory neurons from hairs on the cerci project to the last abdominal ganglion. The four giant interneurons, and the intersegmental and local interneurons that respond to air currents directed at the cerci have cell bodies in the last abdominal ganglion. Based on Boyan and Ball (1989a) and Boyan *et al.* (1989a, b).

failures that occur in the comparable neurons of the cockroach as the axons narrow to pass through the abdominal ganglia, and thus reliably reach the thoracic ganglia in 6–8 ms (Boyan and Ball, 1989a).

In the cockroach *Periplaneta americana*, the largest axon has a diameter of 45 μm, in the cricket *Acheta domesticus* 28 μm, and in the locust 25 μm. These axons thus represent 6, 8 and 3%, respectively, of the area of the abdominal connective in cross section (Boyan and Ball, 1986). The investment that a locust makes in such neurons is therefore modest compared with that made by a cockroach whose giant interneurons altogether occupy some 12% of the cross-sectional area of an abdominal connective. This suggests that their ability to conduct spikes rapidly from the tip of the abdomen to the thorax, and perhaps beyond, is of crucial importance.

There are probably segmental homologies between giant interneurons in one insect, as in crickets they are known to be formed embryonically from three bilateral clusters of cells that are serially repeated in three of the neuromeres of the terminal abdominal ganglion (Jacobs and Murphey, 1987). There may also be homologies between these giant interneurons in different insects, based on their similar morphology, the grouping of their axons in either dorsal or ventral tracts in the connectives, and their response properties. The differences in numbers may simply reflect what is defined as a giant based on the diameter of their axons relative to others in the connectives. The interneurons have nevertheless been given different names in different insects (see Tables 10.1–10.3). Thus, in locusts they are simply called GI1-4, in cockroaches GI1-7, and in crickets they are named according to the neuromere that contains their cell body. In cockroaches, the giant interneurons are divided into ventral (GI1-4) and dorsal groups (GI5-7) according to the position of their axons in the abdominal connectives (Table 10.3). This subdivision has led to their being considered as distinct functional groups with an inevitable bias on the interpretation of their actions. Whatever the precise number of neurons, and the validity of the subdivisions, it is clear that the information about the stimuli detected by the receptors on the cerci is carried in only a few parallel channels, and yet this can be decoded successfully to ensure a motor response in contexts that may be life threatening.

10.4.2. Response properties of the giant interneurons

The giant interneurons in locusts, cockroaches and crickets all respond with a burst of spikes to an air current directed at a cercus, but whether they also receive inputs from hairs or other receptors on other parts of the abdomen has been largely unexplored. Each interneuron is excited directly by the sensory neurons from arrays of hairs on the cerci, with some 4–5 ms elapsing from the detection of an air current by a sensory receptor to the generation of a spike in an interneuron. The connections between the hair afferents and the interneurons are direct and in the cockroach are cholinergic (Sattelle, 1985).

In a locust, the single sensory neuron from one hair diverges to excite more than one interneuron. Interneurons GI1 and 3 are excited equally by both cerci, and GI2 and 4 more strongly by the cercus contralateral to their cell bodies. GI1 responds with a phasic burst of spikes to an air current caused by an initial input from the hair sensory neurons and a later input from local interneurons excited in parallel by the sensory neurons and possibly other giants (Boyan and Ball, 1989c). The other interneurons give a more sustained burst of spikes. Some of the interneurons on one side of the ganglion are also linked to each other synaptically; GI4 makes an excitatory connection with GI2; GI3 with GI1; and GI1 excites GI4 by a pathway that must involve other interneurons (Fig. 10.3). There are no synaptic connections between the left and the right giant interneurons.

All seven pairs of giant interneurons in cockroaches respond to air currents with mean frequencies of spikes that can be as high as 400 Hz, and with instantaneous frequencies for the initial spikes approaching 900 Hz. These are therefore well above the values at which spikes become blocked at the regions of low safety factor for spike propagation as the axons pass through the abdominal ganglia (Spira *et al.*, 1976). There may therefore be unreliable transmission of the spike sequences to the thorax, but it is unclear whether this happens with natural stimuli.

The response of a giant interneuron in terms of the stimulus threshold and the amount of adaptation varies between the different interneurons. At one extreme are those that adapt rapidly, so that their response consists of only a brief burst of a few spikes, and at the other are those which give more prolonged bursts of many spikes. In crickets, the phasic interneurons such as 9-1a (lateral giant) are sensitive to rapidly changing signals and code acceleration with an air particle threshold of about 0.6 $m.s^{-2}$ (Kanou and Shimozawa, 1984). The more tonic interneurons such as 10-3a are velocity sensitive, particularly to low frequencies of stimuli, with thresholds of about 30 $\mu m.s^{-1}$ at 5 Hz, while intermediate interneurons such as 8-1a (medial giant) are acceleration sensitive and are relatively insensitive to slowly changing stimuli, with thresholds of about 10 $mm.s^{-1}$ at 5Hz. In part, these different response properties result from the different patterns of connections made by the sensory neurons from hairs with different properties. This combination of response properties should enable a small moving object to be distinguished from the air currents generated by the cricket's own movements and the necessary information to be coded in the small number of available channels.

10.4.3. Coding of the direction of air currents by the giant interneurons

In locusts, the uneven distribution of the hairs on the cerci and the fact that they all have the same preferred directional sensitivity suggests that the different interneurons have little directional information to code. By contrast, in cockroaches, the directional information provided by the orientation of the cercal receptors is preserved in five out of the seven giant interneurons on each side of the body (Fig. 10.4). The directional sensitivity of these interneurons arises from three sources.

First, sensory neurons from particular arrays of receptors with certain preferred orientations make connections of different strength (Daley and Camhi, 1988). GI1 and GI2, for example, both receive inputs from receptors on all nine rows on the cercus ipsilateral to their axons, but two particular rows provide stronger inputs to GI2 than they do to GI1, and this may explain their different spike responses to air currents of different directions. Hairs within a column may also show gradients in the amplitudes of EPSPs that they evoke in interneurons, in much the same way as

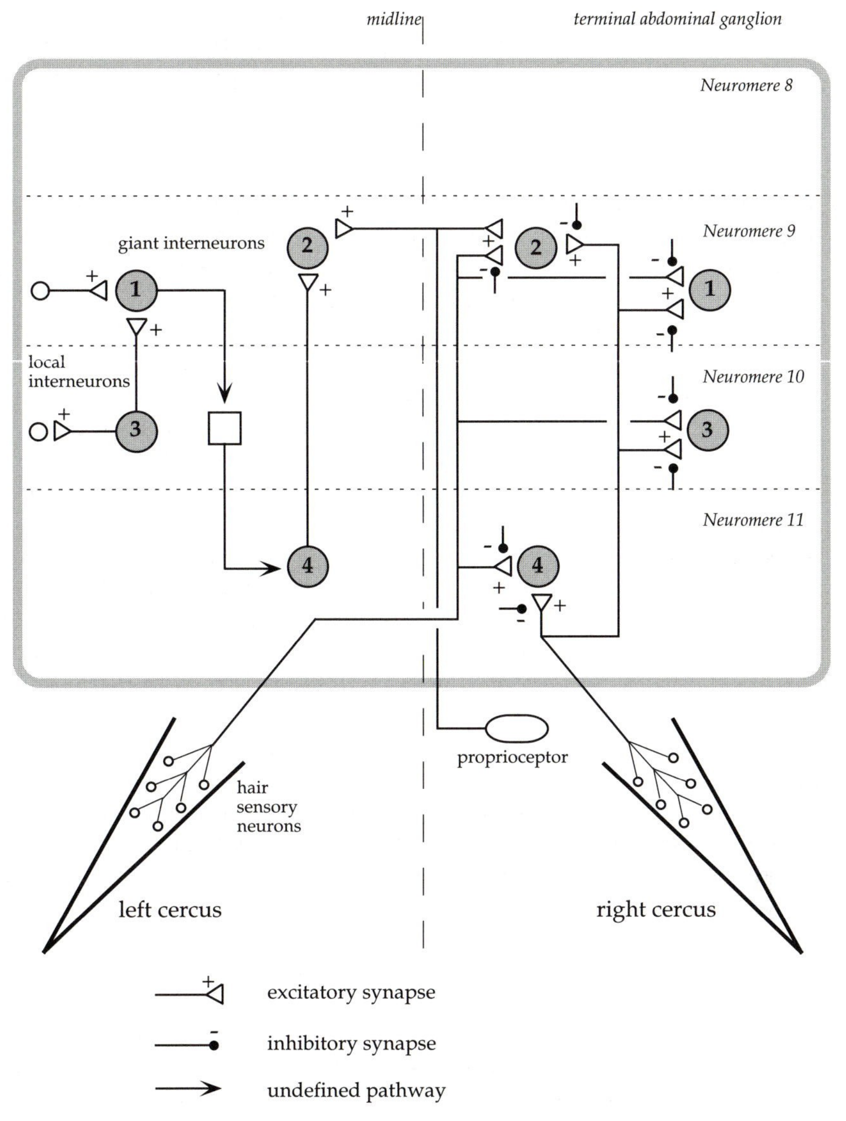

Fig. 10.3 Connections made by sensory neurons from cercal hairs that respond to air currents and interneurons in the last abdominal ganglion of a locust. The four giant interneurons are labelled 1–4. Also shown are the connections made by an unidentified proprioceptor at the base of a cercus. The box indicates that giant interneuron 1 affects giant interneuron 4 by an indirect pathway involving unidentified neurons.

do sensory neurons from hairs on the legs in their connections with local interneurons (*see Chapter 7*).

Second, hairs with particular orientations may make specific connections with particular interneurons. GI3, which does not respond to air currents from the rear,

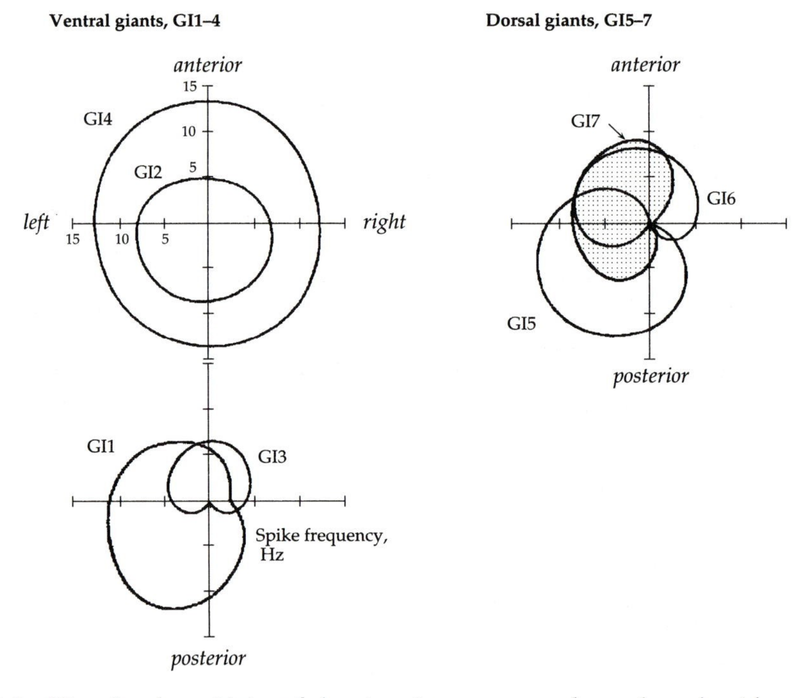

Fig. 10.4 Directional sensitivity of the giant interneurons of a cockroach with axons in the left connective to air currents directed at the cerci. The graphs are polar plots of the frequency of spikes to air currents from different directions. The interneurons are divided into ventral giants (GI1-4), and dorsal giants (GI5-7). Based on Westin *et al.* (1977).

receives no inputs from the columns of hairs that respond to air currents from the rear (Hamon *et al.*, 1994). Inhibitory connections from local interneurons may also sharpen the responses caused by the direct excitatory connections from the hair afferents.

Third, the directionality of an interneuron results from the arrangement of its branches in the terminal abdominal ganglion relative to the spatial map formed by the terminals of the sensory neurons from hairs on the cerci, in much the same way as for the spiking local interneurons in the metathoracic ganglion and their receptive fields on a leg of a locust (*see Chapter 7*). Interneuron 10-3a in crickets, for example, has three main branches in different areas of the neuropil that contains the terminals of sensory neurons with different directional sensitivities, so that its overall response is the nonlinear sum of the inputs to these three branches. Removal of individual branches changes the response properties of the interneuron in the way predicted from the relation of those branches relative to the afferent map (Jacobs and Miller, 1985). Spatial overlap of the branches of the sensory and interneurons is not the only factor, because the relative weighting given to different sensory inputs will also depend on their position relative to the spike-initiating site of the interneuron (Jacobs *et al.*, 1986).

Despite these different, if overlapping, directional sensitivities of the giant interneurons, an air current from the front to either the left or the right side evokes spikes in all interneurons, implying a collective coding of the direction. An air current from one side evokes more spikes with a shorter latency and with some differences in their temporal patterning in the connective ipsilateral to the stimulus, but the difference in the total number of spikes to air currents from different directions is small (Camhi and Levy, 1989). At its simplest, therefore, the code for directionality could be based on the greater number of spikes in a given time in the whole population of giant interneurons on one side, the outputs of which would then sum at the motor circuitry to produce a greater turn to that side (Camhi and Levy, 1988). An alternative code based on the first spikes to arrive in the thorax seems less likely, even though interneurons with axons on the same side as the stimulus do give the first spike, and the spikes in the ventral giant interneurons arrive at the thorax some 5 ms ahead of those in the dorsal giant interneurons (Westin *et al.*, 1977). Experimentally increasing the number of spikes in one of the giant interneurons well above that normally seen indicates that the number of spikes rather than the first to arrive in the thorax is the important feature of the code (Liebenthal *et al.*, 1994). For example, if during an air current from the right, GI1 in the left connective is stimulated so that it produces more spikes than its counterpart in the right connective, then the cockroach will incorrectly turn to the right and into the air current. Merely altering the onset of the spikes does not alter the direction of turning.

Not all of the timing differences between dorsal and ventral giant interneurons can be explained by their differing conduction velocities, but instead there must be differences in the time taken to initiate spikes in the terminal abdominal ganglion. It is not known what the source of this delay could be if, as is generally supposed, the sensory neurons from the hair afferents connect directly with the giant interneurons. If one of the ventral giant interneurons on the left is killed by injection of pronase (Comer, 1985), then more incorrect turns are made to the left than by an intact animal. More mistakes are made the more giant interneurons are killed. Killing GI1 has the most pronounced effect, but the mistakes never exceed 60% as compared with the 100% mistakes that are made if one cercus is covered, or one abdominal connective is cut. This indicates that some directionality is coded by GI1-3, but that other interneurons must also be involved, and implies that direction emerges from a complex decoding of the spike patterns in many parallel channels. Nevertheless, the processing may only need to ensure that the cockroach turns away from a stimulus, and the angle of the turn may not need to be defined accurately.

10.5. OTHER ASCENDING INTERNEURONS CODE INFORMATION FROM THE CERCI

In addition to the giant interneurons there are other intersegmental interneurons in the terminal abdominal ganglion, some with similar response properties, similar shapes (Fig. 10.2), and with conduction velocities approaching those of the giant interneurons, that suggest they may either be segmental homologues, or that they developed from the same cluster of precursor cells (Boyan *et al.*, 1989b). In crickets, it is estimated that there are about 70 pairs of interneurons with ascending axons in addition to the giant interneurons (Baba *et al.*, 1995). In locusts, some eight of these interneurons are characterised by their responses to tactile stimulation of hairs on the ovipositors of females and to air currents directed at the cerci and ovipositiors (Kalogianni, 1995). In crickets, as many as 16 so-called nongiant interneurons also respond to air currents to the cerci (Baba *et al.*, 1991), of which some may be those that also respond to sound (Kämper, 1984). For some of these interneurons it is plainly an arbitrary decision to designate them as nongiant interneurons when they appear to be part of a continuum of sizes. They provide a parallel pathway for signals about tactile stimuli and air currents, although they may integrate signals from receptors over a wider area of the posterior abdomen. The destination of their outputs is largely unknown, but there is no reason to suppose that all are involved in effecting the same thoracic motor outputs as the giant interneurons. Indeed, some of the ascending interneurons with mechanosensory inputs from this part of the abdomen are more clearly involved in controlling the movements of the ovipositor and coordinating the rhythms of egg laying.

10.6. LOCAL INTERNEURONS CODE INFORMATION FROM THE CERCI

Much processing of the sensory signals from the cerci of locusts, cockroaches and crickets occurs in the terminal abdominal ganglion through the action of spiking and nonspiking local interneurons that act in parallel to the signals that are fed directly to the giant interneurons. It has been suggested that there may be as many as 1000 local interneurons in this fused ganglion of a cricket (Baba *et al.*, 1995). The processes of these local interneurons are confined to the terminal ganglion and even to one neuromere, whereas the branches of others may spread over more than one neuromere (Fig. 10.2). Most have branches in both halves of the ganglion, and are thus obvious candidates for processing the signals from the two cerci. These bilateral branches can be linked by up to three commissural processes (Kondoh *et al.*, 1993) (Fig. 10.5) that pose fascinating but unanswered questions about the internal propagation and integration of signals. Many local interneurons of similar shape can be recognised in different insects; for example, interneuron 5 in the locust (Boyan *et al.*, 1989b), neuron 12 in the preying mantis (Boyan and Ball, 1986), local neuron 1 (LN1) in the cricket (Kobashi and Yamaguchi, 1984) and the LDS (local directionally sensitive) neuron of the crayfish (Reichert *et al.*, 1982).

Most of these interneurons that have been characterised so far are normally nonspiking, or generate only the occasional spike when they are held hyperpolarised (Bodnar *et al.*, 1991). In cave crickets (*Phaeophilacris spectrum*), however, interneurons that resemble in shape certain of the nonspiking interneurons in the locust produce bursts of spikes in response to air particle movements (Heidelbach

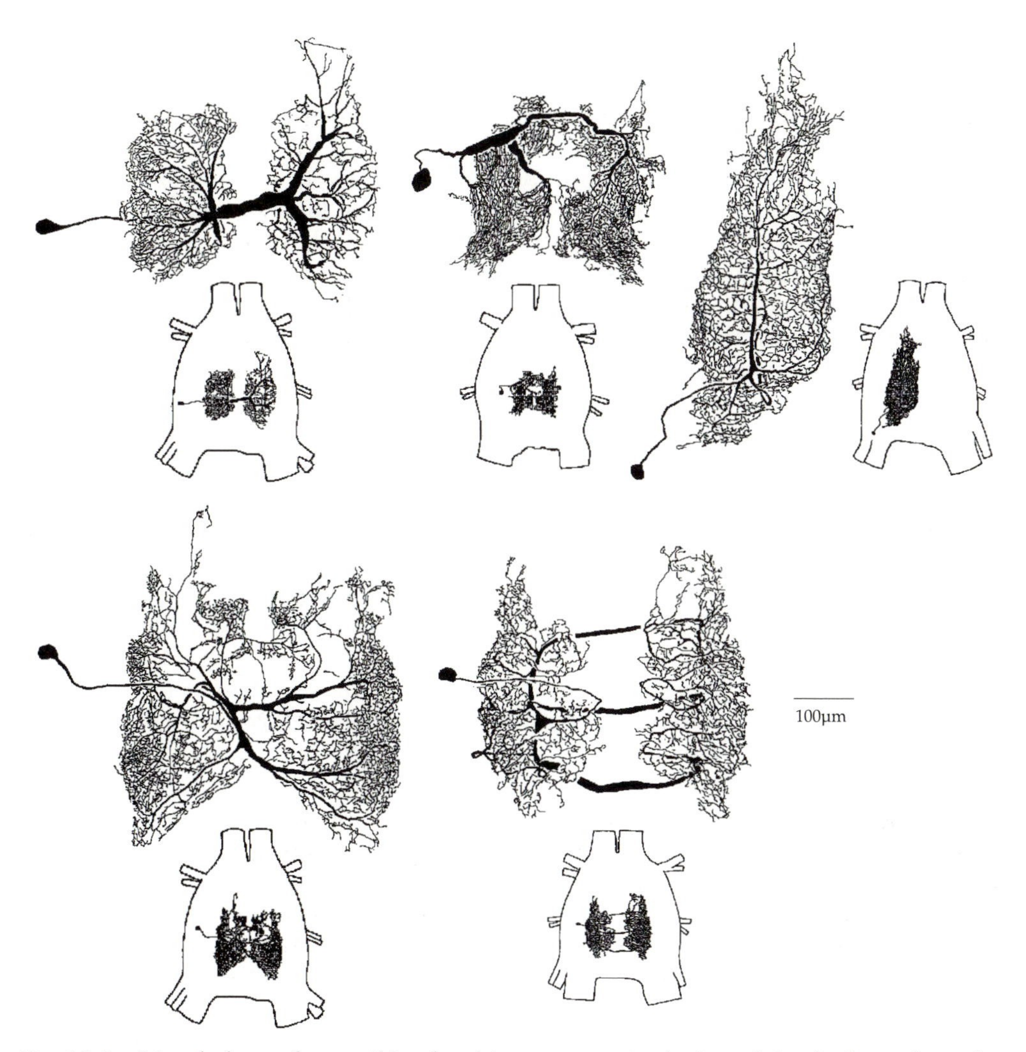

Fig. 10.5 Morphology of nonspiking local interneurons in the last abdominal ganglion of a cockroach. Each of the five interneurons shown responds to air currents directed at the cerci. Each interneuron is shown within the outline of the last abdominal ganglion and enlarged to show the intricate array of fine branches. All but one have branches in both halves of the ganglion, some linked by separate commissural processes. Based on Kondoh *et al.* (1993).

and Dambach, 1991). Similarly, in the cricket *Gryllus bimaculatus*, both spiking and nonspiking interneurons are found that respond to air currents directed at the cerci (Baba *et al.*, 1995). In locusts, the nonspiking local interneurons receive direct inputs from the sensory neurons of one or both cerci (Boyan *et al.*, 1989b) and are thought to be instrumental in augmenting the more sustained responses of the intersegmental interneurons and in shaping their responses to air currents from different directions. In crickets, the nonspiking interneurons also receive direct inputs from cercal sensory neurons but most of the spiking local interneurons apparently do not. This organisation therefore represents a curious reversal in the role of the spiking and nonspiking interneurons as compared with those that process the mechanosensory signals from the legs (*see Chapter* 7).

The nine types of nonspiking local interneurons characterised so far in the terminal abdominal ganglion of cockroaches can be grouped into three physiological classes on the basis of their responses to white noise-modulated air currents (Kondoh *et al.*, 1991b, 1993). First are interneurons with a power spectrum of 70–90 Hz that are both velocity and direction sensitive. Their responses could result from excitatory and inhibitory signals from two subsets of sensory receptors with opposite directional sensitivities. The excitatory input could be direct and the inhibitory input through other interneurons. Second are interneurons with a power spectrum that falls rapidly above 20 Hz but which are directionally sensitive. Their input could come from one subset of receptors with the same directional sensitivity. Third, are nonlinear interneurons that are omnidirectional and that could receive inputs from hairs of any directional sensitivity. The first two groups could give information about the air currents along two coordinates, whereas the third could give information about their velocity.

The connections that these local interneurons make with each other and with the giant interneurons have scarcely been explored. In crickets, photoinactivation of a local interneuron reduces the spike response of a giant interneuron (10-3a) to an air current from the preferred direction of the local interneuron, provided that the branches of the two interneurons overlap (Bodnar, 1993). The overall directional sensitivity of the giant is not changed, indicating either that the direct connections of the sensory neurons predominate, or that several local interneurons shape its receptive field. Depolarising a local interneuron with current can cause a giant to depolarise and spike, but the evidence is insufficient to determine whether the effect is synaptically mediated or the connection is direct. It must thus remain a supposition that the local interneurons are responsible for shaping the responses of the giant interneurons, and there remains no evidence as yet to invoke them to explain the inhibition of the giant interneurons that is caused by inputs from receptors of particular orientations on particular cerci. An analysis of the properties and connections of the local interneurons is crucial for further understanding of the processing of the sensory signals in the terminal abdominal ganglion and the coding of information in the giant (and nongiant) interneurons.

10.7. PRESERVING SENSITIVITY DURING VOLUNTARY MOVEMENTS

The high sensitivity of the receptors that enables them to detect air currents generated by predators means that they will inevitably be stimulated by the air currents that are generated when the insect itself moves. Such continued stimulation could lead to a progressive reduction in the responsiveness of the receptors, thus making them unable to detect externally generated and important signals, and making them unfitted to respond when they are most needed. It might even lead to self-stimulation through activation of the output connections of the giant interneurons. All animals must preserve the responsiveness of their most sensitive receptors during voluntary movements, and it may also be imperative under some conditions for an animal to distinguish between a stimulus that results from its own movements and one that is generated externally. There are several solutions to this problem but a common one is for neurons controlling the execution of the voluntary movement to presynaptically inhibit the terminals of the sensory neurons that would be stimulated. The result is to reduce the efficacy of transmission from the sensory neurons to their target neurons while preserving their sensitivity by preventing a reduction in their stores of transmitter, or their ability to release transmitter (*see Chapter 7*).

10.7.1. Presynaptic inhibition of the sensory terminals

The terminals of the cercal sensory neurons of a locust are depolarised by a synaptic input, both during active movements of the cerci and when they are passively deflected by a current of air. This input reduces the excitability of the terminals and the amplitude of their spikes by as much as 50% and this is thought to reduce the efficacy of their transmission to interneurons in the terminal abdominal ganglion (Boyan, 1988). The depolarisations, which are probably inhibitory, can occur at the same time as spikes in an unidentified proprioceptor at the base of the cercus that codes movements of the cercus and which makes direct excitatory connections with both GI2 interneurons. The depolarisation of the hair sensory neurons is thought to be caused by the sensory neuron from this proprioceptor acting on an interposed interneuron so that the effectiveness of the hairs would be reduced during an active movement and during an air current that was strong enough to move a cercus passively. There is, however, no evidence that the proprioceptive afferent causes either directly or indirectly depolarisation of the sensory neurons that are sensitive to air currents, and its connections with postsynaptic neurons such as GI2 makes it impossible to assess whether the transmission from the sensory neurons sensitive to air currents has really been reduced. An alternative explanation that is consistent with the experimental data is that the depolarisation results from the activation of interposed interneurons by other sensory neurons sensitive to air currents excited by the same stimulus. This would act as a gain control mechanism preventing saturation of a postsynaptic response, in which the action of one sensory neuron would be placed in the context of the network of actions of the other sensory neurons activated by the same stimulus. This mechanism would thus be parallel with that operating for the sensory neurons in the femoral chordotonal organ (*see Chapter 8*). This interpretation is also consistent with the interactions that occur between the cercal sensory neurons of the cockroach (Blagburn and Sattelle, 1987). First instar cockroaches have just two hairs on each cercus and the spikes in one sensory neuron presynaptically inhibit the other. The spikes excite an interposed interneuron that affects a K^+ current in the second sensory neuron, reducing the amplitude of its spikes and the EPSPs it produces in a giant interneuron (Blagburn and Sattelle, 1987). The result is to accentuate the direction of an air current coded by each sensory neuron.

A second mechanism for distinguishing air currents generated as a result of its own movements from those generated externally might lie in the characteristics of the air currents themselves. For example, acceleration of the air seems to be an important feature in evoking a response from the cercal hairs, and it is possible that differences in acceleration properties of air currents from different directions indicate whether or not it is generated by an external source (Plummer and Camhi, 1981).

In addition to these neural mechanisms, the cercal nerves in the cockroach may be mechanically deformed when they are pressed against the cuticle as the cerci are actively moved medially by about 60° during flight, thus blocking the spikes of the hair sensory neurons before they can reach the terminal ganglion (Goldstein and Camhi, 1988; Libersat and Camhi, 1988; Libersat *et al.*, 1987). During escape running the cerci are not moved medially so that the sensory block does not occur. This nonsynaptic inhibitory mechanism lacks the specificity of the neural mechanisms but can reduce substantially (by up to 40%) the number of sensory spikes, and consequently the number of spikes generated by both the ventral and dorsal giant interneurons in response to an air current. The effect may protect the sensitivity of the central neurons so that they do not become habituated to the inevitable air currents that occur during flight. The possibility of similar blocks of other nerves as they pass through joints would be worth exploring, but of course they must not disturb the spikes in the motor and neuromodulatory neurons if locomotion is to be successful.

10.7.2. Changing the responsiveness of the giant interneurons

The responsiveness of the giant interneurons may change during locomotion either as a result of the sensory input they receive, or of the signals sent in parallel to them by the neurons generating the movements. The interpretation of these changes in sensitivity depends on the role that the different interneurons might play in initiating or influencing particular movements.

When a cockroach is flying, some of the giant interneurons are depolarised rhythmically at the wingbeat frequency by bursts of spikes in axons descending from the thorax, by sensory input caused by the vibrations of the cerci at the wingbeat frequency, and by the air currents over the cerci that are generated by the movements of the wings. Active movements of the cerci that occur during flight are monitored by stretch receptors, the sensory neurons of which synapse on particular giant interneurons and may desensitise their responses to air currents (Bernard, 1987). By contrast, the ventral giant interneurons of cockroaches do not spike despite receiving a synaptic input at the wingbeat period, and even though the air movements over the cerci are above the threshold that will excite them in a nonflying animal. These differential effects occur despite the overall reduction in the spikes of sensory neurons from receptors on the cerci, and must indicate that the excitability of the dorsal but not the ventral giant interneurons is restored by inputs from descending neurons that spike in time with the wing beats.

When a cockroach is walking, the dorsal giant interneurons spike at each step, while the ventral giant interneurons are silent and their sensitivity is somewhat decreased (Delcomyn and Daley, 1979; Daley and Delcomyn, 1980a,b), although they are still able to respond to applied air currents (Camhi and Nolen, 1981). The level of excitation and inhibition of the two groups of giant interneurons increases with the speed of walking. The major source of these inputs is a central command from the thorax associated with the production of the rhythmic walking movements, with an additional but smaller input from the cercal receptors excited during the movements. A peripheral component can be shown by gluing an isolated abdomen of one cockroach to another cockroach, whereupon the spikes in the giant interneurons of the attached abdomen are modulated by the walking movements of the carrier cockroach (Orida and Josephson, 1978). Covering the cerci during walking also reduces the effects on the giant interneurons, particularly at faster walking speeds. The changes in responsiveness of both groups of giant interneurons nevertheless still occur in a deafferented preparation that generates a fictive locomotory pattern (*see Chapter 8*). Thus, both a central command and peripheral feedback must contribute to the changed excitability of the giant interneurons during walking, with the result that the sensitivity of the dorsal giant interneurons is increased while that of the ventral giant interneurons is decreased.

10.8. GIANT INTERNEURONS AND THE INITIATION OF BEHAVIOUR

The giant interneurons and the cercal receptors that feed them are implicated in the initiation of distinct movements. In locusts, stimulation of the cerci may initiate jumping. In cockroaches, an air current directed at the cerci causes turning and running when the feet are in contact with the ground, but flying when they are not. These responses can be mimicked by stimulation, either individually or in pairs, of particular giant interneurons. Only the dorsal giant interneurons, however, can initiate flying, with stimulation of two evoking more prolonged sequences provided the feet are not in contact with the ground (in these stimulation experiments the legs were actually removed). The corollary of this observation would be that the ventral giant interneurons are responsible for the turns that occur during walking or running, but there is much confusion and little direct evidence over whether this is one of their actions. In crickets, the ventral giant interneurons receive input from the short acceleration-sensitive hairs which are 100 times less sensitive to air movements than are the longer velocity-sensitive filiform hairs (Shimozawa and Kanou, 1984) which connect with the dorsal giant interneurons that evoke flight. This might suggest that the threshold for inducing flying is lower than that for running, or that the subdivision of function between the dorsal and ventral giant interneurons is less distinct than is supposed.

Table 10.4. Behavioural actions of the giant interneurons in cockroaches

Behaviour	Ventral giant interneurons GI1-4	Dorsal giant interneurons GI5-7
Walking	Silent Respond to air currents	Spike at each step Respond to air currents
Flying	? Response to air currents reduced	Spike at each wingbeat Response to air currents the same
Jumping	Excite an interneuron that, in turn, excites one of the crucial motor neurons involved in jumping	?
Electrical stimulation	GI1 can evoke weak and labile excitation of some leg motor neurons Can sometimes cause turning movements	Can initiate flying Can evoke spikes in coxal depressor and levator motor neurons

10.8.1. Do the giant interneurons initiate jumping?

The evidence for an involvement of the giant interneurons in initiating a jump of a locust is at best equivocal and stems largely from the observation that air currents directed at the cerci can sometimes initiate such movements, in much the same way that any unusual and threatening stimulus can. The inference is that the fastest conducting pathways should carry the signals if survival is paramount. There is, however, no convincing evidence yet available for direct connections between the giant interneurons and any of the motor neurons or thoracic interneurons involved in producing a jump. GI1 and 3 may excite a mesothoracic interneuron (714) which, in turn, makes an excitatory connection with the fast extensor tibiae motor neuron of a hind leg (Boyan and Ball, 1989b), but a role for this thoracic interneuron in jumping has yet to be demonstrated. It would seem sensible if highly tuned receptors such as those on the cerci were able to influence jumping, but the need to use the fastest conducting pathways is less obvious, given the time (up to 500ms) that the leg muscles require to generate the necessary force to execute a jump (*see Chapter 9*). This suggests the involvement of many parallel pathways that may include both the giant and other nongiant interneurons that process mechanosensory signals from the abdomen.

10.8.2. Do the giant interneurons initiate flying?

An air current directed at the cerci of a cockroach can initiate flying. Removal of a cercus may cause flight to become asymmetrical, suggesting that the cerci may act as equilibrium receptors, and once it is in progress can lead to abnormal steering

(Fraser, 1977) although this suggestion has not been analysed further. An air current directed to the cerci of a locust during flight can also change the relative timing of the motor neurons in a way that is normally associated with a turning response. Electrical stimulation of a cercal nerve produces excitatory and inhibitory synaptic potentials in many thoracic interneurons and flight motor neurons with a range of delays.

The inference drawn from all these observations is that the initiating signals are carried from the terminal abdominal ganglion by interneurons that receive inputs from the hair sensory neurons, and of course the obvious candidates are the giant interneurons. Indeed, during flight the membrane potential of GI2 of a locust oscillates in the flight rhythm, perhaps because the air movements caused by the wing beats are detected by the cercal hairs. If either GI2 or GI4 is depolarised with current, then spikes in some flight motor neurons are induced, but without the rhythmicity characteristic of flight (Boyan and Ball, 1989b). The input from these interneurons can, however, reinforce a flight pattern that is already being produced (Boyan *et al.*, 1986). GI1 does make apparently monosynaptic excitatory connections with mesothoracic interneuron 302 that spikes during elevation of the front wings. Some of the 404 flight interneurons (*see Chapter 11*), that can themselves initiate flying are depolarised by an air current to the cerci, but whether the giant interneurons convey these signals is not known. It would seem, therefore, that GI1, and perhaps some of the other giant interneurons, can provide an input to the thoracic circuitry for flight that could either initiate the rhythm or help in sustaining it. This is also supported by the fact that, in cockroaches, stimulation of a dorsal giant at high frequency can raise the frequency of the wing beats by 1 Hz, whereas stimulation of a ventral giant terminates the flight sequence (Libersat, 1992). The spikes of the dorsal giant interneurons that occur in time with the wing beats may thus contribute to the maintenance of flight. In cockroaches, flight can also be initiated by stimulation of individual dorsal giant interneurons (Ritzmann *et al.*, 1982) at frequencies higher than seen when natural stimuli evoke flight. When the cerci are removed, cockroaches are unable to produce evasive manoeuvres to an air current (Ganihar *et al.*, 1994). Stimulating a single dorsal giant at 330 Hz increases the number of spikes produced by a subalar motor neuron at each wing beat and alters the timing between the onset of the subalar spikes on the two sides of the body (Libersat, 1994a). The effect of such changes may be to cause the wings to twist and thus to alter the steering. Some of the giant interneurons may therefore be involved in flight steering, taking the cockroach away from an air current that may indicate danger. There is even a suggestion that the cercal receptors and the giant interneurons could signal the approach of a bat, but if their sensitivity is set this high then they should also respond to the normal turbulence in the air.

10.8.3. Do the giant interneurons initiate turning and running?

Locusts show little in the way of oriented walking or running in response to air currents directed at the cerci. By contrast, a cockroach that is either standing or walking responds to an approaching predator, such as a toad, by pivoting about the posterior part of its body and then running in the opposite direction (Fig. 10.6) (Camhi and Levy, 1988). Whether or not the cockroach will respond to these stimuli with an escape movement depends on some measure of its behavioural state; it is less likely to escape when quiescent, perhaps then using immobility to avoid predators that strike at moving targets, when grooming or when it is in the dark. In an escape movement, all the legs are coordinated and moved in a way that depends on their initial starting positions and on the direction of the stimulus, thus implying the need for considerable processing in the thoracic ganglia. The turn involves five, or even all six legs performing a stance phase simultaneously, but exerting different forces on the ground, as determined by the direction of the stimulus. The transition to the tripod gait of running is achieved within the first few steps. Crickets will also run for a short distance, performing 5–10 steps in response to a short air current, but to repeated stimuli show short bursts of running interspersed with stationary periods (Gras and Hörner, 1992). The initial turn is often absent in crickets so that 20–30% of the movements show no orientation away from the stimulus.

It takes a standing cockroach some 50 ms (range 30–90 ms) to start to enact the complex sequence of movements involved in turning away from an air current directed at its cerci. The spikes of the sensory neurons from receptors at the tip of a cercus will take about 5ms to reach the terminal abdominal ganglion, the spikes in the largest diameter ventral giant interneurons will take 3–4 ms to travel from the terminal abdominal ganglion to the metathoracic ganglion, and the motor spikes will take a further 5ms to reach the muscles. Thus, 13–14 ms is taken in conducting signals so that most of the time must be absorbed by the processing within the central nervous system and the production of the appropriate patterns of spikes at all stages, and the time taken by the muscles to produce the movement. Given the long processing time, the savings effected by using the fast conduction of the giant interneurons is relatively small. During slow walking, however, an air current

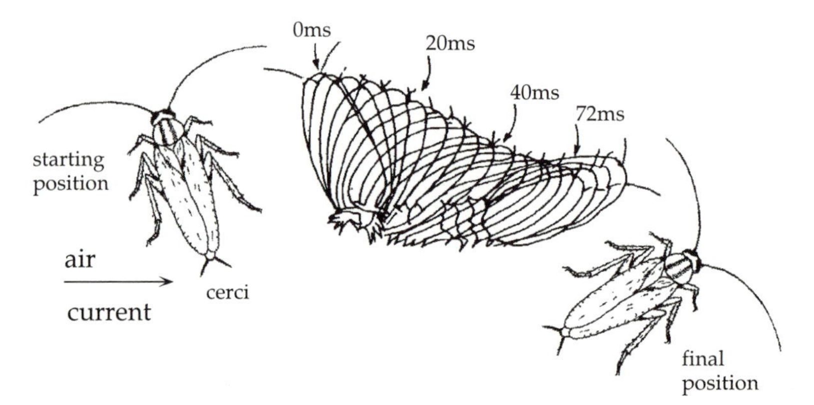

Fig. 10.6 Evasive turning movements of a freely moving cockroach in response to an air current directed at its left cercus. The outlines are drawings from successive frames (separated by 4 ms) of a film. The turn takes the cockroach away from the offending stimulus. Based on Camhi and Levy (1988).

directed at a cercus can accelerate the leg movements within 12–20 ms (Camhi and Nolen, 1981) so that only the first spikes in the giant interneurons could be involved in initiating these changes.

Analyses have increasingly revealed both the intricacies of the movements needed for effective escape and the underlying neural control mechanisms, but it has proved surprisingly hard to establish whether the giant interneurons play an exclusive role, and then to determine the pathways by which they, or perhaps other interneurons, activate the leg motor neurons. Only of late has come the acceptance that many interneurons must be acting in parallel, in part because the actions of individual giant interneurons fail to explain the movements, and also because it is clear that essentially the same movements can be elicited by tactile stimuli to many different parts of the body, and by different modalities of stimuli.

10.8.3.1. *Generating a motor output*

In cockroaches, stimulation of an individual **dorsal** giant interneuron (GI5-7) evokes spikes in some of the slow motor neurons innervating coxal levator and depressor muscles of the legs (Ritzmann and Pollack, 1981). The motor responses are enhanced and the latency to their first spike is reduced if the dorsal giant interneurons are stimulated in pairs, suggesting a summation of their outputs on the motor circuitry, even though each may excite a particular subset of motor neurons. Only one of the larger diameter **ventral** giant interneurons (GI1) can evoke even a weak and labile response in leg motor neurons, although its effects are enhanced by simultaneous stimulation of another ventral giant that on its own produces no motor effect at all (Ritzmann, 1981). Occasional stimulation of this giant can cause turning movements away from the source of an air current (Liebenthal *et al.*, 1994). These results would seem to be at odds with the observation that it is the **ventral** giant interneurons (GI1-4) that are activated by air currents and cause turning movements whilst the insect is walking (Camhi and Nolen, 1981). Moreover, the spikes of the larger ventral giant interneurons arrive at the thorax some 5 ms before those of the dorsal giant interneurons and so would be expected to initiate the first motor response. None of the motor responses evoked by stimulating the giant interneurons directly is as strong or as rapid as that caused by an air current directed at the cerci. This may reflect the reduced excitability of the dissected insects in which these measurements have to be made, but more likely emphasises that normally the signals are carried in many parallel giant, and nongiant pathways. None of the giant interneurons connects directly with any of the leg motor neurons, and instead their motor effects are exerted through connections with sets of thoracic interneurons.

10.8.3.2. *Connections with thoracic interneurons*

The ventral giant interneurons of cockroaches directly excite particular intersegmental interneurons that have cell bodies in the meso- and metathoracic ganglia and axons that project anteriorly, posteriorly, or in both directions (Fig. 10.7) (Ritzmann and Pollack, 1988; Westin *et al.*, 1988). One of these interneurons (Lambda cell in Fig. 10.7) has an axon 25 μm in diameter and would thus distribute

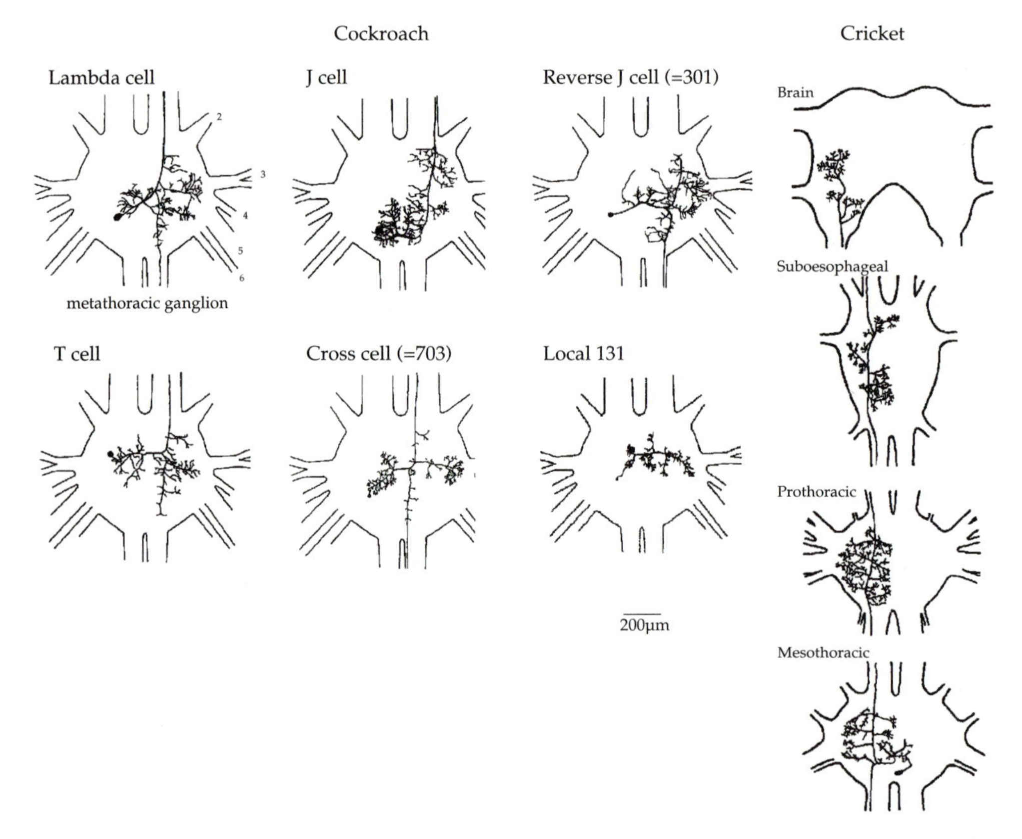

Fig. 10.7 Morphology of metathoracic interneurons in the cockroach. These interneurons receive inputs from particular giant interneurons and many other neurons, and make output connections with interneurons and motor neurons involved in controlling the movements of the legs. Based on Ritzmann and Pollack (1986), Westin *et al.* (1988) and Hörner (1992).

the motor command among the thoracic ganglia faster than could any of the giant interneurons, the axons of which narrow considerably in the thorax so that they are smaller than the axon of this thoracic interneuron. Each of the 13 intersegmental interneurons that have been identified so far has a characteristic shape and a cell body in either a ventral or a dorsal group. Each of the thoracic interneurons receives inputs from particular combinations of ventral giant interneurons so that they either respond best to an air current from a particular direction, or from any direction (Ritzmann *et al.*, 1991). Just one interneuron has been found to receive inputs from dorsal but not ventral giant interneurons. In addition to these inputs from giant interneurons, the intersegmental interneurons also receive inputs from other neurons carrying signals about many different modalities of stimuli. It would seem appropriate for any escape response to be influenced by sensory signals from the head and from elsewhere, and indeed, cockroaches with descending input removed by lesion of a cervical connective make more incorrect turns than do normal animals and run for less long after the stimulus (Keegan and Comer, 1993). In addition, some local DUM interneurons (Pollack *et al.*, 1988) (*see Chapter 3*), and one other local interneuron also receive direct inputs from the ventral giant interneurons. Presumably, more connections with local interneurons in particular, and probably other intersegmental interneurons, await discovery.

The intersegmental interneurons of the dorsal group make excitatory connections, and those in the ventral group make inhibitory connections either with local interneurons, or with motor neurons of the segment containing their cell body or in an adjacent segment to which their axons project (Ritzmann and Pollack, 1990). The local neurons are surprisingly undefined as to whether they are spiking or nonspiking, or whether their motor effects are mediated by direct connections with the motor neurons. Thus, the signals in the giant interneurons are distributed to a diverse population of thoracic interneurons where they are integrated with signals from the head and possibly from leg proprioceptors before being relayed in parallel to local interneurons and to the motor neurons. These intersegmental, thoracic interneurons are thus seen as key elements in the pathways that control turning and running movements during escape. In crickets, however, members of a similar group of interneurons that also receive inputs from the giant fibres are inhibited some 30 ms before a movement of the legs can be seen in response to an air current directed at the cerci (Hörner, 1992). It is questionable, therefore, whether these interneurons could be involved in initiating the first movements of the legs. During subsequent movements the spikes of these interneurons are also suppressed in a way that depends on the velocity of movement, so that during a rapid movement the suppression is total.

Interneurons descending from the brain are excited before a movement is initiated, with the implication that the projections of the giant interneurons to the brain are responsible for their activation. Only the ventral giant interneurons project to the brain, while the dorsal giant interneurons are thought to end in the thorax. No connections of the giant interneurons are known with neurons in the brain and only a few fragmentary effects of their inputs are described. It seems likely that the mechanosensory information that they carry is brought together with the variety of sensory signals received by the head before influencing the output of the many descending interneurons that have branches in the tritocerebrum. To describe it in these vague terms is to reveal the depths of ignorance of the processing that occurs in these regions of the brain. Nevertheless, such an interpretation has the attraction of drawing into the pathways the otherwise unexplained brain projections of the ventral giant interneurons. Perhaps it is also necessary to revive the suggestion that the outputs of the ventral giant interneurons in the brain might lead to movements of the antennae (Dagan and Parnas, 1970) or other appendages of the head.

The current conclusion has to be that our knowledge of the giant interneurons and their connections with the thoracic neurons is insufficiently detailed to explain turning and running escape behaviour, and how a directional response that is appropriate to the current positions of the legs could be produced. The connections of the giant interneurons are, unfortunately, not described in a way that would allow them to be placed into the circuitry that is known to organise the movements of the legs (*see Chapters 8* and *7*). This should be remedied to make use of the enormous information that is available for locusts, at least, on the thoracic interneurons. Moreover, curiously unexplained, and largely unaddressed, is that stimulation of dorsal but only occasionally of one of the ventral giant interneurons causes turning, but that it is the connections of the ventral and not the dorsal giant interneurons with the thoracic interneurons that have been analysed to offer an explanation of escape turning. Furthermore, it is the ventral giant interneurons that are inhibited during walking and which could not therefore control an escape-like movement if the cockroach is already moving.

10.8.4. Future questions

To make further progress there are some underlying assumptions that need to be challenged and much more that we need to understand. Is it appropriate to divide the giant interneurons into two groups and assume that they have different functional roles, simply on the basis of the position of their axons in the connectives? Is it correct to interpret all connections and effects in terms of escape movements? What sorts of connections are made by individual giant interneurons with the populations of local neurons that have already been characterised? With what thoracic neurons do the fast-conducting dorsal giant interneurons connect? How is the responsiveness of the circuitry set by neuromodulators? Some clues to possible new directions are already available. For example, the ability of the dorsal giant interneurons to evoke flight when the legs are removed might suggest that outputs with the flight circuitry should be sought, as has already been successfully achieved for some of the giant interneurons in the locust. The diversity of the giant interneurons and the need to treat them as individuals is indicated by GI4, which is normally designated as a ventral giant from the position of its axon but has some physiological properties more akin to the dorsal giant interneurons (Libersat, 1994b). This recognition may hopefully be extended to include an analysis of the other ascending interneurons when considering the processing that accompanies different behaviour. It is salutary to have to conclude that, after almost 50 years of studying the giant interneurons in cockroaches, there are so few firm conclusions that can be drawn about their role in initiating escape-like behaviour.

11 Flying

11.1. FLYING MOVEMENTS

If there was one development in evolution that enabled insects to adopt the huge diversity of lifestyles that they now enjoy, and which explains why they account for more than 70% of known species of animals, it was the emergence of flight. Locusts are dramatically good at flying for long periods and are renowned for the enormous swarms that they can form and the ensuing damage they can cause when feeding. They are reliably reported to fly continuously for 40 h, and are quite capable of covering 100 km in a few hours at altitudes of 1 km or more, refuelling by eating 1–1.5 times their body mass, then resuming their flight. While the integrity and cohesion of the swarms and the distances covered are clearly influenced by the wind, great distances can be covered as when, most recently in 1988, the remnants of swarms originally formed in North Africa appeared in the Caribbean. To understand how these formidable feats of flight can be achieved, we need to analyse the mechanics of the wing movements, the muscles that move the wings, the neural mechanisms that generate the motor patterns, and the control mechanisms that enable a locust to stay airborne in a fluctuating environment and manoeuvre itself towards its goal. Like many other insects in diverse groups, some species of grasshoppers have lost the ability to fly and these changes are useful in illuminating the specialised mechanisms used for flight.

11.1.1. Initiating flight

Locusts often start to fly by jumping into the air (*see Chapter* 9). A decision to jump is not always accompanied by a decision to fly, but flight can result from the consequence of a jump setting up an air current to the head and a loss of tarsal contact with the ground. When in large numbers it is probably the contact with each other that leads to increased excitability, and when coupled with a lack of food leads to the swarms taking to the air. Many sensory clues can lead to this initial response;

these include a looming visual stimulus, a shadow, a disruptive sound, or an air current detected by hairs on the head or the cerci. A wind stimulus to the head coupled with a loss of contact of the tarsi with the ground are the experimental stimuli most commonly used to evoke flight.

11.1.2. Posture

During horizontal flight, the body is inclined upwards by about 7°, the abdomen is raised, the antennae are pointed directly forwards into the wind stream, and all the legs adopt a characteristic posture (Cooter, 1973). The front legs are folded tightly against the body with the tarsi pointing downwards and backwards, the middle legs are extended posteriorly, and the hind legs are pointed backwards with the tibia either flexed against the femur, or fully extended for use in steering. When coming in to land, the legs adopt a 'spread eagled' landing posture in which the front legs are held out to the side and forwards, the middle legs at right angles to the body, and the hind legs pointed backwards and extended at the femoro-tibial joint. When flying in swarms, the spacing between individuals is probably sufficient to prevent direct interference between the wing movements of neighbours, but the turbulence created should influence their flight. In the laboratory, when the spacing between two individuals is about 35 cm, and is therefore still greater than the spacing in some swarms, the locust at the rear tries to couple its wing beat frequency to the locust in front by detecting this turbulence with wind-sensitive hairs on its head (Kutsch *et al.*, 1994).

11.1.3. Frequency of wing movements

A locust flies by flapping its two pairs of wings at frequencies of about 20Hz to give an air speed of about 3.5 $m.s^{-1}$, with one wing beat moving it forward some 0.2 m. In free flight, the wing beat frequency can rise to 22 Hz or sometimes more, and the air speed to 4.6 $m.s^{-1}$ (Baker *et al.*, 1981). The exact frequency of the wing beats depends on the individual, its sex and its age. Just after the final moult to adulthood, the wings are fully functional for the first time but the cuticle is soft, the wing beat frequency is low and flight is not sustained. Over the next 3 weeks the wing beat frequency increases gradually as the cuticle progressively hardens, largely because of maturational changes in the neural, muscular and mechanical machinery (Kutsch, 1971). Once flight is well established in a particular individual, the frequency usually remains constant, at least when flying in a controlled environment, and is not a parameter that is much changed. The frequency varies by a small amount with the power that is produced and falls slightly after the first few minutes of a sustained flight. The frequency also changes by 0.3 Hz for each degree change in ambient temperature (Foster and Robertson, 1992). Normally, locusts can only fly at air temperatures above 24°C and below 50°C, but the action of their muscles raises the thoracic temperature by some 6°C. Temperature also has a similar effect on the flight motor pattern that is produced by a deafferented locust, so that together these observations indicate that temperature has many possible sites of action, not least the central nervous system itself. In different grasshoppers and locusts the frequency of the wing beats is inversely proportional to body size, so that in some smaller species it can be as high as 60 Hz.

11.1.4. Coupling of the two pairs of wings

The two pairs of wings move at the same frequency but about 30° out of phase with each other so that the hind wings reach their fully elevated position some 7 ms before the front wings, and their fully depressed position some 4ms before (Fig. 11.1) (Wilson and Weis-Fogh, 1962).

11.1.5. The wing beat cycle

The basic movements of a wing consist of repetitive up (elevation) and down (depression) movements, with a twisting upwards of the leading edge of the wing (supination) during the upstroke, and during the downstroke a twisting downwards of the leading edge (pronation) and a bending downwards of the trailing edge of a front wing that starts as it moves below the horizontal. About 80% of the lift is generated by the downstroke movements and, of this, some 70% is produced by the larger but more flexible hind wings. This is a much simplified description of what actually happens but the analyses of the neural mechanisms rarely take us into the more sophisticated considerations of flight mechanics and aerodynamics.

In regular tethered flight, the upstroke and the downstroke of a wing are of about equal duration (Weis-Fogh, 1956a), but in free flight the ratio between them varies from locust to locust, and even within the wing beats of one locust. The ratio of upstroke to downstroke can reach 1: 1.8 depending on whether the locust is flying straight, climbing or descending (Fig. 11.2) (Baker and Cooter, 1979a, b). Each front wing moves through an angle of about 100° in one wing beat, but the movements of a hind wing can approach 155°, with the result that the left and right wings can meet at the top of the upstroke and bend at the bottom of the downstroke.

11.1.6. Flapping and gliding

In free flight, flapping movements of the wings may give way to short periods when a locust merely glides on its outstretched wings (Baker and Cooter, 1979a). The hind wings stop moving as they begin their downstroke (20° down) and the front wings stop at the same time, which means that, because of the lag in their movements, they have just reached the top of their upstroke. The wings remain stationary for as long as 300 ms, during which time the attitude of the body may change and the locust may begin to lose height, before they resume beating by first moving downwards. These gliding movements may be used for a controlled descent at a forward velocity lower than could be achieved by flapping.

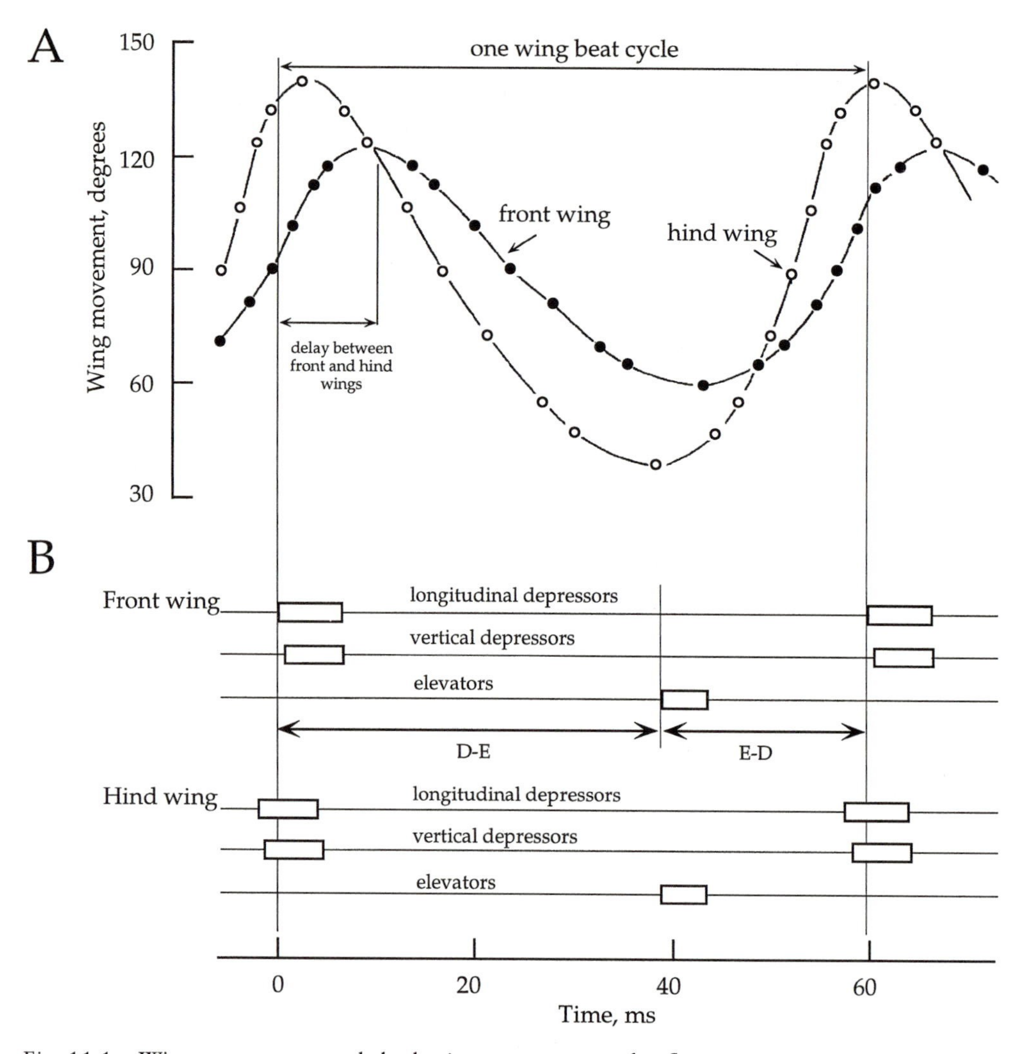

Fig. 11.1 Wing movements and the basic motor pattern for flying. A. Movements of a front wing precede that of a hind wing during flight. The time from the start of depression to the start of elevation (D–E) and from the start of elevation to the start of depression (E–D) are indicated. B. The action of the main groups of muscles (on one side of the body) that move the wings during flight. Based on Weis-Fogh (1956a).

11.1.7. How regular are the wing movements in flight?

Most of the analyses of the wing movements have used a locust glued by the hard cuticle of its neck (pronotum), or ventral thorax, to a flight balance that monitors the lift it generates, placed in a laminar flow of warm air in a wind tunnel. The visual environment is either constant with an artificial horizon, or there is a moving pattern

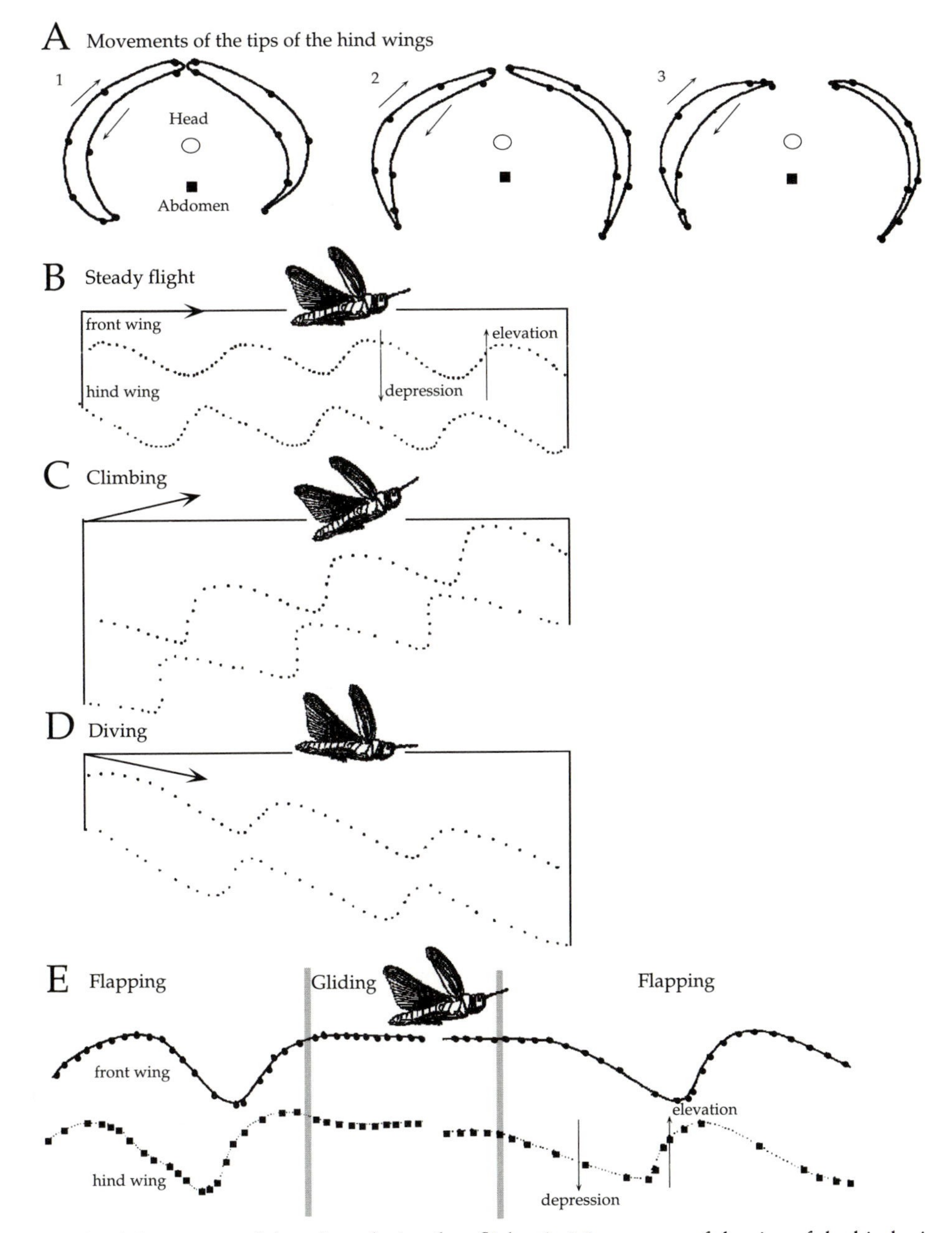

Fig. 11.2 Movements of the wings during free flight. A. Movements of the tips of the hind wings during three cycles of the wing beat. The locust is viewed head-on with the position of the head and the tip of the abdomen marked. B–D. Relationship between the movements of the front and hind wings during steady flight (B), when climbing (C) and when diving (D). Each of the points is 2.5ms apart. E. Periods of gliding when the wings are held elevated can be interspersed with flapping flight. Based on Baker and Cooter (1979a, b).

of black and white stripes that simulates forward movement over the ground. All the movements are thus open loop, in that any changes that the locust produces have no effect on the stimuli that it receives. By contrast, observing the wing movements during free flight emphasises that the flight motor pattern is a reflection of the conditions in which it is performed. The movements can be regular and produced with little variation for hours in the controlled conditions of a wind tunnel with a constant flow of air and an unchanging visual environment. In free flight under natural conditions, however, virtually all parameters of the wing movements can vary to keep the locust airborne in a highly unstable medium, and on its desired course, which may require navigation around hazards. Most analyses of the neural mechanisms which produce flight make use of the regularity of the rhythm under controlled conditions, which is a sensible strategy provided it does not lead to conclusions that are unable to incorporate the wide range of variability that is true flight.

11.2. THE MACHINERY FOR FLYING

11.2.1. The wings

The front wings on the mesothoracic segment are narrow, thick and stiff whereas the hind wings on the metathoracic segment are broader and stiff along the leading edge, but with the remainder thin, flexible and pleated. In a nonflying animal, the thin hind wings are folded along these pleats so that they are completely covered and protected by the thicker front wings. The complex series of hollow radiating veins in each wing serve as conduits for the distribution of tracheae, nerves to the sense organs, and for haemolymph which is necessary to maintain hydration of the wing membrane and hence its compliance and toughness. The veins also stiffen each wing by their arrangement at the peaks and troughs of corrugations, and by being linked by smaller cross veins to form a lattice girder-like structure of small cells that has a high rigidity for its low weight. The membrane between the veins acts as a stressed skin and limts deformation of the framework of veins. The stiffness of the veins, the degree of corrugation and the amount of cross-linking varies in different regions of a wing, so allowing different amounts of deformation. The wings taper, which keeps their centre of mass close to their articulation with the thorax, but means that the tips are less rigid and can therefore bend. A consequence of this overall design of a wing, which is rare in human technology, is that the wings do not remain rigid as they are flapped, but that the hind wings, in particular, can be changed in shape either by the action of particular muscles, or by the inertial and aerodynamic forces. The three-dimensional shape of a wing is important in generating force during flight, not least because it must generate thrust to move the insect forwards, and more upward than downward force to keep it aloft. The pleats in a hind wing open rather like a fan as it moves forwards during the downstroke, thus pulling the outer edge inwards and compressing the radial veins. The result is to increase the camber of a hind wing and its area, and thus its effectiveness as an aerofoil, but to prevent the thin trailing edge from fluttering. On the upstroke, the pleats retract so that the area of the wing and its camber are reduced, with the consequence that it generates less force. Lift is always perpendicular to the path of the wing and drag is parallel to it, so that considered in this way, the upstroke can provide little lift. The primary, active way of changing the shape of a wing is to twist it, although changes in camber, and the stroke plane of the wing, as determined by the attitude of the body, also contribute to changing its effectiveness as an aerofoil. The muscles, however, insert only at the hinge, so that any changes in shape must be transmitted by relative movements of the rigid veins and thickened membranes, especially to the fastest moving tips that have the greatest aerodynamic influence. For example, the angle of attack of a front wing is controlled by axial movements that radiate towards its tip along lines of flexion (Wootton, 1992).

The aerodynamics of flight in the locust are complex and variable. They involve several phenomena that are not found in aircraft with fixed wings, and while they enable locusts to fly well and must be a prime reason for their success, this complexity must be reflected in the organisation of the central nervous system and its sensory feedback mechanisms.

11.2.1.1. *Haemolymph supply to the wings*

The supply of haemolymph to a pair of wings is aided by pulsatile organs in the thorax close to the bases of the wings that act as suction pumps. They are encased dorsally by a dome-shaped swelling of the tergum to form a scutellum from which a pair of corrugated tubes lead to the posterior wing veins (Krenn and Pass, 1994). Just below this dome is a small subscutellar chamber separated from the main thoracic haemolymph by a horizontal septum and below this is the dorsal vessel, or heart, which is enlarged with thick muscles on the dorsal side. Its contractions cause it to arch forward and flatten, reduce the pressure in the subscutellar chamber and thus suck in haemolymph from the wings. The haemolymph then enters the dorsal vessel through paired openings (ostia), and from there can be expelled into the haemolymph of the thorax through exhalent openings. These specialisations of the wing circulatory system develop only as the wings themselves are formed, so that they are apparent only from the fourth instar onwards (Krenn, 1993).

11.2.2. The muscles

Each wing is moved by 10 muscles innervated by a specific set of motor neurons (Figs 11.3 and 11.4). The thorax is bilaterally symmetrical so the two sets of muscles and motor neurons moving the pair of wings on one segment are the same, and those moving the front and hind wings are homologous. For each wing, the muscles are divisible into three groups (Table 11.1):

Direct muscles insert directly onto the wing, or the associated sclerites.
Indirect muscles cause the wings to move by distorting the shape of the thorax.
Accessory muscles modify the effects of the other muscles and cause twisting movements of a wing.

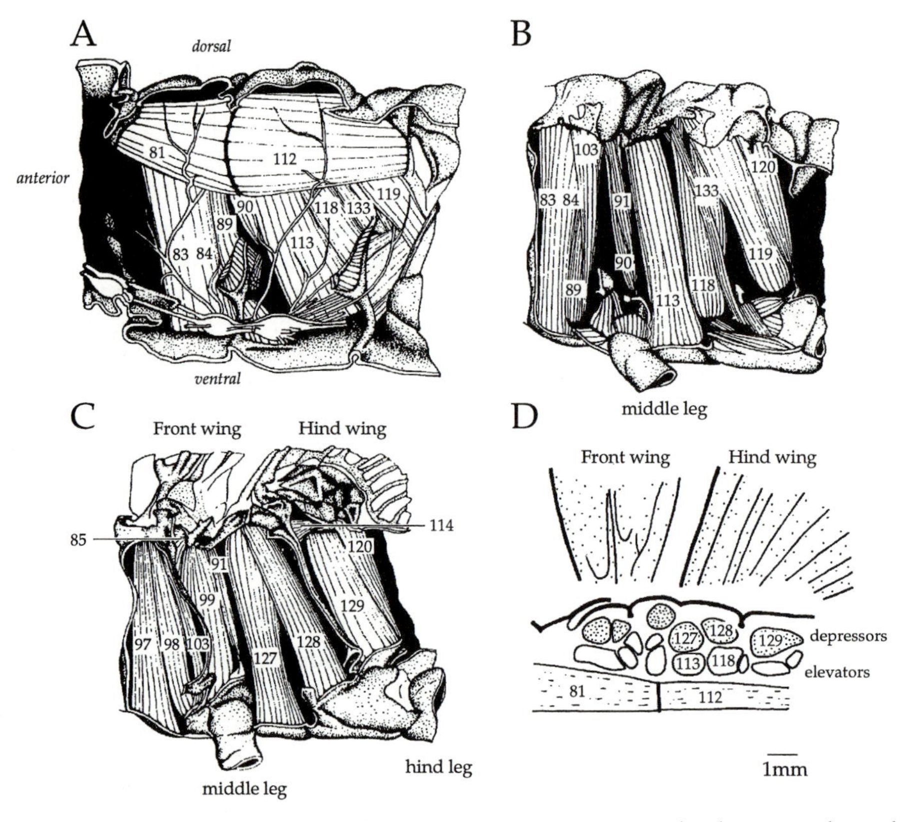

Fig. 11.3 Thoracic muscles that move the wings. A–C. Longitudinal section through the midline of the thorax. A. The ventral thoracic ganglion and the nerves innervating the most medial dorso-ventral muscles and the dorsal longitudinal muscles. B. The dorso-ventral elevator muscles. C. The dorso-ventral depressor muscles. D. Diagram of a horizontal section through the thorax to show the relative positions of muscles that move the wings on one side; elevators are shown as open profiles, depressors as shaded ones. The muscles are numbered after Snodgrass (1929); see Table 11.1 for full listing of the muscles. A–C are based on Sandeman (1961). D is from Möhl and Zarnack (1977).

A wing is moved upwards (elevated) by contractions of a set of elevator muscles arranged vertically in the thorax. It is moved downwards (depressed) by a combination of elasticity and the contraction of both vertically and horizontally arranged muscles. The vertical depressor muscles are located more laterally in the thorax than the vertical elevators, and act directly on the wings, whereas the horizontally arranged depressor muscles, the dorsal longitudinals, act indirectly on the wings. Some of the vertical depressor muscles (mostly the basalars) also control the pronation of a wing during the downstroke, but the supination (a twisting upwards of the leading edge of the wing) during the upstroke results from aerodynamic forces that are opposed by the contractions of a small pleuroaxillary

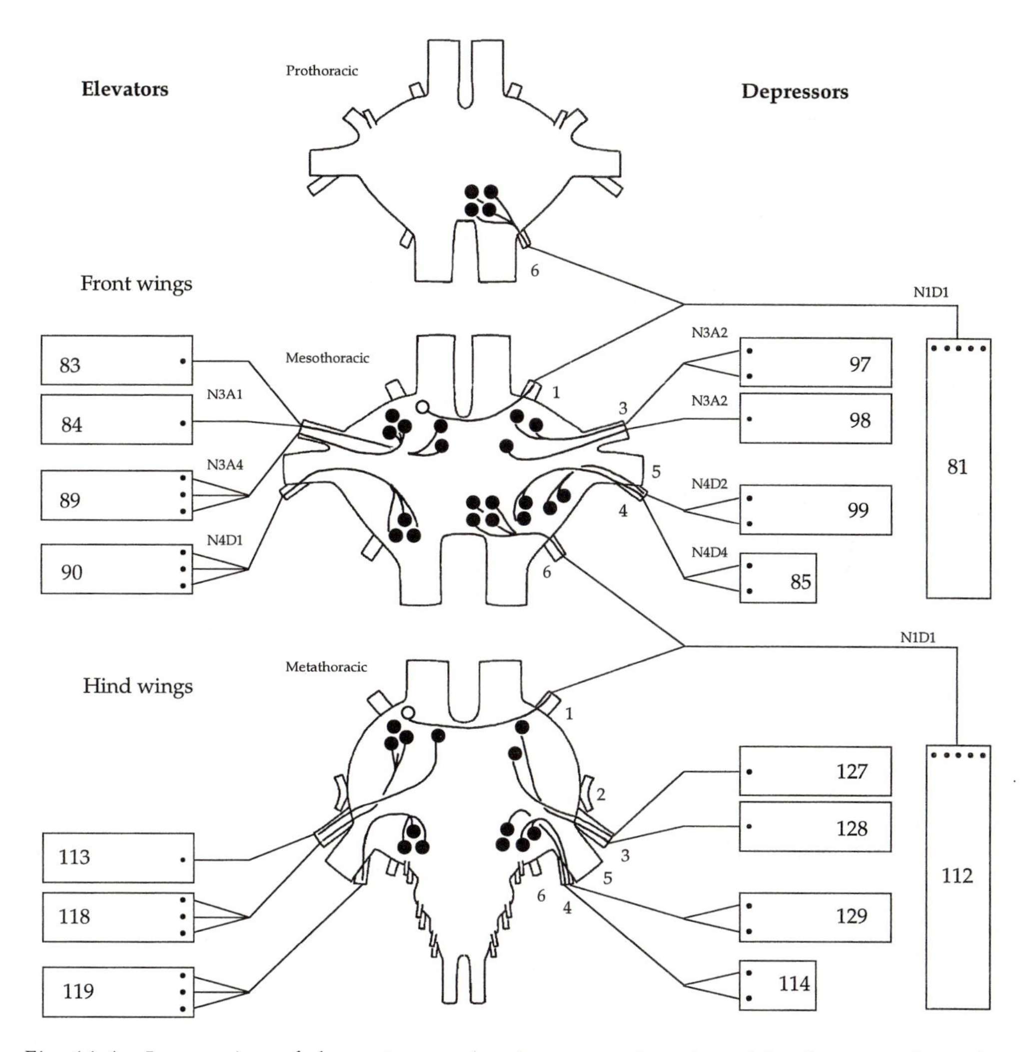

Fig. 11.4 Innervation of the main muscles that move the wings. The drawings show the position of the cell bodies of the motor neurons in the thoracic ganglia and the nerves in which their axons run. The muscles are represented by rectangles with the dots indicating the number of motor neurons that innervate them (see also Table 11.1). Elevator muscles and their motor neurons are on the left, depressors on the right. Based on Burrows (1975a).

muscle (Pfau and Nachtigall, 1981). The pleuroaxillary muscle also opposes the pronation (a twisting downwards of the leading edge) caused by the direct depressor muscles during the downstroke, and can therefore increase the angle of attack of a wing in both phases of its movements.

All these muscles are activated neurogenically and thus operate in the same way as normal skeletal muscle; a single spike in a motor neuron produces a single twitch contraction and a movement of the wing. These flight muscles are sometimes called

Table 11.1. Muscles of the wings: their action and innervation

Muscle name	Muscle number		Nerve	Action		Number of motor neurons	Motor neuron group
	Front wing	Hind wing		Flying	Walking		
Direct depressors							
First basalar	97	127	3A	Depressor and pronator	–	1/2	3
Second basalar*	98	128	3A	Depressor and pronator	–	1/3	1
Subalar*	99	129	4D3	Depressor and supinator	–	2	7
Indirect depressors							
Dorsal longitudinal	81	112	1	Depressor	–	5	–
Elevators							
First tergosternal	83	113	3A	Elevator	–	1	2
Second tergosternal	84	–	3A	Elevator	–	1	–
Anterior tergocoxal (remotor)*	89	118	3A	Elevator	Lifts anterior edge of coxa upwards. 118 not used in walking. Not known if 89 is used in walking	3	1/2
First posterior tergocoxal (remotor 1)*	90	119	4D1	Elevator	Lifts posterior edge of coxa upwards. 119 not used in walking. Not known if 90 is used in walking	3	7
Second posterior tergocoxal (remotor 2)*	91	120	4D2	Elevator	Lifts posterior edge of coxa upwards	91, 2 large 2 small: 120, 2 large 4 small including CI1	7

Table 11.1. Continued

Muscle name	Muscle number		Nerve	Action		Number of motor neurons	Motor neuron group
	Front wing	Hind wing		Flying	Walking		
Tergotrochanteral*	103c	133c	3C	Elevator	Extends trochanter/ femur	1	4
Accessory muscles							
Pleuroaxillary (pleuro–alar)	85	114	4D4	Decreases pronation Unfolds wings	–	2 + CI1	7

* Suggested as bifunctional muscles in flying and walking, but only 120 may act in this way.
For definitions of motor neuron groups see Table 3.2 in *Chapter 3*.

synchronous or nonfibrillar simply to distinguish them from the myogenically activated, asynchronous or fibrillar muscle used to power the high wing beat frequencies of many small insects such as flies. The force that the muscle produces is controlled mainly by the number and frequency of spikes that a motor neuron produces during each wing beat, because the small number of motor neurons innervating a particular muscle does not allow much scope for recruitment. A muscle fibre receives many terminals along its length from a particular motor neuron, and in muscles innervated by more than one motor neuron each fibre is also likely to be polyneuronally innervated. In most flight muscles, however, the distribution of the different motor neurons is segregated, notably in the dorsal longitudinal muscles, so that distinct motor units are formed.

The duration of a twitch contraction elicited by a single motor spike is about 50 ms at 25°C but only 25 ms at 40°C (Neville and Weis-Fogh, 1963). Locusts can start to fly at 25°C, at which time the contraction of any one muscle will occupy the whole of the wing beat cycle, so that the contractions of the elevators and depressors will overlap. This will be inefficient as far as moving the wings is concerned and will dissipate energy as heat, thus reducing the twitch duration in a passive temperature-regulatory mechanism while increasing efficiency. The release of octopamine that occurs during the first few minutes of flight should increase the force produced by the muscles, but at the same time will decrease the duration of their contractions (Whim and Evans, 1988). The latter effect should thus reduce the overlap in contractions and therefore the amount of heat that is generated. During sustained flight, the work performed by the muscles maintains the thoracic temperature at about 6°C above the ambient temperature.

This design of the muscular system contrasts with that in insects such as flies, whose flight muscles are activated myogenically, and in which the frequency of the wing beats is largely determined by the resonant properties of the thorax. The motor spikes in a fly control the power output but not the patterning of the movements, and thus their role is to prime the muscles, which are then induced to contract by the stretch that results from the contraction of other muscles distorting the shape of the thorax.

11.2.3. Bifunctional muscles

Some of the muscles that move the wings, and by inference some of the motor neurons that innervate them, are thought to be used in both flying and walking, and hence are called bifunctional. The idea is that when the wings are folded and locked into place, contractions of these muscles move the middle or hind legs through their insertions on the coxae, but when the wings are unfolded their contractions then contribute to the movements of the wings (Wilson, 1962). The legs are not moved rhythmically during flight in time with the wing movements because muscles that would act as antagonists in moving a coxa are activated simultaneously. If the wings are fixed in the closed position and wind is blown at the head, the flight muscles attempt to move the wings in a flight motor pattern, and then the legs move in time with this rhythm.

Six muscles of both the front and hind wings were suggested to be bifunctional (marked with * in Table 11.1) and although the evidence for their action during both flying and walking was always small, the acceptance of this idea now appears to be widespread. The original myogram recordings by D.M. Wilson do not substantiate all the conclusions he drew, and no subsequent recordings have shown that the second basalar or subalar muscles of the hind wings are active in walking. Intracellular recordings from some motor neurons of supposedly bifunctional muscles of the front wings show activity during both tethered walking and tethered flying (Ramirez and Pearson, 1988). By contrast, myogram recordings from supposedly bifunctional muscles of a hind wing and a hind leg show that only muscle 120 contributes to walking in a variety of attitudes (horizontal, vertical and upside-down) (Duch and Pflüger, 1995). Muscle 120 is unusual for a vertical flight muscle in being innervated by six motor neurons and it appears that different motor neurons are active in flight and in walking. The distribution of these motor neurons in the muscle is not yet known, but the muscle itself seems to consists of two parts. Following autotomy of a hind leg, only part of this muscle degenerates while the rest remains intact, suggesting that different parts may have a different pattern of innervation (Clinton and Arbas, 1994). In flightless grasshoppers (e.g. *Barytettix psolus* or *Romalea microptera*), the muscles that would normally only be used in flying are greatly reduced or even absent, although the motor neurons still innervate glia and basement membranes at the expected site of these muscles (Arbas and Tolbert, 1986). A testable idea, therefore, is that the motor neurons of muscle 120 are divided into two pools, one activated in flying and the other in walking. Certainly, the interneurons that have so far been identified and that synapse with motor neurons to these muscles, spike only in one rhythm (Ramirez and Pearson,

1988). The conclusion to be drawn from the partial data so far available is that only metathoracic muscle 120 is bifunctional. For the mesothoracic muscles, the issue could be resolved by making recordings from the muscles of the supposedly bifunctional muscles during free walking in different attitudes.

11.2.4. The motor neurons

11.2.4.1. *Excitatory motor neurons*

Each muscle is innervated by a small number (1-6) of excitatory motor neurons (Fig. 11.4 and Table 11.1). They all produce twitch contractions and can be considered to be fast motor neurons, so that, in contrast to the leg muscles, there are no motor neurons that produce slow contractions. Only gliding flight, when the wings are outstretched and are not flapped, requires the maintenance of one position, but it is uncertain how this is produced. Each front wing is controlled by a total of 26 excitatory motor neurons and a hind wing by 28 excitatory motor neurons, so that there are only 108 output channels carrying all the motor information to move the wings. The motor neurons innervating a particular muscle have a characteristic shape that allows them to be identified in different locusts. All have enormous arborisations in neuropil on the same side of the ganglion as the muscle they innervate (Fig. 11.5). The fine neurites of these motor neurons are in dorsal and lateral regions of the neuropil and are thus somewhat distinct from the neurites of the leg motor neurons, which are more ventral. The cell bodies are located in the ipsilateral side of the ganglion of the same segment as the wing they innervate. Thus, the muscles of the front wings are mostly innervated by motor neurons from the mesothoracic ganglion and those of the hind wing by motor neurons in the metathoracic ganglion. The only exceptions are the dorsal longitudinal motor neurons; one is in the contralateral side of the same segmental ganglion as the muscle it innervates, whereas the remaining four are in the ipsilateral side of the next anterior ganglion. Thus, a dorsal longitudinal muscle of a front wing is innervated by one motor neuron with a contralateral cell body in the mesothoracic ganglion and by four with ipsilateral cell bodies in the prothoracic ganglion. The pattern is repeated for a hind wing dorsal longitudinal muscle with their cell bodies in the meso- and metathoracic ganglia, respectively.

Like all motor neurons in the locust, those to the muscles that move the wings have intrinsic membrane properties that determine the way they respond to synaptic inputs from sensory neurons and interneurons. They are not simply passive follower neurons used to distribute signals to the distant muscles but can shape the pattern of spikes that they generate (*see Chapter 3*). The motor neurons that are part of the pool that innervates a particular muscle are not electrically coupled to each other. They do, however, receive many synaptic inputs in common, so that they tend to spike at about the same time.

11.2.4.2. *Inhibitory motor neurons*

Correlated with this organisation of fast muscle fibres and fast motor neurons is the absence of any inhibitory innervation to the power-producing muscles. This therefore

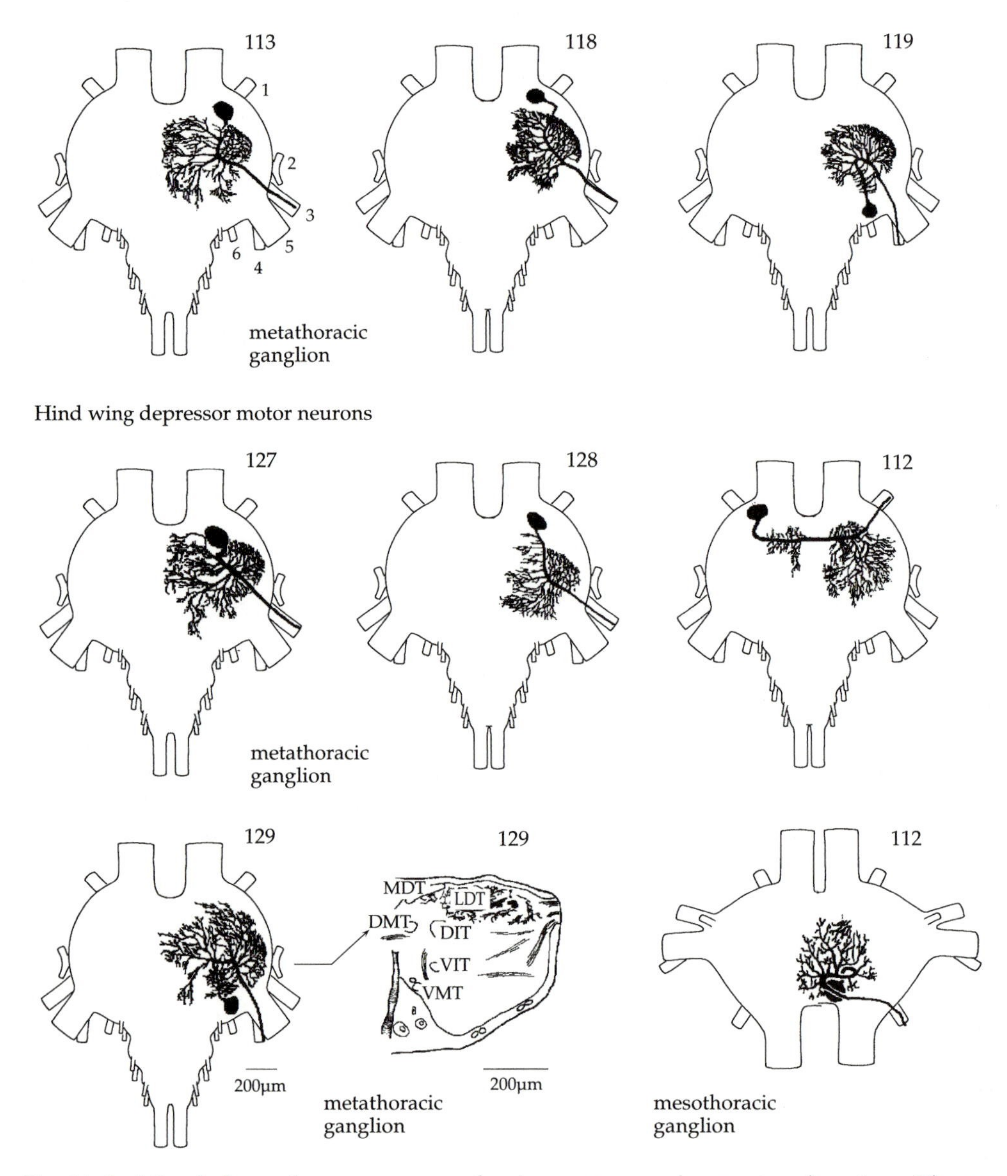

Fig. 11.5 Morphology of motor neurons that innervate muscles moving the wings. The top row shows drawings of motor neurons innervating elevator muscles of the right hind wing following physiological identification and staining by intracellular injection of dye, the bottom two rows show depressor motor neurons. One motor neuron innervating the dorsal longitudinal muscle (112) has its cell body in the contralateral half of the metathoracic ganglion, and the other four (only one is shown) have cell bodies in the ipsilateral half of the mesothoracic ganglion. A transverse section through a motor neuron to muscle 129 shows that its branches are in the most dorsal areas of neuropil. Based on Hedwig and Pearson (1984).

contrasts strongly with the organisation of the leg muscles, most of which are innervated by at least one common inhibitory motor neuron. The muscle fibres appear to be of the fast type, so the absence of inhibitory innervation makes sense if their role is to influence the force produced by fibres that contract slowly. The pleuroaxillary muscles are the only exception to this rule of organisation as they are innervated by branches of CI1, but they control the twisting of the wings and are not power producers. Muscle 120 (second posterior tergocoxal, a wing elevator) is also innervated by CI1 but this may be the part of the muscle that is used in walking and not in flight (*see section 11.2.3*). The dorsal longitudinal muscles of larvae, whose wings are not yet developed, are innervated by one of the common inhibitors that supplies other thoracic muscles (these are distinct neurons from those innervating the leg muscles). In the adult, the dorsal longitudinal muscles lose this innervation but it persists to supply some of the tiny larval thoracic longitudinal muscles that are still present in the adult. All of this implies that the power for flight is generated by sets of excitatory motor neurons and that the inhibitory motor neurons play no part in controlling either the amount or the time course of the force produced.

11.2.4.3. *Neurosecretory neurons*

Many, perhaps all, of the flight muscles are supplied by dorsal unpaired median (DUM) neurons (*see Chapter 3*) from either the meso- or metathoracic ganglia that branch to supply the same sets of muscles on both sides of the locust. The dorsal longitudinal muscles in one segment are supplied by a single DUM neuron called DUMDL, and the pleuroaxillary muscles by a branch of a DUM3,4,5 neuron. By contrast, nine DUM3,4 neurons supply the subalar and posterior tergocoxal muscles (Kutsch and Schneider, 1987). For the other flight muscles, no DUM supply has been identified, but there are DUM neurons whose targets have yet to be found. It would clearly be worthwhile to determine which muscles are supplied by these neurons and to see if the actions of the muscles correlates in any way with the pattern of supply. At the moment, there is no explanation for the substantial supply to the subalar muscles but the apparent absence of a supply to the basalars.

11.3. THE MOTOR PATTERN FOR FLYING

The small number of motor neurons that innervate each muscle, and the few spikes that they produce at each wing beat, allows the patterns of spikes in individual motor neurons to be recorded by fine wires implanted into the muscles that do not impede the normal movements of the wings in tethered flight. Such techniques in the hands of some skilled experimenters have been used to define the complete motor score for flight. For example, the sequence in which depressor motor neurons are activated is known from an impressive series of experiments by Zarnack and Möhl (1977) in which they recorded simultaneously with 14 sets of electrodes from flight muscles (Fig. 11.6). The normal flight motor pattern consists of three basic features.

1. The elevators and the depressors of one wing spike in alternating bursts with a period of about 50 ms (20 Hz).
2. The elevators and then the depressors of the left and right wings of one pair spike at the same time so that both wings move synchronously.
3. The depressors of the hind wings spike some 5ms before their counterparts of the front wings so that the movements of the hind wings always lead the front wings.

The main essence of this rhythmic pattern is described by the latencies between the first spike in an Elevator and the first spike in a Depressor (E–D latency), and between the first spike in a depressor and the first spike in the next cycle of elevator activity (D–E latency) (Fig. 11.1). In straight flight, the D–E latency is longer and occupies about 55% of the wing beat cycle, whereas the E–D latency is shorter and occupies some 45% of the cycle. As the duration of the wing beat cycle increases, the duration of both latencies increases so that the phase of a depressor spike in the elevator cycle remains constant at about 0.55 (Waldron, 1967; Stevenson and Kutsch, 1987). This correlates with the movements of the wings, in which the downstroke is usually longer than the upstroke.

The elevator motor neurons seem to act as an equivalent group so that their spikes to the different elevator muscles occur at about the same time and there is no obvious sequence to their action. By contrast, in tethered flight, the depressor motor neurons to one wing spike at different times relative to each other over a period of some 10 ms, and an individual motor neuron may spike at different times on successive cycles of the wing beat (Möhl, 1985a). These differences in timing, particularly of the basalars, are sufficient to influence the movements of a wing and its pronation. Moreover, the timing relationships between the depressor motor neurons to the left and right wings of a pair, and between the front and hind wings, may show considerable changes in their timing and even some inherent asymmetries. The changes in the timing of all the depressor motor neurons are linked in complex and nonrandom arrangements, but are nevertheless a consequence of tethered flight as they disappear during free flight (Möhl, 1988). The apparent artefact arises not from the way an individual locust might be mounted, or from its surrounding environment, but from the fact that it is flying in open loop conditions with the various exteroceptors unable to close this loop (*see section 11.7*). Asymmetries in the number of spikes to different muscles can nevertheless be expressed as a turning tendency to one side when there are no visual clues (Wilson, 1968). The difference between the sequencing of the elevators and the depressor motor neurons may have its origin in the contribution that some of the depressor muscles play in controlling the supination and pronation of a wing.

11.3.1. Does the motor pattern depend on sensory feedback?

The integral nature of the sensory signals in the flight motor pattern is readily shown by grasping one front wing of a flying locust and then moving that wing at a frequency slightly different from that of the other wings moving in normal flight

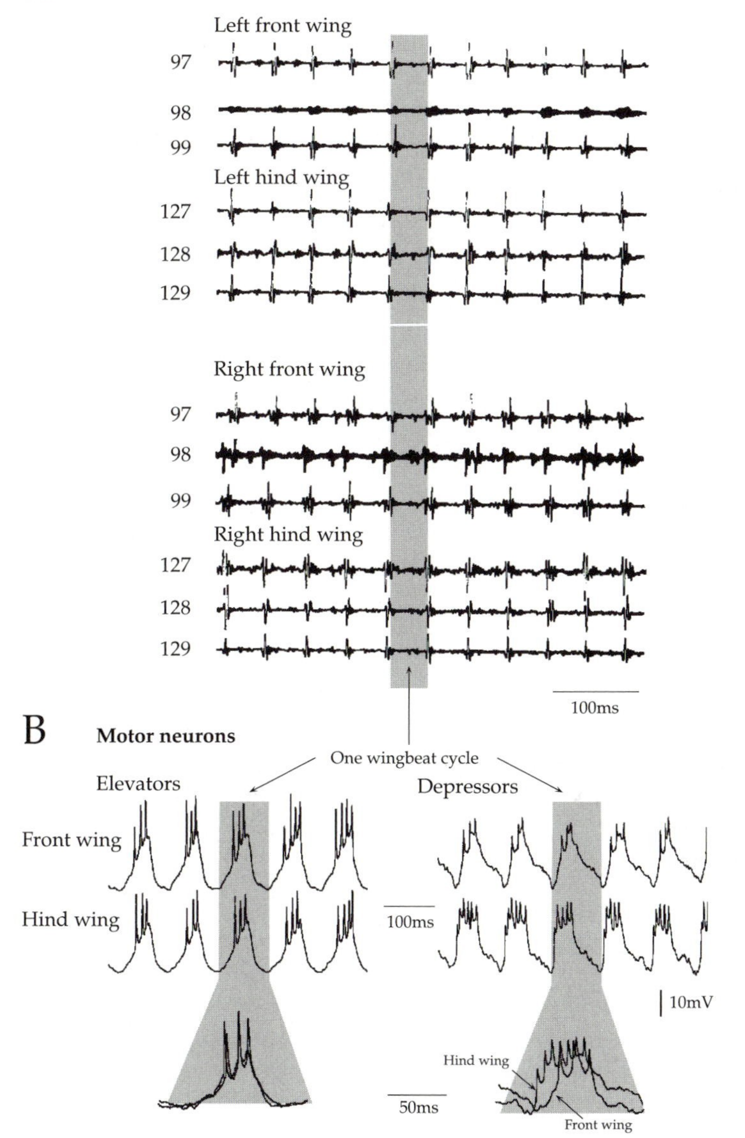

Fig. 11.6 Action of wing muscles and motor neurons during flight. A. Action of the dorso-ventral depressor muscles of the four wings recorded simultaneously during tethered flight. B. Action of front and hind wing elevator and depressor motor neurons on the same side of the body during the flight motor pattern. Spikes in front and hind elevators occur at the same time, but spikes of hind wing depressors occur before those in front wing depressors. The shaded areas indicate one wing beat cycle. A is based on Zarnack and Möhl (1977); B on Hedwig and Pearson (1984).

(Wendler, 1974). Provided that the imposed movements are close to the natural flight frequency, the movements of all the wings become locked to the stimulus. Sensory signals from the wing that is being moved must be responsible for the alterations in the movements of the other wings. To test the contribution of the sensory signals from the wings and the wing hinges that occur at each wing beat to the generation of the flight motor pattern, D.M. Wilson (1961) first cut the wings and then eliminated their sensory input to the central nervous system. The frequency of the wing beats is not changed by reducing the length of the wings, as it is in myogenic insects, but it is reduced by about half when the nerves (N1) from all the wings that contain sensory neurons are cut. Flights are then of short duration, but still show the basic pattern, including alternation of elevators and depressors, twisting of the wings, and delays between the movements of the front and hind wings. This pattern can still be produced in a preparation that consists only of the head, thoracic nerve cord and tracheae and in which the dorso-ventral muscles are all cut in half transversely, so that no wings or legs are present. Furthermore, if all the flight muscles are removed, motor spikes can still be produced in a slow rhythm typical of flight. This slow rhythm can be elicited by wind stimuli to the head, or by electrical stimulation of a connective between the suboesophageal and prothoracic ganglion with randomly occurring pulses that have no phase relationships with the output (Wilson and Wyman, 1965). The input is thus unpatterned but the output of motor spikes is patterned, thus indicating that the central nervous system is capable of generating the pattern without repetitive timing clues from the periphery.

These historic experiments by D.M. Wilson led him to conclude that the basic elements of the flight motor pattern are organised within the central nervous system and do not need any sensory feedback for their execution. This was a conclusive demonstration of what came to be called a **central pattern generator** for a complex movement. Nevertheless, Wilson (1961) was clear that sensory feedback was an essential element: '*The existence of an innate pattern does not preclude the possibility that peripheral loops are* ***also sufficient*** *to produce the same pattern, but important as those feedbacks must be for the setting of the frequency of cycling, for accurate timing of contraction cycle and movement cycle, and for fine adjustments for control manoeuvres they act on top of what is already determined by the central nervous structure and function.*' He was also clear that the concept of a pattern generator should not be taken too literally: ' . . *the vague concept of a "flight control centre" is supported neither in the anatomical sense nor in the sense of a group of hypothetical neurons which are uniquely concerned with integrating input with a certain pattern and passing this on to "motor centres".*'

A huge range of receptors associated with the wings and their articulation with the thorax continuously monitor the movements and positions of the wings, and the forces that are generated. The extensive inflow of signals means that although a motor pattern, which resembles the pattern during free flight in most major features, can be produced by the central nervous system in the absence of sensory feedback, it is more sensible to treat the central and peripheral components as part of an integrated whole, rather than as separable components. A locust will only produce sustained free flight when there is sensory feedback. The implication is that the

central pattern generator supplies a basic rhythmicity whose frequency and patterning can be substantially altered by sensory feedback to produce the motor spike sequence that is often called a **motor programme**. The function of the central networks is therefore to integrate the sensory signals with the inherent membrane and network properties of the central neurons and generate an adaptive motor pattern that can generate an appropriate sequence of movements.

Experiments (and their interpretation) have more recently sought to determine whether the flight motor pattern expressed when there is no sensory feedback differs from that in intact animals. The three basic features of the flight motor pattern listed above remain when the animal is deafferented, despite the increase in the period of the rhythm and the production of more spikes by each motor neuron during each cycle. These changes may result from a lack of input from the wing receptors at each cycle of the wing beat, but they can be partially restored by octopamine (Stevenson and Kutsch, 1987). What other features survive deafferentation is more problematical and essentially unresolved by conflicting results. In tethered animals flying the right way up, the durations of the D–E and E–D latencies increase with period, as in the intact animal, although the variability is much greater. Moreover, the same changes in these latencies with period are seen in the rhythm produced by an isolated meso- and metathoracic ganglion complex when treated with 0.1 M octopamine. This means that the phase of an elevator in the depressor cycle remains constant in these situations. In tethered animals flying upside-down in a wind stream, however, the D–E latency remains constant as the period changes so that consequently the phase changes (Pearson and Wolf, 1987). This result was also obtained in a locust tethered in a normal orientation, and in fictive flight in an inverted and highly dissected animal (Pearson and Wolf, 1989). The fused thoracic ganglia of the moth *Manduca sexta* that are still attached to an intact head, but otherwise isolated from the body can produce fictive flight motor patterns in which the motor neurons show either a constant phase relationship, or a constant latency relationship. The expression of a particular pattern does not therefore depend on sensory feedback (or its absence) from the wings, but on properties of the central nervous system. It should be possible to resolve these observational differences, but little is likely to be gained from a resolution of the formal problem of 'central *versus* peripheral' control of the flight motor pattern. Clearly, the sensory input supplied by the huge range of receptors associated with the wings is an integral part of the flight mechanism and this knowledge is sufficient for further analyses of mechanisms.

11.4. WHERE IS THE MOTOR PATTERN GENERATED?

The organisation of the muscles and the motor neurons into repeating sets might suggest that the rhythmic flight motor pattern is also generated by repeating sets of neurons. On this basis, there would be a total of four networks distributed so that there is one in each half of the meso- and metathoracic ganglia, but linked together to produce a coordinated movement. This presupposes a strict relationship between the segmental organisation of the appendages and a segmental organisation of all the neurons controlling their movements. The alternative organisation envisages one network that generates the rhythm with its components distributed in different parts of the nervous system.

11.4.1. Lesion experiments

Lesion experiments have sought to distinguish between these possibilities by establishing whether each thoracic ganglion, or even each half of a ganglion, can produce a flight motor pattern, or whether the integrity of the thoracic ganglia is necessary and therefore that the neuronal network for flight is distributed. It should come as no surprise to find that the outcome of these sorts of experiments is ambiguous and can be interpreted to favour either organisation.

Wilson (1961) gradually pared down the central nervous system to determine which parts were essential for the production of a flight motor pattern and found that a locust can still generate sufficient lift to support its body weight during flight if the brain, suboesophageal and abdominal ganglia are removed. This pinpoints the thoracic ganglia as the site where the flight motor pattern is generated.

11.4.1.1. *Cutting connectives between ganglia*

If the connectives between the pro- and mesothoracic ganglia are cut, flight can only be sustained for short periods, and although the pattern remains essentially the same as in an intact animal, the frequency is low (Wolf and Pearson, 1987). If the mesothoracic ganglion is separated from the metathoracic ganglion by cutting the meso-metathoracic connectives, the front wings can still be flapped if sensory feedback from them is present. The pattern is, however, very different from normal flight in that the wing beat frequency is low, the number of spikes produced by each motor neuron is more variable, and the relationships between the elevator and depressor phases are different. If the mesothoracic ganglion is isolated so that it receives no feedback from the front wings, rhythmic activity is rarely produced in the front wing motor neurons. Similarly, rhythmic activity rarely occurs in hind wing motor neurons when the metathoracic ganglion is isolated. An isolated metathoracic, but not an isolated mesothoracic ganglion can, however, generate a flight motor pattern when octopamine is bath applied or injected into the neuropil (Stevenson and Kutsch, 1987).

11.4.1.2. *Cutting ganglia in half longitudinally*

More drastic lesions can be performed by opening the thorax, cutting a thoracic ganglion in half longitudinally, gluing the wound in the cuticle and then testing flight ability 3h to 15days later (Ronacher *et al.*, 1988). If the mesothoracic ganglion is cut then most of the operated locusts can still fly normally and can also walk and kick. When tethered or free, they will fly with a motor pattern that shows all the characteristics of normal flight, with the only apparent differences being that the locust takes longer to open its wings and flies for shorter periods. The lack of information flow between the two halves of the ganglion is not compensated for by sensory inflow, because the pattern is not much perturbed by cutting the nerves from the wings, save that the frequency of the wing beats is then reduced by half. By contrast, locusts in which the metathoracic ganglion has been bisected cannot fly freely and only a small number are able to beat their wings in an unstable rhythm. When both meso- and metathoracic ganglia are cut, the locusts cannot fly.

The synaptic input to the motor neurons in the mesothoracic ganglion that drives them in the flight rhythm continues even if the ganglion is bisected longitudinally, indicating that signals in the commissures are not essential (Wolf *et al.*, 1988). If, however, the metathoracic ganglion is bisected in a similar way, there remains only a poor indication of a rhythmic input to the motor neurons with none of the normal characteristics. If both the meso- and metathoracic ganglia are cut then no rhythmic input is delivered to the motor neurons. These experiments demonstrate the more dominant role of the metathoracic ganglion in generating the flight motor pattern and support the idea that the rhythm results from the action of a distributed network of interneurons with many parallel pathways.

Similar experiments on song production in crickets and stridulation in grasshoppers also point to the distributed nature of the neurons controlling these movements. Longitudinal section of the pro- or mesothoracic ganglion has no effect on the wing movements of a cricket during singing, but section of the metathoracic ganglion disrupts the song. Sectioning the mesothoracic ganglion and one meso-metathoracic connective leaves the wings on the sectioned side unable to move, but the wings on the intact side move normally. In grasshoppers, longitudinal section of the metathoracic ganglion in species with a slow rhythm of movements does not disrupt the coordination between the two hind legs, but in species with a faster rhythm they then move independently. The coordination may, however, depend on the element of the song being considered.

These skilful and careful experiments nevertheless still do not distinguish between the alternative interpretations about the origin of the flight motor pattern. The first hypothesis is that each half-ganglion has the ability to generate a rhythm but with a distributed network of coordinating interneurons ensuring that they normally work together. None of these experiments clearly demonstrate the ability of any isolated part of the central nervous system, other than the metathoracic ganglion (which consists of four neuromeres), to produce a flight motor pattern. The second hypothesis is that the network generating the pattern is distributed with many parallel pathways, so that a lesion impairs the function of only a small proportion of these, allowing isolated parts to express a rhythmic pattern with some features of the normal pattern. This is the most parsimonious interpretation of the data that have been reported and is the one I favour. Thus, after 20 years of experiments on the component neurons, we come almost full circle back to the proposal made by Waldron (1967) before any analysis at this level was undertaken, that '*there is a single functionally integrated flight system which extends through many ganglia*'.

This concept is supported by the anatomy and distribution of the interneurons and by the pattern of connections that they make with each other and with the flight motor neurons. Many of the interneurons are apparently segmentally homologous, and occur in repeating, bilaterally arranged sets (they always occur as symmetrical pairs) that are distributed in five neuromeres (Robertson *et al.*, 1982; Robertson and Pearson, 1983; Pearson and Robertson, 1987; Ramirez and Pearson, 1988). The 401, 501, 503 and 504 interneurons are, for example, repeated in the metathoracic and A1-3 neuromeres, the 503 interneurons in the A1-3 neuromeres and the 401 interneurons in the metathoracic and A1-2 neuromeres. These sets of interneurons are, however, not duplicated for each of the two pairs of wings, and moreover there are individual interneurons for which no apparent homologues have yet been found. For example, the 301 and 701 interneurons in the mesothoracic ganglion are apparently not repeated in any of the neuromeres of the metathoracic ganglion, and conversely, the 501 interneurons of the metathoracic ganglion are not repeated in the mesothoracic ganglion. Of course, some of these differences may simply reflect a failure to find the interneurons in the different ganglia, or a failure to recognise them as homologues if their properties are different. The individual meso- and metathoracic ganglia can thus each be viewed as the repository for most of the motor neurons that innervate a particular pair of wings, but the interneuronal network for generating the motor pattern must be viewed as distributed.

11.5. INTERNEURONS AND THE GENERATION OF THE MOTOR PATTERN FOR FLYING

The simplest theoretical explanation for the generation of the flight motor pattern envisaged that it could result from connections between the motor neurons themselves (Wilson, 1966), but such mechanisms are improbable because if the wings are forcibly held in an elevated position during flight, rhythmical bursts of spikes continue in depressor muscles in the absence of any activity in elevators (Waldron, 1967). Moreover, intracellular recording from the motor neurons shows that the appropriate connections are not present (there is no reciprocal coupling between antagonists, or coupling between synergists) and instead reveals that they are driven in specific patterns by synaptic inputs that must be generated by interneurons. The search for the mechanisms responsible for generating flight has thus focused on interneurons, the signals that they receive from sensory neurons and the output that they deliver to the motor neurons.

The actions of interneurons that are involved in generating and controlling the flight motor pattern have been analysed in a highly dissected locust that nevertheless produces a pattern that resembles flight and is called **fictive flight**. The locust is mounted dorsal side up but has its wings, legs and viscera removed, and nerves 3, 4 and 5 that innervate the thorax and the middle and hind legs cut (Robertson and Pearson, 1982). Sometimes, the connectives to the abdomen, and the nerves (N1) that innervate the wings and the dorsal longitudinal muscles are also cut. Thus, in the most reduced preparation, the flight muscles are denervated and there is no sensory feedback from the wings, the thorax or the legs. The flight motor pattern that can be recorded in such preparations usually only lasts for a few seconds, or at most a few tens of seconds, so that often the rhythm has not reached a fully stable state and is either starting or stopping. A promising preparation that has yet to reveal its full potential is in a moth *Manduca sexta*, where the head and attached thoracic ganglion isolated from the rest of the body will generate a flight motor pattern at less than half the normal frequency for a few hours when treated with the octopamine agonist chlordimeform (Wendler and Suder, 1994). The pattern is not expressed continuously all this time, but for periods of minutes so that it is stable. Movements of the visual environment induce changes in the pattern that are similar to those that would occur during steering responses.

What is the evidence that the motor pattern recorded in interneurons or motor neurons of such preparations is a representation of the one used during normal flight? It resembles the normal flight motor pattern in four ways. First, it is elicited by wind stimuli to the head that initially excite the elevator motor neurons. Second, elevator and depressor motor neurons of one wing spike rhythmically in a pattern that has a frequency of about half that of normal flight (approximately 10 Hz instead of 20 Hz), but the homologous motor neurons to the two wings of one segment spike synchronously. Third, there is a shorter delay from the start of elevator activity to the start of depressor activity (E–D latency) within one cycle than there is from the start of depressor activity to the start of the next cycle of elevator activity (D–E latency). The former stays relatively constant as the period of the cycle increases, while the latter increases. The relationship between these two phases of the rhythm is thus similar to that in walking and in breathing (*see Chapters 8 and 12*). Fourth, the spikes in the hind wing motor neurons lead those in the front wing motor neurons by about 5ms, or about the same delay as in normal flight, although the frequency of the rhythm is 50% less.

11.5.1. Morphology of the interneurons

Some 85 morphological types of interneurons have now been characterised that participate in the flight motor pattern, 46 in the mesothoracic ganglia (see Rowell and Reichert, 1991) and 39 in the various neuromeres of the metathoracic ganglion (Figs 11.7–11.10). No equivalent interneurons are described in the prothoracic ganglion, even though it contains flight motor neurons. It is unclear what the total number of interneurons that contribute to flight will be, whether they can be separated by this description from many of the other interneurons, or whether there really are more in the mesothoracic than the metathoracic ganglion. Most of these interneurons are intersegmental interneurons with axons projecting either anteriorly, posteriorly or in both directions. Surprisingly few local interneurons have been identified. The interneurons are given identification numbers (*see Chapter 3*), implicit in which, but nevertheless still an assumption, is that there is only one pair of interneurons with a particular number. For some of the interneurons, this assumption seems valid in the face of the many recordings and stainings that have been made; for others, known only from a few recordings, the validity remains open; and for still others, there are clearly several interneurons with a similar morphology that are lumped together under the same identification number, although this is not made explicit. This is a continuing and sometimes insuperable problem in the identification of interneurons as more is learnt about them in subsequent

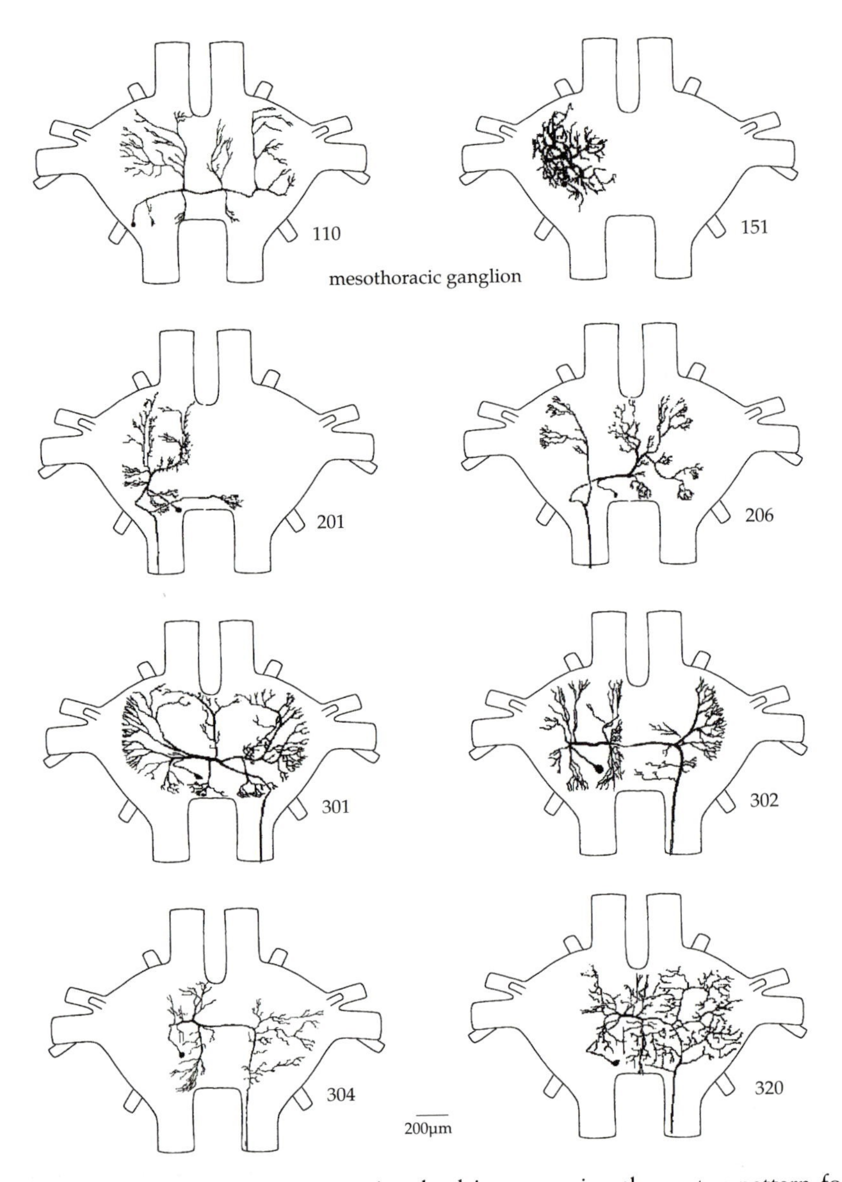

Fig. 11.7 Morphology of interneurons involved in generating the motor pattern for flying. The neurons are numbered according to the scheme in Fig. 3.3 (*see Chapter 3*). The top row shows two local interneurons (100 group interneurons) in the mesothoracic ganglion; the second row shows two interneurons with descending axons ipsilateral to their cell bodies (200 group interneurons); the bottom two rows show four interneurons with descending axons contralateral to their cell bodies (300 group interneurons). The drawings here and in Figs 11.8 and 11.9 are based on the following papers; Pearson and Robertson (1987); Ramirez and Pearson (1988); Robertson and Wisniowski, (1988); Robertson and Pearson (1985a); Pearson *et al.* (1985b); Wolf and Pearson (1989).

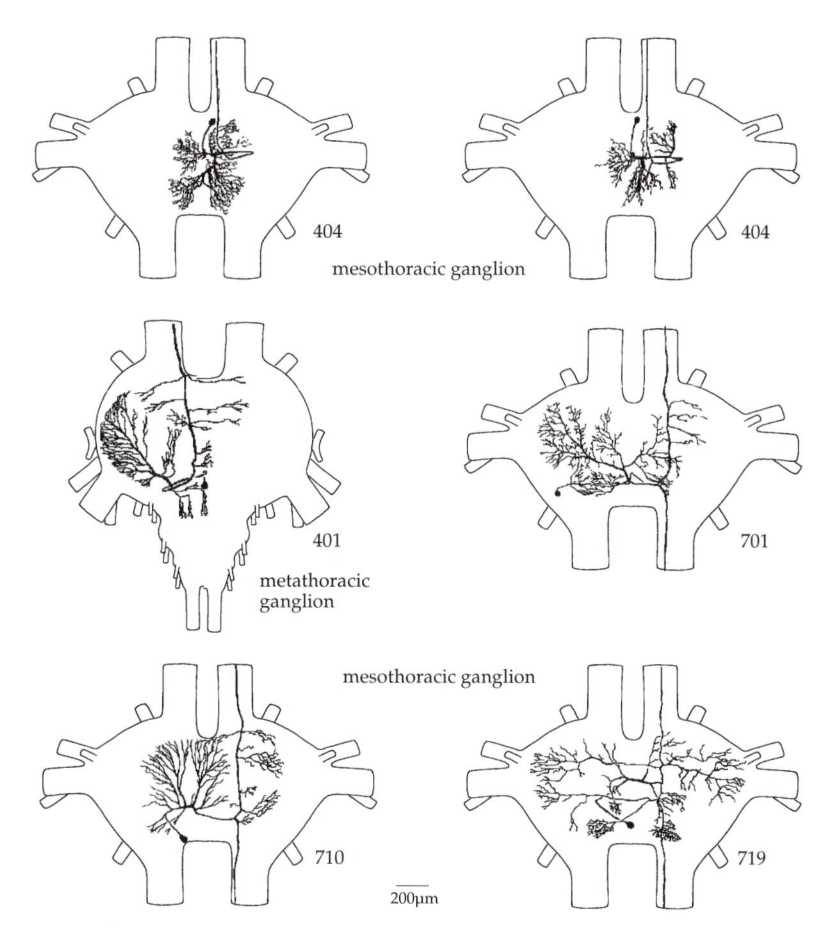

Fig. 11.8 Morphology of interneurons involved in generating the motor pattern for flying. The drawings show interneurons in the meso- and metathoracic ganglia that have ascending axons ipsilateral to their cell bodies (400 group interneurons), and interneurons with both ascending and descending axons contralateral to their cell bodies (700 group interneurons).

experiments. Do interneurons with slight differences in their properties, when examined in different animals, represent the normal variation that is found even for a particular interneuron, or do they represent different members of a similar class of interneuron? These possibilities have yet to be tested adequately for the interneurons involved in flight.

11.5.2. Classifying the interneurons

Various attempts have been made to classify the interneurons according to their particular actions as revealed in recordings during fictive flight. The usual groupings and names that have been used are:

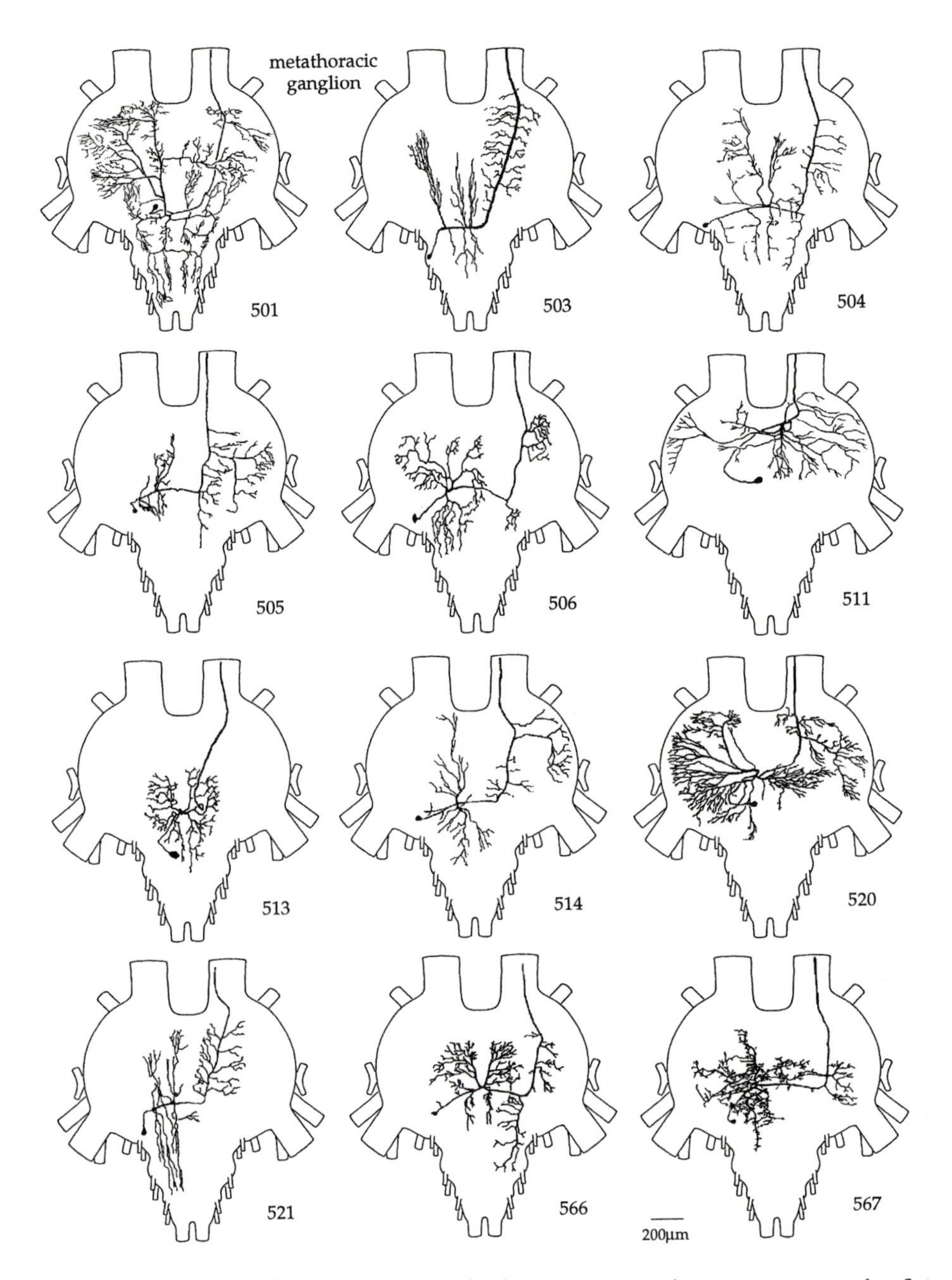

Fig. 11.9 Morphology of interneurons involved in generating the motor pattern for flying. The drawings show interneurons in the metathoracic ganglion with ascending axons contralateral to their cell bodies (500 group interneurons).

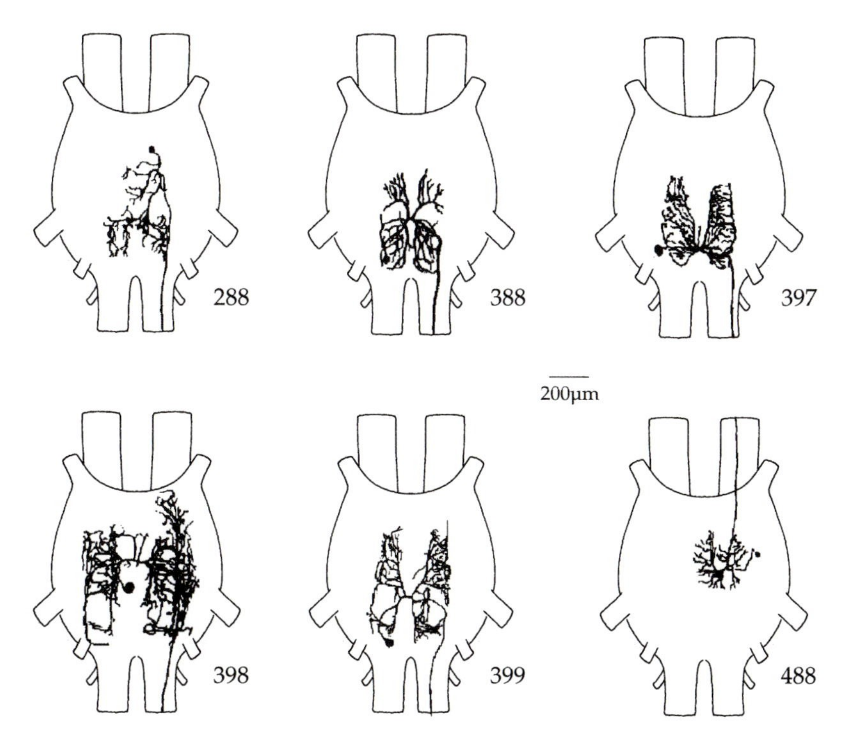

Fig. 11.10 Morphology of interneurons in the suboesophageal ganglion that are involved in generating the motor pattern for flying. Based on Ramirez (1988).

1. Flight-initiating interneurons.
2. Interneurons that generate the rhythmic pattern.
3. Interneurons that modulate the rhythmic pattern.
4. Corollary discharge interneurons that distribute the pattern to other networks.
5. Premotor interneurons that synapse directly with the motor neurons.
6. Reflex-integrating interneurons that perform the first step in processing the sensory feedback.

These well-intentioned attempts to seek order from a huge array of information have always foundered because an interneuron does not do just one thing. For example, an interneuron can be designated as a pattern-generating interneuron if it is able to reset the rhythm when its membrane potential is manipulated with injected current, and if it does not connect directly with the motor neurons. Similarly, an interneuron can be designated as a premotor interneuron if it makes direct synapses with a motor neuron but if it does not reset the rhythm. Interneurons designated as having pattern-generating functions do, however, connect directly with motor neurons, and interneurons designated as premotor interneurons can reset the rhythm. These schemes based on actions are further complicated because the failure of an interneuron to reset the rhythm does not necessarily indicate that it is not involved, but may simply be a reflection of the number of neurons active at any one time. Even the most extensive recordings give only a partial picture of the actions of a neuron during a particular set of circumstances and we still know few of their input and output connections and little of their membrane characteristics. Were this desirable knowledge to be available it would still seem best to steer clear of these functional terms and stick to a neutral terminology such as the use of numbers and live with the more trivial problems that attend the use of these.

11.5.3. Driving the motor neurons

In fictive flight, the membrane potential of the motor neurons oscillates by as much as 25mV in the flight motor pattern, superimposed on a maintained shift in membrane potential, with excitatory synaptic inputs contributing to the depolarisations and inhibitory ones to the repolarisations. These large changes in membrane potential indicate that they are caused by the summation of inputs from many interneurons (and often sensory neurons as well), each of which contributes synaptic potentials of only a few millivolts. The result of this patterned synaptic input is that each motor neuron spikes once or a few times at a particular phase of the cycle.

The motor neurons to the elevator muscles all seem to act in a similar way in that they are depolarised in two phases; first, a slow depolarisation that occurs after the end of the depolarising phase in the depressor motor neurons, and second, a more rapid depolarisation that follows with a delay that depends on the cycle period (Fig. 11.6) (Hedwig and Pearson, 1984). The repolarisation then occurs rapidly as the depressors are depolarised. These changes in membrane potential occur simultaneously in front and hind wing elevator motor neurons. By contrast, the depressors do not act as a uniform group and show different patterns depending on the muscle they innervate and whether it controls a front or a hind wing. They are depolarised immediately after the rapid depolarisation of the elevators, with the depolarisation of hind wing motor neurons preceding that of the front wing motor neurons by 5–15 ms (Fig. 11.6). There is thus a short and relatively fixed delay from the end of the elevator spikes to the start of the depressor spikes (E–D latency), but a longer and more variable delay from the end of the depressor spikes to the start of the next cycle of spikes in the elevators (D–E latency). This suggests that a basic unit of activity is the depolarisation of an elevator followed by the depolarisation of a depressor.

During fictive flight, motor neurons to the same muscle receive many synaptic inputs in common that cause them to spike at about the same time because there are no direct electrical or synaptic connections between them (Robertson, 1990). Simultaneous synaptic driving also causes the motor neurons to the same muscles of the left and right wings of one pair to spike at about the same time so that the wings are moved synchronously.

The synaptic drive to the motor neurons can be ascribed to many interneurons that originate in the mesothoracic ganglion and the four neuromeres of the metathoracic ganglion. In general, the interganglionic connections are better known because it is easier to make the paired recordings needed to determine connections from neurons in separate ganglia (Fig. 11.11) It seems reasonable to presume that the mesothoracic interneurons also connect with mesothoracic motor neurons and that interneurons in all the neuromeres of the metathoracic ganglion also connect with metathoracic motor neurons. Some interneurons provide excitatory inputs and others inhibitory ones. Some of the interneurons that make inhibitory connections probably use GABA but others that are also inhibitory do not (Robertson and Wisniowski, 1988). By correlating the shapes of some of these interneurons with their output connections it has been suggested that particular features of the anatomy of an interneuron can be used to predict the polarity of its output connections; interneurons with a cell body near the midline should make inhibitory connections and those with more lateral cell bodies should make excitatory connections (Pearson and Robertson, 1987). While the correlation may hold for a restricted set of interneurons, it does not extend to the other known interneurons within the same ganglia.

11.5.4. Interconnections between the interneurons

None of the interneurons identified so far appears to have an innate ability to oscillate at the wing beat frequency, so that the rhythm must result from the network of interactions between a large number of interneurons. Most produce a burst of 5–10 spikes at instantaneous frequencies of up to 250 Hz in time with either the elevation or depression phases of the wing beat cycle (Fig. 11.12). Understanding how these interneurons are connected is a powerful, but not an exclusive way of unravelling the intricacies of a network and providing insights into how a motor pattern might be generated. Nevertheless, although many interneurons that are active in flight are known, few interconnections between them have been established (Fig. 11.13).

A few connections have been determined by recordings from pairs of interneurons, so that it can be said with some certainty that they are direct. For example, interneuron 501 makes an inhibitory connection with 301, and 504 makes an excitatory connection with 301. More outputs of the interneurons are known from injecting current into one and observing the resulting, usually delayed, effects in others, but in a complex interconnected network such effects are to be expected and it is then impossible to know the pathway by which they are mediated. Interneurons 301, 302 and 401 all produce delayed changes in the membrane potentials of other interneurons but the connections are not direct. If each interneuron makes widespread connections, as is suspected for those with inhibitory outputs or for interneurons such as 201 which is thought to connect with all depressor motor neurons (Robertson, 1986), then the task of further analyses will be difficult and unrewarding. Perhaps this explains why this approach has apparently stopped. Can any of the attributes of the network be explained by what we have learnt to date about the way the interneurons are interconnected?

Making sense of the connections has centred on possible mechanisms that could lead to the alternating pattern of spikes that occur first in the elevator motor neurons, followed at a constant latency by spikes in the depressors. Particular attention has been focused on the connections of interneurons 301 and 501

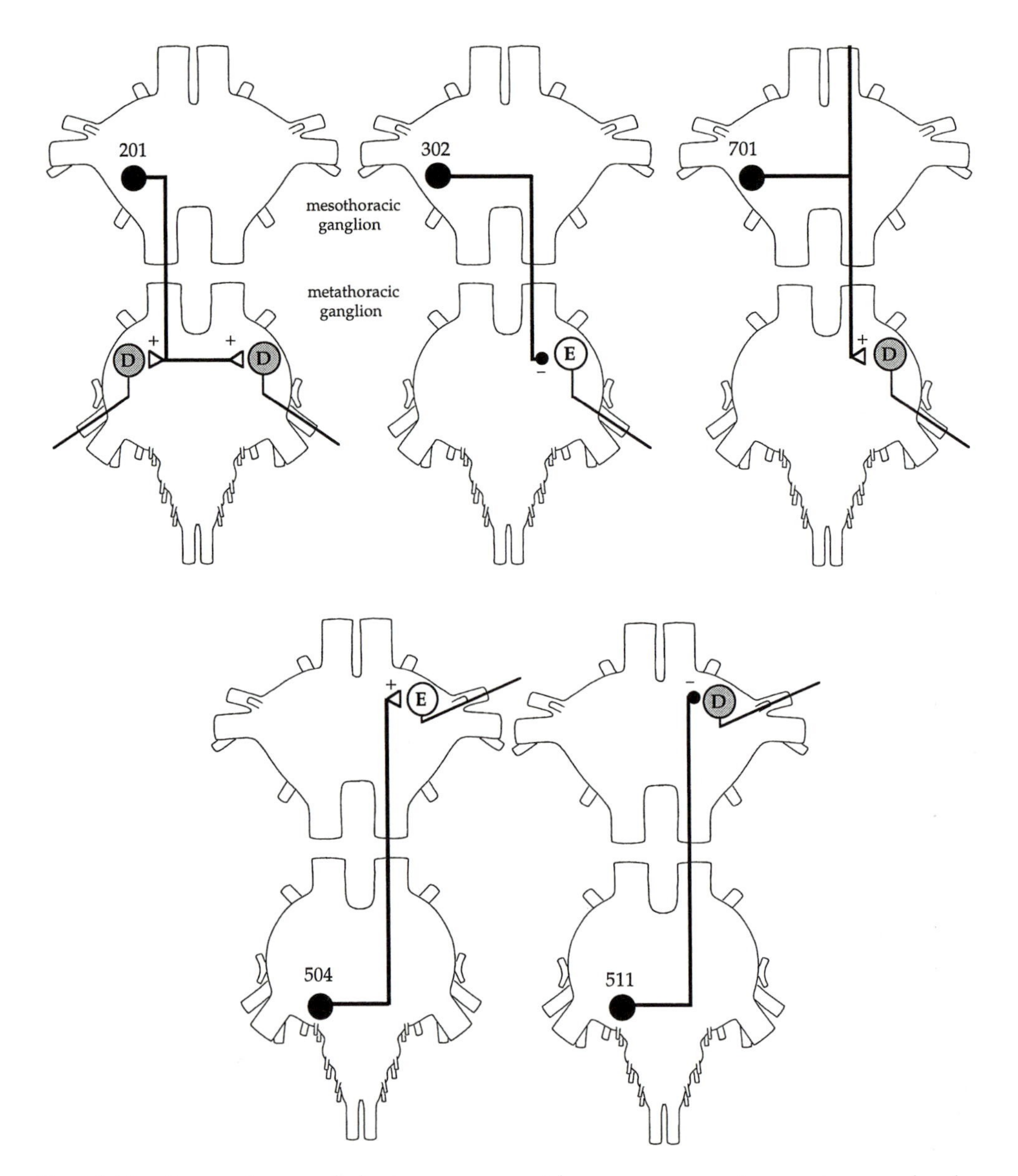

Fig. 11.11 Connections made by interneurons with motor neurons innervating muscles that elevate (E) or depress (D) the wings in flying. The diagrams illustrate the basic pattern of connections found so far, but do not include the connections made by local interneurons.

(Fig. 11.13), probably because current injected into 301 and not any of the other interneurons that are known can advance or delay the timing of the next cycle of the flight motor pattern (Robertson and Pearson, 1985a). A spike in interneuron 501, which in flight normally occurs during depression, evokes an IPSP directly in

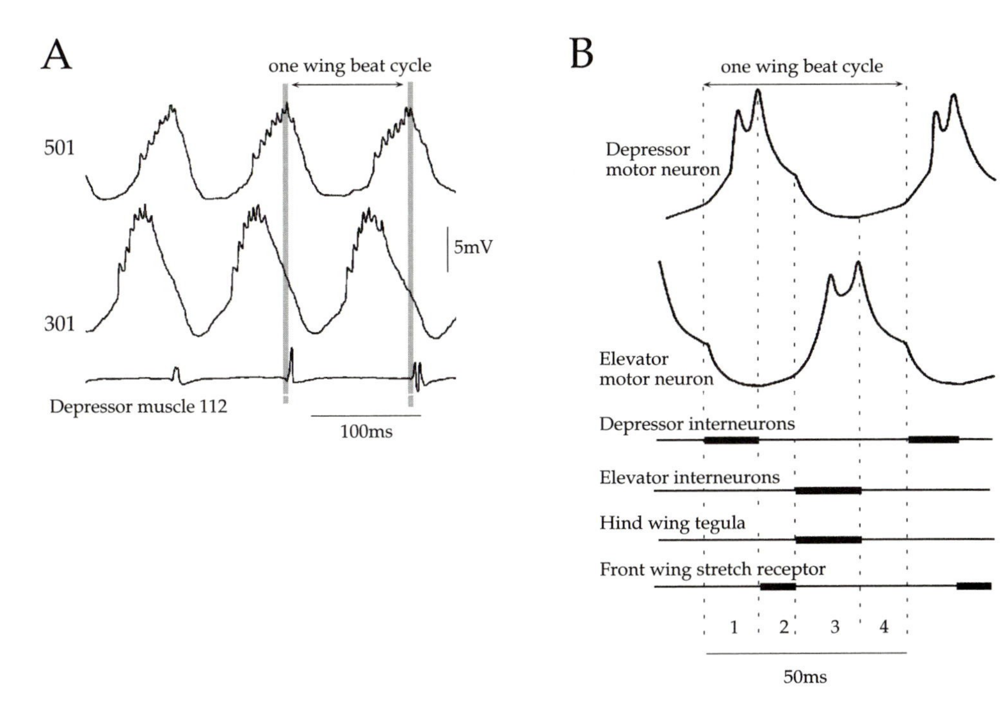

Fig. 11.12 Action of interneurons during the motor pattern for flying. A. Interneuron 301 is depolarised and generates spikes during the elevation phase of the wing beat cycle, whereas interneuron 501 is depolarised and spikes during depression. B. Four components of the synaptic drive to the motor neurons related to the action of sensory neurons from the hind wing tegula and front wing stretch receptor. A is based on Robertson (1988), B on Pearson and Ramirez (1990).

interneuron 301, whereas a spike in 301, which normally occurs before the depressor motor neurons start to spike, evokes a depolarising potential in 501 through a longer pathway. The potential in 501 is probably caused by disinhibition of at least one intervening interneuron that is not identified. For this pathway to work, several complex events have to be presumed. The inhibitory connection of 301 with the intervening interneuron would have to suppress the tonic release of inhibitory transmitter from this interneuron onto 501. Moreover, if the pathway is disynaptic, it requires that the intervening interneuron would have to release transmitter in a graded fashion according to the level of its membrane potential. This would mean that it would have to be a nonspiking interneuron, but these interneurons have yet to be implicated in the generation of the flight motor pattern. If it were a spiking interneuron then it would either have to release transmitter in response to its spikes, or independently in response to graded changes in its membrane potential, but there is only suggestive evidence that some flight interneurons can sometimes release transmitter in this way (Robertson and Reye, 1988). The pathway involving interneurons 301 and 501 could generate alternating bursts of spikes by the following sequence of events, but has not yet been demonstrated to do so.

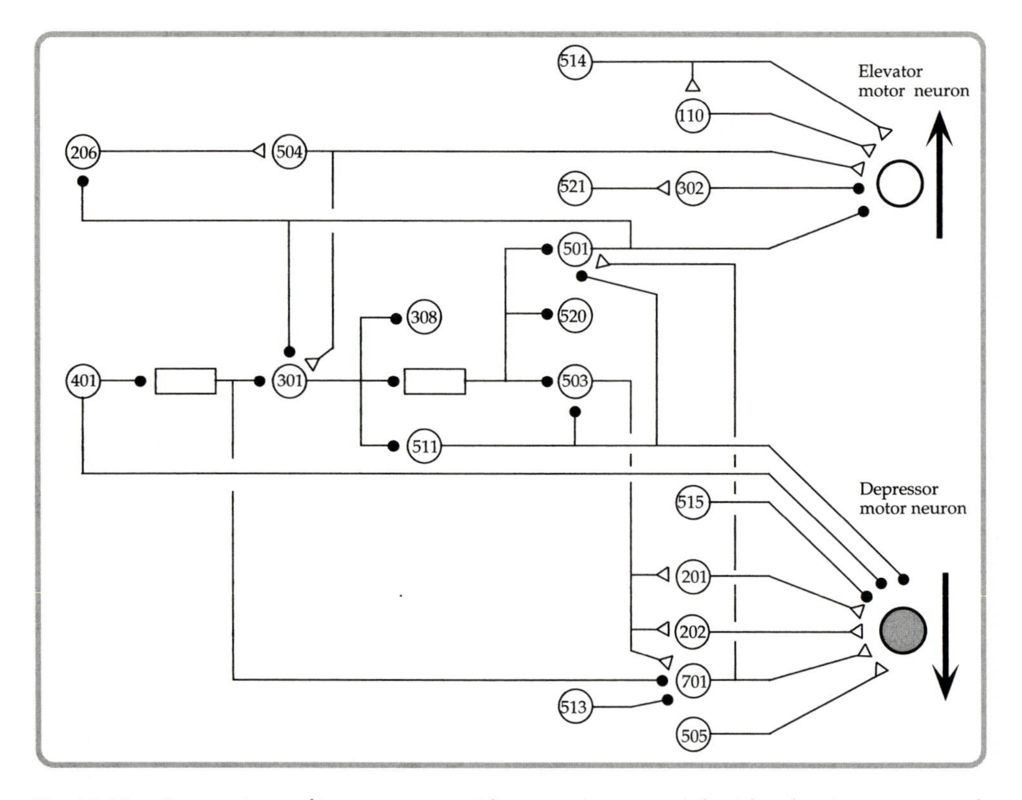

Fig. 11.13 Connections of interneurons with motor neurons and with other interneurons that may underlie the generation of the motor pattern for flying. Based on Robertson and Pearson (1985b). Excitatory connections are indicated by open triangles, inhibitory ones by filled circles. The rectangles indicate unknown pathways by which one interneuron may affect another.

1. The spikes in 301 evoke spikes in 501 after a delay.
2. The spikes in 501 then feed back to inhibit 301.
3. When 301 is inhibited, the inhibition of 501 will resume and it will therefore also stop spiking.
4. 301 now spikes again as it is no longer inhibited by 501.

Support for such a scheme comes from experiments where tonic spikes that are induced in 301 by injected current can sometimes cause bursting in 501, but blocking the delayed excitation with picrotoxin does not abolish the ability of the whole network to produce oscillations. Moreover, longitudinal section of the mesothoracic ganglion at its midline must severely damage the 301 interneurons and yet an apparently normal flight motor pattern is still generated (Ronacher *et al.*, 1988). These results show how difficult it is to ascribe particular roles to a small set of interneurons and to a few connections amongst the many, especially when they are not known in the required detail. In addition to these connections with each other, interneurons 301 and 501 are also embedded in a web of connections with other interneurons, which surely implies that many pathways must operate in parallel.

From these still imperfect details of the connections between interneurons we can extract some partial explanations of how the rhythmic drive is delivered to the sets of motor neurons (Fig. 11.13).

1. The fixed delay from the elevator to the depressor motor spikes could result from 504 directly exciting both the elevator motor neurons and interneuron 301, which then causes a delayed excitation of the depressor motor neurons *via* interneurons 503 and 201.
2. The constant duration of the burst of depressor spikes could result if 301 indirectly excites them and at the same time removes the inhibition from them by 511 thus delimiting a set period when they can spike.
3. Wind to the head will excite 206. This then excites 504 which, in turn, excites elevator motor neurons and interneuron 301. The excitation of 301 will lead to a delayed excitation of the depressor motor neurons as in steps 1 and 2, and an inhibition of the elevator motor neurons through interneuron 501.

This painstaking analysis of the connections made by the interneurons has provided some glimmers of what the mechanisms are but little insight into the guiding design principles of the network, for there are no operating rules that can be extracted from what is known so far. It is not possible to decide whether there are networks in each half of the meso- and metathoracic ganglia that can generate the rhythm coordinated by distributed interneurons, or whether all the pattern-generating interneurons are distributed. Too little is known about the properties and connections of the interneurons, and the schemes as extracted above tend to push the limited data too far. What is surprisingly missing from the proposed networks is a contribution from local interneurons (both spiking and nonspiking), given their widespread and critical roles in the networks that organise movements of the legs (*see Chapters 8 and 7*).

A continuation of this approach could provide deeper insights by addressing and then analysing selected issues in the control of flight. It seems reasonable to suppose that a good proportion of the interneurons has already been identified and that access to some of them at least could become routine, but the challenge is to design tests that will reveal their part in flight without trying to ascribe to them a particular action. Reluctance to pursue this important approach is pervaded by the fear that a network with so many interneurons will not be understandable, but my plea is not for the elaboration of an unsatisfying circuit diagram, but for analyses of particular sets of connections, and synaptic and membrane properties that will enable testable predictions to be made.

11.5.5. Physiological properties of the interneurons

By itself, more details of the connectivity of the interneurons would not provide a satisfactory explanation of how the flight motor pattern is produced. More quantitative information must be obtained on the properties of the synaptic connections and their dependence on time and the presence of neuromodulators, and the cellular properties of the interneurons and the motor neurons. We know surprisingly little about the properties of the components of the flight system, save that some of the interneurons can show plateau potentials (*see Chapters 3 and 5*). We can extrapolate from what we know of the properties of other insect neurons, or even neurons in general, about some likely mechanisms, but this is no substitute for an analysis of the mechanisms that actually exist and of the context in which they are likely to be expressed and have an effect.

11.5.6. Flight-initiating interneurons

Flight is often a means of escape, launched by jumping, so it is not surprising to find that it can be initiated by a number of different stimuli, such as wind to the head or to the cerci, and by visual and auditory stimuli (*see Chapters 9 and 10*). Interneurons have therefore been sought which could provide the link between these diverse releasing stimuli and the interneurons that generate the flight movements. The problem is again to ascribe such an action to an individual interneuron, or even a group of interneurons, when many other interneurons are active in parallel.

On each side of the mesothoracic ganglion is a group of 3–5 interneurons (the 404 interneurons) that have distinctive morphologies (Fig. 11.8) and that may be able to initiate and maintain flight (Pearson *et al*., 1985b). The group may consist of subsets based on their differing responses to different sets of sensory receptors; some are excited by touching the abdomen and receive inputs from unspecified wing receptors, whereas others are inhibited by inputs from the abdomen and receive no input from the wing receptors. Within the mesothoracic ganglion, each forms numerous branches that have both input and output synapses (Watson and Burrows, 1983) before the axon loops to ascend anteriorly to unknown destinations.

In fictive flight induced by a wind stimulus to the head, and in which sensory feedback from the wings is abolished, one of these interneurons initially spikes at a high frequency and then continues to spike tonically during a flight sequence. None of these interneurons receives an input from the interneurons generating the rhythm, but the pattern of inputs that they receive from unidentified sense organs on a wing suggests that in normal flight they would spike rhythmically. If, however, flight is induced by a wind stimulus to the abdomen, some at least of these interneurons are hyperpolarised and spike only sporadically, and therefore could then play no role in either initiating or maintaining flight. This further indicates that these interneurons do not act as a uniform population.

Depolarising current injected into one of these interneurons to evoke spikes of higher frequency than are normally observed at the start of flight can sometimes initiate a flight motor pattern. Furthermore, the probability of eliciting flight by a wind stimulus to the head can be reduced if one of the interneurons is hyperpolarised to prevent spikes. These results suggest that the interneurons may normally need to act together to initiate flight, but this argument is weakened if it is accepted that the processing of information from different sets of sensory receptors is subdivided amongst different members of this small group; such a subdivision would leave few interneurons in each set able to act in concert.

The output connections made by these interneurons are not known. They have few if any axonal branches in the mesothoracic ganglia and the extent of their axonal projections in more anterior ganglia is not known. They cannot connect directly with metathoracic motor neurons, are unlikely to connect with mesothoracic motor neurons and have been shown not to connect with prothoracic flight motor neurons (Watson and Burrows, 1983). Similarly, they cannot connect directly with the majority of flight interneurons in the metathoracic ganglion complex. They could interact locally with mesothoracic interneurons through the output synapses that occur intermingled with their input synapses on their neurites. If so, then a role has to be sought for the anterior projections of the axon. The inference then would be that interneurons in the brain or suboesophageal ganglion are activated so that the initiation of flight results from a complex series of interneuronal looped pathways. The action of these interneurons is thus suggestive of some role in the processing of the descending or ascending sensory information coded by intersegmental interneurons and the networks that generate flight. Under some circumstances they may be able to initiate flight, but they are not necessary for initiation to occur.

Interneurons have also been identified in all three neuromeres of the suboesophageal ganglion that participate in flight and may play some part in initiating it (Fig. 11.10) (Ramirez, 1988). Some, like interneurons 288 and 388 are excited indirectly by a 404 interneuron and spike tonically during fictive flight, whereas interneuron 488 also spikes tonically but excites 404 indirectly through other interneurons in the brain or suboesophageal ganglion, and can, when depolarised with injected current, sometimes initiate a few cycles of flight. Interneuron 397 directly excites a front wing elevator motor neuron, but although excited at the start of flight, it is then inhibited. The contribution of these interneurons to flight further emphasises the widespread nature of the pathways that are involved.

Flight can also be initiated by wind stimuli to the cerci and in both locusts and cockroaches some of the giant interneurons are implicated in the transmission of this information to the flight motor circuitry in the thorax (*see Chapter 10*).

11.6. SENSORY EFFECTS ON FLYING

Flying through the unpredictable and turbulent medium that is the air is a hazardous undertaking requiring continuous adjustments of wing movements if a locust is to remain airborne, stable flight is to be maintained, and the particular goal of a flight achieved. Continuous sensory monitoring of the repetitive movements of the light and fragile wings that can occur for long periods is essential to avoid potentially damaging strains in them and to ensure stable flight. Moreover, as in all other movements, the performance of the muscles and the moving parts must be monitored by sensory receptors, because it cannot be assumed that they will always work with the same reliability and linearity. To fly successfully, therefore, a locust must first initiate flight in response to the appropriate sensory signals, control the direction and velocity of its flight path in response to exteroceptive signals, ensure the correct and most efficient execution of wing movements at each wing beat as a result of proprioceptive signals, and finally stabilise its flight about the three major axes of its body (pitch, roll and yaw) through a combination of all these sensory signals.

Most parts of the body, from the head to the abdomen and including all the legs, also participate in specific movements associated with flight, and experience vibrations as the wings are flapped. It would seem likely that a large proportion of the sensory receptors over most parts of the body will be stimulated during flight, and not just those associated with the wings. The widespread consequences of the wing movements will therefore tend to blur any easy distinction between a mechanoreceptor that acts as an extero- or as a proprioceptor, so that it is difficult to divide the sensory effects on flight into these two categories. The proprioceptors of the wings will certainly be excited at each wing beat, but the consequences of these movements may also modulate the stimuli that excite the exteroceptors.

The proprioceptors are usually considered to be those directly associated with the joint between a wing and the thorax, the most extensively studied being the wing stretch receptors, and the receptors in and on the tegulae. The contribution of receptors in the thorax itself, such as the chordotonal organs innervated by nerve 2 (*see Chapter 8*), has been almost entirely overlooked, even though these receptors must also be stimulated rhythmically at each wing beat by the distortions of the thorax caused by the contractions of the flight muscles. They could thus act as true proprioceptors for flight movements.

The task in analysing the sensory neurons from receptors associated with the wings is to establish the sorts of information they provide to the central neurons, and the effects they can have on the flight movements. This information then has to be integrated with that from receptors on the head and elsewhere on the body, and with the signals that are generated by the connections between the central neurons themselves. It is essential at all times to avoid the tendency to ascribe greater importance to those sensory neurons that are easy to study than to those that are less accessible.

11.6.1. Stretch receptors

The stretch receptors at the hinges of the wings have long been regarded as important elements in generating and controlling the flight motor pattern. This has largely stemmed from the large size of their axons which makes their spikes prominent in recordings from the wing nerves (N1). Prominence is then readily associated with importance. While this may have overemphasised their contribution relative to other wing sense organs, they do spike during each wing beat and provide rhythmical information about each elevation movement. When stimulated at the correct phase of the wing beat cycle, the combined input from the stretch receptors of two wings can raise the frequency of the flight motor pattern, entrain it to the frequency of stimulation, and reset the rhythm. The sensory neurons from each of the four stretch receptors connect with motor neurons and interneurons involved in producing the motor pattern for flight.

11.6.1.1. *Structure*

A wing stretch receptor is a single sensory neuron with its cell body embedded in an elastic strand that is slung from an internal cuticular wall of the thorax to a sclerite in the wing articulation (Fig. 11.14). As the wing is elevated, the strand is stretched and the dendrites of the sensory neuron initiate spikes that are conducted towards the central nervous system. The axon of a stretch receptor is up to 10 μm in diameter and is therefore amongst the largest of all those of sensory neurons from a wing. As it approaches the central nervous system, the axon of a receptor from a front wing branches to enter both the prothoracic and mesothoracic ganglia, and the axon from a hind wing enters the meso- and metathoracic ganglia. An axon from a front wing receptor also extends to the metathoracic ganglia through the ipsilateral posterior connective. The stretch receptors are the only sensory neurons associated with the wings to have this pattern of axonal branches. Within the ganglia the axon branches to form an elaborate but characteristic array of fine branches (Fig. 11.14) in regions of neuropil to which the motor and interneurons involved with flight also project.

11.6.1.2. *Action*

In tethered flight, a stretch receptor produces up to 15-20 spikes during each wing beat, depending on the extent of elevation, with instantaneous frequencies of more than 500 Hz and hence a minimum interval of less than 2.0 ms (Möhl, 1985b). Changes in the wing movements are signalled by alterations in the number of spikes, their instantaneous frequency, the duration of the burst of spikes, and the phase of the burst in the wing beat cycle (Fig. 11.14). When the receptor is isolated mechanically from the hinge joint and stretched, it becomes clear that it is coding the amplitude, and not the rate of change (velocity), of stretch of its elastic strand (Pfau *et al.*, 1989). It is thus a tonic receptor signalling the position of the wing.

Most of the experiments that have sought to simulate wing movements with imposed movements have failed to reproduce the high frequencies seen in normal flight. Part of the problem is the difficulty of reproducing the natural movements of the wing articulation and the active tension of the muscles, but a large factor is the sensitivity of the receptor to temperature. Movements of 20 Hz imposed at temperatures approaching those of the thorax in flight (for example, 32°C) produce frequencies comparable to those in normal flight. This sensitivity to temperature will, however, mean that the signals from the stretch receptors at the start of flight could give ambiguous information about the positions of the wings until the temperature of the thorax has stabilised.

The spikes of a stretch receptor occur at a constant phase of the wing beat cycle, with the carrier frequency of the signal being the frequency of the wing beats, and the signal itself being the burst of spikes. Such an arrangement of an ever present carrier frequency means that there is no threshold and that any change in the movements of a wing can be signalled immediately by a considerable phase shift of the spikes. To assess the phase of the stretch receptor signal, however, the output pattern of the motor spikes would need to be compared with the resulting sensory signals, and this can perhaps be accomplished by the connections that a stretch receptor makes with central neurons. It is doubtful whether the central nervous system does this, but instead is more likely to interpret the signal in the context of the immediate motor pattern.

11.6.1.3. *Synaptic connections with motor neurons*

A stretch receptor of a front wing synapses directly on most of the depressor motor neurons which control that front wing, and in a similar way, a stretch receptor of a hind wing synapses directly on most of the depressor motor neurons which move that hind wing (Burrows, 1975a). All the physiological evidence indicates that the connections are direct, excitatory and mediated by the release of a chemical transmitter that is probably acetylcholine. In keeping with the anatomical projections that are restricted to one side of the central nervous system, no connections are made with motor neurons moving the contralateral wing of the same segment. Connections are, however, made with the depressor motor neurons of the ipsilateral wing in the adjacent segment: a front wing stretch receptor synapses on hind wing depressor motor neurons, and a hind wing receptor on ipsilateral front wing motor neurons, although the amplitude of the EPSP evoked in the motor neurons of distant segments is smaller. This means that a front wing receptor makes synaptic connections in all three thoracic ganglia, and a hind wing receptor in two. Each physiological connection of a stretch receptor with a motor neuron is probably represented by many anatomical contacts, with estimates from the synapses seen with the electron microscope placing this number as high as 600 (Peters *et al.*, 1985).

What effects do these excitatory synaptic connections have? The direct inputs do not make the motor neurons spike when they are delivered in an animal that is not flying, even if the frequency of stimulation approximates that generated by a stretch receptor in flight. On this basis, a continuing but unsubstantiated interpretation is that the connections are weak and will be ineffective during flight. The 'strength' of a synaptic connection relates only to the context in which the observation is made, and to the summed input from all other sources that impinges on the target neuron at that time, and cannot therefore be extrapolated easily to other contexts. It is thus

Fig. 11.14 The stretch receptor at a wing hinge and its effects on neurons controlling the movements of the wings. A. Morphology of thoracic skeleton and the position of the stretch receptor. The boxed area is magnified on the right to show the insertion of the single stretch receptor sensory neuron and the many sensory neurons of the nearby chordotonal organ. B. During tethered flight, the stretch receptor spikes at each upstroke, with larger amplitude elevations evoking more spikes. C. Projections of the sensory neuron from the left front wing and right hind wing into the meso- and metathoracic ganglia. D. Connections of a hind wing stretch receptor with ipsilateral hind wing motor neurons. E. Effects of spikes in a stretch receptor on the action of flight interneurons. When the input is added an interneuron spikes more often at each wing beat and the waveform of depolarisation in an elevator motor neuron is altered. A is based on Pfau (1983), B on Möhl (1985b), E on Reye and Pearson (1987) and Pearson and Ramirez (1990).

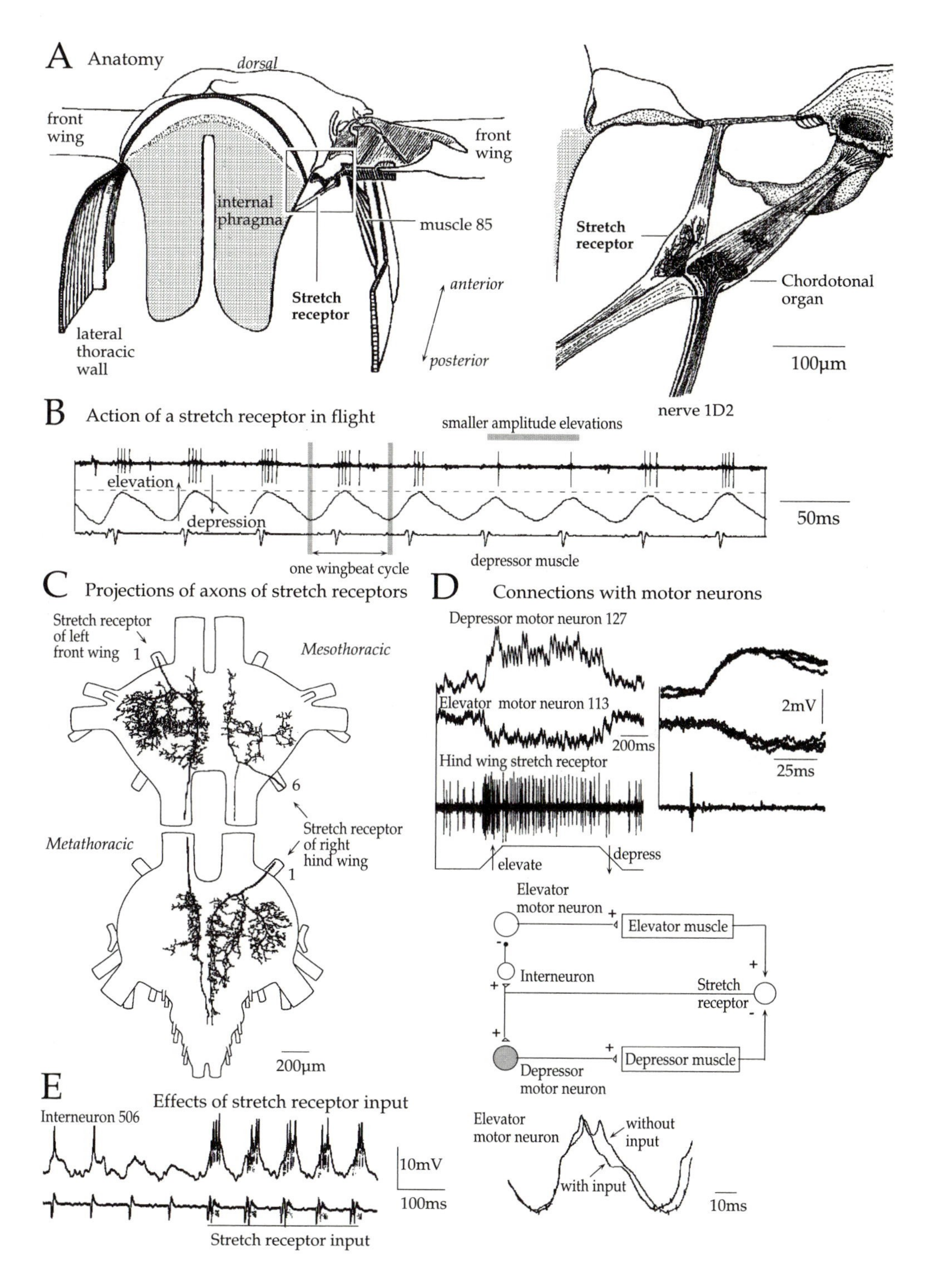
A Anatomy
dorsal
front wing
front wing
internal phragma
muscle 85
Stretch receptor
lateral thoracic wall
anterior
posterior
Stretch receptor
Chordotonal organ
100μm
nerve 1D2
B Action of a stretch receptor in flight
smaller amplitude elevations
elevation
depression
50ms
one wingbeat cycle
depressor muscle
C Projections of axons of stretch receptors
Stretch receptor of left front wing 1
Mesothoracic
6
Stretch receptor of right hind wing 1
Metathoracic
200μm
D Connections with motor neurons
Depressor motor neuron 127
Elevator motor neuron 113
200ms
Hind wing stretch receptor
2mV
25ms
elevate
depress
Elevator motor neuron
Elevator muscle
Interneuron
Stretch receptor
Depressor muscle
Depressor motor neuron
Elevator motor neuron
without input
with input
10ms
E Effects of stretch receptor input
Interneuron 506
10mV
100ms
Stretch receptor input

inappropriate to dismiss these connections as 'weak' simply because they are subthreshold for eliciting spikes under one set of conditions. The connections exist and would be expected to change the probability that a motor neuron will spike under particular circumstances, most notably in flight, unless they are actively reduced by processes such as presynaptic inhibition (*see Chapter 7*).

In parallel to these excitatory connections with depressor motor neurons, the spikes in a stretch receptor also evoke inhibitory potentials in elevator motor neurons of its own wing. These IPSPs occur with a longer latency than the EPSPs in depressors and are caused by interposed interneurons, which reverse the sign of the sensory signal. The connection with these interneurons must be reliable and of high gain as there can be a close linkage between the sensory spikes and the IPSPs in the elevator motor neurons (Fig. 11.14). The excitatory connections with the depressor motor neurons and the inhibitory connections with the elevators form a simple negative feedback loop that could regulate the extent of elevation movements; the more the wing is elevated, the greater the spike response of the receptor and the greater the excitation of the depressors and inhibition of the elevators.

Similar patterns of connections occur with the flight motor neurons of *Locusta migratoria*, with small differences, perhaps explained by the fact that the intersegmental connections were not sought using signal averaging; the failure to find IPSPs in elevators may be explained by the different gain of the connection with the interneuron(s) that reverse the sign of the stretch receptor input (Reye and Pearson, 1987).

11.6.1.4. *Synaptic connections with interneurons*

In addition to the widespread connections with motor neurons, a stretch receptor also synapses with many interneurons involved in flight (Reye and Pearson, 1987). Of the 21 interneurons found with direct inputs from a stretch receptor, 16 spike in time with depression and 14 receive convergent inputs from more than one stretch receptor, but only one of these interneurons is capable of resetting the flight motor pattern. The most common form of convergence is from the front and hind wing on one side of the body. This convergence onto the interneurons suggests that any influence on the movement of an individual wing may have to rely more on the connections that are made with the motor neurons of that wing than with the interneurons.

The stretch receptors have a varied effect on the pattern of spikes in the interneurons. In some, the number of spikes at each cycle is increased and the duration of the period of spiking is increased, while in others that are depolarised strongly in time with the flight motor pattern, the effect is not detectable. In interneurons that normally spike in time with elevation, the stretch receptor input evokes an additional burst of spikes in time with depression.

11.6.1.5. *Effects on the flight motor pattern*

Intracellular recordings from depressor motor neurons during the flight motor pattern show that the spikes from front wing stretch receptors arrive in the central nervous system at a time when the depressor motor neurons are repolarising and could therefore sustain but not initiate their activity (Pearson and Ramirez, 1990). In these experiments, however, the thoracic temperature is low so that the conduction velocity of the spikes in the axons of the stretch receptor will also be low, as will the number and frequency of spikes. In freely flying animals, however, the elevation of the wings is greater and the thoracic temperature is higher so that the stretch receptors should produce more spikes at higher frequencies and these should occur earlier relative to the spikes in the depressor motor neurons.

Stimulation to mimic the stretch receptor input slows the rate of repolarisation of the depressors, decreases the level of hyperpolarisation that is reached in the elevation phase of the cycle, and increases the rate of rise of the next depolarisation. In the elevator motor neurons, the duration of the depolarisation close to its peak is reduced. One consequence of these effects is to increase the frequency of the flight motor pattern through an antagonism in the depressors between an excitatory drive from the stretch receptors and an inhibitory drive from interneurons. Any delay in the onset of depressor spikes will allow the wing to elevate further. This would evoke more spikes in the stretch receptors whose central connections would oppose the repolarisation of the depressors and advance the onset of the next cycle of depressor spikes.

Initial experiments suggested that stimulation of a stretch receptor would raise the frequency of the flight motor pattern by about 25% and with a time constant of some 2 s (Wilson and Gettrup, 1963; Wilson and Wyman, 1965). This effect was obtained in completely deafferented animals by simultaneously stimulating nerves from the front wing and hind wing on one side of the body, either with frequencies much higher than that of the rhythm itself, or with random stimuli. The results were interpreted in terms of the action of the stretch receptor alone, although other receptors must also have been stimulated that would normally not spike at the same time. It is therefore difficult to equate these effects with a normal sensory inflow, or to attribute them directly to the stretch receptor. If the stretch receptors of the two front wings are stimulated selectively so that they are the only wing receptors to be excited (stimuli are applied to N6 of the prothoracic ganglion), the frequency of the flight motor pattern in a dissected animal is raised (Pearson *et al.*, 1983; Reye and Pearson, 1988). This effect still occurs if the animal is deafferented, but the period of the rhythm is 5–10 ms shorter if feedback is permitted from the movements of the stumps of the hind wings. The effect is critically dependent on the timing of the stimuli relative to the spikes in depressor motor neurons, indicating that the input is exerting a cyclically phase-dependent reaction. Single stimuli also have a weak resetting effect on the flight motor pattern. Furthermore, stimulation will also, within certain boundaries, entrain the flight motor pattern to the imposed frequency of stimulation. Stimuli to both front wing stretch receptors that differ in frequency from the 10 Hz flight motor pattern by 4–6 Hz (or 40–50%) will cause entrainment if feedback from movements of the hind wings is permitted (Pearson *et al.*, 1983). The entrainment range is, however, reduced to 1 Hz (10%) if feedback from the hind wing is abolished (Reye and Pearson, 1988). Similar effects are produced by stimulating nerves from the hind wings that contain the axons of the stretch receptors, but no entrainment is produced by stimulating a single stretch receptor, or by simultaneously stimulating the stretch receptors from the front and hind wings on the same side of the body. The implication is that the network of neurons producing the flight motor pattern must receive input from both sides of the body if sensory influences such as these are to be expressed in changes of the motor output, but adjustments of the movements of an individual wing in an intact animal must still be possible.

During tethered flight in intact animals, variations in the relative timing of the spikes of a stretch receptor during the wing beat cycle, the number of spikes at each wing beat, and the instantaneous frequency of those spikes, all correlate with the kinematics of the wing movements, as illustrated by two examples (Möhl and Neumann, 1983). First, during a turn to the right, the left basalar motor neuron spikes later in the wing beat cycle whereas the left stretch receptor spikes earlier and with more spikes at higher instantaneous frequencies. Here, the increased excitation of the stretch receptor is attempting to pull the basalar spikes forward to an earlier time in the wing beat cycle, so stabilising the yaw movement by a negative feedback loop. Second, if the depressor spikes are delayed in one wing beat cycle then the wing will stay elevated for a longer time and this will be signalled by a longer burst of stretch receptor spikes. These signals will correct the motor pattern at the next wing beat by a negative feedback loop that ensures that the depressors will be more strongly excited and thus spike earlier and the elevators will be more strongly inhibited and thus spike later.

These results emphasise the importance of the phasic information provided by the rhythmical pattern of spikes from the stretch receptors and provide no evidence that this input has a tonic role. The results are therefore in accordance with the behavioural observations when the flight movements are disturbed, and with expectations from the known patterns of connections of the stretch receptors with interneurons and motor neurons that are involved in producing the flight motor pattern.

11.6.2. Tegula

A tegula is a knob-shaped membranous protuberance from the anterior of a wing articulation that is 0.6 mm long at the front wings and 1.0 mm long at the hind wings (Fig. 11.15) (Kutsch *et al.*, 1980). On the posterior surface is a patch of about 45 hairs approximately 45 μm long that are brushed by the wing as it moves downwards. Inside the tegula is a chordotonal organ that contains some 30 sensory neurons embedded in a strand slung from its dorsal wall. The proximity of the tegula to the basalar sclerite suggests that the chordotonal organ might monitor the contractions of the basalar muscles during the downstroke. All these receptors are innervated by N1C1a and have axons ranging in diameter from 1 to 10 μm (Altman *et al.*, 1978). During flight, spikes in this nerve begin just after the start of the downstroke, precede the spikes generated by elevator motor neurons, and may continue for much of the downstroke (Fig. 11.15) (Neumann, 1985). Their role is thus to signal the impending end of the downstroke and their action may be to ensure that the next elevation phase follows promptly, or is even advanced, so that the appropriate amount of lift is generated.

Sensory neurons from both the hairs and the chordotonal organ of the tegula spike during the downstroke, but their relative contributions to the movements of a wing are not known, so that effects of the tegula on the flight motor pattern are thus attributable only to its complete array of receptors. If there are more spikes in these receptors then pronation of a wing is greater, and if there are fewer then pronation is less. Similarly, the projections of the axons of these receptors into the central nervous system have not been separated (Fig. 11.15). Most project to dorsal regions of neuropil, as do the stretch receptors, and many have branches that extend to more than one thoracic ganglion (Tyrer and Altman, 1974). So far, it seems that the tegulae of the hind wings have a more important influence on flight than do the ones associated with the front wings, but why this should be so is unexplained.

11.6.2.1. *Synaptic connections with motor neurons*

Electrical stimulation of nerve N1C1a of a hind wing distinguishes two populations of afferents on the basis of their different conduction velocities, but it is not known whether these populations represent the hairs and the chordotonal afferents, or two populations of chordotonal afferents. Nevertheless, sensory neurons in both populations make direct, excitatory connections with hind wing elevator motor neurons which control movements of the wing that is stimulated (Pearson and Wolf, 1988) (Fig. 11.15); those afferents with the faster conduction velocity appear to connect with all elevator motor neurons that have been examined. The connections are powerful enough to make the motor neurons spike to a single volley stimulation, but normally the afferents would not spike simultaneously in this way. The input from these sensory neurons makes the elevator motor neurons spike earlier in the wing beat cycle than they would in a deafferented animal, in which the timing is derived solely from the central connections between neurons. Longer latency excitatory connections are also made with contralateral hind wing elevators, and with both ipsilateral and contralateral front wing elevators. The longer latency and the compound nature of the evoked potential indicate that an intervening interneuron spikes a few times, or that several interneurons acting in parallel are involved in these pathways. Inhibitory responses are elicited in depressor motor neurons controlling all four wings, which suggests the involvement of interneurons that reverse the sign of the afferent signal. The sensory neurons from the tegulae of the front wings make the same pattern of connections and again are powerful enough in a locust that is not flying to evoke spikes in elevator motor neurons (Burrows, 1976).

11.6.2.2. *Synaptic connections with interneurons*

Some of the interneurons distributing the signals from the tegula afferents have been identified from the many that receive either a direct or an indirect input from these sensory neurons. The general effect is to excite interneurons which, in turn, excite elevators, and to excite those which inhibit depressors (Fig. 11.15). For example, interneuron 566 receives input from the tegula of the left and right hind wings and

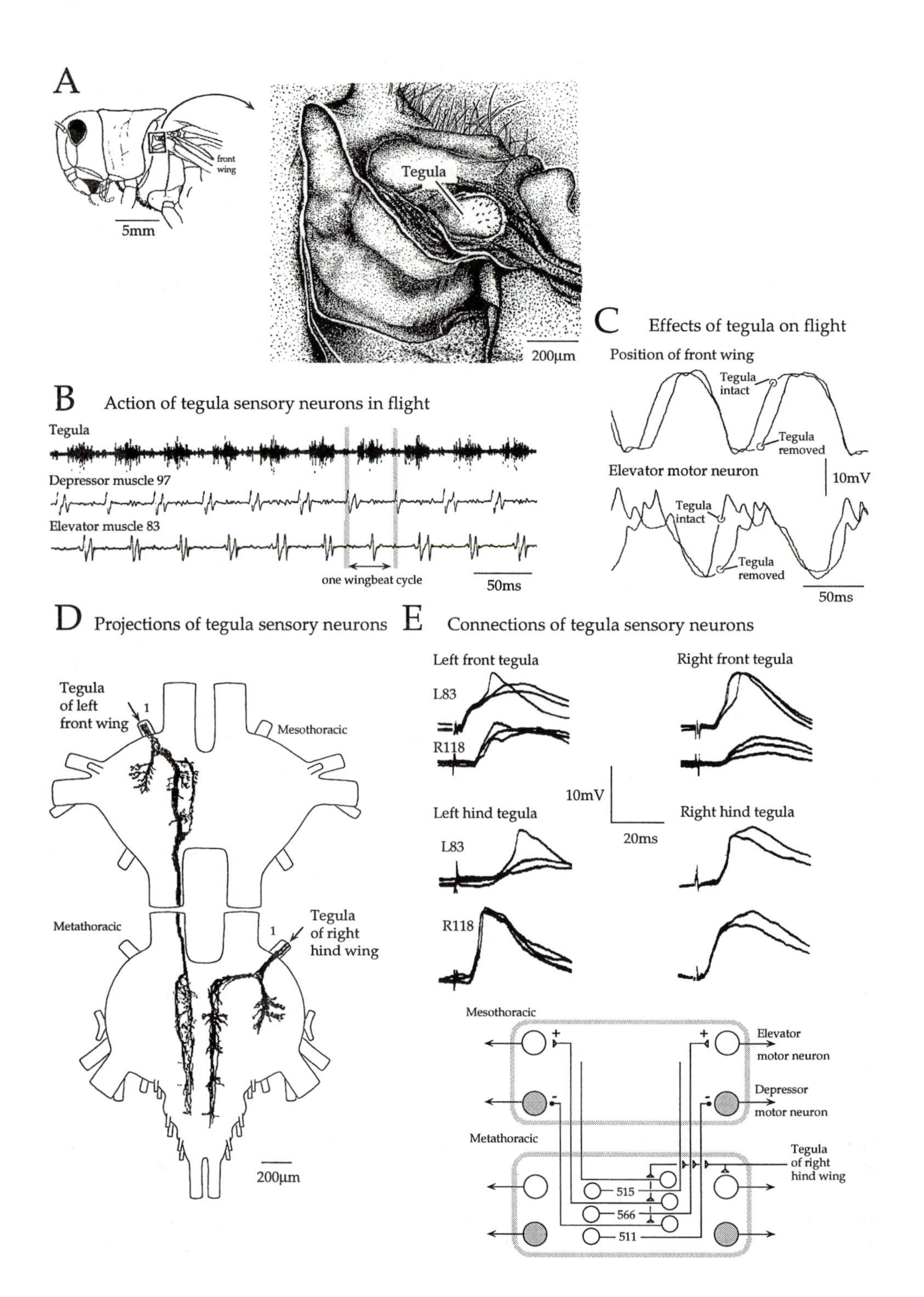
A
front wing
5mm
Tegula
200μm
B Action of tegula sensory neurons in flight
Tegula
Depressor muscle 97
Elevator muscle 83
one wingbeat cycle
50ms
C Effects of tegula on flight
Position of front wing
Tegula intact
Tegula removed
Elevator motor neuron
10mV
Tegula intact
Tegula removed
50ms
D Projections of tegula sensory neurons
Tegula of left front wing
1
Mesothoracic
Metathoracic
1
Tegula of right hind wing
200μm
E Connections of tegula sensory neurons
Left front tegula
Right front tegula
L83
R118
10mV
20ms
Left hind tegula
Right hind tegula
L83
R118
Mesothoracic
Elevator motor neuron
Depressor motor neuron
Metathoracic
Tegula of right hind wing
515
566
511

makes excitatory connections with elevator motor neurons in the mesothoracic ganglion contralateral to its soma. In a deafferented animal expressing the flight motor pattern, the membrane potential of this interneuron is depolarised rhythmically due to connections between central interneurons, but does not spike (Pearson and Wolf, 1988). When sensory feedback is present, however, bursts of spikes are produced at the start of each wing elevation, the additional synaptic input presumably caused by tegula afferents, but also aided by the activation of currents that lead to plateau potentials (Ramirez and Pearson, 1993) (*see Chapter 3*). Interneurons 511 and 515 also receive a similar pattern of inputs from the tegula afferents but make inhibitory connections with mesothoracic depressors. Abolishing the spikes in these interneurons that are evoked by stimulation of the tegula reduces but does not abolish the synaptic potentials in the mesothoracic motor neurons. This confirms the suggestion that several interneurons in parallel distribute the tegula signals: interneurons 504 and 514 are probable candidates for this role in exciting the elevators. None of the connections that contribute to the excitation or inhibition of motor neurons in the same segment as the stimulated tegula have yet been identified, although it is possible that the same interneurons are involved.

The widespread connections of the sensory neurons with the interneurons is, in part, a reflection of the widespread distribution of their terminals but is also a reflection of the interconnections amongst the interneurons themselves, and the consequent ability of a powerful input to propagate around the network.

11.6.2.3. *Effects on the flight motor pattern*

Removal of the tegula on both hind wings increases the interval between the start of spikes in depressors and the start of spikes in the elevators (D-E latency) by up to 50%, increases the duration of the elevator activity, and reduces the wing beat frequency by up to 20% (Pearson and Wolf, 1988). Electrical stimulation of the nerve containing the sensory neurons from the tegula resets the wing beat to the elevation phase. By contrast, however, removing the tegula on both front wings has little effect on the flight motor pattern (Büschges and Pearson, 1991). This is a surprising result, given the efficacy of the connections that the sensory neurons from the front wing tegulae make with motor neurons of all four wings.

Fig. 11.15 The tegula receptors and their effects on neurons controlling the movements of the wings. A. The position of the tegula associated with a front wing. The pronotum is deflected forward to reveal the articulation of the wing. The boxed area is enlarged to show the small protuberance that is the tegula. B. Action of the sensory neurons from the tegula during tethered flight. C. Effects of spikes in sensory neurons from the tegula on the movements of the wings and the action of an elevator motor neuron. The depolarisation of the motor neuron is advanced when tegula spikes are present. D. Projections of the sensory neurons from the tegulae of the left front wing and right hind wing into the meso- and metathoracic ganglia. E. Connections of sensory neurons from the tegulae of all the wings with motor neurons and interneurons involved in generating the motor pattern for flying. A and B are based on Neumann (1985), C on Wolf (1993), D on Büschges *et al.* (1993).

The depolarisation of an elevator motor neuron during the flight motor pattern occurs in two stages: an initial rapid depolarisation and a later, more variable depolarisation. The initial rapid depolarisation is lost when the tegulae are removed, but can be restored by electrical stimulation of the nerves from both hind wing tegulae at the time in the wing beat cycle when they would normally spike (Pearson and Wolf, 1988; Wolf, 1993). Stimulation at other times during the wing beat cycle resets the rhythm by initiating a rapid depolarisation in elevator motor neurons and terminating depressor spikes. All this suggests that the sensory effects are a powerful determinant of their actions on the elevator motor neurons, but that their influences on the depressor motor neurons are less profound. They confirm the original suggestion of Burrows (1976) that input from these receptors is important in initiating the spikes in the elevator motor neurons. An earlier activation of the elevators will advance the timing of the upstroke of the wing and prevent it dwelling too long at the fully depressed position where it is likely to generate negative lift and also drag (Wolf, 1993). Removal of the tegula of the same wing confirms this role by abolishing the early depolarisation of the elevators, delaying the onset of elevation of the front wings by as much as 13ms, and consequently reducing the amount of lift that is produced. The tegulae therefore form part of a feedback mechanism that regulates the timing of elevation in order to regulate the amount of lift that is generated (Zarnack, 1982). Further actions are, however, to be expected when the roles of the front wing tegulae are elucidated and when the relative contribution of the chordotonal organ is understood. For example, do some of its receptors provide information about the velocity of the wing movements during depression?

11.6.3. Campaniform sensilla

The wings are subjected to many distorting forces when they are moved through the air. Moreover, they are not rigid structures and are deliberately twisted to change the angle of attack and hence the amount of lift that is produced during flight. The strains set up by the twisting movements are monitored by campaniform sensilla, which are concentrated into groups at the base of the wings along the anterior veins, and scattered as individuals along the primary veins and in the membranes between them (Fig. 11.16). The front and hind wings each have a group of 60–70 sensilla on the ventral surface of the subcostal vein with their long axes oriented parallel to the long axes of the wings (Gettrup, 1966). The front wings have two additional proximal groups of sensilla with their long axes arranged like a fan. The number of campaniform sensilla and the complexity of their distribution probably means that different features of wing distortions are monitored by different arrays. Destroying the front wing receptors interferes with the ability to rapidly correct angular deviations about all three axes of the body, so that flight becomes unstable and unsustainable. If the receptors on the hind wings are destroyed, a locust can still fly in a well coordinated fashion, but the twisting of the front wings is abolished and with it the ability to maintain constant lift.

The campaniform sensilla produce bursts of spikes at each wing beat, beginning just before the start of the downstroke and ending at the start of the upstroke, so that they signal distortions of the wings caused by the aerodynamic lift and thrust forces that are primarily generated during this depression phase of the wing movements. They are excited by supination of a wing and inhibited by pronation (Fig. 11.16) giving a phasic response to each movement and a prolonged but adapting response if the wing is held in a new position (Elson, 1987a). In one set of experiments (Elson, 1987a), forcibly bending a front wing excited elevator motor neurons and inhibited depressors. In a second set of experiments (Horsmann and Wendler, 1985), the opposite effect was produced when one wing was moved, but the same effects when both front wings were bent simultaneously. This surprising contradiction in the effects of the campaniform sensilla on the flight motor neurons

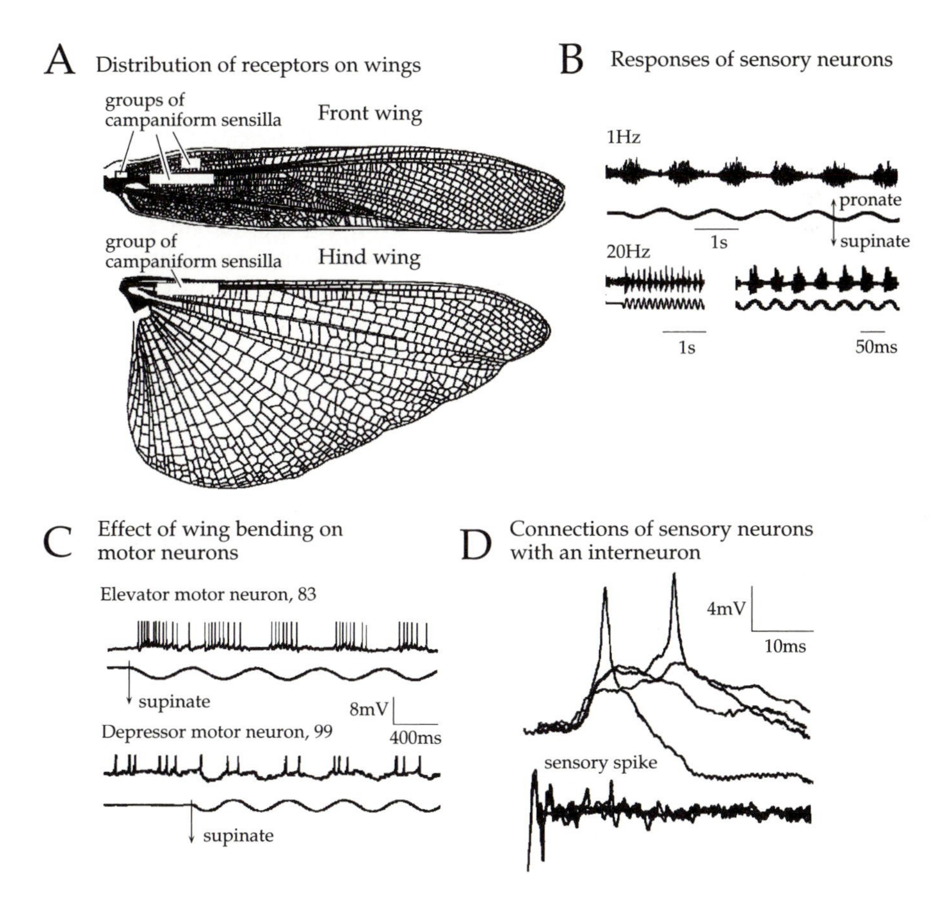

Fig. 11.16 Campaniform sensilla on the wings and their connections with neurons involved in producing the motor pattern for flying. A. Distribution of receptors on a front and a hind wing. B. Spikes are evoked by imposed supinations of a wing at different frequencies. C. Supination of a wing leads to spikes in an elevator motor neuron and hyperpolarisation of a depressor motor neuron. D. Spikes in sensory neurons cause a depolarisation and spikes in an interneuron involved in generating the motor pattern for flying. Based on Albert *et al.* (1976) and Elson (1987a,b).

presumably arises from the excitation of different sets of receptors in the two experiments. At best, the bending movements will inevitably also excite many of the hairs and bristles on a wing and may even excite receptors of the wing hinge (tegula, stretch receptor and chordotonal organ). This issue is not resolved because the connections of the sensory neurons from particular campaniform sensilla with the motor neurons and interneurons have not been elucidated. The excitatory effects on the motor neurons of the same wing could be direct, but the longer latency inhibition must involve interposed interneurons, both to explain the delay and to reverse the sign of the signal. Similarly, the effects on motor neurons of other wings that result from stimulating the campaniform sensilla of a front wing must also be caused by interneurons, as the sensory neurons project only to the ipsilateral half of their own segmental ganglion. A few of the interneurons that could distribute these signals to the neurons controlling the other wings have been identified (Fig. 11.16) (Elson, 1987b). An intersegmental interneuron in the metathoracic ganglion, for example, receives a direct synaptic input from sensory neurons of campaniform sensilla on one hind wing and has an axon that projects to the contralateral side of the mesothoracic ganglion. It could therefore be involved in controlling the lift generated by the front wings in response to input from the hind wing campaniform sensilla. This interneuron also receives signals from the brain about changes in light to the medial ocellus and to the ocellus on the side of its head contralateral to its cell body, and about movements of the head relative to the body, so that it could be integrating the sort of signals that are generated during a deviation from a desired course. It does not, however, synapse directly on any flight motor neurons, so that its effects on wing twisting must be mediated through further sets of interneurons.

These actions of at least some of the campaniform sensilla and their connections with the neurons generating and controlling flight mean that they will exert a maximal excitatory effect on the elevator motor neurons just as they start to spike at the bottom of a downstroke, and an inhibitory effect on the depressors with a longer latency so that it occurs before they begin to spike at the top of the upstroke. The increased input to the elevators will make them spike earlier on the same cycle of movement as the sensory signals are generated, while the spikes in the depressors will be delayed on subsequent cycles. The result is essentially a negative feedback loop, advancing the upstroke in the face of signals indicating increased twisting, or loading, that has resulted either from increased depressor activity or from external disturbances. By monitoring active twisting of the wing during depression, the campaniform sensilla may also set the profile of the wing by their effects on the pleuroaxillary motor neurons, and thus control the angle of attack; the less the wing is pronated during the downstroke, the greater the amount of lift that will be produced. The excitatory effects are powerful enough to evoke motor spikes, even in a locust that is not flying, and these evoked motor spikes follow repetitive mechanical stimulation of the receptors at the wing beat frequency (Wendler, 1978b). Repetitive mechanical stimulation of the receptors can also entrain the flight motor pattern to the imposed stimulus and can also reset the rhythm (Horsmann and Wendler, 1985).

The front and hind wings do not, however, contribute equally to the control of lift. As the body angle increases, the front wings are twisted so that their angle of attack remains constant, whereas the hind wings are twisted less so that their angle of attack also increases and will thus slow flight speed (Gettrup and Wilson,1964). The twisting of the front wings would keep their lift constant were it not for the reduced flight speed caused by the hind wings. The reduced lift of the front wings will thus be subtracted from the increased lift of the hind wings.

11.6.4. Chordotonal organs

A chordotonal organ at the hinge of each wing shares its origin with the stretch receptor on an internal wall of the thorax and inserts on a sclerite in the membrane of the wing joint (Fig. 11.14). It contains some 21 receptor cells that have axons of 2.5–4 μm in diameter in N1D2 (Gettrup, 1962). This nerve contains some 400 axons (Altman *et al.*, 1978) including the one from the stretch receptor, and many smaller axons from hairs on the thorax. Individual sensory neurons from a hind wing chordotonal organ project to the medial ventral association centre (mVAC) in the metathoracic ganglion and also have a branch to the mesothoracic ganglion. Sensory neurons from a front wing chordotonal organ project to both the meso- and the metathoracic ganglion and have a small branch to the prothoracic ganglion (Fig. 11.17).

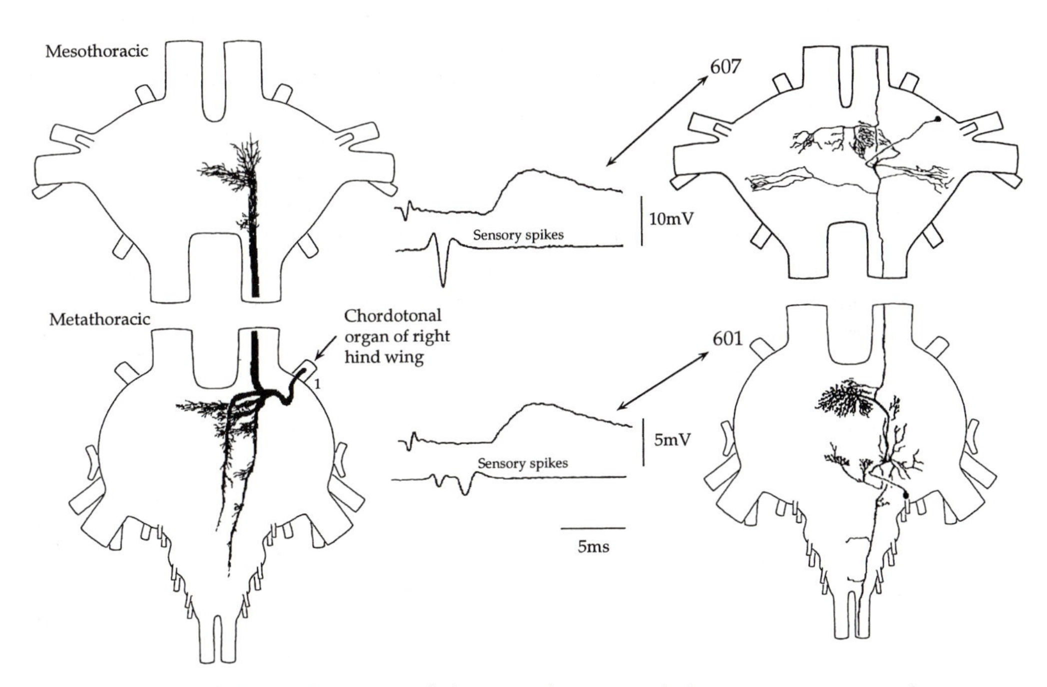

Fig. 11.17 Chordotonal organs of the wing hinges and their connections with neurons involved in generating the motor pattern for flying. The projections of sensory neurons from the chordotonal organ of the right hind wing are shown on the left. The recordings in the middle show the EPSPs evoked in interneurons by spikes in sensory neurons from a chordotonal organ of a hind wing. The interneurons from which these recordings were made are drawn on the right. Based on Pearson *et al.* (1989).

The summed spike discharge of all the chordotonal sensory neurons is maximal to sounds of about 3 kHz (Pearson *et al.*, 1989), but the tuning of individual neurons is not known. Interpretation of the responses of chordotonal organs is notoriously difficult as many, including even the femoral chordotonal organ in the leg (*see Chapter 8*), will respond to sound under the conditions necessary to record their sensory spikes. Many of the sensory neurons are also excited by movements of the wings, spiking maximally midway between the burst of spikes of the depressor motor neurons, but this response could be due to the sound that is generated by the wings as they are moved through the air. Indeed, the sensory neurons are not influenced by the static position of the wing and thus presumably provide no positional information. The flight motor pattern and the frequency of the rhythm in a deafferented preparation are not altered by evoking spikes in the chordotonal afferents, or by pulses of low frequency sound to which the sensory neurons respond best. The sensory neurons do not connect with flight motor neurons, or with flight interneurons (for example interneuron 511), but instead they connect with interneurons that also process sound and low frequency vibration from the leg. Direct connections are implicated with interneurons 601 (= TN1 of Romer and Marquart, 1984), 607 and 608 as a result of electrical stimulation of N1D2, but no correlation is demonstrated between individual sensory spikes and synaptic potentials in these interneurons.

The conclusion is that these sensory neurons have no influence over the flight motor pattern. Why then are these chordotonal organs located at the wing hinge if they are not coding a stimulus of relevance to the wing beat movements? There are two possibilities. First, subtle effects might be produced on neurons other than those surveyed. If these chordotonal organs detect low frequency sound or even vibration, it is not clear why they are needed in addition to the tympanum and the subgenual organ. In the moth *Actias luna* the wing hinge chordotonal organ consists of just three neurons, but again there is uncertainty as to its function, with suggestions that it may monitor wing movements during flight or simply the vibrations that are produced when warming up for flight (Yack and Roots, 1992). Second, the organs may have nothing to do with flight and may simply be serial homologues of those in the abdomen which form the ear in A1 and the pleural chordotonal organs in A2-7 (Meier and Reichert, 1990). Such a statement still begs the question of their function.

11.6.5. Bristles (hairs) on the wings

The dorsal and, to a lesser extent, the ventral surfaces of the wings bear an array of hairs. In the grasshopper *Melanoplus sanguinipes*, the dorsal surface of a front wing has some 850 and a hind wing some 400 hair receptors ranging in length from less that 15 μm to more than 250 μm (Albert *et al.*, 1976), whereas in *Locusta migratoria* there are estimated to be more than 12 000 hairs and bristles on both surfaces (Altman *et al.*, 1978). The function of these receptors during flying has not been tested, but it seems reasonable to suppose that the longer hairs could act as directional air movement detectors, or air pressure detectors. The shorter hairs are less likely to be displaced by air currents, but their arrangement along the main veins may indicate that they can signal distortions or folding of the wing. The small diameter of their axons and the concomitant slow conduction velocity of their spikes probably means that they cannot exert an influence within one cycle of the wing beat, but would instead have to act cumulatively over many cycles. When the wings are folded in a locust that is not flying, tactile stimulation of these receptors can lead to reflex movements of the legs.

11.6.6. Hair plates

A hair plate on the anterior surface of each of the first cervical sclerites in the neck, where they engage the back of the head, is stimulated by many movements, including rotation of the head (Goodman, 1959). Normally, locusts will fly so as to maintain a maximal amount of light on the dorsal part of the compound eyes, a so-called dorsal light reaction in which the head is first moved into the appropriate orientation and then the body is moved into alignment with the head. The ability of the locust to maintain this orientation is reduced by ablation of this hair plate.

11.7. MOTOR PLASTICITY

It is common to assume that repetitive motor patterns such as those of flight are rather rigid and not capable of adjustments to any long-term changes in sensory information or the effects of experience. Just how false these assumptions are is shown dramatically by the long-term changes that occur in response to sensory loss, and by the inherent plasticity in the flight motor pattern that can be demonstrated during normal flight.

11.7.1. Recovery of the flight motor pattern after loss of the tegulae

The organisation of the flight motor pattern is sufficiently plastic that it can recover from the disruptions caused by removal of the tegulae. If the tegulae of the two front wings are removed there is little disruption of the flight motor pattern. If the tegulae of all four wings are removed then over a 2 week period, the flight motor pattern gradually recovers, but the recovery is not complete. If the tegulae of the two hind wings are removed then recovery is complete in 80% of animals (Büschges and Pearson, 1991; Büschges *et al.*, 1992a,b). In these fully recovered animals, electrical stimulation of the front wing tegulae, which has little effect in normal animals, can act like the input from the hind wing tegulae and reset the flight motor pattern. Moreover, if the front wing tegulae are then destroyed, a further disruption of the flight motor pattern results that resembles the initial changes caused by removal of the hind wing tegulae. The suggestion is that when the hind wing tegulae are destroyed, their functional role is replaced by the front wing tegulae, which have similar patterns of branches in the central nervous system and similar patterns of connections with interneurons and motor neurons. It must therefore follow that the partial recovery seen when all tegulae are destroyed is explained by their role being subsumed by other neurons. Four changes occur in the pathways from the tegula sensory neurons when functional recovery is apparently complete. First, sensory neurons from the front wing tegulae make more reliable connections with interneurons that participate in flight. Interneuron 566, for example, receives inputs from these sensory neurons in all recovered animals, but in only 58% of normal animals. Second, the amplitude of the compound EPSP in an interneuron caused by electrical stimulation of the nerve innervating a front wing tegula is larger in recovered than in normal animals, probably because more sensory neurons make synaptic connections. Third, in recovered animals, there are more extensive branches in the metathoracic ganglion from sensory neurons of a front wing tegula. Fourth, interneurons receiving inputs from the front wing tegulae may also have more extensive branches in recovered animals. All this suggests, that freed from the competition that normally exists with the hind wing tegulae, the branches of the front wing sensory neurons can expand and take over the functional role usually established by the hind wing sensory neurons.

11.7.2. Establishing a preferred flight motor pattern

A function of the interactions of the proprioceptive feedback with central neurons may be to define a preferred motor pattern (Möhl, 1985a). The problem that the central nervous system faces is in interpreting the meaning of the sensory feedback. Without some sort of reference baseline it would be unable to tell whether the sensory information it receives was indicating that the wing movements were correct or incorrect. Different sensory information would therefore have to update this stable pattern if it were to cause changes in the motor pattern. If such an organisation exists then it implies plasticity in the central nervous system because the baseline pattern must also be capable of change to allow for alterations in the body that occur during development and maturation, and to allow for unexpected damage and changes in performance. The existence of such plasticity of the flight motor system can be shown in two ways (Möhl, 1988), thereby lending considerable support to the idea that this is indeed a function of the proprioceptive feedback.

First, under open loop conditions (tethered flight), the flight motor pattern at a particular time depends on the previous pattern (Fig. 11.18). For example, imposed changes of yaw alter the time differences between the spikes in depressor motor neurons to a left and a right wing during each wing beat that are responsible for the corrective steering reactions. When the locust subsequently flies directly ahead into a wind stimulus, the time differences between these spikes now depend on the preceding direction and angle of yaw. Thus, the flight motor pattern at a particular time is not set solely by the immediate sensory information, but is also a function of the preceding experience and thus retains some characteristics of its immediate past.

Second, under closed loop conditions, a locust is able to regulate the yaw angle at which it flies into a wind stream. The locust is tethered but can actively determine the position of its body in the yaw plane according to the differences in the timing of spikes of two depressor motor neurons to the left and right hind wings. These timing

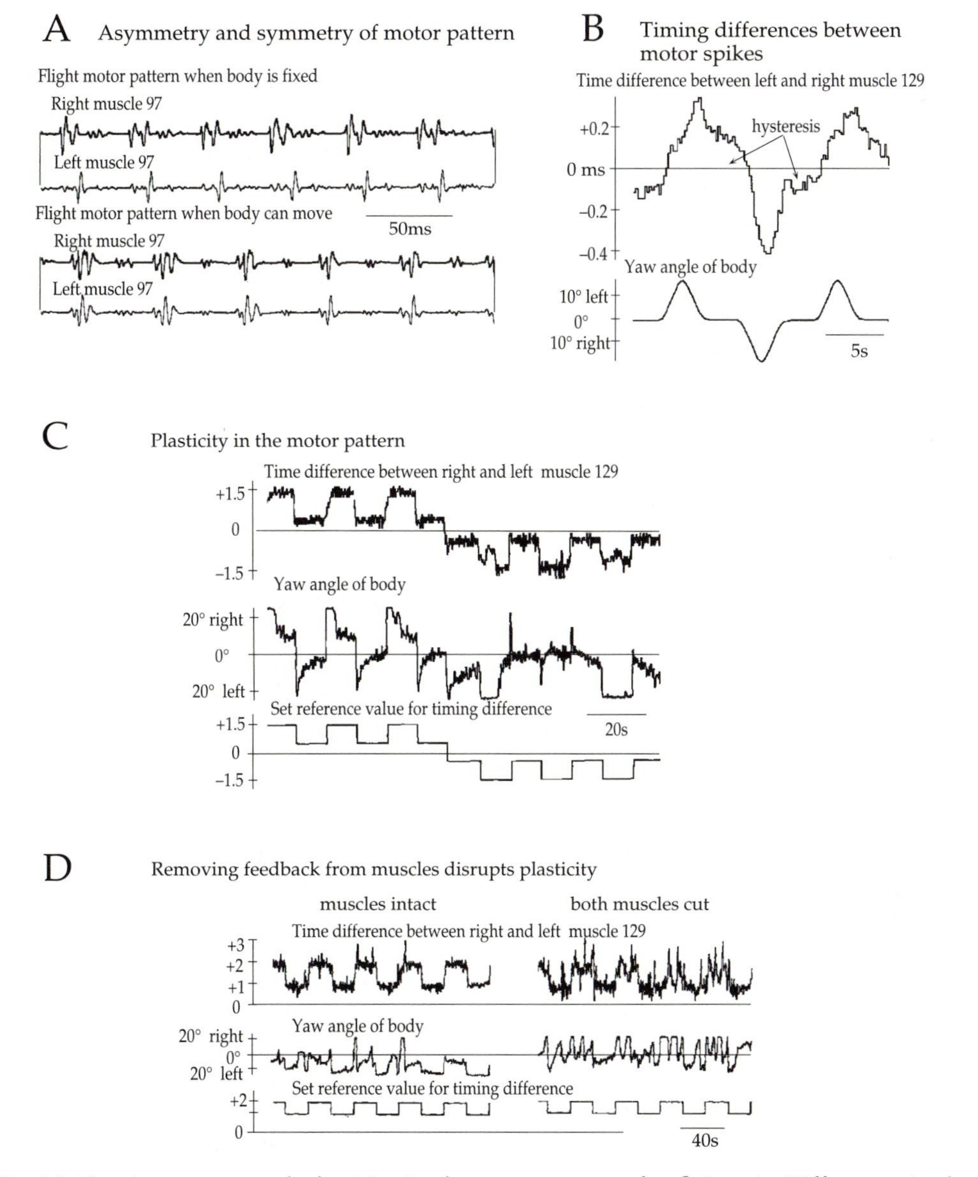

Fig. 11.18 Asymmetry and plasticity in the motor pattern for flying. A. Differences in the timing of motor spikes to front wing muscles 97 (basalar) of the left and right sides when the body is fixed or allowed to move. B. Timing differences to muscles 129 (subalar) of the left and right hind wings during imposed yaw movements of the body. C. Plasticity of the motor pattern as exemplified by the timing differences between spikes to hind wing muscles 129. The pattern is able to adjust to imposed changes in the timing differences. D. The ability of the motor pattern to change is disrupted when feedback is changed by cutting particular muscles. A–C are based on Möhl (1988), D on Möhl (1993).

signals are fed back artificially to alter the yaw angle of the body, and at the same time the experimenter can intervene in this feedback loop by setting an arbitrary timing difference that is required for straight flight. In these circumstances, a locust can adjust the timing difference between the motor spikes to alter the angle of its body so that it flies directly into the wind. The timing of the spikes to other pairs of muscles shows little adjustment to match the stimulus, but if the muscles selected for the timing signals are switched, then the newly selected pair (now in closed loop) adjust their timing to the stimulus and the old pair (now in open loop) no longer do so (Möhl, 1989). If the muscles of the pair selected for the timing measurements are cut, so that they no longer contribute to the movements of the wings, the adjustments to a changed stimulus are now slower and weaker (Möhl, 1993). If other intact muscles are then selected for the timing signals, adjustments to changed stimuli can now produce stable steering.

The implication from these experiments is that a locust can recognise the muscle pair that is in closed loop, and that it can recognise from the inflow of proprioceptive signals the contribution of an individual muscle to the movements of a wing. This runs counter to the accumulated picture of the functional organisation of the flight system. The locust may, however, be able to select the appropriate muscle pair by trial and error in much the same way as *Drosophila* may do when using visual clues for flight stabilisation (Heisenberg and Wolf, 1988). Thus, the natural variation that occurs in the timing of the different muscle pairs would allow the locust to explore and select those combinations of time shifts that produce the most effective steering responses during flight. Those selected would then be more sensitive to yaw stimuli than the other possible pairs. The mechanism would have to compare the efferent signal sent to the muscle with the proprioceptive signal caused by the muscle contraction and with the changed exteroceptive signals from the head as it turns to face a different wind direction. The command to produce a yaw movement is fed to the muscles and their action results in a movement that stimulates exteroceptors and proprioceptors. A copy of this efferent signal could be compared with the sensory signals, and if they are coincident could strengthen synapses in the pathway feeding back to the networks generating the motor commands. This mechanism would thus keep a desired course constant in the face of disturbing influences and, moreover, would improve its performance based on experience gained during flight. Within such a control loop, the main problem will be to restrict the progressive increase in strength of the synapses so that the gain does not exceed 1.0 and the loop therefore become unstable. This could be achieved by delaying the sensory feedback for one cycle of the wing beat.

It is hard to imagine how the contribution of an individual muscle could be extracted from the signals provided by the receptors at the wing hinge, which merely give information about the trajectory of a wing. It might, however, be provided by the many thoracic chordotonal organs and other receptors that are innervated by nerve 2. Some, at least, of these thoracic receptors must signal the contractions of the dorso-ventral thoracic muscles that power the movements of the wings, but their possible contribution to the flight motor pattern has not been explored. It is also a widely held assumption that receptors are not associated with individual flight muscles, but there is not much evidence to support this, so that the application of modern staining techniques would be informative. It seems entirely possible that the contractions of the different flight muscles could be signalled by different patterns of spikes in the various thoracic chordotonal organs and that the central nervous system could extract from this pattern the contribution of an individual muscle.

The important idea to emerge is that the flight motor pattern is organised not just according to the prevailing sensory input, but that this input has a persistent effect which can change the subsequent pattern that is used. The flight motor pattern of an individual locust is thus shaped and optimised by its own experiences, making use of processes with long time constants first to select and then to adopt a particular sequencing of the motor spikes to particular muscles that make up the complete flight motor pattern. This is surely an implication that must be absorbed into our thinking about other motor patterns.

11.7.3. Implications for the concept of a central pattern generator

The experiments on flight steering show that proprioceptive feedback is important in establishing the regular features of a movement and what can be predicted, and that this may require the contribution of many different sorts of receptors. Once a regular movement has been 'learnt', it can then be performed with progressively less reliance on the feedback so that the objective of the design can be seen as one that minimises dependence on feedback. Initially, much feedback is required to enable an animal to reduce the eventual dependence on feedback. There will, however, still be many occasions when feedback is necessary to monitor the unpredictable. Thus, a spectrum of movements exists which has different requirements for feedback. At one extreme are movements that are produced quickly and in rapid succession in a stable environment and which must rely on feedforward central commands because there is insufficient time for feedback. At the other extreme are those that are produced slowly and in a complex environment, and that require much feedback. Such extremes are represented by a cockroach running rapidly over a flat surface and a stick insect walking slowly over rough terrain. In between would be a flying locust which, although its movements are produced rapidly, nevertheless moves through an unstable and unpredictable environment.

The central pattern generator in this scheme can be thought of as a **predictor** of events. It can, for example, foresee that depressor motor neurons of the wings must be activated at a certain time following the elevators when the flight is stable and in the absence of signals indicating a deviation. The regular feedback from such regular movements may of course contribute to the generation of the pattern. Such a design principle is faster, more reliable and more robust than if feedback were necessary to initiate every movement in a sequence and explains why central commands are important for the generation of so many different types of movements. Such a scheme could also explain why so many sensory neurons receive presynaptic inputs at the time when they respond to a movement with their best spike response. This presynaptic input can thus be seen as the output of the predictor reducing the efficacy of sensory signals that it is expecting. Recordings from the terminals of sensory neurons from the wings during the flight motor pattern would be useful to test further the generality of this concept.

11.8. STEERING MOVEMENTS

It is obvious that a flying locust must adjust the movements of its wings, and perhaps other parts of its body, so that it can steer. The steering mechanisms must continually correct for small deviations from the intended flight path to cope with the nonlinearities in its neural and muscular machinery and the unpredictability of the environment, and at the same time must allow for the execution of voluntary turns. Such responses will alter the orientation of the body about the three main axes of the body (Fig. 11.19):

pitch a vertical movement of the body so that the head (and the rest of the body) either points downwards or upwards.

roll a movement of the body about its long axis so that the one side moves either downwards or upwards

yaw a movement of the body in the horizontal plane so that the head veers to the left or the right instead of pointing directly ahead into the wind

The steering responses are of two main types.

Correctional steering movements can be likened to the action of an autopilot in an aircraft, and derive from two needs. First, to correct for nonlinearities or even errors in the performance of the motor circuitry. Second, to stabilise movements against changes in the turbulence of the air, and to stabilise the visual surroundings in an optomotor response that minimises the movement of the visual world over the eyes. In essence, these movements are caused by unintentional deviations from course. Nearly all the mechanisms of steering that have been revealed so far relate to how locusts correct for changes in these sorts of sensory inputs. The control mechanisms must be flexible, not least to permit the second type of movements.

Intentional steering movements are those which lead to the avoidance of an object seen approaching on the flight path, and to stimuli such as the pulses of ultrasound emitted by a predatory bat (Robert and Rowell, 1992b). Such a strategy is particularly necessary when flying at air speeds of 4.6 $m.s^{-1}$ in swarms and separated from neighbours by as little as 300 mm, and yet its success is clear because collisions are avoided. The strategy used is to initiate an avoidance response when a looming object subtends an angle of about 10° (Robertson and Johnson, 1993). Other steering movements are made without obvious reference to an external stimulus and might therefore be called voluntary. To make these movements requires that a locust act against the stabilising effects of the correctional steering responses. If a freely flying locust deviates to the right from its flight path, either intentionally or unintentionally, the visual world will appear to move to the left. The unintended movement will be countered by an optomotor response acting as a correctional steering response, whereas the visual signals caused by the intended movement must be allowed so that the desired steering can occur. The locust must therefore either

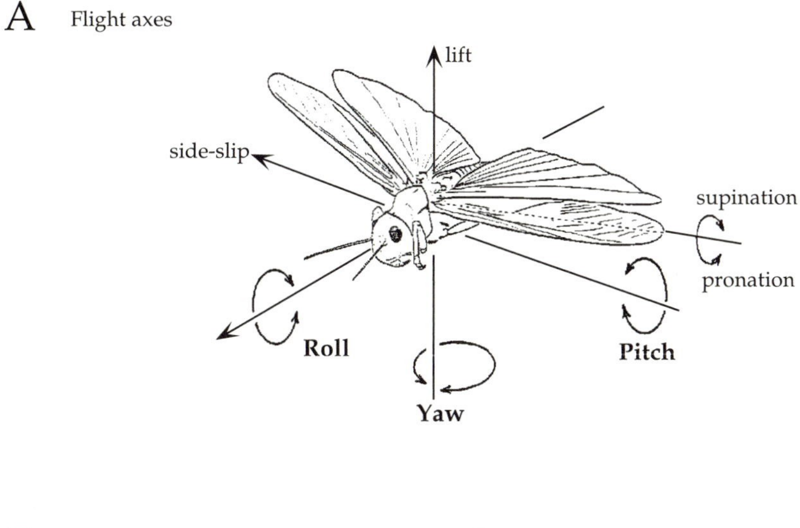

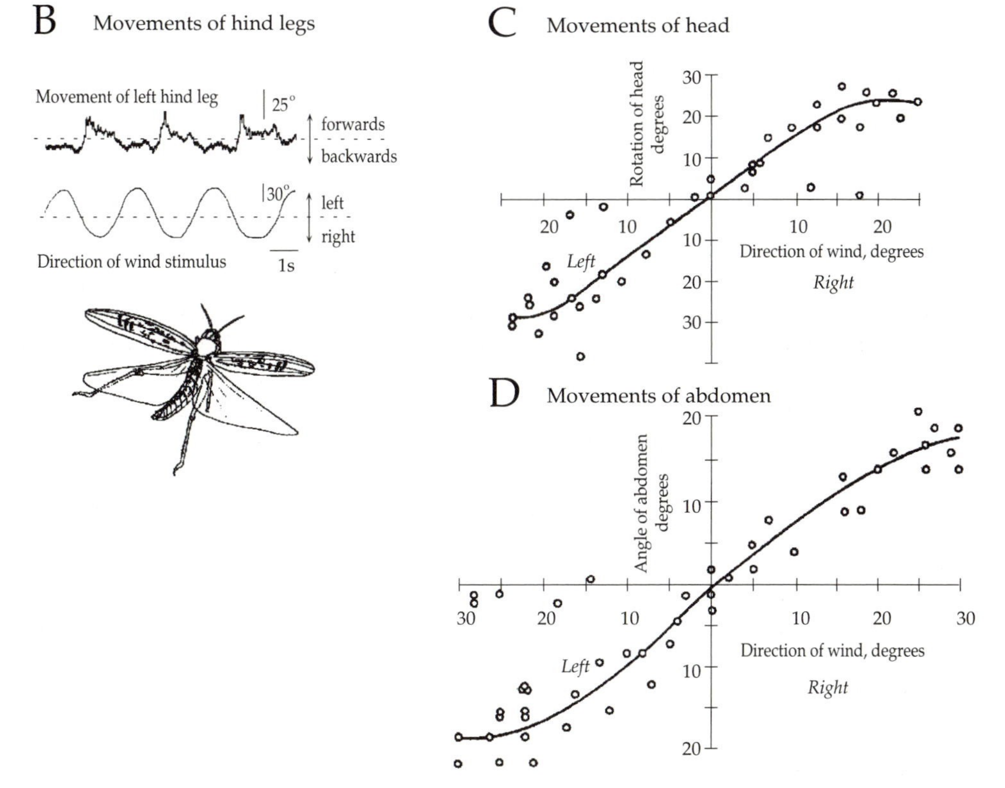

Fig. 11.19 Movements of the body that are used in steering during flight. A. The main axes for flight, the forces generated, and the twisting movements of the wings. Movements of the hind legs (B), head (C) and abdomen (D) during steering movements in response to changes in the direction of the wind. B is based on Arbas (1986), C, D on Camhi (1970).

modulate the channels that are used for correctional steering, or exert control through parallel pathways that override the correctional signals. Most of the current evidence suggests that the visual signals are either suppressed or ignored as a result of a simultaneous feedback or corollary discharge from the motor commands.

Steering the body while in flight can involve several changes in the movements of the wings, the abdomen and the legs, and virtually anything that can have an aerodynamic effect. The movements can be performed individually or in various combinations.

1. Phase shifts of the wings relative to each other within one wing beat cycle.
2. Changes in the amplitude of a wing beat.
3. Changes in the angle of attack of a wing.
4. Ruddering movements of the abdomen, either up and down, or left and right.
5. Ruddering with the hind legs, which are extended.
6. Movements of the neck are also associated with steering but these have no aerodynamic effect, although they may change the gain of the steering response and provide proprioceptive input from neck receptors.

11.8.1. Movements of the wings and changes in the flight motor pattern

Movements of the wings are the primary mechanism for producing steering movements, but the relationship between these changes and the changes in the underlying flight motor pattern is nevertheless complex. Most of the data about steering concern the actions of the muscles in tethered locusts that are flying in response to imposed deviations of their body, their visual surroundings, or the direction of wind stimuli. There is little information on the form of the wing movements, and the assumption is largely that certain patterns of muscle activity will be translated into certain changes in the attitude of the wings and wing movements. The motor output to the wing muscles can be altered in three ways.

First, by changing the number of motor neurons that spike and thus that contribute to the force generated by the contraction of a particular muscle. This mechanism is of limited usefulness because only a small number (1–6) of motor neurons innervate each muscle, and for those such as the first basalar that are innervated by a single motor neuron the choice is whether to contribute to the wing movements or not.

Second, by changing the number of times that each motor neuron spikes during one cycle of the wing beat and the interval between these spikes. Again, this mechanism is limited because motor neurons spike only a few times (typically 1–3 times) at each wing beat.

Third, by changing the phase relationships of the motor spikes to different muscles within one cycle of the wing beat.

Changing the number of motor neurons that are active and the number of times that they spike will alter the power output of a muscle but this can only be expressed as changes in the wing beat frequency, the amplitude of the wing beat, or the angle of attack (loading). Alterations in the frequency of wing beats are not used in steering, and amplitude and loading depend on phase shifts in the timing of the motor spikes. Thus, the first two methods are secondary to the third, so that the timing of the first spikes becomes the most sensitive variable. Moreover, use of the first two methods will inevitably produce a stepwise change in the force whereas the timing relationships can be varied continuously and by small fractions of the total wing beat cycle (Zarnack and Möhl, 1977). These time shifts between spikes of motor neurons to the same muscles on different sides of the body, or between different muscles on the same side, occur when the body itself rolls, or when the visual horizon is shifted (Thüring, 1986). They are proportional to the displacement of the body and can be produced within 1–2 wing beats, so that a 20° displacement causes a shift of some 6 ms in the timing of depressor motor spikes (Zarnack and Möhl, 1977). Differences as small as 0.5 ms can be produced by smaller deviations and yet are still manifested as changes in the wing movements. During an imposed **yaw** movement of the body to the left, the basalar (depressor) motor neurons of the front and hind wings on the left side spike earlier and the subalars (also depressors) later in the wing beat cycle, while those on the right-hand side show the opposite shifts. During an upward **pitch** of the head, basalar motor neurons spike earlier and the front wing subalar later. During a **roll** movement to the left, the right front wing depressor motor neurons spike earlier and the left ones later, producing changes in the flight torque that should lead to correction of an induced deviation. A combination of these changes in the motor pattern sometimes results in changes in the amplitude of the wing beats and in their phasing. On the side that rolls down, the wing beat is advanced in its cycle and decreased in amplitude, with the opposite changes occurring in the wings on the side that rolls upwards (Thüring, 1986).

These changes in the movements of the wings will inevitably alter the proprioceptive feedback from the wings and therefore a steering response may have to act in opposition to the proprioceptive feedback delivered to the central nervous system by the wing receptors. One role for the proprioceptors may be to confirm the expected performance of the motor system and could therefore prevent manoeuvres that are likely to be hazardous while allowing those that are likely to be more adaptive.

11.8.2. Movements of the head, legs and abdomen

Movements of the wings change the torque that is developed, whereas movements of the abdomen and legs change the drag, shift the centre of mass, and change the moment of inertia of the body (Fig. 11.19). When making a turn to the right or receiving a changed wind stimulus to the right side of the head, the head rotates to the right, the right hind leg is moved forwards and away from the body, the left hind leg is moved backwards and towards the body, and the abdomen is curled to the right to act like a rudder, mostly by contraction of muscles in its anterior segments. The movements are proportional to the changes in the direction of the wind stimulus and could therefore stabilise flight against changes in yaw, an interpretation that is reinforced by the finding that none of these movements can be induced unless the locust is flying. The muscles that move the hind legs in these movements are driven in time with the flight motor pattern (Dugard, 1967), so that a coxal abductor (muscle 126, see Tables 8.2 and 8.3 in *Chapter 8*) is depolarised during elevation and an anterior coxal rotator (muscle 121) is depolarised during depression (Lorez, 1995). The result may be a slight vibration of the legs at the wing beat frequency.

The effects are produced rapidly, so that a changed wind stimulus to the head, detected largely by hairs in field 1, induces changes in the torque generated by the wings, and movements of the legs and abdomen within 60ms, but movements of the head always start some 30 ms later (Robert and Rowell, 1992a). This sequence of movements is the same under open loop conditions when the movements have no effect on the exteroceptive input, and under closed loop conditions when they do. Abdominal ruddering movements are smaller under closed loop conditions. The sequence of movements of these different parts of the body indicates that they are each induced in parallel by a wind stimulus to the head. If the head is rotated artificially through 180° then the movements of the abdomen are reversed. Moreover, in correctional steering the head moves in the same direction as the turn whereas in intentional steering the head is moved away from the stimulus, but both movements occur after the movements of the abdomen (Robert and Rowell, 1992b). The movements thus do not depend on serial proprioceptive feedback from each other for their induction or support. Nevertheless, fixing the head to the body leads to a poorer steering response, and forced movements of the head, detected by the hair plates and a myochordotonal organ associated with muscle 54 in the neck, cause slower compensatory movements of the abdomen. Proprioceptive signals generated by head movements can thus modulate the steering commands sent to the thorax, even though they do not themselves initiate steering movements.

11.8.2.1. *Controlling the abdomen*

Normally, the abdomen is held in line with the thorax with a slight upward curl at the tip but if the wind velocity is progressively reduced then the abdomen is progressively lowered (depressed) within 500 ms of the change and the pronation of the fore wings is increased (Camhi, 1969a; 1970). The movements of the abdomen will increase the lift of the posterior part of the body, and the twisting of the front wings will reduce the lift of the anterior part, so that together they will make the locust dive and thus restore its flight speed. The reaction should prevent stalling if the flight speed falls below a critical value.

During flight, the abdomen vibrates vertically in time with the movements of the wings because the motor neurons that innervate abdominal muscles spike in bursts at the wing beat rhythm, and not simply because it is linked mechanically to the vibrating thorax (Camhi and Hinkle, 1972). Wind stimuli to the right side of the head increase the number of spikes that are delivered in each burst to muscles on the right side of the abdomen, and decrease those to muscles on the left, so that the abdomen curls to the right in responses that are proportional to the angle of the wind. The rhythmic drive to the abdominal motor neurons must originate from the networks generating the flight motor pattern and could be conveyed to them by some of the flight interneurons that have axons descending from the metathoracic ganglion (Figs 11.7 and 11.8). One effect is that the flight motor pattern and the ventilatory rhythm are combined in the output of some of the abdominal motor neurons. The coordination between the two rhythms is such that the ventilatory spikes in motor neurons to dorso-ventral muscles are subdivided into the faster flight motor pattern and alternate with those in the dorsal longitudinal muscles to give an alternating pattern similar to that in the elevator and depressor muscles of the wings (Kutsch *et al.*, 1991).

11.8.3. Avoidance steering in response to sound

While they are flying, many insects steer away from sources of ultrasound that might indicate the approach of potentially predatory bats. *Locusta migratoria* has directional hearing, with receptors for high frequency sounds, that provides an early warning of a source of ultrasound above 10 kHz and sound pressure levels above 45 dB, so that a bat will be detected when it is about 20 m away. By contrast, the echolocating system of a bat will probably only be able to detect the locust when it is within 5 m, so that there is a spatial safety factor of 4 in favour of the locust. This gives the locust about 2–5 s in which to complete a successful evasion response from a bat flying at a speed of 10–14 $m.s^{-1}$. Within 65 ms, or about one wing beat, sound pulses cause ruddering movements of the abdomen that generate a yaw torque, and an increase in the wing beat frequency by about 15% (from 17 to 20 Hz) (Robert, 1989). The abdomen is moved away from the sound source while the head is turned towards it. A number of interneurons that are sensitive to sounds between 5 and 15 kHz cause appropriate steering movements of the abdomen, and to a lesser extent the head, when they are stimulated electrically, but normally these interneurons would act as a group (Baader, 1991). Both the sensitivity of the hearing and the rapidity of the steering response enable the flight course to be altered and the locust to steer away from the potential predator. The steering response is specific for particular sound frequencies and can be suppressed by a simultaneous tone of lower frequency. This is necessary if a locust is to use the same interneurons to enable it to turn towards an attractive auditory stimulus and to escape from a potentially dangerous one. Field observations are now needed to complement this laboratory-based demonstration of avoidance behaviour.

11.9. RECEPTORS INVOLVED IN STEERING

A combination of visual, wind and aerodynamic sensors are used by a locust to give information about any changes that might occur in its flight path. Gravity itself would be difficult to use, as without other clues it cannot be distinguished from the inertial forces generated by flight. Some receptors on the cerci of certain cockroaches do give information about gravity, but it is unclear whether a locust is able to monitor the weight of its own body, which will change with age, recent feeding and the stage in the egg laying cycle of females.

Most of the receptors detecting exteroceptive stimuli are on the head, with the main receptors being the compound eyes, the ocelli, the antennae and numerous wind-sensitive hairs. The stimuli that they detect will therefore not be directly linked to the wing beats, although the vibrations of the thorax may cause the head to bob up and down and the turbulence created by the wings may modulate the air that flows over the wind-sensitive receptors on the head and antennae. Moreover, the position of these receptors is not fixed relative to the wings because the head can move in optomotor responses that attempt to minimise slippage of the visual world over the retina. During steering movements, a locust must therefore take account of the exteroceptive stimuli impinging on its head and proprioceptive signals from receptors in its neck that signal the position of the head relative to the body. These must then be integrated with the rhythmic signals from the receptors on the wings and at the wing hinges.

11.9.1. Ocelli

An ocellus acts as a luminosity detector with a wide visual field because the optics of its single lens cannot form a focused image at the level of its receptors (Wilson, 1978). Moreover, the receptors themselves respond preferentially to the ultraviolet component of the spectrum, which is greater in the sky than on the ground. Together, therefore, the three ocelli, one median at the front of the head and two lateral, just anterior to each of the compound eyes, can act as horizon detectors. If the body pitches, then the illumination of the median ocellus will change, and if the body rolls the relative illumination of the left and right lateral ocelli will be altered. This information can be signalled rapidly by these sensors as it appears to be free of the form, colour and motion processing that is performed by the compound eyes, and then fed to fast-conducting interneurons for transmission to the thoracic ganglia where the appropriate changes in the flight motor pattern can be instigated.

Some complex interactions between the signals resulting from wing movements and the visual signals from the ocelli may nevertheless occur in the plexus associated with the ocellar photoreceptors (Kondo, 1978). During flight in the dragonfly, efferent neurons in the ocellar nerves spike in time with the wing beats, and forced movements of the wings produce patterns of efferent spikes that are similar to those recorded in a nerve from a wing. Afferent neurons from the ocelli also spike in time with the wing beats, even in the dark. This suggests that the processing of visual signals by the ocelli may be altered by the movements of the wings.

11.9.2. Compound eyes

The compound eyes also respond to changes in the orientation of the horizon and their integrated output may be fed to some of the same interneurons carrying ocellar and wind information, or carried in separate channels to the thoracic ganglia. They also, of course, provide more complex analyses of the visual world, but in addition provide inputs that stabilise the visual world through the optomotor response.

If a locust is shown a moving pattern (usually alternate black and white stripes), it will track the movements by changes in the movements of its head and/or body, in an optomotor response that partly stabilises its visual world. It does not fixate the pattern, but instead there is a slippage of the pattern over the eye. In the fly, the visual signals generated by such stimuli are processed by neurons in the optic ganglia and the outcome of this processing is seen in sets of visual interneurons in the lobula plate that respond preferentially to a movement in a particular direction of a large area of the visual field. The net result is that when an unintentional turn is made during flight, this creates a visual flow in the opposite direction which will then be opposed. This control nevertheless means that an intentional turn must be made in the face of this conflicting visual signal. By feeding back the turning torque that is generated in flight to control the movements of the visual world, tethered flying locusts can successfully stabilise their visual panorama, the horizon, or even a single feature such as a stripe (Robert, 1988).

11.9.3. Head hairs

A wind stimulus to the front of the head is detected by hairs arranged in five bilateral fields (Fig. 11.20) that are essential for the maintenance of sustained flight (Weis-Fogh, 1949). If a wind stimulus is given to one side of the head, the locust will turn until it flies directly into the wind and the hairs on the head are stimulated symmetrically. If these hair fields are covered, then it is more difficult to initiate flight with a wind stimulus, and once initiated, the flight is usually of a shorter duration, occurs at a reduced speed, and the wing beat frequency is reduced because there is a longer delay between the elevator and depressor muscle activity. Moreover, the locust is no longer able to perform complex manoeuvres. It has not proved possible to ascribe a particular behavioural role to a particular field of hairs and there seems little point in trying to push the attributions further. What can be said is that signals generated by these hairs can stabilise flight movements which involve some translation in the yaw plane. The signals from these hairs may also enable larvae that cannot fly to orient with respect to the wind, and this may be important in directing long marches.

In *Schistocerca gregaria*, there are about 200 hairs in these fields with shafts up to 300 μm long. The hairs within one group point in the same direction (Smola, 1970), and are particularly sensitive to wind from the opposite direction (Camhi, 1969b). About the same number of shorter hairs are present that are not normally stimulated by air currents. Sensory neurons from hairs in all the fields make similar patterns of branches in the ipsilateral tritocerebrum, but in the suboesophageal ganglion hairs from field 3 project ipsilaterally and the others contralaterally. Only a few of the sensory neurons from fields 1–3 project to the prothoracic ganglion, but, in contrast, most of the sensory neurons from fields 4 and 5 form dense tufts of branches close to the midline of the prothoracic ganglion and some may even project to the mesothoracic ganglion.

The sensory neurons from head hairs can thus deliver signals about wind stimuli directly to the circuitry in the thoracic ganglia that generates and controls flight, but the only known output of these distant projections is with the axonal branches of the A4I1 interneurons (*see section 11.10.4*) in the prothoracic ganglion. The information is also distributed in parallel by interneurons in the brain that receive direct inputs

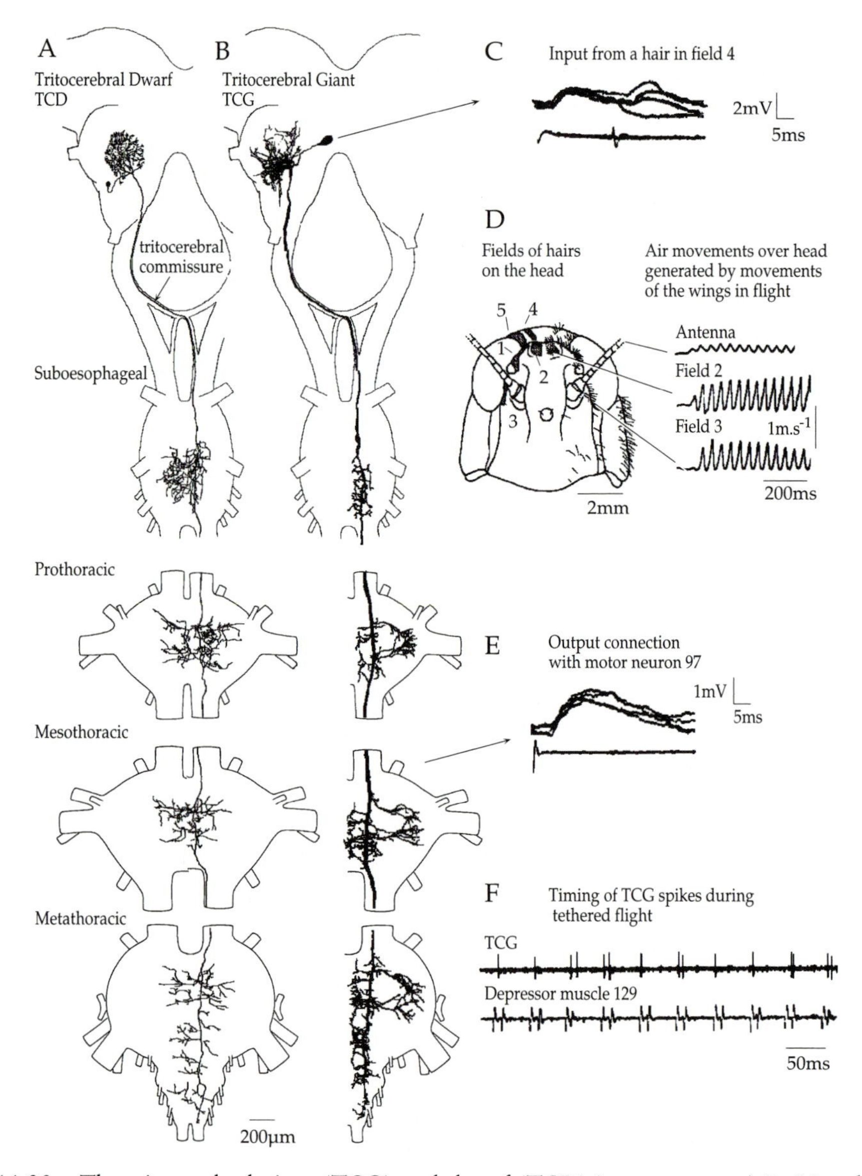

Fig. 11.20 The tritocerebral giant (TCG) and dwarf (TCD) interneurons. A,B. Morphology of the interneurons. C. Sensory neurons from hairs in field 4 on the head make direct excitatory connections with the TCG. D. The fields of hairs on the head and the airflow over them and the antennae generated by the movements of the wings in tethered flight. E. The TCG makes an excitatory connection with the motor neuron innervating depressor muscle 97 of a front wing. F. A TCG interneuron spikes at each cycle of the wing beat during tethered flight. A is based on Tyrer *et al.* (1988), B on Bacon and Tyrer (1978), C, F on Bacon and Möhl (1983), D on Weis-Fogh (1956a) (drawing of hair fields), and Horsmann *et al.* (1983) (flow over head), E on Bacon and Tyrer (1979).

from the primary sensory neurons. There are likely to be many such neurons processing and distributing the information from wind receptors on the head, and indeed, some of them have already been identified. While it is important that the properties and connections of the individual interneurons are known, it is essential to bear in mind that they are only one element in many parallel channels and a behavioural response is likely to result only from the cooperative interaction of many interneurons.

11.10. INTERNEURONS INVOLVED IN STEERING

11.10.1. Estimating the number

The large number of body parts that participate in steering means that the brain and as many as nine other neuromeres are involved, from those in the suboesophageal ganglion to those in the last abdominal ganglion. Estimates of the number of neurons involved in flight steering vary enormously. One estimate suggests that perhaps 50 pairs of intersegmental interneurons carrying information from the head, or from posterior sense organs such as those on the cerci, 342 pairs of premotor interneurons and 378 pairs of motor neurons are involved, excluding those neurons responsible for sensory processing and for generating the flight motor pattern itself. This estimate is an extrapolation based on the number of neurons in the mesothoracic ganglion thought to be involved in steering [see Rowell and Reichert (1991) for a catalogue of 28 of these interneurons] and assumes that a similar number of homologous neurons are involved in the other ganglia. It is unlikely, however, that the same complement of interneurons is present in each ganglion, and we know that the same, or even the homologous neurons, do not necessarily make the same connections in each ganglion. It is also artificial to separate sensory processing and motor control when it is clear that the same neurons can be involved in both. Nevertheless, the estimates emphasise the widespread involvement of many neurons in these movements, perhaps as many as 10% of those outside the brain, and that a single function cannot be ascribed to a single neuron, because each will participate in many different actions together with many other neurons.

Any description of particular interneurons tends to emphasise their importance at the expense of other interneurons that have yet to be discovered. Indeed, there is a strong temptation to explain behavioural observations in terms of the actions of the neurons that we know, even when it is clear that only a partial view is so far available. The description of the few interneurons that are well characterised and which follows should therefore be read with these caveats firmly in mind.

11.10.2. TCG (tritocerebral giant)

11.10.2.1. *Structure*

The tritocerebral giants (TCGs) are a prominent pair (one on each side of the midline) of interneurons that receive direct inputs from head hairs and make direct synaptic connections with flight motor neurons (Fig. 11.20). Each interneuron has a 70μm diameter cell body in the medial part of the tritocerebrum of the brain and many laterally projecting branches, but its diagnostic feature is the route taken by its axon; it leaves the brain in the ipsilateral connective before crossing to the contralateral connective in the small tritocerebral commissure. This commissure contains the axons of the TCG and the much smaller axon of the TCD (tritocerebral dwarf) interneuron. The axon of the TCG then projects at least as far as the abdominal neuromeres of the metathoracic ganglion in tract MDT (see Table 2.2 in *Chapter 2* for a description of the longitudinal tracts), forming a series of ipsilateral branches in the dorsal neuropil in each of the intervening ganglia.

11.10.2.2. *Response properties*

A TCG interneuron is excited by wind on the head and by movements of the ipsilateral antenna. The sensory neurons from head hairs in ipsilateral fields 2, 4, 5 and the posterior part of field 1 make direct excitatory connections, but the receptors providing the input from the antenna are not identified (Bacon and Möhl, 1983).

In a flying locust, a TCG interneuron spikes once or twice as the wings reach the top of their upstroke (Möhl and Bacon, 1983), because the sensory neurons presynaptic to it respond to the rhythmic turbulences of 1.0 $m.s^{-1}$ in the air currents around the head. These are created mostly by the wing movements, and by the nodding movements of the head that change the wind direction by 5.5° at particular groups of head hairs (Horsmann *et al.*, 1983). An imposed yaw movement during flight produces reciprocal responses in the two TCG interneurons in such a way that, for example, a yaw to the right is signalled by a greater number of spikes at each wing beat in a TCG with its cell body on the right-hand side of the brain (Fig. 11.20). These changes are accompanied by a shift in the timing of the depressor motor neurons on the left side of the body to which the axon of the right TCG projects. Electrical stimulation of a TCG can mimic some of these time shifts of the wing motor neurons, but they are not explained by its few known output connections with flight motor neurons such as the fore wing basalar (depressor) motor neuron (Bacon and Tyrer, 1979). TCG branches in other ganglia suggest that many more output connections have yet to be revealed.

In a nonflying locust, a TCG produces a burst of spikes at the sudden onset of a wind stimulus to the head, suggesting that it could signal changes in the attitude of the head relative to a wind stream. Its response to an imposed yaw is, however, smaller when the locust is flying than when it is not flying (Möhl and Bacon, 1983). When flying, the velocity of the air currents impinging on the head is modulated in time with the wing movements, and experimental modulation of these currents within 3 Hz of the flight frequency forces the flight motor pattern to adopt a distinct phase relative to the stimulus (Horsmann *et al.*, 1983). Similarly, electrical stimulation of the head hairs during flight advances the start of the spikes in elevator or depressor motor neurons depending on the phase of the wing beat cycle (Bacon and Möhl, 1983). Part of these adjustments may be caused by a TCG acting in concert with other wind-sensitive interneurons.

11.10.2.3. *Behavioural role*

These results suggest that a TCG is involved in stabilising the yaw component of flight movements, and emphasise the importance of sensory input for the maintenance of flight. The TCG may also have other actions, because when the locust jumps and launches itself into the air (*see Chapter* 9), a high frequency burst of spikes occurs in the TCG (Bicker and Pearson, 1983). Electrical stimulation of the TCG with the same pattern of impulses can initiate flight. If the commissural connectives are cut, and hence the axons of both TCG neurons, flight can still be elicited by wind on the head, but the latency to the first elevator motor spike and thus to wing opening is increased. This suggests that the TCG can initiate flight to a wind stimulus but that many pathways normally operate in parallel.

11.10.3. TCD (tritocerebral dwarf)

This neuron also has its axon in the tritocerebral commissure, although it is much smaller (5 μm compared to 20 μm) than that of the TCG axon (Tyrer *et al.*, 1988). The cell body and branches occur in the same region of the tritocerebrum as the TCG and its axon follows a similar course to the metathoracic ganglia, but its branches cross the midline whereas those of the TCG are ipsilateral. It is immunoreactive for GABA. Its inputs are only roughly characterised and few are recognised that give a reliable response. It spikes when hairs on the thorax, abdomen, eyes, and those in field 3 on the head are touched, when the ipsilateral antenna is moved, and when the ambient illumination is reduced. None of its output connections are known, so that its role in behaviour can only be surmised. It is thus only by association with the TCG that its action could be supposed to be in controlling steering.

11.10.4. A4I1 interneuron

The two A4I1 interneurons, named after the location of their bilaterally paired cell bodies in the first unfused abdominal ganglion (A4), have axons that project to the brain and yet process signals from wind-sensitive hairs on the head and neck (Fig. 11.21). They make output synapses with pleuroaxillary motor neurons (Table 11.1), and therefore have all the appropriate properties to be involved in influencing flight steering.

11.10.4.1. *Structure*

The axon of an A4I1 interneuron projects in the connective contralateral to its cell body to the posterior part of the dorsal deutocerebrum, forming a series of short branches within each of the intervening ganglia. The diameter varies from 5 to 20 μm along its length, and it gives rise to its most extensive arrays of branches in the prothoracic ganglion (in the VAC neuropil) where input synapses predominate and the processing of the wind-sensitive signals takes place (Watson and Pflüger, 1989). Only input synapses are found in the ganglion containing the cell body, but

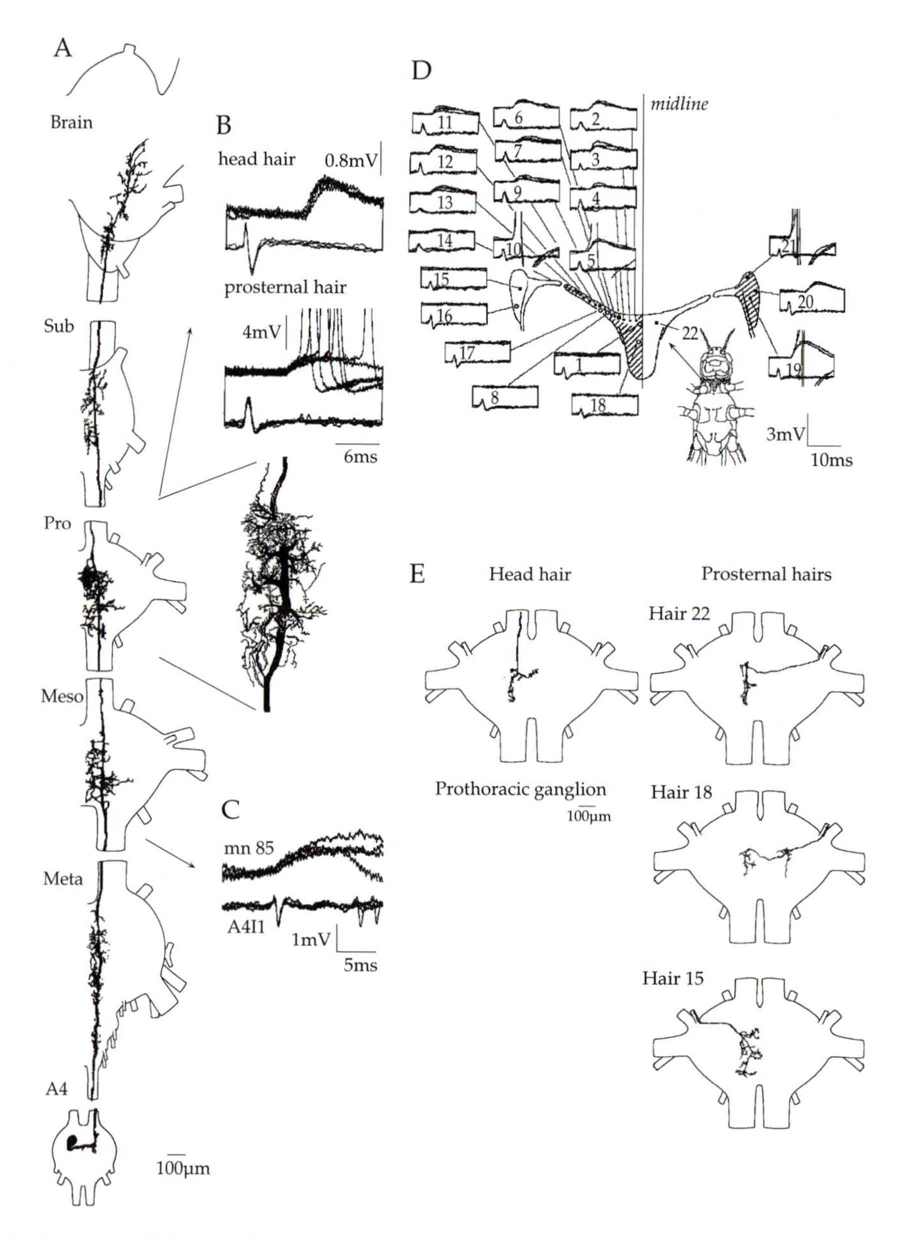

Fig. 11.21 Morphology and connections of interneuron A4I1. A. Morphology. The cell body is in abdominal ganglion A4 and its axon projects to the brain. There is a large array of branches in the prothoracic ganglion where it receives synapses (B) from sensory neurons of particular hairs on the head and the prosternum. C. Excitatory connections are made with motor neurons innervating muscle 85 of a front wing. D. Effectiveness of the array of hairs on the prosternum in generating EPSPs in an interneuron. E. Projections of sensory neurons from head and prosternal hairs into the prothoracic ganglion. A is based on Pflüger (1984), the inset on Watson and Pflüger (1989), B,D,E on Pflüger and Burrows (1990) and C on Burrows and Pflüger (1992).

input and output synapses are intermingled in the other thoracic ganglia. Spikes are initiated in the prothoracic ganglion by inputs from mechanosensory hairs, and close to the cell body by inputs from unidentified receptors, with the result that they can travel anteriorly and posteriorly in the thorax. It is not known whether the input synapses in other ganglia can also initiate spikes, or merely modulate the effectiveness of spikes initiated elsewhere at their local output synapses. Most spikes seem to originate from the prothoracic site.

11.10.4.2. *Response properties*

An A4I1 interneuron responds primarily to wind stimuli detected by arrays of highly sensitive hair receptors in field 1 (Fig. 11.20) on the side of the head ipsilateral to its axon, on the pronotum (dorsal part of the neck), and on two regions of the prosternum (ventral part of the neck), one on the lateral edge and ipsilateral to the axon, the other medial but contralateral (Pflüger, 1984). The sensory neurons from all these hairs project to the prothoracic ganglion where they make direct excitatory synapses on the branches of the interneuron in ventral neuropil. Hairs on the lateral region of the prosternum project to the ipsilateral neuropil, but those in the medial region project contralaterally, so that the inputs from both are brought together on the same interneuron. Some hairs at the midline have bilateral projections and connect with both interneurons. In early instars, hairs in the medial region also have bilateral projections and connect with both interneurons, although the largest amplitude EPSPs are generated in the contralateral interneuron (Pflüger *et al.*, 1994). During postembryonic development, the ipsilateral branches and the connections with the ipsilateral interneuron are lost.

The effectiveness of the synaptic input from these sensory neurons depends on the type of receptor from which they originate and their position, therefore paralleling the processing of mechanosensory signals from exteroceptive hairs on the legs (*see Chapter 7*). The most powerful inputs, defined by the amplitude of the EPSPs and their ability to make the interneuron spike, are made by the afferents from the long (500–600 μm) filiform hairs (Pflüger and Burrows, 1990). The gain of many of these synapses is high, so that single sensory spikes reliably evoke spikes in an interneuron. The shorter filiform hairs generate smaller amplitude EPSPs, and the thicker, short hairs do not connect. The effectiveness of an input from a filiform hair also correlates with the birth date of the receptor in the various larval stages, the longest hairs with the most powerful effect being born first in the first instar.

The long filiform hairs on the prosternum generate a maximal spike response to a 1° displacement (Klee and Thurm, 1986), which translates to a sensitivity to wind speeds as low as 30 $mm.s^{-1}$. This sensitivity is maintained in the spike response of the interneuron (Pflüger and Tautz, 1982), presumably as a result of the summation of many inputs through high gain synapses. One interneuron can thus signal small changes in wind velocity, and the different pattern of spikes in both interneurons to wind from different directions may allow information about wind direction to be extracted by the next neurons down the line. Just what sort of information is given by these interneurons during free flight is unknown because it is hard to establish what the air flow over the most sensitive hairs on the prosternum will be. It might be expected that the high sensitivity receptors would become desensitised quickly, but the posture of the front legs may channel turbulent air over the prosternum when the flight heading or the wind direction changes, thus preventing them from adapting.

11.10.4.3. *Behavioural role*

An A4I1 interneuron connects directly with the two motor neurons that innervate the pleuroaxillary muscle of the ipsilateral fore wing, and indirectly excites the two hind wing pleuroaxillary motor neurons and some motor neurons innervating abdominal muscles (Burrows and Pflüger, 1992). Other output connections are made with a mesothoracic interneuron whose axon projects to the metathoracic ganglion and which is possibly the source of the excitation of the hind wing pleuroaxillary motor neurons, and with a motor neuron that innervates an abdominal muscle. Both the direct and indirect effects on the pleuroaxillary motor neurons are powerful enough to make them spike, even in a locust that is not flying. These sparse and specific output connections of these interneurons suggest that they are involved in the control of steering during flight.

Contractions of the small pleuroaxillary muscles in a nonflying animal alter the angular setting of a wing (Heukamp, 1984) and during flight should decrease its pronation so that the angle of attack and the amount of lift produced during a downstroke are increased (Pfau, 1977, 1978; Wolf, 1990b). These actions are in addition to the role they play in folding the wings at the end of a flight (Snodgrass, 1929; Pringle, 1968). During straight, tethered flight, a fore wing pleuroaxillary motor neuron spikes once or twice during each downstroke (Elson and Pflüger, 1986) and during an imposed roll, the spikes of a motor neuron on the side that is rolled down are advanced, the number of spikes at each wing beat is increased, and the second motor neuron is recruited. The result should be an earlier and more powerful contraction, and thus an earlier increase in the amount of lift produced on this side. A symmetrical input to the right and left motor neurons from the A4I1 interneurons should therefore contribute to the maintenance of steady flight, but an asymmetrical input will result in different amounts of lift on the two sides. A role for these motor neurons, separate from the production of power for flight, is also indicated, because their spikes are not coupled in a preferred phase to the flight motor pattern during fictive flight in the absence of sensory feedback. This suggests that they are not driven by the central network of interneurons producing the flight motor pattern. A further effect of the interneurons may be to control ruddering movements of the abdomen by their effects on motor neurons to abdominal muscles. As a consequence of this organisation, some steering signals may not be processed by the flight motor network because they have direct access to the control of certain motor neurons.

11.10.5. Cooperative action of descending interneurons

Any multimodal sensory stimulation to the head will activate many receptors and their signals will be combined and processed in various parts of the brain. The integrated output of this processing will then be carried in parallel interneuronal pathways to the flight motor circuitry in the thorax. Each interneuron codes a particular pattern of sensory stimuli and makes a specific pattern of synaptic output connections. Some of these input and output patterns of connections may overlap; both the TCG (Bacon and Tyrer, 1978) and A4I1 respond to wind stimuli, and both the DCMD (*see Chapter 9*) (Simmons, 1980) and A4I1 connect with pleuroaxillary motor neurons. This means that the steering response of a locust can be interpreted only in terms of the cooperative action of the pathways that are active and not simply in terms of the action of an individual interneuron. We do not yet know whether other projection interneurons that signal deviation from a flight course synapse on the pleuroaxillary motor neurons, although some do synapse on flight interneurons and motor neurons, by which action they could then control steering movements (Bacon and Tyrer, 1979; Rowell and Pearson, 1983; Reichert and Rowell, 1985; Rowell and Reichert, 1991). For example, the TCG can alter the timing of the fore wing first basalar and subalar motor neurons in the flight motor pattern (Bacon and Möhl, 1983). This concept of neurons receiving overlapping sets of sensory signals and making overlapping patterns of output connections is particularly important for those interneurons which receive visual and wind inputs, to the extent that these neurons can be thought to be deviation detectors.

11.10.6. Deviation detector interneurons

Some of the brain interneurons with axons that project to the thoracic ganglia are excited by particular combinations of visual stimuli to the head, suggesting that the best interpretation of their action is that they act as deviation detectors signalling a particular change in the orientation of the body about one of its three axes (Figs 11.22 and 11.23 and Table 11.2). Four bilateral pairs of these interneurons are known in some detail, but many other neurons with similar arrays of sensory inputs are probably also involved and await characterisation. They may have eluded recognition so far because they are smaller or have responses that are indistinguishable from the established types. The problem of identification of these interneurons is compounded by the fact that their response properties may change depending on the recent history of stimuli to which they have been exposed. It has to be recognised that such neurons may even change their responses as a result of the presentation of the very stimuli that must be used to characterise them. Moreover, it is always difficult to know whether the adequate stimulus has been tested, and the history of such investigations tells us that revisions are constantly necessary in the face of newly exposed facets of neuronal properties.

11.10.6.1. *Structure*

All of the interneurons in Table 11.2, and many of the other descending interneurons involved in flight, have their cell bodies in the protocerebrum of the brain and axons that project to the abdominal neuromeres in the metathoracic ganglion and perhaps beyond (Figs 11.22 and 11.23). In the brain there are extensive branches in the

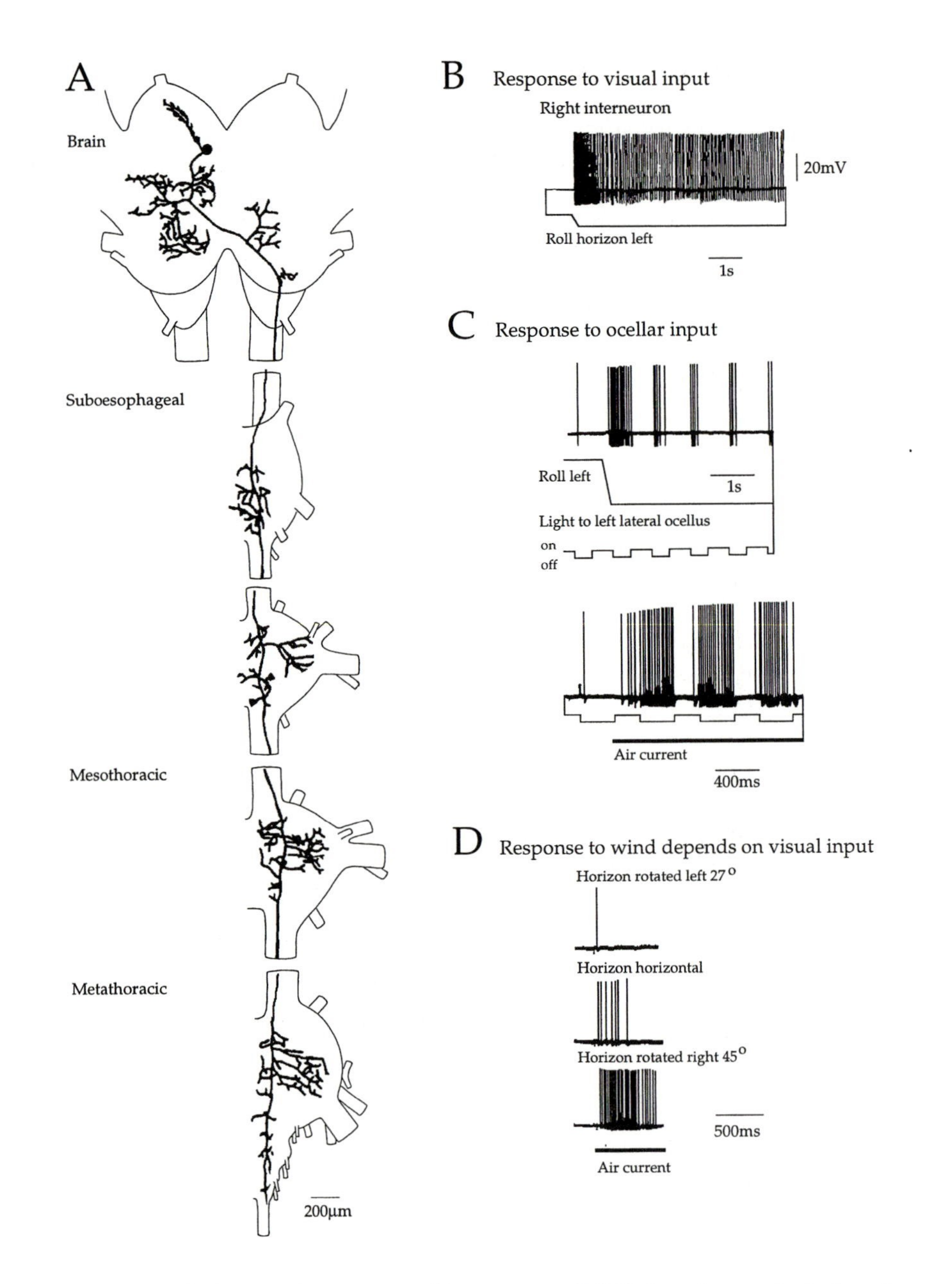

Fig. 11.22 Morphology and responses of interneuron DNC that detects deviations from the flight path. A. Morphology. The axon runs in the connective contralateral to the cell body in the protocerebrum. B. Spikes are evoked in the interneuron with its axon in the right connective when the horizon is rolled to the left. C. The response to a light input to the left ocellus is modified by movement of the horizon and by an air current to the front of the head. D. The response to an air current directed at the head depends on the position of the horizon. A is based on Griss and Rowell (1986), B–D on Rowell and Reichert (1986).

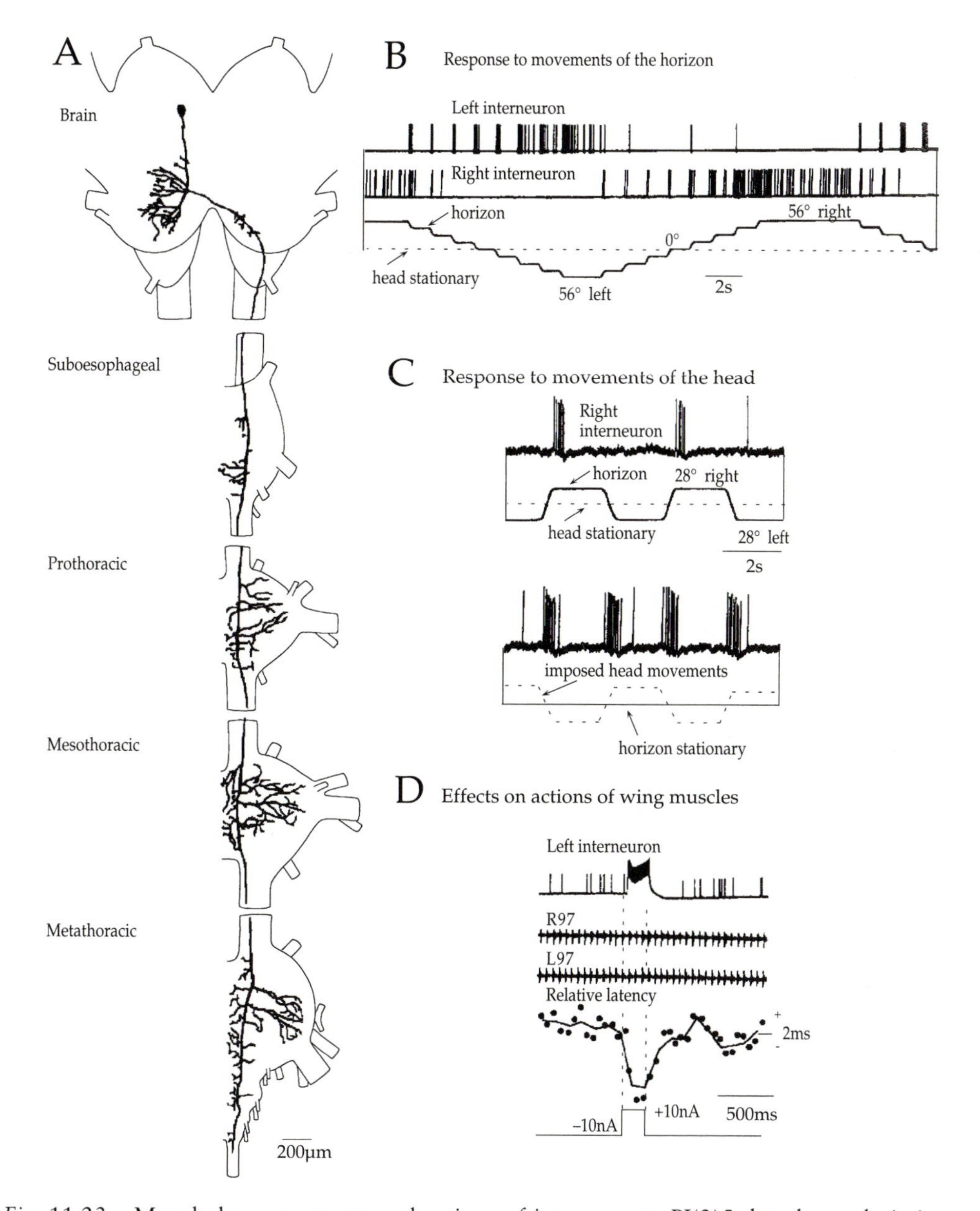

Fig. 11.23 Morphology, responses and actions of interneurons PI(2)5 that detect deviations from the flight path. A. Morphology. The axons run in the connective contralateral to the cell body in the protocerebrum. B. Responses of both interneurons to roll movements of the horizon. C. Responses to movements of the horizon with the head stationary, and to movements of the head when the horizon is stationary. D. A burst of spikes induced by intracellular injection of current into one interneuron changes the timing of spikes in the left and right basalar muscles 97 of the front wings. A–C are based on Hensler (1988), D on Hensler and Rowell (1990).

Table 11.2. Brain interneurons that could act as deviation detectors during flight

Sensory inputs	DNI descending interneuron ipsilateral	DNM descending interneuron median	DNC descending interneuron contralateral	PI(2)5 Pars intercerebralis interneuron (2)5
Ocellus	Ipsilateral	Median	Contralateral	no input
Compound eye	Roll down ipsilateral Pitch down Yaw ipsilateral	Pitch down	Roll down contralateral Yaw contralateral	Roll down ipsilateral Pitch up Yaw ipsilateral
Wind-sensitive head hairs	Yaw ipsilateral	Pitch down	Yaw contralateral	Inhibitory input
Neck proprioceptors	?	?	Excitatory input	Excitatory input
Combined deviation signal	Diving banked turn, ipsilateral	Dive	Banked turn contralateral	Climbing banked turn, ipsilateral

The names **DNI, DNM** and **DNC** refer to the ocellus that provides an excitatory input; ipsilateral and contralateral are relative to the axon of the interneuron. The name **PI(2)5** stems from the original anatomical description of this and other interneurons in a small group in the brain (Williams, 1975).

protocerebrum, with DNI and DNC having prominent branches in the lateral ocellar tract. DNI and DNM have axons ipsilateral to their somata, while the other two interneurons have contralateral axons. Branches are formed in the dorsal neuropil of all ganglia through which their axon passes, with only those of DNM extending to both sides. In the thoracic ganglia the branching patterns of the neurons are very similar, making the use of these branches hazardous for identification.

11.10.6.2. *Response properties*

Each interneuron receives a particular set of convergent signals from the compound eyes, an ocellus, wind-sensitive hairs and perhaps neck proprioceptors, which means that each responds to movements of the locust in space. Amongst the population of interneurons there is considerable overlap in their sensory responses, so that many parallel lines of signals are fed to the circuitry generating flight movements.

The input from an ocellus occurs when the light is turned off and is sufficient to generate one or a few spikes that arrive at a thoracic ganglion within 12 ms of a stimulus at 35°C (Reichert and Rowell, 1985). An input from the compound eyes gives information about large field movements in a particular direction, such as are provided by movements of the horizon. Overall, the best spike response is to a movement that takes the locust away from its preferred orientation relative to the horizon. These visual responses may be even more specific in that they depend on the initial starting position of the horizon, so that the closer the locust is oriented towards the preferred position for the interneuron, the greater will be the response. The input from the wind-sensitive hairs is generated by air currents at the normal velocity of flight. None of these interneurons is excited rhythmically by the interneurons that generate the flight motor pattern, but their overall responsiveness may be changed during flight.

The sensory inputs to these interneurons are thus complementary, but instead of simply summing, they interact in such a way that one modality alters the responsiveness to another in a graded fashion (Rowell and Reichert, 1986). The best response of an interneuron to a wind stimulus occurs at the preferred position of the horizon, and is less effective at other orientations (Fig. 11.22). Similarly, the illumination of an ocellus can alter the response of the visual input from the compound eyes, and for interneurons that receive inputs from neck proprioceptors there are further interactions between head movements and the sensory stimuli (Fig. 11.23). The sensory neurons of the neck proprioceptors project only to the prothoracic ganglion, so that their effects on these brain interneurons must be caused by other interneurons that convey the sensory signals to the brain. A characterisation of particular interneurons illustrates these features of the sensory coding.

The DNC interneuron (Fig. 11.22) with its axon in the left connective is excited when the compound eyes detect that the horizon has moved downwards to the right, and when the light input to the right lateral ocellus is reduced (light off stimulus) (Reichert *et al.*, 1985). The combination of these stimuli indicates a roll to the right. Similarly, this interneuron is excited by inputs from the compound eyes signalling a horizontal rotation of the visual world, and by inputs from head hairs signalling a wind input to the right side of the head. The combination of these stimuli indicates a yaw to the left. The interactions between these stimuli are such that illumination of the right ocellus during a roll to the right, when it would normally be darkened, inhibits the excitation from the compound eyes and from the wind-sensitive hairs. Similarly, the position of the horizon affects the response to wind stimuli in such a way that it is greatest at the position that also gives the greatest visual response. The response to wind also increases as ambient light decreases.

In interneuron PI(2)5 (Fig. 11.23), forcibly rolling the head in the dark produces only a weak response. Moving the head in either direction against a stationary visual pattern produces a response, whereas moving the visual field gives a response in only one direction. The response to an imposed head movement that supplies a stimulus in the preferred visual direction is greater than the response to a visual movement itself, suggesting a summation of inputs. If the head is held in different positions, then the response to movements of the visual horizon is altered.

Other interneurons that may act as deviation detectors can be grouped into four types (Hensler, 1992). First, those that signal position of the head but ignore visual input. Second, those that code for horizontal visual movements but ignore neck proprioceptive input. Third, those that receive visual and proprioceptive inputs that sum during normal deviation responses. Fourth, those in which deviation causes opposing visual and proprioceptive signals so that head movements are ignored and course deviations are recognised.

11.10.6.3. *Behavioural role*

To cause a steering change, the outputs of the interneurons need to alter the actions of motor neurons controlling movements of the head, wings, legs or abdomen. Intracellular stimulation of some of these individual interneurons can cause such changes. Stimulation of DNC or PI(2)5 causes shifts in the timing of the motor spikes to flight muscles within one wing beat cycle that are comparable to those that normally occur during flight (Hensler and Rowell, 1990). Similarly, stimulation of PI(2)5 causes the head to roll down (Hensler, 1992), whereas stimulation of other interneurons can evoke the appropriate ruddering movements of the abdomen (Baader, 1990). Some interneurons with cell bodies in the metathoracic ganglion that have either descending or ascending axons and that respond to visual stimuli may also cause steering movements by altering the movements of the abdomen, the head and the wings (Baader, 1990). The exteroceptive signals conveyed to the thorax by the descending interneurons from the brain are also fed back to control movements of the head in coordination with the ruddering movements of the abdomen and twisting movements of the wings.

To be effective, the stimulation of one of these interneurons must evoke spikes at a higher frequency and for a longer time than would normally occur during flight, indicating that normally a group of descending interneurons acting in parallel is responsible for effecting a particular change. It is a tempting further extrapolation to suggest therefore that no individual interneuron is vitally important but some interneurons may be more effective than others in particular circumstances. Clearly, parallel processing, summation of inputs from many sources, and flexibility are the important design features of the control mechanisms.

11.10.6.4. *Synaptic connections with motor neurons and interneurons*

To understand how the interneurons cause changes in the movements, it is necessary to know the pattern of connections that they make with the motor neurons and other interneurons that, in turn, drive the motor neurons. The required detail is, however, not available to determine whether the motor effects of the interneurons that would be predicted from their sensory response properties are realised by the pattern of connections with the flight neurons. Some of the interneurons do make direct connections with flight motor neurons and direct parallel connections with thoracic interneurons, and some of these thoracic interneurons are presynaptic to motor neurons (Reichert and Rowell, 1985; Hensler, 1992). It is assumed, but not shown, that the connections with the motor neurons are widespread and not restricted to the sorts of subsets that would be expected to adjust movements of the wings. Some interneurons are said to make connections with elevators, others with depressors, some to make excitatory connections, others inhibitory ones, but these patterns are not ascribed to a particular interneuron.

The thoracic interneurons that receive input from the descending interneurons are diverse, but all the ones examined so far are spiking interneurons and have cell bodies in the mesothoracic ganglion. It is assumed that similar connections are made with the homologous neurons in other thoracic ganglia. Some are local interneurons with processes restricted to this ganglion, while others have axons that project to different thoracic ganglia. For only a few of these interneurons is the input from a particular descending interneuron shown to be direct, and again their pattern of outputs to motor neurons is not known in enough detail to evaluate their contribution to an adjustment of a wing movement. During fictive flight, most of these interneurons are driven in the flight motor pattern by a synaptic drive from other interneurons.

11.11. MATCHING SIGNALS FROM HEAD RECEPTORS TO THE WING BEATS

The signals from the receptors on the head must be delivered to the motor neurons at a time when they can exert an effect most likely to bring about the desired flight manoeuvre. Such a coupling may also be necessary for the movements of the head (Hensler, 1990). Some coupling between the head and the thorax is purely mechanical because the rhythmic movements of the wings cause the head to bob up and down. These movements are sufficient to modulate the wind that impinges on the head hairs and antennae in time with the wing movements. Most of the coupling must, however, be brought about by the connections made by the descending interneurons with thoracic interneurons and motor neurons. The following design scheme has been proposed to explain the coupling (Reichert and Rowell, 1985; Reichert *et al.*, 1985). The connections with the motor neurons are considered to be weak on the basis of their small amplitude relative to the amplitude of membrane oscillations that underlie flight, and their inability to evoke spikes in a locust that is not flying. This design makes sense as stimuli received when walking will not then lead to contractions of the flight muscles. The connections with the thoracic interneurons and the connections that these interneurons make, in turn, with the motor neurons are considered to be stronger. Moreover, as the thoracic interneurons are depolarised and hyperpolarised rhythmically in the flight motor pattern, signals from the descending interneurons will only be effective in altering the output to the flight motor neurons when they occur during the depolarising phase of the thoracic interneurons. On this basis, the thoracic interneurons are thought to act as a gate, linking the exteroceptive signals to a particular phase of the wing beat cycle as determined by the timing of their depolarising phase in the flight motor pattern. Summation of sensory information delivered by the descending interneurons from the brain with the oscillatory drive from other flight interneurons in these thoracic interneurons and motor neurons means that the exteroceptive signals are expressed only in the appropriate context and at the appropriate phase of the wing beat cycle. The output of the thoracic interneurons could then advance, or delay, the participation of a motor neuron in the wing beat, or could recruit new motor neurons to the flight motor pattern.

It is not clear why the motor neurons themselves could not act as the gate; their membrane potential shows rhythmic changes in time with the wing beat so that an input from the descending interneurons would be effective only at a particular phase, just as is supposed for the thoracic interneurons. An interposed set of interneurons carrying out this gating function might, nevertheless, ensure that fewer unwanted spikes are generated in the motor neurons. Imposed roll movements during a flight motor pattern produce profound changes in the inputs to the flight motor neurons, but have little effect on the interneurons thought to generate the pattern (e.g. interneuron 511) (Fig. 11.24) (Reichert and Rowell, 1989). This suggests that steering is mediated through interneurons that are postsynaptic to such stable pattern-generating interneurons, but presynaptic to the motor neurons. It has not been demonstrated, however, that inputs from a descending interneuron increase the number of spikes produced by a thoracic interneuron during a particular phase of the wing beat cycle, or that the motor output is altered by this supposed 'gating' action. Some descending interneurons can shift the timing of the motor spikes but it is not known whether this effect involves sets of thoracic interneurons. Furthermore, no role is proposed for the direct connections with the motor neurons, which are

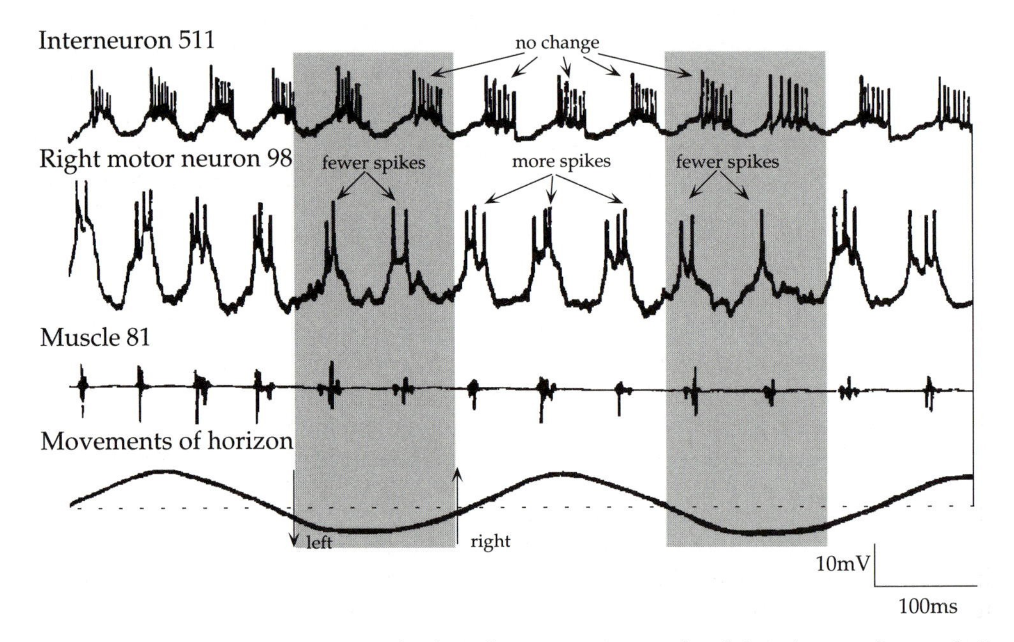

Fig. 11.24 Some interneurons involved in the motor pattern for flying do not change their activity during steering. The horizon is rolled to the left (shaded areas) and the spikes in a motor neuron to front wing basalar muscle 98 are decreased at each cycle, and increased as the horizon is rolled to the right. At the same time, there is no change in the pattern of spikes in interneuron 511. Based on Reichert and Rowell (1989).

simply advanced as another example, like those of the wing stretch receptors, of direct sensory input to motor neurons being subthreshold. Experimental observations nevertheless must always be placed in a behavioural framework, and the description of an input has to be related to the context in which it is observed. It has to be a basic tenet of these types of analyses that an input to a neuron will have some effect in some circumstances, unless it is to be dismissed as an evolutionary relic (*see section 11.14*). Another line of action for these thoracic interneurons that has not been considered is that the oscillatory drive in time with the wing beats allows them to sample sensory signals only intermittently.

11.12. CONTROL OF FLIGHT SPEED AND ALTITUDE

11.12.1. Flight speed

Speed is sensed by airflow receptors on the antennae, which signal the movement of the body relative to the air, and by the eyes, which signal movement relative to the ground. The airflow receptors also provide stability and allow the efficient use of the wing movements for developing lift and forward propulsion, but without the information from the eyes the locust would not know whether it was making any forward progress relative to the ground. Honeybees, for example, adjust their flight speed against a variable wind in an attempt to keep their speed over the ground constant (Heran and Lindauer, 1963). Just how insects measure their speed over the ground using visual signals is largely unknown but must involve the optomotor reaction. A first requirement must surely be that the altitude is both known and controlled, because the closer to the ground the locust is, the faster will be the apparent movement. The antennae, however, are the site of the main detectors of wind speed.

11.12.1.1. *Antennae*

The antennae are about 15 mm long and are usually held pointing horizontally forward at an angle of 50° to the long axis of the body. Each is capable of independent movement under the control of two pairs of muscles, both consisting of about 65 fibres, which move the two basal segments; one pair moves the scape and the second pair the pedicel, but the many articulated segments of the distal flagellum can only be moved passively. The two tentorio-scapal muscles in the head levate or depress the scape about its articulation with the head, so causing the antenna to move up and down. They are innervated by four and six motor neurons, respectively, including a common inhibitor, that have their axons in a branch of the lateral tegumentary nerve from the brain. The two muscles within the scape itself move the pedicel medially and laterally, and are each innervated by five motor neurons with axons in a branch of the antennal nerve. All the cell bodies and neurites of the motor neurons are in the deutocerebrum. Cell bodies in the same region as those of the motor neurons show proctolin-like immunoreactivity, as do nerve terminals on all the muscles (Bauer, 1991), and in crickets double labelling shows that many of the antennal motor neurons have proctolin-like immunoreactivity (Bartos *et al.*, 1994). Contractions of these muscles and their levels of basal tension are increased by proctolin, so that a combination of these two observations suggests that proctolin may be coreleased with a fast-acting neurotransmitter, presumed to be glutamate.

Air currents from directly ahead that initiate flight first cause passive lateral movements of both antennae that are subsequently followed by active medial movements in resistance reflexes maintained by increased contractions of the median scapal muscles. The deflections of the anntennae are detected by internal mechanoreceptors predominantly at their base, consisting of campaniform sensilla, a chordotonal organ and Johnston's organ, which itself contains several hundred sensory neurons. The feedback loop thus created results in the adoption of a flight posture in which the angles of the antennae relative to the long axis of the body are reduced so that they point more directly forwards. During continuous flight, this feedback loop pulls the antennae further forward as the wind speed increases, so that the torque exerted is proportional to the velocity of the air currents (Gewecke and Heinzel, 1980). If the mechanoreceptors are eliminated then the antennae show no compensatory movements with changes in air currents. The control loop is therefore a mechanism that allows both the proprioceptors and the exteroceptors to provide information about the velocity of air currents and hence about air speed, by continually matching the operating range of the proprioceptors to the velocity of the air currents. At the same time, the movements of the wings set up air currents around the antennae that cause them to vibrate in time with the wing beats. Nodding movements of the head and general vibrations of the body may also contribute to the vibrations of the antennae (Heinzel and Gewecke, 1987) and may enable receptors in Johnston's organ to respond to the static changes in antennal position when deflected by different air speeds.

When a tethered locust is flying in a wind tunnel, the wing beat frequency starts at a high level and then gradually declines to a lower and more sustained level. Immobilising the proximal joints of the antennae slows this decline in frequency and thus results in a greater sustained flight speed (Gewecke, 1975). The antennae also reduce the stroke amplitude of both the front and hind wings as the wind velocity increases so that lift is reduced, and when they are immobilised the stroke amplitude increases. In contrast, immobilising the patches of hairs on the head causes a rapid decline in the wing beat frequency to a lower level than normal so that flight speed is reduced. There are thus two opposing systems controlling flight speed through changes in wing beat frequency, with the antennal receptors normally throttling back flight speed. Both sets of receptors ensure that a locust uses an economical flight speed in terms of the energy expended for distance flown through winds of different velocities. The antennal receptors, however, have no effect on controlling the direction of flight, whereas the hair receptors on the head cause changes in the torque generated by the wings that compensate for an asymmetric wind stimulus (Gewecke and Phillipen, 1978).

The antennae, unlike those in bees, crickets and lobsters, do not track moving objects, but they do show avoidance and resistance reflexes. Touching an antenna results in a movement away from the source of the stimulus that can sometimes be followed by 2–10 small oscillations at 8–12 Hz (Saager and Gewecke, 1989). These reflexes are thus opposite to those that occur in response to air currents during flight.

11.12.2. Flight altitude

Locusts flying during the day generally maintain altitudes that can approach 1km, but towards dusk they fly at only a few metres above the ground, suggesting a need to keep in visual contact with the ground. Nevertheless, on moonlit nights they might be able to use visual clues to detect a drift caused by a cross-wind of only a few metres per second, even when flying at altitudes of 2 km (Riley *et al.*, 1988). It may not be necessary to measure the actual distance but to maintain a height, so that the ground has a certain contrast, or so that landmarks can indicate relative changes in height. This will inevitably lead to ambiguities that depend on the weather, but may explain why insects crash into featureless landscapes such as calm water or a snow field.

11.13. DEVELOPMENT OF FLIGHT

Locusts are able to fly only when they are adults because larvae do not possess fully formed and functional wings. Nevertheless, the neural and muscular components develop embryonically and postembryonically, so they are present in larvae. This means that many of the neural components for flight are present in advance of the structures that they control, and therefore raises the question of whether they are connected together appropriately to produce a flight pattern in advance of being able to move the wings.

11.13.1. Changes in the flight machinery

Moveable wings appear for the first time at the final moult from the fifth instar larva to the adult. In instars 1-3, there are small patches of cuticle, called wing pads, where the wings will eventually be formed, and in instars 4-5 these develop into independent, but immoveable wing buds. The motor neurons, the flight muscles themselves and some of the sensory neurons associated with the wing hinge are formed embryonically and can function in the larval instars, although contraction of the muscles can only distort the shape of the thorax.

11.13.1.1. *Motor neurons*

The motor neurons of the wing muscles are all formed in the embryo and the basic morphology of their branches in the central nervous system is recognisable in the early instars. This basic but characteristic shape of an individual motor neuron is then embellished by the addition of fine branches. Synaptic connections with the muscles are established at least by the second instar, although at this stage the muscles are still rudimentary.

11.13.1.2. *Sensory neurons*

Sensory neurons involved in flight comprise those on the body, which may have functions other than the control of flight, and those on the wings, which can only develop as the wings themselves are formed. There are, nevertheless, postembryonic changes in both populations. The number of hairs on the head that detect air currents and supply inputs to many descending interneurons increases throughout the larval instars (Svidersky, 1969). Sensory neurons associated with receptors on the wings are also formed progressively during the larval instars as the structure of the wings themselves develops (Altman *et al.*, 1978). The nerve (N1C1) that innervates a wing and a tegula contains fewer than 50 axons in the first instar but this number increases gradually to about 200 by the third instar, and then abruptly to almost 900 at the fourth instar as the wing buds are first formed. In the young adult, the number increases to 2000, largely as the result of the addition of small diameter profiles, before being reduced to 1000 in mature adults (Table 11.3). The increase is associated with the ingrowth of axons from receptors on the wing, but the reduction in the adult has not been linked to degeneration of axons as the result of failure to establish synaptic contacts with central neurons in competition with other axons. The final tally contains some 70 axons from the tegula with the rest from wing receptors. In N1D1, which innervates receptors at the wing hinge, there is also an increase in the number of axons to the final moult, probably from the hair fields, but then no subsequent reduction.

The sensory neuron of each wing stretch receptor, by contrast, can first be identified in a 40% development embryo, when its peripheral cell body starts to produce an axon that grows towards the central nervous system (Heathcote, 1981) (see Fig. 4.7). In a 60% embryo, the sensory neuron becomes able to produce action potentials but the basic pattern of arborisations in the central nervous system is not formed until 80% development. Further maturation postembryonically involves the growth of fine branches from the basic structure that is laid down in the embryo. Synaptic connections with depressor motor neurons are established at least as early as the third instar (Heathcote, 1980) so that a functional circuit is established before the wings are formed and before they can be moved. The stretch receptor in the larval instars might therefore function to signal movements in the thorax, such as those caused by ventilatory movements (Möss, 1971).

11.13.1.3. *Interneurons*

Few of the interneurons involved in flight have been followed through embryonic or postembryonic development, but it is assumed that all are formed embryonically and

Table 11.3. Developmental changes in the numbers of axons in nerves supplying the wings

Nerve	First instar	Third instar	Fourth instar	Fifth instar	Young adult	Mature adult
N1C1	<50	<200	900	1300	2000	1000
N1D2	50	100	150	200	400	400

achieve their adult pattern of branching and connections by the time of hatching. This supposition still allows for postembryonic elaboration and tuning of branching patterns and connections. The TCG interneuron, for example, has all the features of its adult shape in the first instar larva (Kutsch and Hemmer, 1994), but the latency of its response to wind stimuli decreases and the number of spikes it gives increases during the larval instars, perhaps because of an increased number of synapses from the increasing number of hairs on the head. The potential for this interneuron to elicit flight is thus present well in advance of the capability of the mechanical machinery to produce flight. The development of DCMD (Bentley and Toroian-Raymond, 1981) and the giant interneurons with inputs from cercal hair receptors is also essentially complete by hatching, with only a further elaboration of branches occurring postembryonically.

11.13.2. Development of the flight motor pattern

The early development of many of the neural components of flight, but the late development of the moveable wings raises the question of when the flight motor pattern is first assembled and when it can first be expressed. Does a locust show behaviour associated with flight before it is able to fly and is a motor pattern that has features allowing it to be called a flight motor pattern produced before real flight is possible? If so, do the behaviour and the motor pattern develop during the progressive developmental stages and mature in the adult?

11.13.2.1. *When does a flight motor pattern first appear?*

If second instar larvae are suspended in an air stream, the front legs adopt a folded posture similar to that of adults when flying, and fifth instar larvae adopt the complete adult flight posture (Cooter, 1973). While this may indicate the premature development of a posture that is of no use until further developmental changes have taken place, it may also represent some elements of the posture that is used when jumping, a behaviour that larvae perform regularly (*see Chapter* 2). Adult flightless grasshoppers (e.g. *Barytettix psolus*) will also assume a characteristic posture with their legs when suspended in a wind stream, and will move their hind legs in ways appropriate for steering when the direction of the wind changes (Arbas, 1983).

Early larvae held in a wind stream for several minutes eventually show sequences of motor spikes to their flight muscles that lead to co-contraction of the elevators and depressors, but there is no indication of a rhythmic pattern (Kutsch, 1985). In fifth instar larvae, the sequences of spikes become more regular, lead to alternating contractions of the elevators and depressors, and can show a rhythmicity with a period of about 10 Hz. In newly emerged adults, the pattern is more regular but still of the same period, with clear alternation between elevators and depressors of one wing. It cannot, however, be sustained for more than a few seconds and lacks the delay between the spikes in front and hind wing motor neurons that would lead to the correct phasing of the wing movements. Thus, some semblance of a flight motor pattern seems to be present and can be expressed before it is used behaviourally.

Injection of octopamine into the metathoracic ganglion of dissected larvae as early as the first instar can release a pattern that resembles that of normal flight in the alternation of the elevator and depressor motor neurons, the duration of the D-E and E-D latencies and their dependence on the period of the rhythm (Fig. 11.25) (Stevenson and Kutsch, 1988). The period of the rhythm is similar to that in deafferented adults. This indicates that the central connections which are necessary for the production of a

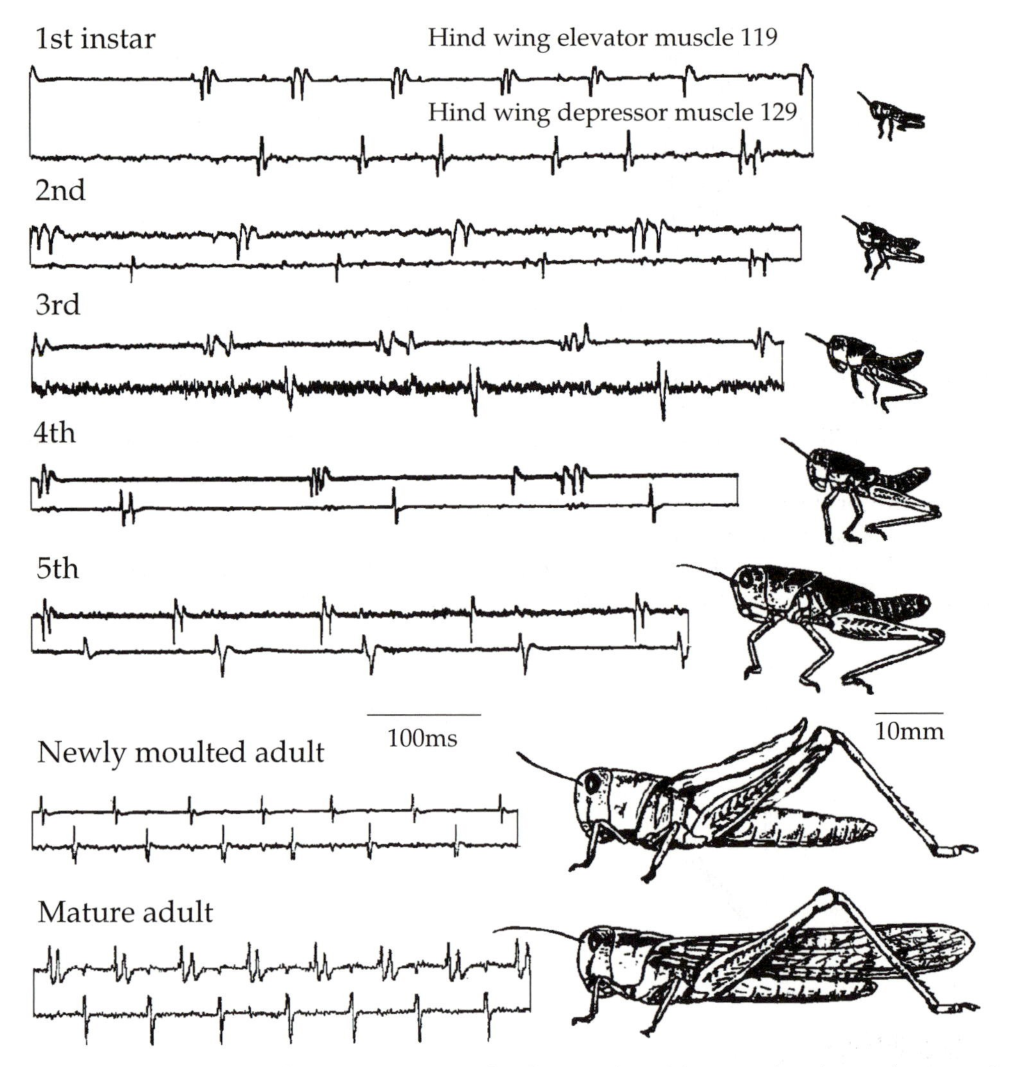

Fig. 11.25 Expression of the motor pattern for flying in larval locusts after the application of octopamine. Recordings are made from an elevator (119) and a depressor (129) of a hind wing in the five larval instars, and in newly moulted and mature adults. In the early larval instars the pattern can only be expressed in the presence of octopamine. Based on Stevenson and Kutsch (1988).

basic pattern for flight are established in the embryo, but that the full expression of the flight motor pattern normally awaits the development of moveable wings and the concomitant sensory feedback that accompanies their movement.

11.13.2.2. *Does the pattern change during adult life?*

Over the first 3 weeks of adulthood, the body mass of a locust increases and the cuticle stiffens but the only variable of the flight motor pattern that alters to match these changes is the frequency of the wing beats, which almost doubles (Fig. 11.26) at an exponential rate that depends on the temperature (*Locusta migratoria*, Kutsch, 1971, 1973). Within a few hours of the final moult, a locust adopts a flight posture when placed in a wind stream, and can flap its wings in a coordinated way, but because the wings are still soft they tend to crumple and therefore generate little lift. By the second day of adulthood, locusts can maintain their altitude in free flight (Kutsch and Gewecke, 1979), but with a wing beat frequency of only about half that of mature adults. Both the wing stroke angles and the stroke plane are by now fully established. The time spent in the air and the vigour of the flight sequences then increase day by day, and migratory flights are undertaken in the second week. The cause of this increase in the wing beat frequency, which seems to be characteristic of many, if not all insects, has been hard to pin down, but four factors that do not appear to influence it can be stated.

First, it still occurs if use of the wings is prevented by waxing them in the folded position.

Second, it does not depend on flight experience as it occurs with the same time course in locusts prevented from flying and those allowed a daily training flight.

Third, it does not depend on changes in sensory feedback as an increase still occurs in locusts deprived of sensory feedback from wing hinge receptors from the

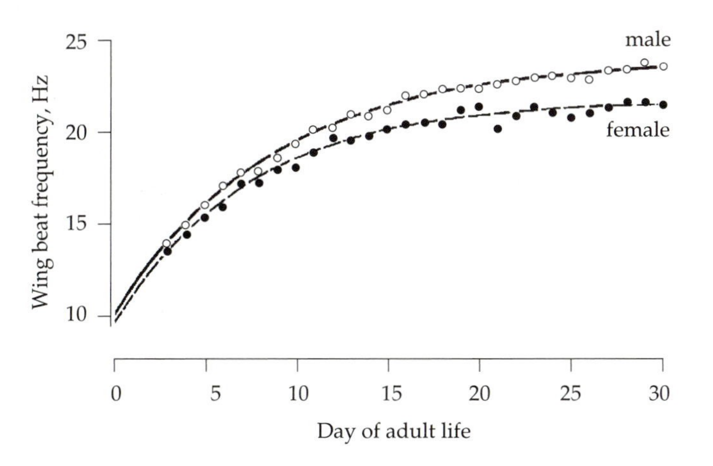

Fig. 11.26 Changes in the frequency of wing beats during maturation of adult locusts. The frequency increases from about 10 Hz on the first day of adult life to more than 20 Hz when mature. Initially, both males and females have the same frequency, but eventually males have the higher frequency. Based on Kutsch (1973).

day after the final moult. These deafferented locusts can fly within a week of the operation and the changes in their wing beat frequency parallel, and reach final values close to, those in normal animals. This interpretation assumes that no plastic changes occur as a result of the lesions (*see section 11.6.2*).

Fourth, it does not depend on the level of juvenile hormone in the haemolymph, because, although this rises in an adult at a rate that almost parallels the increase in wing beat frequency, destroying the corpora allatal cells that secrete it using precocene (*see section 4.3.2*) does not prevent the rise in frequency (Kutsch and Stevenson, 1984). After each larval moult, there is a transient increase in the level of juvenile hormone in the haemolymph. The absence of this hormone in fifth instar larvae ensures that the next moult will be to an adult. If precocene is applied to fourth instar larvae, precocious adultiform fifth instars are formed that are about the same size as normal fifth instar larvae would be, but now actively flap their miniature wings. Azadirachtin, from the Indian neem tree, prevents moulting so that when applied to fifth instar larvae they remain as permanent larvae and fail to moult into adults. A flight motor pattern can be more readily elicited from these animals as they age, but the frequency remains at 10 Hz.

These negative results indicate that there must be changes in the performance of some of the neurons involved in flight as the locust matures. Few that could account for the maturation of the pattern have, however, been found. The wing stretch receptor produces more spikes at each imposed movement as maturation progresses and the rate of adaptation to a step displacement of the wing also increases (Gray and Robertson, 1994). At the same time, the diameter of its axon increases so that its spikes reach the central nervous system earlier, but the EPSPs that they generate in most interneurons remain of constant form, even though the branching pattern of the interneurons may increase (Gee and Robertson, 1994). The changing response properties could reflect changes in the mechanics of the wing hinge as the cuticle hardens but could also be caused by intrinsic changes in the neuron like those in the diameter of its axon, and the interplay between these two factors may change during maturation. If more spikes are delivered earlier in the wing beat cycle for an equivalent movement of the wing, then the timing of the depressor phase of the wing beat cycle will be advanced, so raising the overall frequency of the wing beats.

The mechanism(s) responsible for the gradual increase in wing beat frequency thus remains unresolved but could result from maturation of the properties of the neurons and their connections in the central nervous system, perhaps under the influence of circulating amines and peptides and, of course, in conjunction with the changes listed above.

11.14. EVOLUTION OF FLIGHT MECHANISMS

The acquisition of flight by insects was a profound evolutionary development that occurred some 330–400 million years ago at the end of the Devonian (Wootton, 1985). The finding that interneurons involved with the generation of the flight motor pattern are distributed in five neuromeres (T2,3 and A1-3) even though only two segments have wings, has led to speculation as to what this arrangement might tell us about the evolution of flight (Robertson *et al.*, 1982). Just how flight arose has sparked speculation for almost 200 years (reviewed most recently by Ellington, 1991), but the current theories can broadly be reduced to two, although both suppose that it arose just once.

First the **Paranotal Lobe** theory (Müller, 1875; Crampton, 1916) suggests that wings evolved from expansions of the thoracic terga that originally functioned to protect gills or spiracles. They then became adapted for use in display and temperature regulation before being used in flight, initially in gliding only, but then, with the secondary derivation of an articulation, in flapping flight.

Second, the now more widely accepted **Pleural Appendage** theory (Oken, 1811; Wigglesworth, 1963; Kukalova-Peck, 1978, 1983) suggests that wings evolved from articulated and moveable (by the same muscles that move the gills of present day mayflies) pleural appendages (exites of basal leg segments), that were repeated along the thoracic and abdominal segments and that these have been progressively reduced to the one or two pairs of wings on present insects. The fossil record indicates that juvenile Paleoptera from the Palaeozoic period did indeed have articulated thoracic and abdominal winglets (Kingsolver and Koehl, 1994), but the record is very incomplete and there is no equivalent of an *Archaeopteryx* (Wootton and Ellington, 1982). These winglets may have been used by terrestrial insects for courtship displays, as heat exchangers, or by aquatic insects for gaseous exchange, as fins for swimming, or for generating a ventilatory current and thereby increasing ventilatory efficiency, finally becoming adapted for flight when these insects became terrestrial. An origin from thermoregulators is favoured by experimental models of insects with pro-wings (winglets) which suggest that for all sizes of body there is a wing length below which there is a thermoregulatory but not aerodynamic advantage (Kingsolver and Koehl, 1985), but other treatments of these models suggest that even the smallest of winglets would help to decrease the terminal velocity of a fall (Ellington, 1991). If the wing length increases, the thermoregulatory advantages decrease as the aerodynamic advantages increase, so allowing a smooth transition of function. For small insects moving at low speed, small winglets would have little aerodynamic advantage, but once the winglets are above a certain size then their aerodynamic contribution would be felt. For larger insects that will inevitably fall at faster speeds, even small winglets confer an aerodynamic advantage.

Both theories suppose that the original pro-wings allowed an insect falling from a tree to control its altitude by parachuting, then by gliding and finally by powered flight. The crucial factor is that these insects were sufficiently large that even small winglets would help to stabilise a fall, then to reduce the angle of descent and thus prolong gliding. Twisting movements of the articulated winglets could then be used to control gliding performance. Further advances would involve the addition of simple flapping movements that could further prolong a flight and the introduction of twisting movements to these flapping movements that could add lift.

A possible intermediate stage in the evolution of these winglets could have been their use in skimming along the surface of water (Marden and Kramer, 1994). Modern stoneflies progress in this way by flapping their hairy wings, but keeping their tarsi in contact with the water so that aerodynamic forces produce forward thrust but do not have to be large enough to support the weight of the body. The larger the wings, the more adroit the stoneflies are at skimming. Such a mechanism would allow the progressive evolution of the gill appendages first as devices for rowing across the surface of the water, then skimming and finally generating enough power to leave the surface that supports the weight of the body and take to the air. This might allow selection during evolution for the production of thrust, but not for support of the body during flying, which is essential.

11.14.1. Interneurons and the evolution of flight

The widespread distribution of neurons involved in flight lends some weight to the pleural appendage theory, and suggests that these neurons may originally have been used to control the movements of the many articulated appendages. Evolution must therefore have acted to reduce the number of appendages but to retain the use of the distributed interneurons for the control of the remaining ones. The concentration of the T3 and A1-3 neuromeres into one ganglion may have been a further refinement to improve flight ability, by reducing conduction times between the controlling elements. This hypothesis is, however, weak on two grounds. First, many insects that do not have this fusion of ganglia are nevertheless good fliers. Second, it is not known whether the same sets of interneurons are repeated in the other, unfused abdominal ganglia.

Whatever the contribution that neurobiology can make to this problem, it is worthwhile to reflect that the nervous system has evolved and that the precursor upon which evolution acted may have produced behaviour that is quite different from that required by today's representatives. The current design of the nervous system for a particular task may not therefore be the optimal solution that an engineer would use if designing from scratch, because natural selection can operate only on what already exists. Moreover, there may be more than one neural solution to a behavioural problem and yet evolution will act only on the overt behaviour that is produced (Dumont and Robertson, 1986). The extrapolation of this argument can, however, attribute any ill-understood, or puzzling feature of the nervous system to a consequence of evolution and therefore attribute it no functional role. The argument should therefore be used cautiously and sparingly.

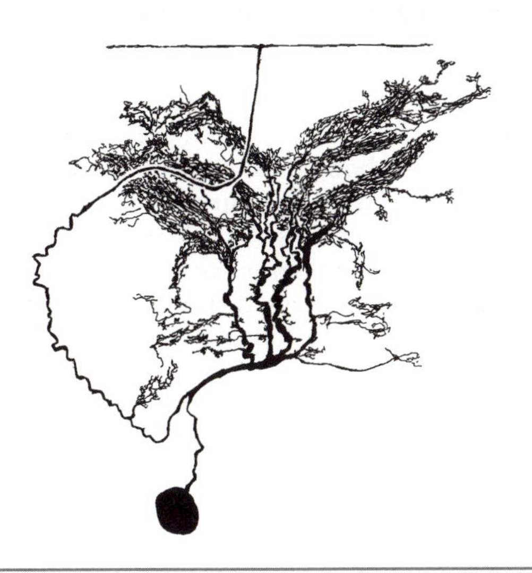

Breathing

12.1. BREATHING MOVEMENTS

The neural control of breathing or ventilatory movements has attracted much attention over the years because of the persistence and reliability of the underlying motor rhythm. The persistence derives from the obvious need to provide a continuous exchange of gases between the tissues of the body and the surrounding air. The drive maintaining the rhythm must thus come either from receptors that monitor the levels of carbon dioxide in the inspired gases or from some monitor of its level in the tissues. Probably both effects occur together but the receptors responsible are not well identified. Nevertheless, the ventilatory motor pattern can be recorded as readily from an isolated central nervous system as it can from an intact locust.

The study of ventilation in insects stems from the early days of electrophysiological analyses of the central nervous system. In some of the first observations of spikes ever made from any nervous system, rhythmic bursts were recorded from an isolated nerve cord removed from the water beetle *Dytiscus marginalis* and these were assumed to be the underlying cause of ventilatory movements (Adrian, 1931). The period of this rhythm generated by the abdominal nervous system could be altered by removing more anterior parts of the nervous system. These simple observations established that a motor rhythm could be generated in a particular region of the central nervous system, and that it could then be influenced by descending commands from the brain. Subsequently it has been shown that the centrally generated rhythm can also be influenced by sensory feedback resulting from the movements themselves, or from external influences. The most detailed information on the neural control of ventilation in insects has been obtained from the locust.

12.1.1. Pattern of movements

Air is circulated throughout the body in a network of tubes called tracheae which convey gases directly to the tissues of the body so that the haemolymph is relegated to only a minor role in gaseous exchange. Movement of the air in this network is caused by rhythmical pumping movements of the abdomen consisting of expiratory and inspiratory phases. The air can enter or leave the body through paired openings, called spiracles, in the cuticle of most segments, whose aperture is under direct muscular control. In normal ventilation, air is drawn into the body during inspiration through the open four pairs of spiracles on anterior segments while the spiracles on posterior segments remain closed. The pattern is then reversed during expiration so that the air is expelled through the posterior spiracles which are now open while the anterior ones are closed, thus creating a flow of air in the body.

The persistence of the breathing rhythm under a wide range of circumstances often leads to the false assumption that the pattern is consistent, and capable of change only in the frequency with which it is expressed. This is reinforced by the restrictions placed on insects that are examined experimentally and which tend to show a more consistent pattern. In unrestrained locusts, however, the breathing movements can vary widely in form, pattern and frequency. The frequency can be as low as 0.08 Hz when no locomotion is occurring and can even be punctuated by long pauses lasting 30 min, but can rise 50-fold to 4 Hz after vigorous activity such as flight. Ventilatory movements are essential to inflate the body and, in particular, the wings during moulting, and the day after the moult ventilation can still be intermittent. In very young or very old locusts that are quiescent, ventilation can also be intermittent (Hamilton, 1964), and in the large grasshopper *Romalea microptera* ventilation can be markedly cyclical, with the period strongly affected by temperature (Quinlan and Hadley, 1993). In other insects, the patterns of ventilatory movements can vary enormously and often depend on the current behaviour. At rest, some cockroaches (*Byrsotria fumigata*) produce only a few pumping strokes at a frequency of 5 min^{-1} separated by quiescent periods of up to 7 min, so that they ventilate for only 4 min in every hour.

12.1.2. Oxygen consumption

The extensive network of tracheae and associated air sacs contain some 0.5–1.0 ml of air, with almost 100–150 µl of this in the large thoracic air sacs. At rest, about 40 litres of air per kg body weight is pumped every hour, but at maximum rates this can rise steeply so that 50 µl is pumped at each ventilatory stroke by a 2 g adult locust (Weis-Fogh, 1967). A locust at rest will consume about 0.63 litres of oxygen.$kg^{-1}.h^{-1}$, or 18 µl.min^{-1} (Krogh and Weis-Fogh, 1951), but when flying there is a dramatic 24-fold increase. Increasing the abdominal pumping movements can only contribute a 4–5-fold increase, so that additional ventilatory mechanisms must then come into play. The flight muscles themselves contribute by distorting the shape of the thorax.

12.1.3. Water loss

When the spiracles are open they provide a direct route for water loss from the body, and thus for all insects and particularly those like locusts that live in hot deserts, it would seem that there may need to be compromises between the beneficial effects of opening the spiracles to permit gaseous exchange and the detrimental effects of dehydration. Often, the patterns of intermittent (discontinuous) ventilation have been seen as a way of reducing water loss through the open spiracles, but in *Romalea microptera* at least, transpiratory water loss during ventilation accounts at most for only 4% of the total water loss from the body. Furthermore, intermittent ventilation only occurs in *Romalea microptera* that are not moving, so that its usefulness in active behaviour must be restricted. The conclusion in these grasshoppers is that opening the spiracles to allow exchange of gases does not contribute significantly to water loss. In insects where cuticular water loss is less then clearly water loss during ventilation assumes more significant proportions, and discontinuous ventilation becomes a method of limiting this (Lighton, 1994).

12.1.4. Behavioural constraints on breathing

Various behavioural considerations, including those just described, set constraints on the neural mechanisms that control ventilation; the central nervous system must deliver a reliable pattern of motor spikes to generate the movements, but one that can be changed rapidly and that can also operate over a wide range to meet the fluctuating demands of behaviour. Nevertheless, the usual pattern of ventilation that is examined for its neural control is typical of the hyperventilation that occurs after strenuous activity or when the animal is stressed by high concentrations of carbon dioxide in the air. The abdominal movements are then strong, regular and rapid with the activity in motor neurons innervating expiratory and inspiratory muscles alternating with some precision. Analysis of these specific patterns is valid for the objective of understanding how the basic rhythm is produced, provided it is recognised that in normal behaviour the ventilatory movements can vary widely.

12.2. THE MACHINERY FOR BREATHING

12.2.1. Network of tracheae

The spiracles are linked to large tracheae that are stiffened by helically arranged strips of cuticle so that they do not collapse yet remain flexible. These tracheae then branch extensively to form finer and finer tracheoles that penetrate all the tissues and even individual cells. Some penetrate the basal lamina surrounding cells of the midgut so that they establish direct contact with epithelial cells of the midgut. The tracheoblast cells that contain the terminal tracheoles have many fine branches that can reach out to supply new cells as the insect grows. The tracheoles have ends that are filled with fluid and in which the gaseous exchange with individual cells takes place. The network of tubes thus ends blindly in the target tissues, so that gaseous movement is tidal in the larger tracheae and final exchange is by diffusion at the tips of the tracheoles. In the long tracheae that supply the legs, much of the movement of the gases must be by diffusion, unless it is aided by contraction of the muscles in locomotion or by the myogenic contractions of a small group of fibres in the extensor tibiae muscle of the hind leg (*see Chapter 5*). So extensive are the branches of these tracheoles that in the muscles and the central nervous system no cell is more than 5–10 μm from the end of a tracheole and hence from the site of gaseous exchange (Longley and Edwards, 1979; Burrows, 1980b) (Fig. 12.1). The importance of gaseous exchange in the central nervous system is indicated by the extensive tracheation of the thoracic ganglia which are supplied directly by short branches from a pair of large diameter tracheae running the length of the body. In some insects, the proper functioning of neurons fails rapidly when the tracheae to the central nervous system are interrupted. The tracheae also form a network of air sacs, particularly in the thorax and close to the spiracles, which inflate and collapse with each cycle of ventilation. They therefore increase the volume of air in the tracheal system and smooth the changes in pressure that occur with each cycle of the rhythmic pumping movements of the abdomen.

This elaborate and complex system for the rapid exchange of gases with the tissues of the body also has the potential for the rapid dissemination of insecticides and baculoviruses within the body (Engelhard *et al.*, 1994). The basal lamina of most cells provides a physical barrier to particles even as small as viruses, but intracellular invasion through the tracheoblasts circumvents this barrier and provides a means of spread within the body through the interdigitating tracheal epidermal cells that form a network of channels containing lymph.

12.2.2. The abdominal pump

The pump that drives the air around the body, and that sucks it in through anterior spiracles and forces it out through posterior spiracles is powered by contractions of the abdominal muscles. Expiration provides the most power but inspiration also involves active muscular contraction in addition to elastic recoil. Although the abdomen consists of 11 segments, only the middle ones actively contribute to the ventilatory movements because segments 1 and 2 lack dorso-ventral inspiratory muscles, while segments 9–11 are modified to bear the genitalia. The ganglia of segments 1–3 are fused to the metathoracic ganglion and those of segments 8–11 are fused to each other (*see Chapter 2*). The movements of each segment involve the action of 13 muscles (Fig. 12.2 and Table 12.1) which contract rhythmically to generate the pumping movements that circulate the air around the body (Fig. 12.3). The dorso-ventral muscles lift the sternites upwards and into the body cavity whereas the two inspiratory muscles in each segment push them downwards and outwards (Hustert, 1975). The abdomen is thus alternately contracted and expanded so that air is forced through the longitudinal tracheae. The other large muscles in the abdomen, the dorsal longitudinal muscles, contract in the ventilatory rhythm only when ventilation is stressed, when they cause telescoping movements of the abdomen. The thin sheet of muscle that forms the ventral diaphragm, running across the abdomen in the horizontal plane just dorsal

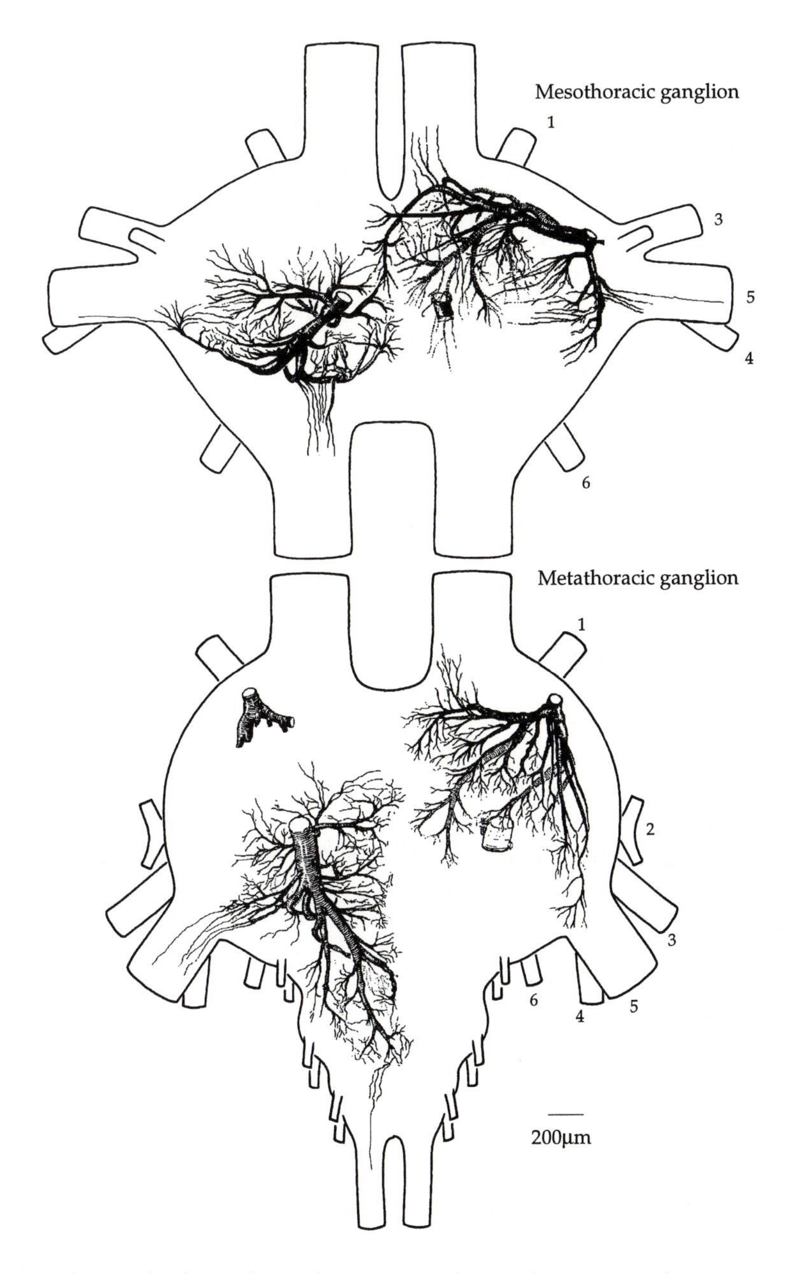

Fig. 12.1 The tracheal supply to the meso- and metathoracic ganglia. Four tracheae enter each ganglion and divide to form an extensive network of tracheoles that ensure that no neuron is more then 5 μm from the site of gaseous exchange. The branches of the posterior trachea on the left and the anterior one on the right are drawn after they had been injected with cobalt and developed in the same way as that used to reveal the shapes of neurons. Based on Burrows (1980b).

Table 12.1. Abdominal muscles involved in breathing (segments A2-5)

Movement	Muscle	Segment			
		A2	A3	A4	A5
Expiration	Dorso-ventral	157/8	175	190	205
		159	176	191	206
		157/8	178	193	208
		161/2	179	194	209
	Spiracle opener	165	180	195	210
Inspiration	Dorso-ventral	–	177	192	207
	Longitudinal	156	174	189	204
	Spiracle closer	166	181	196	211
Auxiliary expiration	Dorsal longitudinal	149–155	167–173	182–188	197–203

to the chain of ganglia, contracts in time with inspiration but this is probably not its only function.

12.2.2.1. *Motor supply of the abdominal pump*

The muscles that provide the force for the ventilatory movements are innervated by the abdominal ganglia. Each of the unfused abdominal ganglia (A4-7) has two pairs of lateral nerves and a single median nerve.

Lateral nerve 1 (**N1**) contains the axons of about 30 motor neurons, 24 with cell bodies on one or other side of the ganglion, and six with axons leaving in the ipsilateral anterior connective to their cell bodies in the next anterior ganglion, some of which innervate dorsal longitudinal muscles.

Lateral nerve 2 (**N2**) contains the axons of some 13 motor neurons with cell bodies on either side of the ganglion that innervate dorso-ventral muscles of the same segment. These motor neurons spike during expiration (Fig. 12.3).

The **median nerve** contains the axons of four motor neurons with cell bodies at the midline of the ganglion that innervate spiracular and inspiratory muscles in the next posterior segment. These motor neurons spike during inspiration.

12.2.2.2. *Morphology of the motor neurons*

The motor neurons have a variety of shapes, with axons either ipsilateral or contralateral to the main areas of neurites and to their cell bodies (Fig. 12.4). Some have axons in the connectives that run to the next ganglion from which they emerge in one of its lateral nerves to innervate their muscles. They do not form branches in the ganglion distant from their cell body. Unlike all the major leg motor neurons, they can have branches in both halves of a ganglion containing their cell bodies,

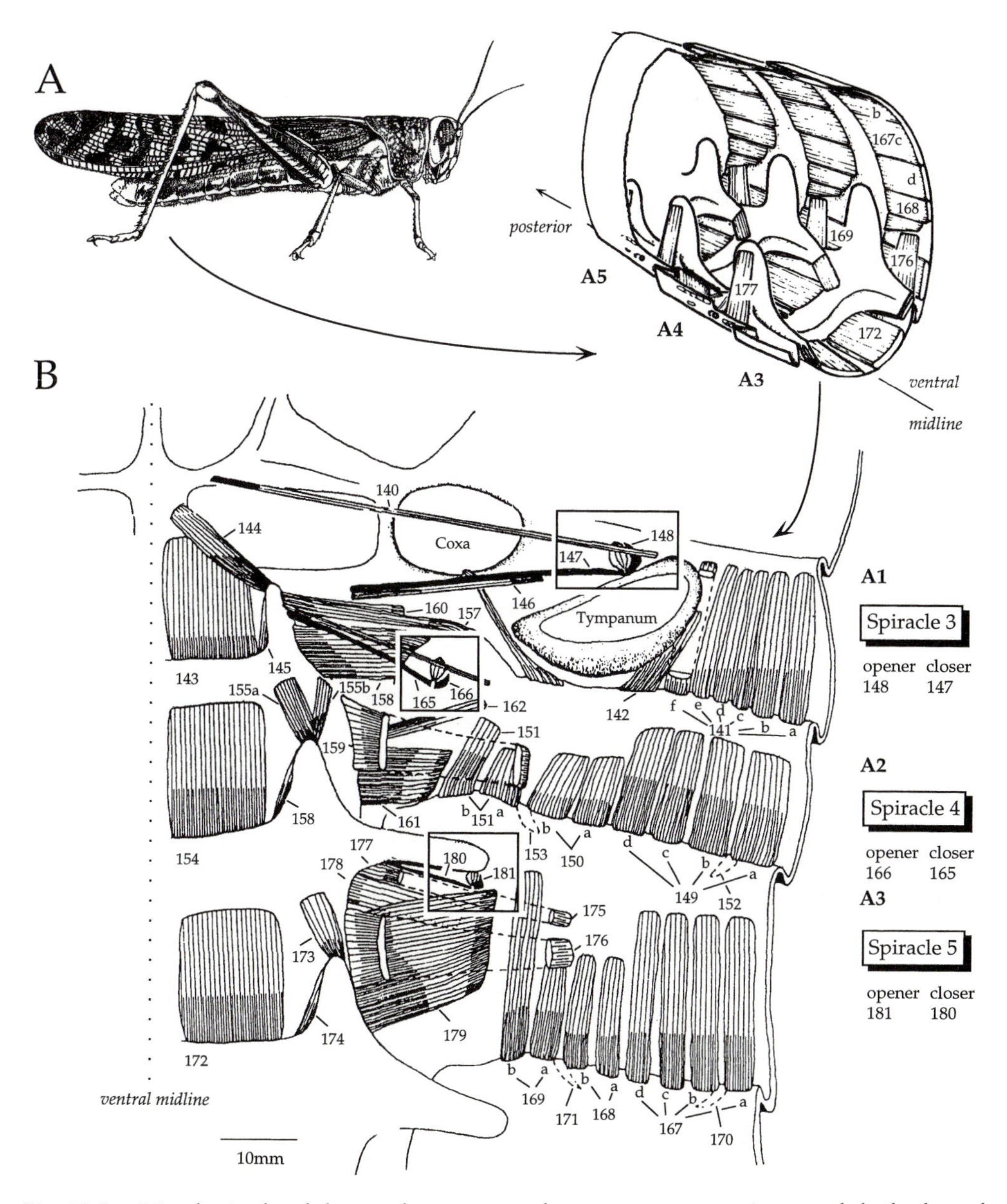

Fig. 12.2 Muscles in the abdomen that generate the power to pump air around the body and allow it to enter and leave through the spiracles. A. The general arrangement of the muscles in a cut-away view of the anterior part (segments A3-5) of the abdomen. B. Details of the muscles in the right half of segments A1-3 of the abdomen. The segments are laid flat with the ventral midline on the left and the dorsal midline on the right. The muscles are numbered according to Snodgrass (1929) (see Table 12.1 for their actions). A is based on Hustert (1975), B on Hill-Venning (1988).

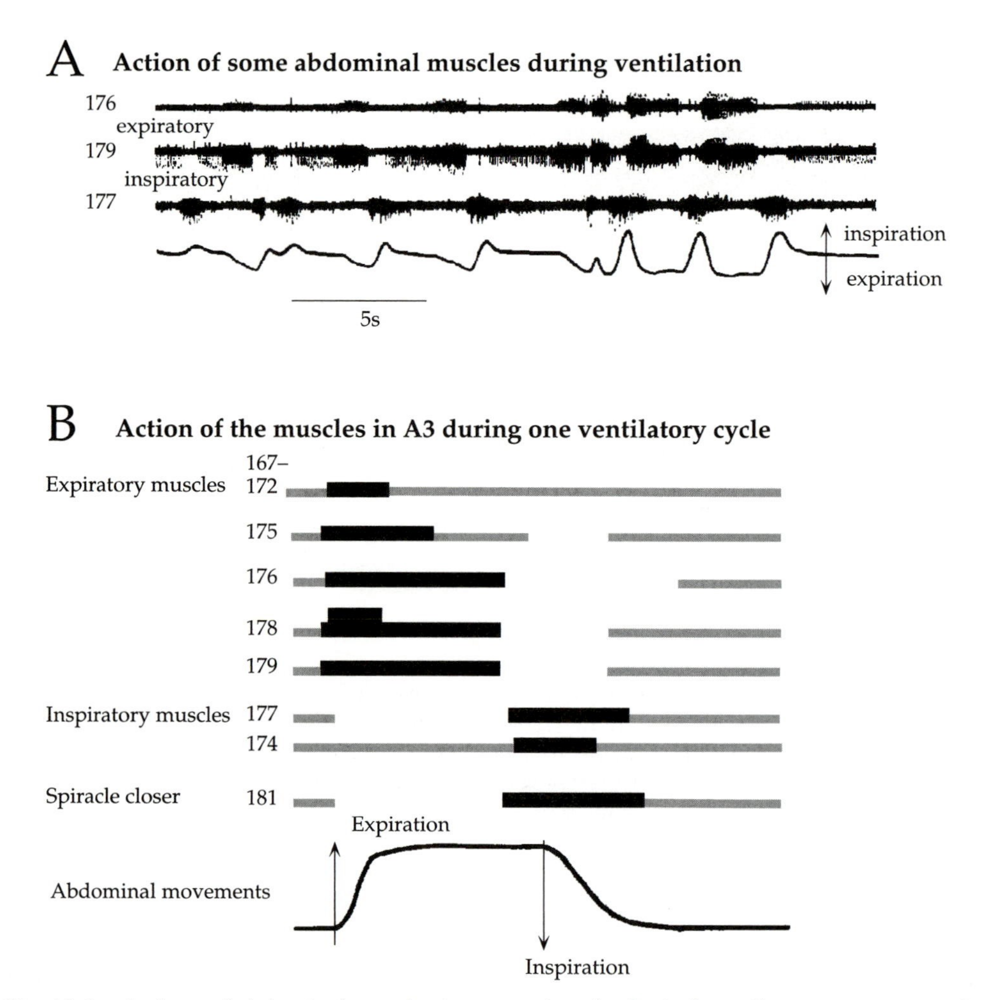

Fig. 12.3 Actions of abdominal muscles in generating rhythmical ventilatory movements. A. The actions of two expiratory (176, 179) and one inspiratory (177) dorso-ventral muscle of segment A3 during ventilation in a freely moving locust. The sequence starts with a series of slow movements followed by a few cycles of fast and deep ventilation. B. The actions of the main expiratory, inspiratory and spiracular muscles in segment A3 during one ventilatory cycle. The solid black bars indicate when the motor neurons to these muscles spike most rapidly, the grey bars when spikes may also occur. Based on Hustert (1975).

implying the need to ensure synchrony of action by the two sides of the abdomen during ventilatory movements. The muscle fibres of the ventral diaphragm in one segment are innervated by two pairs of motor neurons, one from the segmental ganglion and one from the next anterior ganglion (Peters, 1977). The fields of innervation of these motor neurons overlap so that each innervates two segments and their axons enter the muscle from the paramedian nerve formed by the anastomosis of lateral nerves of adjacent ganglia.

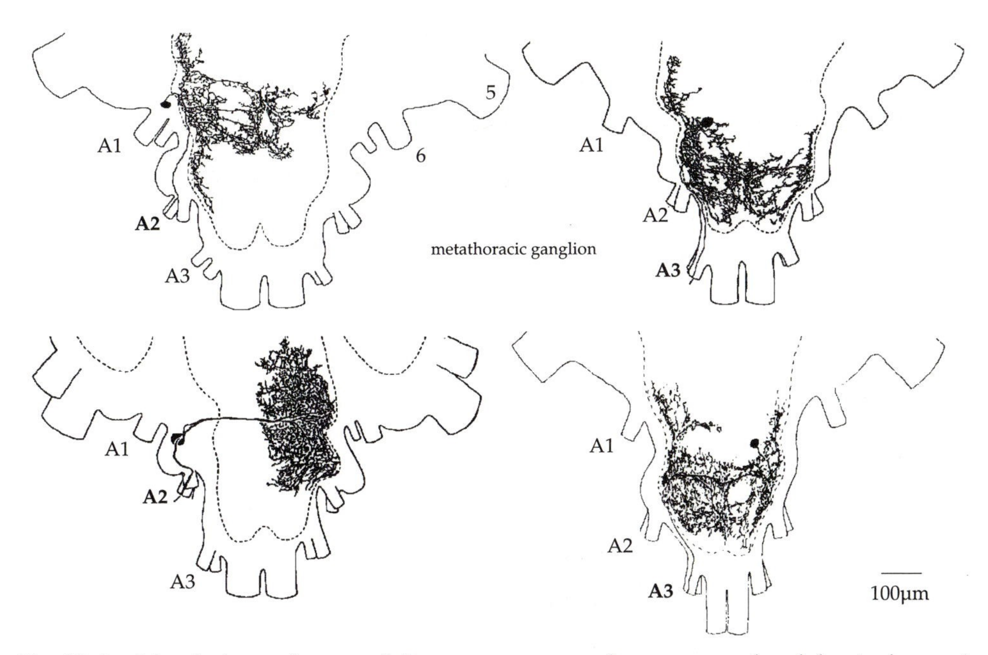

Fig. 12.4 Morphology of some of the motor neurons that generate the abdominal pumping movements of ventilation. All these motor neurons have their cell bodies in abdominal neuromeres (A1-3) that are fused to the metathoracic neuromere to form the metathoracic ganglion. Each of the motor neurons was stained by the intracellular injection of dye and then drawn. Based on Hill-Venning (1988).

12.3. THE MOTOR PATTERN FOR BREATHING

Of the two phases of breathing, inspiration is usually the shorter and stays relatively constant as the period of the rhythm increases. By contrast, expiration is of more variable duration and changes with cycle duration (Fig. 12.5). This asymmetry of the two phases of the rhythm is thus similar to that in walking (*see Chapter 8*) and flying (*see Chapter 11*) and suggests a general design feature of networks that generate rhythmic motor patterns.

In a freely moving locust ventilating deeply, the motor spikes that indicate the start of expiration begin at about the same time in all segments of the abdomen, so that there is no discernible sequence of contractions and the abdomen acts as a unit. In shallower ventilation, a sequence of contractions from front to rear can sometimes be detected because there are delays of 80–400 ms between the activation of adjacent segments (Hustert, 1975), and some ganglia may even occasionally fail to contribute to the rhythm so that the amplitude of the contraction by that segment is reduced.

In a restrained locust, similar patterns of coordination also occur (Lewis *et al.*, 1973). Usually all the segments of the abdomen contract together because each abdominal ganglion produces bursts of motor impulses at about the same time and

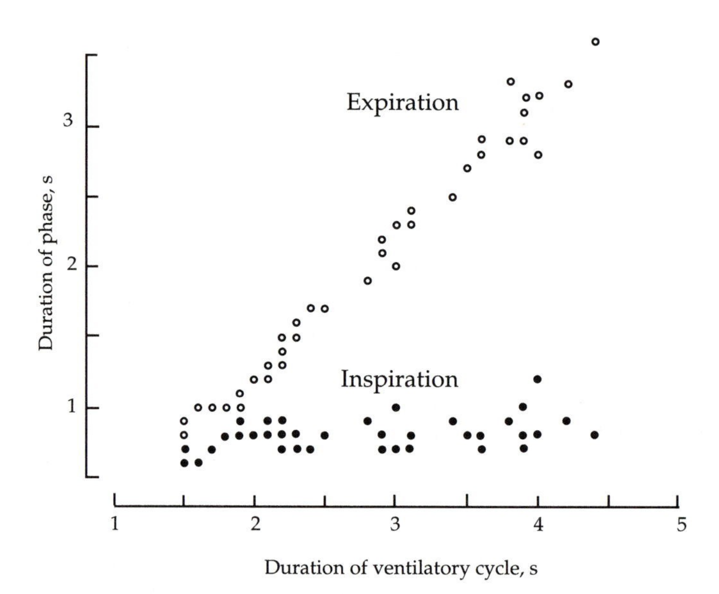

Fig. 12.5 The pattern of ventilatory movements. As the frequency of ventilation changes, the duration of the inspiratory phase stays relatively constant, while the duration of expiration changes. Compare this pattern with the swing and stance phases of walking (see Figs 8.1 and 8.3 in *Chapter 8*) and the elevator and depressor phases of the wing beat cycle (see Fig. 11.2 in *Chapter 11*). Based on Lewis *et al.* (1973).

at the ventilatory frequency set by the metathoracic ganglion complex. In slow ventilation, the contractions of the different segments do not occur in any particular sequence, so that either the most anterior or the most posterior ganglion may lead in the production of its motor spikes. There is certainly no metachronal wave passing either anteriorly or posteriorly, nor is a preferred order to the outputs of all the ganglia maintained in one animal. In faster ventilation there is still much variation in the order in which the ganglia are active, but the anterior ganglia tend to lead. In these circumstances there is a delay of 10–15 ms between adjacent segments and a total delay of 50–100 ms between A1 and A8. This pattern of coordination implies either that interneurons send signals posteriorly and that differing thresholds of the neurons in each ganglion result in their outputs being produced at variable times following the descending signals, or that sensory feedback occurs between neighbouring segments. When the ventilation is faster, the descending information may be stronger (more spikes and at higher frequencies), and the intraganglionic thresholds lower, so that the contraction of the whole abdomen more closely reflects the timing of the descending signals. The estimated conduction velocity of the interneurons conveying this information (1–1.3 $m.s^{-1}$; Lewis *et al.*, 1973) can, however, account for only some 30 ms of the delay between A1 and A8. The remainder must reside in the local interactions that occur in the ganglia. Some of the interneurons thought to be responsible for the coordination have been identified in the metathoracic ganglion, lending support to the idea that this is the dominant pacemaker for ventilation. It is not known, however, whether an isolated chain of abdominal ganglia can produce a coordinated output or whether any of these ganglia contribute signals capable of effecting intersegmental coordination.

12.3.1. Where is the motor pattern generated?

Attempts to define the site from which the ventilatory rhythm originates have been predicated by the assumption that a pattern generator resides in a discrete part of the nervous system. This reasoning has relied on lesion experiments to provide supporting evidence, in which particular groups of ganglia, or even individual ganglia, are physically isolated from the rest of the CNS. Such procedures identify a ganglion (or group of fused neuromeres) that is able to sustain a rhythm, and if this ganglion has a rhythm with the highest frequency, then it is interpreted to be the dominant pacemaker and the origin of the overall ventilatory rhythm (Table 12.2). In reality, however, the rhythm could originate from the interactions between neurons distributed in several ganglia and the isolation of individual ganglia may simply show that local interactions in some are sufficient to generate a rhythm but in others they are not. The expression of a rhythm by a ganglion, or the failure to express a rhythm, may say little about the organisation of the neurons because the lesions may have different effects on different ganglia depending on the number and types of neurons whose axons are severed. It is easy to see that these sorts of experiments may lead to inappropriate interpretations. The idea that the rhythm could equally well result from the interaction between neurons distributed in many different parts of the nervous system has only recently gained credibility.

In locusts, lesion experiments indicate that the metathoracic ganglion complex containing the metathoracic and first three abdominal neuromeres (*see Chapter* 2) is the source of the rhythm. Removing the brain reduces the frequency of ventilation, but subsequent removal of the suboesophageal, pro- or mesothoracic ganglia has little further effect. An isolated metathoracic ganglion will produce a rhythm that has the same structure as that in an intact locust, but none of the other thoracic ganglia or the suboesophageal ganglion when isolated, or when linked together, are able to generate a ventilatory rhythm. Any of the free abdominal ganglia (A4-8) can, when isolated, also produce a rhythm, but the frequency is always low and the pattern is abnormal; expiration is short and alternates with prolonged inspiratory periods (Lewis *et al.*, 1973). Isolating the metathoracic ganglion complex is not, however, equivalent to isolating one of the unfused abdominal ganglia, as the interconnections between four ganglia still remain (the metathoracic and A1-3). An isolated chain of the four unfused abdominal ganglia (A4-8) is, however, unable to generate a rhythm comparable to that produced when the metathoracic ganglion is present. On this basis, it has become accepted that the metathoracic ganglion of the locust (or one of the neuromeres within this fused mass) is the dominant or primary oscillator that imposes its rhythms on the secondary oscillators in each of the abdominal ganglia. It should follow from this inference that the segments of the abdomen are activated in a sequence as the result of commands issued by this particular ganglion.

When similar lesion experiments are made in other insects, different ganglia are implicated as the source of the rhythm, but the interpretations are made more complex because different numbers of abdominal ganglia can be fused to the metathoracic ganglion (Table 12.2).

12.4. DOES THE PATTERN DEPEND ON SENSORY FEEDBACK?

An isolated nerve cord deprived of all sensory feedback will generate a breathing rhythm, and even a single isolated metathoracic or an abdominal ganglion can produce a similar if slower rhythm. The conclusion drawn is that deafferentation reduces the frequency of the rhythm (Lewis *et al.*, 1973), but that it is not needed for the expression of the rhythm. Undoubtedly, a ventilatory rhythm can be produced in the complete absence of sensory feedback, but the effects of the many abdominal proprioceptors, such as chordotonal organs and muscle receptors, on the patterning of the different phases of the movements and on the shaping of the motor output, have not been examined in sufficient detail to evaluate their normal role in the control mechanisms. Some of these receptors are clearly activated by each cycle of the ventilatory movements in a cockroach (Farley *et al.*, 1967) and could potentially influence many aspects of the pattern. Spanning the folds between the dorsal and ventral plates (tergites and sternites) at the sides of each segment of a locust are chordotonal organs, and attached to certain intersegmental muscles are stretch receptors (Slifer and Finlayson, 1956). The chordotonal organs signal changes of one segment relative to another, whereas the stretch receptors monitor stretch in their associated muscle and the telescoping movements of the abdomen. The sensory neurons from both the proprioceptive chordotonal organs and stretch receptors, and also the exteroceptive hairs, project to several ganglia so that information from the different segments will overlap (Hustert, 1978). These projections therefore provide

Table 12.2. Source of the breathing rhythm in different insects

Insect	Ganglion	Reference
Locust, *Schistocerca gregaria*	T3	Miller (1960)
Cockroach, *Periplaneta americana*	T3	Farley *et al.* (1967)
Cockroach, *Blaberus craniifera*	T3, A1 or A2	Case (1961)
Cockroach, *Byrsotria fumigata*	A1	Myers and Retzlaff (1963)
Dobsonfly larvae, *Corydalus cornotus*	A2 or 3	Fitch and Kammer (1982)
Dragonfly larvae, *Aeschna juncea, Anax imperator, Anax parthenope julius*	A8	Mill and Hughes (1966); Komatsu (1982)

the anatomical basis for the connections that would be necessary to ensure the coordinated movements of the abdomen.

Even exteroceptors on the abdomen can have a powerful effect on abdominal movements, for when they are touched a local contraction of the abdomen occurs and the ventilatory rhythm stops for a few seconds (Hustert, 1975). In other insects, signals from exteroceptors can entrain the rhythm (Farley and Case, 1968), or reset it if they occur at a particular time during the cycle (Fitch and Kammer, 1982). It is to be expected that sensory feedback will play an important role in regulating the ventilatory movements, particularly in the face of changes in the abdomen correlated with the moulting cycle, the amount of food in the gut, or the presence of eggs in a female.

12.5. CONTROL OF THE MOTOR NEURONS

12.5.1. Patterns of motor spikes during a breathing cycle

In any of the abdominal ganglia, the expiratory bursts of spikes of motor neurons with axons in N2 show little overlap with the inspiratory bursts of spikes in the median nerve. Typically, some 10 ms separate the end of expiration and the start of inspiration, but a longer period of some 80 ms separates the end of inspiration and the start of expiration. None of the motor neurons that participate in ventilation show any intrinsic rhythmicity and none are coupled to each other either by direct electrical or chemical synapses. Furthermore, their spikes do not activate pathways that can alter the pattern of spikes in other ventilatory motor neurons. Thus, when current is injected into a motor neuron, the frequency of its spikes can be altered but there is no effect on the ventilatory pattern expressed in other motor neurons. Moreover, there is too much variation between the end of one phase of the rhythm and the start of the next for simple reciprocal coupling between the two pools of motor neurons to be responsible for the switch. The motor neurons themselves are thus incapable of generating a ventilatory rhythm so that their rhythmic pattern of spikes must be imposed by a synaptic drive from interneurons.

12.5.2. Synaptic drive to the motor neurons

The synaptic drive determines whether a motor neuron is recruited to the breathing rhythm and whether it spikes during inspiration or expiration, its pattern of spikes, and in large part, the frequency code of those spikes. It can consist of at least four different patterns. First, there may be excitation alone during the phase when spikes are produced and an apparent absence of a synaptic input during the opposite phase when no spikes are produced. Second, there may be excitation during one phase and inhibition during the other. Third, there may be a mixture of excitation and inhibition during both inspiration and expiration, with one type dominating during a particular phase. An inhibitory input at the end of one phase may result in spikes at a higher frequency through postinhibitory rebound at the start of the next. The spikes at the end of one phase may stop because of a lack of excitation, or may be terminated by an active inhibition. Fourth, the synaptic inputs during one phase may be patterned, so lending a substructure to the pattern of spikes in a motor neuron. The motor neurons are recruited to the rhythm in accordance with their size, at least as far as it applies to the diameter of their axons (Hinkle and Camhi, 1972), so that some fast motor neurons do not spike during ventilation at rest, but still receive a synaptic drive in time with ventilation (Burrows, 1974). As ventilation becomes more stressed, or if their membrane potential is reduced by a depolarising current, these motor neurons will spike and thus add further force to the abdominal contractions.

12.6. INTERNEURONS AND THE GENERATION OF THE BREATHING RHYTHM

The essential nature of the synaptic input to the motor neurons means that knowledge of the action of these presynaptic interneurons is necessary for an understanding of how the rhythm is generated. Success in identifying these interneurons has been limited, despite the obviously attractive features of ventilation for analysis; the rhythm is reliable, can be expressed by a single ganglion, and the actions of the participating motor neurons are known. So far, twelve interneurons involved in ventilation have been identified (Table 12.3 and Fig. 12.6), but of these only one (577) (*see section 12.8.2*) is known to be directly presynaptic to motor neurons. Moreover, no interactions between interneurons have been found that might begin to explain how the rhythm is produced. Part of the problem results from the small size of the interneurons, their wide distribution in the suboesophageal,

Table 12.3. Interneurons involved in breathing

Interneuron	Position of cell body	Axon	Spikes during	Effects of current		Able to reset rhythm
				depol	hyper	
326	A1	contra/descend	expiration	+	–	yes
327	A1	contra/descend	expiration	+	–	yes
328	A1	contra/descend	expiration	+	–	yes
329	A1	contra/descend	expiration	+	–	yes
606	A1	ipsi/ascend	expiration	–	+	
516	A1	contra/ascend	inspiration	+		
577, 578	A1	contra/ascend	inspiration			
720	T2	contra/asc+desc	expiration		+	
377	Suboes.	contra/descend	no rhythm	+		
725	A4	contra/asc+desc	expiration	+		
unnamed	A1	contra/desc	expiration	+		

+ ventilation rate is increased
– ventilation rate is decreased
Suboes. = suboesophageal ganglion
Data from Pearson (1980); Ramirez and Pearson (1989a); Burrows (1982a)

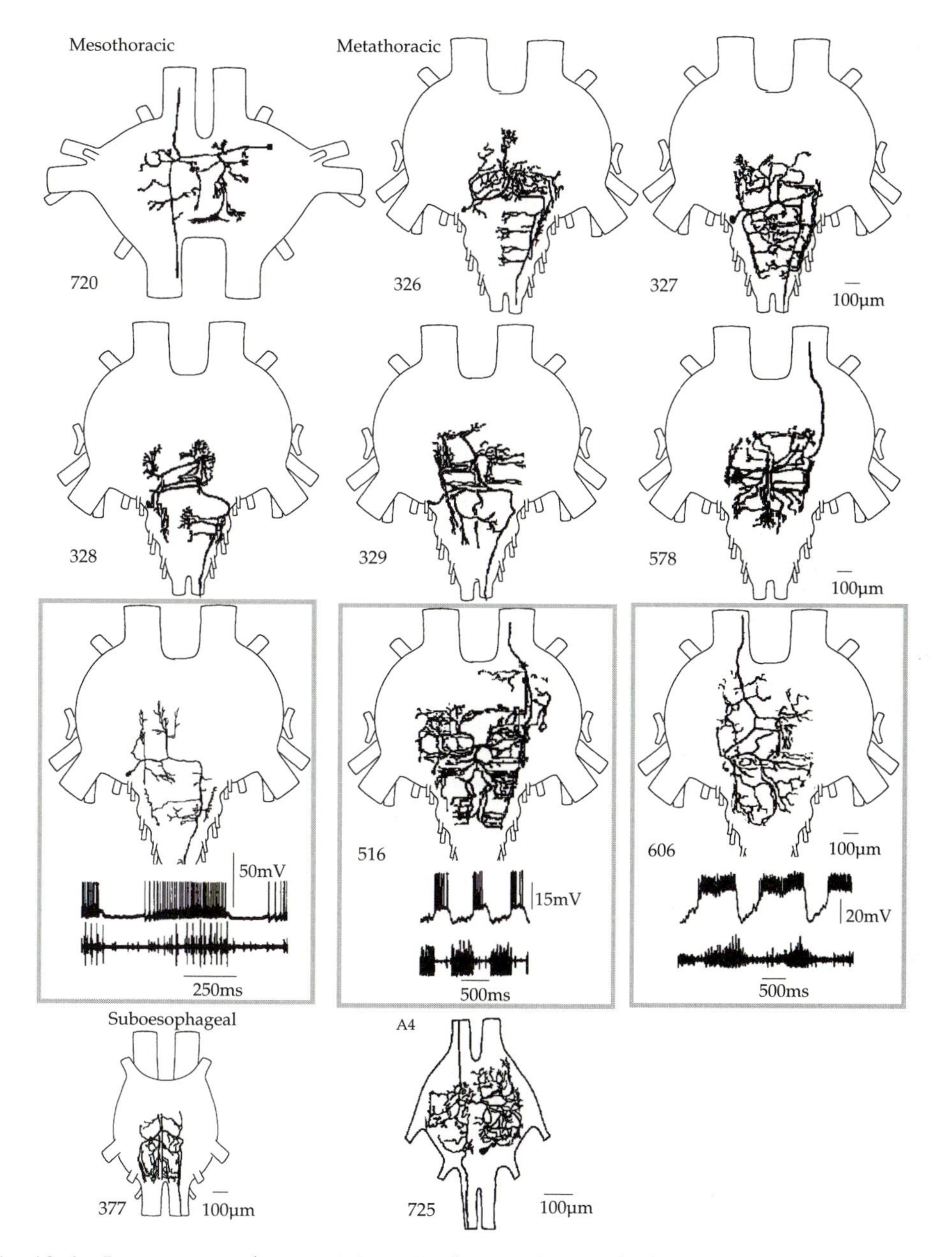

Fig. 12.6 Interneurons that participate in the ventilatory rhythm. Eight interneurons in the metathoracic ganglion are shown, and one each in the suboesophageal, A4 and the mesothoracic ganglion. The patterns of spikes during ventilation of three of these interneurons (enclosed in boxes) are illustrated relative to expiratory bursts of spikes recorded extracellularly from nerves innervating expiratory muscles. Each of the interneurons was stained by the intracellular injection of dye and then drawn. Based on Pearson (1980) and Ramirez and Pearson (1989a,b).

mesothoracic, metathoracic and abdominal ganglia 1–4, and the difficulty in distinguishing them from other neurons showing a ventilatory rhythm, unless their morphology is revealed by intracellular staining. Many neurons, such as those innervating leg muscles, often show a ventilatory rhythm in their synaptic inputs or spike outputs, without these events subserving any apparent behavioural function, and certainly no action that is causal to the generation of the ventilatory rhythm. Interneurons have been attributed to the network responsible for generating ventilation if their membrane potential fluctuates in time with the rhythm, and if current injected into them shifts the phase of the next cycle of the rhythm (i.e. the rhythm is reset) or changes the frequency of the rhythm. These criteria presuppose that manipulating the membrane potential of an individual interneuron will have an observable effect on the whole pattern and would cause to be overlooked those interneurons that must act with others to produce such effects.

The majority of interneurons revealed so far spike during expiration and increase the frequency of ventilation by a small amount when they are depolarised with a steady current. Some can reset the rhythm but others cannot. Most have axons in the connective contralateral to their cell bodies that either run anteriorly or posteriorly, but sometimes branch to do both. For example, one unnamed interneuron has an axon contralateral to its soma in A1 that runs to the last abdominal ganglion (Fig. 12.6) (Pearson, 1980). It produces an accelerating burst of spikes during expiration and, when depolarised with injected current, increases the rate of ventilation. It has no inherent rhythmicity and is made to spike by an excitatory synaptic input that has many of the characteristics of that to spiracle motor neurons and interneurons (*see section 12.8.2*). The output connections of this interneuron are unknown, in common with all but one of the above interneurons, but it would appear suitable for some role in coordinating the output of the ventilatory motor neurons in the chain of abdominal ganglia.

The preponderance of interneurons involved in ventilation that have cell bodies in A1 suggests that at least some of the main elements of the primary ventilatory pacemaker may reside here, provided that the small sample is not giving a biased picture. Nevertheless, the existence of interneurons in A4 suggests either that the network for producing ventilation is distributed in many ganglia, or that there is feedback from secondary pacemakers to the primary one. The former interpretation is preferred because it is reinforced by the presence of interneurons in the mesothoracic and suboesophageal ganglia, suggesting that they are part of the network and are not simply responsible for the wider distribution of the rhythm. The silence of the suboesophageal interneuron during normal ventilation may simply indicate that it is responsible for recruiting and coordinating the activity of motor neurons responsible for movements of the head and prothorax to help with increased ventilation.

12.7. DESCENDING CONTROL

The rate and pattern of ventilation must be fitted to the behaviour of the animal and this implies that those regions of the nervous system, most notably the brain and suboesophageal ganglion, which may initiate the appropriate behaviour will also affect ventilation. In crickets, electrical stimulation of particular regions in the protocerebrum of the brain can change the rate of ventilation (Huber, 1960). Similarly, locally raising the temperature of the brain or suboesophageal ganglion by 10°C, but not the metathoracic or A4 ganglia, increases the ventilation rate by as much as 90% (Janiszewski *et al.*, 1988). While these experiments suggest the presence of neurons in these regions that affect ventilation, they must be interpreted with caution because changes in ventilation accompany alterations in most behaviour. To show that the effects are specific requires that the actual interneurons be identified. In the suboesophageal ganglion of crickets, two pairs of identified interneurons can reset the ventilatory rhythm or reduce its frequency (Otto and Janiszewski, 1989). They are driven in the ventilatory rhythm by inputs from the metathoracic ganglion so that they spike continuously during one phase (probably expiration) and are inhibited during the other phase (inspiration). They are also excited by wind stimuli to the antennae and the cerci and by the sound of the song as detected by the cercal receptors (*see Chapter 10*). If the insect is singing, then they are also inhibited in time with this rhythm with a single spike in time with each of the pulses of sound (Otto and Hennig, 1993). Depolarising one of these interneurons reduces the frequency of the chirps in the song. The presence of these interneurons that can both alter the ventilatory rhythm and participate in this rhythm demonstrates the distributed nature of the network that is responsible for the control of ventilation. It also emphasises the close interrelationship between ventilation and different movements such as singing in a behaving animal, and suggests that these interactions result from the participation of some neurons common to the different movements.

12.8. CONTROLLING THE ENTRY AND EXIT OF AIR THROUGH THE SPIRACLES

The opening and closing movements of the spiracles allow air to enter and leave the body, and the sequence in which the spiracles in different segments are moved allows a directional flow of air within the body. A locust has ten pairs of spiracles, two pairs in the thorax and eight pairs in the abdomen (Fig. 12.7). Each has two valves that control the aperture of the spiracle and hence determine the phase of the ventilatory cycle when air may enter or leave. The movements of these valves are typically controlled by an opener and a closer muscle that receive rhythmic patterns of impulses in motor neurons linked to the ventilatory rhythm. In addition, however, the extent and duration of the movements can be influenced by local concentrations of carbon dioxide at the spiracle that may act directly on the spiracular muscles. No proprioceptors have been described that could monitor the movements of the spiracles and no cycle-by-cycle sensory control of these movements is known.

Spiracles 1–4 open during inspiration, while the more posterior abdominal ones (5–10) are closed. During expiration, the pattern is reversed so that a posteriorly directed flow is set up in the body, with the air leaving through the posterior spiracles

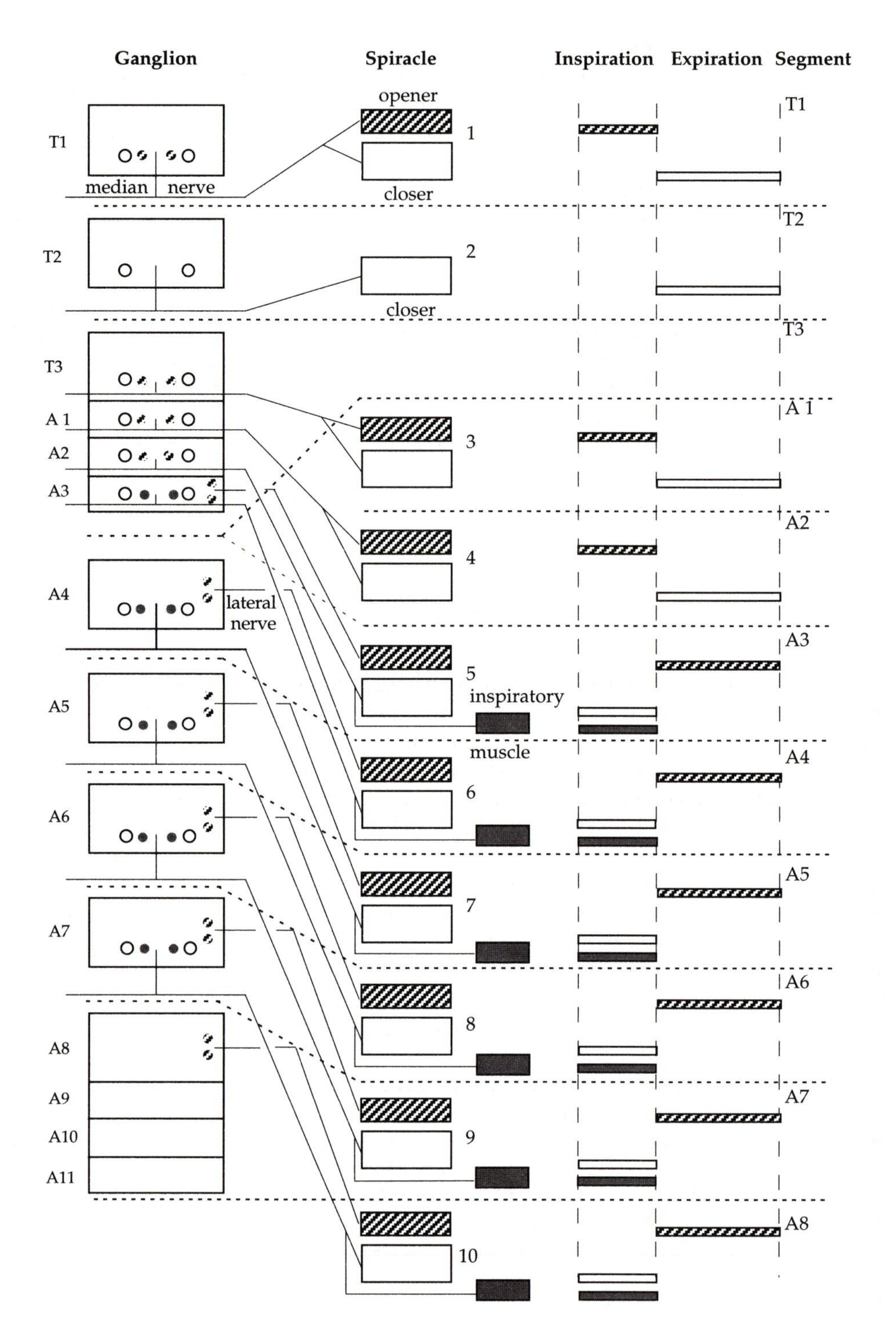
Ganglion
Spiracle
Inspiration
Expiration
Segment
opener
closer
median
nerve
lateral
nerve
inspiratory
muscle
T1
T2
T3
A 1
A2
A3
A4
A5
A6
A7
A8
A9
A10
A11
1
2
3
4
5
6
7
8
9
10

(Lee, 1925). At low frequencies of ventilation, even fewer spiracles may contribute to the movement of the air; it may simply enter through spiracles 1, 2 and 4, and leave by spiracles 10 while the others remain closed. At a ventilatory rate of 2 Hz, the open phase of the inspiratory spiracles 1–4 occupies no more that 20% of the cycle and that of the expiratory spiracles 5–10%.

The movements of the spiracles are always described as being linked to the ventilatory rhythm rather than being an integral part of it, implying that under some circumstances they can be uncoupled from the ventilatory rhythm. In the locust, the movements of the spiracles are apparently always tightly coupled to the ventilatory rhythm, but in some cockroaches the linkage can be varied. *Gromphadorhina portentosa* can produce hissing sounds by expelling air through particular abdominal spiracles (Nelson, 1979), and in *Blaberus giganteus* changes in the sequence in which the spiracles are moved are thought to reverse the direction of airflow within the body (Miller, 1973). When ventilation is fast, spiracle 1 closes during expiration while spiracle 10 opens, with those on both sides of the body working symmetrically to give a posterior flow of air. By contrast, when ventilation is slow, spiracle 1 opens for much of the cycle and spiracle 10 on one side of the body now opens during inspiration to give an anterior flow of air.

12.8.1. Actions of the individual spiracles

12.8.1.1. *Spiracle 1*

Each of the two spiracles on the prothoracic segment (spiracle 1) is moved by two muscles, an opener that contracts during inspiration and a closer that contracts during expiration. Both muscles are innervated by excitatory motor neurons with cell bodies in the prothoracic ganglion and with axons in the unpaired median nerve that divide to innervate the muscles of both the left and the right spiracle. None of the spiracular muscles are innervated by inhibitory motor neurons, although in other insects they may be. The unpaired median nerve thus contains the axons of four motor neurons but also contains many smaller axons from neurons with a neurosecretory function (*see Chapter 5*). This unusual pattern of motor innervation means that the spiracles on the left and right sides of one segment are both innervated by the same motor neurons. It is a characteristic feature of all the spiracles that at least some of their innervation is provided by neurons with bifurcating axons in the median nerves (Fig. 12.8). These spiracular motor neurons

---

Fig. 12.7 Innervation of the 10 pairs of spiracles in the thorax and abdomen and the timing of the actions of their motor neurons during the ventilatory rhythm. The chain of segmental ganglia (T1-3 and A1-8) are represented on the left, the muscles belonging to the right half of the body in the centre, and the timing of the motor spikes on the right. The motor neurons with axons in the median nerve innervate muscles on both sides, but those with axons in a lateral nerve innervate muscles only one side. Closer motor neurons, closer muscles and their axons are indicated by open circles and rectangles, openers by hatched symbols and inspiratory motor neurons and muscles by shaded symbols.

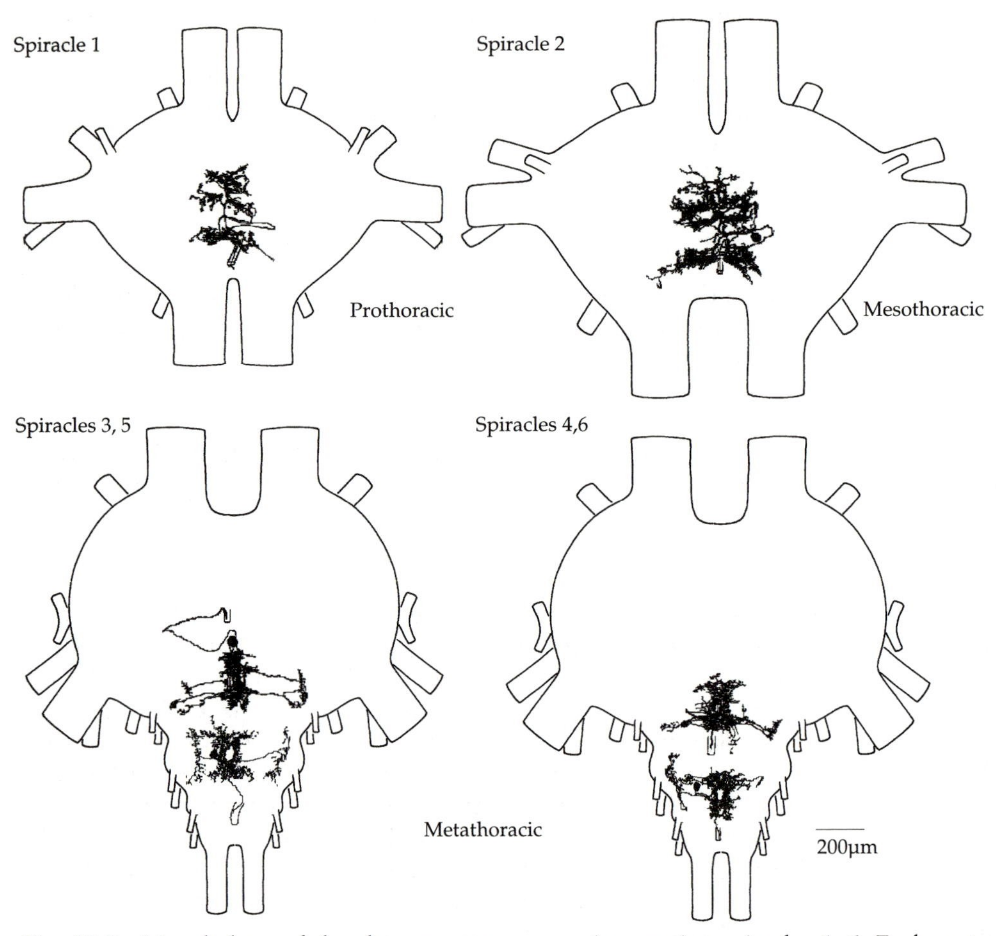

Fig. 12.8 Morphology of the closer motor neurons innervating spiracles 1–6. Each motor neuron has an axon in the median nerve that divides within this nerve to innervate spiracles on both sides of the body. Based on Burrows (1982b) and Hill-Venning (1988).

with axons in the median nerves have branches in both the left and the right halves of the neuropil, suggesting a symmetrical arrangement of their inputs. The only other muscles innervated in this way are two neck muscles in the cricket that may be the segmental homologues of the thoracic spiracular opener and closer motor muscles (Bartos and Honegger, 1992). These small dorso-ventral muscles are innervated by motor neurons with cell bodies in the suboesophageal ganglion and axons in the median nerve from that ganglion.

The spiracular opener muscle contracts during inspiration against a cuticular spring that stores energy until the closer has relaxed. It is innervated by two motor neurons with small diameter axons that conduct spikes to the muscles at a velocity of about 0.4 $m.s^{-1}$. A third motor neuron with an axon in N1 also joins the median nerve close to the opener muscle, and spikes during the opposite phase of ventilation, expiration. This has led to the suggestion that it might be an inhibitor, but there is no evidence that it actually innervates the spiracle, although it certainly spikes in a ventilatory rhythm. The closer, which contracts during expiration, is also innervated by two motor neurons, but these have larger diameter axons that conduct spikes at about 1 $m.s^{-1}$ (Miller, 1965, 1966).

12.8.1.2. *Spiracle 2*

The spiracles in the mesothoracic segment lack opener muscles and are therefore closed by contraction of the closer muscle and, when it relaxes, open due to the elasticity in the articulation of the valves. The closer muscle, the structure of which may be typical of other spiracular muscles, consists of an outer layer of about 20 thicker fibres, 30–65 µm in diameter and with Z bands 3–6 µm apart, and a core of 50 thinner fibres, 10–15 µm in diameter and with Z bands at 6–9 µm (Hoyle, 1959). Each of the fibres is innervated by two median nerve motor neurons from the mesothoracic ganglion. These have been characterised as fast and slow, respectively, according to the junctional potentials they evoke in the muscle fibres and the contractions that result (Hoyle, 1959), but the evidence for this distinction is slight and they are best regarded as equivalent. Additional control is exerted by changes in the level of carbon dioxide in the tracheae near one of these spiracles. Increases lower the amplitude of the synaptic potentials in the muscle fibres and the force produced by the whole muscle (Hoyle, 1960), with the stimulus being the increased level of carbon dioxide itself and not the change in pH that may accompany it. The result is that the valve will open, even when the closer motor neurons are spiking. The movements of this spiracle are thus influenced both by the commands carried in the sequences of motor impulses generated by the central nervous system, and by the local effects of carbon dioxide on the operation of the closer muscle. This would allow some independence of action by the two spiracles in the face of a common motor drive, and similar mechanisms operating at other spiracles may allow independent action between segments.

The mesothoracic closer muscles are also supplied by an octopaminergic DUM neuron from the metathoracic ganglion that also supplies some other thoracic muscles (Bräunig *et al.*, 1994). They also have octopamine receptors which, when activated, can elevate the intracellular levels of cyclic AMP (Swales *et al.*, 1992). The median nerve itself may also be associated with peripheral neurosecretory cells that show immunoreactivity to certain peptides (Myers and Evans, 1985a,b). This suggests that the force produced by these muscles, and perhaps those of the other spiracles, can be modulated in ways similar to that elucidated for the extensor tibiae muscle of a hind leg (*see Chapter 5*), and that these possible effects must be considered as essential elements in controlling the exchange of gases.

12.8.1.3. *Spiracles 3 and 4*

Spiracles 3 and 4 on abdominal segments 1 and 2 have basically the same arrangement of muscles and motor neurons as spiracle 1. The cell bodies of the motor neurons are in the metathoracic and A1 neuromeres, respectively, of the metathoracic ganglion.

12.8.1.4. *Spiracles 5–10*

These spiracles in the abdomen have both opener and closer muscles but their pattern of innervation is different to the more anterior spiracles, and their action is reversed, so that the opener muscles contract during expiration and the closers during inspiration. The closer muscles of the spiracles on the left and right sides of one segment are innervated by two motor neurons with axons in the median nerve, but these motor neurons now spike during inspiration in contrast to those that innervate more anterior spiracles which spike during expiration. The opener muscles of the two spiracles are innervated separately by two motor neurons with axons in lateral nerve 1 of the next posterior ganglion. These motor neurons spike during expiration. Thus, spiracles 5, for example, have their closer muscles innervated by the same two median nerve motor neurons from ganglion A2, but their opener muscles are each separately innervated by two motor neurons with axons in lateral nerves 1 of A3. The separate innervation of the left and right opener muscles of the spiracles in one segment suggests that independent action is possible under central control, in addition to any local effects on the muscles themselves. The remaining two median nerve motor neurons now innervate a dorso-ventral inspiratory muscle (one of only two inspiratory muscles in each segment) on the left and the right side of the abdomen instead of spiracular muscles (Fig. 12.7).

The closer muscles of all the spiracles are therefore innervated by median nerve motor neurons, but there is a reversal in their action in the anterior and posterior parts of the body so that those that innervate spiracles 1–4 spike during expiration, whereas those that innervate spiracles 5–10 spike during inspiration. Only the opener muscles of spiracles 1–4 are innervated by median nerve motor neurons and these spike during inspiration, whereas the opener muscles of spiracles 5–10 are innervated by separate paired neurons with axons in lateral nerves that spike during expiration. The remaining median nerve motor neurons in the abdomen innervate inspiratory muscles and, like their homologues in the thorax, spike during inspiration. This switch in the innervation patterns of the spiracles and in the actions of some of the median nerve motor neurons correlates with the normal behavioural role of these different spiracles in directing the flow of air within the body, and represents a further example of the different behavioural roles that apparently homologous neurons can play in different segments of the body.

12.8.2. Motor commands to the spiracles

12.8.2.1. *Closer motor neurons*

The synaptic control of the closer motor neurons of spiracles 1–4 is quite simple: two interneurons make them spike during expiration and two interneurons suppress their spikes during inspiration. The motor neurons have no inherent rhythmicity so that without these synaptic inputs they just spike tonically at a variable but low frequency.

Each pair of motor neurons that innervate the left and right closer muscles of spiracles 1–4 produce bursts of spikes at the same time during expiration and are inhibited during inspiration (Fig. 12.9). The patterns of spikes in each of these motor neurons are closely matched, but the spikes are not phase locked and there is no electrical or direct synaptic coupling between them (Burrows, 1982b). These features are particularly apparent during ventilatory pauses, when both motor neurons spike tonically. The spikes in one motor neuron then drift relative to those in the other and thus set up a beating pattern in extracellular recordings from the median nerve or from the muscle fibres. During the expiratory bursts, similar beating patterns may also occur. A further substructure to the pattern is also apparent during expiration, when the spikes occur in groups of two or more at short intervals separated by longer intervals of 50 ms. This patterning is caused by a patterned synaptic input in which EPSPs occur in groups. Electrical stimulation of the median nerve with pairs of pulses produces greater muscular tension than if the stimulation is unpatterned (Hoyle, 1959), so that the normal sequences of motor impulses may represent a way of making use of a possible pattern sensitivity of the neuromuscular junction to produce greater amounts of force. The valves of the spiracles often show small

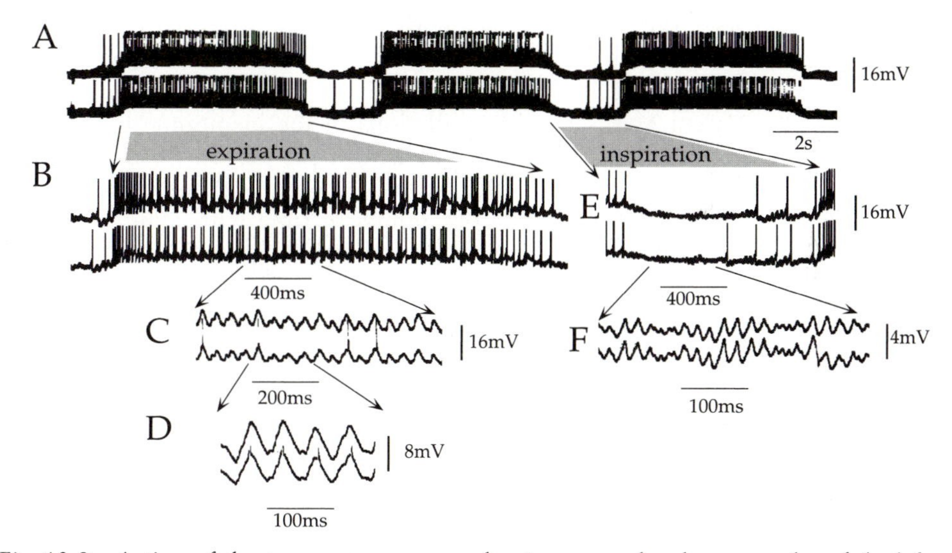

Fig. 12.9 Action of the two motor neurons that innervate the closer muscles of the left and right spiracles 2 in the mesothoracic segment. A. Both motor neurons spike at the same time during expiration. B. The spikes in both motor neurons occur in a similar pattern that consists of groups of 2–3 spikes separated by longer intervals. C, D. The excitatory synaptic drive that underlies these spikes is matched in the two neurons. E. During inspiration, the motor neurons are inhibited. F. The inhibitory synaptic input to the two motor neurons is matched in the same way as the excitatory input during expiration. The synaptic drive indicates that both motor neurons are excited by the same interneurons during expiration, and inhibited by common interneurons during inspiration. From Burrows (1982b).

fluttering movements that could well be caused by these patterns of spikes but the functional significance of these movements is unknown.

The EPSPs that cause the spikes in the closer motor neurons of spiracles 1–4 during expiration are matched in both motor neurons that innervate one spiracle (Fig. 12.9), and in the motor neurons that innervate different spiracles (Burrows, 1975b). The simplest interpretation is that they are caused by the same pair of interneurons (or a closely coupled set of interneurons), one with an axon in the left connective, the other with an axon in the right connective, that link all these spiracles. The two interneurons, which have not been identified but about which much can be inferred, must have their cell bodies in the metathoracic ganglion complex. They are not synaptically coupled to each other, but are driven by common synaptic inputs, they must spike during expiration, and the interplay between their spikes must give rise to the periodicity in the evoked motor spikes. They also make more widespread connections with thoracic motor neurons involved in the auxiliary head movements of stressed ventilation. Normally, their input to these motor neurons is subthreshold.

The ventilatory action of these interneurons is thus to coordinate the expiratory movements of the closer muscles of the thoracic spiracles, to shape the patterns of spikes in their motor neurons by superimposing a substructure to the bursts, and to recruit motor neurons for auxiliary ventilatory movements. Less easy to explain is the surprising fact that they also connect with many (at least 30) other motor neurons, particularly those that innervate the elevator muscles which move the two sets of wings in flight (Fig. 12.10) (Burrows, 1975c). The depolarisation in these neurons is again apparent as a series of oscillations in the membrane potential at intervals of 50 ms, equivalent to the normal wing beat period in flight. Thus, spiracular closer motor neurons and wing elevator motor neurons in any of the thoracic ganglia receive matching excitatory inputs in time with expiration. The input to these flight motor neurons is usually subthreshold in a locust that is not flying, but under appropriate conditions will make them spike at intervals that are multiples of the flight rhythm, and in bursts that are in time with ventilation. Whether this input represents any manifestation of the flight rhythm is not established, but the reason for its existence needs to be explained. Similarly, there is a need to explain why flight motor neurons should be driven in time with the ventilatory rhythm when, in a nonflying animal, they contribute no force to the ventilatory movements, although they may possibly compress the thoracic air sacs. The widespread connections of these interneurons must imply a neural and behavioural requirement for the close integration and interaction between the different rhythms that the locust produces. It may indicate the efficient use of energy so that several actions performed at once are matched to the resonant frequency of the thorax.

The motor spikes stop at the end of expiration because the excitatory synaptic drive declines and because an inhibitory drive begins (Fig. 12.9). The inhibitory synaptic potentials are matched in the two motor neurons of one segment and in neurons of different segments, again implying a common source. As found for the EPSPs during expiration, so the IPSPs during inspiration tend to occur in groups

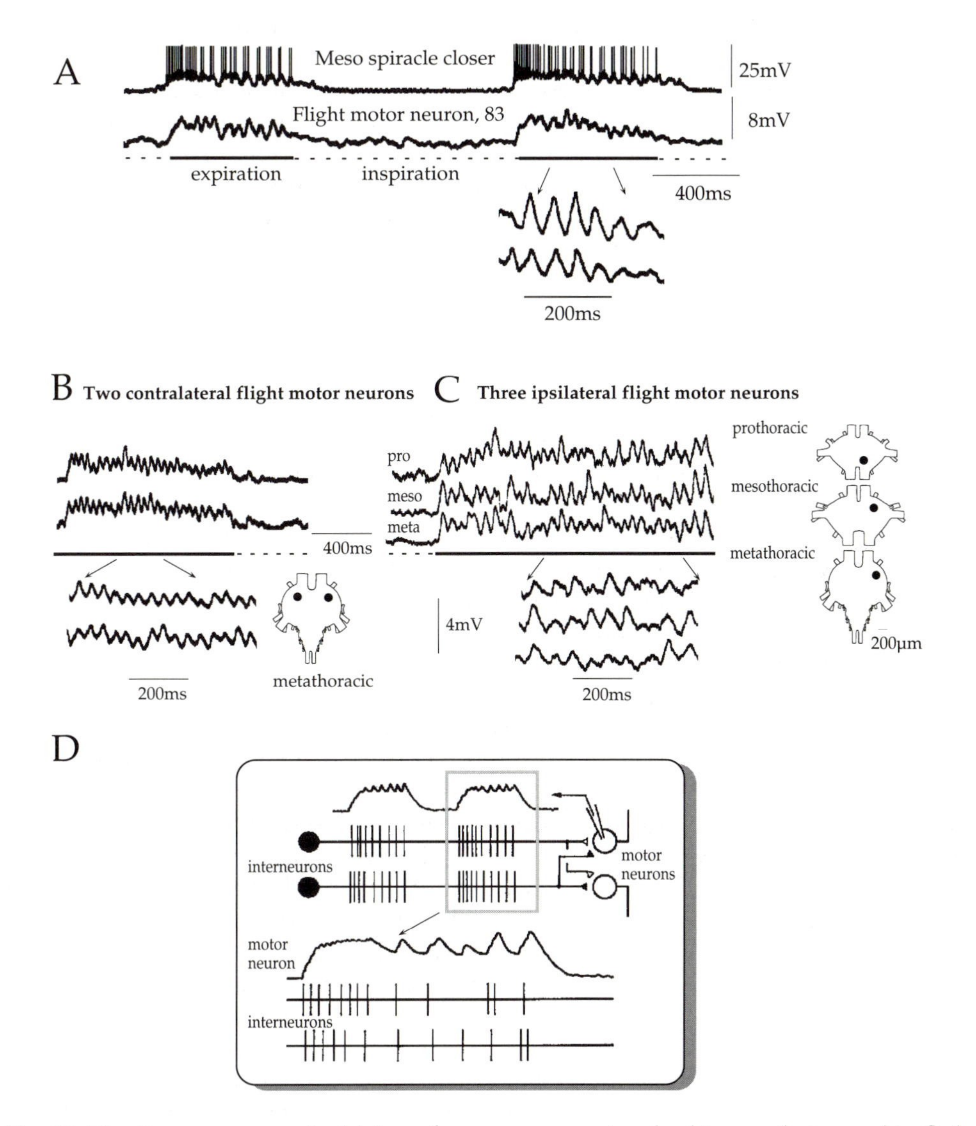

Fig. 12.10 Common synaptic driving of motor neurons involved in ventilation and in flight. A. When a mesothoracic spiracle closer motor neuron spikes during expiration, a motor neuron innervating a front wing elevator muscle (tergosternal, 83: see Table 11.1) is depolarised. The underlying EPSPs in the two motor neurons are matched, indicating that they are caused by the same interneurons. B. The motor neurons to hind wing elevator muscles (tergosternal, 113) on the left and right sides of the body receive a common excitatory synaptic input during expiration. C. Three motor neurons in three different thoracic ganglia that innervate flight muscles on the same side of the body, all receive a common excitatory synaptic depolarisation during expiration. D. A summary diagram of the actions of two interneurons that could explain the depolarisations in ventilatory and flight motor neurons during expiration. Based on Burrows (1975b,c).

caused by beating between the spikes of the common presynaptic interneurons. The two interneurons causing these inputs have axons that ascend to the prothoracic ganglion in the connective contralateral to their cell bodies in the metathoracic ganglion complex (Fig. 12.11). They begin to spike towards the end of expiration, continue at a constant frequency throughout inspiration and are then abruptly inhibited at the start of the next expiration. Manipulating the spikes in these interneurons does not alter the period of the ventilatory rhythm, but does alter the structure of the bursts of spikes in the closer motor neurons. Depolarising an

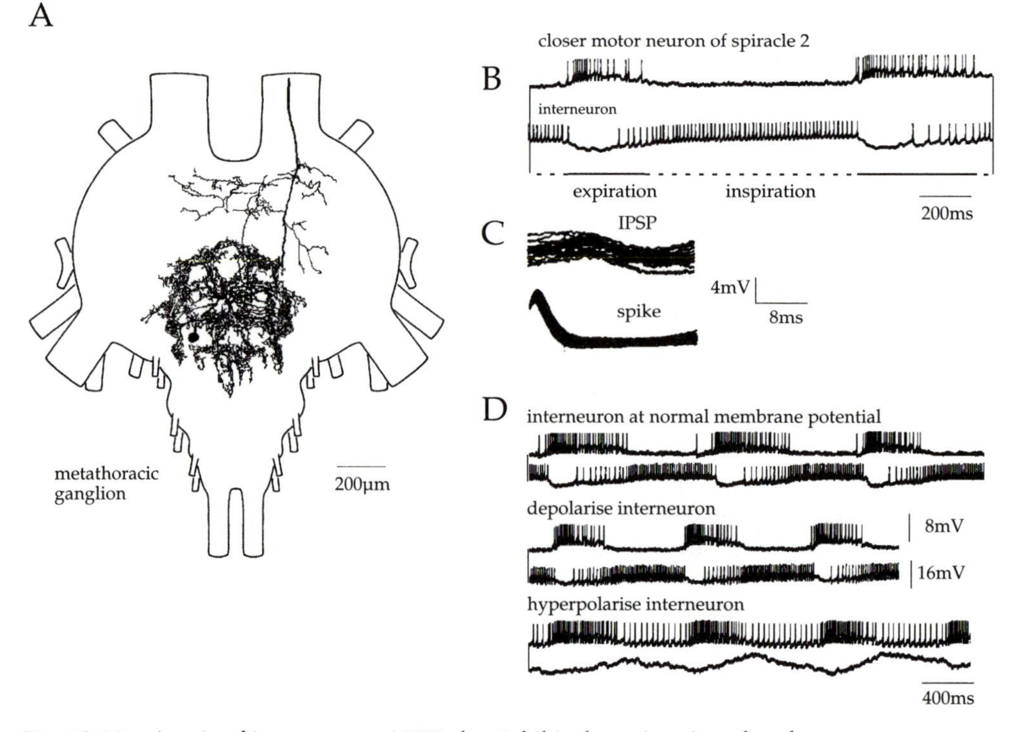

Fig. 12.11 A pair of interneurons (577) that inhibit thoracic spiracular closer motor neurons during inspiration. A. The morphology of one of these interneurons that has its cell body in the metathoracic ganglion and an axon in the contralateral anterior connective. B. It spikes during inspiration when a closer motor neuron of spiracle 2 is inhibited. C. Each spike in the interneuron is followed by an IPSP in the motor neuron, indicating a direct connection. D. Changing the pattern of spikes in the interneuron with applied d.c. current alters the pattern of spikes in a closer motor neuron. Normally, only an occasional spike occurs in the motor neuron during inspiration. When the interneuron is depolarised so that it spikes faster, none occur, but when it is hyperpolarised so that it does not spike, many motor spikes occur. Some inhibition still persists in the motor neuron that is explained by the action of the bilateral homologue of the impaled interneuron. These different actions of the interneuron would allow either a complete opening of the spiracles during inspiration, a partial opening, or would prevent opening altogether. Based on Burrows (1982a).

interneuron results in a more abrupt start and end to the closer bursts, whereas hyperpolarising allows the closer spikes to continue at a reduced frequency during inspiration. The applied hyperpolarisation also reveals that the interneurons themselves are driven synaptically in the ventilatory rhythm and that without this drive they show no inherent rhythmicity. They are inhibited by a synaptic input during expiration, but during inspiration produce spikes from a smoothly rising membrane potential. The spikes could thus arise from the release of inhibition allowing the membrane potential to gradually exceed and remain above spike threshold, from sustained inputs from nonspiking interneurons, or from inputs from spiking interneurons that are not registered as discrete potentials at the recording site.

12.8.2.2. *Opener/inspiratory motor neurons*

The control of the opener motor neurons also appears to be quite simple, with two interneurons inhibiting them during expiration, and rebound from this inhibition, perhaps with the addition of some excitatory input, allowing them to spike during inspiration (Fig. 12.12). At the beginning of inspiration they are depolarised and spike throughout inspiration at a steady frequency until they are abruptly hyperpolarised at the start of expiration. The two motor neurons innervating the opener muscles of one pair of spiracles or the inspiratory muscles of one segment, are not directly coupled to each other by chemical or electrical synapses. The synaptic drive thus means that although the overall burst of spikes occurs at the same time in each, the individual spikes occur at slightly different times and frequencies. No clear synaptic input underlies the pattern of spikes during inspiration: the depolarisation starts slowly at the beginning of inspiration, is maintained throughout inspiration and then declines sharply at the start of expiration. The drive may thus be explained in the same way as for the interneurons that inhibit the spiracle closer motor neurons. By contrast, a clear inhibitory synaptic input suppresses the spikes during expiration and is matched in the two motor neurons of one segment, implying that it is derived from a common source. These synaptic potentials also occur in groups, setting up oscillations in the membrane potential that closely match the depolarising oscillations that occur at the same time in the closer motor neurons.

The ventilatory pattern in the thoracic spiracles and their coordination in a segmental sequence is thus determined by only a small group of interneurons. Four interneurons are sufficient to explain the actions of the closer motor neurons, and a further two can explain most of the actions of the openers. Moreover, the strong correlation between the IPSPs in the openers and the EPSPs in the closers during expiration indicates that the premotor interneurons must themselves be closely coupled, probably by the same presynaptic interneurons. While the detail on the control of the abdominal spiracles is, as yet, not so extensive, the control mechanisms appear to be similar, so that the coordination of all the segments in the ventilatory rhythm is quite simple. None of the interneurons that synapse on the spiracular motor neurons seem, however, to be involved in generating the ventilatory rhythm, but instead their role seems to be to distribute this rhythm and produce a

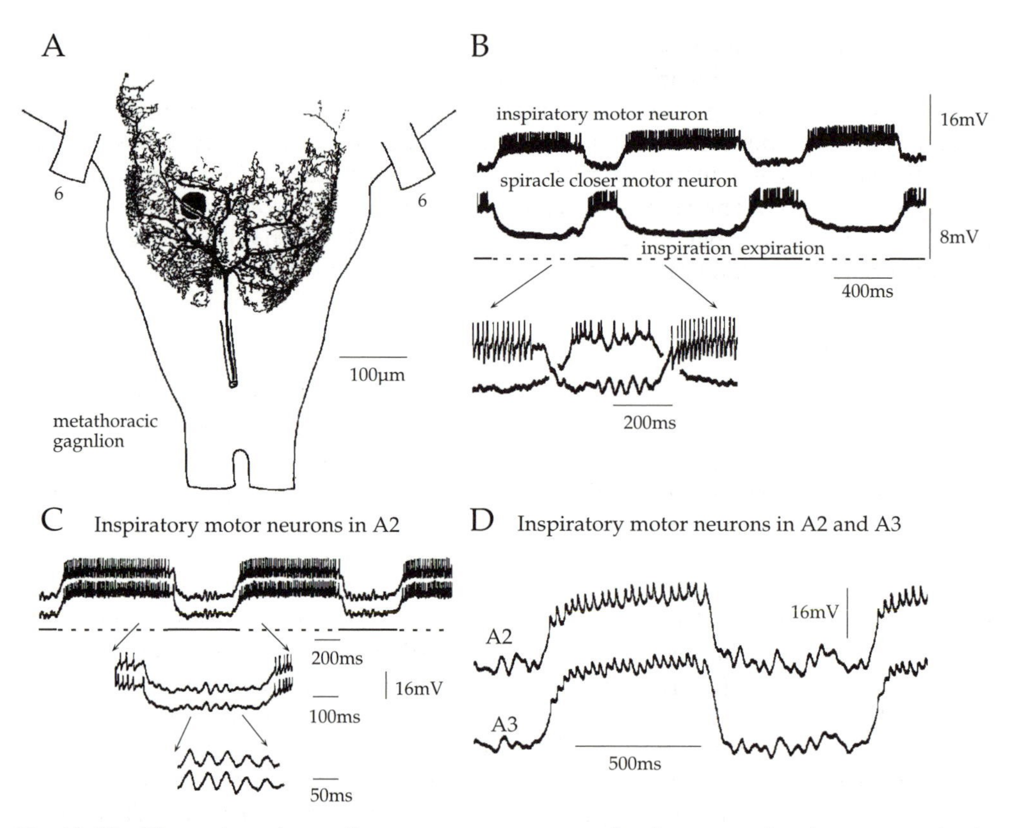

Fig. 12.12 The action of a median nerve motor neuron that innervates inspiratory muscles in the abdomen. In contrast to the median nerve spiracular closer motor neurons in the thorax, this motor neuron spikes during inspiration. A. Morphology of the motor neuron. B. Action during ventilation. It spikes during inspiration, whereas a spiracle closer motor neuron spikes during expiration. C. The two median nerve inspiratory motor neurons of one segment show patterns of spikes that are closely matched. During expiration their IPSPs are matched, indicating that they are driven by the same interneurons. D. The patterns of spikes during inspiration and IPSPs during expiration are similar, even in motor neurons of adjacent segments. A–C are based on Burrows (1982b). D is based on Hill-Venning (1988).

coordinated motor output in different parts of the body. Our knowledge of the way the rhythm is generated will remain rudimentary until the interneurons generating the pattern are characterised.

12.9. INTERACTIONS BETWEEN BREATHING AND OTHER MOTOR PATTERNS

Breathing is a prerequisite for the production of metabolic energy, so it is natural to expect that the ventilatory rhythm might set limits on the expression of behaviour, and that, in turn, the behaviour might modify the ventilatory movements. These sorts of interactions between breathing and movements such as flying, walking, running, swallowing and chewing are seen in many animals and presumably represent an adaptation for the efficient use of muscles and hence a saving of energy. In insects, the three movements that interact most strongly with breathing are flying, walking and singing (stridulating). The close association between these motor patterns has led to the suggestion that the pattern generators for these different movements are linked (Kutsch, 1969) to the extent that they may share common neurons.

12.9.1. Breathing and flying

In locusts that are flying, the flight muscles help in moving air through the body to meet the huge requirement for oxygen. This organisation implies that the ventilatory and flight rhythms must become linked. The meso- and metathoracic spiracles remain open continuously during flight and the tracheae on each side of the thorax are almost completely isolated from each other and from the rest of the tracheae in the abdomen. The movements of the thorax accompanying the wing movements provide an air exchange mechanism that is superimposed on that of the abdominal pump and which directly ventilates the main tracheae supplying the power-producing flight muscles and the thoracic air sacs. They nevertheless exchange only some 7 μl of air at each wing stroke (equivalent to some 900 $ml.min^{-1}$) and therefore only a small fraction of the volume of the thoracic tracheae and air sacs (Weis-Fogh, 1964). At the same time, the prothoracic spiracles provide an air supply to the head, and the abdominal spiracles open and close in time with ventilation, but with a greater proportion of their ventilatory cycle now spent in the open position. This again implies flexibility in the pattern that controls and coordinates the movements of the spiracles, or that the contractions of the thoracic flight muscles merely cause a mechanical override.

What then happens to ventilation when flight starts? Initially, the pumping movements of the abdomen are not apparent, probably because of an increase in the stiffness of the abdomen caused by the activation of many of its muscles. After a few seconds of flight, the rhythmic pumping movements of the abdomen reappear, and the spike patterns of many of the ventilatory motor neurons now assume a flight rhythm as well, so that the abdomen is moved in time with the wings, and with the slower ventilatory rhythm superimposed (Camhi and Hinkle, 1972). The rhythm in the ventilatory interneurons persists throughout the transition to flight, but is initially reset by inhibition of expiratory neurons and excitation of inspiratory neurons, before continuing at an increased frequency during sustained flight (Ramirez and Pearson, 1989b). For example, the closer motor neurons of spiracle 4 and expiratory interneurons 326–329 and 606 are first inhibited, while inspiratory interneurons 516 and 578 increase the frequency of their spikes. Interneuron 577 spikes continuously during flight, so that its inhibitory connections with the thoracic spiracle closer motor neurons will ensure that these spiracles are more likely to stay open. Some of the neurons that express only the ventilatory rhythm in their spike patterns before flight begins, express both the ventilatory and the flight pattern once it starts. This is a further indication of the ability of neurons to adopt different rhythms in different circumstances that has been most elegantly shown for neurons which control the various movements of the guts of crabs and lobsters (Harris-Warrick *et al.*, 1992).

A further demand on the abdomen is made during flight in that it must act as a rudder to aid in the control of steering. Many of the motor neurons contributing to ventilatory movements must thus be biased to move the abdomen up or down, or to the left or right. Some of the interneurons that are rhythmically active during ventilation can cause the abdomen to bend in ways appropriate for steering, but these interneurons are not driven by the usual visual signals that initiate steering during flight (Baader, 1990). This must imply that there are other interneurons still to be revealed.

These interactions between flying and breathing also occur in birds, where ventilation is usually, but not obligatorily, coupled to the wing beat. For example, in pigeons, the wing movements during free flight are coupled 1: 1 to ventilation and thus clearly aid the ventilation of the lungs (Berger *et al.*, 1970), while in other birds as many as 11 different forms of coordination can occur. The coupling results, in part, from sensory feedback, as passive movements of the wings can entrain the ventilatory rhythm (Funk *et al.*, 1992), and also from interactions between neurons in the central nervous system, as the fictive rhythms are still coupled in a bird unable to move. This means that in free flight, sensory feedback can modify the central linkage between the two rhythms.

12.9.2. Breathing and walking

Interactions between the walking and breathing patterns have long been proposed, even to the extent that they may be the expression of a single pattern-generating network, but the linkage during normal walking has yet to be expounded. The reason for the linkage in insects, as in many other animals, is that the locomotory movements set mechanical constraints on the effectiveness of the ventilatory movements in exchanging gases. The need for ventilation may also constrain, or even dictate, the locomotory movements. Before walking begins in the locust, the abdomen is lifted and is then held in a more elevated position, indicating that at least some of the motor neurons to muscles controlling movements of the abdomen are driven by the same neurons that initiate and control walking. Two statements can be made about the coupling between breathing and walking. First, that descending interneurons from the suboesophageal ganglion of crickets which spike in time with ventilation are modulated during walking (Hörner, 1992). Second, that the rhythmic motor activity expressed by isolated ganglia is coupled to ventilation in such a way that one phase, which may correspond to the swing phase of normal walking, is always excluded from the inspiratory phase of ventilation (Ryckebusch and Laurent, 1993; *see Chapter 8*). It might be expected that there would be a close coupling between more vigorous locomotory movements such as kicking, jumping, swimming, and breathing but these have not been explored.

There are many precedents for these sorts of interactions in other animals. In decerebrate cats with no sensory feedback, fictive walking movements elicited by stimulation of the mesencephalon locomotor region can be coupled 1: 1 by central pathways to fictive ventilatory movements, and the ratio of the coupling can be varied by changes in arterial carbon dioxide concentration (Persegol *et al.*, 1988; Kawahara *et al.*, 1989). During normal locomotion, wallabies take one breath per hop (Baudinette *et al.*, 1987), horses, dogs and hares all take one breath for each stride when cantering and galloping, and in humans breathing and the step cycle in running can be coupled in various ratios (Bramble and Carrier, 1983). In race horses, this linkage between the two rhythms could result in large part from the movements that occur when the back flexes (Young *et al.*, 1992). Thus, there are both central and peripheral feedback mechanisms that link these two rhythms.

In mammals, there are further strong linkages between breathing and the rhythms of chewing and swallowing movements that would not necessarily be expected of insects with their arrangement of tracheae. In humans, most swallowing movements occur during expiration, and in cats fictive swallowing and breathing are coupled 1: 1 so that swallowing always occurs at the transition from one phase of ventilation to the other (Dick *et al.*, 1993).

12.9.3. Breathing, hearing and singing

The sound detecting systems of locusts and many other insects depend on the specialisation of particular tracheae or air sacs, but as these are regularly inflated and deflated in time with the ventilatory rhythm, a strong relationship between the performance of auditory neurons and the ventilatory rhythm is to be expected.

The ears of the locust are positioned between air sacs in the first abdominal segment in such a way that the membranes of the two tympana move in by 30 μm and out by 90 μm in time with the ventilatory movements. Any changes in the stiffness of the tympanal membrane will alter the vibrations induced by sounds and hence the response of the receptors, which can detect nanometer movements of the membrane. Ventilatory movements enhance the high frequencies of a sound pattern by about 7 dB SPL and reduce the low frequencies by some 15 dB SPL, thus altering the spectral composition of the sound that is received (Meyer and Hedwig, 1995). These changes are also reflected in the summed spikes of the sensory neurons responding to the sound. This fluctuation, in time with ventilation, of the signals sent to the central nervous system by the sensory neurons, will, unless some compensation is added, be further reflected in the responses of the central interneurons that decode these signals, and ultimately in the behaviour that they evoke. The apparent price that the locust has to pay for having highly sensitive auditory receptors is that they will be contaminated by signals generated by movements such as ventilation. This is a common problem with high sensitivity receptors and is usually solved by gating their signals at some stage in their processing as, for example, occurs in the integration of signals from wind receptors on the cerci (*see Chapter 10*).

For insects that produce sound, there is the additional problem of the interaction of both the sound emitters and the sound detectors with the ventilatory rhythm. When a male cricket sings, the movements of the fore wings that produce the pulses of sound (chirps) are synchronised with the expiratory movements of the abdomen (Kutsch, 1969). Certain interneurons from the suboesophageal ganglion then spike in patterns linked to both ventilatory and song rhythms and are able to influence both those rhythms (Otto and Hennig, 1993). The ventilatory movements of a female grasshopper are suppressed when she hears the song of a male, presumably so that hearing is not distorted. The suppression is initiated only by the song of a conspecific and not by white noise pulses of sound. A similar suppression of ventilation occurs for a few seconds after the male has emitted a sound pulse so that any response by other conspecifics can be detected without distortion.

12.9.4. Breathing and moulting

A locust periodically softens and sheds its hard exoskeleton so that it may grow. The ventilatory system is intimately involved in the shedding of the old cuticle and the inflation of the new one, together with the air that is swallowed to inflate the gut. The whole process of moulting involves a complex pattern of movements, particularly of the abdomen, in repeated sequences of contractions. The head and thorax emerge first from the old cuticle and then the locust hangs from the tip of the abdomen while the hind legs are extricated (Hughes, 1980a). After the emergence of the abdomen, the locust hangs from its prothoracic legs and begins to expand the new soft exoskeleton. Both the later stages of the emergence behaviour and the expansion of the new exoskeleton involve rhythmical contractions of the abdomen in two superimposed rhythms involving some of the same sets of motor neurons (Fig. 12.13) (Hughes, 1980b; Elliott, 1982); a slow expiratory contraction that is maintained for up to 60s and which alternates with several cycles of a faster rhythm, or a single, long and deep inspiration lasting 2–3 s. During the long contractions, a continuous pressure is applied to the inflated gut and closed tracheal system that can stretch the cuticle. Exchange of gases can occur only during the fast rhythm when the spiracles are opened and closed. Increasing the level of carbon dioxide in the air decreases the period of the slow rhythm of maintained contractions and increases the frequency of the fast ventilatory cycles. The same rhythms also occur in the neck muscles that are normally used only in stressed ventilation. The spiracles open and close in time with the fast rhythm, with the thoracic spiracles opening and the abdominal ones closing during the fast inspirations, but remain closed during the slow compression phase. Furthermore, during wing expansion, which can last for 30min, wing elevator, but not depressor, muscles contract in time with the slow rhythm and finally, as the wings are folded to their closed position, the pleuroaxillary muscles (85 and 114, *see Chapter 11*) become rhythmically active.

12.9.5. Breathing and egg laying

The movements of the spiracles are also coordinated with the digging movements of the abdomen that a female uses to lay her eggs in the ground. The eggs are laid in a sandy substratum to avoid desiccation and predation, and to do this the abdomen must stretch from its normal length of about 3 cm to about 13 cm. This is achieved

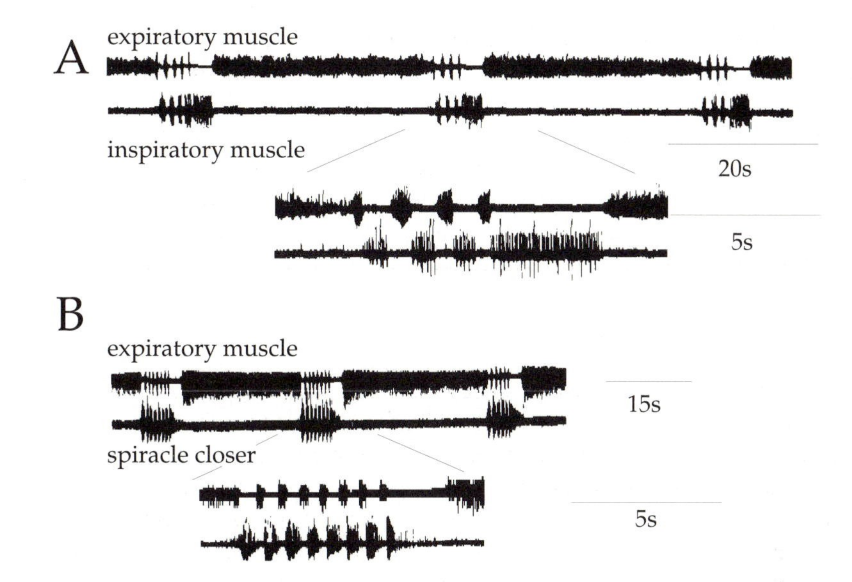

Fig. 12.13 Action of abdominal muscles during moulting. A. Action of expiratory and inspiratory muscle in the abdomen during the period of moulting when the wings are being expanded. A slow rhythm of deep inspirations and expirations is punctuated by a faster rhythm of inspiratory and expiratory activity. B. Spiracle closer muscles are active during the pauses in expiratory muscle activity and show a series of contractions with a faster rhythmicity that alternate with the spikes in the expiratory muscle. Based on Elliott (1982).

by unfolding the overlapping segments of the normally telescoped abdomen and by stretching the intersegmental membrane, particularly between segments 4–5 and 5–6, by as much as 10 times. These movements inevitably mean that the intersegmental muscles (dorsal longitudinals) will also be elongated by 10 times their contracted length during digging. These muscles can supercontract at resting lengths because they have perforated Z discs that allow the thick filaments of adjacent sarcomeres to overlap, and can still contract at superextended lengths to restore the abdomen to its normal shape, because the Z discs fragment (Jorgensen and Rice, 1983). During these movements, the volume of air in the tracheal system increases by 120%, but the extension is caused more by the pull exerted by the digging movements of the ovipositor than by internal pressure, which remains much the same (Vincent and Wood, 1972). The maintenance of intra-abdominal pressure is important for the rigidity of the abdomen during digging and egg laying, and to achieve this the abdominal spiracles close to retain the air within the abdomen. Again, this implies flexibility in the control of the different spiracles and in the coordination of ventilatory movements with other movements that use the same groups of muscles. The enormous stretching of the abdomen must mean that the nerve cord is also stretched, but the consequences of this on its processing and transmission of signals are not explored.

12.10. THE SPIRACLES ARE ALSO USED FOR DEFENSIVE SECRETIONS

The spiracles of some grasshoppers and cockroaches are used to store and then forcibly expel defensive secretions. In the grasshopper *Romalea guttata*, a series of coiled tracheae close to the metathoracic spiracles are partially filled with a complex secretion consisting of phenols and quinones that is produced by associated, thick glandular epithelia (Whitman *et al.*, 1991). Normally, the metathoracic spiracles (spiracles 3) remain shut to prevent evaporation of the stored secretion, so that they and their associated coiled tracheae play no part in normal ventilation. When molested, the metathoracic spiracles open while all the others close, the abdomen contracts dorso-ventrally and longitudinally, raising the pressure in the tracheae and forcing air and the secretion simultaneously out of the left and right metathoracic spiracles with a hissing sound. The valves of the spiracles flutter to narrow the aperture so that the secretion is ejected as a spray that can reach as far as 20 cm. The ventilatory movements then resume, with the valves of the metathoracic spiracles remaining open so that the remaining secretion mixes with the expelled air to form foam at the mouth of the spiracles. The behaviour thus uses odour and sound as the main long-distance repellents, and a noxious contact repellent to deal with more persistent predators. For this behaviour, the movements of the metathoracic spiracles must be uncoupled from the normal ventilatory movements so that they open only when the secretion is to be released. It thus implies a further complexity in the control mechanisms of the spiracles, but nothing is known of these mechanisms.

Glossary

This is a list of definitions of the more unusual terms that are used and as they apply specifically to the locust. Words or phrases in **bold** in the definitions are defined elsewhere in their own right.

5-HT. 5-hydroxytryptamine, often called **serotonin**. Contained in only a few central neurons that have extensive arborisations, and in neurons of the **satellite nervous system** that form release sites along certain nerves. Acts as a neuromodulator by blocking particular K^+ channels.

A4I1. An identified interneuron (I) with its cell body in abdominal ganglion 4 (A4), and arbitrarily given the number 1. It receives direct synaptic inputs from hairs on the ventral region of the neck that are sensitive to low velocities of air movements and is probably involved in controlling steering during flight by virtue of its output connections with motor neurons innervating the pleuroaxillary muscle of a wing.

ACh. Acetylcholine. Probably the major transmitter of mechanosensory neurons. Probably also used by many interneurons but this use is not established.

Acheta domesticus. Domestic cricket.

Adipokinetic hormones. A large family of peptides the most well known of which are AKH-I and AKH-II. Others are known by names such as red pigment concentrating hormone and neurohormone D (myotropin I). About 20 members of this family have been sequenced so far. In the locust, AKH-I and -II and at least three other peptides (adipokinetic precursor-related peptides, APRPs) are synthesised in the glandular cells of the **corpora cardiaca.**

Air sac. A thin walled, expandable extension of the **tracheae** that inflates and deflates with each cycle of the ventilatory rhythm so that it can smooth the changes in pressure.

AKH. See **adipokinetic hormones.**

Alternating tripod. The normal **gait** used by a locust and many other insects during walking, in which three legs are always supporting the weight of the body by their contact with the ground. The sequence of movements involves the middle leg on one side moving in time with the front and hind legs of the opposite side.

Amacrine. A neuron that lacks an axon and which has branches restricted to one region of the central nervous system. The name could describe either a nonspiking or a spiking local interneuron, but is not commonly used in these contexts.

Anastomosis. The fusion of two nerves from the same or adjacent ganglia, found most commonly in the abdomen and forming the ladder-like arrangement of the paired **paramedian** and lateral heart nerves, and in the thorax between N1 of one ganglion and N6 of the next posterior ganglion.

Antagonist. A description of a muscle that works in opposition to another muscle in moving a particular joint. Thus, a muscle that extends a joint will often work as an antagonist to a muscle that flexes the same joint. Sometimes, however, these muscles need to work as **synergists.** The term is thus not an immutable description of the actions of muscles, but refers to their actions in a particular behavioural context.

Antennal heart. An accessory muscular pump at the base of an antenna which aids in the flow of **haemolymph** into that antenna.

Antidromic. A spike (action potential) which is conducted in the opposite direction to the normal (**orthodromic**) direction. This definition means that such spikes are normally only evoked experimentally, but in neurons with distinct spike-initiating zones it is used to describe the two way traffic of the spikes along an axon.

Apodeme. Cuticular structure onto which muscle fibres insert, equivalent therefore to a tendon of a vertebrate. It is also the structure, sometimes called a ligament, to which the **dendrites** of sensory neurons in some chordotonal organs insert. The apodeme is an extension of cuticle, often from the adjacent section of an appendage, so that the movements of the joint are transmitted to sensory neurons.

Assistance reflex. An adjustment of a movement to a stimulus that results in an enhancement of that movement. It is thus the reverse of a **resistance reflex.** The same stimuli can elicit either response depending on the behavioural context.

Azadirachtin. A natural substance extracted from the Indian neem tree (*Azadirachta indica*) that prevents feeding and moulting. Fifth instar larvae treated with azadirachtin remain as larvae and fail to **moult** into adults that can fly. It may disrupt the synthesis or release of ecdysone. (See **precocene.**)

Band. A large accumulation of larval locusts moving as a group, progressing by marching and hopping, and feeding prodigiously as they go.

Basiconic sensilla. Sensory receptors with shafts protruding from the surface of the cuticle that are generally shorter than those of **trichoid sensilla,** set in a membranous socket, and which are innervated by several sensory neurons. They can serve a purely olfactory function, as on the antennae where the different sensory neurons respond to different odours. On the body, limbs and mouthparts, they generally function as contact chemoreceptors, although they may still respond to certain odours, but have one neuron that is a mechanoreceptor.

Bistable membrane properties. See **plateau potential.**

Brunner's organ. A small, soft tubercle on the ventral surface of the femur of a hind leg; it is not present on the other legs. It is associated with a group of tactile hairs and **campaniform sensilla.** When the tibia is fully flexed the tubercle is distorted and may deflect the hairs. No function has been ascribed to this structure.

C neuron. An intersegmental interneuron with its cell body in the mesothoracic ganglion and an axon projecting to the metathoracic ganglion, given the functional name cocking interneuron because of its previously supposed action during kicking and jumping.

***Calliphora erythrocephala*.** Blowfly.

Campaniform sensillum. Sensory receptor set in the cuticle, in which the **dendrite** of a single sensory neuron is attached to a dome of flexible cuticle in an oval depression surrounded by the normal hard cuticle. It responds to strains and can thus act either as a **proprioceptor,** monitoring the forces generated by the muscles, or as an **exteroceptor**

monitoring externally applied forces. May occur individually on many parts of the body and limbs, and as groups in specific places.

Catch property. Strictly, this should be called 'catch-like property' to distinguish it clearly from the catch state of certain molluscan muscles. It refers to the **hysteresis** in the force generated by a muscle that depends on the past history of its motor impulses. The force that is generated by a particular frequency of motor spikes will be greater if it is preceded by a higher frequency of spikes than if preceded by a lower frequency.

CCAP. Crustacean cardioactive peptide.

CCK. Cholecystokinin.

Central complex. The central region of the brain where fibres pass from one hemisphere of the brain to the other and which consists of the protocerebral bridge, the upper and lower divisions of the central body, and paired noduli. A role in visual processing is implicated on the basis of pathways from the compound eyes and ocelli and it receives a large input from neurons that show **serotonin**-like **immunoreactivity**, which may indicate that its processing is highly modifiable.

Central pattern generator. A group of central neurons that, by virtue of their intrinsic membrane properties and the connections that they form with each other, are able to produce the basics of a motor pattern without reference to the sensory feedback that would normally be generated by such movements. The basics of many motor patterns such as those underlying flying and breathing can be produced by the central nervous system that is isolated from all sensory feedback. The motor programme, defined as the action of particular motor neurons, muscles or joints, results from the interplay between the output of these central neurons with sensory information fed back from these movements. The idea of central pattern generators has led to the unfortunate idea that separate groups of neurons are responsible for the production of different patterns, and that they represent a distinct entity in the central nervous system separable from other neurons. The reality is that neurons participate in many computations and may even contribute to the generation of different patterns in different circumstances. See **circuit, motor programme, network**.

Cerci. Paired appendages of the last abdominal segment that have many hair **sensilla** that are highly sensitive to air currents, and other receptors. These appendages can be moved.

Chordotonal organ. Sensory receptor with no obvious external cuticular structure that is slung from one point on the body to another, often in such a way that they can respond to movements of a joint. It consists of three types of cell: bipolar ciliated sensory neurons, a **scolopale** cell which envelops the distal region of the sensory neuron by secreting the fibrous scolopale to form a **scolopidium**, and a support cell. In each organ there may be several to many sensory neurons that insert on a ligament moved by joints or other parts of the body, and which therefore serve a proprioceptive function (e.g. FeCO at the femoro-tibial joint of a leg). They can also act as vibration (**subgenual** organ) or sound receptors (**tympanum**).

CI1,2,3. Three inhibitory motor neurons that each innervate several muscles of a leg, so that they are called **common inhibitory** motor neurons and numbered 1–3.

Circuit. A term often used interchangeably with **network** to define the neurons processing particular sensory stimuli, or with **central pattern generator** to define those neurons contributing to the production of a particular movement. A local circuit results from interactions between neurons in a restricted region of the central nervous system, such as a ganglion. As for central pattern generator, a circuit should not be considered as a separable entity from other neurons in the central nervous system; the contribution of particular neurons to a circuit can vary in different circumstances.

Commissure. Bundle of neuronal processes running between the two sides of the central nervous system. In the segmental ganglia, many of the commissures are discrete and are therefore named, and contain the processes of identified neurons.

Commissures in segmental ganglia

Dorsal commissures	DCI–DCVI
Ventral commissures	VCI, VCII
Supramedian commissure	SMC
Posterior ventral commissure	PVC

Common inhibitor. Motor neurons with an inhibitory action on muscles. Each motor neuron innervates several muscles. See **CI1,2,3**.

Corollary discharge. A group or pattern of spikes in an interneuron or population of interneurons that corresponds to the expression of some feature of a motor response that is being generated. The interneurons showing the corollary discharge are not thought to be primarily responsible for the generation of the motor pattern. It may simply be a way of describing the response of an interneuron when its action is unclear, or it may imply that certain signals are being fed back or fed forward to other known components of the networks involved in the production of the movement.

Corpora allata. Paired neurosecretory structures behind the brain that secrete **juvenile hormone** (JH3), which plays an important role in development and in maturation. They are each linked to the **corpora cardiaca** by NCAI and to the suboesophageal ganglion by NCAII (also called suboesophageal N3). Each half of the corpora allata is innervated by 17 neurons, 13 with ipsilateral cell bodies and axons, and four with bilateral axons projecting in NCCII to the corpora cardiaca and then by NCAI to the ipsilateral corpora allata.

Corpora cardiaca. Paired secretory structures just behind the brain and in front of the oesophagus innervated directly by two nerves from the brain and indirectly by a third (NCCI-III). The storage lobes contain the terminals of neurosecretory cells originating in the brain. The glandular lobes consist of intrinsic neurosecretory cells that synthesise, store and release into the **haemolymph** peptides, most notably the **adipokinetic hormones** (AKH).

Coxa. Section of the proximal part of the leg of a locust between the thorax and the **trochanter**. The articulation with the thorax is complex and allows movement in three dimensions under the control of six muscle groups arranged in three pairs. It contains two of the parts of the depressor muscle and the levator muscles that move the trochanter about the hinge joint of the coxa with the trochanter. In the hind leg, the fusion of the trochanter and **femur** means that the femur is also moved by the action of these muscles. See also **trochantin, tibia, tarsus, unguis.**

DCMD. Descending contralateral movement detector. An interneuron with its cell body in the **protocerebrum** of the brain and an axon in the contralateral connective to the thorax that codes the approach of an object along the axis of the eye.

Decticus albifrons. A tettigonid from southern Europe.

Dendrite. A term derived from vertebrates which means a branch from a cell body that bears input synapses. Strictly, the term has no meaning in insects because there are no such branches. Sometimes, however, it is used to describe the fine branches in the **neuropil**, but a better term is **neurite** or simply a branch. In sensory neurons, it is a process from the cell body that, for example, inserts into the shaft of a hair or onto the **apodeme** of a **chordotonal organ.**

Deutocerebrum. A posterior region of the brain, often called the antennal lobe, that is the main processing region for the primary olfactory signals. It is indented where the gut protrudes between the connectives, so that it is separated ventrally into left and right parts. Each part of the deutocerebrum consists of a dome-shaped anterior (ventral) lobe that protrudes on either side of the oesophagus, and a smaller dorsal lobe, with the antennal nerve entering laterally. Cortically arranged cell bodies of neurons are clustered into particular groups and **neuropil** areas are concentrated into spherical **glomeruli**. See **protocerebrum, tritocerebrum.**

Diadic. A synaptic arrangement, as seen with an electron microscope, in which a presynaptic process is associated with two postsynaptic processes. This contrasts with a **monodic** arrangement in which there is only one postsynaptic process.

Drosophila melanogaster. Fruit fly.

DUM neuron. Dorsal unpaired median neuron with a cell body in a group at the dorsal midline of the segmental ganglia and which is probably the progeny of the median **neuroblast** (with possibly a few neurons formed by other neuroblasts). There are at least three general classes.

1. *Efferent DUM neurons* have axons in particular nerves on both sides of a ganglion and run to muscles and other effectors on both sides of the body. Most, if not all, contain **octopamine.**
2. *Local DUM neurons* have bilateral processes that are contained within one **neuromere,** and may function as typical interneurons with **GABA** as their possible transmitter.
3. *DUM interneurons* have axons that project to other parts of the nervous system.

DUMETi. An efferent, dorsal unpaired median neuron (**DUM neuron**) that supplies the extensor tibiae muscles of a segmental pair of legs. It modulates the contractions caused by the excitatory motor neurons (see **SETi, FETi**) by releasing **octopamine.**

Equalising time constant. Describes the passive distribution of charge to non-isopotential regions of a neuron, and is usually much smaller than the **membrane time constant.**

Exteroceptor. A receptor on the surface of the body and appendages that responds to signals in the environment of a particular modality. Many of these receptors signal stimuli at a distance (e.g. vision, smell and sound) while others signal contact of the limbs or body with an external object, or with other parts of its own body (e.g. touch, mechanoreception). Some of these mechanoreceptors can thus act as **proprioceptors,** especially those that are placed at joints or at particular points of strain or stress in the cuticle.

Fascicle. A bundle of fibres that group together and run parallel to each other in peripheral nerves or in tracts and commissures of the developing central nervous system. Many fascicles contribute to each nerve, tract or commissure and newly developing neurons join a particular fascicle, indicating that its constituent fibres have unique labels.

Femur. Section of the leg between the **trochanter** and the **tibia.** It contains the extensor and flexor tibiae muscles that operate the hinge joint with the **tibia,** and part of the retractor unguis muscle. In the hind leg, the femur is greatly enlarged to contain the large extensor muscle involved in jumping and kicking, has a hexagonal profile, and is marked externally by a herring-bone pattern formed by the insertion points of the extensor muscle. See also **coxa, tarsus, unguis.**

FETi. Fast motor neuron innervating the extensor tibiae muscle of a leg. See also **SETi, DUMETi.**

Fibrillar muscle. Muscle that is used to move the wings of insects such as flies and bees that have high wing beat frequencies when flying. The muscles do not act directly on the wings but instead distort the shape of the thorax. Contractions result from a muscle being stretched and not directly from its activation by motor spikes. Consequently, the pattern of wing movements is not a direct reflection of the pattern of spikes in the motor neurons. The motor spikes alter the power output. Locusts have conventional muscles to move their wings so that the wing movements are a direct reflection of the pattern of spikes in the motor neurons.

Fictive. A term that has come to describe a motor pattern produced by a nervous system from which sensory feedback has been eliminated. The pattern is thus produced without any of the proprioceptive or other sensory cues that would occur during normal behaviour. To understand what the fictive pattern might represent it is necessary to define the features of a normal motor pattern and then compare them with those of the fictive pattern. The fictive patterns expressed in motor neurons innervating muscles that move the wings seem to represent flight, but those in motor neurons innervating leg muscles could represent several movements, of which walking is only one. Even if the relationships seem clear, there may still remain doubts about whether the pattern is being produced in the way that it normally would be when sensory information is present.

Filopodium. An extension from the growth cone of a developing neuron that samples its immediate environment and at successive choice points makes decisions about the direction in which to extend. Each growth cone can have a number of filopodia with diameters of less than 0.5 μm extending up to 100 μm. A particular filopodium advances at rates of 1–2 $\mu m.h^{-1}$ and can either persist for more than 10 min or can be retracted within 1 min. Internal second messenger signals are transmitted from the tip of a filopodium to the actomyosin filaments at the base to amplify small differences in extrinsic clues. This ensures that the growth cone does not simply advance in a way that is determined by the sum of the adhesive forces acting on all its filopodia.

Flight balance. A device used to support a locust during **tethered** flight and to measure the forces that it is producing. Such balances vary in the sophistication of their design with some merely allowing lift to be measured and others rotational torques. Some allow the phase relationships between selected motor neurons to be fed back and control horizontal turning torque in response to horizontal movement of the visual world. The contribution of the flight balance used to the motor pattern that is observed must be recognised, and will impose differences from the motor pattern that is used during free flight where all feedback loops are closed.

Frontal ganglion. A ganglion of some 100 neurons on the dorsal surface of the pharynx anterior to the brain, linked to the **tritocerebrum** by paired frontal connectives, and by a recurrent nerve to the **hypocerebral ganglion.** It is one of the **stomatogastric ganglia.**

GABA. γ-Aminobutyric acid, a putative transmitter of a few inhibitory motor neurons, and of many other neurons in the central nervous system. It is associated with an inhibitory effect on most neurons mediated by receptors with properties different from vertebrate $GABA_A$ and $GABA_B$ receptors. The action is usually through an increase in chloride conductance that can be blocked by **picrotoxin.**

Gait. The sequence in the which the six legs are moved during walking. Typically, the legs are moved in an alternating tripod gait, with three of the legs on the ground while the other three legs are moved forwards.

Ganglion mother cell. Progeny of the asymmetric division of a **neuroblast** in the embryonic nervous system. Each ganglion mother cell divides symmetrically just once to form cells that differentiate as neurons.

GI. Abbreviation for **giant interneuron.**

Giant interneuron. An interneuron that is substantially larger than other neurons in the nervous system. In insects, this relative term is applied only to a group of **intersegmental interneurons** that receive inputs from wind-sensitive hairs on the **cerci.** The term refers to the diameter of their axons in the abdominal nervous system.

Glioblast. A precursor cell in an embryo with properties similar to a **neuroblast** except that its progeny are glial cells instead of neurons.

Glomerulus. Spherical **neuropil** area characteristic of olfactory processing in the **deutocerebrum.** Primary sensory neurons project directly to these regions and make synaptic contacts with **local** and **projection interneurons.**

Gregarious phase. Locusts with a particular morphological appearance and behaviour that results from them living in crowded conditions, as opposed to those living in isolated conditions (see **solitary phase**). The larvae are black and yellow and develop more quickly than solitary ones. The adults have a broader head and larger compound eyes than solitary ones, and fly readily during the day rather than at night. The phase differences are under hormonal control.

Gryllus campestris. Field cricket.

Guidepost cell. A cell in an embryo that may provide one of the clues enabling a pioneer neuron and its followers to establish the route of a peripheral nerve or the tracts and commissures within the developing central nervous system.

Hair row. An array of hairs that are shorter and stiffer than other **trichoid sensilla** (tactile hairs). Probably serve a proprioceptive function signalling changes in joint movements and positions.

Haemolymph. Fluid (with a volume of about 200 μl in an adult locust) bathing the tissues and which carries neuromodulators and neurohormones to their targets. It plays little part in gaseous exchange as this is carried out by the network of **tracheae** and **tracheoles.** The circulation is largely open through various sinuses but is propelled by the rhythmic beating of the tubular heart. Flow to the hind legs may be aided by a **myogenic** rhythmicity of a small bundle of fibres in the extensor tibiae muscles, and to the wings and antennae by pulsatile organs.

Hair plates. Concentrations of short **trichoid sensilla** in small patches close to the articulations of the joints. They are stimulated when the joint is close to one of its extremes and thus act as **proprioceptors.**

Hemimetabolous. Hemimetabolous insects, such as flies and moths, emerge from the egg as larvae that are of quite different appearance to an adult. After a period as freely moving larvae, they pupate. Adults eventually emerge from the pupae. See **holometabolous.**

HEPES. *N*-2-Hydroxyethylpiperazine-*N*′-2-ethanesulphonic acid, used as a buffer in some Ringer solutions.

Holometabolous. Holometablous insects, such as the locust, that undergo a period of embryonic growth and development within the egg before hatching as an insect that has most of the characteristics of an adult, but is much smaller. There then follows a postembryonic period of growth and moulting through five larval stages, or **instars,** of progressively increasing size before a final **moult** to the adult. See **hemimetabolous.**

Homeotic genes. The loss of one of these genes results in the transformation of a segment or part of a segment into the pattern characteristic of another. They do not appear to affect segment boundaries. Many of these genes occur in two clusters.

1. *Bithorax cluster* containing, for example, *Ultrabithorax* and *bithoraxoid.*
2. *Antennapedia complex* containing, for example, *Antennapedia*, *Deformed* and *Sex combs reduced.*

These genes are expressed both in the embryonic nervous system and body, with particular ones expressed most strongly in a particular region consisting of the posterior part of one segment and the anterior part of the next posterior segment, a **parasegment.** See also **segmentation genes.**

Homosynaptic depression. Depression of the effects of a particular neuron on its postsynaptic neurons as a result of its preceding spikes. This effect contrasts with heterosynaptic depression caused by spikes in other neurons.

Hopper. See **larva.**

Hypocerebral ganglion. A small unpaired ganglion, sometimes called the **occipital ganglion,** behind the brain, dorsal to the pharynx and ventral to the **corpora cardiaca,** linked anteriorly by the nervus recurrens to the frontal ganglion, dorsally to the corpora cardiaca, and posteriorly by the paired oesophageal nerves. Its neurons are undescribed.

Hysteresis. Nonlinearity in the force generated by a muscle, or in the membrane potential or spike frequency of a neuron as a result of the history of action of that muscle, or of other neurons. In muscle, it is most readily expressed as a **catch-like property.**

Immunoreactivity. Staining with antibodies raised against specific antigens that is circumspectly described as antigen-like immunoreactivity. The caution is necessary because of the uncertainty of just what the antibody is binding to.

Ingluvial ganglion. One of a pair of ganglia on either side of the crop, each containing perhaps 60 neurons that arise from the oesophageal nerves from the **hypocerebral ganglion.** The ganglia innervate the anterior portion of the gut and are part of the **stomatogastric ganglia.**

Instar. See **larva.**

Integrating segment. A term loosely applied to the expanded section of the **primary neurite** of a neuron as it courses through the **neuropil** and from which many of the fine **neurites** or branches arise. It is a misleading functional term, for there is little evidence that this branch performs more integration than any of the other branches. In those neurons that spike, the **spike-initiating zone** is generally thought to be somewhere along its length. The neutral term **neuropilar segment** is to be preferred.

Intersegmental interneuron. An interneuron (sometimes called a **projection interneuron** or a **principal interneuron**) with a cell body in one ganglion, or one part of the brain, and an axon that projects to another ganglion or to another distinct region of the brain. These neurons contrast with **local interneurons** that have all their branches restricted to one part of the nervous system (often to a single segmental ganglion).

Junction(al) potential. The potential change evoked in a muscle fibre by the release of transmitter from an excitatory or inhibitory motor neuron. The term is synonymous with synaptic potential used to describe the potential change in a neuron resulting from the release of transmitter from another neuron.

Juvenile hormone. Hormone secreted by the **corpora allata** that prevents moulting in larvae and the maturation of the ovaries in adults. Its action is antagonised by **precocene.**

Kenyon cell. Small interneurons in the **mushroom bodies** of the **protocerebrum** named after their discoverer. They are thought to be involved in olfactory processing and learning.

Lamina. First (distal) **neuropil** region of an **optic lobe** that receives inputs in the form of graded depolarising signals from the axons of the **retinula cells** and generates graded hyperpolarising output signals in lamina monopolar neurons. The lamina consists of an array of cartridges that equal the number of **ommatidia.** Within each cartridge, synaptic interactions occur between the input neurons (the retinula cells), intrinsic neurons that run between adjacent cartridges, feedback neurons from the medulla, and the output neurons (the lamina monopolar neurons). Each cartridge may be electrically isolated from its neighbours by glial cell wrappings.

Larva. One of a number of stages in the postembryonic development of **holometabolous** insects such as the locust. A locust has five larval stages or **instars** which progressively increase in size at each **moult.** They all resemble the adult locust but are sexually immature and lack functional wings and particular arrays of sensory neurons. Also called **nymph,** or **hopper** after their characteristic mode of progression. In **hemimetabolous** insects, the larva is the postembryonic stage that emerges from the egg and that then pupates.

Lobula. The third proximal region of an **optic lobe** where convergence starts to occur so that the number of repeating arrays of neurons is less than the number of **ommatidia.** Many of the neurons are large and have outputs in the central regions of the brain or have axons that extend to the other optic lobe. Neurons here either have wide **receptive fields** and provide information that is necessary for the stabilisation of the visual world, or small receptive fields and provide information about the movement of small objects in the visual world.

Local circuit. See **central pattern generator, circuit, network, pathway.**

Local interneuron. A neuron that may lack an axon or has only a short axon and whose branches are restricted to one **neuromere.** In the segmental chain of ganglia, it can be applied strictly to mean a neuron whose processes are restricted to one ganglion. In the brain, the term becomes less useful because distinctions between regions may not be so anatomically clear and it may then define an interneuron that projects from one region to another.

Locusta migratoria migratoriodes. Migratory locust from Africa.

M neuron. An intersegmental interneuron with its cell body in the metathoracic ganglion and given the functional name multimodal neuron because it responds to more than one modality of sensory stimulus. Many interneurons are, however, affected by stimuli of different modalities. It is thought to be involved in controlling the kicking and jumping movements of the hind legs.

Malpighian tubules. A series of about 200 long, narrow, blind-ending tubules that arise close to the junction of the midgut and hind gut and lie in the **haemolymph** of the abdominal cavity. Their secretory epithelium is involved in the transport of fluids and the production of urine, and thereby the regulation of the fluid and ionic balance of the haemolymph.

Manduca sexta. Tobacco hornworm moth from North America.

Median nerve. An unpaired nerve arising from the dorsal midline of the thoracic and abdominal ganglia. It contains the axons of motor neurons that branch to supply spiracular muscles on the left and right sides of the body and the axons of neurons containing peptides that end in **neurohaemal** swellings of this nerve. In the prothoracic and abdominal ganglia there are both anterior and posterior median nerves which **anastomose** in the abdomen to form a chain.

Medulla. The second **neuropil** region of an **optic lobe,** in which the repeating arrays of neurons still match the number of **ommatidia.** The number of neuron types is greatly increased over those present in the **lamina** (first distal neuropil layer). Some of these neurons signal aspects of movement in the visual world and the orientation of polarised light.

Melanoplus sanguinipes. Rocky mountain locust from North America.

Membrane time constant. Describes the time taken for a voltage to decay to 63% of its original amplitude when the membrane is in a passive state. (See **equalising time constant.**)

Midline precursor cell. One of a small group of precursor cells at the midline of the developing nervous system. Some are recognised as **ganglion mother cells** of the median neuroblast. Others are distinct from the **neuroblasts** in that they arise from different ectodermal cells and divide symmetrically.

Monodic. A synaptic arrangement, as seen with an electron microscope, in which a presynaptic process is associated with a single postsynaptic process. This contrasts with a **diadic** arrangement of two postsynaptic neurons.

Motor pool. Set of motor neurons that innervate a particular muscle. Each motor neuron in a pool may innervate a particular part of the muscle, or more commonly the innervation overlaps so that a muscle fibre is innervated by more than one excitatory motor neuron.

Motor programme. The pattern of spikes in motor neurons (and other efferent neurons), muscle contractions, or joint movements that underlie a particular movement. In many of these movements a **central pattern generator** may supply a basic rhythmicity whose frequency and patterning can be substantially altered to produce the sequence of motor spikes that is observed as the motor programme.

Moult. A periodic shedding of the cuticle that enables a locust to increase in size through five larval **instars**, and which is under hormonal control. The movements that cause the old cuticle to split and the locust to emerge and inflate its soft body are aided by the abdominal muscles pumping air around the **tracheae.**

Multipolar receptor. A receptor cell at the joints that acts in parallel to the **strand receptors** and **chordotonal organs.** Usually occurs in small groups of 1–3 but the information it might provide in addition to that from the more numerous cells of the chordotonal organs is not understood.

Muscle receptor organ. A sensory receptor associated with a muscle whose contractions can potentially set the sensitivity of the receptor cells. One such receptor in the **coxa** of a leg (coxo-trochanteral muscle receptor organ, CxTrMRO) consists of a single sensory cell associated with a bundle of three muscle fibres innervated by a single motor neuron. The **dendrites** of the sensory neuron insert in series only on the collagenous insertion of the receptor muscle, so that they should be affected by both force and length changes in the receptor muscle. It is the only receptor in the leg for which specific efferent control has so far been demonstrated. See also **tension receptor.**

Mushroom bodies. Region of the anterior **protocerebrum** that derives its name from a superficial resemblance to an inverted lamellated mushroom. Involved in the processing of olfactory and probably other modalities of sensory information.

Myogenic. A contraction of a muscle that results from intrinsic properties of its fibres and/or interactions between those fibres. The contractions, which are often rhythmical, do not depend on spikes in the motor neurons that innervate the muscle, although these spikes may alter the contractions. Only a few skeletal muscles, such as a small part of the extensor tibiae muscle of a hind leg, show such contractions, but they are common in muscles of the gut and oviducts. See also **neurogenic.**

Myogram. An electrical recording of the activity of a muscle usually recorded with pairs of small diameter wires. The activity reflects the response of the muscle fibres to spikes in motor neurons, although sometimes the motor spikes themselves can be recorded. The use of thin and light wires allows recordings to be made during unimpeded natural movements. See **neurogram.**

Network. See **central pattern generator** and **circuit.**

Neurite. Branch of a neuron within the **neuropil,** variously called a process, an arborisation, or simply a branch, but sometimes erroneously called a **dendrite.** The single process emerging from the cell body of a neuron is called the **primary neurite,** and this may expand to form the so-called **integrating segment,** from which many smaller branches or neurites emerge.

Neuroblast. A cell that emerges from the sheet of ectoderm of an embryo and enlarges before embarking on a series of asymmetric divisions to form **ganglion mother cells.** These then divide symmetrically to form two cells that differentiate into neurons. Each segment contains a fixed number of neuroblasts which form distinct lineages of neurons. Only a few neuroblasts in **holometabolous** insects such as the locust remain mitotic in postembryonic development. See also **glioblast.**

Neurogenic. A contraction of a muscle produced as a result of the action of spikes in the motor neurons that innervate it. Skeletal muscles typically contract in this way, although a few can also contract **myogenically**. See also **neurogenic genes.**

Neurogenic genes. The spatial and temporal sequence of **neuroblast** formation is controlled by the neurogenic genes such as *Notch*. Loss of a neurogenic gene results in all cells in the ventral strip of ectoderm becoming neuroblasts, whereas loss of the proneural *achaete-scute* complex can result in embryos that lack only a single neuroblast but which then show much neuronal degeneration.

Neurogram. An electrical recording of potential changes in many neurons in the central nervous system usually made with electrodes larger than those used for intracellular recording or for extracellular recording of spikes. Sometimes, small wires have been inserted into the tracheae that invade the central areas of the CNS, but this technique has yet to be exploited fully. See **myogram.**

Neurohaemal organ. Typically a swollen enlargement of a **median nerve** containing the terminals of central neurons and the cell bodies of other neurons that release a variety of neurosecretory or neuromodulatory substances. The release is probably into the **haemolymph.**

Neuromere. Part of the central nervous system belonging to one segment of the body. For much of the body, the organisation of the nervous system reflects its segmentation, with one segmental ganglion representing the neuromere of a particular segment. However, in the brain, suboesophageal, metathoracic and last abdominal ganglion, several neuromeres are fused. In the brain of the adult, the segmental origins of the different regions are not obvious.

Neuropil (neuropile). The region in the central nervous system where most of the synaptic interactions between neurons take place. It is a fibrous region inside the cortex of cell bodies and contains no cell bodies itself. It is penetrated by longitudinal and vertical tracts and by commissures. Prominent regions of neuropil in segmental ganglia are named.

Neuropil regions in segmental ganglia

Ventral neuropils	
Ventral association centre	VAC
anterior	aVAC
lateral	lVAC
medial	mVAC (aRT)
ventralmost	vVAC
Dorsal neuropils	
Lateral association centre	LAC
anterior	aLAC
posterior	pLAC

Neuropilar segment. The expanded **primary neurite** of a neuron as it courses through the **neuropil** and gives rise to many fine **neurites** or branches. Sometimes, the functional term **integrating segment** is also used but the implication behind this term may be misleading.

Nonspiking interneuron. An interneuron that does not produce action potentials even during the performance of apparently normal behaviour and which cannot be made to spike by experimentally activating a particular input **pathway**, or by the intracellular injection of depolarising current. Under some extreme and unphysiological experimental conditions these neurons can generate small spikes, but there is no evidence that these are normally used for intracellular or intercellular communication. All the nonspiking interneurons that are known are local interneurons with their branches restricted to one ganglion, or one region of the central nervous system. The term nonspiking is thus a pragmatic one, based on what is normally observed. The important feature of these neurons is their ability to communicate with other neurons with graded signals and without the intervention of spikes.

Nymph. See **larva.**

Occipital ganglion. See **hypocerebral ganglion.**

Octopamine. *p*-Hydroxyphenylethanolamine, a neuromodulatory substance first isolated from an octopus, and contained in a number of central neurons of the locust, most notably the efferent **DUM neurons**. There are two isomers of octopamine: D-octopamine the natural isomer occurs as the minus form, D(-)octopamine; L-octopamine occurs as the plus form, L(+)octopamine. DL-octopamine is thus a mixture of the two isomers. Sometimes, when the isomers present are not known, octopamine is called *p*-octopamine to indicate that the hydroxyl group is parallel to the chain.

Ommatidium (ommatidia). One of the several thousand repeating optical and neural units in the compound eye. Each ommatidium consists of a hexagonally shaped corneal lens and eight **retinula cells** that contain the visual pigment, and may be screened from neighbouring ommatidia by pigment-containing cells.

Optic lobe. Region of the **protocerebrum** below the compound eye forming a prominent lateral lobe. It is chiefly concerned with the processing of visual signals and its most distal two regions, the **lamina** and **medulla** contain as many repeating elements as there are **ommatidia** in the eye. Convergence occurs in the third and most proximal region, the **lobula.**

Optomotor response. A movement of the head, or of both the head and the whole body, in response to a movement of the visual world, that attempts to stabilise the visual image on the retina. Experimentally, it is usually induced by the rotation of a pattern of black and white stripes around the animal.

Orthodromic. The direction in which a spike is normally conducted in a neuron away from the **spike-initiating zone**. It contrasts with **antidromic** spikes that can be induced experimentally, or which describe spikes in neurons with distinct spike-initiating zones.

Paramedian nerve. A ladder-like arrangement of paired nerves running parallel to the main nerve cord in the abdomen and formed by **anastomoses** of the unpaired **median nerve** with lateral nerves of the abdominal ganglia.

Parasegment. A unit of the nervous system (primarily in the embryo) consisting of the posterior part or compartment of one segment and the anterior compartment of the next posterior segment.

Paraventricular ganglion. Largely undescribed ganglion associated with the foregut that is part of the **stomatogastric ganglia.**

Pathway. A general term used to describe the route taken by signals within the nervous system. *The pathway from sensory to motor neurons* could imply that the sensory neurons make direct connections with the motor neurons, or that several layers of interneurons intervene. See **central pattern generator, circuit, network.**

PDH. Pigment dispersing hormone.

Perineurium. Thin specialised layer of glia beneath the connective tissue sheath of the central nervous system that provides an ionic barrier, essentially acting like a blood-brain barrier, so that the ionic concentrations at the surface of the neurons are quite different from those in the **haemolymph.**

Periplaneta americana. American cockroach.

Pheromone. A chemical released by an insect that affects the behaviour of another insect, usually of the same species, over a distance. Many pheromones are used as sexual attractants.

Photoinactivation. A method for killing a whole neuron or selected parts of an individual neuron by the injection of a photoactive dye and its subsequent activation by a bright light of the appropriate wavelength. Thus, if the dye Lucifer Yellow is injected into a neuron, the neuron will be killed if the whole preparation is illuminated with a bright blue light, but if a laser is used to focus the light to a specific region then a **neurite** or an axon can be severed from the rest of the neuron.

Picrotoxin. A substance that is used to block the action of **GABA** by its action on chloride channels.

Pilocarpine. A cholinergic, muscarinic agonist, most notable for its ability to induce rhythmic motor activity in isolated thoracic ganglia.

Pioneer neuron. A neuron in an embryo that is the first to chart the route of a peripheral nerve to the central nervous system, or to establish the path of a tract or commissure within the central nervous system.

Plateau potential. A prolonged depolarisation in a neuron caused by a brief depolarisation above a certain voltage threshold. It results from the activation of a voltage-dependent inward current that in some neurons may be carried by Ca^{2+} and which imparts a region of negative slope in the current–voltage relationship. In some neurons, these effects can be generated by synaptic inputs alone, but in others they occur only in the presence of neuromodulators. This intrinsic membrane property means that a neuron will respond in a nonlinear way to the synaptic inputs that it receives depending on whether the voltage changes are sufficient to activate or deactivate these currents. A neuron can thus exist at two stable states; one at resting potential, the other at a depolarised plateau. Such neurons are thus often said to have **bistable membrane properties.**

Pleomorphic. Synaptic vesicles with an irregular, rather flattened or ovoid shape. They are often associated with neurons that also show **GABA**-like immunoreactivity, but the existence of one of these features does not necessarily herald the presence of the other.

Polyneuronal. The innervation of an individual muscle fibre by more than one motor neuron.

Precocene. A natural plant product extracted from *Ageratum houstonianum* (commonly used as a summer bedding plant) that has an anti-**juvenile hormone** action. Application to a larval stage results in precocious moulting to the next stage. Two compounds, precocene 1 and 2, are known. (See **azadirachtin.**)

Primary neurite. The single process that emerges from the cell body. Within the **neuropil** this process often expands in diameter, and is then sometimes called the **integrating segment.** From the primary neurite arise the many fine branches or **neurites** which ramify in the neuropil.

Principal interneuron. See **intersegmental interneuron.**

Proctodeum. The hind gut, which like the foregut or **stomodeum** is formed from ectoderm and is thus lined by a layer of cuticle that is shed at each **moult**. It consists of the ileum and rectum. The midgut is, however, formed from endoderm.

Projection interneuron. See **intersegmental interneuron.**

Promotor. The name used to describe the anterior tergocoxal muscles (muscle 62 of the prothorax, 89 of the mesothorax and 118 of the metathorax) and their supposed action in lifting the anterior edge of the **coxa** upwards. These muscles in the meso- and metathorax have been considered to be bifunctional in that they are active in elevating the wing during flying and moving the coxa during walking. Evidence that the metathoracic muscle 118 is used in walking is, however, lacking. See **remotor**.

Pronation. Twisting movement of a wing during flight in which the anterior edge is depressed and the posterior part is elevated about the long axis of the wing. The effect is a reduction in the amount of lift that the wing produces. The subalar muscle normally causes pronation accompanying depression, and this can be opposed by contraction of the pleuroaxillary muscle, thus restoring the lift generating capability of the wing. See **supination**.

Pronotum. Rigid, highly sclerotised region of the exoskeleton on the dorsal and lateral surfaces of the body behind the head (neck) .

Proprioceptor. A receptor (usually internal) that may signal the movements of a joint, the force generated by a muscle, and the distortions of the cuticle that result from changes in the loading on the body, or from the contractions of the muscles. These effects are thus usually generated by the actions of the locust itself, but the same receptors will also signal externally imposed movements of the joint or deformations of the exoskeleton. Some proprioceptors may thus at times act like **exteroceptors**.

Protocerebrum. The anterior part of the brain that consists of prominent areas such as the **optic lobes, mushroom bodies** and **central complex**. Generally considered the region for higher order processing where signals from different modalities are brought together and where learning occurs. See **deutocerebrum, tritocerebrum**.

Protraction. See **swing phase**.

PTTH. Prothoracicotropic hormone.

Quisqualate. A potent agonist of a particular glutamate receptor characterised on muscle membrane that gates cations and causes a depolarisation. It is extracted with difficulty from the seeds of *Quisqualis indica*, the Rangoon creeper, a semi-evergreen climbing plant.

Range fractionation. Subdivision of the coding of a sensory stimulus among a group of sensory neurons. For example, in the femoral **chordotonal organ** of a leg, some sensory neurons respond to flexion and others to extension, and within the groups that respond to one direction, individual sensory neurons respond to different velocities of joint movements and/or to different ranges of joint angles.

Receptive field. The volume of space, or the area on the surface of the body, that leads to the excitation (or inhibition) of a neuron by a particular modality of sensory stimulus. For example, the **DCMD** interneuron is excited by a moving stimulus anywhere in the visual field of the compound eye contralateral to its axons. A spiking **local interneuron** of the midline group in a thoracic ganglion receives direct synaptic inputs from sensory neurons of mechanoreceptors on specific regions of a leg and these define its receptive field. Receptive fields of multimodal neurons are composed of several regions and, by definition, of more than one modality.

Remotor. The name used to describe the first and second posterior tergocoxal muscles (63, 64 of the prothorax; 90, 91 of the mesothorax; and 119, 120 of the metathorax) and their supposed action in lifting the posterior edge of the **coxa** upwards. These muscles in the meso- and metathorax have been considered to be bifunctional in that they are active in elevating the wing during flying and moving the coxa during walking. In the metathorax, it seems that different sets of motor neurons that innervate muscle 120 are active in the two movements, and that 119 is active only in flying. See **promotor**.

Resistance reflex. A postural adjustment in which an imposed movement is met by increased force in muscles generating the opposing movements. For example, an imposed flexion of the **tibia** of a leg is met by increased excitation of excitatory extensor tibiae motor neurons and inhibition of excitatory flexor tibiae motor neurons so that greater extension force is developed. The effect may spread to other joints of the leg so that stability is enhanced. The gain of the motor response to the same stimulus can change in different circumstances and may even reverse in sign to give an **assistance reflex**, particularly during locomotion.

Retinula cell. A receptor cell in the compound eyes that contains visual pigment and is responsible for the conversion of light into a graded depolarising electrical signal that is conveyed by its axon to the **lamina** of an **optic lobe.**

Retraction. See **stance phase.**

Retrocerebral complex. Group of paired and unpaired ganglia close to the brain and suboesophageal ganglion. The paired ganglia are the **corpora cardiaca** and **corpora allata** and have a secretory function; the unpaired ganglia are the **stomatogastric ganglia.**

Ringer solutions. The following Ringer solutions have been used for analyses of the central nervous system and for peripheral nerves and muscles.

Author(s)	Tissue	NaCl	KCl	$CaCl_2$	$MgCl_2$	Buffer		Sucrose	pH
		in mmol.l^{-1} (mM)							
Hoyle (1953)	nerve	140	10	2	2	$NaHCO_3$ NaH_2PO_4	4 6	–	6.8
Usherwood and Grundfest (1965)	skeletal muscle (used subsequently by many others for recording from intact CNS)	140	10	4	–	$NaHCO_3$ NaH_2PO_4	4 6	–	6.8
Clements and May (1974)	skeletal muscle	140	10	2	–	NaH_2PO_4 Na_2HPO_4	4 6	90	6.8
Cuthbert and Evans (1989)	heart muscle	150	–	4	–	$KHCO_3$ KH_2PO_4	4 6	90	6.8
Dubas (1991)	desheathed ganglion	155	3	4	–	Hepes	10	25	6.8
Wolf and Laurent (1994)	intact CNS and isolated sheathed ganglia	150	5	5	1	$NaHCO_3$ Hepes	4 6.3	–	7.1
Robertson and Pearson (1982)	recording from intact CNS, particularly during flight	146	10	3	–	MOPS	5	–	6.8

Sarcoplasmic reticulum. An internal series of membranes in muscle fibres that are contacted by the **T tubule** invaginations of the plasma membrane. Ca^{2+} may be stored and released from it to activate the contractile machinery.

Satellite nervous system. A series of superficial neurohaemal release sites formed by the terminals of seven pairs of neurons in the suboesophageal ganglion along particular nerves of this ganglion. These neurons contain **5-HT.**

Schistocerca gregaria. African desert locust.

Sclerite. A small piece of cuticle that may articulate with a main segment of a limb and which is often embedded in the joint membrane. Muscle fibres attach to it. Sclerites are most obvious at the articulation of a wing with the thorax.

Scolopidium. An internal mechanoreceptor **sensillum** with 1–3 associated bipolar sensory neurons that have ciliated **dendrites.** It is enveloped by a scolopale cell that secretes a fibrous, barrel-shaped sleeve called the scolopale around the cilia. The whole is suspended from the cuticle by an attachment cell. These structures linked in various numbers are characteristic of **chordotonal organs.**

Segmentation genes. These genes interact to establish the segmentation of the embryo and to control the identity of the cells within each segment. They can be divided into three groups based on the defects seen in mutants lacking a particular gene.

1. *Zygotic gap* genes produce a deletion between contiguous segments
2. *Pair-rule* genes such as *fushi tarazu* and *even-skipped* loci produce homologous defects in alternate segments.
3. *Segment polarity* genes such as *engrailed* are needed for pattern formation within each segment.

See also **homeotic genes.**

Semilunar process. A black, hard and brittle region of cuticle at the femoro-tibial joint of a hind leg of a locust that is bowed during the contraction of the extensor and flexor tibiae muscles that precedes jumping and kicking. The distortion of the cuticle stores energy which, when released suddenly, contributes to the rapid extension of the **tibia.**

Sensillum (sensilla). A sensory structure (receptor) protruding from, or embedded in, the cuticle consisting of four cell types; the **trichogen,** or hair-forming cell, the **tormogen,** or socket-forming cell, a neuron, and a neurilemma cell. See **trichoid, basiconic,** and **campaniform sensilla.**

Serotonin. See **5-HT.**

SETi. Slow motor neuron innervating the extensor tibiae muscle of a leg. See also **FETi, DUMETi.**

SN1, 2. Salivary neurons 1 and 2. SN1 has its cell body in the labial **neuromere** of the suboesophageal ganglion, contains **dopamine** and has an axon ipsilateral to its soma. SN2 has its cell body in the mandibular neuromere of the suboesophageal ganglion, contains **5-HT** and **GABA** and has an axon contralateral to its soma.

Solitary phase. These are locusts with a particular morphological appearance and behaviour that results from them living in isolated conditions, as opposed to those living in crowded conditions (see **gregarious phase**). The larvae are green or yellow-green, and develop more slowly than gregarious ones. Solitary adults fly only at night but lay more eggs than gregarious adults. The phase differences are under hormonal control.

Soma (somata). The cell body of a neuron. Most sensory neurons have cell bodies in the periphery associated with particular sensory structures, the exception being **strand receptors.** A few neurosecretory neurons have cell bodies associated with peripheral nerves. All other neurons have cell bodies in the cortex of the brain or a ganglion. They have no synapses, even though receptors for various potential neurotransmitters and neuromodulators may be present. Synaptic potentials and spikes are therefore reflected into a cell body from sites in the **neuropil.** They are usually electrically inexcitable (but not **DUM neurons**). Each cell body is enveloped by glial cells which form many invaginations close to the exit of the single **primary neurite.** Groups of cell bodies are bundled together by further glial wrappings and this organisation may reflect an origin from a common **neuroblast**(s).

Spike-initiating zone. The region of membrane in a spiking neuron where the integrated synaptic input initiates a spike that is then transmitted **orthodromically** along the axon. It is generally thought that the zone can shift along a small length of membrane and is not a discrete site.

Spiracle. An opening in the cuticle connecting with **tracheae** and through which air can be exchanged. The aperture is controlled by the action of muscles. The spiracles are paired structures on most segments and open and close with the ventilatory rhythm.

Stance phase. Part of the step cycle of an individual leg during walking that is often called **retraction**, the support phase, or the power stroke, in which the foot is in contact with the ground and propels the body forwards (or backwards). The duration of this phase varies with the speed of walking. The stance phase alternates with the **swing phase** (**protraction**).

Stomatogastric ganglia. A group of small aggregations of neurons into ganglia that are closely associated with the brain, suboesophageal ganglion, **corpora cardiaca** and **corpora allata** in the **retrocerebral complex**. They consist of the **frontal**, **hypocerebral** (**occipital**) and **ingluvial** (**ventricular**), and paraventricular ganglia. They would appear to be responsible for controlling some of the movements of the gut, but their actions are not known in detail.

Stomodeum. The foregut, which like the hindgut or **proctodeum** is formed from ectoderm and thus is lined by a layer of cuticle that is shed at each **moult**. It consists of the pharynx, oesophagus and crop. The midgut, however, is formed from endoderm.

Strand receptor. Sensory cell associated with a strand of connective tissue in the periphery that responds to movements of a joint. These sensory cells are set apart from all other sensory neurons in that their cell bodies (**somata**) are in the central nervous system and not at the receptor itself.

Stretch receptor. Sensory structure often consisting of a single neuron associated with a strand of connective tissue (e.g. at a wing hinge), or with a receptor muscle (e.g. **coxa** or abdomen). Gives information about length changes and hence about the movements of a joint.

Subgenual organ. A sense organ consisting of a group of **chordotonal sensilla** in the **tibia** that respond to vibrations transmitted through the ground. In cockroaches they may also respond to sound. In locusts, their function is largely unexplored.

Supination. Movement of the wing during flight in which the anterior edge is elevated and the posterior part is depressed about the long axis of the wing. The effect will be to increase the amount of lift that a wing can produce. See **pronation**.

Swarm. Vast aggregation of adult locusts (**gregarious phase**) either in flight or feeding.

Swing phase. Part of the step cycle of an individual leg during walking that is often called **protraction**, the recovery phase, or the return stroke, in which the leg is swung forwards and the foot is off the ground. The duration of this phase varies little with the speed of walking. The swing phase alternates with the **stance phase** (**retraction**).

Synergist. A description of a muscle that is working with another muscle, in contrast to one that is opposing the action of another muscle and therefore acting as an **antagonist**. Synergy in the actions of muscles is common at complex joints where a movement results only from the balance of actions of a set of muscles.

T tubules. Invaginations of the plasma membrane of muscle fibres that allow ionic changes associated with excitation and inhibition to be carried deep within a fibre. Contacts are made with an internal system of membranes called the **sarcoplasmic reticulum** from which Ca^{2+} may be released to activate the contractile machinery.

Tarsus. Section of the distal part of the leg between the **tibia** and the **unguis**, that forms the greater part of the foot. It is moved by muscles in the tibia but contains no muscles itself. The ventral surface has several pads packed with sensory receptors. See also **coxa, trochanter, femur**.

TCD. Tritocerebral dwarf neuron with its cell body in the **tritocerebrum** of the brain and an axon that passes along the tritocerebral commissure before joining the contralateral connective.

TCG. Tritocerebral giant neuron with its cell body in the **tritocerebrum** of the brain and an axon that passes along the tritocerebral commissure before joining the contralateral connective.

Tegula. A sensory structure associated with a wing that consists of exteroceptive hairs and an internal **chordotonal organ**. Both signal the downstroke of a wing and provide an important synaptic input to wing elevator motor neurons.

Tension receptor. Single multipolar sensory neuron embedded in muscles that responds primarily to the isometric force generated by a particular muscle rather than the movement of a joint or an appendage that may result. So far known only in the flexor tibiae muscles of the middle and hind legs.

Tethered (walking or flying). A description of an experimental arrangement to study walking or flying, in which the movements of the locust are restricted to allow the actions of neurons and muscles to be analysed. The term is not specific and describes a number of different experimental arrangements. In tethered walking, the locust may be glued to a rigid holder and allowed to grasp a light ball which it can then spin as it walks, or it may be attached to a balance so that it has to support its body weight as it turns a servo-controlled ball, or walks on a tread-wheel. In some experiments on stick insects, the left and right legs walk on separate tread-wheels whose speed can be varied independently. In other arrangements the orientation of the insect may vary, so that it may walk upside-down inside a tread-wheel against which it has to push by an amount equivalent to its own body weight. In tethered flying, the support may again be either rigid or a **flight balance**, which typically requires the insect to lift its own body weight. More sophisticated, but highly unstable balances allow more degrees of freedom of movement.

Tibia. Section of the leg between the **femur** and the **tarsus**. It is light and tubular in construction, containing two parts of the retractor unguis muscle proximally, and the levator and depressor tarsi muscles distally that operate the hinge joint with the tarsus. The dorsal surface of the tibia of a hind leg has two rows of spines; the most distal two pairs of spines can be moved passively, but the remainder are fixed. By contrast, the front and middle legs have passively moveable spines on the ventral surface. See also **coxa, trochanter, unguis**.

Tormogen. The cell associated with a sensory **sensillum** that forms the socket of the receptor. See **Trichogen**.

Trachea(e). A tube, often reinforced by helical cuticular strips, that is part of a complex network conveying gases from the **spiracles** to the tissues. In the body, it normally has a silvery appearance because it is filled with air. The tracheae are connected to **air sacs** that inflate with each cycle of the ventilatory rhythm. Within the tissues, tracheae branch into fine blind-ending tubes called **tracheoles**.

Tracheole. Fine, fluid-filled and blind-ending branches of the **tracheae** which are responsible for gaseous exchange with tissues.

Tract. A bundle of axons that are aggregated together as they pass through a particular region of the central nervous system. The connectives as they pass through the thoracic and abdominal ganglia split up into nine prominent longitudinal tracts. Smaller tracts mark the entry of the axons from the peripheral nerves and the accumulation of small groups of axons and processes that run vertically or obliquely.

Tracts in segmental ganglia

Longitudinal tracts	
Dorsal intermediate tract	DIT
Dorsal median tract	DMT
Median dorsal tract	MDT
Median ventral tract	MVT
Lateral dorsal tract	LDT
Lateral ventral tract	LVT
Ventral intermediate tract	VIT
Ventral lateral tract	VLT
Ventral median tract	VMT
Vertical and oblique tracts	
C tract	CT
I tract	IT
Oblique tract	OT
Perpendicular tract	PT
Ring tract	RT
T tract	TT
Deep DUM tract	DDT
Superficial DUM tract	SDT

Trichogen. The cell associated with a sensory **sensillum** that forms the hair of the receptor. See **Tormogen.**

Trichoid sensilla. Sensory receptors with a shaft of variable length that protrudes from the cuticle. They are set in a membranous socket, and are usually innervated by one (or sometimes two) sensory neurons. The stiffness and thickness of the shaft, and its mounting in its socket, determine whether its sensory neuron responds to air movements or to tactile stimuli. See also **basiconic sensilla.**

Tritocerebrum. Middle region of the brain between the **protocerebrum** anteriorly and the **deutocerebrum** posteriorly.

Trochanter. Section of the proximal part of the leg between the **coxa** and the **femur**. In the hind leg it is fused to the femur, so that no movement about the joint is possible, and no muscles are present that could move the femur about it. In the front and middle legs, some movement is possible and a single muscle is present (the reductor femora). See also **trochantin, tibia, tarsus, unguis.**

Trochantin. Small **sclerite** embedded in the arthrodial membrane at the **coxa** on which the coxal promotor muscle inserts. See also **trochanter, femur, tibia, tarsus, unguis.**

Trophospongium. An association between glial cells and a neuron, typically at the point where the **primary neurite** emerges from the cell body. At this point, the glial cells form numerous invaginations into the neuron and the neuron may send processes into the glia so that there are reciprocal interdigitations. Based on the Greek word *Trophos* meaning 'one who feeds'.

Tympanum (tympana). Auditory receptor consisting of a thin cuticular membrane, **chordotonal organ**, and closely associated **tracheae**. In locusts, the tympana are on the abdomen just behind the hind legs, but in crickets they are in the tibia of the front legs. The number of **sensilla** varies from two in some moths to more than 1500 in some cicadas.

Unguis. Terminal section of the leg, often called the claw. It is moved by a retractor unguis muscle that has three parts, one in the **femur** and two in the proximal **tibia** that all attach to the same **apodeme** (tendon) that runs through the femur, tibia and **tarsus** to insert on the **unguis.** The action of this muscle will curl the unguis and should increase traction when the tarsus is placed on the ground or during climbing. See also **coxa, trochanter.**

Ventral diaphragm. A horizontal septum in the abdomen just dorsal to the nerve cord separating the sinus (perineurial) around the nerve cord from that (perivisceral) around the viscera. It consists of a thin sheet of muscle that may contract in time with the inspiratory cycle of breathing.

Wind tunnel. A device for inducing sustained flight in a locust that is **tethered.** The airflow is arranged to be warm and laminar and the locust is presented with a visual environment that may consist simply of a horizon or may be more structured. The locust may be rigidly mounted or may be attached to a **flight balance** that can measure various forces and torques.

References

Adams, M.E. and O'Shea, M. (1983). Peptide cotransmitter at a neuromuscular junction. Science (NY), 221: 286–288.

Adams, M.E., Bishop, C.A. and O'Shea, M. (1989). Functional consequences of peptide cotransmission in arthropod muscle. Am. Zoologist, 29: 1231–1330.

Adrian, E. D. (1931). Potential changes in the isolated nervous system of *Dytiscus marginalis*. J. Physiol., 72: 132–151.

Agricola, H.-J. and Bräunig, P. (1995). Comparative aspects of peptidergic signaling pathways in the nervous systems of arthropods. In '*The nervous systems of invertebrates: an evolutionary and comparative approach*'. pp. 303–327. Ed. Breidbach, O. and Kutsch, W. Birkhäuser Verlag, Basel.

Albert, P.J., Zacharuk, R.Y. and Wong, L. (1976). Structure, innervation and distribution of sensilla on the wings of a grasshopper. Can. J. Zool., 54: 1542–1553.

Alexander, R.McN. (1981). Factors of safety in the structure of animals. Sci. Prog. Oxford, 67: 109–130.

Alexandrowicz, J.S. (1951). Muscle receptor organs in the abdomen of *Homarus vulgaris* and *Palinurus vulgaris*. Q. J. Microsc. Sci., 92: 163–199.

Ali, D.W., Orchard, I. and Lange, A.B. (1993). The aminergic control of locust (*Locusta migratoria*) salivary glands: Evidence for dopaminergic and serotonergic innervation. J. Insect Physiol., 39: 623–632.

Allgäuer, C. and Honegger, H.W. (1993). The antennal motor system of crickets: modulation of muscle contractions by a common inhibitor, DUM neurons, and proctolin. J. Comp. Physiol., 173A: 485–494.

Aloe, L. and Levi-Montalcini, R. (1972). *In vitro* analysis of the frontal and ingluvial ganglia from nymphal specimens of the cockroach *Periplaneta americana*. Brain Res., 44: 147–163.

Alsop, D.W. (1978). Comparative analysis of the intrinsic leg musculature of the American cockroach, *Periplaneta americana*. J. Morphol., 158: 199–242.

Altman, J.S. and Kien, J. (1987). Functional organisation of the suboesophageal ganglion in arthropods. In '*Arthropod brain: its evolution, development, structure and functions*' pp. 265–301. Ed. Gupta, A.P. Wiley, New York.

Altman, J.S., Anselment, E. and Kutsch, W. (1978). Postembryonic development of an insect sensory system: ingrowth of axons from hindwing sense organs in *Locusta migratoria*. Proc. R. Soc. Lond. B, 202: 497–516.

Alvarez-Buyalla, A (1990). Mechanism of neurogenesis in adult avian brain. Experientia, 46: 948–955.
Ammermüller, J., Oltrogge, M. and Janssen-Bienhold, U. (1994). Neurotensin-like immunoreactivity in locust supraesophageal ganglion and optic lobes. Brain Res., 636: 40–48.
Anderson, H. (1978). Postembryonic development of the visual system of the locust *Schistocerca gregaria*. I. Patterns of growth and developmental interactions in the retina and optic lobes. J. Embryol. Exp. Morphol., 45: 55–83.
Anderson, H. and Tucker, R.P. (1988). Pioneer neurones use basal lamina as a substratum for outgrowth in the embryonic grasshopper. Development, 104: 601–608.
Arakawa, S., Gogayne, J.D., McCombie, W.R., Urquart, D.A., Hall, L.M., Fraser, C.M. and Venter, J.C. (1990). Cloning, localisation and permanent expression of a *Drosophila* octopamine receptor. Neuron, 2: 343–354.
Arbas, E.A. (1983). Aerial manoeuvring reflexes in flightless grasshoppers. J. Exp. Biol., 107: 509–513.
Arbas, E.A. (1986). Control of hindlimb posture by wind-sensitive hairs and antennae during locust flight. J. Comp. Physiol., 159A: 849–857.
Arbas, E.A. and Tolbert, L.P. (1986). Presynaptic terminals persist following degeneration of 'flight' muscle during development of a flightless grasshopper. J. Neurobiol., 17: 627–636.
Armet-Kibel, C., Meinertzhagen, I.A. and Dowling, J.E. (1977). Cellular and synaptic organisation in the lamina of the dragonfly *Sympetrum rubicundulum*. Proc. R. Soc. Lond. B, 196: 385–413.
Atwood, H.L. and Hoyle, G. (1965). A further study of the paradox phenomenon of crustacean muscle. J. Physiol., 181: 225–234.
Aubele, E. and Klemm, N. (1977). Origin, destination and mapping of tritocerebral neurons of locust. Cell Tissue Res., 178: 199–219.
Baader, A. (1990). The posture of the abdomen during locust flight: regulation by steering and ventilatory interneurones. J. Exp. Biol., 151: 109–131.
Baader, A. (1991). Auditory interneurons in locusts produce directional head and abdomen movements. J. Comp. Physiol., 169A: 87–100.
Baba, Y., Hirota, K. and Yamaguchi, T. (1991). Morphology and response properties of wind-sensitive non-giant interneurons in the terminal abdominal ganglion of the cricket. Zool. Sci., 8: 437–445.
Baba, Y., Hirota, K., Shimozawa, T. and Yamaguchi, T. (1995). Differing afferent connections of spiking and nonspiking wind-sensitive local interneurons in the terminal abdominal ganglion of the cricket *Gryllus bimaculatus*. J. Comp. Physiol., 176A: 17–30.
Bacon, J. and Möhl, B. (1983). The tritocerebral commissure giant (TCG) wind-sensitive interneurone in the locust. I. Its activity in straight flight. J. Comp. Physiol., 150A: 439–452.
Bacon, J. and Tyrer, N.M. (1978). The tritocerebral giant (TCG): a bimodal interneurone in the locust, *Schistocerca gregaria*. J. Comp. Physiol., 126A: 317–325.
Bacon, J. and Tyrer, M. (1979). Wind interneurone input to flight motor neurones in the locust, *Schistocerca gregaria*. Naturwissenschaften, 66: 116.
Bacon, J.P. and Altman, J.S. (1977). A silver intensification method for cobalt-filled neurones in wholemount preparations. Brain Res., 138: 359–363.
Bacon, J.P. and Murphey, R.K. (1984). Receptive fields of cricket giant interneurones are related to their dendritic structure. J. Physiol., 352: 601–623.
Bai, D. and Sattelle, D.B. (1995). A $GABA_B$ receptor on an identified insect motor neurone. J. Exp. Biol., 198: 889–894.

Baines, R.A. and Bacon, J.P. (1994). Pharmacological analysis of the cholinergic input to the locust VPLI neuron from an extraocular photoreceptor system. J. Neurophysiol., 72: 2864–2874.

Baines, R.A., Lange, A.B. and Downer, R.G.H. (1990). Proctolin in the innervation of the locust mandibular closer muscle modulates contractions through the elevation of inositol trisphosphate. J. Comp. Neurol., 297: 479–486.

Baines, R.A., Thompson, K.S.J., Rayne, R.C. and Bacon, J.P. (1995). Analysis of the peptide content of the locust vasopressin-like immunoreactive (VPLI) neurons. Peptides, 16: 799–807.

Baird, D.H., Schalet, A.P. and Wyman, R.J. (1990). The *Passover* locus in *Drosophila melanogaster*: complex complementation and different effects on the giant fiber neural pathway. Genetics, 126: 1045–1059.

Baird, D.H., Koto, M. and Wyman, R.J. (1993). Dendritic reduction in *Passover*, a *Drosophila* mutant with a defective giant fiber neuronal pathway. J. Neurobiol., 24: 971–984.

Baker, P.S. and Cooter, R.J. (1979a). The natural flight of the migratory locust, *Locusta migratoria* L. I. Wing movements. J. Comp. Physiol., 131A: 79–87.

Baker, P.S. and Cooter, R.J. (1979b). The natural flight of the migratory locust, *Locusta migratoria* L. II. Gliding. J. Comp. Physiol., 131: 89–94.

Baker, P.S., Gewecke, M. and Cooter, R.J. (1981). The natural flight of the migratory locust *Locusta migratoria*. III. Wing beat frequency, flight speed and attitude. J. Comp. Physiol., 141A: 233–237.

Ball, E.E. and Goodman, C.S. (1985a). Muscle development in the grasshopper embryo. II. Syncytial origin of the extensor tibiae muscle pioneers. Dev. Biol., 111: 399–416.

Ball, E.E. and Goodman, C.S. (1985b). Muscle development in the grasshopper embryo. III. Sequential origin of the flexor tibiae muscle pioneers. Dev. Biol., 111: 417–424.

Ball, E.E., Ho, R.K. and Goodman, C.S. (1985). Development of neuromuscular specificity in the grasshopper embryo: guidance of motoneuron growth cones by muscle pioneers. J. Neurosci., 5: 1808–1819.

Ballard, J.W.O., Olsen, G.J., Faith, D.P., Odgers, W.A., Rowell, D.M. and Atkinson, P.W. (1992). Evidence from 12S ribosomal RNA sequences that onychophorans are modified arthropods. Science (NY), 258: 1345–1348.

Baroni Urbani, C., Boyan, G.S., Blarer, A., Billen, J. and Musthak Ali, T.M. (1994). A novel mechanism for jumping in the Indian ant *Harpegnathos saltator* (Jerdon) (Formicidae, Ponerinae). Experientia, 50: 63–71.

Bartos, M. and Honegger, H.W. (1992). Complex innervation of three neck muscles by motor and dorsal unpaired median neurons in crickets. Cell Tissue Res., 267: 399–406.

Bartos, M., Allgäuer, C., Eckert, M. and Honegger, H.W. (1994). The antennal motor system of crickets: proctolin in slow and fast motor neurons as revealed by double labelling. Eur. J. Neurosci., 6: 825–836.

Bässler, U. (1967). Zur Regelung der Stellung des Femur-Tibia-Gelenkes bei der Stabheuschrecke *Carausius morosus* in der Ruhe und im Lauf. Kybernetik, 4: 18–26.

Bässler, U. (1977). Sensory control of leg movement in the stick insect *Carausius morosus*. Biol. Cybernetics, 25: 61–72.

Bässler, U. (1993). The walking- (and searching-) pattern generator of stick insects, a modular system composed of reflex chains and endogenous oscillators. Biol. Cybernetics, 69: 305–317.

Bässler, U., Hofmann, T. and Schuch, U. (1986). Assisting components within a resistance reflex of the stick insect, *Cuniculina impigra*. Physiol. Entomol., 11: 359–366.

Bastiani, M.J. and Goodman, C.S. (1984). Neuronal growth cones: specific interactions mediated by filopodial insertion and induction of coated vesicles. Proc. Natl Acad. Sci. USA, 81: 1849–1853.

Bastiani, M.J. and Goodman, C.S. (1986). Guidance of neuronal growth cones in the grasshopper embryo. III. Recognition of specific glial pathways. J. Neurosci., 6: 3542–3551.

Bastiani, M., Pearson, K.G. and Goodman, C.S. (1984). From embryonic fascicles to adult tracts: organization of neuropile from a developmental perspective. J. Exp. Biol., 112: 45–64.

Bastiani, M.J., du Lac, S. and Goodman, C.S. (1986). Guidance of growth cones in grasshopper embryo. I. Recognition of a specific axonal pathway by the pCC neuron. J. Neurosci., 6: 3518–3531.

Bastiani, M.J., De Couet, H.G., Quinn, J.M.A., Karlstrom, R.O., Kotrla, K., Goodman, C.S. and Ball, E.E. (1992). Position-specific expression of the annulin protein during grasshopper embryogenesis. Dev. Biol., 154: 129–142.

Bate, C.M. (1976a). Embryogenesis of an insect nervous system. I. A map of the thoracic and abdominal neuroblasts in *Locusta migratoria*. J. Embryol. Exp. Morphol., 35: 107–123.

Bate, C. M. (1976b). Pioneer neurones in an insect embryo. Nature (Lond.), 260: 54–56.

Bate, C.M. and Grunewald, E.B. (1981). Embryogenesis of an insect nervous system II: a second class of neuron precursor cells and the origin of the intersegmental connectives. J. Embryol. Exp. Morphol., 61: 317–330.

Bate, M., Goodman, C.S. and Spitzer, N.C. (1981). Embryonic development of identified neurons: segment-specific differences in the H cell homologues. J. Neurosci., 1: 103–106.

Baudinette, R.V., Gannon, B.J., Runciman, W.B., Wells, S. and Love, J.B. (1987). Do cardiorespiratory frequencies show entrainment with hopping in the Tammar wallaby? J. Exp. Biol., 129: 251–263.

Bauer, C.K. (1991). Modulatory action of proctolin in the locust (*Locusta migratoria*) antennal motor system. J. Insect Physiol., 37: 663–673.

Belanger, J.H. and Orchard, I. (1993). The locust ovipositor opener muscle: Proctolinergic central and peripheral neuromodulation in a centrally driven motor system. J. Exp. Biol., 174: 343–362.

Bellah, K.L., Fitch, G.K. and Kammer, A.E. (1984). A central action of octopamine on ventilation frequency in *Corydalus cornutus*. J. Exp. Zool., 231: 289–292.

Bennet-Clark, H.C. (1975). The energetics of the jump of the locust *Schistocerca gregaria*. J. Exp. Biol., 63: 53–83.

Bennet-Clark, H.C. and Lucey, E.C.A. (1967). The jump of the flea: a study of the energetics and a model of the mechanism. J. Exp. Biol., 47: 59–76.

Benson, J.A. (1993). The electrophysiological pharmacology of neurotransmitter receptors on locust neuronal somata. In '*Comparative molecular neurobiology*'. pp. 390–413. Ed. Pichon, Y. Birkhäuser Verlag, Basel.

Bentley, D. and Caudy, M. (1983). Pioneer axons lose directed growth after selective killing of guidepost cells. Nature (Lond.), 304: 62–65.

Bentley, D. and Toroian-Raymond, A. (1981). Embryonic and postembryonic morphogenesis of a grasshopper interneuron. J. Comp. Neurol., 201: 507–518.

Bentley, D. and Toroian-Raymond, A. (1986). Disoriented pathfinding by pioneer neurone growth cones deprived of filopodia by cytochalasin treatment. Nature (Lond.), 323: 712–715.

Bentley, D. and Toroian-Raymond, A. (1989). Pre-axogenesis migration of afferent pioneer cells in the grasshopper embryo. J. Exp. Zool., 251: 217–223.

Bentley, D.R., Keshishian, H., Shankland, M. and Toroian-Raymond, A. (1979). Quantitative staging of embryonic development of the grasshopper, *Schistocerca nitens*. J. Embryol. Exp. Morphol., 54: 47–74.

Berger, M., Roy, O.Z. and Hart, J.S. (1970). The co-ordination between respiration and wing beats in birds. Z. vergl. Physiol., 66: 190–200.

Bermudez, I., Beadle, D.J. and Benson, J.A. (1992). Multiple serotonin-activated currents in isolated, neuronal somata from locust thoracic ganglia. J. Exp. Biol., 165: 43–60.

Bernard, J. (1987). Effectiveness of the cercal chordotonal inhibitory organ in the cockroach. Synaptic activity during imposed cercal movements. Comp. Biochem. Physiol., 87A: 53–56.

Bernays, E.A. (1971). The vermiform larva of *Schistocerca gregaria* (Forskål): form and activity (Insecta: Orthoptera). Z. Morphol. Tiere., 70: 183–200.

Bernays, E.A. (1972). The muscles of newly hatched *Schistocerca gregaria* larvae and their possible functions in hatching, digging and ecdysial movements (Insecta: Acrididae). J. Zool., 166: 141–158.

Bicker, G. and Pearson, K.G. (1983). Initiation of flight by an identified wind sensitive neurone (TCG) in the locust. J. Exp. Biol., 104: 289–293.

Bicker, G., Schafer, S., Ottersen, O.P. and Storm-Mathisen, J. (1988). Glutamate-like immunoreactivity in identified neuronal populations of insect nervous systems. J. Neurosci., 8: 2108–2122.

Bieber, M. and Fuldner, D. (1979). Brain growth during the adult stage of a holometablous insect. Naturwissenschaften, 66: 426.

Bishop, C.A. and O'Shea, M. (1982). Neuropeptide proctolin immunocytochemical mapping of neurons in the central nervous system of the cockroach. J. Comp. Neurol., 207: 223–238.

Blagburn, J.M. and Sattelle, D.B. (1987). Presynaptic depolarization mediates presynaptic inhibition at a synapse between an identified mechanosensory neurone and giant interneurone 3 in the first instar cockroach, *Periplaneta americana*. J. Exp. Biol., 127: 135–157.

Blagburn, M., Beadle, D.J. and Sattelle, D.B. (1985). Development of synapses between identified sensory neurones and giant interneurones in the cockroach *Periplaneta americana*. J. Embryol. Exp. Morphol, 86: 227–246.

Blaschko, H., Cattell, M. and Kahn, J.L. (1931). On the nature of the two types of response in the neuromuscular system of the crustacean claw. J. Physiol., 73: 25–35.

Bodnar, D.A. (1993). Excitatory influence of wind-sensitive local interneurons on an ascending interneuron in the cricket cercal sensory system. J. Comp. Physiol., 172A: 641–651.

Bodnar, D.A., Miller, J.P. and Jacobs, G.A. (1991). Anatomy and physiology of identified wind-sensitive local interneurons in the cricket cercal sensory system. J. Comp. Physiol., 168A: 553–564.

Bogdany, F.J. (1978). Linkage of learning signals in honey bee orientation. Behav. Ecol. Sociobiol., 3: 323–336.

Böhm, H. and Schildberger, K. (1992). Brain neurones involved in the control of walking in the cricket *Gryllus bimaculatus*. J. Exp. Biol., 166: 113–130.

Bowers, W.S. (1981). How anti-juvenile hormones work. Am. Zoologist, 21: 737–742.

Boyan, G. (1992). Common synaptic drive to segmentally homologous interneurons in the locust. J. Comp. Neurol., 321: 544–554.

Boyan, G., Williams, L. and Meier, T. (1993). Organization of the commissural fibers in the adult brain of the locust. J. Comp. Neurol., 332: 358–377.

Boyan, G.S. (1988). Presynaptic inhibition of identified wind-sensitive afferents in the cercal system of the locust. J. Neurosci. 8: 2748–2757.

Boyan, G.S. and Altman, J.S. (1985). The suboesophageal ganglion: a 'missing link' in the auditory pathway of the locust. J. Comp. Physiol., 156A: 413–428.

Boyan, G.S. and Ball, E.E. (1986). Wind-sensitive interneurones in the terminal ganglion of praying mantids. J. Comp. Physiol., 159A: 773–789.

Boyan, G.S. and Ball, E.E. (1989a). The wind-sensitive cercal receptor/giant interneurone system of the locust, *Locusta migratoria*. II. Physiology of the giant interneurones. J. Comp. Physiol., 165A: 511–521.

Boyan, G.S. and Ball, E.E. (1989b). The wind-sensitive cercal receptor/giant interneurone system of the locust, *Locusta migratoria*. III. Cercal activation of thoracic motor pathways. J. Comp. Physiol., 165A: 523–537.

Boyan, G.S. and Ball, E.E. (1989c). Parallel inputs shape the response of a giant interneurone in the cercal system of the locust. J. Insect Physiol., 35: 305–312.

Boyan, G.S., Ashman, S. and Ball, E.E. (1986). Initiation and modulation of flight by a single giant interneuron in the cercal system of the locust. Naturwissenschaften, 73: 272–274.

Boyan, G.S., Williams, J.L.D. and Ball, E.E. (1989a). The wind-sensitive cercal receptor/giant interneurone system of the locust, *Locusta migratoria*. I. Anatomy of the system. J. Comp. Physiol., 165A: 495–510.

Boyan, G.S., Williams, J.L.D. and Ball, E.E. (1989b). The wind-sensitive cercal receptor/giant interneurone system of the locust, *Locusta migratoria*. IV. The non-giant interneurones. J. Comp. Physiol., 165A: 539–552.

Bramble, D.M. and Carrier, D.R. (1983). Running and breathing in mammals. Science (NY), 219: 251–256.

Bräunig, P. (1982). The peripheral and central nervous organization of the locust coxo-trochanteral joint. J. Neurobiol., 13: 413–433.

Bräunig, P. (1985). Strand receptors associated with the femoral chordotonal organs of locust legs. J. Exp. Biol. 116: 331–341.

Bräunig, P. (1987). The satellite nervous system – an extensive neurohemal network in the locust head. J. Comp. Physiol., 160A: 69–77.

Bräunig, P. (1988). Identification of a single prothoracic 'dorsal unpaired median' (DUM) neuron supplying locust mouthpart nerves. J. Comp. Physiol., 163A: 835–840.

Bräunig, P. (1990a). The morphology of suboesophageal ganglion cells innervating the nervus corporis cardiaci III of the locust. Cell Tissue Res., 260: 95–108.

Bräunig, P. (1990b). The mandibular ganglion – a new peripheral ganglion of the locust. J. Exp. Biol., 148: 313–324.

Bräunig, P. (1991a). Suboesophageal DUM neurons innervate the principal neuropiles of the locust brain. Phil. Trans. R. Soc. Lond. 332B: 221–240.

Bräunig, P. (1991b). A suboesophageal ganglion cell innervates heart and retrocerebral glandular complex in the locust. J. Exp. Biol., 156 :567–582.

Bräunig, P. and Hustert, R. (1985a). Actions and interactions of proprioceptors of the locust hind leg coxo-trochanteral joint. I. Afferent responses in relation to joint position and movement. J. Comp. Physiol., 157A: 73–82.

Bräunig, P. and Hustert, R. (1985b). Actions and interactions of proprioceptors of the locust hind leg coxo-trochanteral joint. II. Influence on the motor system. J. Comp. Physiol., 157A: 83–87.

Bräunig, P., Hustert, R. and Pflüger, H.-J. (1981). Distribution and specific central projections of mechanoreceptors in the thorax and proximal leg joints of locusts. I. Morphology, location and innervation of internal proprioceptors of pro- and metathorax and their central projections. Cell Tissue Res., 216: 57–77.

Bräunig, P., Cahill, M.A. and Hustert, R. (1986). The coxo-trochanteral muscle receptor organ of locusts. Dendritic tubular bodies in a non-ciliated insect mechanoreceptive neuron. Cell Tissue Res., 243: 517–524.

Bräunig, P., Stevenson,P.A. and Evans, P.D. (1994). A locust octopamine-immunoreactive dorsal unparied median neuron forming terminal networks on sympathetic nerves. J. Exp. Biol., 192: 225–238.

Breer, H. and Heilgenberg, H. (1985). Neurochemistry of GABAergic activities in the central nervous system of *Locusta migratoria*. J. Comp. Physiol., 157A: 343–354.

Broadie, K., Sink, H., Van Vactor, D., Fambrough, D., Whitington, P.M., Bate, M. and Goodman, C.S. (1993). From growth cone to synapse: the life history of the RP3 motor neuron. Development, 119 : 227–238.

Brogan, R.T. and Pitman, R.M. (1981). Axonal regeneration in an identified insect motoneurone. J. Physiol., 319: 34P–35P.

Brown, R.H.J. (1967). Mechanism of locust jumping. Nature (Lond.) 214: 939.

Brown, S.D., Johnson, F. and Bottjer, S.W. (1993). Neurogenesis in adult canary telencephalon is independent of gonadal hormone levels. J. Neurosci., 13: 2024–2032.

Brunn, D.E. and Dean, J. (1994). Intersegmental and local interneurons in the metathorax of the stick insect *Carausius morosus* that monitor middle leg position. J. Neurophysiol., 72: 1208–1219.

Brunner von Wattenwyl, C. (1880). Neues Organ bei Acridiern. Verh. Zool. Bot. Ges. Wien, 29: 26–27.

Buchner, E., Buchner, S., Burg, M.G., Hofbauer, A., Pak, W.L. and Pollack, I. (1993). Histamine is a major mechanosensory neurotransmitter candidate in *Drosophila melanogaster*. Cell Tissue Res., 273: 119–125.

Burkhardt, W. and Braitenberg, V. (1976). Some peculiar synaptic complexes in the first visual ganglion of the fly, *Musca domestica*. Cell Tissue Res., 173: 287–308.

Burns, M.D. (1973). The control of walking in Orthoptera. I. Leg movements in normal walking. J. Exp. Biol., 58: 45–58.

Burns, M.D. (1974). Structure and physiology of the locust femoral chordotonal organ. J. Insect Physiol., 20: 1319–1339.

Burns, M.D. and Usherwood, P.N.R. (1978). Mechanical properties of locust extensor tibiae muscles. Comp. Biochem. Physiol., 61A: 85–95.

Burns, M.D. and Usherwood, P.N.R. (1979). The control of walking in Orthoptera. II. Motor neurone activity in normal free-walking animals. J. Exp. Biol., 79: 69–98.

Burrows, M. (1969). The mechanics and neural control of the prey capture strike of the mantid shrimps *Squilla* and *Hemisquilla*. Z. vergl. Physiol., 62: 361–381.

Burrows, M. (1974). Modes of activation of motoneurons controlling the ventilatory movements of the locust abdomen. Phil. Trans. R. Soc. Lond. B, 269: 29–48.

Burrows, M. (1975a). Monosynaptic connexions between wing stretch receptors and flight motoneurones of the locust. J. Exp. Biol., 62: 189–219.

Burrows, M. (1975b). Co-ordinating interneurones of the locust which convey two patterns of motor commands: their connexions with ventilatory motoneurones. J. Exp. Biol., 63: 735–753.

Burrows, M. (1975c). Co-ordinating interneurones of the locust which convey two patterns of motor commands: their connexions with flight motoneurones. J. Exp. Biol., 63: 713–733.

Burrows, M. (1976). The influence of sensory inflow on the flight system of the locust. In '*Perspectives in experimental biology. Vol. II Zoology*'. pp. 399–409. Ed. Spencer-Davies, P. Pergamon, Oxford.

Burrows, M. (1979a). Synaptic potentials effect the release of transmitter from locust nonspiking interneurons. Science (NY), 204: 81–83.

Burrows, M. (1979b). Graded synaptic transmission between local pre-motor interneurons of the locust. J. Neurophysiol. 42: 1108–1123.

Burrows, M. (1980a). The control of sets of motoneurones by local interneurones in the locust. J. Physiol. 298: 213–233.

Burrows, M. (1980b). The tracheal supply to the central nervous system of the locust. Proc. R. Soc. Lond. B, 207: 63–78.

Burrows, M. (1982a). Interneurones co-ordinating the ventilatory movements of the thoracic spiracles in the locust. J. Exp. Biol., 97; 385–400.

Burrows, M. (1982b). The physiology and morphology of median nerve motor neurones in the metathoracic ganglion of the locust. J. Exp. Biol., 96: 325–341.

Burrows, M. (1983). Local interneurones and the control of movement in insects. In '*Behavioural physiology and neuroethology: roots and growing points*'. pp. 26–41. Eds. Huber, F. and Markl, H. Springer-Verlag, Berlin.

Burrows, M. (1985). The processing of mechanosensory information by spiking local interneurones in the locust. J. Neurophysiol., 54: 463–478.

Burrows, M. (1987). Inhibitory interactions between spiking and nonspiking local interneurones in the locust. J. Neurosci., 7: 3282–3292.

Burrows, M. (1988). Responses of spiking local interneurones in the locust to proprioceptive signals from the femoral chordotonal organ. J. Comp. Physiol., 164A: 207–217.

Burrows, M. (1989). Effects of temperature on a central synapse between identified motor neurones in the locust. J. Comp. Physiol., 165A: 687–695.

Burrows, M. (1992). Reliability and effectiveness of transmission from exteroceptive sensory neurons and spiking local interneurons in the locust. J. Neurosci., 12: 1477–1489.

Burrows, M. (1995). Motor patterns during kicking movements in the locust. J. Comp. Physiol., 176A: 289–305.

Burrows, M. and Laurent, G. (1993). Synaptic potentials in the central terminals of locust proprioceptive afferents generated by other afferents from the same sense organ. J. Neurosci., 13: 808–819.

Burrows, M. and Matheson, T. (1994). A presynaptic gain control mechanism among sensory neurons of a locust leg proprioceptor. J. Neurosci., 14: 272–282.

Burrows, M. and Newland, P.L. (1993). Correlation between the receptive fields of locust interneurons, their dendritic morphology, and the central projections of mechanosensory neurons. J. Comp. Neurol., 329: 412–426.

Burrows, M. and Newland, P.L. (1994). Convergence of mechanosensory afferents from different classes of exteroceptors onto spiking local interneurons in the locust. J. Neurosci., 14: 3341–3350.

Burrows, M. and Pflüger, H.-J. (1986). Processing by local interneurones of mechanosensory signals involved in a leg reflex of the locust. J. Neurosci., 6: 2764–2777.

Burrows, M. and Pflüger, H.-J. (1988). Positive feedback loops from proprioceptors involved in leg movements of the locust. J. Comp. Physiol., 163A: 425–440.

Burrows, M. and Pflüger, H.-J. (1992). Output connections of a wind sensitive interneurone with motor neurones innervating flight steering muscles in the locust. J. Comp. Physiol., 171A: 437–446.

Burrows, M. and Pflüger, H.-J. (1995). Action of locust neuromodulatory neurons is coupled to specific motor patterns. J. Neurophysiol., 74: 347–357.

Burrows, M. and Rowell, C.H.F. (1973). Connections between descending visual interneurons and metathoracic motoneurons in the locust. J. Comp. Physiol., 85: 221–234.

Burrows, M. and Siegler, M.V.S. (1978). Graded synaptic transmission between local interneurones and motoneurones in the metathoracic ganglion of the locust. J. Physiol., 285: 231–255.

Burrows, M. and Siegler, M.V.S. (1982). Spiking local interneurons mediate local reflexes. Science (NY), 217: 650–652.

Burrows, M. and Siegler, M.V.S. (1984). The morphological diversity and receptive fields of spiking local interneurons in the locust metathoracic ganglion. J. Comp. Neurol., 224: 483–508.

Burrows, M. and Siegler, M.V.S. (1985). The organization of receptive fields of spiking local interneurones in the locust with inputs from hair afferents. J. Neurophysiol., 53: 1147–1157.

Burrows, M. and Watkins, B.L. (1986). Spiking local interneurones in the mesothoracic ganglion of the locust: homologies with metathoracic interneurones. J. Comp. Neurol., 245: 29–40.

Burrows, M., Laurent, G.J. and Field, L.H. (1988). Proprioceptive inputs to nonspiking local interneurones contribute to local reflexes of a locust hindleg. J. Neurosci., 8: 3085–3093.

Burrows, M., Watson, A.H.D. and Brunn, D.E. (1989). Physiological and ultrastructural characterization of a central synaptic connection between identified motor neurones in the locust. Eur. J. Neurosci., 1: 111–126.

Büschges, A. (1990). Nonspiking pathways in a joint-control loop of the stick insect *Carausius morosus*. J. Exp. Biol., 151: 133–160.

Büschges, A. and Pearson, K.G. (1991). Adaptive modifications in the flight system of the locust after the removal of wing proprioceptors. J. Exp. Biol., 157: 313–333.

Büschges, A. and Schmitz, J. (1991). Nonspiking pathways antagonize the resistance reflex in the thoraco-coxal joint of stick insects. J. Neurobiol., 22: 224–237.

Büschges, A., Ramirez, J.-M. and Pearson, K.G. (1992a). Reorganization of sensory regulation of locust flight after partial deafferentation. J. Neurobiol., 23: 31–43.

Büschges, A., Ramirez, J.-M., Driesang, R. and Pearson, K.G. (1992b). Connections of the forewing tegulae in the locust flight system and their modification following partial deafferentation. J. Neurobiol., 23: 44–60.

Büschges, A., Kittmann, R. and Ramirez, J.-M. (1993). Octopamine effects mimic state-dependent changes in a proprioceptive feedback system. J. Neurobiol., 24: 598–610.

Büschges, A., Kittmann, R. and Schmitz, J. (1994). Identified nonspiking interneurons in leg reflexes and during walking in the stick insect. J. Comp. Physiol., 174A: 685–700.

Büschges, A., Schmitz, J. and Bässler, U. (1995). Rhythmic patterns in the thoracic nerve cord of the stick insect induced by pilocarpine. J. Exp. Biol., 198: 435–456.

Callaway, J.C. and Stuart, A.E. (1989). Biochemical and physiological evidence that histamine is the transmitter of barnacle photoreceptors. Visual Neurosci., 3: 311–325.

Callec, J.J., Guillet, J.C., Pichon, Y. and Boistel, J. (1971). Further studies on synaptic transmission in insects. II. Relations between sensory information and its synaptic integration at the level of a single giant axon in the cockroach. J. Exp. Biol., 55: 123–149.

Camhi, J. (1969a). Locust wind receptors. III. Contribution to flight initiation and lift control. J. Exp. Biol., 50: 363–373.

Camhi, J. (1969b). Locust wind receptors. I. Transducer mechanics and sensory response. J. Exp. Biol., 50: 335–348.

Camhi, J.M. (1970). Yaw-correcting postural changes in locusts. J. Exp. Biol., 52: 519–532.

Camhi, J.M. (1980). The escape system of the cockroach. Scient. Am., 243: 158–172.

Camhi, J.M. and Hinkle, M. (1972). Attentiveness to sensory stimuli: central control in locusts. Science (NY), 175: 550–553.

Camhi, J.M. and Levy, A. (1988). Organization of a complex movement: fixed and variable components of the cockroach escape behavior. J. Comp. Physiol., 163A: 317–328.
Camhi, J.M. and Levy, A. (1989). The code for stimulus direction in a cell assembly in the cockroach. J. Comp. Physiol., 165A: 83–88.
Camhi, J.M. and Nolen, T.G. (1981). Properties of the escape system of cockroaches during walking. J. Comp. Physiol., 142A: 339–346.
Campbell, H.R., Thompson, K.J. and Siegler, M.V.S. (1995). Neurons of the median neuroblast lineage of the grasshopper: a population study of the efferent DUM neurons. J. Comp. Neurol., 358: 541–551.
Candy, D.J. (1978). The regulation of flight muscle metabolism by octopamine and other compounds. Insect Biochem., 8: 177–181.
Carlson, A.D. and Jalenak, M. (1986). Release of octopamine from the photomotor neurones of the larval firefly lanterns. J. Exp. Biol., 122: 453–457.
Casagrand, J.L. and Ritzmann, R.E. (1992). Biogenic amines modulate synaptic transmission between identified giant interneurons and thoracic interneurons in the escape system of the cockroach. J. Neurobiol., 23: 644–655.
Case, J.F. (1961). Organization of the cockroach respiratory center. Biol. Bull., 121: 385.
Caudy, M. and Bentley, D. R. (1986). Pioneer growth cone morphologies reveal proximal increases in substrate affinity with leg segments of grasshopper embryos. J. Neurosci., 6: 364–379.
Cayre, M., Strambi, C. and Strambi, A. (1994). Neurogenesis in an adult insect brain and its hormonal control. Nature (Lond.), 368: 57–59.
Chang, W.S., Kerikawa, K., Allen, K. and Bentley, D. (1992). Disruption of pioneer growth cone guidance *in vivo* by removal of glycosyl-phosphatidylinositol-anchored cell surface proteins. Development, 114: 507–519.
Chen, Y., Veenstra, J.A., Davis, N.T. and Hagedorn, H.H. (1994). A comparative study of leucokinin-immunoreactive neurons in insects. Cell Tissue Res., 276: 69–83.
Chiba, A., Shepherd, D. and Murphey, R.K. (1988). Synaptic rearrangement during postembryonic development in the cricket. Science (NY) 240:901–905.
Chiba, A., Kämper, G. and Murphey, R.K. (1992). Response properties of interneurons of the cricket cercal sensory system are conserved in spite of changes in peripheral receptors during maturation. J. Exp. Biol., 164: 205–226.
Chrachri, A. and Clarac, F. (1990). Fictive locomotion in the fourth thoracic ganglion of the crayfish, *Procambarus clarkii.* J. Neurosci., 10: 707–719.
Christensen, T.A. and Carlson, A.D. (1982). The neurophysiology of firefly luminescence: direct activation through four bifurcating (DUM) neurones. J. Comp. Physiol., 148A: 503–514.
Christensen, T.A., Sherman, T.G., McCaman, R.E. and Carlson, A.D. (1983). Presence of octopamine in firefly photomotor neurones. Neuroscience, 9: 183–189.
Claasen, D.E. and Kammer, A.E. (1986). Effects of octopamine, dopamine and serotonin on production of flight motor output by thoracic ganglia of *Manduca sexta*. J. Neurobiol., 17: 1–14.
Claiborne, B.J. and Selverston, A.I. (1984). Histamine as a transmitter in the stomatogastric nervous system of the spiny lobster. J. Neurosci., 4: 708–721.
Clarke, K.U. and Langley, P.A. (1963). Studies on the initiation of growth and moulting in *Locusta migratoria migratorioides* R. & F. II. The role of the stomatogastric nervous system. J. Insect Physiol., 9: 363–373.
Clements, A.N. and May, T.E. (1974). Studies on locust neuromuscular physiology in relation to glutamic acid. J. Exp. Biol., 60: 673–705.

Clinton A.S. and Arbas E.A. (1994). Neuromuscular transmission during trans-neuronally induced atrophy. Beitrage 22. *Göttinger Neurobiologentagung* Vol. II. p. 133. Ed. Elsner, N. and Breer, H. Georg Thieme Verlag, Stuttgart.

Coast, G.M., Rayne, R.C., Hayes, T.K., Mallet, A.I., Thompson, K.S.J. and Bacon, J.P. (1993). A comparison of the effects of two putative diuretic hormones from *Locusta migratoria* on isolated locust Malpighian tubules. J. Exp. Biol., 175: 1–14.

Cochrane, D.G., Elder, H.Y. and Usherwood, P.N.R. (1972). Physiology and ultrastructure of phasic and tonic skeletal muscle fibres in the locust, *Schistocerca gregaria*. J. Cell Sci., 10: 419–441.

Coillot, J.P. (1974). Analyse du codage d'un mouvement périodique par des récepteurs à l'étirement d'un insecte. J. Insect Physiol., 20: 1101–1116.

Coillot, J.P. and Boistel, J. (1968). Localisation et description des récepteurs a l'étirement au niveau de l'articulation tibio-fémorale de la patte sauteuse du criquet, *Schistocerca gregaria*. J. Insect Physiol., 14: 1661–1667.

Coillot, J.P. and Boistel, J. (1969). Etude de l'activitie electrique propagee de récepteurs a l'etirement de la patte metathoracique du criquet, *Schistocerca gregaria*. J. Insect Physiol., 15: 1449–1470.

Collins, C. and Miller, T. (1977). Studies on the action of biogenic amines on cockroach heart. J. Exp. Biol., 67: 1–15.

Collins, J.J. and Stewart, I. (1993). Hexapod gaits and coupled nonlinear oscillator models. Biol. Cybernetics, 68: 287–298.

Collins, J.J. and Stewart, I. (1994). A group-theoretic approach to rings of coupled biological oscillators. Biol. Cybernetics, 71: 95–103.

Comer, C.M. (1985). Analyzing cockroach escape behavior with lesions of individual giant interneurons. Brain Res., 335: 342–346.

Comer, C.M., Mara, E., Murphy, K.A., Getman, M. and Mungy, M.C. (1994). Multisensory control of escape in the cockroach *Periplaneta americana*. II. Patterns of touch-evoked behavior. J. Comp. Physiol., 174A: 13–26.

Condic, M.L. and Bentley, D. (1989a). Removal of the basal lamina *in vivo* reveals growth cone-basal lamina adhesive interactions and axonal tension in grasshopper embryos. J. Neurosci., 9: 2678–2686.

Condic, M.L. and Bentley, D. (1989b). Pioneer growth cone adhesion *in vivo* to boundary cells and neurons after enzymatic removal of basal lamina in grasshopper embryos. J. Neurosci., 9: 2687–2696.

Condron, B.G. and Zinn, K. (1994). The grasshopper median neuroblast is a multipotent progenitor cell that generates glia and neurons in distinct temporal phases. J. Neurosci., 14: 5766–5777.

Condron, B.G. and Zinn, K. (1995). Activation of cAMP-dependent protein kinase triggers a glial-to-neuronal cell-fate switch in an insect neuroblast lineage. Curr. Biol., 5: 51–61.

Condron, B.G., Patel, N.H. and Zinn, K. (1994). *engrailed* controls glial/neuronal cell fate decisions at the midline of the central nervous system. Neuron, 13: 541–554.

Cook, D.J. and Milligan, J.V. (1972). Electrophysiology and histology of the medial neurosecretory cells in adult male cockroaches, *Periplaneta americana*. J. Insect Physiol., 18: 1197–1214.

Cook, P.M. (1951). Observations on giant fibres of the nervous system of *Locusta migratoria*. Q. J. Microsc. Sci., 92: 297–305.

Cooter, R.J. (1973). Flight and landing posture in hoppers of *Schistocerca gregaria* (Forskål). Acrida, 2: 307–317.

Copenhaver, P.F. and Truman, J.W. (1986). Identification of the cerebral neurosecretory cells that contain eclosion hormone in the moth *Manduca sexta*. J. Neurosci., 6: 1738–1747.

Coss, R.G., Brandon, J.G. and Globus, A. (1980). Changes in morphology of dendrite spines on honeybee calycal interneurons associated with cumulative nursing and foraging experiences. Brain Res., 192: 49–59.

Crampton, G.C. (1916). The orders and relationships of Apterygotan insects. J. NY Ent. Soc., 24: 267–301.

Crossman, A.R., Kerkut, G.A. and Walker, R.J. (1971a). Axon pathways of electrically excitable cell bodies in the insect nervous system. J. Physiol., 218: 55–56.

Crossman, A.R., Kerkut, G.A., Pitman, R.M. and Walker, R.J. (1971b). Electrically excitable nerve cell bodies in the central ganglia of two insect species *Periplaneta americana* and *Schistocerca gregaria*. Comp. Biochem. Physiol. 40A: 579–594.

Crossman, A.R., Kerkut, G.A. and Walker, R.J. (1972). Electrophysiological studies on the axon pathways of specified nerve cells in the central ganglia of two insect species, *Periplaneta americana* and *Schistocerca gregaria*. Comp. Biochem. Physiol., 43A: 393–415.

Cruse, H. (1976a). The control of body position in the stick insect (*Carausius morosus*), when walking over uneven terrain. Biol. Cybernetics, 24: 25–33.

Cruse, H. (1976b). The function of the legs in the free walking stick insect *Carausius morosus*. J. Comp. Physiol., 112A: 235–262.

Cruse, H. (1979a). A new model describing the coordination patterns of the legs of a walking stick insect. Biol. Cybernetics, 32: 107–113.

Cruse, H. (1979b). The control of the anterior extreme position of a walking insect *Carausius morosus*. Physiol. Entomol., 4: 121–124.

Cruse, H. (1980a). A quantitative model of walking incorporating central and peripheral influences. I. The control of the individual leg. Biol. Cybernetics, 37: 131–136.

Cruse, H. (1980b). A quantitative model of walking incorporating central and peripheral influences. II. The connections between the different legs. Biol. Cybernetics, 37: 137–144.

Cruse, H. (1985). Coactivating influences between neighbouring legs in walking insects. J. Exp. Biol., 114: 513–519.

Cruse, H. (1990). What mechanisms coordinate leg movement in walking arthropods. Trends Neurosci., 13: 15–21.

Cruse, H. and Epstein, S. (1982). Peripheral influences on the movement of the legs in a walking insect *Carausius morosus*. J. Exp. Biol., 101: 161–170.

Cruse, H. and Pflüger, H.-J. (1981). Is the position of the femur-tibia joint under feedback control in the walking stick insect? II. Electrophysiological recordings. J. Exp. Biol., 92: 97–107.

Cruse, H. and Saxler, G. (1980). Oscillations in force in the standing legs of a walking insect (*Carausius morosus*). Biol. Cybernetics, 36: 159–163.

Cruse, H., Dean, J. and Suilmann, M. (1984). The contributions of diverse sense organs to the control of leg movement by a walking insect. J. Comp. Physiol., 154A: 695–705.

Cruse, H., Riemenschneider, D. and Stammer, W. (1989). Control of body position of a stick insect standing on an uneven surface. Biol. Cybernetics, 61: 71–77.

Cruse, H., Schmitz, J., Braun, U. and Schweins, A. (1993). Control of body height in a stick insect walking on a treadwheel. J. Exp. Biol., 181: 141–155.

Cruse, H., Bartling, C., Cymbalyuk, G., Dean, J. and Dreifert, M. (1995). A modular artificial neural net for controlling a six-legged walking system. Biol. Cybernetics, 72: 421–430.

Cuthbert, B.A. and Evans, P.D. (1989). A comparison of the effects of FMRFamide-like peptides on locust heart and skeletal muscle. J. Exp. Biol., 144: 395–414.

Cuttle, M.F., Hevers, W., Laughlin, S.B. and Hardie, R.C. (1995). Diurnal modulation of photoreceptor potassium conductance in the locust. J. Comp. Physiol., 176A: 307–316.
Cymborowski, B. and Korf, H.-W. (1995). Immunocytochemical demonstration of S-antigen (arrestin) in the brain of the blowfly *Calliphora vicina*. Cell Tissue Res., 279: 109–114.
Dagan, D. and Camhi, J.M. (1979). Responses to wind recorded from the cercal nerve of the cockroach *Periplaneta americana*. II. Directional selectivity of the sensory neurons innervating single columns of filiform hairs. J. Comp. Physiol., 133A: 103–110.
Dagan, D. and Parnas, I. (1970). Giant fibre and small fibre pathways involved in the evasive response of the cockroach, *Periplaneta americana*. J. Exp. Biol., 52: 313–324.
Dagan, D. and Volman, S. (1982). Sensory basis for directional wind detection in first instar cockroaches, *Periplaneta americana*. J. Comp. Physiol., 147A: 471–478.
Daley, D.L. and Camhi, J.M. (1988). Connectivity pattern of the cercal-to-giant interneuron system of the American cockroach. J. Neurophysiol., 60: 1350–1368.
Daley, D. and Delcomyn, F. (1980a). Modulation of the excitability of cockroach giant interneurons during walking. I. Simultaneous excitation and inhibition. J. Comp. Physiol., 138A: 231–240.
Daley, D. and Delcomyn, F. (1980b). Modulation of the excitability of cockroach giant interneurons during walking. II. Central and peripheral components. J. Comp. Physiol., 138A: 241–251.
Dambach, M., Rausche, H.-G. and Wendler, G. (1983). Proprioceptive feedback influences the calling song of the field cricket. Naturwissenschaften, 70: 417–418.
David, J.A. and Pitman, R.M. (1993). The pharmacology of α-bungarotoxin-resistant acetylcholine receptors on an identified cockroach motoneurone. J. Comp. Physiol., 172A: 359–368.
Davis, G.W. and Murphey, R.K. (1993). A role for postsynaptic neurons in determining presynaptic release properties in the cricket CNS: evidence for retrograde control of facilitation. J. Neurosci., 13: 3827–3838.
Davis, J.P.L. and Pitman, R.M. (1991). Characterization of receptors mediating the actions of dopamine on an identified inhibitory motoneurone of the cockroach. J. Exp. Biol., 155: 203–217.
Davis, N.T. and Hildebrand, J.G. (1992). Vasopressin-immunoreactive neurons and neurohemal systems in cockroaches and mantids. J. Comp. Neurol., 320: 381–393.
Dean, J. and Cruse, H. (1986). Evidence for the control of velocity as well as position in leg protraction and retraction by the stick insect. Exp. Brain Res., 15: 263–274.
Dean, J. and Wendler, G. (1982). Stick insects walking on a wheel: pertubations induced by obstructions of leg protraction. J. Comp. Physiol., 148A: 195–207.
Debrodt, B. and Bässler, U. (1989). Motor neurones of the flexor tibiae muscle in phasmids. Zool. Jb. Physiol., 93: 481–494.
Delcomyn, F. (1971a). The locomotion of the cockroach *Periplaneta americana*. J. Exp. Biol., 54: 443–452.
Delcomyn, F. (1971b). The effect of limb amputation on locomotion in the cockroach *Periplaneta americana*. J. Exp. Biol., 54: 453–469.
Delcomyn, F. (1973). Motor activity during walking in the cockroach *Periplaneta americana*. II. Tethered walking. J. Exp. Biol., 59: 643–654.
Delcomyn, F. (1991a). Perturbation of the motor system in freely walking cockroaches. I. Rear leg amputation and the timing of motor activity in leg muscles. J. Exp. Biol., 156: 483–502.
Delcomyn, F. (1991b). Activity and directional sensitivity of leg campaniform sensilla in a stick insect. J. Comp. Physiol., 168A: 113–119.

Delcomyn, F. (1993). The walking of cockroaches – deceptive simplicity. In '*Biological neural networks in invertebrate neuroethology and robotics*'. pp. 21–41. Ed. Beer, R.D., Ritzman, R.E. and McKenna, T. Academic Press, Boston.

Delcomyn, F. and Daley, D. (1979). Central excitation of cockroach giant interneurons during walking. J. Comp. Physiol., 130A: 39–48.

Delcomyn, F. and Usherwood, P.N.R. (1973). Motor activity during walking in the cockroach *Periplaneta americana*. I. Free walking. J. Exp. Biol., 59: 629–642.

Dethier, V.G. (1972). Sensitivity of the contact chemoreceptors of the blowfly to vapours. Proc. Natl. Acad. Sci. USA, 69: 2189–2192.

Dethier, V.G. (1976). *The hungry fly*. Harvard University Press, Cambridge, MA.

Dick, T.E., Oku, Y., Romaniuk, J.R. and Cherniack, N.S. (1993). Interaction between central pattern generators for breathing and swallowing in the cat. J. Physiol., 465: 715–730.

Dircksen, H. and Homberg, U. (1995). Crustacean cardioactive peptide-immunoreactive neurons innervating brain neuropils, retrocerebral complex and stomatogastric nervous system of the locust, *Locusta migratoria*. Cell Tissue Res., 279: 495–515.

Dircksen, H. Müller, A. and Keller, R. (1991). Crustacean cardioactive peptide in the nervous system of the locust, *Locusta migratoria*: an immunocytochemical study on the ventral nerve cord and peripheral innervation. Cell Tissue Res., 263: 439–457.

Doe, C.Q. (1992). Molecular markers for identified neuroblasts and ganglion mother cells in the *Drosophila* central nervous system. Development, 116: 855–863.

Doe, C.Q. and Goodman, C.S. (1985). Early events in insect neurogenesis. I. Development and segmental differences in the pattern of neuronal precursor cells. Dev. Biol., 111: 193–205.

Doe, C.Q., Kuwada, J.Y. and Goodman, C.S. (1985). From epithelium to neuroblasts to neurons: the role of cell interactions and cell lineage during insect neurogenesis. Phil. Trans. R. Soc. Lond. B., 312: 67–81.

Doe, C.Q., Hiromi, Y., Gehring, W.J. and Goodman, C.S. (1988a). Expression and function of the segmentation gene *fushi tarazu* during *Drosophila* neurogenesis. Science (NY), 239: 170–175.

Doe, C.Q., Smouse, D. and Goodman, C.S. (1988b). Control of neuronal fate by the *Drosophila* segmentation gene *even-skipped*. Nature (Lond.), 333: 376–378.

Doe, C.Q., Chu-Lagraff, Q., Wright, D.M. and Scott, M.P. (1991). The *prospero* gene specifies cell fates in the *Drosophila* nervous system. Cell, 65: 451–464.

Downer, R.G.H., Hiripi, L. and Juhos, S. (1993). Characterization of the tyraminergic system in the central nervous system of the locust, *Locusta migratoria migratoides*. Neurochem. Res., 18: 1245–1248.

Dresden, D. and Nijenhuis, E. (1958). Fiber analysis of the nerves of the second thoracic leg in *Periplaneta americana*. Proc. K. Ned. Akad. Wet. Ser. C, 61: 213–233.

Driesang, R.B. and Büschges, A. (1993). The neural basis of catalepsy in the stick insect. IV. Properties of nonspiking interneurons. J. Comp. Physiol., 173A: 445–454.

du Lac, S., Bastiani, M.J. and Goodman, C.S. (1986). Guidance of neuronal growth cones in the grasshopper embryo. II. Recognition of a specific axonal pathway by the aCC neuron. J. Neurosci., 6: 3532–3541.

Dubas, F. (1991). Actions of putative amino acid neurotransmitters on the neuropile arborizations of locust flight motoneurones. J. Exp. Biol. 155: 337–356.

Duch, C. and Pflüger, H.-J. (1995). Motor patterns for horizontal and upside-down walking and vertical climbing in the locust. J. Exp. Biol., 198: 1963–1976.

Dugard, J.J. (1967). Directional change in flying locusts. J. Insect Physiol., 13: 1055–1063.

Dumont, J.P.C. and Robertson R.M. (1986). Neuronal circuits: an evolutionary perspective. Science (NY), 233: 849–853.

Dunphy, G.B. and Downer, R.G.H. (1994). Octopamine, a modulator of the haemocytic nodulation response of non-immune *Galleria mellonella* larvae. J. Insect Physiol., 40: 267–272.

Ebens, A.J., Garren, H., Cheyette, B.N.R. and Zipursky, S.L. (1993). The *Drosophila* anachronism locus: a glycoprotein secreted by glia inhibits neuroblast proliferation. Cell, 74: 15–27.

Edwards, J.S., Chen, S.-W and Berns, M.W. (1981). Cercal sensory development following laser microlesions of embryonic apical cells in *Acheta domesticus*. J. Neurosci., 1: 250–258.

Eisemann, C.H., Jorgensen, W.K., Merritt, D.J., Rice, M.J., Cribb, B.W., Webb, P.D. and Zalucki, M.P. (1984). Do insects feel pain? – A biological view. Experientia, 40: 164–167.

El Manira, A. and Clarac, F. (1994). Presynaptic inhibition is mediated by histamine and GABA in the crustacean escape reaction. J. Neurophysiol., 71: 1088–1095.

El Manira, A., Cattaert, D., Wallén, P., DiCaprio, R.A. and Clarac, F. (1993). Electrical coupling of mechanoreceptor afferents in the crayfish: a possible mechanism for enhancement of sensory signal transmission. J. Neurophysiol., 69: 2248–2251.

Elia, A.J. and Orchard, I. (1995). Peptidergic innervation of leg muscles of the cockroach, *Periplaneta americana* (L.), and a possible role in modulation of muscle contraction. J. Comp. Physiol., 176A: 425–435.

Elias, M.S. and Evans, P.D. (1983). Histamine in the insect nervous system: distribution, synthesis and metabolism. J. Neurochem., 41: 562–568.

Ellington, C.P. (1991). Aerodynamics and the origin of insect flight. Adv. Insect Physiol., 23: 171–210.

Elliott, C.J.H. (1982). Neurophysiological analysis of locust behaviour during ecdysis: the slow rhythm underlying expansion. J. Insect Physiol., 28: 53–60.

Elphick, M.R., Green, I.C. and O'Shea, M. (1993). Nitric oxide synthesis and action in an invertebrate brain. Brain Res., 619: 344–346.

Elphick, M.R., Rayne, R.C., Riveros-Moreno, V., Moncada, S. and O'Shea, M. (1995). Nitric oxide synthase in locust olfactory interneurones. J. Exp. Biol., 193: 821–829.

Elson, R. and Pflüger, H.-J. (1986). The activity of a steering muscle in flying locusts. J. Exp. Biol., 120: 421–441.

Elson, R.C. (1987a). Flight motor neurone reflexes driven by strain-sensitive wing mechanoreceptors in the locust. J. Comp Physiol., 161A: 747–760.

Elson, R.C. (1987b). Integration of wing proprioceptive and descending exteroceptive sensory inputs by thoracic interneurones of the locust. J. Exp. Biol., 128: 193–217.

Elson, R.C., Sillar, K.T. and Bush, B.M.H. (1992). Identified proprioceptive afferents and motor rhythm entrainment in the crayfish walking system. J. Neurophysiol., 67: 530–546.

Emptage, N.J. (1991). *The neural integration of mechanosensory information in the desert locust*. PhD Thesis, University of Cambridge, UK.

Emson, P.C., Burrows, M. and Fonnum, F. (1974). Levels of glutamate decarboxylase, choline acetyl transferase and acetylcholine esterase in identified motor neurones of the locust. J. Neurobiol., 5: 33–42.

Engelhard, E.K., Kam-Morgan, L.N.W., Washburn, J.O. and Volkman, L.E. (1994). The insect tracheal system: a conduit for the systemic spread of *Autographa californica* M nuclear polyhedrosis virus. Proc. Natl. Acad. Sci. USA, 91: 3224–3227.

Erber, J., Kloppenburg, P. and Scheidler, A. (1993). Neuromodulation by serotonin and octopamine in the honeybee: behaviour, neuroanatomy and electrophysiology. Experientia, 49: 1073–1083.

Evans, M.E.G. (1972). The jump of the click beetle (Coleoptera-Elateridae) – a preliminary study. J. Zool. Lond., 167: 319–336.

Evans, M.E.G. (1973). The jump of the click beetle (Coleoptera:Elateridae) – energetics and mechanics. J. Zool. Lond., 169: 181–194.

Evans, P.D. (1981). Multiple receptor types for octopamine in the locust. J. Physiol., 318: 99–122.

Evans, P.D. and Cournil, I. (1990). Co-localization of FLRF-like and vasopressin-like immunoreactivity in a single pair of sexually dimorphic neurones in the nervous system of the locust. J. Comp. Neurol., 292: 331–348.

Evans, P.D. and Myers, C.M. (1986). The modulatory actions of FMRFamide and related peptides on locust skeletal muscle. J. Exp. Biol., 126: 403–422.

Evans, P.D. and O'Shea, M. (1978). The identification of an octopaminergic neurone and the modulation of a myogenic rhythm in the locust. J. Exp. Biol., 73: 235–260.

Evans, P.D. and Siegler, M.V.S. (1982). Octopamine mediated relaxation of maintained and catch tension in locust skeletal muscle. J. Physiol., 324: 93–112.

Farley, R.D. and Case, J.F. (1968). Sensory modulation of ventilatory pacemaker output in the cockroach *Periplaneta americana*. J. Insect Physiol., 14: 591–601.

Farley, R.D., Case, J.F. and Roeder, K.D. (1967). Pacemaker for tracheal ventilation in the cockroach, *Periplaneta americana* (L.). J. Insect Physiol., 13: 1713–1728.

Ferber, M. and Pflüger, H.-J. (1990). Bilaterally projecting neurones in pregenital abdominal ganglia of the locust: anatomy and peripheral targets. J. Comp. Neurol., 302: 447–460.

Ferber, M. and Pflüger, H.-J. (1992). An identified dorsal unpaired median neurone and bilaterally projecting neurones exhibiting bovine pancreatic polypeptide-like/FMRFamide-like immunoreactivity in abdominal ganglia of the migratory locust. Cell Tissue Res., 267: 85–98.

Field, L.H. (1991). Mechanism for range fractionation in chordotonal organs of *Locust migratoria* (L.) and *Valanga* sp. (Orthoptera: Acrididae). Int. J. Insect Morphol. Embryol., 20: 25–39.

Field, L.H. and Burrows, M. (1982). Reflex effects of the femoral chordotonal organ upon leg motor neurones of the locust. J. Exp. Biol., 101: 265–285.

Field, L.H. and Coles, M.M.L. (1994). The position-dependent nature of postural resistance reflexes in the locust. J. Exp. Biol., 188: 65–88.

Field, L.H. and Pflüger, H.-J. (1989). The femoral chordotonal organ: a bifunctional orthopteran (*Locusta migratoria*) sense organ? Comp. Biochem. Physiol., 93A: 729–743.

Field, L.H. and Rind, F.C. (1981). A single insect chordotonal organ mediates inter- and intra-segmental leg reflexes. Comp. Biochem. Physiol., 68A, 99–102.

Field, L.H., Meyer, M.R. and Edwards, J.S. (1994). Selective expression of glionexin, a glial glycoprotein, in insect mechanoreceptors. J. Neurobiol., 25: 1017–1028.

Fifield, S.M. and Finlayson, L.H. (1978). Peripheral neurons and peripheral neurosecretion in the stick insect. Proc. R. Soc. Lond. B, 200: 63–85.

Finlayson, L.H. and Osborne, M.P. (1968). Peripheral neurosecretory cells in the stick insect (*Carausius morosus*) and the blowfly (*Phormia terrae-novae*). J. Insect Physiol., 14: 1793–1801.

Fitch, G.K. and Kammer, A.E. (1982). Modulation of the ventilatory rhythm of the hellgrammite *Corydalus cornutus* by mechanosensory input. J. Comp. Physiol., 149A: 423–434.

Forssberg, H. (1979). Stumbling corrective reaction: a phase-dependent compensatory reaction during locomotion. J. Neurophysiol. 42: 936–953.

Foster, J.A. and Robertson, R.M. (1992). Temperature dependency of wing-beat frequency in intact and deafferented locusts. J. Exp. Biol., 162: 295–312.

Foth, E. and Bässler, U. (1985). Leg movement of stick insects walking with five legs on a treadwheel and with one leg on a motor-driven belt. I. General results and 1 : 1-coordination. Biol. Cybernetics, 51: 313–318.

Foth, E. and Graham, D. (1983). Influence of loading parallel to the body axis on the walking coordination of an insect. I. Ipsilateral changes. Biol. Cybernetics, 47: 17–23.

Fraser, P.J. (1977). Cercal ablation modifies tethered flight behaviour of cockroach. Nature (Lond.), 268: 523–524.

French, A.S. and Torkkeli, P.H. (1994). The time course of sensory adaptation in the cockroach tactile spine. Neurosci. Lett., 178: 147–150.

French, A.S., Sanders, E.J., Duszyk, E., Prasad, S., Torkkeli, P.H., Haskins, J. and Murphy, R.A. (1993). Immunocytochemical localization of sodium channels in an insect central nervous system using a site-directed antibody. J. Neurobiol., 24: 939–948.

Fridman-Cohen, S. and Pener, M.P. (1980). Precocenes induce effect of juvenile hormone excess in *Locusta migratoria*. Nature (Lond.), 286: 711–713.

Full, R.J. and Tu, M.S. (1991). Mechanics of a rapid running insect: two-, four- and six-legged locomotion. J. Exp. Biol., 156 :215–231.

Funk, G.D., Milsom, W.K. and Steeves, J.D. (1992). Coordination of wingbeat and respiration in the Canada goose. I. Passive wing flapping. J. Appl. Physiol., 73: 1014–1024.

Furth, D.G., Traub, W. and Harpaz, I. (1983). What makes *Blepharida* jump? A structural study of the metafemoral spring of a flea beetle. J. Exp. Zool., 227: 43–47.

Fuzeau-Braesch, S. and David, J.-C. (1978). Étude du taux d'octopamine chez *Locusta migratoria* (Insecte: Orthoptère) comparaison entre insectes grégaires, solitaires et traités au gaz carbonique. C.r. Acad. Sci. Paris, 286D: 697–699.

Gabriel, J.M. (1985). The development of the locust jump mechanism. I. Allometric growth and its effect on the jumping performance. J. Exp. Biol., 118: 313–326.

Ganihar, D., Libersat, F., Wendler, G. and Camhi, J.M. (1994). Wind-evoked evasive responses in flying cockroaches. J. Comp. Physiol., 175A: 49–65.

Gee, C.E. and Robertson, R.M. (1994). Effects of maturation on synaptic potentials in the locust flight system. J. Comp. Physiol., 175A: 437–447.

Gettrup, E. (1962). Thoracic proprioceptors in the flight system of locusts. Nature (Lond.), 193: 498–499.

Gettrup, E. (1966). Sensory regulation of wing twisting in locusts. J. Exp. Biol., 44: 1–16.

Gettrup, E. and Wilson, D.M. (1964). The lift control reaction of flying locusts. J. Exp. Biol., 41: 183–190.

Gewecke, M. (1975). The influence of air current sense organs on the flight behaviour of *Locusta migratoria*. J. Comp. Physiol., 103A: 79–95.

Gewecke, M. and Heinzel, H.-G. (1980). Aerodynamic and mechanical properties of the antennae as air-current sense organs in *Locusta migratoria*. I. Static characterisitics. J. Comp. Physiol., 139A: 357–366.

Gewecke, M. and Phillipen, J. (1978). Control of the horizontal flight-course by air-current sense organs in *Locusta migratoria*. Physiol. Entomol., 3: 43–52.

Gibson, G.A.P. (1986). Mesothoracic skeletomusculature and mechanics of flight and jumping in Eupelminae (Hymenoptera, Chalcidoidea: Eupelmidae). Can. Entomol., 118: 691–728.

Godden, D.H. and Graham, D. (1984). A preparation of the stick insect *Carausius morosus* for intracellular recording from identified neurones during walking. Physiol. Entomol., 9: 275–286.

Goettel, M.S., Johnson, D.L. and Douglas Inglis, G. (1995). The role of fungi in the control of grasshoppers. Can. J. Bot., 73: 571–575.

Goldstein, R.S. and Camhi, J.M. (1988). Modulation of activity in sensory neurons and wind-sensitive interneurons by cercal displacement in the cockroach. J. Comp. Physiol., 163A: 479–487.

Goldstein, R.S. and Camhi, J.M. (1991). Different effects of the biogenic amines dopamine, serotonin and octopamine on the thoracic and abdominal portions of the escape circuit in the cockroach. J. Comp. Physiol., 168A: 103–112.

Goodman, C.S. (1982). Embryonic development of identified neurons in the grasshopper. In '*Neuronal development*'. pp. 171–211. Ed. Spitzer, N.C. Plenum Press, New York.

Goodman, C.S. and Bate, M. (1981). Neuronal development in the grasshopper. Trends Neurosci., 4: 163–169.

Goodman, C.S. and Heitler, W.J. (1979). Electrical properties of insect neurones with spiking and non-spiking somata: normal, axotomised and colchicine treated neurones. J. Exp. Biol., 83: 95–121.

Goodman, C.S. and Spitzer, N.C. (1979). Embryonic development of identified neurones: differentiation from neuroblast to neurone. Nature (Lond.) 280: 208–214.

Goodman, C.S. and Spitzer, N.C. (1981). The development of electrical properties of identified neurones in grasshopper embryos. J. Physiol., 313: 385–403.

Goodman, C.S., Pearson, K.G. and Spitzer, N.C. (1980). Electrical excitability: a spectrum of properties in the progeny of a single embryonic neuroblast. Proc. Natl. Acad. Sci. USA, 77: 1676–1680.

Goodman, C.S., Bate, M. and Spitzer, N.C. (1981). Embryonic development of identified neurons: origin and transformation of the H cell. J. Neurosci., 1: 94–102.

Goodman, C.S., Raper, J.A., Ho, R. and Chang, S. (1982). Pathfinding by neuronal growth cones in grasshopper embryos. In '*Cytochemical methods in neuroanatomy*'. pp. 461–494. A.R. Liss, New York.

Goodman, L.J. (1959). Hair plates on the first cervical sclerites of the Orthoptera. Nature (Lond.), 183: 1106–1107.

Goosey, M.W. and Candy, D.J. (1980). The release and removal of octopamine by tissues of the locust *Schistocerca americana gregaria*. Insect Biochem., 12: 681–685.

Gosbee, J.L., Milligan, J.V. and Smallman, B.N. (1968). Neural properties of the protocerebral neurosecretory cells of the adult cockroach *Periplaneta americana*. J. Insect Physiol., 14: 1785–1792.

Gould, J.L. and Gould, C.G. (1982). The insect mind: physics or metaphysics? In '*Animal mind – human mind*'. pp. 269–297. Ed. Griffin, D.R. Springer Verlag, Berlin.

Graham, D. (1972). A behavioural analysis of the temporal organization of walking movements in the first instar and adult stick insect *Carausius morosus*. J. Comp. Physiol., 81A: 23–52.

Graham, D. (1977). The effect of amputation and leg restraint on the free walking coordination of the stick insect *Carausius morosus*. J. Comp. Physiol., 116A: 91–116.

Graham, D. (1978). Unusual step pattens in the free walking grasshopper *Neoconocephalus robustus*. I. General features of the step pattern. J. Exp. Biol., 73: 147–157.

Graham, D. (1979). Effects of circum-oesophageal lesion on the behaviour of the stick insect *Carausius morosus*. I. Cyclic behaviour patterns. Biol. Cybernetics, 32: 139–145.

Graham, D. (1981).Walking kinetics of the stick insect using a low-inertia, counter-balanced pair of independent treadwheels. Biol. Cybernetics, 40: 49–57.

Graham, D. (1985). Influence of coxa-thorax joint receptors on retractor motor output during walking in *Carausius morosus*. J. Exp. Biol., 114: 131–139.

Graham, D. and Bässler, U. (1981). Effects of afference sign reversal on motor activity in walking stick insects (*Carausius morosus*). J. Exp. Biol., 91: 179–193.

Graham, D. and Cruse, H. (1981). Coordinated walking of stick insects on a mercury surface. J. Exp. Biol., 92: 229–241.

Gramoll, S. and Elsner, N. (1987). Morphology of local 'stridulatory' interneurons in the metathoracic ganglion of the Acridid grasshopper *Omocestus viridulus* L. J. Comp. Neurol., 263: 593–606.

Gras, H. and Hörner, M. (1992). Wind-evoked escape running of the cricket *Gryllus bimaculatus*. I. Behavioural analysis. J. Exp. Biol., 171: 189–214.

Gras, H., Hörner, M., Runge, L. and Schurmann, F.-W. (1990). Prothoracic DUM neurons of the cricket *Gryllus bimaculatus* – responses to natural stimuli and activity in walking behavior. J. Comp. Physiol., 166A: 901–914.

Graubard, K. (1978). Synaptic transmission without action potentials: input-output properties of a non-spiking presynaptic neuron. J. Neurophysiol., 41: 1014–1025.

Graubard, K., Raper, J.A. and Hartline, D.K. (1980). Graded synaptic transmission between spiking neurons. Proc. Natl. Acad. Sci. USA, 77: 3733–3735.

Gray, J.R. and Robertson, R.M. (1994). Activity of the forewing stretch receptor in immature and mature adult locusts. J. Comp. Physiol., 175A: 425–435.

Gregory, G.E. (1974). Neuroanatomy of the mesothoracic ganglion of the cockroach *Periplaneta americana* (L.). The roots of the peripheral nerves. Phil. Trans. R. Soc. Lond. B, 267: 421–465.

Gregory, G.E. (1984). Neuroanatomy of the mesothoracic ganglion of the cockroach *Periplaneta americana* (L.) II. Median neuron cell body groups. Phil. Trans. R. Soc. Lond. B, 306: 191–218.

Grillner, S. and Rossignol, S. (1978). On the initiation of the swing phase of locomotion in chronic spinal cats. Brain Res., 146: 269–277.

Griss, C. and Rowell, C.H.F. (1986). Three descending interneurons reporting deviation from course in the locust. I. Anatomy. J. Comp. Physiol., 158A: 765–774.

Grolleau, F. and Lapied, B. (1994). Transient Na^+-activated K^+ current in beating pacemaker-isolated adult insect neurosecretory cells (DUM neurones). Neurosci. Lett., 167: 46–50.

Grolleau, F. and Lapied, B. (1995). Separation and identification of multiple potassium currents regulating the pacemaker activity of insect neurosecretory cells (DUM neurons). J. Neurophysiol., 73: 160–171.

Gronenberg, W., Tautz, J. and Holldobler, B. (1993). Fast trap jaws and giant neurons in the ant *Odontomachus*. Science (NY), 262: 561–563.

Gundel, M. and Penzlin, H. (1978). The neuronal connections of the frontal ganglion of the cockroach *Periplaneta americana*. Cell Tissue Res., 193: 353–371.

Gwilliam, G.F. and Burrows, M. (1980). Electrical characteristics of the membrane of an identified insect motor neurone. J. Exp. Biol., 86: 49–61.

Gymer, A. and Edwards, J.S. (1967). The development of the insect nervous system. I. An analysis of postembryonic growth in the terminal ganglion of *Acheta domesticus*. J. Morphol., 123: 191–198.

Gynther, I.C. and Pearson, K.G. (1986). Intracellular recordings from interneurones and motoneurones during bilateral kicks in the locust: implications for mechanisms controlling the jump. J. Exp. Biol., 122: 323–343.

Gynther, I.C. and Pearson, K.G. (1989). An evaluation of the role of identified interneurons in triggering kicks and jumps in the locust. J. Neurophysiol., 61: 45–57.

Hale, J.P. and Burrows, M. (1985). Innervation patterns of inhibitory motor neurones in the thorax of the locust. J. Exp. Biol., 117: 401–413.

Halter, D.A., Urban, J., Rickert, C., Ner, S.S., Ito, K., Travers, A.A. and Technau, G.M. (1995). The homeobox gene *repo* is required for the differentiation and maintenance of glia function in the embryonic nervous system of *Drosophila melanogaster*. Development, 121: 317–332.

Hamilton, A.G. (1964). The occurrence of periodic output or discontinuous discharge of carbon dioxide by male desert locusts (*Schistocerca gregaria* Forskål) measured by infrared gas analyser. Proc. R. Soc. Lond. B, 160: 373–395.

Hammer, M. (1993). An identified neuron mediates the unconditioned stimulus in associative olfactory learning in honeybees. Nature (Lond.), 366: 59–63.

Hamon, A. and Guillet, J.-C. (1994). Some electrical properties of the cercal anemoreceptors of the cockroach, *Periplaneta americana*. Comp. Biochem. Physiol., 107A: 357–368.

Hamon, A., Guillet, J.C. and Callec, J.J. (1994). Patterns of monosynaptic input to the giant interneurons 1–3 in the cercal system of the adult cockroach. J. Comp. Physiol., 174A: 91–102.

Hancox, J.C. and Pitman, R.M. (1991). Plateau potentials drive axonal impulse bursts in insect motoneurons. Proc. R. Soc. Lond. B, 244: 33–38.

Hancox, J.C. and Pitman, R.M. (1992). A time-dependent excitability change in the soma of an identified insect motoneurone. J. Exp. Biol., 162: 251–263.

Hancox, J.C. and Pitman, R.M. (1993). Plateau potentials in an insect motoneurone can be driven by synaptic stimulation. J. Exp. Biol., 176: 307–310.

Hardie, R.C. (1987). Is histamine a neurotransmitter in insect photoreceptors? J .Comp. Physiol., 161A: 201–213.

Hardie, R.C. (1988). Effects of antagonists on putative histamine receptors in the first visual neuropile of the housefly (*Musca domestica*). J. Exp. Biol., 138: 221–241.

Hardie, R.C. (1989). A histamine-activated chloride channel involved in neurotransmission at a photoreceptor synapse. Nature (Lond.), 339: 704–706.

Harrelson, A.L. and Goodman, C.S. (1988). Growth cone guidance in insects: fasciclin II is a member of the immunoglobulin superfamily. Science (NY), 242: 700–708.

Harris-Warrick, R., Marder, E., Selverston, A.I. and Moulins, M. (Eds) (1992) '*Dynamic biological networks: the stomatogastric nervous system*'. MIT press, Cambridge, MA.

Hartenstein, V., Rudloff, E. and Campos-Ortega, J.A. (1987). The pattern of proliferation of the neuroblasts in the wild-type embryo of *Drosophila melanogaster*. Roux's Arch. Dev. Biol., 196: 473–485.

Hatsopoulos, N.G., Burrows, M. and Laurent, G. (1995). Hysteresis reduction in proprioception using presynaptic shunting inhibition. J. Neurophysiol., 73: 1031–1042.

Heathcote, R.D. (1980). Physiological development of a monosynaptic connection involved in an adult insect behavior. J. Comp. Neurol., 191: 155–166.

Heathcote, R.D. (1981). Differentiation of an identified sensory neuron (SR) and associated structures (CTO) in grasshopper embryos. J. Comp. Neurol., 202: 1–18.

Hedwig, B. and Pearson, K.G. (1984). Patterns of synaptic input to identified flight motoneurons in the locust. J. Comp. Physiol., 154A: 745–760.

Heidelbach, J. and Dambach, M. (1991). Processing of wing flick-generated air-vortex signals in the African cave cricket *Phaeophilacris spectrum*. Naturwissenschaften, 78: 277–278.

Heinzel, H.-G. and Dambach, M. (1987). Travelling air vortex rings as potential communication signals in a cricket. J. Comp. Physiol., 160A: 79–88.

Heinzel, H.-G. and Gewecke, M. (1987). Aerodynamic and mechanical properties of the antennae as air-current sense organs in *Locusta migratoria*. II. Dynamic characteristics. J. Comp Physiol., 161A: 671–680.

Heisenberg, M. and Wolf, R. (1988). Reafferent control of optomotor yaw torque in *Drosophila melanogaster*. J. Comp. Physiol., 163A: 373–388.

Heitler, W.J. (1974). The locust jump. Specialisations of the metathoracic femoral-tibial joint. J. Comp. Physiol., 89: 93–104.

Heitler, W.J. (1977). The locust jump. III. Structural specializations of the metathoracic tibiae. J. Exp. Biol., 67: 29–36.

Heitler, W.J. (1978). Coupled motoneurones are part of the crayfish swimmeret central oscillator. Nature (Lond.), 275: 231–234.

Heitler, W.J. (1983). Suppression of a locust visual interneurone (DCMD) during defensive kicking. J. Exp. Biol., 104: 203–215.

Heitler, W.J. (1995). Quasi-reversible photo-axotomy used to investigate the role of extensor muscle tension in controlling the kick motor programme of grasshoppers. Eur. J. Neurosci., 7: 981–992.

Heitler, W.J. and Bräunig, P. (1988). The role of fast extensor motor activity in the locust kick reconsidered. J. Exp. Biol., 136: 289–309.

Heitler, W.J. and Burrows, M. (1977). The locust jump. II. Neural circuits of the motor programme. J. Exp. Biol., 66: 221–241.

Heitler, W.J. and Goodman, C.S. (1978). Multiple sites of spike initiation in a bifurcating locust neurone. J. Exp. Biol., 76: 63–84.

Heitler, W.J., Goodman, C.S. and Fraser Rowell, C.H. (1977). The effects of temperature on the threshold of identified neurons in the locust. J. Comp. Physiol., 117A: 163–182.

Hekimi, S., Fischer-Lougheed, J. and O'Shea, M. (1991). Regulation of peptide stoichiometry in neurosecretory cells. J. Neurosci., 11: 3246–3256.

Hengstenberg, R. (1977). Spike responses of 'non-spiking' visual interneurone. Nature (Lond.), 212: 1242–1245.

Hensler, K. (1988). The pars intercerebralis neurone PI(2)5 of locusts: convergent processing of inputs reporting head movements and deviations from straight flight. J. Exp. Biol., 140: 511–533.

Hensler, K. (1990). Neural control of optomotor head rolling in locusts. Naturwissenschaften, 77 35–37.

Hensler, K. (1992). Neuronal control of course deviation and head movements in locusts. I. Descending deviation detectors. J. Comp Physiol., 171A: 257–271.

Hensler, K. and Rowell, C.H.F. (1990). Control of optomotor responses by descending deviation detector neurones in intact flying locusts. J. Exp. Biol., 149: 191–205.

Heran, H. and Lindauer, M. (1963). Windkompensation und Seitenwindkorrektur der Bienen beim Flug über Wasser. Z. vergl. Physiol., 47: 39–55.

Hertel, W. and Penzlin, H. (1992). Function and modulation of the antennal heart of *Periplaneta americana* (L.). Acta Biol. Hung., 43: 113–125.

Hertel, W., Pass, G. and Penzlin, H. (1985). Electrophysiological investigation of the antennal heart of *Periplaneta americana* and its reactions to proctolin. J. Insect Physiol., 31: 563–572.

Heukamp, U. (1984). Sensory regulation of the pleuroalary muscles of the migratory locust. Naturwissenschaften, 71: 481–482.

Hewes, R.S. and Truman, J.W. (1994). Steroid regulation of excitability in identified insect neurosecretory cells. J. Neurosci., 14: 1812–1819.

Hill-Venning, C. (1988). *The neural control of ventilation in the desert locust* Schistocerca gregaria *Forskål*. PhD Thesis, University of Cambridge, UK.

Hinkle, M. and Camhi, J.M. (1972). Locust motoneurons: bursting activity correlated with axon diameter. Science (NY), 175: 553–556.

Hofmann, T., Koch, U.T. and Bässler, U. (1985). Physiology of the femoral chordotonal organ in the stick insect, *Cuniculina impigra*. J. Exp. Biol., 114: 207–223.

Holman, G.M., Nachman, R.J. and Wright, M.S. (1990). Insect neuropeptides. Annu. Rev. Entomol., 35: 201–217.

Homberg, U. (1991). Neuroarchitecture of the central complex in the brain of the locust *Schistocerca gregaria* and *S. americana* as revealed by serotonin immunocytochemistry. J. Comp. Physiol., 303A: 245–254.

Homberg, U. (1994). Flight-correlated activity changes in neurons of the lateral accessory lobes in the brain of the locust *Schistocerca gregaria*. J. Comp. Physiol., 175A: 597–610.

Homberg, U. and Hildebrand, J.G. (1991). Histamine-immunoreactive neurons in the midbrain and suboesophageal ganglion of the sphinx moth *Manduca sexta*. J. Comp. Neurol., 307: 647–657.

Homberg, U., Kingan, T.G. and Hildebrand, J.G. (1987). Immunocytochemistry of GABA in the brain and suboesophageal ganglion of *Manduca sexta*. Cell Tissue Res., 248: 1–24.

Homberg, U., Würden, S., Dircksen, H. and Rao, K.R. (1991). Comparative anatomy of pigment-dispersing hormone-immunoreactive neurons in the brain of orthopteroid insects. Cell Tissue Res., 266: 343–357.

Hörner, M. (1992). Wind-evoked escape running of the cricket *Gryllus bimaculatus*. II. Neurophysiological analysis. J. Exp. Biol., 171: 215–245.

Horseman, G., Hartmann, R., Virant-Doberlet, M., Loher, W. and Huber, F. (1994). Nervous control of juvenile hormone biosynthesis in *Locusta migratoria*. Proc. Natl. Acad. Sci. USA, 91: 2960–2964.

Horsmann, U. and Wendler, G. (1985). The role of a fast wing reflex in locust flight. In '*Insect locomotion*'. pp. 157–166. Eds Gewecke, M. and Wendler, G. Paul Parey, Berlin.

Horsmann, U., Heinzel, H.G. and Wendler, G. (1983). The phasic influence of self-generated air current modulations on the locust flight motor. J. Comp. Physiol., 150A: 427–438.

Howes, E.A., Armett-Kibel, C. and Smith, P.J.S. (1993). A blood-derived attachment factor enhances the *in vitro* growth of two glial cell types from adult cockroach. Glia, 8: 33–41.

Hoyle, G. (1953). Potassium ions and insect nerve muscle. J. Exp. Biol., 30: 121–135.

Hoyle, G. (1955). Neuromuscular mechanisms of a locust skeletal muscle. Proc. R. Soc. Lond. B, 143: 343–367.

Hoyle, G. (1959). The neuromuscular mechanism of an insect spiracular muscle. J. Insect Physiol., 3: 378–394.

Hoyle, G. (1960). The action of carbon dioxide on an insect spiracular muscle. J. Insect Physiol., 4: 63–79.

Hoyle, G. (1966). Functioning of the inhibitory conditioning axon innervating insect muscles. J. Exp. Biol., 44: 429–453.

Hoyle, G. (1974). A function for neurons (DUM) neurosecretory on skeketal muscle of insects. J. Exp. Zool., 189: 401–406.

Hoyle, G. (1975). Evidence that insect dorsal unpaired median (DUM) neurones are octopaminergic. J. Exp. Zool., 193: 425–431.

Hoyle, G. (1978a). The dorsal unpaired median neurons of the locust metathoracic ganglion. J. Neurobiol., 9: 43–57.

Hoyle, G. (1978b). Distribution of nerve and muscle fibre types in locust jumping muscle. J. Exp. Biol., 73: 203–233.

Hoyle, G. (1986). Glial cells of an insect ganglion. J. Comp. Neurol., 246: 85–103.

Hoyle, G. and Burrows, M. (1973). Neural mechanisms underlying behavior in the locust *Schistocerca gregaria*. I. Physiology of identified motorneurons in the metathoracic ganglion. J. Neurobiol., 4: 3–41.

Hoyle, G. and Dagan, D. (1978). Physiological characteristics and reflex activation of DUM (octopaminergic) neurones of locust metathoracic ganglion. J. Neurobiol., 9: 59–79.

Hoyle, G. and O'Shea, M. (1974). Intrinsic rhythmic contractions in insect skeletal muscle. J. Exp. Zool., 189: 407–412.

Hoyle, G., Dagan, D., Moberly, B. and Colquhoun, W. (1974). Dorsal unpaired median insect neurons make neurosecretory endings on skeletal muscle. J. Exp. Zool., 187: 159–165.

Hoyle, G., Colquhoun, W. and Williams, M. (1980). Fine structure of an octopaminergic neuron and its terminals. J. Neurobiol., 11: 103–126.

Huber, F. (1960). Experimentelle Untersuchungen zur nervösen Atmungsregulation der Orthopteren (Saltatoria: Gryllidae). Z. vergl. Physiol., 43: 358–391.

Hughes, T.D. (1980a). The imaginal ecdysis of the desert locust *Schistocerca gregaria*. I. A description of the behaviour. Physiol. Entomol., 5: 47–54.

Hughes, T.D. (1980b). The imaginal ecdysis of the desert locust *Schistocerca gregaria*. II. Motor activity underlying the pre-emergence and emergence behaviour. Physiol. Entomol., 5: 55–71.

Hustert, R. (1975). Neuromuscular coordination and proprioceptive control of rhythmical abdominal ventilation in intact *Locusta migratoria migratorioides*. J. Comp. Physiol., 97A: 159–179.

Hustert, R. (1978). Segmental and interganglionic projections from primary fibers of insect mechanoreceptors. Cell Tissue Res., 194: 337–351.

Hustert, R. (1983). Proprioceptor responses and convergence of proprioceptive influence on motoneurones in the mesothoracic thoraco-coxal joint of locusts. J. Comp. Physiol., 150A: 77–86.

Hustert, R. (1985). Multisegmental integration and divergence of afferent information from single tactile hairs in a cricket. J. Exp. Biol., 118: 209–227.

Hustert, R. and Gnatzy, W. (1995). The motor program for defensive kicking in crickets: performance and neural control. J. Exp. Biol., 198: 1275–1283.

Hustert, R. and Topel, U. (1986). Location and major postembryonic changes of identified 5-HT-immunoreactive neurones in the terminal ganglion of a cricket (*Acheta domesticus*). Cell Tissue Res., 245: 615–621.

Hustert, R., Pflüger, H.-J. and Bräunig, P. (1981). Distribution and specific central projections of mechanoreceptors in the thorax and proximal leg joints of locusts. III. The external mechanoreceptors: the campaniform sensilla. Cell Tissue Res., 216: 97–111.

Jacobs, J.R. and Goodman, C.S. (1989). Embryonic development of axon pathways in the *Drosophila* CNS. I. A glial scaffold appears before the first growth cones. J. Neurosci., 9: 2402–2411.

Jacobs, G.A. and Miller, J.P. (1985). Functional properties of individual neuronal branches isolated *in situ* by laser photoinactivation. Science (NY), 228: 344–346.

Jacobs, G.A. and Murphey, R.K. (1987). Segmental origins of the cricket giant interneuron system. J. Comp. Neurol., 265: 145–157.

Jacobs, G.A. and Nevin, R. (1991). Anatomical relationships between sensory afferent arborizations in the cricket cercal system. Anat. Rec., 231: 563–572.

Jacobs, G.A., Miller, J.A. and Murphey, R.K. (1986). Integrative mechanisms controlling directional sensitivity of an identified sensory interneuron. J. Neurosci., 6: 2298–2311.

Jagota, A. and Habibulla, M. (1992). The frontal ganglionic system: cauterization effects on serotonin circadian rhythms of the cockroach corpora allata and corpora cardiaca. Insect Biochem., 22: 747–755.

Janiszewski, J., Kosecka-Janiszewska, U. and Otto, D. (1988). Changes in rate of abdominal ventilatory pumping induced by warming individual ganglia in the male cricket *Gryllus bimaculatus* (De Geer). J. Thermal. Biol., 13: 185–188.

Jay, D.G. and Keshishian, H. (1990). Laser inactivation of fasciclin I disrupts axon adhesion of grasshopper pioneer neurons. Nature (Lond.), 348: 548–550.

Johnson, B. (1966). Fine structure of the lateral cardiac nerves of the cockroach. J. Insect Physiol., 12: 645–653.

Jorgensen, W.K. and Rice, M.J. (1983). Superextension and supercontraction in locust ovipositior muscles. J. Insect Physiol., 29: 437–448.

Kaczmarek, L.K. and Levitan, I.B. (1987). *Neuromodulation.* Oxford University Press, New York, NY.

Kalogianni, E. (1995). Morphology and physiology of abdominal projection interneurons in the locust with mechanosensory inputs from ovipositor hair receptors. J. Comp. Neurol., 366: 656–673.

Kalogianni, E. and Pflüger, H.-J. (1992). The identification of motor and unpaired median neurones innervating the locust oviduct. J. Exp. Biol., 168: 177–198.

Kalogianni, E. and Theophilidis, G. (1993). Centrally generated rhythmic activity and modulatory function of the oviductal dorsal unpaired median (DUM) neurones in two orthopteran species (*Calliptamus* sp. and *Decticus albifrons*). J. Exp. Biol., 174: 123–138.

Kämper, G. (1984). Abdominal ascending interneurons in crickets: responses to sound at the 30-Hz calling-song frequency. J. Comp. Physiol. 155: 507–520.

Kämper, G. and Dambach, M. (1981). Response of the cercus-to-giant interneuron system in crickets to species-specific song. J. Comp. Physiol., 141A: 311–317.

Kanou, M. and Shimozawa, T. (1984). A threshold analysis of cricket cercal interneurons by an alternating air-current stimulus. J. Comp. Physiol., 154A: 357–365.

Katz, S.L. and Gosline, J.M. (1992). Ontogenetic scaling and mechanical behaviour of the tibiae of the African desert locust (*Schistocerca gregaria*). J. Exp. Biol., 168: 125–150.

Katz, S.L. and Gosline, J.M. (1993). Ontogenetic scaling of jump performance in the African desert locust (*Schistocerca gregaria*). J. Exp. Biol., 177: 81–111.

Kawahara, K., Nakazono, Y., Yamauchi, Y. and Miyamoto, Y. (1989). Coupling between respiratory and locomotor rhythms during fictive locomotion in decerebrate cats. Neurosci. Lett., 103: 326–3332.

Keegan, A.P. and Comer, C.M. (1993). The wind-elicited escape response of cockroaches (*Periplaneta americana*) is influenced by lesions rostral to the escape circuit. Brain Res., 620: 310–316.

Kendall, M.D. (1970). The anatomy of the tarsi of *Schistocerca gregaria* Forskål. Z. Zellforsch., 109: 112–137.

Kent, K.S. and Hildebrand, J.G. (1987). Cephalic sensory pathways in the central nervous system of larval *Manduca sexta* (Lepidoptera: Sphingidae). Phil. Trans. R. Soc. Lond. B. 315: 1–36.

Kent, K.S. and Levine, R.B. (1988). Neural control of leg movements in a metamorphic insect: persistence of larval leg motor neurons to innervate the adult legs of *Manduca sexta.* J. Comp. Neurol., 276: 30–43.

Keshishian, H. (1980). The origin and morphogenesis of pioneer neurons in the grasshopper metathoracic leg. Dev. Biol., 80: 388–397.

Keshishian, H. and Bentley, D. (1983a). Embryogenesis of peripheral nerve pathways in grasshopper legs. I. The initial nerve pathways to the CNS. Dev. Biol., 96: 98–102.

Keshishian, H. and Bentley, D. (1983b). Embryogenesis of peripheral nerve pathways in grasshopper legs. II. The major nerve routes. Dev. Biol., 96: 103–115.

Keshishian, H. and Bentley, D. (1983c). Embryogenesis of peripheral nerve pathways in grasshopper legs. III. Development without pioneer neurons. Dev. Biol., 96: 116–124.

Keshishian, H., and O'Shea, M. (1985a). The acquisition and expression of a peptidergic phenotype in the grasshopper embryo. J. Neurosci., 5: 1005–1015.

Keshishian, H., and O'Shea, M. (1985b). The distribution of a peptide neurotransmitter in the postembryonic grasshopper central nervous system. J. Neurosci., 5: 992–1004.

Kien, J. (1980). Mechanisms of motor control by plurisegmental interneurons in locusts. J. Comp. Physiol., 140A: 303–320.

Kien, J. (1983). The initiation and maintenance of walking in the locust: an alternative to the command neuron concept. Proc. R. Soc. Lond. B, 219: 137–174.

King, D.G. and Wyman, R.T. (1980). Anatomy of the giant fibre pathway in *Drosophila*. Three thoracic components of the pathway. J. Neurocytol., 1: 753–770.

Kingsolver, J.G. and Koehl, M.A.R. (1985). Aerodynamics, thermoregulation, and the evolution of insect wings: differential scaling and evolutionary change. Evolution, 39: 488–504.

Kingsolver, J.G. and Koehl, M.A.R. (1994). Selective factors in the evolution of insect wings. Annu. Rev. Entomol., 39: 425–451.

Kirby, P., Beck, R. and Clarke, K.U. (1984). The stomatogastric nervous sytem of the house cricket *Acheta domesticus* L. I. The anatomy of the system and innervation of the gut. J. Morphol., 180: 81–103.

Kirschenbaum, S.R. and O'Shea, M. (1993). Postembryonic proliferation of neuroendocrine cells expressing adipokinetic hormone peptides in the corpora cardiaca of the locust. Development, 118: 1181–1190.

Kiss, T., Varanka, I. and Benedekzy, I. (1984). Neuromuscular transmission in the visceral muscle of locust oviduct. Neuroscience, 12: 309–322.

Kittmann, R., Dean, J. and Schmitz, J. (1991). An atlas of the thoracic ganglia in the stick insect, *Carausius morosus*. Phil. Trans. R. Soc. Lond. B, 331: 101–121.

Klee, S. and Thurm, U. (1986). Mechanorezeptive Fadenhaar-Sensillen mit zwei Sinneszellen von *Locusta migratoria* L. Ver. Dtsch Zool. Ges. 79: 220.

Klemm, N. and Sundler, F. (1983). The organisation of catecholamine and serotonin-immunoreactive fibres in the corpora pedunculata of the desert locust, *Schistocerca gregaria* Forsk. (Insecta, Orthoptera). Neurosci. Lett. 36: 13–17.

Klemm, N., Hustert, R., Cantera, R. and Nässel, D.R. (1986). Neurons reactive to antibodies against serotonin in the stomatogastric nervous system and the alimentary canal of locust and crickets (Orthoptera, Insecta). Neuroscience, 17: 247–261.

Kloppenburg, P. and Hildebrand, J.G. (1995). Neuromodulation by 5-hydroxytrptamine in the antennal lobe of the sphinx moth *Manduca sexta*. J. Exp. Biol., 198: 603–611.

Klose, M. and Bentley, D. (1989). Transient pioneer neurons are essential for formation of an embryonic peripheral nerve. Science (NY), 245: 982–984.

Knechtel, W.K. von. (1938). Über die Wanderheuschrecke in Rumänien. Bull. entomol. Res., 29: 175–183.

Kobashi, M. and Yamaguchi, T. (1984). Local non-spiking interneurons in the cercus-to-giant interneuron system of crickets. Naturwissenschaften, 71: 154–156.

Kolodkin, A.L., Matthes, D.J., O'Connor, T.P., Patel, N.H., Admon, A., Bentley, D. and Goodman, C.S. (1992). Fasciclin IV: sequence, expression, and function during growth cone guidance in the grasshopper embryo. Neuron, 9: 831–845.

Komatsu, A. (1982). Respiratory nervous activity in the isolated nerve cord of the larval dragonfly, and location of the respiratory oscillator. Physiol. Entomol., 7: 183–191.

Kondo, H. (1978). Efferent system of lateral ocellus in dragonfly – its relationships with ocellar afferent units, compound eyes and wing sensory system. J. Comp. Physiol., 125A: 341–349.

Kondoh, Y., Morishita, H., Arima, T., Okuma, J. and Hasegawa, Y. (1991a). White noise analysis of graded response in a wind-sensitive, nonspiking interneuron of the cockroach. J. Comp. Physiol., 168A: 429–443.

Kondoh, Y., Arima, T., Okuma, J. and Hasegawa, Y. (1991b). Filter characteristics of cercal afferents in the cockroach. J. Comp. Physiol. 169A: 653–662.

Kondoh, Y., Arima, T., Okuma, J. and Hasegawa, Y. (1993). Response dynamics and directional properties of nonspiking local interneurons in the cockroach cercal system. J. Neurosci., 13: 2287–2305.

Kondoh, Y., Okuma, J. and Newland, P.L. (1995). Dynamics of neurons controlling movements of a locust hind leg: Wiener kernel analysis of the responses of proprioceptive afferents. J. Neurophysiol., 73: 1829–1842.

Konings, P.N.M., Vullings, H.G.B., Siebinga, R., Diederen, J.H.B. and Jansen, W.F. (1988). Serotonin-immunoreactive neurones in the brain of *Locusta migratoria* innervating the corpus cardiacum. Cell Tissue Res., 254: 147–153.

Konings, P.N.M., Vullings, H.G.B., Kok, O.J.M., Diederen, J.H.B. and Jansen, W.F. (1989a). The innervation of the corpus cardiacum of *Locusta migratoria*: a neuroanatomical study with the use of Lucifer Yellow. Cell Tissue Res., 258: 301–308.

Konings, P.N.M., Vullings, H.G.B., Van Gemert, W.M.J.B., DeLeeuw, R., Diederen, J.H.B. and Jansen, W.F. (1989b). Octopamine-binding sites in the brain of *Locusta migratoria* L. Eur. J. Pharmacol., 35: 519–524.

Krauthamer, V. and Fourtner, C.R. (1978). Locomotory activity in the extensor and flexor tibiae of the cockroach *Periplaneta americana*. J. Insect Physiol., 24: 813–819.

Krenn, H.W. (1993). Postembryonic development of accessory wing circulatory organs in *Locusta migratoria* (Orthoptera: Acrididae). Zool. Anz., 230: 227–236.

Krenn, H.W. and Pass, G. (1994). Morphological diversity and phylogenetic analysis of wing circulatory organs in insects, part I: non-Holometabola. Zoology, 98: 7–22.

Krogh, A. and Weis-Fogh, T. (1951). The respiratory exchange of the desert locust (*Schistocerca gregaria*) before, during and after flight. J. Exp. Biol., 28: 344–357.

Kuenzi, F. and Burrows, M. (1995). Central connections of sensory neurones from a hair plate proprioceptor in the thoraco-coxal joint of the locust. J. Exp. Biol., 198: 1589–1601.

Kukalova-Peck, J. (1978). Origin and evolution of insect wings and their relation to metamorphosis, as documented in the fossil record. J. Morphol., 156: 52–126.

Kukalova-Peck, J. (1983). Origin of the insect wing and wing articulation from the arthropodan leg. Can. J. Zool., 61: 1618–1669.

Kutsch, W. (1969). Neuromuskuläre Aktivität bei verschiedenen Verhaltensweisen von drei Grillenarten. Z. vergl. Physiol., 63: 335–378.

Kutsch, W. (1971). The development of the flight pattern in the desert Locust *Schistocera gregaria*. Z. vergl. Physiol., 74: 156–168.

Kutsch, W. (1973). The influence of age and culture-temperature on the wing-beat frequency of the migratory locust *Locusta migratoria*. J. Insect Physiol., 19: 763–772.

Kutsch, W. (1985). Pre-imaginal flight motor pattern in *Locusta*. J. Insect Physiol., 31: 581–586.

Kutsch, W. (1989). Formation of the receptor system in the hind leg of the locust embryo. Roux's Arch. Dev. Biol. 198: 39–47.

Kutsch, W. and Gewecke, M. (1979). Development of flight behaviour in maturing adults of *Locusta migratoria*. II: Aerodynamic parameters. J. Insect Physiol., 25: 299–304.

Kutsch, W., and Hemmer, W. (1994). Ontogenetic studies of flight initiation in *Locusta migratoria*: flight muscle activity. J. Insect Physiol., 40: 519–525.

Kutsch, W. and Schneider, H. (1987). Histological characterization of neurones innervating functionally different muscles of *Locusta*. J. Comp. Neurol., 261: 515–528.

Kutsch, W. and Stevenson, P. (1984). Manipulation of the endocrine system of *Locusta* and the development of the flight motor pattern. J. Comp. Physiol., 155A: 129–138.

Kutsch, W., Hanloser, H. and Reinecke, M. (1980). Light and electron-microscopic analysis of a complex sensory organ: the tegula of *Locusta migratoria*. Cell Tissue Res., 210: 461–478.

Kutsch, W., Neubauer, K. and Krämer, H. (1991). Flight motor pattern in abdominal segments of locusts. J. Exp. Biol., 156: 629–635.

Kutsch, W., Camhi, J. and Sumbre, G. (1994). Close encounters among flying locusts produce wing-beat coupling. J. Comp. Physiol., 174A: 643–649.

Lagueux, M., Sall, C. and Hoffmann, J.A. (1981). Ecdysteroids during embryogensis in *Locusta migratoria*. Am. Zoologist, 21: 715–726.

Lange, A.B. and Orchard, I. (1984). Dorsal unpaired median neurones and ventral bilaterally paired neurones, project to a visceral muscle in an insect. J. Neurobiol., 15: 441–453.

Lange, A.B. and Tsang, P.K.C. (1993). Biochemical and physiological effects of octopamine and selected octopamine agonists on the oviducts of *Locusta migratoria*. J. Insect Physiol., 39: 393–400.

Lange, A., Orchard, I. and Loughton, B.G. (1984). Neural inhibition of egg-laying in the locust, *Locusta migratoria*. J. Insect Physiol., 30: 271–278.

Lange, A.B., Orchard, I. and Adams, M.E. (1986). Peptidergic innervation of insect reproductive tissue – The association of proctolin with oviduct visceral musculature. J. Comp. Neurol., 254: 279–286.

Lange, A.B., Orchard, I. and Te Brugge, V.A. (1991). Evidence for the involvement of a SchistoFLRF-amide-like peptide in the neural control of locust oviduct. J. Comp. Physiol., 168A: 383–391.

Lange, A.B., Chan, K.K. and Stay, B. (1993). Effect of allatostatin and proctolin on antennal pulsatile organ and hindgut muscle in the cockroach, *Diploptera punctata*. Arch. Insect Biochem. Physiol., 24: 79–92.

Lapied, B., Malecot, C.O. and Pelhate, M. (1989). Ionic species involved in the electrical activity of single adult aminergic neurons islolated from the sixth abdominal ganglion of the cockroach *Periplaneta americana*. J. Exp. Biol., 144: 535–549.

Lapied, B., Malecot, C.O. and Pelhate, M. (1990). Patch-clamp study of the properties of the sodium current in cockroach single isolated adult aminergic neurones. J. Exp. Biol., 151: 387–403.

Lapied, B., Tribut, F. and Hue, B. (1992). Effects of McN-A-343 on insect neurosecretory cells: evidence for muscarinic-like receptor subtypes. Neurosci. Lett., 139: 165–168.

Laughon, A. and Scott, M.P. (1984). Sequence of a *Drosophila* segmentation gene: protein homology with DNA-binding proteins. Nature (Lond.), 310: 25–31.

Laurent, G. (1986). Thoracic intersegmental interneurones in the locust with mechanoreceptive inputs from a leg. J. Comp. Physiol., 159A: 171–186.

Laurent, G. (1987). The morphology of a population of thoracic intersegmental interneurones in the locust. J. Comp. Neurol., 256: 412–429.

Laurent, G. (1988). Local circuits underlying excitation and inhibition of intersegmental interneurones in the locust. J. Comp. Physiol., 162A: 145–157.

Laurent, G. (1990). Voltage-dependent nonlinearities in the membrane of locust nonspiking local interneurons, and their significance for synaptic integration. J. Neurosci., 10: 2268–2280.

Laurent, G. (1991). Evidence for voltage-activated outward currents in the neuropilar membrane of locust nonspiking local interneurons. J. Neurosci., 11: 1713–1726.

Laurent, G. (1993). A dendritic gain control mechanism in axonless neurons of the locust, *Schistocerca americana*. J. Physiol. (Lond.), 470: 45–54.

Laurent, G. and Burrows, M. (1988). A population of ascending intersegmental interneurones in the locust with mechanosensory inputs from a hind leg. J. Comp. Neurol. 275: 1–12.

Laurent, G. and Burrows, M. (1989a). Intersegmental interneurones can control the gain of reflexes in adjacent segments by their action on nonspiking local interneurones. J. Neurosci., 9: 3030–3039.

Laurent, G. and Burrows, M. (1989b). Distribution of intersegmental inputs to nonspiking local interneurones and motor neurones in the locust. J. Neurosci., 9: 3019–3029.

Laurent, G. and Sivaramakrishnan, A. (1992). Single local interneurons in the locust make central synapses with different properties of transmitter release on distinct postsynaptic neurons. J. Neurosci., 12: 2370–2380.

Laurent, G., Seymour-Laurent, K.J. and Johnson, K. (1993). Dendritic excitability and a voltage-gated calcium current in locust nonspiking local interneurons. J. Neurophysiol., 69: 1484–1498.

Laurent, G.J. and Hustert, R. (1988). Motor neuronal receptive fields delimit patterns of activity during locomotion of the locust. J. Neurosci., 8: 4349–4366.

Lee, M.O. (1925). On the mechanism of respiration in certain Orthoptera. J. Exp. Zool., 41: 125–254.

Leitch, B. and Laurent, G. (1993). Distribution of GABAergic synaptic terminals on the dendrites of locust spiking local interneurones. J. Comp. Neurol., 337: 461–470.

Leitch, B., Laurent, G. and Shepherd, D. (1992). Embryonic development of synapses on spiking local interneurones in locust. J. Comp. Neurol., 324: 213–236.

Leitch, B., Shepherd, D. and Laurent, G. (1995). Morphogenesis of the branching pattern of a group of spiking local interneurons in relation to the organization of embryonic sensory neuropils in locust. Phil. Trans. R. Soc. Lond. B., 349: 433–447.

Lewis, G.W., Miller, P.L. and Mills, P.S. (1973). Neuro-muscular mechanisms of abdominal pumping in the locust. J. Exp. Biol., 59: 149–168.

Libersat, F. (1992). Modulation of flight by the giant interneurons of the cockroach. J. Comp. Physiol., 170A: 379–392.

Libersat, F. (1994a). The dorsal giant interneurons mediate evasive behavior in flying cockroaches. J. Exp. Biol., 197: 405–411.

Libersat, F. (1994b). Physiological properties of an identified giant interneuron (GI4) as related to the escape and flight circuitries of the cockroach *Periplaneta americana*. J. Insect Physiol., 40: 431–438.

Libersat, F. and Camhi, J.M. (1988). Control of cercal position during flight in the cockroach: a mechanism for regulating sensory feedback. J. Exp. Biol., 136: 483–488.

Libersat, F., Goldstein, R.S. and Camhi, J.M. (1987). Nonsynaptic regulation of sensory activity during movement in cockroaches. Proc. Natl. Acad. Sci. USA, 84: 8150–8154.

Libersat, F., Levy, A. and Camhi, J.M. (1989). Multiple feedback loops in the flying cockroach: excitation of the dorsal and inhibition of the ventral giant interneurons. J. Comp. Physiol., 165A: 651–668.

Liebenthal, E., Uhlmann, O. and Camhi, J.M. (1994). Critical parameters of the spike trains in a cell assembly: coding of turn direction by the giant interneurons of the cockroach. J. Comp. Physiol., 174A: 281–296.

Lighton, J.R.B. (1994). Discontinuous ventilation in terrestrial insects. Physiol. Zool., 67: 142–162.

Longley, A. and Edwards, J.S. (1979). Tracheation of abdominal ganglia and cerci in the house cricket *Acheta domesticus* (Orthoptera, Gryllidae). J. Morphol., 159: 233–244.

Lorez, M. (1995). Neural control of hindleg steering in flight in the locust. J. Exp. Biol., 198: 869–875.

Lundquist, C.T., Clottens, F.L., Holman, G.M., Riehm, J.P., Bonkale, W. and Nässel, D.R. (1994). Locustatachykinin immunoreactivity in the blowfly central nervous system and intestine. J. Comp. Neurol., 341: 225–240.

Lutz, E.M. and Tyrer, N.M. (1988). Immunohistochemical localization of serotonin and choline acetyltransferase in sensory neurones of the locust. J. Comp. Neurol., 267: 335–342.

Macfarlane, R.G., Midgley, J.M., Watson, D.G. and Evans, P.D. (1990). Identification and quantitation of phenylalanine, tyrosine and dihydroxyphenylalanine in the thoracic nervous system of the locust *Schistocerca gregaria*, by gas chromatography-negative-ion chemical ionisation mass spectrometry. J. Chromatogr. Biomed. Appl., 532: 1–11.

Macmillan, D.L. and Kien, J. (1983). Intra- and intersegmental pathways active during walking in the locust. Proc. R. Soc. Lond. B, 218: 287–308.

Maddrell, S.H.P. (1966). Nervous control of the mechanical properties of the abdominal wall at feeding in *Rhodnius*. J. Exp. Biol., 44: 59–68.

Maitland, D.P. (1992). Locomotion by jumping in the Mediterranean fruit-fly larva *Ceratitis capitata*. Nature (Lond.), 355: 159–161.

Malamud, J.G., Mizisin, A.P. and Josephson, R.K. (1988). The effects of octopamine on contraction kinetics and power output of a locust flight muscle. J. Comp. Physiol., 162A: 827–835.

Marchand, A.R. and Leibrock, C.S. (1994). Functional aspects of central electrical coupling in mechanoreceptor afferents of crayfish. Brain Res., 667: 98–106.

Marden, J.H. and Kramer, M.G. (1994). Surface-skimming stoneflies: a possible intermediate stage in insect flight evolution. Science (NY) 266: 427–430.

Marquart, V. (1985). Local interneurons mediating excitation and inhibition onto ascending neurons in the auditory pathway of grasshoppers. Naturwissenschaften, 72: 42–44.

Mason, C.A. (1973). New features of the brain-retrocerebral complex of the locust *Schistocerca vaga* (Scudder). Z. Zellforsch., 141: 19–32.

Matheson, T. (1990). Responses and location of neurones in the locust metathoracic femoral chordotonal organ. J. Comp. Physiol., 166A: 915–927.

Matheson, T. (1992). Morphology of the central projections of physiologically characterised neurones from the locust metathoracic femoral chordotonal organ. J. Comp. Physiol., 170A: 101–120.

Matheson, T. and Field, L.H. (1990). Innervation of the metathoracic femoral chordotonal organ of *Locusta migratoria*. Cell Tissue Res., 259: 551–560.

Matheson, T. and Field, L.H. (1995). An elaborate tension receptor system highlights sensory complexity in the hind leg of the locust. J. Exp. Biol., 198: 1673–1689.

Maulik, S. (1929). On the structure of the hind femur in halticine beetles. Proc. Zool. Soc. Lond., 2: 305–308.

May, T.E., Brown, B.E. and Clements, A.N. (1979). Experimental studies upon a bundle of tonic fibres in the locust extensor tibalis muscle. J. Insect Physiol., 25: 169–181.

Meier, T. and Reichert, H. (1990). Embryonic development and evolutionary origin of the orthoptera auditory organs. J. Neurobiol., 21: 592–610.

Meier, T., Therianos, S., Zacharias, D. and Reichert, H. (1993). Developmental expression of TERM-1 glycoprotein on growth cones and terminal arbors of individual identified neurons in the grasshopper. J. Neurosci., 13: 1498–1510.

Mendelson, M. (1971). Oscillator neurons in crustacean ganglion. Science (NY), 171: 1170–1173.

Menzel, R. and Erber, J. (1978). Learning and memory in bees. Scient. Am., 239: 102–110.

Mercer, A.R., Hayashi, J.H. and Hildebrand, J.G. (1995). Modulatory effects of 5-hydroxytryptamine on voltage-activated currents in cultured antennal lobe neurones of the sphinx moth *Manduca sexta*. J. Exp. Biol., 198: 613–627.

Mestre, J. (1988). *Les Acridiens des formations herbeuses d'Afrique de l'ouest*. Cirad-Prifas, Paris.

Meurant, K., Sernia, C. and Rembold, H. (1994). The effects of azadirachtin A on the morphology of the ring complex of *Lucilia cuprina* (Wied) larvae (Diptera: Insecta). Cell Tissue Res., 275: 247–254.

Meyer, D. and Walcott, B. (1979). Differences in the responsiveness of identified motoneurons in the cockroach: role in the motor program for stepping. Brain Res., 178: 600–605.

Meyer, J. and Hedwig, B. (1995). The influence of tracheal pressure changes on the response of the tympanal membrane and auditory receptors in the locust *Locusta migratoria*. J. Exp. Biol., 198: 1327–1339.

Mill, P.J. and Hughes, G.M. (1966). The nervous control of ventilation in dragonfly larvae. J. Exp. Biol., 44: 297–316.

Miller, P.L. (1960). Respiration in the desert locust. I. The control of ventilation. J. Exp. Biol., 37: 224–236.

Miller, P.L. (1965). The central nervous control of respiratory movements. In '*Physiology of the insect central nervous system*'. p.141–155. Eds. Treherne, J.E. and Beament, J.W.L. Academic Press, New York.

Miller, P.L. (1966). The regulation of breathing in insects. Adv. Insect Physiol., 3: 279–354.

Miller, P.L. (1973). Spatial and temporal changes in the coupling of cockroach spiracles to ventilation. J. Exp. Biol., 59: 137–148.

Miller, T. and Usherwood, P.N.R. (1971). Studies of cardio-regulation in the cockroach, *Periplaneta americana*. J. Exp. Biol., 54: 329–348.

Möhl, B. (1985a). The role of proprioception in locust flight control. I. Asymmetry and coupling within the time pattern of motor units. J. Comp. Physiol., 156A: 93–101.

Möhl, B. (1985b). The role of proprioception in locust flight control. II. Information signalled by forewing stretch receptors during flight. J. Comp. Physiol., 156A: 103–116.

Möhl, B. (1988). Short-term learning during flight control in *Locusta migratoria*. J. Comp. Physiol., 163A: 803–812.

Möhl, B. (1989). 'Biological noise' and plasticity of sensorimotor pathways in the locust flight system. J. Comp. Physiol., 166A: 75–82.

Möhl, B. (1993). The role of proprioception for motor learning in locust flight. J. Comp. Physiol., 172A: 325–332.

Möhl, B. and Bacon, J. (1983). The tritocerebral commissure giant (TCG) wind-sensitive interneurone in the locust. II. Directional sensitivity and role in flight stabilisation. J. Comp. Physiol., 150A: 453–465.

Möhl, B. and Neumann, L. (1983). Peripheral feedback-mechanisms in the locust flight system. In '*BIONA-report 2: Insektenflug II*'. pp. 81–87. Ed. Nachtigall, W. Gustav Fischer Verlag, Stuttgart.

Möhl. B. and Zarnack, W. (1977). Flight steering by means of time shifts in the activity of the direct downstroke muscles in the locust. Fortschr. Zool., 24: 333–339.

Morton, D.B. and Evans, P.D. (1983). Octopamine distribution in solitarious and gregarious forms of the locust *Schistocerca americana gregaria*. Insect Biochem., 13: 177–183.

Morton, D.B. and Evans, P.D. (1984). Octopamine release from an identified neurone in the locust. J. Exp. Biol., 113: 269–287.

Möss, D. (1971). Sinnesorgane im Bereich des Flügels der Feldgrille (*Gryllus campestris* L.) und ihre Bedeutung für die Kontrolle der Singbewegung und die Einstellung der Flügellage. Z. vergl. Physiol., 73: 53–83.

Mücke, A. (1991). Innervation pattern and sensory supply of the midleg of *Schistocerca gregaria* (Insecta, Orthopteroidea). Zoomorphology 110: 175–187.

Müller, A.R., Wolf, H., Galler, S. and Rathmayer, W. (1992). Correlation of electrophysiological, histochemical, and mechanical properties in fibres of the coxa rotator muscle of the locust, *Locusta migratoria*. J. Comp. Physiol., 162B: 5–15.

Müller, F. (1875). Beiträge zur Kenntnis der Termiten. Z. Naturwissenschaften, (Jena), 9: 241–264.

Müller, U. (1994). Ca^{2+}/calmodulin-dependent nitric oxide synthase in *Apis mellifera* and *Drosophila melanogaster*. Eur. J. Neurosci., 6: 1362–1370.

Müller, U. and Bicker, G. (1994). Calcium-activated release of nitric oxide and cellular distribution of nitric oxide-synthesising neurons in the nervous system of the locust. J. Neurosci., 14: 7521–7528.

Murphey, R.K. (1981). The structure and development of a somatotopic map in crickets: the cercal afferent projection. Dev. Biol., 88: 236–246.

Murphey, R.K., Possidente, D., Pollack, G. and Merritt, D.J. (1989). Modality-specific axonal projections in the CNS of the flies *Phormia* and *Drosophila*. J. Comp. Neurol., 290: 185–200.

Myers, C. and Ball, E.E. (1987). Comparative development of the extensor and flexor tibiae muscles in the legs of the locust, *Locusta migratoria*. Development, 101: 351–361.

Myers, C.M. and Evans, P.D. (1985a). The distribution of bovine pancreatic polypeptide/FMRFamide-like immunoreactivity in the ventral nervous system of the locust. J. Comp. Neurol. 234: 1–16.

Myers, C.M. and Evans, P.D. (1985b). An FMRFamide antiserum differentiates between populations of antigens in the ventral nervous sytem of the locust, *Schistocerca gregaria*. Cell Tissue Res., 242: 109–114.

Myers, C.M. and Evans, P.D. (1987). An FMRFamide antiserum differentiates between populations of antigens in the brain and retrocerebral complex of the locust, *Schistocerca gregaria*. Cell Tissue Res., 250: 93–99.

Myers, C.M. and Evans, P.D. (1988). Peripheral neurosecretory cells on the thoracic median nerves of the locust *Schistocerca gregaria*. J. Morphol., 195: 45–58.

Myers, P.Z. and Bastiani, M.J. (1993). Growth cone dynamics during the migration of an identified commissural growth cone. J. Neurosci., 13: 127–143.

Myers, T.H. and Retzlaff, E. (1963). Localization and action of the respiratory centre of the Cuban burrowing cockroach. J. Insect Physiol., 9: 607–614.

Nagayama, T. (1989). Morphology of a new population of spiking local interneurones in the locust metathoracic ganglion. J. Comp. Neurol., 283: 189–211.

Nagayama, T. (1990). The organisation of receptive fields of an antero-medial group of spiking local interneurones in the locust with exteroceptive inputs from the legs. J. Comp. Physiol., 166A: 471–476.

Nagayama, T. and Burrows, M. (1990). Input and output connections of an anteromedial group of spiking local interneurones in the metathoracic ganglion of the locust. J. Neurosci., 10: 785–794.

Nashner, L.M. (1976). Adapting reflexes controlling the human posture. Exp. Brain Res., 26: 59–72.

Nässel, D.R. (1993). Neuropeptides in the insect brain: a review. Cell Tissue Res., 273: 1–29.

Nässel, D.R. and Elekes, K. (1984). Ultrastructural demonstration of serotonin-immunoreactivity in the nervous system of an insect (*Calliphora erythrocephala*). Neurosci. Lett., 48: 203–210.

Nässel, D.R. and Elekes, K. (1992). Aminergic neurons in the brain of blowflies and *Drosophila*: dopamine- and tyrosine hydroxylase-immunoreactive neurons and their relationship with putative histaminergic neurons. Cell Tissue Res., 267: 147–167.

Nässel, D.R. and O'Shea, M. (1987). Proctolin-like immunoreactive neurons in the blowfly central nervous system. J. Comp. Neurol., 265: 437–454.

Nässel, D.R., Holmqvist, M.H., Hardie, R.C., Hakanson, R. and Sundler, F. (1988). Histamine-like immunoreactivity in photoreceptors of the compound eyes and ocelli of flies. Cell Tissue Res., 253: 639–646.

Nässel, D.R., Pirvola, U. and Panula, P. (1990). Histaminelike immunoreactive neurons innervating putative neurohaemal areas and central neuropil in the thoraco-abdominal ganglia of the flies *Drosophila* and *Calliphora*. J. Comp. Neurol., 297: 525–536.

Nathanson, J.A. (1979). Octopamine receptors, adenosine 3′5′- monophosphate, and neural control of firefly flashing. Science (NY) 203: 65–68.

Nathanson, J.A. and Greengard, P. (1973). Octopamine-sensitive adenylate cyclase: evidence for a biological role of octopamine in nervous tissue. Science (NY), 180: 308–310.

Nelson, M.C. (1979). Sound production in the cockroach *Gromphadorhina portentosa*: the sound-producing apparatus. J. Comp. Physiol., 132A: 27–38.

Neumann, L. (1985). Experiments on tegula function for flight coordination in the locust. In '*Insect locomotion*'. pp. 149–156. Eds Gewecke, M. and Wendler, G. Paul Parey Verlag, Hamburg.

Neville, A.C. and Weis-Fogh, T. (1963). The effect of temperature on locust flight muscle. J. Exp. Biol., 40: 111–121.

Newland, P.L. (1990). Morphology of a population of mechanosensory ascending interneurones in the metathoracic ganglion of the locust. J. Comp. Neurol., 299: 242–260.

Newland, P.L. (1991a). Physiological properties of afferents from tactile hairs on the hindlegs of the locust. J. Exp. Biol., 155: 487–503.

Newland, P.L. (1991b). Morphology and somatotopic organisation of the central projections of afferents from tactile hairs on the hind leg of the locust. J. Comp. Neurol., 312: 493–508.

Newland, P.L. and Burrows, M. (1994). Processing of mechanosensory information from gustatory receptors on a hind leg of the locust. J. Comp. Physiol., 174A: 399–410.

Newland, P.L., Watkins, B., Emptage, N.J. and Nagayama, T. (1995). The structure, response properties and development of a hair plate on the mesothoracic leg of the locust. J. Exp. Biol., 198: 2397–2404.

Nicklaus, R. (1965). Die Erregung einzelner Fadenhaare von *Periplaneta americana* in Abhängigkeit von der Größe und Richtung der Auslenkung. Z. vergl. Physiol., 50: 331–362.

Nicol, D. and Meinertzhagen, I.A. (1982). Regulation in the number of fly photoreceptor synapses: the effects of alterations in the number of presynaptic cells. J. Comp Neurol., 207: 45–60.

Nijhout, H.F. (1977). Control of antennal hair erection in male mosquitoes. Biol. Bull., 153: 591–603.

Norman, A.P. (1995). Adaptive changes in locust kicking and jumping behaviour during development. J. Exp. Biol., 198: 1341–1350.

Nurnberger, A., Rapus, J., Eckert, M. and Penzlin, H. (1993). Taurine-like immunoreactivity in octopaminergic neurones of the cockroach, *Periplaneta americana* (L.). Histochemistry, 100: 285–292.

O'Dell, D.A. and Watkins, B.L. (1988). The development of GABA-like immunoreactivity in the thoracic ganglia of the locust *Schistocerca gregaria*. Cell Tissue Res., 254: 635–646.

O'Shea, M. (1975). Two sites of spike initiation in a bimodal interneuron. Brain Res., 96: 93–98.

O'Shea, M. and Adams, M.E. (1981). Pentapeptide (proctolin) associated with an identified neuron. Science, 213: 567–569.

O'Shea, M. and Adams, M. (1986). Proctolin: from 'gut factor' to model neuropeptide. Adv. Insect Physiol., 19: 1–28.

O'Shea, M. and Evans, P.D. (1979). Potentiation of neuromuscular transmission by an octopaminergic neurone in the locust. J. Exp. Biol., 79: 169–190.

O'Shea, M. and Rayne, R.C. (1992). Adipokinetic hormones: cell and molecular biology. Experientia, 48: 430–438.

O'Shea, M., Rowell, C.H.F. and Williams, J.L.D. (1974). The anatomy of a locust visual interneurone; the descending contralateral movement detector. J. Exp. Biol., 60: 1–12.

O'Shea, M., Adams, M.E. and Bishop, C.A. (1982). Identification of proctolin-containing neurons. Fedn. Proc., 41: 2940–2947.

O'Shea, M., Hekimi, S., Witten, J. and Worden, M.K. (1988). Funtions of aminergic and peptidergic skeletal motoneurones in insects. In '*Neurohormones in invertebrates*'. pp. 159–172. Eds Thorndyke, M.C. and Goldsworthy, G.J. Cambridge University Press, Cambridge.

Oertel, D., Linberg, K.A. and Case, J.F. (1975). Ultrastructure of the larval firefly light organ as related to control of light emission. Cell Tissue Res., 164: 27–44.

Oken, L. (1811). *Lehrbuch der Naturphilosophie*. Friedrich Fromman, Jena.

Oldfield, B.P. (1982). Tonotopic organisation of auditory receptors in Tettigoniidae (Orthoptera: Ensifera). J. Comp. Physiol., 147A: 461–469.

Orchard, I. and Finlayson, L.H. (1977). Electrically excitable neurosecretory cell bodies in the periphery of the stick insect *Carausius morosus*. Experientia, 33: 226–228.

Orchard, I. and Lange, A.B. (1985). Evidence for octopaminergic modulation of an insect visceral muscle. J. Neurobiol., 16: 171–181.

Orchard, I. and Lange, A.B. (1986). Neuromuscular transmission in an insect visceral muscle. J. Neurobiol., 17: 359–372.

Orchard, I. and Osborne, M.P. (1977). The effects of cations upon the action potentials recorded from neurohaemal tissue of the stick insect. J. Comp. Physiol., 118A: 1–12.

Orchard, I., Lange, A.B., Cook, H. and Ramirez, J.-M. (1989). A subpopulation of dorsal unpaired median neurons of the blood-feeding insect *Rhodnius prolixus* displays serotonin-like immunoreactivity. J. Comp. Neurol., 289: 118–128.

Orida, N. and Josephson, R.K. (1978). Peripheral control of responsiveness to auditory stimuli in giant fibres of crickets and cockroaches. J. Exp. Biol., 72: 153–164.

Orona, E. and Ache, B.W. (1992). Physiological and pharmacological evidence for histamine as a neurotransmitter in the olfactory CNS of the spiny lobster. Brain Res., 590: 136–143.

Otto, D. and Hennig, R.M. (1993). Interneurons descending from the cricket subesophageal ganglion control stridulation and ventilation. Naturwissenschaften, 80: 36–38.

Otto, D. and Janiszewski, J. (1989). Interneurones originating in the suboesophageal ganglion that control ventilation in two cricket species: effects of the interneurones (SD-AE neurones) on the motor output. J. Insect Physiol., 35: 483–491.

Paemen, L., Schoofs, L. and DeLoof, A. (1992). Localisation of Lom-AG-myotropin I-like substances in the male reproductive and nervous tissue of the locust, *Locusta migratoria.* Cell Tissue Res., 268: 91–97.

Palka, J., Levine, R. and Schubiger, M. (1977). The cercus-to-giant interneuron system of crickets. 1: Some attributes of the sensory cells. J. Comp. Physiol., 119A: 267–283.

Panganiban, G., Nagy, L. and Carroll, S.B. (1994). The role of the *Distal-less* gene in the development and evolution of insect limbs. Curr. Biol., 4: 671–675.

Parker, D. (1994). Glutamatergic transmission between antagonistic motor neurons in the locust. J. Comp. Physiol., 175A: 737–748.

Parker, D. (1995a). Serotonergic modulation of locust motor neurons. J. Neurophysiol., 73: 923–932.

Parker, D. (1995b). Octopaminergic modulation of locust motor neurones. J. Comp. Physiol., 178A: 243–252.

Parker, D. and Newland, P.L. (1995). Cholinergic synaptic transmission between proprioceptive afferents and a hind leg motor neuron in the locust. J. Neurophysiol., 73: 586–594.

Patel, N.H., Kornberg, T.B. and Goodman, C.S. (1989). Expression of *engrailed* during segmentation in grasshopper and crayfish. Development, 107: 201–212.

Pearson, K.G. (1972). Central programming and reflex control of walking in the cockroach. J. Exp. Biol., 56: 173–193.

Pearson, K.G. (1980). Burst generation in coordinating interneurons of the ventilatory system of the locust. J. Comp. Physiol., 137A: 305–313.

Pearson, K.G. (1983). Neural circuits for jumping in the locust. J. Physiol. (Paris), 78: 765–771.

Pearson, K.G. and Fourtner, C.R. (1975). Nonspiking interneurons in walking system of the cockroach. J. Neurophysiol., 38: 33–52.

Pearson, K.G. and Franklin, R. (1984). Characteristics of leg movement patterns of coordination in locusts walking on rough terrain. Int. J. Robotics Res., 3: 102–112.

Pearson, K.G. and Goodman, C.S. (1979). Correlation of variability in structure with variability in synaptic connections of an identified interneuron in locusts. J. Comp. Physiol., 184A: 141–166.

Pearson, K.G. and Goodman, C.S. (1981). Presynaptic inhibition of transmission from identified interneurons in the locust central nervous system. J. Neurophysiol., 45: 501–515.

Pearson, K.G. and Iles, J.F. (1970). Discharge patterns of coxal levator and depressor motoneurones of the cockroach, *Periplaneta americana*. J. Exp. Biol., 52: 139–165.

Pearson, K.G. and Iles, J.F. (1973). Nervous mechanisms underlying intersegmental coordination of leg movements during walking in the cockroach. J. Exp. Biol., 58: 725–744.

Pearson, K.G. and Ramirez, J.M. (1990). Influence of input from the forewing stretch receptors on motoneurones in flying locusts. J. Exp. Biol., 151: 317–340.

Pearson, K.G. and Robertson, R.M. (1981). Interneurons coactivating hind leg flexor and extensor motoneurons in the locust. J. Comp. Physiol., 144A: 391–400.

Pearson, K.G. and Robertson, R.M. (1987). Structure predicts synaptic function of two classes of interneurons in the thoracic ganglia of *Locusta migratoria*. Cell Tissue Res., 250: 105–114.

Pearson, K.G. and Wolf, H. (1987). Comparison of motor patterns in the intact and deafferented flight system of the locust. I. Electromyographic analysis. J. Comp. Physiol., 160A: 259–268.

Pearson, K.G. and Wolf, H. (1988). Connections of hindwing tegulae with flight neurons in the locust, *Locusta migratoria*. J. Exp. Biol., 135: 381–409.

Pearson, K.G. and Wolf, H. (1989). Timing of forewing elevator activity during flight in the locust. J. Comp. Physiol., 165A: 217–228.

Pearson, K.G., Wong, R.K.S. and Fourtner, C.R. (1976). Connexions between hair-plate afferents and motoneurones in the cockroach leg. J. Exp. Biol., 64: 251–266.

Pearson, K.G., Heitler, W.J. and Steeves, J.D. (1980). Triggering of locust jump by multimodal inhibitory interneurons. J. Neurophysiol., 43: 257–278.

Pearson, K.G., Reye, D.N. and Robertson, R.M. (1983). Phase-dependent influences of wing stretch receptors on flight rhythm in the locust. J. Neurophysiol., 49: 1168–1181.

Pearson, K.G., Boyan, G.S., Bastiani, M. and Goodman, C.S. (1985a). Heterogeneous properties of segmentally homologous interneurons in the ventral nerve cord of locusts. J. Comp. Neurol., 233: 133–145.

Pearson, K.G., Reye, D.N., Parsons, D.W. and Bicker, G. (1985b). Flight-initiating interneurones in the locust. J. Neurophysiol., 53: 910–925.

Pearson, K.G., Gynther, I.C. and Heitler, W.J. (1986). Coupling of flight initiation to the jump in locusts. J. Comp. Physiol., 158A: 81–89.

Pearson, K.G., Hedwig, B. and Wolf, H. (1989). Are the hindwing chordotonal organs elements of the locust flight pattern generator? J. Exp. Biol., 144: 235–255.

Pener, M.P. (1991). Locust phase polymorphism and its endocrine control. Adv. Insect Physiol., 23: 1–79.

Persegol, L., Jordan, M., Viala, D. and Fernandez, C. (1988). Evidence for central entrainment of the medullary respiratory pattern by the locomotor pattern in the rabbit. Exp. Brain Res., 71: 153–162.

Peters, B.H., Altman, J.S. and Tyrer, M. (1985). Synaptic connections between the hindwing stretch receptor and flight motor neurones in the locust revealed by double cobalt labelling for electron microscopy. J. Comp. Neurol., 233: 269–284.

Peters, B.H., Römer, H. and Marquart, V. (1986). Spatial segregation of synaptic inputs and outputs in a locust auditory interneurone. J. Comp. Neurol., 254: 34–50.

Peters, M. (1977). Innervation of the ventral diaphragm of the locust (*Locusta migratoria*). J. Exp. Biol., 69: 23–32.

Peterson, B.A. and Weeks, J.C. (1988). Somatotopic mapping of sensory neurons innervating mechanosensory hairs on the larval prolegs of *Manduca sexta*. J. Comp. Neurol., 275: 128–144.

Pfau, H.K. (1977). Zur Morphologie und Funktion des Vorderflügels und Vorderflügelgelenks von *Locusta migratoria* (L.). Fortschr. Zool., 24: 341–345.

Pfau, H.K. (1978). Funktionsanatomische Aspekte des Insektenflugs. Zool. Jb. (Anat.), 99: 99–108.

Pfau, H.K. (1983). Mechanik und sensorische Kontrolle der Flügel-Pronation und -Supination. In '*BIONA-report 1: Insektenflug I*'. pp. 61–77. Ed. Nachtigall, W. Gustav Fischer Verlag, Stuttgart.

Pfau, H.K. and Nachtigall, W. (1981). Der Voderflügel großer Heuschrecken als Luftkrafterzeuger. II. Zusammenspiel von Muskeln und Gelenkmechanik bei der Einstellung der Flügelgeometrie. J. Comp. Physiol., 142A: 135–140.

Pfau, H.K., Koch, U.T. and Möhl, B. (1989). Temperature dependence and response characteristics of the isolated wing hinge stretch receptor in the locust. J. Comp. Physiol., 165A: 247–252.

Pflüger, H.-J. (1977). The control of the rocking movements of the phasmid *Carausius morosus*. J. Comp. Physiol., 120A: 181–202.

Pflüger, H.-J. (1980). Central nervous projections of sternal trichoid sensilla in locusts. Naturwissenschaften, 67: 316–317.

Pflüger, H.-J. (1984). The large fourth abdominal intersegmental interneuron: a new type of wind-sensitive ventral cord interneuron in locusts. J. Comp. Neurol., 222: 343–357.

Pflüger, H.-J. and Burrows, M. (1978). Locusts use the same basic motor pattern in swimming as in jumping and kicking. J. Exp. Biol., 75: 81–93.
Pflüger, H.-J. and Burrows, M. (1987). A strand receptor with a central cell body synapses upon spiking local interneurones in the locust. J. Comp. Physiol., 160A: 295–304.
Pflüger, H.-J. and Burrows, M. (1990). Synaptic connections of different strength between wind-sensitive hairs and an identified projection interneuron in the locust. Eur. J. Neurosci., 2: 1040–1050.
Pflüger, H.-J. and Tautz, J. (1982). Air movement sensitive hairs and interneurons in *Locusta migratoria*. J. Comp. Physiol., 145A: 369–380.
Pflüger, H.-J. and Watson, A.H.D. (1988). Structure and distribution of Dorsal Unpaired Median (DUM) neurones in the abdominal nerve cord of male and female locusts. J. Comp. Neurol., 268: 329–345
Pflüger, H.-J. and Watson, A.H.D. (1995). GABA and glutamate-like immunoreactivity at synapses received by dorsal unpaired median neurones in the abdominal nerve cord of the locust. Cell Tissue Res., 280: 325–333.
Pflüger, H.-J., Bräunig, P. and Hustert, R. (1981). Distribution and specific central projections of mechanoreceptors in the thorax and proximal leg joints of locusts. II. The external mechanoreceptors: hair plates and tactile hairs. Cell Tissue Res., 216: 79–96.
Pflüger, H.-J., Bräunig, P. and Hustert, R. (1988). The organization of mechanosensory neuropiles in locust thoracic ganglia. Phil. Trans. R. Soc. Lond. B, 321: 1–26.
Pflüger, H.-J., Witten, J.L. and Levine, R.B. (1993). Fate of abdominal ventral unpaired median cells during metamorphosis of the hawkmoth, *Manduca sexta*. J. Comp. Neurol., 335: 508–522.
Pflüger, H.-J., Hurdelbrink, S., Czjzek, A. and Burrows, M. (1994). Activity-dependent structural dynamics of insect sensory fibers. J. Neurosci., 14: 6946–6955.
Phillips, C.E. (1980). An arthropod muscle innervated by nine excitatory motor neurons. J. Exp. Biol., 88: 249–258.
Pitman, R.M. (1975). The ionic dependence of action potentials induced by colchicine in an insect motorneurone cell body. J. Physiol., 247: 511–520.
Pitman, R.M. (1979). Intracellular citrate or externally applied TEA ions produce calcium-dependent action potentials in an insect motorneurone cell body. J. Physiol., 291: 327–337.
Pitman, R.M. (1988). Delayed effects of anoxia upon the electrical properties of an identified cockroach motoneurone. J. Exp. Biol., 135: 95–108.
Pitman, R.M., Tweedle, C.D. and Cohen, M.J. (1972a). Electrical responses of insect central neurons: augmentation by nerve section or colchicine. Science (NY), 178: 507–509.
Pitman, R.M., Tweedle, C.D. and Cohen, M.J. (1972b). Branching of central neurons: intracellular cobalt injection for light and electron microscopy. Science (NY), 176: 412–414.
Plotnikova, S.I. (1969). Effector neurones with several axons in the ventral nerve cord of the Asian grasshopper, *Locusta migratoria*. J. Evol. Biochem. Physiol., 5: 276–278.
Plummer, M.R. and Camhi, J.M. (1981). Discrimination of sensory signals from noise in the escape system of the cockroach: the role of wind acceleration. J. Comp. Physiol., 142: 347–357.
Pollack, A.J., Ritzmann, R.E. and Westin, J. (1988). Activation of DUM cell interneurons by ventral giant interneurons in the cockroach, *Periplaneta americana*. J. Neurobiol., 19: 489–497.
Pollack, I. and Hofbauer, A. (1991). Histamine-like immunoreactivity in the visual system and brain of *Drosophila melanogaster*. Cell Tissue Res., 266: 391–398.
Pond, C.M. (1972). Initiation of flight in unrestrained locusts, *Schistocerca gregaria*. J. Comp. Physiol., 80: 163–178.

Power, M.E. (1948). The thoracico-abdominal nervous system of an adult insect, *Drosophila melanogaster*. J. Comp. Neurol., 88: 347–409.

Pringle, J.W.S. (1940). The reflex mechanism of the insect leg. J. Exp. Biol., 17: 8–17.

Pringle, J.W.S. (1968). Comparative physiology of the flight motor. Adv. Insect Physiol. 5: 163–227.

Prokop, A. and Technau, G.M. (1991). The origin of postembryonic neuroblasts in the ventral nerve cord of *Drosophila melanogaster*. Development, 111: 79–88.

Proux, J.P., Miller, C.A., Li, J.P., Carney, R.L., Girardie, A., Delaage, M. and Schooley, D.A. (1987). Identification of an arginine vasopressin-like diuretic hormone from *Locusta migratoria*. Biochem. Biophys. Res. Commun., 149: 180–186.

Puiroux, J., Pedelaborde, A. and Loughton, B.G. (1993). The effect of proctolin analogues and other peptides on locust oviduct muscle contractions. Peptides, 14: 1103–1109.

Queathem, E. (1991). The ontogeny of grasshopper jumping performance. J. Insect Physiol., 37: 129–138.

Quinlan, M.C. and Hadley, N.F. (1993). Gas exchange, ventilatory patterns, and water loss in two lubber grasshoppers: quantifying cuticular and respiratory transpiration. Physiol. Zool., 66: 628–642.

Raabe, M. (1986). Comparative immunocytochemical study of release sites of insulin, glucagon and AKH-like products in *Locusta migratoria*, *Periplaneta americana*, and *Carausius morosus*. Cell Tissue Res., 245: 267–271.

Radnikow, G. and Bässler, U. (1991). Function of a muscle whose apodeme travels through a joint moved by other muscles: why the retractor unguis muscle in stick insects is tripartite and has no antagonist. J. Exp. Biol., 157: 87–99.

Rall, W. (1981). Functional aspects of neuronal geometry. In '*Neurones without impulses*'. pp. 223–254. Eds Roberts, A. and Bush, B.M.H. Cambridge University Press, Cambridge.

Ramirez, J.M. (1988). Interneurons in the suboesophageal ganglion of the locust associated with flight initiation. J. Comp. Physiol., 162A: 669–686.

Ramirez, J.M. and Orchard, I. (1990). Octopaminergic modulation of the forewing stretch receptor in the locust *Locusta migratoria*. J. Exp. Biol., 149: 255–279.

Ramirez, J.M. and Pearson, K.G. (1988). Generation of motor patterns for walking and flight in motoneurones supplying bifunctional muscles in the locust. J. Neurobiol., 19: 257–282.

Ramirez, J.M. and Pearson, K.G. (1989a). Distribution of intersegmental interneurones that can reset the respiratory rhythm of the locust. J. Exp. Biol., 141: 151–176.

Ramirez, J.M. and Pearson, K.G. (1989b). Alteration of the respiratory system at the onset of locust flight. I. Abdominal pumping. J. Exp. Biol., 142: 401–424.

Ramirez, J.M. and Pearson, K.G. (1991a). Octopamine induces bursting and plateau potentials in insect neurones. Brain Res., 549: 332–337.

Ramirez, J.M. and Pearson, K.G. (1991b). Octopaminergic modulation of interneurons in the flight system of the locust. J. Neurophysiol., 66: 1522–1537.

Ramirez, J.M. and Pearson, K.G. (1993). Alteration of bursting properties in interneurons during locust flight. J. Neurophysiol., 70: 2148–2160.

Ramirez, J.M., Büschges, A. and Kittman, R. (1993). Octopaminergic modulation of the femoral chordotonal organ in the stick insect. J. Comp. Physiol., 173A: 209–219.

Rane, S.G. and Wyse, G.A. (1982). A peripheral sensory-to-motorneuron reflex arc in an arthropod walking leg. Comp. Biochem. Physiol., 73A: 503–512.

Raper, J.A., Bastiani, M. and Goodman, C.S. (1983). Pathfinding by neuronal growth cones in grasshopper embryos. I. Divergent choices made by the growth cones of sibling neurons. J. Neurosci., 3: 20–30.

Raper, J.A., Bastiani, M. and Goodman, C.S. (1984). Pathfinding by neuronal growth cones in grasshopper embryos. IV. The effects of ablating the A and P axons upon the behaviour of the G growth cone. J. Neurosci., 8: 2329–2345.

Rehder, V. (1989). Sensory pathways and motoneurons of the proboscis reflex in the suboesophageal ganglion of the honey bee. J. Comp. Neurol., 279: 499–513.

Rehder, V., Bicker, G. and Hammer, M. (1987). Serotonin-immunoreactive neurons in the antennal lobes and suboesophageal ganglion of the honeybee. Cell Tissue Res., 247: 59–66.

Reichert, H. and Rowell, C.H.F. (1985). Integration of nonphaselocked exteroceptive information in the control of rhythmic flight in the locust. J. Neurophysiol., 53: 1201–1218.

Reichert, H. and Rowell, C.H.F. (1989). Invariance of oscillator interneurone activity during variable motor output by locusts. J. Exp. Biol., 141: 231–239.

Reichert, H., Plummer, M.R., Hagiwara, G., Roth, R.L. and Wine, J.J. (1982). Local interneurons in the terminal abdominal ganglion of the crayfish. J. Comp. Physiol., 149A: 145–162.

Reichert, H., Plummer, M.R. and Wine, J.J. (1983). Identified nonspiking local interneurons mediate nonrecurrent, lateral inhibition of crayfish mechanosensory interneurons. J. Comp. Physiol., 151A: 261–276.

Reichert, H., Rowell, C.H.F. and Griss, C. (1985). Course correction circuitry translates feature detection into behavioural action in locusts. Nature (Lond.), 315: 142–144.

Reingold, S.C. and Camhi, J.M. (1977). A quantitative analysis of rhythmic leg movements during three different behaviors in the cockroach, *Periplaneta americana*. J. Insect Physiol., 23: 1407–1420.

Remy, C. and Girardie, J. (1980). Anatomical organisation of two vasopressin-neurophysin-like neurosecretory cells throughout the nervous system of the migratory locust. Gen. Comp. Endocrinol., 40: 27–35.

Reye, D.N. and Pearson, K.G. (1987). Projections of the wing stretch receptors to central flight neurons in the locust. J. Neurosci., 7: 2476–2487.

Reye, D.N. and Pearson, K.G. (1988). Entrainment of the locust central flight oscillator by wing stretch receptor stimulation. J. Comp. Physiol., 162A: 77–89.

Reynolds, S. E., Taghert, P.H. and Truman, J.W. (1979). Eclosion hormone and bursicon titres and the onset of hormonal responsiveness during the last day of adult development in *Manduca sexta* (L.). J. Exp. Biol., 78: 77–86.

Riley, J.R., Krueger, U., Addison, C.M. and Gewecke, M. (1988). Visual detection of wind-drift by high-flying insects at night: a laboratory study. J. Comp. Physiol., 162A: 793–798.

Rind, F.C. and Simmons, P.J. (1992). Orthopteran DCMD neuron: a reevaluation of responses to moving objectives. I. Selective responses to approaching objects. J. Neurophysiol., 68: 1654–1666.

Ritchie, J.M., Rogart, R.B. and Strichartz, G.R. (1976). A new method for labelling saxitoxin and its binding to non-myelinated fibres of the rabbit vagus, lobster walking leg, and garfish olfactory nerves. J. Physiol., 261: 477–494.

Ritzmann, R.E. (1981). Motor responses to paired stimulation of giant interneurons in the cockroach *Periplaneta americana*. II. The ventral giant interneurons. J. Comp. Physiol., 143A: 71–80.

Ritzmann, R.E. and Pollack, A.J. (1981). Motor responses to paired stimulation of giant interneurons in the cockroach *Periplaneta americana*. I. The dorsal giant interneurons. J. Comp. Physiol., 143A: 61–70.

Ritzmann, R.E. and Pollack, A.J. (1986). Identification of thoracic interneurons that mediate giant interneuron-to-motor pathways in the cockroach. J. Comp. Physiol., 159A: 639–654.

Ritzmann, R.E. and Pollack, A.J. (1988). Wind-activated thoracic interneurons in the cockroach. II. Patterns of connections from ventral giant interneurons. J. Neurobiol., 19: 589–611.

Ritzmann, R.E. and Pollack, A.J. (1990). Parallel motor pathways from thoracic interneurons of the ventral giant interneuron system of the cockroach, *Periplaneta americana*. J. Neurobiol., 21: 1219–1235.

Ritzmann, R.E., Pollack, A.J. and Tobias, M.L. (1982). Flight activity mediated by intracellular stimulation of dorsal giant interneurons of the cockroach *Periplaneta americana*. J. Comp. Physiol., 147A: 313–322.

Ritzmann, R.E., Pollack, A.J., Hudson, S.E. and Hyvonen, A. (1991). Convergence of multi-modal sensory signals at thoracic interneurons of the escape system of the cockroach, *Periplaneta americana*. Brain Res., 563: 175–183.

Robb, S. and Evans, P.D. (1990). FMRFamide-like peptides in the locust: distribution, partial characterization and bioactivity. J. Exp. Biol., 149: 335–360.

Robb, S. and Evans, P.D. (1994). The modulatory effect of SchistoFLRFamide on heart and skeletal muscle in the locust *Schistocerca gregaria*. J. Exp. Biol., 197: 437–442.

Robb, S., Cheek, T.R., Hannan, F.L., Hall, L.M., Midgley, J.M. and Evans, P.D. (1994). Agonist-specific coupling of a cloned *Drosophila* octopamine/tyramine receptor to multiple second messenger systems. EMBO J., 13: 1325–1330.

Robert, D. (1988). Visual steering under closed-loop conditions by flying locusts: flexibility of optomotor response and mechanisms of correctional steering. J. Comp Physiol., 164A: 15–24.

Robert, D. (1989). The auditory behaviour of flying locusts. J. Exp. Biol., 147: 279–301.

Robert, D. and Rowell, C.H.F. (1992a). Locust flight steering. I. Head movements and the organization of correctional manoeuvres. J. Comp. Physiol., 171A: 41–51.

Robert, D. and Rowell, C.H.F. (1992b). Locust flight steering. II. Acoustic avoidance manoeuvres and associated head movements, compared with correctional steering. J. Comp. Physiol., 171A: 53–62.

Robertson, R.M. (1986). Neuronal circuits controlling flight in the locust: central generation of the rhythm. Trends Neurosci., 9: 278–280.

Robertson, R.M. (1988). Insect neurons: synaptic interactions, circuits and the control of behavior. In '*Nervous systems in invertebrates*'. pp. 393–442. Ed. Ali, M.A. Plenum Press, New York.

Robertson, R.M. (1990). Synchronous activity of flight neurons in the mesothoracic ganglion of the locust. J. Comp. Physiol., 167A: 61–70.

Robertson, R.M. and Johnson, A.G. (1993). Retinal image size triggers obstacle avoidance in flying locusts. Naturwissenschaften, 80: 176–178.

Robertson, R.M. and Pearson, K.G. (1982). A preparation for the intracellular analysis of neuronal activity during flight in the locust. J. Comp. Physiol., 146A: 311–320.

Robertson, R.M. and Pearson, K.G. (1983). Interneurons in the flight system of the locust: distribution, connections and resetting properties. J. Comp. Neurol., 215: 33–50.

Robertson, R.M. and Pearson, K.G. (1985a). Neural circuits in the flight system of the locust. J. Neurophysiol., 53: 110–128.

Robertson, R.M. and Pearson, K.G. (1985b). Neural networks controlling locomotion in locusts. In '*Model neural networks and behavior*'. pp. 21–35. Ed. Selverston, A.I., Plenum Press, New York.

Robertson, R.M. and Reye, D.N. (1988). A local circuit interaction in the flight system of the locust. J. Neurosci., 8: 3929–3936.

Robertson, R.M. and Wisniowski, L. (1988). GABA-like immunoreacitivity of identified interneurons in the flight system of the locust, *Locusta migratoria*. Cell Tissue Res., 254: 331–340.

Robertson. R.M., Pearson, K.G. and Reichert, H. (1982). Flight interneurons in the locust and the origin of insect wings. Science (NY), 217: 177–179.

Robinson, N.L. (1982). Anomalous resistance changes following application of the neurotransmitter L-glutamate in insect muscle. J. Comp. Physiol., 148A: 281–285.

Roeder, K.D. (1937). The control of tonus and locomotor activity in the praying mantis (*Mantis religiosa* L.). J. Exp. Zool., 76: 353–374.

Roeder, K.D. (1948). Organization of the ascending giant fiber system of the cockroach (*Periplaneta americana*). J. Exp. Zool., 108: 243–261.

Roeder, T. (1990). Histamine H_1-receptor-like binding sites in the locust nervous tissue. Neurosci. Lett., 116: 331–335.

Roeder, T. (1992). A new octopamine receptor class in locust nervous tissue, the octopamine 3 (OA_3) receptor. Life Sci., 50: 21–28.

Roeder, T. (1994). Biogenic amines and their receptors in insects. Comp. Biochem. Physiol., 107C: 1–12.

Roeder, T. and Nathanson, J.A. (1993). Characterization of insect neuronal octopamine receptors (OA_3 receptors). Neurochem. Res., 18: 921–925.

Roeder, T., Vossfeldt, R. and Gewecke, M. (1993). Pharmacological characterization of the locust neuronal 3H-mianserin binding site, a putative histamine receptor. Comp. Biochem. Physiol., 106C: 503–507.

Römer, H. and Marquart, V. (1984). Morphology and physiology of auditory interneurons in the metathoracic ganglion of the locust. J. Comp. Physiol., 155A: 249–262.

Römer, H., Marquart, V. and Hardt, M. (1988). Organization of a sensory neuropile in the auditory pathway of two groups of orthoptera. J. Comp. Neurol., 275: 201–215.

Ronacher, B., Wolf, H. and Reichert, H. (1988). Locust flight behavior after hemisection of individual thoracic ganglia: evidence for hemiganglionic premotor centers. J. Comp. Physiol., 163A: 749–759.

Rothschild, M., Schlein, Y., Parker, K. and Sternberg, S. (1972). Jump of the oriental rat flea *Xenopsylla cheopsis* (Rothschild). Nature (Lond.), 239: 45–47.

Rowell, C.H.F. (1961). The structure and function of the prothoracic spine of the desert locust, *Schistocerca gregaria* Forskål. J. Exp. Biol. 38: 457–469.

Rowell, C.H.F. (1971). The orthopteran descending movement detector (DMD) neurones: a characterisation and review. Z. vergl. Physiol., 73: 167–194.

Rowell, C.H.F. (1976). The cells of the insect neurosecretory system; constancy, variability, and the concept of the unique identifiable neuron. Adv. Insect Physiol., 12: 63–123.

Rowell, C.H.F. (1989). The taxonomy of invertebrate neurons: a plea for a new field. Trends Neurosci., 12: 169–174

Rowell, C.H.F. and Dorey, A.E. (1967). The number and size of axons in the thoracic connectives of the desert locust *Schistocerca gregaria* Forskål. Z. Zellforschung, 83: 288–294.

Rowell, C.H.F. and Pearson, K.G. (1983). Ocellar input to the flight motor system of the locust: structure and function. J. Exp. Biol., 103: 265–288.

Rowell, C.H.F. and Reichert, H. (1986). Three descending interneurons reporting deviation from course in the locust. II. Physiology. J. Comp. Physiol., 158A: 775–794.

Rowell, C.H.F. and Reichert, H. (1991). Mesothoracic interneurons involved in flight steering in the locust. Tissue Cell, 23: 75–139.

Rudomin, P. (1990). Presynaptic inhibition of muscle spindle and tendon organ afferents in the mammalian spinal cord. Trends Neurosci., 13: 499–505.

Ryckebusch, S. and Laurent, G. (1993). Rhythmic patterns evoked in locust leg motor neurons by the muscarinic agonist pilocarpine. J. Neurophysiol., 69: 1583–1595.

Ryckebusch, S. and Laurent, G. (1994). Interactions between segmental leg central pattern generators during fictive rhythms in the locust. J. Neurophysiol., 72: 2771–2785.

Saager, F. and Gewecke, M. (1989). Antennal reflexes in the desert locust *Schistocerca gregaria*. J. Exp. Biol., 147: 519–532.

Sabry, J.H., O'Connor, T.P., Evans, L., Toroian-Raymond, A., Kirschner, M. and Bentley, D. (1991). Microtubule behavior during guidance of pioneer neuron growth cones *in situ*. J. Cell. Biol., 115: 381–395.

Saint Marie, R.L., Carlson, S.D. and Chi, C. (1984). The glial cells of insects. In '*Insect ultrastructure*' Vol. 2. pp. 453–475. Eds King, R.C. and Akai, H. Plenum Press, New York.

Sandeman, D.C. (1961). Some aspects of the flight mechanism of the migratory locust, *Locusta migratoria migratoroides* R and F. Masters thesis, Department of Zoology, University of Natal, Pietermaritzburg, South Africa.

Sarthy, P.V. (1991). Histamine: a neurotransmitter candidate for *Drosophila* photoreceptors. J. Neurochem., 57: 1757–1768.

Sattelle, D.B. (1985). Acetylcholine receptors. In '*Comprehensive insect physiology, biochemistry and pharmacology*'. Vol 11, pp. 395–434. Eds Kerkut, G.A. and Gilbert, L.I. Pergamon Press, Oxford.

Saudou, F., Boschert, U., Amlaiky, N., Plassat, J.-L. and Hen, R. (1992). A family of *Drosophila* serotonin receptors with distinct intracellular signalling properties and expression patterns. EMBO J., 11: 7–17.

Sbrenna, G. (1971). Postembryonic growth of the ventral nerve cord in *Schistocerca gregaria* Forsk. (Orthoptera: Acrididae). Boll. Zool., 38: 49–74.

Schachtner, J. and Bräunig, P. (1993). The activity pattern of identified neurosecretory cells during feeding behaviour in the locust. J. Exp. Biol., 185: 287–303.

Schachtner, J. and Bräunig, P. (1995). Activity pattern of aminergic nerve cells innervating the salivary glands of the locust *Locusta migratoria*. J. Comp. Physiol., 176A: 491–501.

Schäfer, S. and Bicker, G. (1986). Distribution of GABA-like immunoreactivity in the brain of the honeybee. J. Comp.Neurol., 246: 287–300.

Schäfer, S. and Rehder, V. (1989). Dopamine-like immunoreactivity in the brain and suboesophageal ganglion of the honeybee. J. Comp. Neurol., 280: 43–58.

Schaller, D. (1978). Antennal sensory system of *Periplaneta americana* L. Cell Tissue Res., 191: 121–139.

Schmidt, J. and Rathmayer, W. (1993). Central organization of common inhibitory motoneurons in the locust: role of afferent signals from leg mechanoreceptors. J. Comp. Physiol., 172A: 447–456.

Schmidt, R.F. (1971). Presynaptic inhibition in the vertebrate central nervous system. Ergebn. Physiol., 63: 20–101.

Schmidt-Ott, U. and Technau, G.M. (1992). Expression of *en* and *wg* in the embryonic head and brain of *Drosophila* indicates a refolded band of seven segment remnants. Development, 116: 111–125.

Schmitt, J.B. (1965). Variations in the transverse nerve in the abdominal nervous system of insects. J. NY Entomol. Soc., 73: 144–150.

Schneider, L.E. and Taghert, P.H. (1988). Isolation and characterization of a *Drosophila* gene that encodes multiple peptides related to Phe-Met-Arg-Phe-NH_2 (FRMFamide). Proc. Natl. Acad. Sci. USA, 85: 1993–1997.

Schofield, P.K. and Treherne, J.E. (1986). Octopamine sensitivity of the blood-brain barrier of an insect. J. Exp. Biol., 123: 423–439.

Schofield, P.K., Swales, L.S. and Treherne, J.E. (1984). Potentials associated with the blood-brain barrier of an insect: recordings from identified neuroglia. J. Exp. Biol., 109: 307–318.

Schoofs, L., Jegou, S., Vaudry, H., Verhaert, P. and De Loof, A. (1987). Localization of melanotropin-like peptides in the central nervous system of two insect species, the migratory locust, *Locusta migratoria* and the fleshfly, *Sarcophaga bullata*. Cell Tissue Res., 248: 25–31.

Schoofs, L., Holman, G.M., Hayes, T.K., Nachman, R.J. and DeLoof, A. (1990). Locustatachykinins I and II, two novel insect neuropeptides with homology to peptides of the vertebrate tachykinin family. FEBS Lett., 261: 397–401.

Schoofs, L., Tips, A., Holman, G.M., Nachman, R.J. and DeLoof, A. (1992). Distribution of locustamyotropin-like immunoreactivity in the nervous system of *Locusta migratoria*. Regulat. Peptides, 37: 237–254.

Seabrook, W.D. (1968). The innervation of the terminal abdominal segments (VIII-XI) of the desert locust, *Schistocerca gregaria*. Can. Entomol., 100: 693–715.

Selverston, A.I., Kleindienst, H.-U. and Huber, F. (1985). Synaptic connectivity between cricket auditory interneurons as studied by selective photoinactivation. J. Neurosci., 5: 1283–1292.

Selys-Longchamps, E. de. (1878). Lettre de M. Samuel H. Scudder et observations par M. de Selys-Longchamps sur l'*Acridium peregrinum*. Annls. Soc. entomol. Belge, 21: 5–8.

Sevala, V.M., Sevala, V.L. and Loughton, B.G. (1993). FMRFamide-like activity in the female locust during vitellogenesis. J. Comp. Neurol., 337: 286–294.

Seymour, K.J. (1990). *The neural control of oviposition in the locust* Schistocerca gregaria. PhD Thesis, University of Cambridge, England.

Shanbhag, S.R., Singh, K. and Naresh Singh, R. (1992). Ultrastructure of the femoral chordotonal organs and ther novel synaptic organization in the legs of *Drosophila melanogaster* Meigen (Diptera: Drosophilidae). Int. J. Insect Morphol. Embryol., 21: 311–322.

Shaw, S.R. (1994a). Re-evaluation of the absolute threshold and response mode of the most sensitive known 'vibration' detector, the cockroach's subgenual organ: a cochlea-like displacement threshold and a direct response to sound. J. Neurobiol., 25: 1167–1185.

Shaw, S.R. (1994b). Detection of airborne sound by a cockroach 'vibration detector': a possible missing link in insect auditory evolution. J. Exp. Biol., 193: 13–47.

Shelton, P.M.J., Stephen, R.O., Scott, J.J.A. and Tindall, A.R. (1992). The apodeme complex of the femoral chordotonal organ in the metathoracic leg of the locust, *Schistocerca gregaria*. J. Exp. Biol., 163: 345–358.

Shepherd, D. and Bate, C.M. (1990). Spatial and temporal patterns of neurogenseis in the embryo of the locust *Schistocerca gregaria*. Development, 108: 83–96.

Shepherd, D. and Laurent, G. (1992). Embryonic development of a population of spiking local interneurones in the locust (*Schistocerca gregaria*). J. Comp. Neurol., 319: 438–453.

Shepherd, D., Kämper, G. and Murphey, R.K. (1988). The synaptic origins of receptive field properties in the cricket cercal sensory system. J. Comp. Physiol., 162A: 1–11.

Shimozawa, T. and Kanou, M. (1984). Varieties of filiform hairs: range fractionation by sensory afferents and cercal interneurons of a cricket. J. Comp. Physiol., 155A: 485–493.

Siegert, K.J. and Mordue, W. (1986). Quantification of adipokinetic hormones I and II in the corpora cardiaca of *Schistocerca gregaria* and *Locusta migratoria*. Comp. Biochem. Physiol., 84A: 279–284.

Siegler, M.V.S. (1981a). Posture and history of movement determine membrane potential and synaptic events in nonspiking interneurons and motor neurons of the locust. J. Neurophysiol., 46: 296–309.

Siegler, M.V.S. (1981b). Postural changes alter synaptic interactions between nonspiking interneurons and motor neurons of the locust. J. Neurophysiol., 46: 310–323.

Siegler, M.V.S. (1982). Electrical coupling between supernumerary motor neurones in the locust. J. Exp. Biol., 101: 105–119.

Siegler, M.V.S. and Burrows, M. (1979). The morphology of local non-spiking interneurones in the metathoracic ganglion of the locust. J. Comp. Neurol. 183: 121–148.

Siegler, M.V.S. and Burrows, M. (1983). Spiking local interneurons as primary integrators of mechanosensory information in the locust. J. Neurophysiol., 50: 1281–1295.

Siegler, M.V.S. and Burrows, M. (1984). The morphology of two groups of spiking local interneurones in the metathoracic ganglion of the locust. J. Comp. Neurol., 224: 463–482.

Siegler, M.V.S. and Burrows, M. (1986). Receptive fields of motor neurones underlying local tactile reflexes in the locust. J. Neurosci., 6: 507–513.

Siegler, M.V.S. and Pousman, C.A. (1990a). Motor neurons of grasshopper metathoracic ganglion occur in stereotypic anatomical groups. J. Comp. Neurol., 297: 298–312.

Siegler, M.V.S. and Pousman, C.A. (1990b). Distribution of motor neurons into anatomical groups in the grasshopper metathoracic ganglion. J. Comp. Neurol., 297: 313–327.

Simmons, P.J. (1980). Connexions between a movement-detecting visual interneurone and flight motorneurones of a locust. J. Exp. Biol., 86: 87–97.

Simmons, P.J. and Hardie, R.C. (1988). Evidence that histamine is a neurotransmitter of photoreceptors in the locust ocellus. J. Exp. Biol., 138: 205–219.

Skiebe, P. and Schneider, H. (1994). Allatostatin peptides in the crab stomatogastric nervous system: inhibition of the pyloric motor pattern and distribution of allatostatin-like immunoreactivity. J. Exp. Biol., 194: 195–208.

Skiebe, P., Corrette, B.J. and Wiese, K. (1990). Evidence that histamine is the inhibitory transmitter of the auditory interneuron ON1 of crickets. Neurosci. Lett., 116: 361–366.

Skinner, K. (1985a). The structure of the fourth abdominal ganglion of the crayfish, *Procambarus clarki* (Girard). I. Tracts in the ganglionic core. J. Comp. Neurol., 234: 168–181.

Skinner, K. (1985b). The structure of the fourth abdominal ganglion of the crayfish, *Procambarus clarki* (Girard). II. Synaptic neuropils. J. Comp. Neurol., 234: 182–191.

Skorupski, P. and Hustert, R. (1991). Reflex pathways responsive to depression of the locust coxotrochanteral joint. J. Exp. Biol., 158: 599–605.

Slifer, E.H. (1954). The reaction of a grasshopper to an odorous material held near one of its feet (Orthoptera: Acrididae). Proc. R. Entomol. Soc. Lond. A, 29: 177–179.

Slifer, E.H. (1956). The response of a grasshopper, *Romalea microptera* (Beauvois), to strong odours following amputation of the metathoracic leg at different levels. Proc. R. Entomol. Soc. Lond. A, 31: 95–98.

Slifer, E.H. and Finlayson, L.H. (1956). Muscle receptor organs in grasshoppers and locusts. Q. J. Microsc. Sci., 97: 617–620.

Slifer, E.H. and Uvarov, B.P. (1938). Brunner's organ; a structure found on the jumping legs of grasshoppers (Orthoptera). Proc. R. Entomol. Soc. Lond. A, 13: 111–115.

Smith, P.J.S., Leech, C.A. and Treherne, J.E. (1984). Glial repair in the insect central nervous system: effects of selective glial disruption. J. Neurosci., 4: 2698–2711.

Smola, U. (1970). Untersuchung zur Topographie, Mechanik und Strömungsmechanik der Sinneshaare auf dem Kopf der Wanderheuschrecke *Locusta migratoria*. Z. vergl. Physiol., 67: 382–402.

Snodgrass, R.E. (1928). Morphology and evolution of the insect head and its appendages. Smithson. misc. Collns, 81: 1–158.

Snodgrass, R.E. (1929). The thoracic mechanism of a grasshopper, and its antecedents. Smithson. misc. Collns, 82: 1–111.

Snodgrass, R.E. (1935). *Principles of insect morphology*. McGraw-Hill, New York.

Sobek, L., Eckert, M., Penzlin, H. and Reissmann, S. (1986). Evidence for proctolinergic innervation of the cockroach oviduct. Zool. J. Physiol., 90: 461–466.

Sobel, E.C. (1990). The locust's use of motion parallax to measure distance. J. Comp. Physiol., 167A: 579–588.

Sombati, S. and Hoyle, G. (1984a). Glutamatergic central nervous transmission in locusts. J. Neurobiol. 15: 507–516.

Sombati, S. and Hoyle, G. (1984b). Central nervous sensitization and dishabituation of reflex action in an insect by the neuromodulator octopamine. J. Neurobiol., 15: 455–480.

Sombati, S. and Hoyle, G. (1984c). Generation of specific behaviors in a locust by local release into neuropil of the natural neuromodulator octopamine. J. Neurobiol., 15: 481–506.

Sonetti, D., Ottaviani, E., Bianchi, F., Rodriguez, M., Stefano, M.L., Scharrer, B. and Stefano, G.B. (1994). Microglia in invertebrate ganglia. Proc. Natl. Acad. Sci. USA, 91: 9180–9184.

Spinola, S.M. and Chapman, K.M. (1975). Proprioceptive indentation of the campaniform sensilla of cockroach legs. J. Comp. Physiol., 96A: 257–252.

Spira, M.E., Yarom, Y. and Parnas, I. (1976). Modulation of spike frequency by regions of special axonal geometry and by synaptic inputs. J. Neurophysiol., 39: 882–899.

Spirito, C.P. and Mushrush, D.L. (1979). Interlimb coordination during slow walking in the cockroach. I. Effects of substrate alterations. J. Exp. Biol., 78: 233–243.

Spörhase-Eichmann, U., Vullings, H.G.B., Buijs, R.M., Hörner, M. and Schurmann, F.-W. (1992). Octopamine-immunoreactive neurons in the central nervous system of the cricket, *Gryllus bimaculatus*. Cell Tissue Res., 268: 287–304.

Städler, E. and Hanson, F.E. (1975). Olfactory capabilities of the 'gustatory' chemoreceptors of the tobacco hornworm larvae. J. Comp. Physiol., 104A: 97–102.

Stangier, J., Hilbich, C. and Keller, R. (1989). Occurrence of crustacean cardioactive peptide (CCAP) in the nervous system of an insect, *Locusta migratoria*. J. Comp. Physiol., 159B: 5–11.

Starrat, A.N. and Brown, B.E. (1975). Structure of the pentapeptide proctolin, a proposed neurotransmitter in insects. Life Sci., 17: 1253–1256.

Stay, B., Chan, K.K. and Woodhead, A.P. (1992). Allatostatin-like immunoreactive neurons projecting to the corpora allata of adult *Diploptera punctata*. Cell Tissue Res., 270: 13–23.

Steeves, J.D. and Pearson, K.G. (1982). Proprioceptive gating of inhibitory pathways to hind leg flexor motoneurons in the locust. J. Comp. Physiol., 146A: 507–515.

Stevenson, P.A. and Kutsch, W. (1987). A reconsideration of the central pattern generator concept for locust flight. J. Comp. Physiol., 161A: 115–129.

Stevenson, P.A. and Kutsch, W. (1988). Demonstration of functional connectivity of the flight motor system in all stages of the locust. J. Comp. Physiol., 162A: 247–259.

Stevenson, P.A. and Pflüger, H.-J. (1994). Colocalization of octopamine and FMRFamide related peptide in identified heart projecting (DUM) neurones in the locust revealed by immunocytochemistry. Brain Res., 638: 117–125.

Stevenson, P.A., and Spörhase-Eichmann, U. (1995). Localization of octopaminergic neurones in insects. Comp. Biochem. Physiol., 110A: 203–215.

Stevenson, P.A., Pflüger, H.-J., Eckert, M. and Rapus, J. (1992). Octopamine immunoreactive cell populations in the locust thoracic-abdominal nervous system. J. Comp. Neurol., 315: 382–397.

Stevenson, P.A., Pflüger, H.-J., Eckert, M. and Rapus, J. (1994). Octopamine-like immunoreactive neurones in locust genital ganglia. Cell Tissue Res., 275: 299–308.

Storrer, V.J., Bässler, U. and Mayer, S. (1986). Motor neurons in the meso- and metathoracic ganglia of the stick insect. Zool. J. Physiol., 90: 359–374.

Strausfeld, N.J. (1976). *Atlas of an insect brain.* Springer Verlag, Berlin.

Suder, F. and Wendler, G. (1993). Organisation of the thoracic ganglia of the adult sphinx moth *Manduca sexta* (Insecta, Lepidoptera). Zoomorphol., 113: 103–112.

Sun, X.J., Tolbert, L.P. and Hildebrand, J.G. (1993). Ramification pattern and ultrastructural characteristics of the serotonin-immunoreactive neuron in the antennal lobe of the moth *Manduca sexta*: a laser scanning confocal and electron microscopic study. J. Comp. Neurol., 338: 5–16.

Svidersky, V.L. (1969). Receptors on the forehead of the locust, *Locusta migratoria* in ontogenesis. J. Evol. Biochem. Physiol., 5: 482–490.

Swales, L.S. and Evans, P.D. (1994). Distribution of myomodulin-like immunoreactivity in the adult and developing ventral nervous system of the locust *Schistocerca gregaria.* J. Comp. Neurol., 343: 263–280.

Swales, L.S. and Evans, P.D. (1995a). Distribution of SchistoFLRFamide-like immunoreactivity in the adult ventral nervous system of the locust, *Schistocerca gregaria.* Cell Tissue Res., 281: 339–348.

Swales, L.S. and Evans, P.D. (1995b). Distribution of myomodulin-like immunoreactivity in the brain and retrocerebral complex of the locust, *Schistocerca gregaria.* J. Comp. Neurol., 353: 407–414.

Swales, L.S., Cournil, I. and Evans, P.D. (1992). The innervation of the closer muscle of the mesothoracic spiracle of the locust. Tissue Cell, 24: 547–558.

Taghert, P.H. and Goodman, C.S. (1984). Cell determination and differentiation of identified serotonin-immunoreactive neurons in the grasshopper embryo. J. Neurosci., 4: 989–1000.

Tamarelle, M., Romeuf, M. and Vanderhaegen, J.J. (1988). Immunohistochemical localization of gastrin-cholecystokinin-like material in the central nervous system of the migratory locust. Histochemistry, 89: 201–207.

Tanaka, S. (1993). Hormonal deficiency causing albinism in *Locusta migratoria.* Zool. Sci., 10: 467–471.

Tanaka, S. and Pener, M.P. (1994). A neuropeptide controlling the dark pigmentation in color polymorphism of the migratory locust, *Locusta migratoria.* J. Insect Physiol., 40: 997–1005.

Tanouye, M.A. and Wyman, R.J. (1980). Motor outputs of giant nerve fiber in *Drosophila.* J. Neurophysiol., 44: 405–421.

Tareilus, E., Hanke, W. and Breer, H. (1990). Neuronal acetylcholine receptor channels from insects: a comparative electrophysiological study. J. Comp. Physiol., 167A: 521–526.

Tautz, J., Holldobler, B. and Danker, T. (1994). The ants that jump: different techniques to take off. Zoology, 98: 1–6.

Technau, G.M. (1984). Fiber number in the mushroom bodies of adult *Drosophila melanogaster* depends on age, sex and experience. J. Neurogenetics, 1: 113–126.

ten Cate, J. (1936). Beitrage zur Innervation der Lokomotionsbewegung der Heuschrecke (*Locusta viridissima*). Arch. neerl. Physiol., 21: 562–566.

Theophilidis, G. (1983). A comparative study of the anatomy and innervation of the metathoracic extensor tibia muscle in three orthopteran species. Comp. Biochem. Physiol., 75A, 285–292.

Theophilidis, G. and Burns, M.D. (1979). A muscle tension receptor in the locust leg. J. Comp. Physiol., 131A: 247–254.

Theophilidis, G. and Dimitriadis, V.K. (1990). The structure and innervation of the metathoracic flexor tibiae muscle of two species of orthoptera (Insecta). Comp. Biochem. Physiol., 97A: 583–594.

Thomas, J.B. and Wyman, R.J. (1984). Mutations altering synaptic connectivity between identified neurons in *Drosophila*. J. Neurosci., 4: 530–538.

Thomas, J.B., Bastiani, M.J., Bate, M. and Goodman, C.S. (1984). From grasshopper to *Drosophila*: a common plan for neuronal development. Nature (Lond.), 310: 203–207.

Thomas, J.G. (1965). The abdomen of the female desert locust (*Schistocerca gregaria* Forskål) with special reference to the sense organs. Anti-Locust Bulletin, 42: 1–20.

Thomas, M.B., Wood, S.N. and Lomer, C.J. (1995). Biological control of locusts and grasshoppers using a fungal pathogen: the importance of secondary cycling. Proc. R. Soc. Lond. B., 259: 265–270.

Thomas, M.V. (1984). Voltage-clamp analysis of calcium-mediated potassium conductance in cockroach *Periplaneta americana* central neurones. J. Physiol., 350: 159–178.

Thompson, C.S., Yagi, K.J., Chen, Z.F. and Tobe, S.S. (1990). The effects of octopamine on juvenile hormone biosynthesis, electrophysiology, and cAMP content of the coropora allata of the cockroach *Diploptera punctata*. J. Comp. Physiol., 160B: 241–249.

Thompson, K.J. (1986a). Oviposition digging in the grasshopper. I. Functional anatomy and the motor programme. J. Exp. Biol., 122: 387–411.

Thompson, K.J. (1986b). Oviposition digging in the grasshopper. II. Descending neural control. J. Exp. Biol., 122: 413–425.

Thompson, K.J. and Siegler, M.V.S. (1991). Anatomy and physiology of spiking local and intersegmental interneurons in the median neuroblast lineage of the grasshopper. J. Comp. Neurol., 305: 659–675.

Thompson, K.J. and Siegler, M.V.S. (1993). Development of segment specificity in identified lineages of the grasshopper CNS. J. Neurosci., 13: 3309–3318.

Thompson, K.S.J. and Bacon, J.P. (1991). The vasopressin-like immunoreactive (VPLI) neurons of the locust, *Locusta migratoria*. II. Physiology. J. Comp. Physiol., 168A: 619–630.

Thompson, K.S.J., Tyrer, N.M., May, S.T. and Bacon, J.P. (1991). The vasopressin-like immunoreactive (VPLI) neurons of the locust, *Locusta migratoria*. I. Anatomy. J. Comp. Physiol., 168A: 605–617.

Thompson, K.S.J., Blagburn, J.M., Gibbon, C.R. and Bacon, J.P. (1992). Correlation of filiform hair position with sensory afferent morphology and synaptic connections in the second instar cockroach. J. Comp. Neurol., 320: 213–227.

Thüring, D.A. (1986). Variability of motor output during flight steering in locusts. J. Comp. Physiol., 158A: 653–664.

Ting, L.H., Blickhan, R. and Full, R.J. (1994). Dynamic and static stability in hexapedal runners. J. Exp. Biol., 197: 251–269.

Torkkeli, P.H. and French, A.S. (1994). Characterization of a transient outward current in a rapidly adapting insect mechanosensory neuron. Pflügers Arch., 429: 72–78.

Treherne, J.E. and Schofield, P.K. (1979). Ionic homeostasis of the brain microenvironment in insects. Trends Neurosci., 2: 227–230.

Treherne, J.E. and Schofield, P.K. (1981). Mechanisms of ionic homeostasis in the central nervous system of an insect. J. Exp. Biol., 95: 61–73.

Trimmer, B.A. and Weeks, J.C. (1989). Effects of nicotinic and muscarinic agents on an identified motoneurone and its direct afferent inputs in larval *Manduca sexta*. J. Exp. Biol., 144: 303–337.

Trimmer, B.A. and Weeks, J.C. (1991). Activity-dependent induction of facilitation, depression, and post-tetanic potentiation at an insect central synapse. J. Comp. Physiol. 168A: 27–43.

Trimmer, B.A. and Weeks, J.C. (1993). Muscarinic acetylcholine receptors modulate the excitability of an identified insect motoneuron. J. Neurophysiol., 69: 1821–1836.

Truman, J.W. and Copenhaver, P.F. (1989). The larval eclosion hormone neurones in *Manduca sexta*: identification of the brain-proctodeal neurosecretory system. J. Exp. Biol., 147: 457–470.

Tyrer, N.M. and Altman, J.S. (1974). Motor and sensory neurones in a locust demonstrated using cobalt chloride. J. Comp. Neurol., 157: 117–138.

Tyrer, N.M. and Gregory, G.E. (1982). A guide to the neuroanatomy of locust suboesophageal and thoracic ganglia. Phil. Trans. R. Soc. Lond. B. 297: 91–123.

Tyrer, N.M., Turner, J.D. and Altman, J.S. (1984). Identifiable neurons in the locust central nervous system that react with antibodies to serotonin. J. Comp. Neurol., 227: 313–330.

Tyrer, N.M., Pozza, M.F., Humbel, U., Peters, B.H. and Bacon, J.P. (1988). The tritocerebral commissure 'dwarf' (TCD): a major GABA-immunoreactive descending interneuron in the locust. J. Comp. Physiol., 164A: 141–150.

Tyrer, N.M., Davis, N.T., Arbas, E.A., Thompson, K.S.J. and Bacon, J.P. (1993). Morphology of the vasopressin-like immunoreactive (VPLI) neurons in many species of grasshopper. J. Comp. Neurol., 329: 385–401.

Udolph, G., Prokop, A., Bossing, T. and Technau, G.M. (1993). A common precursor for glia and neurons in the embryonic CNS of *Drosophila* gives rise to segment-specific lineage variants. Development, 118: 765–777.

Ultsch, A., Schuster, C.M., Laube, B., Schloss, P., Schmitt, B. and Betz, H. (1992). Glutamate receptors of *Drosophila melanogaster*: cloning of a kainate-selective subunit expressed in the central nervous system. Proc. Natl. Acad. Sci. USA, 89: 10484–10488.

Usherwood, P.N.R. (1994). Insect glutamate receptors. Adv. Insect Physiol., 24: 309–341.

Usherwood, P.N.R. and Grundfest, H. (1964). Inhibitory postsynaptic potentials in grasshopper muscle. Science (NY), 143: 817–818.

Usherwood, P.N.R. and Grundfest, H. (1965). Peripheral inhibition in skeletal muscle of insects. J. Neurophysiol., 28: 497–518.

Usherwood, P.N.R. and Runion, H.I. (1970). Analysis of the mechanical responses of metathoracic extensor tibiae muscles of free-walking locusts. J. Exp. Biol., 52: 39–58.

Usherwood, P.N.R., Runion, H.I. and Campbell, J.I. (1968). Structure and physiology of a chordotonal organ in the locust leg. J. Exp. Biol., 48: 305–323.

Usherwood, P.N.R., Giles, D. and Suter, C. (1980). Studies of the pharmacology of insect neurones *in vitro*. In '*Insect neurobiology and pesticide action (Neurotox 79)*'. pp. 115–128. Society of Chemical Industry, London.

Veelaert, D., Schoofs, L., Tobe , S.S., Yu, C.G., Vullings, H.G.B., Couillaud, F. and De Loof, A. (1995). Immunological evidence for an allatostatin-like neuropeptide in the central nervous system of *Schistocerca gregaria*, *Locusta migratoria* and *Neobellieria bullata*. Cell Tissue Res., 279: 601–611.

Veenstra, J.A. and Davis, N.T. (1993). Localization of corazonin in the nervous system of the cockroach *Periplaneta americana*. Cell Tissue Res., 274: 57–64.

Vincent, J.F.V. and Wood, S.D.E. (1972). Mechanism of abdominal extension during oviposition in *Locusta*. Nature (Lond.), 235: 167–168.

Virant-Doberlet, M., Horsemann, G., Loher, W. and Huber, F. (1994). Neurons projecting from the brain to the corpora allata in orthopteroid insects: anatomy and physiology. Cell Tissue Res., 277: 39–50.

Wafford, K.A. and Sattelle, D.B. (1989). L-Glutamate receptors on the cell body membrane of an identified insect motor neurone. J. Exp. Biol., 144: 449–462.

Waldron, I. (1967). Mechanisms for the production of the motor output pattern in flying locusts. J. Exp. Biol., 47: 201–212.

Waldrop, B., Christensen, T.A. and Hildebrand, J.G. (1987). GABA-mediated synaptic inhibition of projection neurones in the antennal lobes of the sphinx moth, *Manduca sexta*. J. Comp. Physiol., 161A: 23–32.

Washio, H. and Tanaka, Y. (1992). Some effects of octopamine, proctolin and serotonin on dorsal unpaired median neurones of cockroach (*Periplaneta americana*) thoracic ganglia. J. Insect Physiol., 38: 511–517.

Watkins, B.L. and Burrows, M. (1989). GABA-like immunoreactivity in the suboesophageal ganglion of the locust *Schistocerca gregaria*. Cell Tissue Res., 258: 53–63.

Watkins, B.L., Burrows, M. and Siegler, M.V.S. (1985). The structure of locust non-spiking interneurones in relation to the anatomy of their segmental ganglion. J. Comp. Neurol. 240: 233–255.

Watson, A.H.D. (1984). The dorsal unpaired median neurons of the locust metathoracic ganglion: neuronal structure and diversity, and synapse distribution. J. Neurocytol., 13: 303–327.

Watson, A.H.D. (1986). The distribution of GABA-like immunoreactivity in the thoracic nervous system of the locust *Schistocerca gregaria*. Cell Tissue Res., 246: 331–341.

Watson, A.H.D. (1988). Antibodies against GABA and glutamate label neurones with morphologically distinct synaptic vesicles in locust central nervous system. Neuroscience 26: 33–44.

Watson, A.H.D. (1992a). Presynaptic modulation of sensory afferents in the invertebrate and vertebrate nervous system. Comp. Biochem. Physiol., 103A: 227–239.

Watson, A.H.D. (1992b). The distribution of dopamine-like immunoreactivity in the thoracic and abdominal ganglia of the locust (*Schistocerca gregaria*). Cell Tissue Res., 270: 113–124.

Watson, A.H.D. and Burrows, M. (1982). The ultrastructure of identified locust motor neurones and their synaptic relationships. J. Comp. Neurol., 205: 383–397.

Watson, A.H.D. and Burrows, M. (1983). The morphology, ultrastructure and distribution of synapses on an intersegmental interneurone of the locust. J. Comp. Neurol., 214: 154–169.

Watson, A.H.D. and Burrows, M. (1985). The distribution of synapses on the two fields of neurites of spiking local interneurones in the locust. J. Comp. Neurol., 240: 219–232.

Watson, A.H.D. and Burrows, M. (1987). Immunocytochemical and pharmacological evidence for GABAergic spiking local interneurones in the locust. J. Neurosci., 7: 1741–1751.

Watson, A.H.D. and Burrows, M. (1988). The distribution and morphology of synapses on nonspiking local interneurones in the thoracic nervous system of the locust. J. Comp. Neurol., 272: 605–616.

Watson, A.H.D. and Laurent, G. (1990). GABA-like immunoreactivity in a population of locust intersegmental interneurones and their inputs. J. Comp. Neurol., 302: 761–767.

Watson, A.H.D. and Pflüger, H.-J. (1984). The ultrastructure of prosternal sensory hair afferents within the locust central nervous system. Neuroscience, 11: 269–279.

Watson, A.H.D. and Pflüger, H.-J. (1987). The distribution of GABA-like immunoreactivity in relation to ganglion structure in the abdominal nerve cord of the locust (*Schistocerca gregaria*). Cell Tissue Res., 249: 391–402.

Watson, A.H.D. and Pflüger, H.-J. (1989). Regional specialisation for synaptic input and output on a locust intersegmental interneurone with multiple spike-initiating zones. J. Comp. Neurol., 279: 515–527.

Watson, A.H.D. and Seymour-Laurent, K.J. (1993). The distribution of glutamate-like immunoreactivity in the thoracic and abdominal ganglia of the locust (*Schistocerca gregaria*). Cell Tissue Res., 273: 557–570.

Watson, A.H.D., Burrows, M. and Hale, J.P. (1985). The morphology and ultrastructure of common inhibitory motor neurones in the thorax of the locust. J. Comp. Neurol., 239: 341–359.

Watson, A.H.D., Burrows, M. and Leitch, B. (1993). GABA-immunoreactivity in processes presynaptic to the terminals of afferents from a locust leg proprioceptor. J. Neurocytol., 22: 547–557.

Wedemeyer, S., Roeder, T. and Gewecke, M. (1992). Pharmacological characterization of a 5-HT receptor in locust nervous tissue. Eur. J. Pharmacol., 223: 173–178.

Weeks, J.C. and Jacobs, G.A. (1987). A reflex behavior mediated by monosynaptic connections between hair afferents and motoneurons in the larval tobacco hornworm, *Manduca sexta*. J. Comp. Physiol., 160A: 315–329.

Weis-Fogh, T. (1949). An aerodynamic sense organ stimulating and regulating flight in locusts. Nature (Lond.), 164: 873–874.

Weis-Fogh, T. (1956a). Biology and physics of locust flight. II. Flight performance of the desert locust *Schistocerca gregaria*. Phil. Trans. R. Soc. Lond. B. 239: 459–510.

Weis-Fogh, T. (1956b). The flight of locusts. Scient. Am., 194: 116–124.

Weis-Fogh, T. (1964). Functional design of the tracheal system of flying insects as compared with the avian lung. J. Exp. Biol., 41: 207–227.

Weis-Fogh, T. (1967). Respiration and tracheal ventilation in locusts and other flying insects. J. Exp. Biol., 47: 561–587.

Weiss, M.J. (1981). Structural patterns in the corpora pedunculata of orthoptera: a reduced silver analysis. J. Comp. Neurol., 203: 515–553.

Wendler, G. (1964). Laufen und Stehen der Stabheuschrecke *Carausius morosus*: Sinnesborstenfelder in den Beingelenken als Glieder von Regelkreisen. Z. vergl. Physiol., 48: 198–250.

Wendler, G. (1966). The co-ordination of walking movements in arthropods. Symp. Soc. Exp. Biol., 20: 229–249.

Wendler, G. (1972). Einfluß Erzwungener Flügelbewegungen auf das Motorische Flugmuster von Heuschrecken. Naturwissenschaften, 59: 220.

Wendler, G. (1974). The influence of proprioceptive feedback on locust flight coordination. J. Comp. Physiol., 88A: 173–200.

Wendler, G. (1978a). Erzeugung und Kontrolle koordinierter Bewegungen bei Tieren – Beispiele an Insekten. Kybernetik, 77: 11–34.

Wendler, G. (1978b). The possible role of fast wing reflexes in locust flight. Naturwissenschaften, 65: 65.

Wendler, G. and Suder, F. (1994). Chlordimeform elicits long-term flight in the hawkmoth *Manduca sexta*. *Proceedings of the 22nd Gottingen Neurobiology Conference*, p. 301. Georg Thieme Verlag, Stuttgart.

Wendt, B. and Homberg, U. (1992). Immunocytochemistry of dopamine in the brain of the locust *Schistocerca gregaria*. J. Comp. Neurol., 321: 387–403.

Westin, J. (1979). Responses to wind recorded from the cercal nerve of the cockroach *Periplaneta americana*. I. Response properties of single sensory neurons. J. Comp. Physiol., 133A: 97–102.

Westin, J., Langberg, J.J. and Camhi, J.M. (1977). Responses of giant interneurons of the cockroach *Periplaneta americana* to wind puffs of different directions and velocities. J. Comp. Physiol., 121A: 307–324.

Westin, J., Ritzmann, R.E. and Goddard, D.J. (1988). Wind-activated thoracic interneurons of the cockroach. I. Responses to controlled wind stimuli. J. Neurobiol., 19: 573–588.

Whim, M.D. and Evans, P.D. (1988). Octopaminergic modulation of flight muscle in the locust. J. Exp. Biol., 134: 247–266.

White, P.R. and Chapman, R.F. (1990). Tarsal chemoreception in the polyphagous grasshopper *Schistocerca americana*: behavioural assays, sensilla distributions and electrophysiology. Physiol. Entomol., 15: 105–121.

Whitington, P.M. (1989). The early development of motor axon pathways in the locust embryo: the establishement of the segmental nerves in the thoracic ganglia. Development, 105: 715–721.

Whitington, P.M. and Seifert, E. (1981). Identified neurons in an insect embryo: the pattern of neurons innervating the metathoracic leg of the locust. J. Comp. Neurol., 200: 203–212.

Whitington, P.M., Bate, M., Seifert, E., Ridge, K. and Goodman, C.S. (1982). Survival and differentiation of identified embryonic neurons in the absence of their target muscles. Science (NY), 215: 973–975.

Whitman, D.W., Billen, J.P.J., Alsop, D. and Blum, M.S. (1991). Anatomy, ultrastructure, and functional morphology of the metathoracic tracheal defensive glands of the grasshopper *Romalea guttata*. Can. J. Zool., 69: 2100–2108.

Wicher, D. and Penzlin, H. (1994). Ca^{2+} currents in cockroach neurones: properties and modulation by neurohormone D. NeuroReport 5: 1023–1026.

Wicher, D., Walther, C. and Penzlin, H. (1994). Neurohormone D induces ionic current changes in cockroach central neurones. J. Comp. Physiol., 174A: 507–515.

Wiens, T.J. and Wolf, H. (1993). The inhibitory motoneurons of crayfish thoracic limbs: Identification, structures, and homology with insect common inhibitors. J. Comp. Neurol., 336: 261–278.

Wigglesworth, V.B. (1963). The origin of flight in insects. Proc. R. Entomol. Soc. Lond. Ser. C, 28: 23–32.

Wildman, M.H. and Cannone, A.J. (1991). Interaction between afferent neurones in a crab muscle receptor organ. Brain Res., 565: 175–178.

Willey, R.B. (1961). The morphology of the stomodeal nervous system in *Periplaneta americana* (L.) and other Blattaria. J. Morphol., 108: 219–261.

Williams, J.L.D. (1975). Anatomical studies of the insect central nervous system: a ground plan of the mid brain and an introduction to the central complex in the locust *Schistocerca gregaria* (Orthoptera). J. Zool., 176: 67–86.

Williamson, R. and Burns, M.D. (1978). Multiterminal receptors in the locust mesothoracic leg. J. Insect Physiol., 24: 661–666.

Wilson, D.M. (1961). The central nervous control of flight in a locust. J. Exp. Biol., 38: 471–490.

Wilson, D.M. (1962). Bifunctional muscles in the thorax of grasshoppers. J. Exp. Biol., 39: 669–677.

Wilson, D.M. (1966). Central nervous mechanisms for the generation of rhythmic behaviour in arthropods. Symp. Soc. Exp. Biol., 20: 199–228.

Wilson, D.M. (1968). Inherent asymmetry and reflex modulation of the locust flight motor pattern. J. Exp. Biol., 48: 631–641.

Wilson, D.M. and Gettrup, E. (1963). A stretch reflex controlling wingbeat frequency in grasshoppers. J. Exp. Biol., 40: 171–185.

Wilson, D.M. and Larimer, J.L. (1968). The catch property of ordinary muscle. Proc. Natl. Acad. Sci. USA, 61: 909–916.

Wilson, D.M. and Weis-Fogh, T. (1962). Patterned activity of co-ordinated motor units, studied in flying locusts. J. Exp. Biol., 39: 643–667.

Wilson, D.M. and Wyman, R.J. (1965). Motor output patterns during random and rhythmic stimulation of locust thoracic ganglia. Biophys. J., 5: 121–143.

Wilson, D.M., Smith, D.O. and Dempster, P. (1970). Length and tension hysteresis during sinusoidal and stepfunction stimulation of arthropod muscle. Am. J. Physiol., 219: 916–922.

Wilson, J.A. (1981). Unique, identifiable nonspiking interneurons in the locust mesothoracic ganglion. J. Neurobiol., 12: 353–366.

Wilson, J.A. and Hoyle, G. (1978). Serially homologous neurones as concomitants of functional specialisation. Nature (Lond.), 274: 377–379.

Wilson, J.A. and Phillips, C.E. (1982). Locust local nonspiking interneurons which tonically drive antagonistic motor neurons: physiology, morphology and ultrastructure. J. Comp. Neurol., 204: 21–31.

Wilson, J.A., Phillips, C.E., Adams, M.E. and Huber, F. (1982). Structural comparison of a homologous neuron in Gryllid and Acridid insects. J. Neurobiol., 13: 459–467.

Wilson, M. (1978). The functional organisation of locust ocelli. J. Comp Physiol., 124A: 297–316.

Withers, G.S., Fahrbach, S.E. and Robinson, G.E. (1993). Selective neuroanatomical plasticity and division of labour in the honeybee. Nature (Lond.), 364: 238–240.

Witten, J.L. and O'Shea, M. (1985). Peptidergic innervation of insect skeletal muscle: immunocytochemical observations. J. Comp. Neurol., 242: 93–101.

Witten, J.L., Worden, M.K., Schaffer, M.H. and O'Shea, M. (1984). New classification of insect motoneurons: expression of different peptide transmitters. Neurosci. Abstr., 10: 151.

Witthöft, W. (1967). Absolute Anzahl und Verteilung der Zellen im Hirn der Honigbiene. Z. morphol. Tiere., 61: 160–184.

Wohlers, D.W. and Huber, F. (1985). Topographical organization of the auditory pathway within the prothoracic ganglion of the cricket *Gryllus campestris* L. Cell Tissue Res., 239: 555–565.

Wolf, H. (1990a). Activity patterns of inhibitory motoneurones and their impact on leg movement in tethered walking locusts. J. Exp. Biol., 152: 281–304.

Wolf, H. (1990b). On the function of a locust flight steering muscle and its inhibitory innervation. J. Exp. Biol., 150: 55–80.

Wolf, H. (1992). Reflex modulation in locusts walking on a treadwheel- intracellular recordings from motoneurons. J. Comp. Physiol., 170A: 443–462.

Wolf, H. (1993). The locust tegula: Significance for flight rhythm generation, wing movement control and aerodynamic force production. J. Exp. Biol., 182: 229–253.

Wolf, H. and Burrows, M. (1995). Proprioceptive sensory neurons of a locust leg receive rhythmic presynaptic inhibition during walking. J. Neurosci., 15: 5623–5636.

Wolf, H. and Lang, D.M. (1994). Origin and clonal relationship of common inhibitory motoneurons Cl1 and Cl3 in the locust CNS. J. Neurobiol., 25: 846–864.

Wolf, H. and Laurent, G. (1994). Rhythmic modulation of the responsiveness of locust sensory local interneurons by walking pattern generating networks. J. Neurophysiol., 71: 110–118.

Wolf, H. and Pearson, K.G. (1987). Flight motor patterns recorded in surgically isolated sections of the ventral nerve cord of *Locusta migratoria*. J. Comp. Physiol., 161A: 103–114.

Wolf, H. and Pearson, K.G. (1989). Comparison of motor patterns in the intact and deafferented locust. III. Patterns of interneuronal activity. J. Comp. Physiol., 165A: 61–74.

Wolf, H., Ronacher, B. and Reichert, H. (1988). Patterned synaptic drive to locust flight motoneurons after hemisection of thoracic ganglia. J. Comp. Physiol., 161A: 761–769.

Wong, R.K.S. and Pearson, K.G. (1976). Properties of the trochanteral hair plate and its function in the control of walking in the cockroach. J. Exp. Biol., 64: 233–249.

Wootton, R.J. (1985). The origin of insect flight: where are we now? Antenna, 10: 82–86.

Wootton, R.J. (1992). Functional morphology of insect wings. Annu. Rev. Entomol., 37: 113–140.

Wootton, R.J. and Ellington, C.P. (1982). Biomechanics and the origin of insect flight. In '*Biomechanics in evolution*'. pp. 99–112. Eds. Rayner, J.M.V. and Wootton, R.J. Cambridge University Press, Cambridge.

Worden, M.K., Witten, J.L. and O'Shea, M. (1985). Proctolin is a co-transmitter for the SETi motoneuron. Neurosci. Abstracts, 11: 327.

Yack, J.E. and Roots, B.I. (1992). The metathoracic wing-hinge chordotonal organ of an atympanate moth, *Actias luna* (Lepidoptera, Saturniidae): a light- and electron-microscopic study. Cell Tissue Res., 267: 455–471.

Young, I.S., Alexander, R.M., Woakes, A.J., Butler, P.J. and Anderson, L. (1992). The synchronization of ventilation and locomotion in horses (*Equus caballus*). J. Exp. Biol., 166: 19–31.

Zacharias, D., Williams, J.L.D., Meier, T. and Reichert, H. (1993). Neurogenesis in the insect brain: cellular identification and molecular characterization of brain neuroblasts in the grasshopper embryo. Development, 118: 941–955.

Zarnack, W. (1982). Untersuchungen zum Flug von Wanderheuschrecken - Die Bewegungen, räumlichen Lagebeziehungen sowie Formen und Profile von Vorder- und Hinterflügeln. In '*BIONA-report 1. Insektenflug II*'. pp. 79–102. Ed. Nachtigall, W. Gustav Fischer, Stuttgart.

Zarnack, W. and Möhl, B. (1977). Activity of the direct downstroke flight muscles of *Locusta migratoria* (L.) during steering behavior in flight. I. Patterns of time shift. J. Comp. Physiol., 118A: 215–233.

Zhang, B.G., Torkkeli, P.H. and French, A.S. (1992). Octopamine selectively modifies the slow component of sensory adaptation in an insect mechanoreceptor. Brain Res., 591: 351–355.

Zill, S.N. (1985). Plasticity and proprioception in insects. I. Responses and cellular properties of individual receptors of the locust metathoracic femoral chordotonal organ. J. Exp. Biol., 116: 435–461.

Zill, S.N. (1986). A model of pattern generation of cockroach walking reconsidered. J. Neurobiol., 17: 317–328.

Zill, S.N. and Frazier, S.F. (1990). Responses of locusts in a paradigm which tests postural load compensatory reactions. Brain Res., 535: 1–8.

Zill, S.N. and Moran, D.T. (1981a). The exoskeleton and insect proprioception. I. Responses of tibial campaniform sensilla to external and muscle-generated forces in the American cockroach, *Periplaneta americana*. J. Exp. Biol., 91: 1–24.

Zill, S.N. and Moran, D.T. (1981b). The exoskeleton and insect proprioception. III. Activity of tibial campaniform sensilla during walking in the American cockroach, *Periplaneta americana*. J. Exp. Biol., 94: 57–75.

Zill, S.N. and Moran, D.T. (1982). Suppression of reflex postural tonus: a role of peripheral inhibition in insects. Science (NY), 216: 751–753.

Zill, S.N., Moran, D.T. and Varela, F.G. (1981). The exoskeleton and insect proprioception. II. Reflex effects of tibial campaniform sensilla in the American cockroach, *Periplaneta americana*. J. Exp. Biol., 94: 43–55.

Index

Numbers in **bold** are page references to **Figures**.
Numbers in *italics* are page references to **Tables**.